Advances in Experimental Medicine and Biology

Volume 1484

Advances in Experimental Medicine and Biology provides a platform for scientific contributions in the main disciplines of the biomedicine and the life sciences. This series publishes thematic volumes on contemporary research in the areas of microbiology, immunology, neurosciences, biochemistry, biomedical engineering, genetics, physiology, and cancer research. Covering emerging topics and techniques in basic and clinical science, it brings together clinicians and researchers from various fields.

Advances in Experimental Medicine and Biology has been publishing exceptional works in the field for over 40 years, and is indexed in SCOPUS, Medline (PubMed), EMBASE, Reaxys, EMBiology, the Chemical Abstracts Service (CAS), and Pathway Studio.

2023 CiteScore: 5.9

Parissa Farnia • Ali Akbar Velayati
Jalaledin Ghanavi • Poopak Farnia

Proteins in *Mycobacterium Tuberculosis*

Functions and Therapeutic Advances

Springer

Parissa Farnia
Shahid Beheshti University of Medical Sciences
Mycobacteriology Research Centre (MRC), National Research Institute of Tuberculosis and Lung Diseases (NRITLD)
Tehran, Iran

Ali Akbar Velayati
Shahid Beheshti University of Medical Sciences
Mycobacteriology Research Centre (MRC), National Research Institute of Tuberculosis and Lung Diseases (NRITLD)
Tehran, Iran

Jalaledin Ghanavi
Shahid Beheshti University of Medical Sciences
Mycobacteriology Research Centre (MRC), National Research Institute of Tuberculosis and Lung Diseases (NRITLD)
Tehran, Iran

Poopak Farnia
Shahid Beheshti University of Medical Sciences
Mycobacteriology Research Centre (MRC), National Research Institute of Tuberculosis and Lung Diseases (NRITLD)
Tehran, Iran

ISSN 0065-2598 ISSN 2214-8019 (electronic)
Advances in Experimental Medicine and Biology
ISBN 978-3-031-96882-2 ISBN 978-3-031-96883-9 (eBook)
https://doi.org/10.1007/978-3-031-96883-9

This Springer imprint is published by the registered company Springer Nature Switzerland AG
The registered company address is: Gewerbestrasse 11, 6330 Cham, Switzerland

Understanding Mycobacterium tuberculosis (Mtb) presents a profound scientific challenge. Although this bacterium measures only 1 to 3 microns in length and is invisible to the naked eye, it has an outsized impact on global health. Despite its microscopic scale, Mtb produces approximately 4,000 proteins, organized into diverse functional groups that govern essential processes such as metabolism, replication, virulence, immune evasion, and antibiotic resistance.

What is truly remarkable is the scale and dedication of the global scientific community striving to unravel Mtb's biology. Over the past 140 years, more than half a million researchers worldwide have contributed to improving diagnosis, decoding the bacterium's complex proteome, and advancing strategies for tuberculosis (TB) prevention, control, and eventual eradication. Yet, despite this monumental collective effort, TB remains the leading cause of death from a single infectious agent. This enduring challenge highlights not only the exceptional biological complexity of Mtb but also the persistent difficulties in fully understanding its mechanisms of survival and pathogenicity.

In recognition of these challenges, the authors dedicate this work first to the profound wisdom underlying the creation of the TB bacillus and second to the tireless efforts of researchers around the world. This dedication stands as a tribute to their commitment and as a reminder of the urgent need for continued research. Ultimately, conquering tuberculosis will require sustained innovation, scientific rigor, and the unwavering determination that has driven progress thus far.

Preface

Proteins in *Mycobacterium tuberculosis* are fundamental to the bacterium's survival, pathogenicity, and interaction with the host. This book delves into the intricate protein architecture of *M. tuberculosis*, exploring diverse families that govern cell wall synthesis, virulence, metabolism, and immune modulation. Understanding these proteins is pivotal for uncovering novel targets for drug development, vaccine design, and diagnostics, crucial steps in combating the global tuberculosis epidemic. Advances in proteomic technologies and molecular biology have expanded our insights into the dynamic roles of these proteins, revealing complex regulatory networks and adaptation strategies employed by the bacterium in hostile environments. This volume integrates cutting-edge research on protein function, posttranslational modifications, and secretion systems, providing a comprehensive resource for researchers, clinicians, and students alike. It aims to bridge basic science and clinical application, fostering new avenues for effective tuberculosis control and treatment.

Tehran, Iran

Parissa Farnia
Ali Akbar Velayati
Jalaledin Ghanavi
Poopak Farnia

Contents

1 Tuberculosis: An Ongoing Global Threat

Abstract

Tuberculosis (TB) remains one of the world's most urgent public health challenges, with a substantial global burden that continues to affect millions annually. According to the World Health Organization (WHO) Global Tuberculosis Report 2024, approximately 10.6 million people developed TB in 2023, resulting in 1.6 million deaths. The highest incidence rates persist in low- and middle-income countries, where socioeconomic determinants such as poverty, malnutrition, overcrowded living conditions, and limited access to quality healthcare exacerbate disease transmission and worsen outcomes. The emergence and spread of drug-resistant TB, including multidrug-resistant (MDR-TB), extensively drug-resistant (XDR-TB), and totally drug-resistant (TDR-TB) strains, pose significant challenges to effective treatment and control, contributing to increased mortality and healthcare costs. The coronavirus disease-2019 (COVID-19) pandemic has further disrupted TB services worldwide, causing declines in case detection, delayed diagnoses, and interruptions in treatment, thereby reversing years of progress. Despite the availability of effective vaccines and treatment regimens, gaps in awareness, funding, healthcare infrastructure, and social determinants of health hinder global elimination efforts. Coordinated, multisectoral strategies are essential to address this complex epidemic, improving diagnostic capacity, expanding access to comprehensive treatment, combating stigma, addressing socioeconomic barriers, and investing in research for novel prevention and therapeutic tools. These efforts are critical to achieving the WHO's End TB Strategy targets and ultimately eliminating TB as a public health threat.

Keywords

Mycobacterium tuberculosis · Population dynamics · Age · Gender · Incidence · TB control program

1.1 Introduction

The genus *Mycobacterium* comprises a large and diverse group of Gram-positive, acid-fast organisms within the phylum Actinomycetota, class Actinomycetia, order Mycobacteriales, and family Mycobacteriaceae (Runyon 1959; Lightner and Redman 1998; Hartmans et al. 2006; Gao and Gupta 2012; Magee et al. 2012; Gupta et al. 2018; Kanabalan et al. 2021; Moens et al. 2024; Silva-Pereira et al. 2024). Members of this family inhabit a wide range of environments, including water, soil, and metalworking fluids, and many species are well recognized as human and animal

P. Farnia et al., *Proteins in Mycobacterium Tuberculosis*, Advances in Experimental Medicine and Biology 1484, https://doi.org/10.1007/978-3-031-96883-9_1

pathogens (Kreiss and Cox-Ganser 1997; Fox et al. 1999; Medjahed et al. 2010; Magee et al. 2012; Velayati et al. 2014, 2015a, b; Velayati and Farnia 2019; Rajendran et al. 2021; Zheng et al. 2023; Sakthivadivel et al. 2023; Orgeur et al. 2024). Among these, the most clinically significant are *Mycobacterium tuberculosis* (Mtb) and *Mycobacterium leprae*, causative agents of tuberculosis (TB) and leprosy, respectively (Hansen 1880; Zopf 1883; Kempsell et al. 1992; Gordon et al. 1999; Truman et al. 2008; Brzostek et al. 2009; Falkinham 2015; Tortoli 2012; Vidal et al. 2025). To date, the genus *Mycobacterium* encompasses over 200 species (with 192 validly published with correct names and 13 recognized synonyms) as listed in the "List of Prokaryotic names with Standing in Nomenclature" (LPSN) (Parte et al. 2020; Armstrong and Parrish 2021; Sun et al. 2022; Zhang et al. 2024). In a recent phylogenomic study, Gupta et al. (2018) proposed a major taxonomic revision, recommending the division of the genus *Mycobacterium* into five separate genera based on comprehensive genome-scale phylogenetic analyses and molecular synapomorphies. According to this proposal, the name *Mycobacterium* would be retained only for the clade containing the principal human and animal pathogens, while other species would be reassigned to four newly established genera. This reclassification was intended to clarify evolutionary relationships and enhance the clinical focus on pathogenic species (Gupta et al. 2018). However, Tortoli et al. (2019) offered a more conservative perspective. While acknowledging the genomic evidence for distinct clades within *Mycobacterium*, they prioritized nomenclatural stability and adherence to the rules of bacterial taxonomy. Their genome-based revision recognized many synonymous taxa within the genus but did not fully endorse the division into multiple genera as proposed by Gupta et al. (2018). Instead, Tortoli et al. recommended reclassifications primarily at the species and subspecies levels, maintaining a broader concept of the genus *Mycobacterium* and proposing the synonymization of several species. This approach aimed to balance evolutionary insights with the practical need for taxonomic continuity in clinical and research settings. Understanding the relationships among *Mycobacterium* species is of paramount importance, as the genus includes major opportunistic pathogens capable of causing infections in both humans and animals (Wallace et al. 1992; Portaels et al. 1997; Dailloux et al. 1999; Magee et al. 2012; Gupta et al. 2018; Tortoli 2014; Velayati et al. 2015a, b; Lory 2014; Fedrizzi et al. 2017; Jimeno and Muñoz 2020; Selmane and L'Hadj 2020; Samanta et al. 2023).

Mycobacterium tuberculosis complex (MTBc) comprises a group of closely related species capable of causing tuberculosis (TB) in both humans and animals. Since the identification of *Mycobacterium tuberculosis* as the causative agent of TB in 1882, additional species within this complex have been recognized, expanding our understanding of the pathogens responsible for tuberculosis. Members of the MTBc include *Mycobacterium bovis* (including the vaccine strain *Mycobacterium bovis bacillus Calmette-Guérin* [BCG]) (Lange et al. 2022); *Mycobacterium africanum* (Niemann et al. 2004); *Mycobacterium pinnipedii* (Silva-Pereira et al. 2019; Roe et al. 2019); *Mycobacterium microti* (Smith et al. 2009); *Mycobacterium canettii* (Pfyffer et al. 1998); *Mycobacterium caprae* (Prodinger et al. 2014); *Mycobacterium orygis* (van Ingen et al. 2012); *Mycobacterium mungi* (Alexander et al. 2010); and *Mycobacterium suricattae* (Dippenaar et al. 2015). These species exhibit an extremely high degree of genetic similarity, often exceeding 99%, and are considered variants within a single complex rather than entirely distinct species. The differentiation of MTBc members relies on specific genetic markers and unique genomic regions, which are crucial for distinguishing lineages and elucidating their evolutionary relationships (Brites and Gagneux 2017). Recognizing the MTBc as a unified group is essential for accurate diagnosis, effective treatment, and robust epidemiological surveillance of tuberculosis globally. Within the MTBc, *Mycobacterium tuberculosis* (Mtb) is the principal cause of tuberculosis in humans. TB is an infectious disease that primarily affects the lungs but can also involve other organs such as the kidneys, spine, and brain (Mohammadnabi et al. 2024). Transmission occurs when individuals with active pulmonary or laryngeal TB expel aerosolized droplets containing the bacteria through coughing, sneezing, or talking, which others can then inhale. Upon inhalation, the TB bacilli reach the alveoli in the lungs, where alveolar macrophages phagocytose them. Inside these host cells, the bacteria replicate and eventually cause cell lysis, releasing more bacilli to infect neighboring cells (Wigginton and Kirschner 2001; Barreiro et al. 2012; Cadena et al. 2017; Wessler et al. 2020). The spread of bacteria from one cell to another is a key mechanism in TB pathogenesis, facilitating dissemination within pulmonary tissues and potential entry into the circulatory system. It is estimated that approximately one-quarter of the global population has been infected with TB bacteria. However, only about 5–10% of infected individuals will develop symptomatic TB disease during their lifetime (Rastogi et al. 2001; Parrish et al. 1998; Lillebaek et al. 2002; Esmail et al. 2014; Uden et al. 2017; WHO 2024). Infection with Mtb can result in two distinct outcomes: active TB disease or latent TB infection (LTBI). In active TB, the bacteria multiply and cause clinical symptoms such as persistent cough, chest pain, fever, night sweats, and weight loss, and affected individuals are capable of transmitting the infection to others. In contrast, latent TB infection is characterized by the presence of dormant bacteria in the body, without symptoms or risk of transmission (Tobin and Tristram 2024; WHO 2024). However, LTBI can progress to active TB disease if the host immune system becomes compromised. Understanding the distinction between active and latent TB is critical for the development of effective control strategies and treatment protocols worldwide (Portnoy et al. 2023; Zhang et al. 2025; WHO 2024).

1.2 Tuberculosis from Ancient Origins to a Modern Health Crisis

Tuberculosis (TB) is a disease with a profound and extensive history, having afflicted humanity throughout both recorded history and prehistory. The causative agent, Mtb, is believed to have caused more deaths than any other microbial pathogen. Phylogenetic and genomic studies trace the origins of Mtb to East Africa, where an early progenitor may have infected hominids as far back as three million years ago (Gutierrez et al. 2005). Modern members of the MTBc, including *Mycobacterium tuberculosis*, *Mycobacterium africanum*, *Mycobacterium canettii*, and *Mycobacterium bovis*, are thought to have diverged from a common African ancestor between 35,000 and 15,000 years ago (Kapur et al. 1994; Brosch et al. 2002). The currently circulating strains of Mtb appear to have evolved from a common ancestor approximately 20,000 to 15,000 years ago and are classified into six major lineages, all present in East Africa but with varying global distributions (Gagneux et al. 2006). Compelling evidence supports the co-migration of Mtb with anatomically modern humans during their exodus from Africa around 70,000 years ago. Phylogenetic analyses reveal that the global diversity of Mtb parallels human migration patterns, indicating that Mtb lineages expanded alongside human populations (Comas et al. 2013). Mutation rate analyses suggest that much of the present genetic diversity in Mtb emerged within the last 1000 years (Niemann and Supply 2014). East Africa is thus recognized as the ancestral home of both modern humans and tubercle bacilli. Archaeological evidence documents tuberculosis in ancient Egypt dating back over 3000 years, with characteristic skeletal lesions such as Pott's deformities identified in mummies and depicted in early Egyptian art (Cave 1939; Morse et al. 1964; Zimmerman 1979). Similar evidence of tuberculosis has also been found in the Americas (Daniel 2006). The disease was known in classical antiquity under the term "phthisis," first used in Greek literature around 460 BC, with Hippocrates describing its predilection for young adults (Moonan et al. 2018). The Roman physician Galen, who became the physician to Emperor Marcus Aurelius in 174 AD, documented tuberculosis and suggested treatments such as fresh air, milk, and sea voyages. He described symptoms including fever, sweating, and coughing with blood-stained sputum, although tuberculosis was not a primary focus of his medical texts. Persian physicians made significant advances in the understanding of tuberculosis during the medieval period. Rhazes (865–925 AD) provided one of the earliest clinical descriptions of pulmonary tuberculosis, distinguishing it from other respiratory diseases (Velayati 2023). Avicenna (970–1037 AD) devoted two chapters of his medical encyclopedia, *The Canon of Medicine*, to tuberculosis, detailing its symptoms, causes, and treatments (Saffari and Pakpour 2012; Velayati 2023).

The scientific study of tuberculosis progressed markedly in the early modern period. In 1689, Richard Morton first identified tubercles in the lungs of phthisis patients. Gaspar Laurent Bayle's extensive autopsies in 1810 revealed tubercles in multiple organs (Auerbach 1944). René Laennec's seminal 1819 treatise laid the foundation for modern clinical understanding by describing TB pathology and physical signs (Daniel 2006). The term "tuberculosis" was introduced by Johann Lukas Schönlein in 1839, replacing earlier terms such as "phthisis" and "consumption" (Barberis et al. 2017). Jean-Antoine Villemin demonstrated the infectious nature of TB in 1865, experimentally transmitting the disease to rabbits (Daniel 2006). The discovery of Mtb by Robert Koch in 1882 was a milestone that revolutionized TB diagnosis and treatment. Koch summarized the importance of his findings in a manuscript published in the *Berliner Klinische Wochenschrift* shortly after his announcement: "In the future the fight against this terrible plague of mankind will deal no longer with an undetermined something, but with a tangible parasite, whose living conditions are for the most part known and can be investigated further" (Koch 1932; Blevins and Bronze 2010). Koch's studies on tuberculosis ushered in the bacteriological era in hygiene and public health that led to the "medicalization" of entire societies in the late nineteenth century. Despite fundamental achievements in diagnostic advances during the nineteenth century, therapeutic solutions for tuberculosis were still immature. In the second half of the nineteenth century, it was a common conviction that specific climatic environments could contribute to curing tuberculosis (Friedman et al. 1989; McCarthy 2001; Iseman 2000). Subsequently, the most frequently prescribed remedy for pulmonary forms was a stay in a temperate climate. Unfortunately, the historical record documents that Koch's discovery had no significant consequences in medical practice for over 60 years. The first truly effective antituberculosis drug was streptomycin, reported in 1943 (Schatz et al. 2005). The British Medical Research Council conducted the first large-scale clinical trial of streptomycin in 1948, considered the world's first published drug trial involving random selection of participants. Although many patients were cured during this clinical trial, a substantial proportion experienced relapses, and isolated TB bacilli from these patients showed resistance to streptomycin. At the same time, two new antituberculosis agents, thiacetazone and para-aminosalicylic acid, were introduced, which improved cure rates when combined with streptomycin and reduced antibiotic resistance (Lehmann 1946; British Medical Journal 1950). In 1951, at Sea View Hospital in New York, the clinical trial of the drug isoniazid (INH) was tested and documented a dramatic improvement in clinical outcomes, leading to its widespread use (Robitzek and Selikoff 1952). This was followed by the development of pyrazinamide, cycloserine, ethionamide, rifampin (RF), and ethambutol (Yeager et al. 1952; Hyde 1972; Takayama et al. 1972; Paramasivan et al. 1993; Mitchison 2000; Chakraborty and Rhee 2015; Sotgiu et al. 2015; WHO 2023). Rifampin revolutionized TB treatment due to its high efficacy and ease of administration; however, the introduction of each new drug also led to the emergence of resistant strains (Barberis et al. 2017; Mousavi-Sagharchi et al. 2025). Through multi-country clinical trials led by the British Medical Research Council, a four-drug regimen became recommended for newly diagnosed TB patients. This regimen typically includes isoniazid and rifampin as the backbone, administered for 6–8 months, marking the birth of short-course chemotherapy (Dartois and Rubin 2022). Despite these substantial efforts and progress in treatment protocols, tuberculosis remains among the main challenges that highlight the constant need for attention and innovation in therapeutic strategies.

1.3 The Present State of Tuberculosis

Tuberculosis (TB) has re-emerged as the world's deadliest infectious disease, surpassing coronavirus disease-2019 (COVID-19) in 2023 as the leading cause of infectious mortality. This resurgence is reflected in the World Health Organization's (WHO) report of 8.2 million new TB cases globally in 2023, the highest number recorded since systematic surveillance began in 1995 (WHO 2024; John et al. 2025). The recent rise in TB cases can be partly attributed to the disruptions caused by the COVID-19 pandemic, which severely affected TB control programs worldwide (Daneshmandi et al. 2022).

In 2020, many countries experienced a sharp decline in TB case notifications due to overwhelmed healthcare systems and reduced access to diagnostic and treatment services, with only a partial recovery observed in 2021 (Falzon et al. 2023). This interruption has significantly hindered progress in combating TB, resulting in a global incidence rate decline of only 8.3% between 2015 and 2023, far below the ambitious 50% reduction target set by the WHO End TB Strategy for 2025 (Naidoo and Perumal 2023). Moreover, the pandemic has exacerbated existing disparities in TB detection and care, as low-income countries consistently report higher case notification rates, highlighting persistent inequities in healthcare access and disease surveillance (Koura and Harries 2024). Geographically, the TB burden remains heavily concentrated in the South-East Asia region, which accounts for more than 45% of the global annual incidence. This regional concentration contributes significantly to the global TB epidemic, which continues to grow despite ongoing control efforts. India, in particular, remains the country most affected by TB, responsible for approximately 27% of all global cases, underscoring the need for intensified national and regional interventions (Anirvan 2024). Compounding the challenge of controlling TB is the growing threat of drug-resistant strains. In 2018, rifampicin-resistant (RR) or multidrug-resistant (MDR) TB was identified in nearly 5% of all TB cases worldwide, complicating treatment regimens and threatening to reverse progress made in TB control (MacNeil et al. 2020). Although some advances have been made, such as expanded TB preventive treatment among people living with HIV, these efforts remain insufficient. Achieving global targets for reducing TB incidence and mortality will require more coordinated and intensified actions (MacNeil et al. 2020). Encouragingly, despite the increase in TB cases, mortality rates have shown a modest decline, from 1.32 million deaths in 2022 to 1.25 million in 2023 (WHO 2024). However, this slight improvement should not overshadow the urgent need for sustained global commitment. A further complexity in TB control arises from the increasing recognition of subclinical tuberculosis. Subclinical TB refers to infections detectable by radiographic or microbiological methods but presenting with minimal or no symptoms (Kendall et al. 2021). Individuals with subclinical TB often remain undiagnosed and untreated for extended periods, contributing silently to ongoing transmission within communities. This hidden reservoir of infection complicates efforts to interrupt TB transmission and calls for enhanced diagnostic capabilities. Additionally, the clinical profile of TB patients is evolving, with a rising proportion of extrapulmonary TB cases, older patients, and drug-resistant infections being reported (Vaisman et al. 2020). Unfortunately, current diagnostic tools and treatment protocols are limited in their ability to identify and manage subclinical TB effectively, emphasizing the urgent need for further research and a multidisciplinary approach to address this challenge (Mahmoudi et al. 2024). Thereby, our understanding of TB infection has also advanced beyond the traditional binary classification of latent infection versus active disease. It is now recognized as a dynamic continuum that includes stages such as uninfected, infected, incipient TB, subclinical TB, and symptomatic disease (Migliori et al. 2021). This more nuanced framework has profound implications for TB prevention, diagnosis, and treatment strategies. It highlights the importance of developing more sensitive diagnostic methods and tailored interventions that can detect and treat TB earlier in its disease spectrum, thereby reducing transmission and improving patient outcomes. Ongoing efforts are focused on developing new vaccines, improving diagnostic technologies, and ensuring fair and widespread access to healthcare, particularly for vulnerable and underserved populations (WHO 2024). Despite these advances, tuberculosis remains a persistent and serious global health threat. Recent data emphasize the urgent need for sustained and coordinated action to address major challenges such as drug resistance, disruptions in healthcare services, social inequalities, and the hidden burden of subclinical TB. Only by implementing inclusive, comprehensive, and long-term strategies can the global community hope to effectively control the TB epidemic and prevent further loss of life.

1.4 The Challenge of Drug Resistance in Tuberculosis

Today, based on drug-resistant patterns, TB is classified into major categories that signify the increasing level of resistance to standard treatment regimens, i.e., rifampicin-resistant TB (RR-TB), multidrug-resistant TB (MDR-TB), pre-extensively drug-resistant TB (pre-XDR-TB), extensively drug-resistant TB (XDR-TB), and totally drug-resistant TB (TDR-TB) (CDC 1990, 1991; WHO 1996; Ritacco et al. 1997; Masjedi et al. 2006; Migliori et al. 2008; Velayati et al. 2009, 2013, 2018; Shah et al. 2017; Eldholm et al. 2014; Dean et al. 2022; Menzies et al. 2023; Birhanu et al. 2024; Paul et al. 2025). By convention, multidrug resistance (MDR) is defined as resistance to at least isoniazid (INH) and rifampin (RF). Pre-extensively drug-resistant (pre-XDR) is an MDR-TB strain that is also resistant to either fluoroquinolones or second-line injectable drugs. XDR-TB is TB caused by MDR-TB strains resistant to fluoroquinolone and group A drugs (levofloxacin or moxifloxacin, bedaquiline [BDQ], and linezolid [LZD]). TDR-TB is a strain that does not respond to any available drugs.

The term TDR-TB, described by Velayati et al., has gained the attention of international scientific communication (Velayati et al. 2009). Later, he documented that the term TDR was not chosen solely because of the bacilli's resistance to all available drugs, but also due to complex changes occurring at the genomic and proteomic levels. These changes alter their morphology and enable them to evade the immune system and/or resist any combination of antibiotics (Velayati et al. 2009, 2013; Farnia et al. 2018). Similar dangerous types of strains have also been reported from South Africa and India (Shah et al. 2017; Udwadia et al. 2012). To control the level of resistant cases due to the mismanagement of MDR-TB patients, the WHO effectively started broadening access to MDR-TB care. The key elements of the new WHO Guidelines on MDR-TB treatment include: (I) Second-line treatment is recommended for all RR-TB patients, regardless of whether isoniazid resistance of the mycobacterial isolate is confirmed. (II) A shorter MDR-TB treatment regimen is conditionally recommended for MDR/RR-TB patients, under specific eligibility criteria. (III) Recommendations for the treatment of children with MDR/RR-TB are based on the first-ever individual pediatric patient data meta-analysis on treatment outcomes. (IV) MDR-TB medicines are now differently regrouped, based on updated information about their efficacy and safety. In particular, clofazimine (CFZ) and linezolid are now recommended as core second-line MDR-TB drugs, while p-aminosalicylic acid has been moved to the list of add-on agents. Clarithromycin and other macrolides are no longer included for the treatment of MDR/RR-TB. Delamanid (DLM) may also be used in patients aged 6–17 years. (V) An evidence-informed recommendation on partial resection lung surgery is now included (WHO 2016, 2019, 2020; Dalcolmo et al. 2017; Conradie et al. 2020; Espinosa-Pereiro et al. 2022). Importantly, the recommendation on the "shorter" MDR-TB treatment regimen, which includes bedaquiline, pretomanid, linezolid, and moxifloxacin, referred to as the BPaLM regimen (Lee et al. 2012; WHO 2016, 2019, 2020; James et al. 2024; Jain et al. 2025), has shown significantly improved success rates compared to previous standard regimens (WHO 2016; Sweileh et al. 2016). For patients with pre-XDR-TB, an individualized treatment plan is recommended, often lasting 18 months (WHO 2016; Mase and Chorba 2019; Conradie et al. 2022; WHO 2022a, b; Otto-Knapp et al. 2024). This may include the non-modifiable drug combination of bedaquiline, pretomanid, and linezolid (BPaL) in selected cases where all prerequisites are met (Tweed et al. 2019; Conradie et al. 2022; Patra et al. 2025). In the case of XDR-TB, treatment is more complex due to resistance to most available drugs (Tweed et al. 2019; Conradie et al. 2022; D'Silva et al. 2025). A standardized regimen may include at least four new drugs that have never been used before, ensuring a combination of core drugs with bactericidal and sterilizing activity (Sotgiu et al. 2015; Rendon et al. 2016; Caminero et al. 2017; Falzon et al. 2017). Commonly used drugs in this context include linezolid, bedaquiline, clofazimine, and carbapenems like meropenem-clavulanate (Dooley et al. 2012; WHO 2013; Mase and Chorba 2019; Gao and Zhang 2021; Conradie et al. 2022; WHO 2022a, b; Otto-Knapp et al. 2024). Treatment duration for XDR-TB can extend significantly beyond 18 months, depending on individual patient factors and drug susceptibility (Gandhi et al. 2006; Migliori et al. 2007; WHO 2016; Keshavjee and Farmer 2012; Horne et al. 2025). For TDR-TB, which is resistant to all known anti-TB drugs, treatment options are extremely limited and often require experimental approaches or compassionate use of novel agents under clinical trials (Cegielski 2010; Parida et al. 2015; Caminero et al. 2010). The management of TDR-TB typically involves a highly individualized regimen based on available drugs and patient-specific factors (Keshavjee and Farmer 2012; Gandhi et al. 2006; Migliori et al. 2007). Therefore, it is evident that ongoing research and development are essential to address the various types of drug resistance. This effort is necessary to improve therapeutic options, enhance treatment outcomes, and elevate patient care on a global scale.

1.5 Population Dynamics and Distribution Patterns in Tuberculosis

The World Health Organization (WHO) classifies tuberculosis (TB) incidence rates into three broad categories based on endemicity: low, middle, and high TB burden regions (Dye et al. 2009; Lönnroth and Raviglione 2016; Lei et al. 2023; WHO 2024). This stratification reflects the varying epidemiological landscapes and public health challenges faced by countries worldwide in controlling TB. In low-endemic countries, primarily located in Western Europe and North America, TB incidence rates remain comparatively low relative to global averages. Despite this, these regions continue to be vulnerable to TB outbreaks, mainly due to the presence of sizable immigrant populations originating from high-burden countries and the prevalence of individuals with compromised immune systems, such as those living with HIV/AIDS or other immunosuppressive conditions. For example, in 2023, the United States documented approximately 8000 tuberculosis cases, underscoring its continued presence as a manageable yet ongoing public health concern (Williams 2024; WHO 2024; European Centre for Disease Prevention and Control 2024). To maintain these low incidence rates, comprehensive screening initiatives focusing on high-risk populations, such as immigrants, homeless individuals, and those with compromised immune systems, have been established, complemented by targeted strategies to promote early diagnosis and ensure treatment adherence (Collin et al. 2019; Chakaya et al. 2021; WHO 2024). Nonetheless, significant challenges remain, particularly regarding the timely diagnosis of latent TB infections and ensuring adherence to prolonged treatment regimens, which are critical to preventing disease progression and transmission (Gandhi et al. 2006; Jha et al. 2006; Osei-Wusu et al. 2025a, 2025b; Sazali et al.

2023; Taylor et al. 2025). Middle-endemic countries face more complex challenges, where socioeconomic determinants such as poverty, malnutrition, overcrowding, and inadequate healthcare infrastructure exacerbate the TB epidemic. These factors hinder effective diagnosis, detection, and treatment, contributing to sustained transmission and poor outcomes (Bain et al. 2013; Bhutta et al. 2014; WHO 2024). India and South Africa exemplify this category, with India bearing the highest global TB burden, accounting for nearly one-quarter of all cases worldwide (Bain et al. 2013; Bhutta et al. 2014; WHO 2014a, b, c, 2024). In South Africa, the TB epidemic is further complicated by the high prevalence of HIV co-infection, with approximately 60% of people living with HIV also harboring latent or active TB infection (Abdool Karim et al. 2010; Wouters et al. 2010; Olivier and Luies 2023; WHO 2024). In response, the South African government has adopted integrated care models that combine HIV treatment with TB management, aiming to address the dual burden more effectively (Kalonji and Mahomed 2019; Letang et al. 2020; Wouters et al. 2010; WHO 2024). Despite these advances, persistent barriers such as underfunding of TB programs, social stigma, and limited access to healthcare services continue to impede progress in middle-endemic countries. High-endemic countries, including Nigeria and Indonesia, bear a disproportionate share of the global TB burden. Nigeria alone reported an estimated 450,000 new TB cases in 2023, reflecting the magnitude of the challenge in resource-limited settings (Ugwu et al. 2021; Olaleye et al. 2023; WHO 2024). These nations face multifaceted obstacles, including insufficient healthcare infrastructure, limited availability of diagnostic tools and effective treatment options, and pervasive poverty, all of which contribute to the persistence and spread of TB. In such contexts, community-based interventions have emerged as vital strategies to enhance case detection, improve treatment adherence, and reduce transmission (WHO 2024). Approaches often involve mobilizing local health workers, leveraging community networks, and employing culturally sensitive education campaigns to overcome barriers to care. Overall, the WHO has identified 30 high-burden countries that collectively account for approximately 87% of the global TB burden. Among these, the top five countries, India, Indonesia, China, the Philippines, and Pakistan, represent 56% of all new TB cases reported worldwide (WHO 2024). In Sub-Saharan Africa, countries such as Angola, the Democratic Republic of Congo, Mozambique, Nigeria, South Africa, Somalia, Zambia, and Zimbabwe face severe TB epidemics. These regions confront unique challenges due to the high prevalence of HIV/AIDS co-infection coupled with multidrug-resistant TB (MDR-TB), which complicates disease management and worsens outcomes (Gisso et al. 2022; Obeagu 2023; Yirsaw et al. 2025). In the former Soviet republics, TB remains a significant public health concern, with countries including Ukraine and Kazakhstan exhibiting high burdens of TB and MDR-TB (Dohál et al. 2024; WHO 2024). The ongoing conflict in Ukraine has further exacerbated healthcare challenges, disrupted TB control programs, and hindered access to diagnosis and treatment, thereby increasing the risk of transmission and drug resistance. In most settings, high TB incidence correlates strongly with socioeconomic determinants such as poverty, malnutrition, and weak healthcare systems. The concentration of TB cases in specific geographic and demographic pockets underscores the urgent need for targeted, context-specific interventions. The WHO emphasizes that successful TB prevention, diagnosis, and treatment efforts require sustained and consistent financial investment, alongside political commitment and multisectoral collaboration. Recent outreach initiatives and mobile health clinics have demonstrated efficacy in reaching marginalized and hard-to-reach populations, raising awareness about TB prevention, facilitating early diagnosis, and improving access to treatment (Denkinger et al. 2014; Tumuhimbise et al. 2023; Byrne et al. 2024). These community-centered programs have shown great promise in enhancing treatment adherence and outcomes, particularly in underserved areas where traditional healthcare infrastructure is limited. Overall, the global burden of tuberculosis is unevenly distributed, with low-, middle-, and high-endemic regions facing distinct epidemiological profiles and challenges. Addressing these requires tailored strategies that consider local social, economic, and health system contexts to reduce TB incidence and mortality worldwide effectively.

1.6 Gender and Age-Related Differences in Tuberculosis Incidence and Outcomes

Age and gender are critical determinants shaping the epidemiology, clinical presentation, and outcomes of tuberculosis (TB) worldwide. Globally, TB affects individuals across all ages and genders; yet significant disparities exist that influence disease burden and treatment success. Approximately 56% of TB infections occur in males, a disparity especially pronounced in high-burden countries such as Nigeria, South Africa, and India (WHO 2024). For instance, in India, out of 2.69 million TB cases reported in 2022, around 70% were male (WHO 2024). Similar gender imbalances have been observed in other high-burden countries, where cultural norms and behaviors contribute to higher tuberculosis infection rates among men (Williams 2024). Several factors underlie this gender gap. Biological differences, particularly hormonal variations, play a significant role; estrogen, in particular, has been suggested to exert protective effects against TB infection, which may partly explain the lower incidence observed in females (Nhamoyebonde and Leslie 2014; Ortona et al. 2019; Gupta et al. 2022; Sciarra et al. 2023; Oshi et al. 2024). Estrogen, especially estradiol, influences immune responses by modulating cellular pathways involved in Mtb infection. Studies have demonstrated that estradiol binds to estrogen receptor alpha (ERα) expressed in bronchial epithelial and immune cells, potentially inhibiting bacterial proliferation and enhancing host defense mechanisms. This hormone-driven modulation may contribute to the stronger immunity observed in females compared to males (Gan et al. 2022; Gupta et al. 2022). Moreover, estradiol impacts the balance of pro- and anti-inflammatory cytokines, promoting Th1 immune responses critical for containing Mtb at basal concentrations, which are typical in nonpregnant females. This hormonal effect may enhance macrophage activation and the production of interferon-gamma (IFN-γ) and tumor necrosis factor-alpha (TNF-α), key mediators in TB immunity (Harding and Heaton 2022). While estrogen appears protective, its role is complex and dose-dependent, with some studies indicating that elevated estradiol levels during certain physiological states might reduce protective cell-mediated immunity and exacerbate inflammation, potentially contributing to disease progression in specific contexts (Rey et al. 2017; Dong et al. 2023). Nonetheless, estrogen and estrogen receptor modulators show promise as adjunctive therapies in tuberculosis, warranting further investigation through animal and clinical studies (Garg et al. 2019). Besides biological factors, socioeconomic and behavioral factors, such as higher rates of smoking and alcohol consumption among men, can weaken immune defenses, increasing susceptibility to TB (Narasimhan et al. 2013; Ogbudebe et al. 2023; Osei-Wusu et al. 2025a, 2025b). Access to healthcare also varies by gender and region; while men may have easier access in some cultures, this is not universal and can differ widely by area (Onifade et al. 2010; Krishnan et al. 2014; WHO 2024). Gender disparities extend beyond incidence to treatment outcomes. Women often experience delayed diagnosis due to social and structural barriers (WHO 2021a, b), and even after diagnosis, face challenges in completing treatment. Factors such as stigma, caregiving responsibilities, and lack of familial or healthcare support disproportionately affect women's adherence to TB treatment (WHO 2021a, b, 2024). The social stigma surrounding TB is particularly detrimental to women, deterring timely healthcare seeking and adherence to therapy (Brüggemann et al. 2025). These findings align with studies demonstrating that women may have longer delays before diagnosis and more frequent encounters with lower-quality healthcare providers (Sun et al. 2023). Age is another fundamental factor influencing TB incidence and outcomes. The highest burden is observed in adults aged 15–54 years, but children and older adults also face significant risks (Snow et al. 2018; Vanden Driessche et al. 2013; Carvalho et al. 2018; Laycock et al. 2021; Bai and Ameyaw 2024; WHO 2024). In 2023, approximately 6.0 million TB cases occurred among adults aged 15 and older, while about 1.25 million cases were reported in children aged 0–14 years (WHO 2024). Older adults face increased vulnerability due to immune senescence and comorbidities, which complicate treatment and worsen outcomes (Cruz et al. 2013; Snow et al. 2018; Laycock et al. 2021; Teo et al. 2023; Bai and Ameyaw 2024; WHO 2024). Younger adults aged 15–24 years are particularly susceptible due to increased exposure risks linked to urbanization and social behaviors (Laycock et al. 2021; Bai and Ameyaw 2024; WHO 2022a, b, 2024). Children under 5 years old are at heightened risk for severe TB forms due to immature immune systems (Vanden Driessche et al. 2013; Carvalho et al. 2018; WHO 2022a, b, 2024). In high-burden countries like Nigeria and India, pediatric TB remains underdiagnosed and undertreated, with the WHO estimating around one million children developing TB globally in recent years (Chauhan et al. 2023; WHO 2024). Older adults, increasingly prevalent in many high-burden countries, face unique challenges. Age-related immune decline, comorbidities, mobility issues, and cognitive impairments complicate treatment adherence and outcomes (Negin et al. 2015; Byng-Maddick and Noursadeghi 2016; Ku and Dodd 2019; Olmo-Fontánez and Turner 2022; Teo et al. 2023; Sakthivadivel et al. 2023; Teibo et al. 2024; van Arkel et al. 2024; López-Medrano et al. 2025). These vulnerabilities necessitate tailored approaches for elderly TB patients. Statistical data from highly endemic countries further illustrate these disparities. In India, men accounted for over 70% of new TB infections in 2022, with the highest incidence in the 25–34 age group (WHO 2024). Approximately 10% of cases were children under five (Chauhan et al. 2023; WHO 2024). Nigeria reports one of the highest TB incidence rates globally, with over 219 cases per 100,000 population, males comprising about 60% of reported cases, and most infections occurring between the ages of 15 and 44. Pediatric TB cases are often underreported but are estimated to represent 20% of infections (Ogunniyi et al. 2024; WHO 2024; Chen et al. 2025a, b). South Africa faces a severe TB-HIV co-epidemic, with an incidence rate of approximately 520 per 100,000 people and males constituting nearly two-thirds of cases, predominantly in the 25–34 age group (Kroon et al. 2020; WHO 2024; Akokuwebe et al. 2024). To effectively address these age- and gender-based inequities, comprehensive, population-specific strategies are essential. Community engagement programs aimed at raising awareness of TB prevention and treatment, tailored public health campaigns addressing gender-specific risks, and efforts to improve healthcare access for vulnerable groups, particularly women and children facing stigma and socioeconomic barriers, are critical (Peer et al. 2023; Madaki et al. 2024; WHO 2024).

Integrating TB services with broader healthcare initiatives, such as maternal health and HIV/AIDS programs, can optimize resources and address comorbidities prevalent among high-risk populations (Naidoo et al. 2019; Bulstra et al. 2021; WHO 2024; Maseko et al. 2025). Furthermore, increased investment in research is necessary to develop novel diagnostics, vaccines, and therapeutics to combat emerging challenges, including drug resistance. International collaboration among governments, nongovernmental organizations (NGOs), researchers, and pharmaceutical companies remains vital for effectively sharing knowledge and resources to tackle global TB challenges. Robust surveillance systems are also essential to monitor progress toward TB elimination by ensuring timely and accurate data collection on incidence and treatment outcomes. Overall, significant disparities in TB incidence and outcomes exist across age and gender groups globally. While men generally exhibit higher infection rates, children under five remain particularly vulnerable due to immature immune systems, and older adults face unique challenges related to comorbidities and treatment adherence (Humayun et al. 2022). Addressing these disparities requires multifaceted, tailored approaches that incorporate community engagement, healthcare integration, and sustained research and surveillance efforts to advance global TB control and elimination goals.

1.7 Current Barriers and Future Directions in TB Control

Global efforts to eradicate tuberculosis (TB) face substantial and complex challenges that hinder effective control of this infectious disease. Despite significant advancements in medical science, public health initiatives, and growing awareness, tuberculosis remains a leading cause of mortality and morbidity worldwide, with a disproportionate impact on low- and middle-income countries (Hargreaves et al. 2011; Tanimura et al. 2014; Pradipta et al. 2018; Yang et al. 2025; Ledesma et al. 2024; WHO 2024). The principal challenges and barriers to TB control can be broadly categorized into four domains: biological, healthcare system, socioeconomic, and operational challenges (Hargreaves et al. 2011; Tanimura et al. 2014; Pradipta et al. 2018; WHO 2024). A comprehensive and detailed understanding of the critical aspects of TB is essential for developing effective strategies that address the disease in a holistic manner and improve global public health outcomes.

Among the biological challenges, the emergence and dissemination of multidrug-resistant tuberculosis (MDR-TB), extensively drug-resistant tuberculosis (XDR-TB), and totally drug-resistant tuberculosis (TDR-TB) constitute significant obstacles to effective treatment and disease management (Zumla et al. 2012; Seung et al. 2015; WHO 2024). MDR-TB, in particular, represents a primary global health concern, with approximately 450,000–500,000 new cases reported annually (WHO 2024). The management of drug-resistant TB (DR-TB) requires more complex and prolonged treatment regimens, which are often less effective and associated with more severe adverse effects compared to standard therapies (Kurz et al. 2016; Mancuso et al. 2023; WHO 2024). In many high-burden countries, persistent shortages of second-line anti-TB medications, combined with inadequate treatment infrastructure, further exacerbate challenges in disease control and contribute to the sustained global TB burden. For instance, in South Africa, elevated MDR-TB rates are influenced by factors such as widespread HIV co-infection and limited access to healthcare services (Kroon et al. 2020; Akokuwebe et al. 2024; WHO 2024). Furthermore, the treatment of patients with XDR-TB and TDR-TB is complicated by prolonged delays in treatment initiation, which significantly increase the risk of mortality (Nyasulu et al. 2024; Barteka et al. 2025). Another critical biological challenge lies in the accurate and rapid diagnosis of latent tuberculosis infection (LTBI), especially among individuals co-infected with HIV. This population presents unique diagnostic difficulties due to immunosuppression, which often leads to atypical clinical presentations and reduced sensitivity of conventional diagnostic tests (Hamada et al. 2021; Kapata et al. 2025). Despite advances in diagnostic technologies, detection rates of latent TB in HIV-positive individuals remain suboptimal, hindering the timely initiation of preventive therapy and increasing the risk of progression to active disease. The limitations of current diagnostic methods, including tuberculin skin tests and interferon-gamma release assays, are further compounded by logistical challenges in resource-limited settings where the burden of co-infection is highest. Addressing these diagnostic gaps is essential to improve case detection, reduce transmission, and ultimately control the global TB epidemic, particularly in high-risk populations. In this context, Velayati et al. (2024) identified unique and consistent frameshift mutations in the genes *fbpC1*, *PPE35*, *ald*, and *whiB6* in dormant clinical TB strains cultured in vitro. These mutations were absent during the exponential phase of the same strains. The findings offer promising opportunities for developing molecular diagnostic tests capable of distinguishing between active and latent TB infections; however, further research is required to validate their clinical utility.

Healthcare system challenges constitute a significant barrier to the effective control of TB on a global scale. These challenges are multifaceted and include fragmented leadership and governance structures, insufficient financial resources, and a shortage of adequately trained healthcare personnel. Additionally, poor infrastructure, limited access to healthcare services, particularly in rural and underserved regions, and a scarcity of diagnostic and treatment facilities contribute substantially to underdiagnosis and delays in care (Oga-Omenka et al. 2020; Teibo et al. 2024; Malla et al. 2025). Patients frequently encounter geographic and transportation obstacles that impede timely access to TB testing and treatment. As a result, symptoms of TB are often overlooked or misdiagnosed until the disease has progressed to an advanced stage, leading to a considerable number of undetected cases that perpetuate ongoing transmission within communities. Moreover, inadequate staffing levels within healthcare facilities compromise both patient safety and the overall quality of care delivered (Marahatta et al. 2020). The private healthcare sector, which plays a substantial role in many high-burden countries, often suffers from insufficient equipment, lack of regulatory oversight, and poor adherence to standardized treatment protocols. These deficiencies result in inconsistent quality of care, underreporting of TB cases, increased out-of-pocket costs for patients, and challenges in accurately monitoring TB prevalence and treatment outcomes (Wang et al. 2023; Derakhshani et al. 2024). To address these systemic issues, sustained and coordinated efforts from governments and relevant stakeholders are imperative. This includes increasing funding allocations, demonstrating strong political commitment, and targeting the underlying determinants of TB transmission. Furthermore, continuous training programs for healthcare workers and the implementation of effective retention strategies are essential to enhance the performance and sustainability of TB control programs. In recent years, the COVID-19 pandemic has significantly exacerbated existing challenges within healthcare systems by causing widespread disruption to essential health services globally. This disruption has led to marked declines in tuberculosis (TB) case detection and initiation of treatment (Falzon et al. 2023; Abaynew et al. 2025). The implementation of lockdowns, social distancing measures, and the reallocation of healthcare resources to address the pandemic severely limited the access to diagnostic and treatment services for TB. Consequently, many opportunities for timely diagnosis and intervention were missed, likely contributing to increased TB-related morbidity and mortality (Gulumbe et al. 2025). According to the World Health Organization, there was a 21% reduction in the number of individuals receiving TB treatment in 2020 compared to 2019, reflecting a sharp decrease in case notifications, particularly in countries most severely impacted by the pandemic (WHO 2024). This alarming decline has raised serious concerns regarding a hidden reservoir of undiagnosed and untreated TB cases, which may fuel increased transmission and mortality in the coming years (WHO 2024). Pai et al. (2022) highlight the devastating impact of the pandemic on TB care worldwide and call for urgent recovery strategies focused on restoring and strengthening TB services. They emphasize the importance of leveraging innovations and health system reforms catalyzed by the pandemic to build more resilient TB programs capable of withstanding future systemic shocks. Furthermore, the intersection of COVID-19 and TB has introduced additional clinical complexities (Cioboata et al. 2023). For example, studies from Indonesia, a country with a high TB burden, have documented significant rates of TB co-infection among COVID-19 patients. These co-infections present with clinical manifestations that complicate diagnosis and management, resulting in increased mortality (Muflihah et al. 2024). This dual burden exacerbates disease severity and underscores the critical need for integrated screening and treatment approaches within healthcare settings to mitigate adverse outcomes effectively.

Socioeconomic determinants play a crucial role in shaping the incidence and outcomes of tuberculosis (TB), particularly in low- and middle-income countries. Key factors such as poverty, inadequate housing conditions, malnutrition, low levels of education, unemployment, and overcrowding are strongly associated with an increased risk of TB infection and disease progression (Nalugwa et al. 2025; Gulumbe et al. 2025; Zumla et al. 2025). According to WHO estimates, around 90% of all new tuberculosis cases occur in low- and middle-income countries, where poor living conditions play a significant role in facilitating disease transmission. Overcrowding, especially in urban areas with limited housing, promotes the spread of Mtb by increasing close contact among individuals (WHO 2024). Poor ventilation in homes exacerbates this risk by allowing airborne pathogens to persist longer in enclosed spaces (Gulumbe et al. 2025; Zumla et al. 2025). For example, densely populated slums in Mumbai have been identified as significant TB hotspots due to these factors (Marimuthu 2016; Singal et al. 2025). Malnutrition critically undermines immune function, increasing susceptibility to TB. Numerous studies consistently demonstrate that individuals with a body mass index (BMI) below 18.5 are at a significantly increased risk of developing active tuberculosis compared to those with a normal BMI range. Food insecurity remains a pressing challenge in many high-burden countries, exacerbating malnutrition among vulnerable populations (WHO 2013). Access to healthcare is essential for effective TB management, yet barriers such as high treatment costs, limited transportation options, and insufficient healthcare infrastructure hinder timely diagnosis and treatment initiation. The WHO estimates that approximately one-third of individuals with TB globally remain undiagnosed or untreated due to these obstacles (Appiah et al. 2023). Education plays a crucial role in shaping health outcomes related to TB; higher levels of education are associated with improved health literacy, better health-seeking behaviors, and greater adherence to treatment regimens. Conversely, limited education perpetuates cycles of poverty and poor health, restricting individuals' ability to access information on TB prevention and treatment (Akakpo and Neuerer 2024; Chauhan et al. 2024). Social stigma associated with TB often leads to isolation, particularly among vulnerable groups such as women and children, who may face additional barriers in accessing healthcare services. For example, women with TB in India frequently experience stigmatization from both family and community members, discouraging them from seeking treatment. While various countries have implemented public health programs to reduce stigma, the effectiveness of these interventions varies widely (Medina-Marino et al. 2025). Underdiagnosis and delays in treatment initiation remain significant challenges in TB control. At the individual level, factors such as alcohol addiction, HIV co-infection, and prior TB contact are directly linked to increased TB risk, while at the population level, socioeconomic indicators such as gross domestic product (GDP) per capita and human development index exhibit inverse relationships with TB incidence (Bayigga et al. 2025). Employment status also influences TB rates and treatment success, as unemployment and job instability often lead to financial hardships that impede access to healthcare and adherence to treatment (Ting et al. 2024). Addressing TB, therefore, requires multisectoral collaboration beyond healthcare, involving policymakers, housing authorities, educational institutions, and food security programs. Employment initiatives that foster stable livelihoods contribute to improved health outcomes and support TB control efforts. Considering socioeconomic determinants is vital for shaping sustainable and effective long-term strategies to combat TB globally (WHO 2015, 2024). Through comprehensive public health approaches aimed at improving living conditions, such as enhancing food security, expanding healthcare access, and promoting education and community engagement, significant progress can be made toward reducing the TB burden. These efforts contribute to healthier societies free from preventable diseases like tuberculosis, enabling collective advancement toward shared health goals (Kumah 2025). Ensuring equitable access to resources regardless of background or social challenges is fundamental to this endeavor. Collaborative action is essential to confront the persistent global challenge posed by TB, which continues to affect millions worldwide despite medical advancements.

Operational challenges further complicate TB control by impacting the delivery of effective care. Logistical inefficiencies, including delays in the supply of essential drugs and diagnostic tools, hinder timely treatment initiation and adversely affect patient outcomes. A critical barrier to progress is the insufficient funding allocated to TB research and control programs. Although TB is a preventable and treatable disease, funding levels remain inadequate compared to other global health priorities such as HIV/AIDS and malaria (Getnet et al. 2021; WHO 2024; Zumla et al. 2025). The Global Fund has highlighted that limited financial resources constrain the expansion of diagnostic and treatment services (Global Fund Report 2023). In many low-income countries, national health budgets fall short of covering the costs required for comprehensive TB prevention and treatment. For example, while India has increased its budget for TB control, the available funds remain insufficient to address the disease nationwide effectively. Consequently, overcoming the determinants of TB control necessitates a multifaceted approach that extends beyond the healthcare sector. The Sustainable Development Goals (SDGs) framework offers valuable opportunities for coordinated action addressing the socioeconomic and related factors contributing to the TB burden in high-prevalence countries (Satyanarayana et al. 2020; Ferreira et al. 2023). Successful interventions depend on strong political commitment, continuous healthcare provider training, and active community engagement, underscoring the importance of comprehensive, integrated strategies to advance global TB control and ultimately achieve eradication.

1.8 Advances in Technology and Their Impact on Tuberculosis Control Efforts

Over the past two decades, technological advancements have profoundly transformed the fight against tuberculosis (TB), revolutionizing rapid diagnosis, treatment, and prevention strategies. These innovations have significantly enhanced the detection and control of TB, particularly in countries where the disease continues to pose substantial public health challenges. A breakthrough in tuberculosis management has been the development and widespread implementation of rapid molecular diagnostic tests. These tests significantly reduce turnaround time, enabling earlier detection of Mtb and drug resistance mutations with high sensitivity and specificity, which allows clinicians to initiate appropriate treatment or preventive measures promptly (Nandlal et al. 2022; Chauke et al. 2024; Yayan et al. 2024; Scott et al. 2025). Traditional diagnostic methods, such as sputum smear microscopy and culture examination, often require several weeks to yield results, which delays treatment initiation and increases the risk of ongoing transmission. In contrast, molecular assays like the GeneXpert system, introduced in 2010, have revolutionized TB diagnosis by providing results within a few hours (Williamson et al. 2012; Lawn et al. 2013; WHO 2014a, b, c; Steingart et al. 2014; López-Roa et al. 2021; Gota et al. 2023). The GeneXpert MTB/RIF and GeneXpert Ultra assays (Cepheid, Sunnyvale, CA, USA) detect Mtb infection and rifampin resistance within 1–2 h, facilitating the rapid identification of patients with presumptive multidrug-resistant TB (MDR-TB) (WHO 2014a, b, c, 2020). The deployment of these platforms has demonstrated significant improvements in intermediate clinical outcomes, including reduced time to diagnosis and timely initiation of effective therapy. Complementing these molecular diagnostics are line probe assays (LPAs), such as GenoType MTBDRplus and MTBDRsl (Hain Lifescience GmbH, Nehren, Germany), which provide rapid detection, within 4–5 h, of specific resistance-conferring mutations for selected first- and second-line anti-TB drugs, thereby aiding in the optimization of drug-resistant TB treatment regimens. However, the utility of these genotypic assays in guiding appropriate drug-resistant TB treatment remains limited (Barnard et al. 2012; WHO 2016; Lin et al. 2022). Notably, none of these assays currently detect resistance to newer or repurposed anti-TB drugs such as bedaquiline (BDQ), linezolid (LZD), delamanid (DLM), and clofazimine

(CFZ), which are integral components of the World Health Organization's updated hierarchical classification of drug-resistant TB therapeutics. Although Xpert MTB/RIF and line probe assays (LPAs) significantly reduce diagnostic turnaround times by bypassing culture, their limitation lies in analyzing only a limited number of drugs and gene regions. As a result, a critical gap remains, since no single assay currently provides comprehensive resistance detection across all first-line, second-line, new, and repurposed anti-tuberculosis drugs simultaneously. Next-generation sequencing (NGS), particularly whole genome sequencing (WGS), offers the most comprehensive molecular-based approach for drug susceptibility testing (Walker et al. 2015; Iketleng et al. 2018; Satta et al. 2018; Dookie et al. 2022; Morey-León et al. 2025). WGS enables interrogation of the entire Mtb genome, identifying all mutations potentially conferring drug resistance and thereby facilitating individualized treatment regimens (Witney et al. 2015; Cox et al. 2022). In well-resourced settings, WGS has become the standard of care (Dookie et al. 2022; Arora et al. 2023). The ability of WGS to accurately detect mixed infections, variable drug susceptibility within bacterial populations, transmission dynamics, outbreaks, and individuals with high transmission potential provides significant clinical and public health advantages in tuberculosis management (Ghanavi and Farnia 2025). By offering detailed insights into the genetic diversity and evolution of Mtb strains, WGS enables more precise diagnosis, effective treatment strategies, and improved outbreak control measures, ultimately contributing to better patient outcomes and more efficient disease surveillance. This capability stems from its high-resolution analysis of transmission clusters, identification of drug resistance mutations, and differentiation between relapse and reinfection, thereby enhancing patient care and outbreak management. WGS has become an increasingly affordable tool, with overall costs comparable to other diagnostic tests currently employed for detecting drug-resistant tuberculosis (DR-TB) and conducting cluster analyses. This comprehensive approach offers an "all-in-one" solution, providing detailed information on expected strain drug sensitivity, genetic background, epidemiological data, and potential indicators of laboratory cross-contamination risk (Cirillo et al. 2016; Dookie et al. 2022; Arora et al. 2023; Chekesa et al. 2024). The World Health Organization (WHO) has recommended a set of mutation sites for predicting resistance to common anti-TB drugs using WGS. However, the predictive accuracy and clinical utility of these mutated loci require further validation through extensive studies involving diverse populations and geographic regions. In addition to laboratory-based innovations, advances in digital health technologies have substantially contributed to improving tuberculosis control efforts (Ngwatu et al. 2018; Lee et al. 2023). Beyond diagnostics, new technologies have enhanced data collection and monitoring systems critical for TB management. Electronic health records (EHRs) enable healthcare providers to efficiently track patient progress and facilitate seamless data sharing across healthcare facilities, thus improving continuity and coordination of care (Calderwood et al. 2010; Ratwani 2017; Quinn et al. 2019; Vonnahme et al. 2023). Enhanced surveillance and monitoring systems empower public health officials to track TB incidence trends better and allocate resources strategically to regions with the greatest need. Moreover, technologies that directly engage patients, such as mobile health (mHealth) applications, have become vital tools for improving treatment adherence and patient engagement. These applications provide medication reminders, educational content about tuberculosis, and facilitate communication between patients and healthcare providers. A notable example is India's "mTB" initiative, which utilizes mobile technology to monitor treatment adherence among patients with drug-resistant tuberculosis. This innovative approach supports healthcare systems by improving adherence rates and consequently reducing the incidence of treatment failure (Sodhi et al. 2024). Significant progress has also been made in applying artificial intelligence (AI) and machine learning algorithms to tuberculosis diagnostics

(Qin et al. 2021; Santosh et al. 2022). For instance, the ULTR-AI suite, an AI-powered lung ultrasound tool, has demonstrated superior diagnostic performance compared to human experts, achieving 93% sensitivity and 81% specificity in detecting pulmonary TB. This performance surpasses WHO benchmarks for non-sputum-based triage tests. The technology enables rapid, sputum-free, and accessible diagnosis, particularly benefiting resource-limited settings by providing real-time interpretation through portable, smartphone-connected devices. Such AI-driven innovations reduce operator dependency, accelerate diagnosis, and help decrease patient drop-out rates, thereby strengthening global TB control efforts (Qin et al. 2021; Santosh et al. 2022; Liang et al. 2022; Geric et al. 2023; McClean et al. 2024; Van Rie et al. 2025). Alongside diagnostic advancements, technological progress has significantly contributed to the development of novel treatments for tuberculosis. Recent innovations in anti-TB therapies have been instrumental in addressing the challenges posed by drug-resistant strains of Mtb. Notably, bedaquiline and delamanid have emerged as groundbreaking treatment options for MDR-TB. These drugs operate through mechanisms distinct from conventional anti-TB medications, providing more effective therapeutic alternatives for patients with resistant infections. The introduction of these novel therapies has markedly improved treatment outcomes and played a vital role in reducing mortality associated with drug-resistant forms of TB (Patil et al. 2018, 2025; Mancuso et al. 2023). Another important advancement in molecular TB research is molecular docking, a computational technique widely employed to predict interactions between small molecules, such as potential drugs or natural compounds, and target proteins of Mtb. Molecular docking is pivotal in structure-based drug design, as it estimates the preferred binding orientation and affinity of ligands within active or allosteric sites of target proteins (Nyijime et al. 2025). This approach accelerates the identification of novel drug candidates, elucidates mechanisms of drug resistance, and guides the optimization of therapeutic agents against TB. The docking process involves two fundamental components: a search algorithm that explores numerous possible conformations and orientations of the ligand within the protein's binding pocket, and a scoring function that evaluates each pose by estimating binding energies to rank ligand-target interactions based on predicted stability and likelihood (Rehman et al. 2025). The primary objective is to determine the binding pose with the lowest energy that most accurately represents the biological interaction occurring in vivo. In TB research, molecular docking has been extensively applied to screen large libraries of natural products, synthetic compounds, and FDA-approved drugs against essential Mtb proteins. For example, docking studies targeting InhA, anenoyl-acyl carrier protein (ACP) reductase critical for mycolic acid synthesis, have led to the discovery of novel inhibitors effective against drug-resistant strains. Similarly, DNA gyrase subunits GyrA and GyrB, key enzymes involved in bacterial DNA replication, have been targeted to identify promising leads against fluoroquinolone-resistant TB (Lone et al. 2018; Baptista et al. 2021). Molecular docking also aids in understanding drug resistance mechanisms by modeling how mutations in target proteins alter ligand binding. For instance, resistance to isoniazid frequently arises from mutations in the katG and inhA genes; docking studies visualize how these alterations reduce drug affinity, thereby guiding the design of molecules capable of overcoming resistance (Dar et al. 2024). Target validation and prioritization substantially benefit from molecular docking techniques. With the complete sequencing of the Mtb genome, molecular docking has become a pivotal tool in validating the therapeutic potential of novel targets. Notable examples include Rv0678, a gene associated with resistance to bedaquiline; PknB, a serine/threonine kinase critical for bacterial cell wall synthesis; and MmpL3, a membrane protein involved in lipid transport and cell wall biogenesis. These targets offer promising avenues for the development of new antitubercular agents (Guo et al. 2022). Docking simulations complement experimental data by providing detailed insights into ligand-target interactions, thereby assisting medicinal chemists in focusing their efforts on the most viable therapeutic targets (Muhammed and Aki-Yalcin 2024). Moreover, molecular docking plays a pivotal role in drug repurposing initiatives, wherein existing drugs approved for other diseases are computationally evaluated against Mtb proteins. For example, antimalarial and antiviral compounds have been screened against targets such as the ClpP1P2 protease and ATP synthase, revealing promising candidates that could expedite the development of new TB treatments (Schmitt et al. 2011; Hanscheid et al. 2024). Docking also supports ligand optimization and structure-activity relationship (SAR) studies by visualizing molecular interactions within binding pockets. This enables medicinal chemists to modify compounds strategically to enhance hydrogen bonding, increase hydrophobic contacts, or reduce steric hindrance, thereby systematically improving drug efficacy.

Frequently, docking is integrated with molecular dynamics (MD) simulations, which refine docking poses and assess the stability of protein-ligand complexes over time under physiological conditions. Additional computational tools such as pharmacophore modeling and quantitative structure-activity relationship (QSAR) analyses further augment predictive accuracy and drug design efficiency (Meng et al. 2011; Mao et al. 2021). Despite its significant utility, molecular docking faces challenges including reliance on accurate protein structures, limitations inherent in scoring functions, and simplified treatment of protein flexibility and solvent effects. Nevertheless, advances in artificial intelligence-driven scoring algorithms, improved computational methods, and integration with high-throughput experimental screening continue to mitigate these limitations (Singh et al. 2024). Recent studies exemplify these technological advances. For instance, a high-throughput virtual screening of approximately 460,000 compounds from the National Cancer Institute (NCI) library against the ATP-binding pocket of PknG, a mycobacterial virulence factor that promotes Mtb survival within macrophages, identified novel inhibitor scaffolds including chromene glycosides and benzene sulfonamides. These compounds show considerable promise for combating drug-resistant TB (Macalino et al. 2020). Similarly, reverse docking of 53 natural products against multiple Mtb druggable targets predicted strong interactions with key enzymes such as *PknB* and *DprE1*, thereby guiding subsequent experimental validation (Baptista et al. 2021; Amado et al. 2022). Parallel to these advances in drug discovery, progress in vaccine development has demonstrated promising potential to enhance tuberculosis control efforts. Although the *Bacillus Calmette-Guérin* (BCG) vaccine has been in use for nearly a century, its protective efficacy against pulmonary TB in adults remains limited (Theorupun et al. 2025). Consequently, researchers are investigating new vaccine candidates through various clinical trials aimed at eliciting stronger immune responses against Mtb. The MVA85A vaccine candidate, for example, has shown potential in boosting immunity in individuals previously vaccinated with BCG (Dunachie et al. 2008; Schaible et al. 2017; Srivastava et al. 2023). If proven effective, such novel vaccines could play a critical role in halting TB infections and reducing transmission rates. Incorporating genomics into tuberculosis research has yielded invaluable insights into the biology of Mtb and its interactions with host immune responses. Collectively, these technological developments have significantly enhanced the overall effectiveness of TB control programs by enabling timely case detection, improving diagnostic accuracy, and supporting patient-centered care models. Continued investment in innovation, alongside equitable implementation strategies, remains essential to overcoming the persistent challenges faced by TB-affected communities worldwide (WHO 2024). Despite significant technological advancements in TB control, numerous challenges persist in effectively implementing these innovations across diverse healthcare settings. In countries with fragile healthcare infrastructures, substantial barriers impede the adoption of advanced diagnostic tools and novel treatment regimens. Additionally, difficulties in training healthcare personnel to proficiently utilize new technologies further hinder their successful integration into routine clinical practice.

In conclusion, tuberculosis remains one of the world's deadliest infectious diseases, necessitating urgent, coordinated global action. The WHO and international scientific partners emphasize that ending TB requires scaling up early detection, accurate diagnosis, and effective treatment, with particular focus on drug-resistant TB. Sustained investment in research and innovation is critical to developing new diagnostics, therapeutics, and vaccines that can accelerate progress toward elimination. The global targets outlined in the WHO's End TB Strategy aim for an 80% reduction in TB incidence and a 90% reduction in TB mortality by 2030. Achieving these ambitious goals demands unwavering political commitment, substantially increased funding—currently, only approximately one-quarter of the estimated US$22 billion annual requirement is available—and robust multisectoral collaboration to address the socioeconomic determinants of TB, including poverty, malnutrition, and overcrowding. The setbacks caused by the COVID-19 pandemic have underscored the urgent necessity of building patient-centered, resilient health systems capable of withstanding future public health emergencies. Therefore, alongside scientific and technological advancements, sustained political will and resource mobilization are essential to reduce the global TB burden effectively. The international community must commit to investing in and delivering equitable, high-quality TB care to interrupt transmission chains and ultimately eradicate this ancient yet preventable disease. The message is unequivocal: no more excuses; urgent, coordinated action is imperative to save lives and realize a TB-free world.

Conflict of Interest The authors declare that they have no conflict of interest related to the content of this chapter.

Ethical Consideration This chapter is a review based entirely on data and information obtained from previously published and publicly available sources, all of which have been appropriately cited. The authors conducted no new experimental work involving human participants or animals. Therefore, no additional ethical approval was required. The authors have ensured proper attribution and compliance with ethical standards in the use of all referenced materials.

References

Abaynew Y, Ali A, Taye G (2025) Exploring the impact of the COVID-19 pandemic on the social determinants of tuberculosis in Addis Ababa, Ethiopia: a qualitative study. Discov Soc Sci Health 5:94

Abdool Karim SS, Naidoo K, Grobler A, Padayatchi N, Baxter C, Gray A, Gengiah T, Nair G, Bamber S, Singh A, Khan M, Pienaar J, El-Sadr W, Friedland G, Abdool KQ (2010) Timing of initiation of antiretroviral drugs during tuberculosis therapy. N Engl J Med 362(8):697–706

Akakpo MG, Neuerer M (2024) The relationship between health literacy and health-seeking behavior amongst university students in Ghana: a cross-sectional study. Health Sci Rep 7(5):e2153

Akokuwebe ME, Osuafor GN, Idemudia ES (2024) Prevalence and association of HIV and tuberculosis status in older adults in South Africa: an urgent need to escalate the scientific and political attention to aging and health. Front Public Health 12:1245553

Alexander KA, Laver PN, Michel AL, Williams M, van Helden PD, Warren RM, Gey van Pittius NC (2010) Novel *Mycobacterium tuberculosis* complex pathogen, *M. mungi*. Emerg Infect Dis 16(8):1296–1299

Amado PSM, Woodley C, Cristiano MLS, O'Neill PM (2022) Recent advances of DprE1 inhibitors against *Mycobacterium tuberculosis*: computational analysis of physicochemical and ADMET properties. ACS Omega 7(45):40659–40681

Anirvan P (2024) The dire struggle: India's unfulfilled promise to eliminate tuberculosis. Indian J Med Ethics IX(4):338–339

Appiah MA, Arthur JA, Gborgblorvor D, Asampong E, Kye-Duodu G, Kamau EM, Dako-Gyeke P (2023) Barriers to tuberculosis treatment adherence in high-burden tuberculosis settings in Ashanti region, Ghana: a qualitative study from patient's perspective. BMC Public Health 23(1):1317

Armstrong DT, Parrish N (2021) Current updates on mycobacterial taxonomy, 2018 to 2019. J Clin Microbiol 59(7):e0152820

Arora VK, Jindal SK, Katiyar SK, Behra D, Talwar D, Sarin R, Dhar R, Mehta P, Bhargava S, Singhal P, Joshi S, Tiwaskar M, Nikam C, Chatterjee A, Vora A (2023) Genomic revolution: transforming tuberculosis diagnosis and treatment with the use of whole genome sequencing – a consensus statement. Indian J Tuberc 70(4):383–389

Auerbach O (1944) Acute generalized miliary tuberculosis. Am J Pathol 20(1):121–136. PMID: 19970738; PMCID: PMC2033127

Bai W, Ameyaw EK (2024) Global, regional and national trends in tuberculosis incidence and main risk factors: a study using data from 2000 to 2021. BMC Public Health 24(1):12

Bain LE, Awah PK, Geraldine N, Kindong NP, Sigal Y, Bernard N, Tanjeko AT (2013) Malnutrition in sub-Saharan Africa: burden, causes and prospects. Pan Afr Med J 15:120

Baptista R, Bhowmick S, Shen J, Mur LAJ (2021) Molecular docking suggests the targets of anti-mycobacterial natural products. Molecules 26(2):475

Barberis I, Bragazzi NL, Galluzzo L, Martini M (2017) The history of tuberculosis: from the first historical records to the isolation of Koch's bacillus. J Prev Med Hyg 58(1):E9–E12. PMID: 28515626; PMCID: PMC5432783

Barnard M, Warren R, Gey Van Pittius N, van Helden P, Bosman M, Streicher E, Coetzee G, O'Brien R (2012) Genotype MTBDRsl line probe assay shortens time to diagnosis of extensively drug-resistant tuberculosis in a high-throughput diagnostic laboratory. Am J Respir Crit Care Med 186(12):1298–1305

Barreiro LB, Tailleux L, Pai AA, Gicquel B, Marioni JC, Gilad Y (2012) Deciphering the genetic architecture of variation in the immune response to *Mycobacterium tuberculosis* infection. Proc Natl Acad Sci USA 109(4):1204–1209

Barteka G, Bwayo D, Matovu JKB, Wanume B, Alunyo JP, Sseguya R, Masaba JP, Obbo JS (2025) Treatment outcomes and predictors of success for multidrug resistant tuberculosis (MDR-TB) in Ugandan regional referral hospitals. Sci Rep 15:14144

Bayigga J, Kakai I, Odongpiny EAL, Ddungu A, Semakula L, Nansereko M, Wiltshire CS, Stavia T, Zawedde-Muyanja S (2025) Alcohol use disorder among people diagnosed with tuberculosis in a large urban case-finding project in Central Uganda: prevalence, associated factors and challenges to treatment adherence. Subst Abuse Treat Prev Policy 20(1):10

Bhutta ZA, Sommerfeld J, Lassi ZS, Salam RA, Das JK (2014) Global burden, distribution, and interventions for infectious diseases of poverty. Infect Dis Poverty (3):21

Birhanu MY, Bekele GM, Jemberie SS (2024) Molecular detection of rifampicin-resistant *Mycobacterium tuberculosis* by polymerase chain reaction in Ethiopia: a systematic review and meta-analysis. Front Med (Lausanne) 11:1319845

Blevins SM, Bronze MS (2010) Robert Koch and the 'golden age' of bacteriology. Int J Infect Dis 14(9):e744–e751

Brites D, Gagneux S (2017) The nature and evolution of genomic diversity in the *Mycobacterium tuberculosis* complex. Adv Exp Med Biol 1019:1–26

British Medical Journal (1950) Streptomycin in the treatment of pulmonary tuberculosis: a Medical Research Council investigation. Br Med J 2(4687):769–782

Brosch R, Gordon SV, Marmiesse M, Brodin P, Buchrieser C, Eiglmeier K, Garnier T, Gutierrez C, Hewinson G, Kremer K, Parsons LM, Pym AS, Samper S, van Soolingen D, Cole ST (2002) A new evolutionary scenario for the *Mycobacterium tuberculosis* complex. Proc Natl Acad Sci USA 99(6):3684–3689

Brüggemann R, Schlumberger F, Chinshailo F, Willis M, Kadyrov A, Kalmambetova G, Chen M, Unterkircher SC, Moidunova N, Sydykova A, Fastenau A (2025) Stigmatization and discrimination of female tuberculosis patients in Kyrgyzstan – a phenomenological study. Int J Equity Health 24(1):185

Brzostek A, Pawelczyk J, Rumijowska-Galewicz A, Dziadek B, Dziadek J (2009) *Mycobacterium tuberculosis* is able to accumulate and utilize cholesterol. J Bacteriol 191(21):6584–6591

Bulstra CA, Hontelez JAC, Otto M, Stepanova A, Lamontagne E, Yakusik A, El-Sadr WM, Apollo T, Rabkin M, UNAIDS Expert Group on Integration, Atun R, Bärnighausen T (2021) Integrating HIV services and other health services: a systematic review and meta-analysis. PLoS Med 18(11):e1003836. https://doi.org/10.1371/journal.pmed.1003836. PMID: 34752477; PMCID: PMC8577772

Byng-Maddick R, Noursadeghi M (2016) Does tuberculosis threaten our ageing populations? BMC Infect Dis 16:119

Byrne RL, Wingfield T, Adams ER, Banu S, Bimba JS, Codlin A, Atienzar AC, Garg T, John S, Gurgel RQ, Sander M, Santos VS, Squire SB, Vo LNQ, Creswell J (2024) Finding the missed millions: innovations to bring tuberculosis diagnosis closer to key populations. BMC Glob Public Health 2(1):33

Cadena AM, Fortune SM, Flynn JL (2017) Heterogeneity in tuberculosis. Nat Rev Immunol 17(11):691–702

Calderwood MS, Platt R, Hou X, Malenfant J, Haney G, Kruskal B, Lazarus R, Klompas M (2010) Real-time surveillance for tuberculosis using electronic health record data from an ambulatory practice in eastern Massachusetts. Public Health Rep 125(6):843–850

Caminero JA, Sotgiu G, Zumla A, Migliori GB (2010) Best drug treatment for multidrug-resistant and extensively drug-resistant tuberculosis. Lancet Infect Dis 10(9):621–629

Caminero JA, Cayla JA, García-García JM, García-Pérez FJ, Palacios JJ, Ruiz-Manzano J (2017) Diagnosis and treatment of drug-resistant tuberculosis. Arch Bronconeumol (9):501–509. English, Spanish

Carvalho ACC, Cardoso CAA, Martire TM, Migliori GB, Sant'Anna CC (2018) Epidemiological aspects, clinical manifestations, and prevention of pediatric tuberculosis from the perspective of the end TB strategy. J Bras Pneumol 44(2):134–144

Cave AJE (1939) The evidence for the incidence of tuberculosis in ancient Egypt. Br J Tuberc 33(3):142–152

Cegielski JP (2010) Extensively drug-resistant tuberculosis: "there must be some kind of way out of here". Clin Infect Dis 50(Suppl 3):S195–S200

Centers for Disease Control and Prevention (1990) Initial therapy for tuberculosis in the era of multidrug resistance. MMWR Morb Mortal Wkly Rep 39(46):784–785

Centers for Disease Control and Prevention (1991) Tuberculosis: multidrug-resistant tuberculosis—New York City, 1990–1991. MMWR Morb Mortal Wkly Rep 40(23):388–391

Chakaya J, Khan M, Ntoumi F, Aklillu E, Fatima R, Mwaba P, Kapata N, Mfinanga S, Hasnain SE, Katoto PDMC, Bulabula ANH, Sam-Agudu NA, Nachega JB, Tiberi S, McHugh TD, Abubakar I, Zumla A (2021) Global tuberculosis report 2020 – reflections on the global TB burden, treatment and prevention efforts. Int J Infect Dis 113(Suppl 1):S7–S12

Chakraborty S, Rhee KY (2015) Tuberculosis drug development: history and evolution of the mechanism-based paradigm. Cold Spring Harb Perspect Med 5(8):a021147

Chauhan A, Parmar M, Dash GC, Solanki H, Chauhan S, Sharma J, Sahoo KC, Mahapatra P, Rao R, Kumar R, Rade K, Pati S (2023) The prevalence of tuberculosis infection in India: a systematic review and meta-analysis. Indian J Med Res Feb-157(2&3):135-151:135

Chauhan A, Parmar M, Dash GC, Chauhan S, Sahoo KC, Samantaray K, Sharma J, Mahapatra P, Pati S (2024) Health literacy and tuberculosis control: systematic review and meta-analysis. Bull World Health Organ 102(6):421–431

Chauke SH, Nzuza S, Ombinda-Lemboumba S, Abrahamse H, Dube FS, Mthunzi-Kufa P (2024) Advances in the detection and diagnosis of tuberculosis using optical-based devices. Photodiagn Photodyn Ther 45:103906

Chekesa B, Singh H, Gonzalez-Juarbe N, Vashee S, Wiscovitch-Russo R, Dupont CL, Girma M, Kerro O, Gumi B, Ameni G (2024) Whole-genome sequencing-based genetic diversity, transmission dynamics, and drug-resistant mutations in *Mycobacterium tuberculosis* isolated from extrapulmonary tuberculosis patients in western Ethiopia. Front Public Health 9(12):1399731

Chen Z, Chen X, Chang M, Lu D, Zhang L, Zheng Y (2025a) Age-period-cohort analysis and prediction of tuberculosis trends in China—based on the global burden of disease 2021 data. Front Public Health 13:1512514

Chen Z, Wang T, Du J, Sun L, Wang G, Ni R, An Y, Fan X, Li Y, Guo R, Mao L, Jing W, Shi K, Cheng J, Wang Q, Nie W, Liu H, Liang J, Gong W (2025b) Decoding the WHO global tuberculosis report 2024: a critical analysis of global and Chinese key data. Zoonoses 5:1

Cioboata R, Biciusca V, Olteanu M, Vasile CM (2023) COVID-19 and tuberculosis: unveiling the dual threat and shared solutions perspective. J Clin Med 12(14):4784

Cirillo D, Cabibbe A, de Filippo MR, Trovato A, Simonetti T, Rossolini G, Tortoli E (2016) Use of WGS in *Mycobacterium tuberculosis* routine diagnosis. Int J Mycobacteriol 5(Suppl):1

Collin SM, Wurie F, Muzyamba MC, de Vries G, Lönnroth K, Migliori GB, Abubakar I, Anderson SR, Zenner D (2019) Effectiveness of interventions for reducing TB incidence in countries with low TB incidence: a systematic review of reviews. Eur Respir Rev 28(152):180107

Comas I, Coscolla M, Luo T, Borrell S, Holt KE, Kato-Maeda M, Parkhill J, Malla B, Berg S, Thwaites G, Yeboah-Manu D, Bothamley G, Mei J, Wei L, Bentley S, Harris SR, Niemann S, Diel R, Aseffa A, Gao Q, Young D, Gagneux S (2013) Out-of-Africa migration and Neolithic coexpansion of *Mycobacterium tuberculosis* with modern humans. Nat Genet 45(10):1176–1182

Conradie F, Diacon AH, Ngubane N, Howell P, Everitt D, Crook AM, Mendel CM, Egizi E, Moreira J, Timm J, TD MH, Wills GH, Bateson A, Hunt R, Van Niekerk C, Li M, Olugbosi M, Spigelman M, Nix-TB Trial Team (2020) Treatment of highly drug-resistant pulmonary tuberculosis. N Engl J Med 382(10):893–902

Conradie F, Bagdasaryan TR, Borisov S, Howell P, Mikiashvili L, Ngubane N, Samoilova A, Skornykova S, Tudor E, Variava E, Yablonskiy P, Everitt D, Wills GH, Sun E, Olugbosi M, Egizi E, Li M, Holsta A, Timm J, Bateson A, Crook AM, Fabiane SM, Hunt R, TD MH, Tweed CD, Foraida S, Mendel CM, Spigelman M, ZeNix Trial Team (2022) Bedaquiline-pretomanid-linezolid regimens for drug-resistant tuberculosis. N Engl J Med 387(9):810–823

Cox H, Goig GA, Salaam-Dreyer Z, Dippenaar A, Reuter A, Mohr-Holland E, Daniels J, Cudahy PG, Nicol MP, Borrell S, Reinhard M (2022) Whole-genome sequencing has the potential to improve treatment for rifampicin-resistant tuberculosis in high-burden settings: a retrospective cohort study. J Clin Microbiol 60(3):e02362–e02321

Cruz AT, Hwang KM, Birnbaum GD, Starke JR (2013) Adolescents with tuberculosis: a review of 145 cases. Pediatr Infect Dis J 32(9):937–941

Dailloux M, Laurain C, Weber M, Hartemann P (1999) Water and nontuberculous mycobacteria. Water Res 33(10):2219–2228

Dalcolmo M, Gayoso R, Sotgiu G, D'Ambrosio L, Rocha JL, Borga L, Fandinho F, Braga JU, Galesi VM, Barreira D, Sanchez DA, Dockhorn F, Centis R, Caminero JA, Migliori GB (2017) Effectiveness and safety of clofazimine in multidrug-resistant tuberculosis: a nationwide report from Brazil. Eur Respir J 49(3):1602445

Daneshmandi Z, Pourdowlat G, Rekabi M, Honarpisheh P, Mirzendedel M, Sadati E et al (2022) Coronavirus disease 2019 and *Mycobacterium tuberculosis* reactivation and coinfections: a review of the literature. J Prevent Diagnos Treat Strateg Med 1(2):76–81

Daniel TM (2006) The history of tuberculosis. Respir Med 100(11):1862–1870

Dar NA, Qadir M, Wani SA, Shah WA (2024) Chemical profile, antituberculosis, DSC, and molecular docking studies of Mentha longifolia essential oil. Nat Prod Res 1:1–7

Dartois VA, Rubin EJ (2022) Anti-tuberculosis treatment strategies and drug development: challenges and priorities. Nat Rev Microbiol 20(11):685–701

Dean AS, Tosas Auguet O, Glaziou P, Zignol M, Ismail N, Kasaeva T, Floyd K (2022) 25 years of surveillance of drug-resistant tuberculosis: achievements, challenges, and way forward. Lancet Infect Dis 22(7):e191–e196

Denkinger CM, Schumacher SG, Boehme CC, Dendukuri N, Pai M, Steingart KR (2014) Xpert MTB/RIF assay for the diagnosis of extrapulmonary tuberculosis: a systematic review and meta-analysis. Eur Respir J 44(2):435–446

Derakhshani N, Rezapour R, Azami-Aghdash S, Nafar H, Soleimanpour S, Tahmazi Aghdam E et al (2024[Epub ahead of print]) Factors affecting private sector engagement in achieving universal health coverage: a scoping review. Glob Health Action 17

Dippenaar A, Parsons SDC, Sampson SL, van der Merwe RG, Drewe JA, Abdallah AM et al (2015) Whole genome sequence analysis of *Mycobacterium suricattae*. Tuberculosis (Edinb) 95(6):682–688

Dohál M, Dvořáková V, Šperková M, Pinková M, Ghodousi A, Omrani M, Porvazník I, Rasmussen EM, Škereňová M, Krivošová M, Wallenfels J, Konstantynovska O, Walker TM, Nikolayevskyy V, Cirillo DM, Solovič I, Mokrý J (2024) Tuberculosis in Ukrainian war refugees and migrants in The Czech Republic and Slovakia: a molecular epidemiological study. J Epidemiol Glob Health 14(1):35–44

Dong W, Peng Q, Liu Z, Xie Z, Guo X, Li Y, Chen C (2023) Estrogen plays an important role by influencing the NLRP3 inflammasome. Biomed Pharmacother 167:115554

Dookie N, Khan A, Padayatchi N, Naidoo K (2022) Application of next generation sequencing for diagnosis and clinical management of drug-resistant tuberculosis: updates on recent developments in the field. Front Microbiol 13:775030

Dooley KE, Mitnick CD, Ann DeGroote M, Obuku E, Belitsky V, Hamilton CD, Makhene M, Shah S, Brust JC, Durakovic N, Nuermberger E, Efficacy Subgroup, RESIST-TB (2012) Old drugs, new purpose: retooling existing drugs for optimized treatment of resistant tuberculosis. Clin Infect Dis 55(4):572–581

D'Silva OA, Lancione S, Ananthakrishnan O, Addae A, Shrestha S, Alsdurf H, Thavorn K, Mzizi N, Vasilu A, Kay A, Mandalakas AM, Zwerling AA (2025) The catastrophic cost of TB care: understanding costs incurred by individuals undergoing TB care in low-, middle-, and high-income settings – a systematic review. PLOS Glob Public Health. 5(4):e0004283

Dunachie SJ, Whelan KT, Alder NC, Sander CR, Hill AV, McShane H (2008) Boosting BCG vaccination with MVA85A down-regulates the immunoregulatory cytokine TGF-beta1. Vaccine 26(41):5269–5275

Dye C, Lönnroth K, Jaramillo E, Williams BG, Raviglione M (2009) Trends in tuberculosis incidence and their determinants in 134 countries. Bull World Health Organ 87(9):683–691

Eldholm V, Norheim G, von der Lippe B, Kinander W, Dahle UR, Caugant DA, Mannsåker T, Mengshoel AT, Dyrhol-Riise AM, Balloux F (2014) Evolution of extensively drug-resistant *Mycobacterium tuberculosis* from a susceptible ancestor in a single patient. Genome Biol 15(11):490

Esmail H, Barry CE 3rd, Young DB, Wilkinson RJ (2014) The ongoing challenge of latent tuberculosis. Philos Trans R Soc Lond Ser B Biol Sci 369(1645):20130437

Espinosa-Pereiro J, Sánchez-Montalvá A, Aznar ML, Espiau M (2022) MDR tuberculosis treatment. Medicina (Kaunas) 58(2):188

European Centre for Disease Prevention and Control, World Health Organization Regional Office for Europe (2024) Tuberculosis surveillance and monitoring in Europe 2024–2022 data. Stockholm (ECDC) & Copenhagen (WHO ROE). Available from:

Falkinham JO 3rd. (2015) Environmental sources of nontuberculous mycobacteria. Clin Chest Med 36(1):35–41

Falzon D, Schünemann HJ, Harausz E, González-Angulo L, Lienhardt C, Jaramillo E, Weyer K (2017) World Health Organization treatment guidelines for drug-resistant tuberculosis, 2016 update. Eur Respir J 49(3):1602308

Falzon D, Zignol M, Bastard M, Floyd K, Kasaeva T (2023) The impact of the COVID-19 pandemic on the global tuberculosis epidemic. Front Immunol 29(14):1234785

Farnia P, Farnia P, Ghanavi J, Zhavnerko GK, Poleschuyk NN, Velayati AA (2018) A review on the shape changes in pathogenic bacteria with emphasis on *Mycobacterium tuberculosis*. Biomed Biotechnol Res J 2:242–246

Fedrizzi T, Meehan CJ, Grottola A, Giacobazzi E, Fregni Serpini G, Tagliazucchi S, Fabio A, Bettua C, Bertorelli R, De Sanctis V, Rumpianesi F, Pecorari M, Jousson O, Tortoli E, Segata N (2017) Genomic characterization of nontuberculous mycobacteria. Sci Rep 7:45258

Ferreira MRL, Bonfim RO, Bossonario PA, Maurin VP, Valença ABM, Abreu PD, Andrade RLP, Fronteira I, Monroe AA (2023) Social protection as a right of people affected by tuberculosis: a scoping review and conceptual framework. Infect Dis Poverty 12(1):103

Fox W, Ellard GA, Mitchison DA (1999) Studies on the treatment of tuberculosis undertaken by the British Medical Research Council tuberculosis units, 1946-1986, with relevant subsequent publications. Int J Tuberc Lung Dis 3(10 Suppl 2):S231–S279

Friedman LN, Filderman AE, D'Aquila TG, Reynolds HY (1989) ELISA analysis of BACTEC bottles for the earlier diagnosis of tuberculosis. Am Rev Respir Dis 140(3):668–671

Gagneux S, DeRiemer K, Van T, Kato-Maeda M, de Jong BC, Narayanan S, Nicol M, Niemann S, Kremer

K, Gutierrez MC, Hilty M, Hopewell PC, Small PM (2006) Variable host-pathogen compatibility in *Mycobacterium tuberculosis*. Proc Natl Acad Sci USA 103(8):2869–2873

Gan Y, Hu Q, Li A, Gu L, Guo S (2022) Estradiol inhibits autophagy of *Mycobacterium tuberculosis*-infected 16HBE cells and controls the proliferation of intracellular *Mycobacterium tuberculosis*. Mol Med Rep 25(6):196

Gandhi NR, Moll A, Sturm AW, Pawinski R, Govender T, Lalloo U, Zeller K, Andrews J, Friedland G (2006) Extensively drug-resistant tuberculosis as a cause of death in patients co-infected with tuberculosis and HIV in a rural area of South Africa. Lancet 368(9547):1575–1580

Gao B, Gupta RS (2012) Phylogenetic framework and molecular signatures for the main clades of the phylum Actinobacteria. Microbiol Mol Biol Rev 76(1):66–112

Gao W, Zhang L (2021) Nanomaterials arising amid antibiotic resistance. Nat Rev Microbiol 19(1):5–6

Garg A, Singh B, Sharma R, Singh A (2019) Kumar A, vol 20. Curr Mol Med, Selective estrogen receptor modulators (SERMs): Mechanistic insights against resistant infections, p 19

Geric C, Qin ZZ, Denkinger CM, Kik SV, Marais B, Anjos A, David PM, Ahmad Khan F, Trajman A (2023) The rise of artificial intelligence reading of chest X-rays for enhanced TB diagnosis and elimination. Int J Tuberc Lung Dis 27(5):367–372

Getnet F, Demissie M, Worku A, Gobena T, Tschopp R, Farah AM, Seyoum B (2021) Challenges in delivery of tuberculosis Services in Ethiopian Pastoralist Settings: clues for reforming service models and organizational structures. BMC Health Serv Res 21(1):627

Ghanavi J, Farnia P (2025) Studying the effect of tumor necrosis factor-alpha and tumor necrosis factor gene polymorphisms on the incidence of tuberculosis. Intl J Mycobacteriol 14(2):89–95

Gisso BT, Hordofa MW, Ormago MD (2022 Sep) Prevalence of pulmonary tuberculosis and associated factors among adults living with HIV/AIDS attending public hospitals in Shashamene Town, Oromia region, South Ethiopia. SAGE Open Med 6(10):20503121221122437

Gordon SV, Brosch R, Billault A, Garnier T, Eiglmeier K, Cole ST (1999) Identification of variable regions in the genomes of tubercle bacilli using bacterial artificial chromosome arrays. Mol Microbiol 32(3):643–655

Gota A, Shenoy VP, Kamath A (2023) Evaluating diagnostic utility of geneXpert ultra (*Mycobacterium tuberculosis*/rifampicin), microscopy and liquid culture to isolate *Mycobacterium tuberculosis* and nontuberculous mycobacteria among pulmonary tuberculosis suspects. J Prevent Diagnos Treat Strateg Med 2(4):236–242

Gulumbe BH, Abdulrahim A, Ahmad SK, Lawan KA, Danlami MB (2025) WHO report signals tuberculosis resurgence: addressing systemic failures and revamping control strategies. Decod Infect Transm 3:100044

Guo Q, Bi J, Lin Q, Ye T, Wang Z, Wang Z, Liu L, Zhang G (2022 May) Whole genome sequencing identifies novel mutations associated with Bedaquiline resistance in *Mycobacterium tuberculosis*. Front Cell Infect Microbiol 27(12):807095

Gupta RS, Lo B, Son J (2018 Feb) Phylogenomics and comparative genomic studies robustly support division of the genus mycobacterium into an emended genus mycobacterium and four novel genera. Front Microbiol 13(9):67

Gupta M, Srikrishna G, Klein SL, Bishai WR (2022) Genetic and hormonal mechanisms underlying sex-specific immune responses in tuberculosis. Trends Immunol 43(8):640–656

Gutierrez MC, Brisse S, Brosch R, Fabre M, Omaïs B, Marmiesse M, Supply P, Vincent V (2005) Ancient origin and gene mosaicism of the progenitor of *Mycobacterium tuberculosis*. PLoS Pathog 1(1):e5

Hamada Y, Getahun H, Tadesse BT, Ford N (2021) HIV-associated tuberculosis. Int J STD AIDS 32(9):780–790

Hanscheid T, del Portal R, Luyten C, Hermans SM, Grobusch MP (2024) Repurposing of anti-malarial drugs for the treatment of tuberculosis: realistic strategy or fanciful dead end? Malar J 23(1):132

Hansen GA (1880) Undersøgelser angaaende Spedalskhedens Aarsager. Norsk Mag Laegevidensk 10:1–88

Harding AT, Heaton NS (2022) The impact of estrogens and their receptors on immunity and inflammation during infection. Cancers (Basel) 14(4):909

Hargreaves JR, Boccia D, Evans CA, Adato M, Petticrew M, Porter JD (2011) The social determinants of tuberculosis: from evidence to action. Am J Public Health 101(4):654–662

Hartmans SY, de Bont JA, Stackebrandt ER (2006) The genus Mycobacterium-nonmedical. Prokaryotes 3:889–918

Horne DJ, Inbaraj LR, Zifodya JS, Shapiro AE, Bhaskar A, Church EC, Daniel J, Kay AW, Kreniske J, Lutje V, Narayanan MKS (2025) Web annex B. 1. Low complexity automated nucleic acid amplification tests (LC-aNAATs) for detection of pulmonary TB and resistance to isoniazid/rifampicin in adults and adolescents: a systematic review of diagnostic accuracy. WHO consolidated guidelines on tuberculosis. Module Diagnos 3:1

Humayun M, Chirenda J, Ye W, Mukeredzi I, Mujuru HA, Yang Z (2022) Effect of gender on clinical presentation of tuberculosis (TB) and age-specific risk of TB, and TB-human immunodeficiency virus coinfection. Open Forum Infect Dis 9(10):ofac512

Hyde L (1972) Rifampin in the treatment of pulmonary tuberculosis. Calif Med 117(6):18–21. PMID: 4635395; PMCID: PMC1518748

Iketleng T, Lessells R, Dlamini MT, Mogashoa T, Mupfumi L, Moyo S, Gaseitsiwe S, de Oliveira T (2018) *Mycobacterium tuberculosis* next-generation whole genome sequencing: opportunities and challenges. Tuberc Res Treat:2018, 1298542

Iseman MD (2000) A clinician's guide to tuberculosis. Lippincott Williams & Wilkins, Philadelphia. ISBN: 9780781717496

Jain N, Sharma D, Jain NK (2025) Newer research transforming 24-month treatment of MDR/XDR-TB TO 6 months. Lung India 42(2):140–146

James LP, Klaassen F, Sweeney S, Furin J, Franke MF, Yaesoubi R, Chesov D, Ciobanu N, Codreanu A, Crudu V, Cohen T, Menzies NA (2024) Impact and cost-effectiveness of the 6-month BPaLM regimen for rifampicin-resistant tuberculosis in Moldova: a mathematical modeling analysis. PLoS Med 21(5):e1004401

Jha P, Gajalakshmi V, Gupta PC, Kumar R, Mony P, Dhingra N, Peto R, RGI-CGHR Prospective Study Collaborators (2006) Prospective study of one million deaths in India: rationale, design, and validation results. PLoS Med 3(2):e18

Jimeno VM, Muñoz EC (2020) Diagnosis of atypical mycobacterial and fungal coinfection. Int J Mycobacteriol. 9(4):435–437

John VBJ, Velayutham AK, Bordoloi P (2025) Targeted therapies—a promising cure for TB, the leading deadliest infectious disease worldwide. Int J Sci Technol 15(3):145–152

Kalonji D, Mahomed OH (2019) Health system challenges affecting HIV and tuberculosis integration at primary healthcare clinics in Durban, South Africa. Afr J Prim Health Care Fam Med 11(1):e1–e7

Kanabalan RD, Lee LJ, Lee TY, Chong PP, Hassan L, Ismail R, Chin VK (2021) Human tuberculosis and *Mycobacterium tuberculosis* complex: a review on genetic diversity, pathogenesis and omics approaches in host biomarkers discovery. Microbiol Res 246:126674

Kapata N, Tembo J, Mwaba P, Nabyonga-Orem J, Ntoumi F, McHugh TD, Zumla A (2025) Undiagnosed burden of latent tuberculosis, active tuberculosis and tuberculosis-HIV co-infections in Africa—status quo, needs, priorities, and opportunities. IJID Reg 14(Suppl 2):100585

Kapur V, Whittam TS, Musser JM (1994) Is *Mycobacterium tuberculosis* 15,000 years old? J Infect Dis 170(5):1348–1349

Kempsell KE, Ji YE, Estrada IC, Colston MJ, Cox RA (1992) The nucleotide sequence of the promoter, 16S rRNA and spacer region of the ribosomal RNA operon of *Mycobacterium tuberculosis* and comparison with *Mycobacterium leprae* precursor rRNA. J Gen Microbiol 138(Pt 8):1717–1727

Kendall EA, Shrestha S, Dowdy DW (2021) The epidemiological importance of subclinical tuberculosis. A critical reappraisal. Am J Respir Crit Care Med 203(2):168–174. https://doi.org/10.1164/rccm.202006-2394

Keshavjee S, Farmer PE (2012) Tuberculosis, drug resistance, and the history of modern medicine. N Engl J Med 367(10):931–936

Koch R (1882) The etiology of tuberculosis. Berliner Klinische Wochenschrift 19:221–230

Koch EK (1932) The diary of Elsie Kleiner Koch. Paperback ed. English. ISBN: 9780615242774

Koura KG, Harries AD (2024) The trend of tuberculosis case notification rates from 1995 to 2022 by country income and World Health Organization region. Trop Med Infect Dis 9(12):294. https://doi.org/10.3390/tropicalmed9120294. PMID: 39728821; PMCID: PMC11679434

Kreiss K, Cox-Ganser J (1997) Metalworking fluid-associated hypersensitivity pneumonitis: a workshop summary. Am J Ind Med 32(4):423–432

Krishnan L, Akande T, Shankar AV, McIntire KN, Gounder CR, Gupta A, Yang WT (2014) Gender-related barriers and delays in accessing tuberculosis diagnostic and treatment services: a systematic review of qualitative studies. Tuberc Res Treat. 2014:215059

Kroon EE, Kinnear CJ, Orlova M, Fischinger S, Shin S, Boolay S, Walzl G, Jacobs A, Wilkinson RJ, Alter G, Schurr E, Hoal EG, Möller M (2020) An observational study identifying highly tuberculosis-exposed, HIV-1-positive but persistently TB, tuberculin and IGRA negative persons with M. tuberculosis specific antibodies in Cape Town, South Africa. EBioMedicine 61:103053

Ku CC, Dodd PJ (2019) Forecasting the impact of population ageing on tuberculosis incidence. PLoS One 14(9):e0222937

Kumah A (2025 May) Building community networks and engagement for effective TB case management. Front Public Health 29(13):1576875

Kurz SG, Furin JJ, Bark CM (2016) Drug-resistant tuberculosis: challenges and Progress. Infect Dis Clin N Am 30(2):509–522

Lange C, Aaby P, Behr MA, Donald PR, Kaufmann SHE, Netea MG, Mandalakas AM (2022) 100 years of Mycobacterium bovis bacille Calmette-Guérin. Lancet Infect Dis 22(1):e2–e12

Lawn SD, Mwaba P, Bates M, Piatek A, Alexander H, Marais BJ, Cuevas LE, McHugh TD, Zijenah L, Kapata N, Abubakar I, McNerney R, Hoelscher M, Memish ZA, Migliori GB, Kim P, Maeurer M, Schito M, Zumla A (2013) Advances in tuberculosis diagnostics: the Xpert MTB/RIF assay and future prospects for a point-of-care test. Lancet Infect Dis 13(4):349–361

Laycock KM, Enane LA, Steenhoff AP (2021) Tuberculosis in adolescents and young adults: emerging data on TB transmission and prevention among vulnerable young people. Trop Med Infect Dis 6(3):148

Ledesma JR, Castillo M, Reyes A et al (2024) Tuberculosis collaborators. Global, regional, and national age-specific progress towards the 2020 milestones of the WHO end TB strategy: a systematic analysis for the global burden of disease study 2021. Lancet Infect Dis 24(7):698–725

Lee M, Lee J, Carroll MW, Choi H, Min S, Song T, Via LE, Goldfeder LC, Kang E, Jin B, Park H, Kwak H, Kim H, Jeon HS, Jeong I, Joh JS, Chen RY, Olivier KN, Shaw PA, Follmann D, Song SD, Lee JK, Lee D, Kim CT, Dartois V, Park SK, Cho SN, Barry CE

3rd. (2012) Linezolid for treatment of chronic extensively drug-resistant tuberculosis. N Engl J Med 367(16):1508–1518

Lee S, Rajaguru V, Baek JS, Shin J, Park Y (2023) Digital health interventions to enhance tuberculosis treatment adherence: scoping review. JMIR Mhealth Uhealth 11:e49741

Lehmann J (1946) Para-aminosalicylic acid in the treatment of tuberculosis. Lancet 1(6384):15

Lei Y, Wang J, Wang Y, Xu C (2023) Geographical evolutionary pathway of global tuberculosis incidence trends. BMC Public Health 23(1):755

Letang E, Ellis J, Naidoo K, Casas EC, Sánchez P, Hassan-Moosa R, Cresswell F, Miró JM, García-Basteiro AL (2020) Tuberculosis-HIV co-infection: Progress and challenges after two decades of global antiretroviral treatment roll-out. Arch Bronconeumol 56(7):446–454

Liang S, Ma J, Wang G, Shao J, Li J, Deng H, Wang C, Li W (2022 Jul) The application of artificial intelligence in the diagnosis and drug resistance prediction of pulmonary tuberculosis. Front Med 28(9):935080

Lightner DV, Redman RM (1998) ACUaCUltUre. Aquaculture 164:201–203

Lillebaek T, Andersen AB, Dirksen A, Smith E, Skovgaard LT, Kok-Jensen A (2002) Persistent high incidence of tuberculosis in immigrants in a low-incidence country. Emerg Infect Dis 8:679–684

Lin M, Chen YW, Li YR, Long LJ, Qi LY, Cui TT, Wu SY, Lin JY, Wu T, Yang YC, Yuan WH, Wu GY, Lan QW, Liu JQ, Li YP, Yu ZY, Guo XG (2022) Systematic evaluation of line probe assays for the diagnosis of tuberculosis and drug-resistant tuberculosis. Clin Chim Acta 533:183–218

Lone MY, Kumar SP, Athar M, Jha PC (2018) Exploration of *Mycobacterium tuberculosis* structural proteome: an in-silico approach. J Theoret Biol 439, 14:–23

Lönnroth K, Raviglione M (2016) The WHO'S new end TB strategy in the post-2015 era of the sustainable development goals. Trans R Soc Trop Med Hyg 110(3):148–150

López-Roa P, Martin-Higuera C, Ruiz-Serrano MJ, Toro C, Tato M, Simon M et al (2021) Performance of Xpert MTB/RIF ultra assay on respiratory and extra-respiratory samples in a high-resource setting with a low tuberculosis prevalence. Diagn Microbiol Infect Dis 99(2):115235. https://doi.org/10.1016/j.diagmicrobio.2020.115235

López-Medrano F, Carver PL, Rutjanawech S, Aranha-Camargo LF, Fernandes R, Belga S, Daniels SA, Mueller NJ, Burkhard S, Theodoropoulos NM, Postma DF (2025) Clinical management and outcomes of non-tuberculous mycobacterial infections in solid organ transplant recipients: a multinational case-control study. Transplantation 109(2):e134–e141

Lory S (2014) The Family Mycobacteriaceae. In: Rosenberg E, DeLong EF, Lory S, Stackebrandt E, Thompson F (eds) The prokaryotes. Springer, Berlin/Heidelberg, pp 889–918

Macalino SJ, Billones J, Organo V, Carrillo MC (2020) In Silico strategies in tuberculosis drug discovery. Molecules 25(3):665

MacNeil A, Glaziou P, Sismanidis C, Date A, Maloney S, Floyd K (2020) Global epidemiology of tuberculosis and Progress toward meeting global targets – worldwide, 2018. MMWR Morb Mortal Wkly Rep 69(11):281–285

Madaki AS, Ahmad K, Singh D (2024) IT integration implementation in e-government public sector in developing countries: a systematic literature review and model development. Transform Govern People Process Policy 18(3):451–472

Magee JG, Ward AC, Family III (2012) Mycobacteriaceae Chester 1897, 63AL. In: Bergey's manual of systematic bacteriology: volume 5: the Actinobacteria. Springer, New York, p 312

Mahmoudi S, García MJ, Drain PK (2024) Current approaches for diagnosis of subclinical pulmonary tuberculosis, clinical implications and future perspectives: a scoping review. Expert Rev Clin Immunol 20(7):715–726

Malla E, Shahi S, Forsberg B (2025) Barriers and facilitators to tuberculosis service access for patients in Bajhang, Nepal: a qualitative study based on focus groups with patients, screening camp visitors, community volunteers and medical doctors. J Health Popul Nutr 44:46

Mancuso G, Midiri A, De Gaetano S, Ponzo E, Biondo C (2023) Tackling drug-resistant tuberculosis: new challenges from the old pathogen *Mycobacterium tuberculosis*. Microorganisms 11(9):2277

Mao J, Akhtar J, Zhang X, Sun L, Guan S, Li X, Chen G, Liu J, Jeon HN, Kim MS, No KT, Wang G (2021) Comprehensive strategies of machine-learning-based quantitative structure-activity relationship models. iScience 24(9):103052

Marahatta SB, Yadav RK, Giri D, Lama S, Rijal KR, Mishra SR, Shrestha A, Bhattrai PR, Mahato RK, Adhikari B (2020) Barriers in the access, diagnosis and treatment completion for tuberculosis patients in central and western Nepal: a qualitative study among patients, community members and health care workers. PLoS One 15(1):e0227293

Marimuthu P (2016) Tuberculosis prevalence and socio-economic differentials in the slums of four metropolitan cities of India. Indian J Tuberc 63(3):167–170

Mase SR, Chorba T (2019) Treatment of drug-resistant tuberculosis. Clin Chest Med 40(4):775–795. https://doi.org/10.1016/j.ccm.2019.08.002. PMID: 31731984; PMCID: PMC7000172

Maseko AF, Silumbwe A, Maritim P, Munakampe MN, Chirambo GB, Jacobs C, Zulu JM (2025) Factors that shape the integration of HIV and TB services in Zomba District, Malawi. BMC Health Serv Res 25(1):213. https://doi.org/10.1186/s12913-025-12367-8. PMID: 39915793; PMCID: PMC11800521

Masjedi MR, Farnia P, Sorooch S, Pooramiri MV, Mansoori SD, Zarifi AZ, Akbarvelayati A, Hoffner S

(2006) Extensively drug-resistant tuberculosis: 2 years of surveillance in Iran. Clin Infect Dis 43(7):841–847

McCarthy OR (2001) The key to the sanatoria. J R Soc Med 94(8):413–417

McClean M, Panciu TC, Lange C, Duarte R, Theis F (2024) Artificial intelligence in tuberculosis: a new ally in disease control. Breathe 20(3):240056

Medina-Marino A, de Vos L, Daniels J (2025) Social isolation, social exclusion, and access to mental and tangible resources: mapping the gendered impact of tuberculosis-related stigma among men and women living with tuberculosis in Eastern Cape Province. South Africa BMC Glob Public Health 3:50

Medjahed H, Gaillard JL, Reyrat JM (2010) Mycobacterium abscessus: a new player in the mycobacterial field. Trends Microbiol 18(3):117–123

Meng XY, Zhang HX, Mezei M, Cui M (2011) Molecular docking: a powerful approach for structure-based drug discovery. Curr Comput Aided Drug Des 7(2):146–157

Menzies NA, Allwood BW, Dean AS, Dodd PJ, Houben RMGJ, James LP, Knight GM, Meghji J, Nguyen LN, Rachow A, Schumacher SG, Mirzayev F, Cohen T (2023) Global burden of disease due to rifampicin-resistant tuberculosis: a mathematical modeling analysis. Nat Commun 14(1):6182

Migliori GB, Ortmann J, Girardi E, Besozzi G, Lange C, Cirillo DM, Ferrarese M, De Iaco G, Gori A, Raviglione MC, SMIRA/TBNET Study Group (2007) Extensively drug-resistant tuberculosis, Italy and Germany. Emerg Infect Dis 13(5):780–782

Migliori GB, Lange C, Centis R, Sotgiu G, Mütterlein R, Hoffmann H, Kliiman K, De Iaco G, Lauria FN, Richardson MD, Spanevello A, Cirillo DM, TBNET Study Group (2008) Resistance to second-line injectables and treatment outcomes in multidrug-resistant and extensively drug-resistant tuberculosis cases. Eur Respir J 31(6):1155–1159

Migliori GB, Ong CWM, Petrone L, D'Ambrosio L, Centis R, Goletti D (2021) The definition of tuberculosis infection based on the spectrum of tuberculosis disease. Breathe (Sheff) 17(3):210079

Mitchison DA (2000) Role of individual drugs in the chemotherapy of tuberculosis. Int J Tuberc Lung Dis 4(9):796–806. Erratum in: Int J Tuberc Lung Dis. 2003 7(3):304. PMID: 10985648

Moens C, Bogaerts B, Lorente-Leal V, Vanneste K, De Keersmaecker SCJ, Roosens NHC, Mostin L, Fretin D, Marché S (2024) Genomic comparison between Mycobacterium bovis and Mycobacterium microti and in silico analysis of peptide-based biomarkers for serodiagnosis. Front Vet Sci 11:1446930

Mohammadnabi N, Shamseddin J, Emadi M, Bodaghi AB, Varseh M, Shariati A, Rezaei M, Dastranj M, Farahani A (2024) *Mycobacterium tuberculosis*: The mechanism of pathogenicity, immune responses, and diagnostic challenges. J Clin Lab Anal 38(23):e25122

Moonan PK, Nair SA, Agarwal R, Chadha VK, Dewan PK, Gupta UD, Ho CS et al (2018) Tuberculosis preventive treatment: the next chapter of tuberculosis elimination in India. BMJ Glob Health 3(5):e001135. https://doi.org/10.1136/bmjgh-2018-001135. PMID: 30364389; PMCID: PMC6195150

Morey-León G, Mejía-Ponce PM, Fernández-Cadena JC, García-Moreira E, Andrade-Molina D, Licona-Cassani C, Fresia P, Berná L (2025) Global epidemiology of *Mycobacterium tuberculosis* lineage 4 insights from Ecuadorian genomic data. Sci Rep 15:3823

Morse D, Brothwell DR, Ucko PJ (1964) Tuberculosis in ancient Egypt. Am Rev Respir Dis 90:524–541

Mousavi-Sagharchi SMA, Ghorbani A, Meskini M, Siadat SD (2025) Historical examination of tuberculosis: from ancient affliction to modern challenges. J Infect Public Health 18(3):102649

Muflihah H, Yulianto FA, Rina SE, Ferdiana A, Rahimah SB (2024) Tuberculosis coinfection among COVID-19 patients: clinical presentation and mortality in a tertiary lung Hospital in Indonesia. Int J Mycobacteriol. 13(1):58–64

Muhammed MT, Aki-Yalcin E (2024) Molecular docking: principles, advances, and its applications in drug discovery. Lett Drug Des Discov 21:480–495

Naidoo K, Perumal R (2023) Advances in tuberculosis control during the past decade. Lancet Respir Med 11(4):311–313

Naidoo K, Gengiah S, Singh S, Stillo J, Padayatchi N (2019) Quality of TB care among people living with HIV: gaps and solutions. J Clin Tubercul Other Mycobacter Dis 17:100122

Nalugwa T, Annerstedt KS, Nabwire S, West NS, Kadota JL, Atkins S, Cattamanchi A, Lönnroth K, Katamba A, Shete PB (2025) Identifying mechanisms by which social determinants of health impact TB diagnostic evaluation uptake in Uganda: a qualitative study. Int J Equity Health 24:73

Nandlal L, Perumal R, Naidoo K (2022) Rapid molecular assays for the diagnosis of drug-resistant tuberculosis. Infect Drug Resist 29(15):4971–4984

Narasimhan P, Wood J, Macintyre CR, Mathai D (2013) Risk factors for tuberculosis. Pulm Med 2013:828939

Negin J, Abimbola S, Marais BJ (2015) Tuberculosis among older adults--time to take notice. Int J Infect Dis 32:135–137

Ngwatu BK, Nsengiyumva NP, Oxlade O, Mappin-Kasirer B, Nguyen NL, Jaramillo E, Falzon D, Schwartzman K (2018) Collaborative group on the impact of digital technologies on TB. The impact of digital health technologies on tuberculosis treatment: a systematic review. Eur Respir J 51(1):1701596

Nhamoyebonde S, Leslie A (2014) Biological differences between the sexes and susceptibility to tuberculosis. J Infect Dis (209, Suppl 3):S100–S106

Niemann S, Supply P (2014) Diversity and evolution of *Mycobacterium tuberculosis*: moving to whole-genome-based approaches. Cold Spring Harb Perspect Med 4(12):a021188

Niemann S, Kubica T, Bange FC, Adjei O, Browne EN, Chinbuah MA, Diel R, Gyapong J, Horstmann RD, Joloba ML, Meyer CG, Mugerwa RD, Okwera A, Osei I, Owusu-Darbo E, Schwander SK, Rüsch-Gerdes S (2004) The species Mycobacterium africanum in the

light of new molecular markers. J Clin Microbiol 42(9):3958–3962
Nyasulu PS, Doumbia CO, Ngah V, Togo ACG, Diarra B, Chongwe G (2024) Multidrug-resistant tuberculosis: latest opinions on epidemiology, rapid diagnosis and management. Curr Opin Pulm Med 30(3):217–228
Nyijime TA, Shallangwa GA, Uzairu A, Umar AB, Ibrahim MT, Madkhali HA et al (2025) Computational design, pharmacokinetics, molecular docking and molecular dynamic simulation of novel anti-tubercular inhibitors. In Silico Res Biomed 1:100012
Obeagu EI (2023) Tuberculosis diagnostic and treatment delays among patients in Uganda. Health Sci Rep. 6(11):e1700
Oga-Omenka C, Bada F, Agbaje A, Dakum P, Menzies D, Zarowsky C (2020) Ease and equity of access to free DR-TB services in Nigeria- a qualitative analysis of policies, structures and processes. Int J Equity Health 19(1):221
Ogbudebe C, Jeong D, Odume B, Chukwuogo O, Dim C, Useni S, Okuzu O, Malolan C, Kim D, Nwariaku F, Nwokoye N, Gande S, Nongo D, Eneogu R, Odusote T, Oyelaran S, ChijiokeAkaniro O, Nihalani N, Anyaike C, Gidado M (2023) Identifying hot spots of tuberculosis in Nigeria using an early warning outbreak recognition system: retrospective analysis of implications for active casefinding interventions. JMIR Public Health Surveill 9:e40311
Ogunniyi TJ, Abdulganiyu MO, Issa JB et al (2024) Ending tuberculosis in Nigeria: a priority by 2030. BMJ Glob Health 9:e016820
Olaleye SA, Balogun OS, Adusei-Mensah F (2023) Bibliometric structured review of tuberculosis in Nigeria. Afr Health Sci 23(2):139–160
Olivier C, Luies L (2023) WHO goals and beyond: managing HIV/TB co-infection in South Africa. SN Compr Clin Med 5:251
Olmo-Fontánez AM, Turner J (2022) Tuberculosis in an aging world Pathogens 11(10):1101
Onifade DA, Bayer AM, Montoya R, Haro M, Alva J, Franco J, Sosa R, Valiente B, Valera E, Ford CM, Acosta CD, Evans CA (2010) Gender-related factors influencing tuberculosis control in shantytowns: a qualitative study. BMC Public Health 10:381
Orgeur M, Sous C, Madacki J, Brosch R (2024) Evolution and emergence of *Mycobacterium tuberculosis*. FEMS Microbiol Rev 48(2):fuae006
Ortona E, Pierdominici M, Rider V (2019) Editorial: sex hormones and gender differences in immune responses. Front Immunol 10:1076
Osei-Wusu S, Asare P, Danso EK, Asogun D, Otchere ID, Asante-Poku A, Yeboah-Manu D (2025a) Addressing key risk factors hindering tuberculosis control activities in West Africa – progress in meeting the UN sustainable development goals. IJID Reg. 14(Suppl 2):100594
Osei-Wusu S, Asare P, Danso EK, Asogun D, Otchere ID, Asante-Poku A, Yeboah-Manu D (2025b) Addressing key risk factors hindering tuberculosis control activities in West Africa-progress in meeting the UN sustainable development goals. IJID Regions 14:100594
Oshi DC, Amah PO, Alobu IN, Oshi SN, Brown S, Ukwaja KN (2024) Enhancing knowledge and beliefs: the impact of a gender-transformative training program on tuberculosis care in southern Nigeria. Int J Mycobacteriol. 13(4):394–403
Otto-Knapp R, Edwards S, Kuchukhidze G, Kröger S, Häcker B, Bivol S, Yedilbayev A (2024) Availability of drugs for the treatment of multidrug-resistant/rifampicin-resistant tuberculosis in the World Health Organization European region, October 2023. Euro Surveill 29(17):2400211
Pai M, Kasaeva T, Swaminathan S (2022) Covid-19's devastating effect on tuberculosis care – a path to recovery. N Engl J Med 386:1490
Paramasivan CN, Herbert D, Prabhakar R (1993) Bactericidal action of pulsed exposure to rifampicin, ethambutol, isoniazid & pyrazinamide on *Mycobacterium tuberculosis* in vitro. Indian J Med Res 97:145–150. PMID: 8406638
Parida SK, Axelsson-Robertson R, Rao MV, Singh N, Master I, Lutckii A, Keshavjee S, Andersson J, Zumla A, Maeurer M (2015) Totally drug-resistant tuberculosis and adjunct therapies. J Intern Med 277(4):388–405
Parrish NM, Dick JD, Bishai WR (1998) Mechanisms of latency in *Mycobacterium tuberculosis*. Trends Microbiol 6(3):107–112
Parte AC, Sardà Carbasse J, Meier-Kolthoff JP, Reimer LC, Göker M (2020) List of prokaryotic names with standing in nomenclature (LPSN) moves to the DSMZ. Int J Syst Evol Microbiol 70(11):5607–5612
Patil K, Bagade S, Bonde S, Sharma S, Saraogi G (2018) Recent therapeutic approaches for the management of tuberculosis: challenges and opportunities. Biomed Pharmacother 99:735–745
Patil TS, Nirbhavane P, Aher AA, Gujarathi NA, Khairnar TD, Goyal YS (2025) Future prospects and clinical aspects for the management of tuberculosis. In: Emerging paradigms in delivery Systems for Antitubercular Therapy. Academic Press, pp 371–395
Paul R, Lorenzo F, López B, Alegre MG, Couvin D, Rastogi N, Pérez-Lago L, García de Viedma D, Gamberale A, González N, Palmero D, Altabe S, Simboli N, Yokobori NK (2025) Outbreak caused by multidrug-resistant *Mycobacterium tuberculosis* with unusual combination of resistance mutations, northern Argentina, 2006–2022. Emerg Infect Dis 31(3):601–606
Peer V, Schwartz N, Green MS (2023) Gender differences in tuberculosis incidence rates—a pooled analysis of data from seven high-income countries by age group and time period. Front Public Health 10:997025
Pfyffer GE, Auckenthaler R, van Embden JD, van Soolingen D (1998) Mycobacterium canettii, the smooth variant of *M. tuberculosis*, isolated from a

Swiss patient exposed in Africa. Emerg Infect Dis 4(4):631–634

Portaels F, Aguiar J, Fissette K, Fonteyne PA, De Beenhouwer H, De Rijk P et al (1997) Direct detection and identification of mycobacterium ulcerans in clinical specimens by PCR and oligonucleotide-specific capture plate hybridization. J Clin Microbiol 35(5):1097–1100

Portnoy A, Arcand JL, Clark RA, Weerasuriya CK, Mukandavire C, Bakker R, Patouillard E, Gebreselassie N, Zignol M, Jit M, White RG, Menzies NA (2023) The potential impact of novel tuberculosis vaccine introduction on economic growth in low- and middle-income countries: a modeling study. PLoS Med 20(7):e1004252

Pradipta IS, Forsman LD, Bruchfeld J, Hak E, Alffenaar JW (2018) Risk factors of multidrug-resistant tuberculosis: a global systematic review and meta-analysis. J Infect 77(6):469–478

Prodinger WM, Indra A, Koksalan OK, Kilicaslan Z, Richter E (2014) Mycobacterium caprae infection in humans. Expert Rev Anti-Infect Ther 12(12):1501–1513

Qin ZZ, Ahmed S, Sarker MS, Paul K, Adel ASS, Naheyan T, Barrett R, Banu S, Creswell J (2021) Tuberculosis detection from chest x-rays for triaging in a high tuberculosis-burden setting: an evaluation of five artificial intelligence algorithms. Lancet Digit Health 3(9):e543–e554

Quinn M, Forman J, Harrod M, Winter S, Fowler KE, Krein SL, Gupta A, Saint S, Singh H, Chopra V (2019) Electronic health records, communication, and data sharing: challenges and opportunities for improving the diagnostic process. Diagnosis (Berl) 6(3):241–248

Rajendran P, Padmapriyadarsini C, Mondal R (2021) Nontuberculous mycobacterium: An emerging pathogen: Indian perspective. Int J Mycobacteriol. 10(3):217–227

Rastogi N, Legrand E, Sola C (2001) The mycobacteria: an introduction to nomenclature and pathogenesis. Revue Scientifique Et Technique-Office International Des Epizooties 20(1):21–54

Ratwani R (2017) Electronic health records and improved patient care: opportunities for applied psychology. Curr Dir Psychol Sci 26(4):359–365

Rehman ZU, Alhazmi HA, Albratty M, Algaissi A, Makeen HA, Najmi A et al (2025) Structural insights into PKS13 Thioesterase inhibitors targeting *Mycobacterium tuberculosis* for drug-resistant tuberculosis therapy: identification, stability assessment, and pharmacophore modeling. Results Chem 16:102388

Rendon A, Tiberi S, Scardigli A, D'Ambrosio L, Centis R, Caminero JA, Migliori GB (2016) Classification of drugs to treat multidrug-resistant tuberculosis (MDR-TB): evidence and perspectives. J Thorac Dis 8(10):2666–2671

Rey AD, Mahuad CV, Bozza VV, Bogue C, Farroni MA, Bay ML, Bottasso OA, Besedovsky HO (2017) Endocrine and cytokine responses in humans with pulmonary tuberculosis. Brain Behav Immun 21:171–179

Ritacco V, Di Lonardo M, Reniero A, Ambroggi M, Barrera L, Dambrosi A, Lopez B, Isola N, de Kantor IN (1997) Nosocomial spread of human immunodeficiency virus-related multidrug-resistant tuberculosis in Buenos Aires. J Infect Dis 176(3):637–642

Robitzek EH, Selikoff IJ (1952) Hydrazine derivatives of isonicotinic acid (rimifon marsilid) in the treatment of active progressive caseous-pneumonic tuberculosis; a preliminary report. Am Rev Tuberc 65(4):402–428

Roe WD, Lenting B, Kokosinska A, Hunter S, Duignan PJ, Gartrell B, Rogers L, Collins DM, de Lisle GW, Gedye K, Price-Carter M (2019) Pathology and molecular epidemiology of mycobacterium pinnipedii tuberculosis in native New Zealand marine mammals. PLoS One 14(2):e0212363

Runyon EH (1959) Anonymous mycobacteria in pulmonary disease. Med Clin North Am 43(1):273–290

Saffari M, Pakpour AH (2012) Avicenna's canon of medicine: a look at health, public health, and environmental sanitation. Arch Iran Med 15(12):785–789. PMID: 23199255

Sakthivadivel V, Gaur A, Geetha J (2023) Tuberculosis in elderly population: a cross-sectional comparative study. Int J Mycobacteriol. 12(1):38–42

Samanta J, Mitra S, Chakraborty S, Kumar C, Yashavanth KY, Das S (2023) Pulmonary tuberculosis among patients of tubercular pleural effusion: a single-center experience. Intl J Mycobacteriol 12(2):139–143

Santosh KC, Allu S, Rajaraman S, Antani S (2022) Advances in deep learning for tuberculosis screening using chest X-rays: the last 5 years review. J Med Syst 46(11):82

Satta G, Lipman M, Smith GP, Arnold C, Kon OM, McHugh TD (2018) *Mycobacterium tuberculosis* and whole-genome sequencing: how close are we to unleashing its full potential? Clin Microbiol Infect 24(6):604–609

Satyanarayana S, Thekkur P, Kumar AMV, Lin Y, Dlodlo RA, Khogali M, Zachariah R, Harries AD (2020) An opportunity to END TB: using the sustainable development goals for action on socio-economic determinants of TB in high burden countries in WHO South-East Asia and the Western Pacific regions. Trop Med Infect Dis 5(2):101

Sazali MF, Rahim SSSA, Mohammad AH, Kadir F, Payus AO, Avoi R, Jeffree MS, Omar A, Ibrahim MY, Atil A, Tuah NM, Dapari R, Lansing MG, Rahim AAA, Azhar ZI (2023) Improving tuberculosis medication adherence: The potential of integrating digital technology and health belief model. Tuberc Respir Dis (Seoul) 86(2):82–93

Schaible UE, Linnemann L, Redinger N, Patin EC, Dallenga T (2017) Strategies to improve vaccine efficacy against tuberculosis by targeting innate immunity. Front Immunol 8:1755

Schatz A, Bugie E, Waksman SA (2005) Streptomycin, a substance exhibiting antibiotic activity against gram-positive and gram-negative bacteria. 1944. Clin Orthop Relat Res 437:3–6

Schmitt E, Riwanto M, Sambandamurthy V, Roggo S, Miault C, Zwingelstein C, Krastel P, Noble C, Beer D, Rao S, Au M, Niyomrattanakit P, Lim V, Zheng J, Jeffery D, Pethe K, Camacho L (2011) The natural product Cyclomarin kills *Mycobacterium tuberculosis* by targeting the ClpC1 subunit of the Caseinolytic protease. Angew Chem Int Ed Engl 50(26):5889–5891

Sciarra F, Campolo F, Franceschini E, Carlomagno F, Venneri MA (2023) Gender-specific impact of sex hormones on the immune system. Int J Mol Sci 24(7):6302

Scott GY, Aborode AT, Adesola RO, Benson I, Omulepu I, Ajayi OO, Nibokun EO et al (2025) Diagnostic innovations for tuberculosis in sub-Saharan Africa. Discov Public Health 22:188

Selmane S, L'Hadj M (2020) Epidemiology and clinical characteristics of tuberculosis in Leon Bernard tuberculosis unit in Algeria. Intl J Mycobacteriol 9(3):254–260

Seung KJ, Keshavjee S, Rich ML (2015) Multidrug-resistant tuberculosis and extensively drug-resistant tuberculosis. Cold Spring Harb Perspect Med 5(9):a017863

Shah NS, Auld SC, Brust JC, Mathema B, Ismail N, Moodley P, Mlisana K, Allana S, Campbell A, Mthiyane T, Morris N, Mpangase P, van der Meulen H, Omar SV, Brown TS, Narechania A, Shaskina E, Kapwata T, Kreiswirth B, Gandhi NR (2017) Transmission of extensively drug-resistant tuberculosis in South Africa. N Engl J Med 376(3):243–253

Silva-Pereira TT, Ikuta CY, Zimpel CK, Camargo NCS, de Souza Filho AF, Ferreira Neto JS, Heinemann MB, Guimarães AMS (2019) Genome sequencing of mycobacterium pinnipedii strains: genetic characterization and evidence of superinfection in a south American sea lion (Otaria flavescens). BMC Genomics 20(1):1030

Silva-Pereira TT, Soler-Camargo NC, Guimarães AMS (2024) Diversification of gene content in the *Mycobacterium tuberculosis* complex is determined by phylogenetic and ecological signatures. Microbiol Spectr 12(2):e0228923

Singal S, Modi N, Singal A, Agrawal H (2025) Investigating the prevalence of tuberculosis in urban slums: a pathological cross-sectional approach. Eur J Cardiovasc Med 15(1):198–202

Singh S, Gupta H, Sharma P, Sahi S (2024) Advances in artificial intelligence (AI)-assisted approaches in drug screening. Artif Intellig Chem 2(1):100039

Smith NH, Crawshaw T, Parry J, Birtles RJ (2009) Mycobacterium microti: more diverse than previously thought. J Clin Microbiol 47(8):2551–2559

Snow KJ, Sismanidis C, Denholm J, Sawyer SM, Graham SM (2018) The incidence of tuberculosis among adolescents and young adults: a global estimate. Eur Respir J 51(2):1702352

Sodhi R, Vatsyayan V, Panibatla V, Sayyad K, Williams J, Pattery T, Pal A (2024) Impact of a pilot mHealth intervention on treatment outcomes of TB patients seeking care in the private sector using propensity scores matching-evidence collated from New Delhi, India. PLOS Digit Health 3(9):e0000421

Sotgiu G, D'Ambrosio L, Centis R, Mura I, Castiglia P, Spanevello A, Migliori GB (2015) The multidrug-resistant tuberculosis threat: old problems and new solutions. J Thorac Dis 7(9):E354–E360

Srivastava S, Dey S, Mukhopadhyay S (2023) Vaccines against tuberculosis: where are we now? Vaccine 11(5):1013

Steingart KR, Schiller I, Horne DJ, Pai M, Boehme CC, Dendukuri N (2014) Xpert® MTB/RIF assay for pulmonary tuberculosis and rifampicin resistance in adults. Cochrane Database Syst Rev 14(1):CD009593

Sun Q, Yan J, Liao X, Wang C, Wang C, Jiang G, Dong L, Wang F, Huang H, Wang G, Pan J (2022) Trends and species diversity of non-tuberculous mycobacteria isolated from respiratory samples in northern China, 2014–2021. Front Public Health 10:923968

Sun TY, Hardin J, Nieva HR, Natarajan K, Cheng RF, Ryan P, Elhadad N (2023) Large-scale characterization of gender differences in diagnosis prevalence and time to diagnosis. medRxiv [Preprint] 16:2023. https://doi.org/10.1101/2023.10.12.23296976

Sweileh WM, AbuTaha AS, Sawalha AF, Al-Khalil S, Al-Zyoud SH (2016) Bibliometric analysis of worldwide publications on multi-, extensively, and totally drug-resistant tuberculosis (2006–2015). Multidiscip Respir Med 11:45

Takayama K, Wang L, David HL (1972) Effect of isoniazid on the in vivo mycolic acid synthesis, cell growth, and viability of *Mycobacterium tuberculosis*. Antimicrob Agents Chemother 2(1):29–35

Tanimura T, Jaramillo E, Weil D, Raviglione M, Lönnroth K (2014) Financial burden for tuberculosis patients in low- and middle-income countries: a systematic review. Eur Respir J 43(6):1763–1775

Taylor JW, Wright GEL, Lim LL, Denholm JT (2025) A systematic review of the utility of antituberculosis therapy for presumed tuberculous uveitis. BMC Infect Dis 25:112

Teibo TKA, Andrade RLP, Rosa RJ, de Abreu PD, Olayemi OA, Alves YM, Vaz Tavares RB, da Costa FBP, Dias Moura HS, Ferezin LP, Tártaro AF, de Campos MCT, Ribeiro NM, Berra TZ, Arcêncio RA (2024) Barriers that interfere with access to tuberculosis diagnosis and treatment across countries globally: a systematic review. ACS Infect Dis 10(8):2600–2614

Teo AKJ, Morishita F, Islam T, Viney K, Ong CWM, Kato S, Kim H, Liu Y, Oh KH, Yoshiyama T, Ohkado A, Rahevar K, Kawatsu L, Yanagawa M, Prem K, Yi S, Tran HTG, Marais BJ (2023) Tuberculosis in older adults: challenges and best practices in the Western Pacific region. Lancet Reg Health West Pac 36:100770

Theorupun SC, Massi MN, Nurhasanah A, Syamsuri F, Hatta M, Rasita YD et al (2025) PPE18 and PepA variations in *Mycobacterium tuberculosis* clinical isolates from Makassar, Indonesia: challenges for immune recognition and vaccine development. Intl J Mycobacteriol 14(2):191–200

Ting H, Bozzola M, Ravetti C (2024) Tuberculosis and labour market participation: evidence from South Africa. South Afr J Econ 92:444

Tobin EH, Tristram D (2024) Tuberculosis Overview. In: StatPearls [Internet]. StatPearls Publishing, Treasure Island (FL). [Cited 2025 May?]. Available from: NCBI Bookshelf

Tortoli E (2012) Phylogeny of the genus mycobacterium: many doubts, few certainties. Infect Genet Evol 12(4):827–831

Tortoli E (2014) Microbiological features and clinical relevance of new species of the genus mycobacterium. Clin Microbiol Rev 27(4):727–752

Tortoli E, Meehan CJ, Grottola A, Fregni Serpini G, Fabio A, Trovato A, Pecorari M, Cirillo DM (2019) Genome-based taxonomic revision detects a number of synonymous taxa in the genus mycobacterium. Infect Genet Evol 75:103983

Truman RW, Andrews PK, Robbins NY, Adams LB, Krahenbuhl JL, Gillis TP (2008) Enumeration of mycobacterium leprae using real-time PCR. PLoS Negl Trop Dis 2(11):e328

Tumuhimbise W, Atwine D, Kaggwa F, Musiimenta A (2023) Acceptability and feasibility of a mobile health application for enhancing public private mix for TB care among healthcare Workers in Southwestern Uganda. BMC Digit Health 1(1):9. https://doi.org/10.1186/s44247-023-00009-0. Epub 2023 Mar 3. PMID: 38014370; PMCID: PMC9982777

Tweed CD, Dawson R, Burger DA, Conradie A, Crook AM, Mendel CM, Conradie F, Diacon AH, Ntinginya NE, Everitt DE, Haraka F (2019) Bedaquiline, moxifloxacin, pretomanid, and pyrazinamide during the first 8 weeks of treatment of patients with drug-susceptible or drug-resistant pulmonary tuberculosis: a multicentre, open-label, partially randomised, phase 2b trial. Lancet Respir Med 7(12):1048–1058

Uden L, Barber E, Ford N, Cooke GS (2017) Risk of tuberculosis infection and disease for health care workers: an updated meta-analysis. Open Forum Infect Dis 4(3):ofx137

Udwadia ZF, Amale RA, Ajbani KK, Rodrigues C (2012) Totally drug-resistant tuberculosis in India. Clin Infect Dis 54(4):579–581

Ugwu KO, Agbo MC, Ezeonu IM (2021) Prevalence of tuberculosis, drug-resistant tuberculosis and HIV/TB co-infection in Enugu Nigeria. Afr J Infect Dis 15(2):24–30

Vaisman A, Barry P, Flood J (2020) Assessing complexity among patients with tuberculosis in California, 1993–2016. Open Forum Infect Dis 7(8):ofaa264

van Arkel C, Storms I, Kurver L, Smeenk F, Wielders P, Hoefsloot W, Carpaij N, Boeree MJ, van Crevel R, van Laarhoven A, Magis-Escurra C (2024) Elderly patients with tuberculosis in a low-incidence country: clinical characteristics, inflammation and outcome. J Infect 89(2):106200. https://doi.org/10.1016/j.jinf.2024.106200. Epub 2024 Jun 18. PMID: 38901573

van Ingen J, Rahim Z, Mulder a, Boeree MJ, Simeone R, Brosch R, van Soolingen D. (2012) Characterization of *Mycobacterium orygis* as *M. tuberculosis* complex subspecies. Emerg Infect Dis 18(4):653–655

Van Rie A, Warren R, Meurrens MV, Potgieter T (2025) The role of AI in combatting RR-TB epidemics. Int J Infect Dis 152:107741

Vanden Driessche K, Persson A, Marais BJ, Fink PJ, Urdahl KB (2013) Immune vulnerability of infants to tuberculosis. Clin Dev Immunol 2013:781320. https://doi.org/10.1155/2013/781320. Epub 2013 May 13. PMID: 23762096; PMCID: PMC3666431

Velayati AA (2023) Hospitalization, diagnosis, treatment, and surgery in Iran. In: The lifecycle of medical sciences, 1st edn. Academic. ISBN 9780443153051

Velayati AA, Farnia P (eds) (2019) Nontuberculous mycobacteria (NTM): microbiological, clinical and geographical distribution, 1st edn. Academic, London. Paperback ISBN: 9780128146927. eBook ISBN: 9780128146934

Velayati AA, Masjedi MR, Farnia P, Tabarsi P, Ghanavi J, ZiaZarifi AH, Hoffner SE (2009) Emergence of new forms of totally drug-resistant tuberculosis bacilli: super extensively drug-resistant tuberculosis or totally drug-resistant strains in Iran. Chest 136(2):420–425

Velayati AA, Farnia P, Masjedi MR (2013) Totally drug-resistant tuberculosis (TDR-TB): a debate on global health communities. Int J Mycobacteriol. 2(2):71–72

Velayati AA, Farnia P, Mozafari M, Malekshahian D, Seif S, Rahideh S, Mirsaeidi M (2014) Molecular epidemiology of nontuberculous mycobacteria isolates from clinical and environmental sources of a metropolitan city. PLoS One 9(12):e114428

Velayati AA, Rahideh S, Nezhad ZD, Farnia P, Mirsaeidi M (2015a) Nontuberculous mycobacteria in Middle East: current situation and future challenges. Int J Mycobacteriol 4(1):7–17

Velayati AA, Farnia P, Mozafari M, Mirsaeidi M (2015b) Nontuberculous mycobacteria isolation from clinical and environmental samples in Iran: twenty years of surveillance. Biomed Res Int 2015:254285

Velayati AA, Farnia P, Hoffner S (2018) Drug-resistant *Mycobacterium tuberculosis*: epidemiology and role of morphological alterations. J Glob Antimicrob Resist 12:192–196

Velayati AA, Mitaria S, Farnia P, Farnia P, Ghanavi J, Maleknia M, Murase Y, Hoffner S (2024) Exploring the genetic landscape of *Mycobacterium tuberculosis*: unlocking the differences between latent and active tuberculosis. Int J Mycobacteriol. 13(4):410

Vidal SL, Batista LCF, Lages DDS, Bortolomai BE, Bueno IDC, Niitsuma ENA, Oliveira NGD, Baptista IMFD, Lana FCF (2025) Molecular detection of mycobacterium leprae and the process of infection and illness in contacts of leprosy patients: a systematic review. Trop Med Infect Dis 10(2):32

Vonnahme LA, Raykin J, Jones M, Oakley J, Puro J, Langer A et al (2023) Using electronic health record data to measure the latent tuberculosis infection care Cascade in safety-net primary care clinics. AJPM Focus 2(4):100148

Walker TM, Kohl TA, Omar SV, Hedge J, Del Ojo EC, Bradley P, Iqbal Z, Feuerriegel S, Niehaus KE, Wilson DJ, Clifton DA, Kapatai G, Ip CLC, Bowden R, Drobniewski FA, Allix-Béguec C, Gaudin C, Parkhill J, Diel R, Supply P, Crook DW, Smith EG, Walker AS, Ismail N, Niemann S, Peto TEA, Modernizing Medical Microbiology (MMM) informatics Group (2015) Whole-genome sequencing for prediction of *Mycobacterium tuberculosis* drug susceptibility and resistance: a retrospective cohort study. Lancet Infect Dis 15(10):1193–1202

Wallace RJ, Brown BA, Onyi GO (1992) Skin, soft tissue, and bone infections due to *Mycobacterium chelonae* chelonae: importance of prior corticosteroid therapy, frequency of disseminated infections, and resistance to oral antimicrobials other than clarithromycin. J Infect Dis 166(2):405–412

Wang M, Chen X, Sun Y, Wang Q, Liu G (2023) Functions, advantages and challenges facing private healthcare organisations in China's healthcare system: a qualitative analysis through open-ended questionnaires. BMJ Open 13(6):e069381

Wessler T, Joslyn LR, Borish HJ, Gideon HP, Flynn JL, Kirschner DE, Linderman JJ (2020) A computational model tracks whole-lung *Mycobacterium tuberculosis* infection and predicts factors that inhibit dissemination. PLoS Comput Biol 16(5):e1007280

Wigginton JE, Kirschner D (2001) A model to predict cell-mediated immune regulatory mechanisms during human infection with *Mycobacterium tuberculosis*. J Immunol 166(3):1951–1967

Williams PM (2024) Tuberculosis—United States, 2023. MMWR Morb Mortal Wkly Rep 73:265

Williamson DA, Roberts SA, Bower JE, Vaughan R, Newton S, Lowe O et al (2012) Clinical failures associated with rpoB mutations in phenotypically occult multidrug-resistant *Mycobacterium tuberculosis*. Intl J Tubercul Lung Dis 16(2):216–220

Witney AA, Gould KA, Arnold A, Coleman D, Delgado R, Dhillon J, Pond MJ, Pope CF, Planche TD, Stoker NG, Cosgrove CA, Butcher PD, Harrison TS, Hinds J (2015) Clinical application of whole-genome sequencing to inform treatment for multidrug-resistant tuberculosis cases. J Clin Microbiol 53(5):1473–1483

World Health Organization (1996) Global tuberculosis control: WHO report 1996. World Health Organization, Geneva. Available from: https://apps.who.int/iris/handle/10665/62065

World Health Organization (2013) Nutritional care and support for patients with tuberculosis [internet]. WHO, Geneva. [cited 2025 Jul 9]. Available from: https://www.who.int/publications/i/item/9789241565400

World Health Organization (2014a) Global tuberculosis report 2014 [internet]. WHO, Geneva. [cited 2025 Jul 9]. Available from: https://apps.who.int/iris/handle/10665/137094

World Health Organization (2014b) Xpert MTB/RIF implementation manual: technical and operational 'how-to': practical considerations [internet]. WHO, Geneva. [cited 2025 Jul 9]. Available from: https://apps.who.int/iris/handle/10665/112469

World Health Organization (2014c) Companion handbook to the WHO guidelines for the programmatic management of drug-resistant tuberculosis [internet]. WHO, Geneva. [cited 2025 Jul 9]. Available from: https://www.who.int/publications/i/item/9789241548809

World Health Organization (2015) Implementing the end TB strategy: the essentials [internet]. WHO, Geneva. [cited 2025 Jul 9]. Available from: https://apps.who.int/iris/handle/10665/206499

World Health Organization (2016) The use of molecular line probe assays for the detection of resistance to isoniazid and rifampicin: policy update [internet]. WHO, Geneva. [cited 2025 Jul 9]. Available from: https://www.who.int/publications/i/item/9789241511260

World Health Organization (2019) WHO consolidated guidelines on drug-resistant tuberculosis treatment [internet]. WHO, Geneva. [cited 2025 Jul 9]. Available from: https://iris.who.int/handle/10665/311389

World Health Organization (2020) Consolidated guidelines on tuberculosis treatment (Module 4: drug-resistant TB treatment). World Health Organization, Geneva. Available from: https://www.who.int/publications/i/item/9789240007048

World Health Organization (2021a) WHO consolidated guidelines on tuberculosis: module 3, diagnosis: rapid diagnostics for tuberculosis detection [internet]. WHO, Geneva. [cited 2025 Jul 9]. Available from: https://www.who.int/publications/i/item/9789240037021

World Health Organization (2021b) WHO operational handbook on tuberculosis: module 2, management of tuberculosis in children and adolescents [internet]. WHO, Geneva. [cited 2025 Jul 9]. Available from: https://www.who.int/publications/i/item/9789240036949

World Health Organization (2022a) Drug-resistant tuberculosis treatment, 2022 update [internet]. WHO, Geneva. [cited 2025 Jul 9]. Available from: https://www.who.int/publications/i/item/9789240045061

World Health Organization (2022b) WHO consolidated guidelines on tuberculosis: module 4, treatment—drugs and regimens [internet]. WHO, Geneva. [cited 2025 Jul 9]. Available from: https://www.who.int/publications/i/item/9789240051303

World Health Organization (2023) Global tuberculosis report 2023 [internet]. WHO, Geneva. [cited 2025 Jul 9]. Available from: https://www.who.int/teams/tb-reports/global-tuberculosis-report-2023

World Health Organization (2024) Global tuberculosis report 2024 [internet]. WHO, Geneva. [cited 2025 Jul 9]. Available from: https://www.who.int/teams/tb-reports/global-tuberculosis-report-2024

Wouters E, van Rensburg HC, Meulemans H (2010) The National Strategic Plan of South Africa: what are the prospects of success after the repeated failure of previous AIDS policy? Health Policy Plan 25(3):171–185

Yang Z, Li J, Shen J, Cao H, Wang Y, Hu S, Du Y, Wang Y, Yan Z, Xie L, Li Q, Gomaa SE, Liu S, Li X, Li J (2025)

Recent progress in tuberculosis diagnosis: insights into blood-based biomarkers and emerging technologies. Front Cell Infect Microbiol 8(15):1567592

Yayan J, Franke KJ, Berger M, Windisch W, Rasche K (2024) Early detection of tuberculosis: a systematic review. Pneumonia (Nathan) 16(1):11

Yeager RL, Munroe WG, Dessau FI (1952) Pyrazinamide (aldinamide) in the treatment of pulmonary tuberculosis. Am Rev Tuberc 65(5):523–546. PMID: 14924175

Yirsaw AN, Lakew G, Getachew E, Kasaye MD, Enyew EB, Getachew D, Tareke AA (2025) Health extension workers job satisfaction and associated factors in Ethiopia: a systematic review and meta-analysis. BMC Health Serv Res 25(1):144

Zhang M, Adroub S, Ummels R, Asaad M, Song L, Pain A, Bitter W, Guan Q, Abdallah AM (2024) Comprehensive pan-genome analysis of Mycobacterium marinum: insights into genomic diversity, evolution, and pathogenicity. Sci Rep 14(1):27723

Zhang H, Guan W, Zhou J (2025) Advances in the diagnosis of latent tuberculosis infection. Infect Drug Resist. 18:483–493

Zheng Y, Emam M, Lu D, Tian M, Wang K, Peng X (2023) Analysis of the effect of temperature on tuberculosis incidence by distributed lag non-linear model in Kashgar city, China. Environ Sci Pollut Res 30:11530–11541. https://doi.org/10.1007/s11356-022-22849-5

Zimmerman MR (1979) Pulmonary and osseous tuberculosis in an Egyptian mummy. Bull N Y Acad Med 55(6):604–608. PMID: 380700; PMCID: PMC1807652

Zopf W (1883) Die Spaltpilze, 2nd edn. J.U. Kern's Verlag, Breslau

Zumla A, Raviglione M, Hafner R, von Reyn CF (2012) Tuberculosis. N Engl J Med 368(8):745–755

Zumla A, Sahu S, Ditiu L, Singh U, Park YJ, Yeboah-Manu D, et al (2025) Inequities underlie the alarming resurgence of tuberculosis as the world's top cause of death from an infectious disease: breaking the silence and addressing the underlying root causes. IJID14(Suppl 2):100587. https://doi.org/10.1016/j.ijregi.2025.100587. PMID: 40201557; PMCID: PMC1197369

2 Comprehensive Approaches to Protein Detection and Analysis in *Mycobacterium tuberculosis*

Abstract

Comprehensive strategies for detecting and analyzing proteins in *Mycobacterium tuberculosis* (Mtb) integrate advanced experimental and computational approaches to deepen understanding of the bacterium's biology, pathogenesis, and mechanisms of drug resistance. Recent technological advances have refined genome annotations by identifying novel protein-coding regions and improving existing gene models. Utilizing diverse fractionation techniques combined with mass spectrometry, proteins localized to distinct cellular compartments, including the cell wall, membranes, and cytoplasm, can be profiled, providing critical insights into their spatial organization and functional roles. Comparative proteomic analyses across multiple Mtb strains have uncovered both unique protein variants and conserved elements, shedding light on bacterial adaptation and virulence strategies. Mapping protein interaction networks has revealed essential pathways governing metabolism and survival, highlighting the intricate coordination of proteins that underpin Mtb pathogenicity. Furthermore, the integration of machine learning and bioinformatics tools has significantly advanced the prediction of protein functions, posttranslational modifications, and contributions to drug resistance, thereby enabling the prioritization of promising targets for therapeutic intervention. Collectively, these modern methodologies offer a detailed and dynamic portrait of the Mtb proteome, facilitating the discovery of novel biomarkers, drug targets, and vaccine candidates. This comprehensive knowledge base is vital for guiding the development of improved diagnostic tools, effective treatments, and, ultimately, strategies to control and eradicate tuberculosis.

Keywords

Mycobacterium tuberculosis · Protein detection · Mass spectrometry · Liquid chromatography · SDS-PAGE · Protein profiling

2.1 Introduction

Proteins constitute a fundamental component of the biology and pathogenicity of *Mycobacterium tuberculosis* (Mtb), the causative agent of tuberculosis (TB). A comprehensive investigation of these proteins is imperative for advancing our understanding of the mechanisms by which Mtb induces disease, persists within the host environment, and evades immune defenses. Proteins regulate numerous critical processes throughout the bacterium's life cycle, including virulence, metabolic adaptation, and interactions with host

P. Farnia et al., *Proteins in Mycobacterium Tuberculosis*, Advances in Experimental Medicine and Biology 1484, https://doi.org/10.1007/978-3-031-96883-9_2

cells, thereby representing essential targets for research focused on tuberculosis control and therapeutic intervention (Mészáros et al. 2011a, b; Yazdi et al. 2018; Chai et al. 2018; Amalia et al. 2022; Sreelatha et al. 2023; Sun et al. 2025). A primary rationale for studying Mtb proteins is to elucidate the molecular mechanisms underlying pathogenesis. Proteins synthesized by Mtb directly contribute to its capacity to infect host tissues and cause cellular damage. Through detailed characterization of proteins, it becomes possible to gain insight into how Mtb modulates host immune responses to evade recognition and destruction. This knowledge is critical for understanding the establishment and maintenance of infection, as well as the bacterium's ability to persist despite the host's immune defenses. In addition to their role in pathogenesis, Mtb proteins are pivotal in the discovery and development of novel antimicrobial agents. Many bacterial proteins are indispensable for survival and virulence; thus, targeting these proteins can disrupt essential biological functions. Identification and characterization of such proteins facilitate the rational design of drugs that specifically inhibit these targets, effectively arresting bacterial growth and preventing disease progression (Mohammadnabi et al. 2024; Sun et al. 2025). This approach holds particular significance in the context of the escalating emergence of drug-resistant Mtb strains. Proteins also play a crucial role in enhancing diagnostic methodologies for tuberculosis. Specific proteins secreted by Mtb or released upon bacterial cell death serve as biomarkers detectable in clinical specimens. The exploitation of these protein biomarkers has enabled the development of diagnostic assays with improved sensitivity and specificity, thereby facilitating early and accurate detection of both active and latent tuberculosis infections. For example, immunoassays employing antibodies that target secreted proteins from the Mtb cell wall have proven effective in identifying infected individuals (Nahid et al. 2006; Goletti et al. 2016; Pollock et al. 2021; Ayalew et al. 2024; Temur et al. 2025). A comprehensive understanding of the proteins that trigger protective immune responses is crucial for advancing vaccine development against Mtb. By identifying and characterizing immunogenic proteins, the design of vaccines that induce strong and long-lasting immunity becomes possible. Such vaccines aim to prime the immune system to efficiently recognize and eliminate Mtb, thereby reducing both the risk of infection and the severity of tuberculosis. This focused strategy not only guides the rational development of next-generation vaccines but also enhances the effectiveness and durability of protection, ultimately supporting global efforts to control and eradicate tuberculosis (Lai et al. 2023; Yang et al. 2024; Velayati et al. 2024; An et al. 2025; Barazani et al. 2025; Lei et al. 2025). Moreover, the study of Mtb proteins provides critical insights into the bacterium's immune evasion strategies. Mtb secretes proteins that interact with host immune cells, modulating their functions to prevent adequate immune clearance (Zhuang et al. 2024). Investigating these interactions sheds light on the mechanisms by which Mtb establishes long-term persistence within the host, frequently resulting in latent infection (Chai et al. 2020; Velayati et al. 2024; Abbasnia et al. 2024). Latency poses a significant obstacle in tuberculosis control, as Mtb can remain dormant within the host for extended periods before reactivation occurs (Gideon and Flynn 2011). Protein studies are crucial for understanding how Mtb transitions between dormant and active states, which is essential for developing strategies to prevent tuberculosis relapse. Gaining a deeper understanding of these processes provides valuable insights that can guide the development of interventions aimed at preventing disease recurrence (Gengenbacher and Kaufmann 2012; Yang et al. 2023). Among the various proteins expressed by Mtb, secretory proteins are of particular interest due to their pivotal roles in bacterial virulence and modulation of host immune responses. These secreted proteins have the capacity to localize within diverse host cellular compartments and interact with host molecular targets, thereby altering immune signaling pathways to facilitate bacterial survival within the host environment (Arend et al. 2000; Houben et al. 2014; Yang et al. 2016; Stamm et al. 2019; Khosrobeygi et al. 2021; Pal et al.

2022). Therefore, investigating the catalytic mechanisms underlying protein synthesis in Mtb presents significant opportunities for discovering novel therapeutic targets. A comprehensive molecular understanding of the functions and interactions of proteins involved in these pathways enables the rational design of drugs that selectively inhibit essential components of the protein synthesis machinery. By disrupting these critical processes, such targeted therapies can effectively compromise bacterial viability, offering promising and precise strategies for the treatment of tuberculosis.

2.2 Techniques for Protein Separation and Identification in *Mycobacterium tuberculosis*

Over the past several decades, a multitude of techniques have been developed to separate and identify proteins based on their intrinsic physicochemical properties, such as molecular size, electrical charge, solubility, and other chemical characteristics. These methodologies have evolved significantly, ranging from early classical approaches to modern and advanced techniques, each contributing uniquely to the study of proteins in Mtb (Nagai et al. 1991; Andersen and Brennan 1994; Målen et al. 2007, 2011; Parish and Stoker 2008; Lalvani and Pareek 2010; Saraswat et al. 2013; Stasyk and Hube 2004; Lucas et al. 2021). Understanding these methods is crucial for advancing research into Mtb biology, pathogenicity, and for developing effective diagnostics and therapeutics. One of the earliest and most straightforward methods for protein separation is precipitation. This technique involves the selective aggregation and sedimentation of proteins from crude biological mixtures by the addition of specific precipitating agents. Commonly used agents include ammonium sulfate and trichloroacetic acid (TCA), which induce protein precipitation by altering the solubility of proteins under different ionic conditions (Jiang et al. 2004; Mosavari et al. 2021). Ammonium sulfate precipitation is a classic method used in protein purification. It takes advantage of the fact that proteins have different solubilities in solutions with varying salt concentrations. By gradually increasing the ionic strength of the solution through the addition of ammonium sulfate, proteins become less soluble and precipitate, allowing their separation from other cellular components (Wingfield 2001; Duong-Ly and Gabelli 2014). Following precipitation, proteins are typically collected via centrifugation, washed to remove residual contaminants, and then subjected to further biochemical or structural analyses. This precipitation technique not only serves as a fundamental tool in protein purification but also holds historical significance in tuberculosis research. Robert Koch's discovery of tuberculin, a protein extract derived from heat-killed Mtb cultures, marked a milestone in tuberculosis diagnostics and immunology (Koch 1882; Mazana 2009). Although initially intended as a vaccine or therapeutic agent, tuberculin did not succeed in curing tuberculosis; however, it was later adopted as a diagnostic reagent. The diagnostic potential of tuberculin was first practically demonstrated by Clemens von Pirquet in 1907. He observed that individuals previously exposed to tuberculosis exhibited a localized skin reaction upon application of tuberculin via a cutaneous scratch test, a phenomenon he termed "tuberculin allergy." This reaction served as an early indicator of immune sensitization to Mtb (Martini et al. 2018). Building upon von Pirquet's foundational work, Charles Mantoux refined the diagnostic approach in 1912 by introducing intradermal injection of tuberculin instead of superficial scratching. This modification significantly enhanced the test's accuracy and reliability, establishing the Mantoux test as a cornerstone of tuberculosis diagnosis worldwide. Further improvements in tuberculin preparation came in 1934 when Florence Seibert applied ammonium sulfate precipitation to purify tuberculin. This process substantially reduced contaminating proteins, thereby enhancing the specificity of the diagnostic test. The resulting purified preparation, known as Purified Protein Derivative (PPD), became the standard reagent for tuberculin skin testing (Colover et al. 1962; Kim et al. 1964). In this context, ammonium sulfate precipitation increases the ionic strength of the solution, causing non-specific proteins and impurities to precipitate, which yields a more refined and consistent diagnostic product (Nayak and Acharjya 2012). To ensure consistent and reproducible tuberculin skin test results across diverse populations and geographic regions, efforts to standardize purified protein derivative (PPD) were initiated. These standardization efforts have addressed multiple factors influencing tuberculin potency, such as adsorption to container surfaces, exposure to light, molecular weight distribution, and the route of injection (Landi et al. 1975). Additionally, international comparisons of different PPD batches have been performed, taking into account varying sensitivity levels among populations and recommending specific dosages for routine testing (Guld et al. 1958; Dobbelaer et al. 1977). Subsequent investigations identified the "booster phenomenon," first described by Bass and Serio (1981). This phenomenon describes cases where individuals with prior tuberculosis exposure initially test negative but later convert to a positive result upon retesting after a short interval. To address this, a two-step testing protocol was recommended, involving a second tuberculin test administered 1 to 3 weeks after the first. This method is designed to stimulate a boosted immune response and reduce the likelihood of false-negative results (Bass and Serio 1981). The efficacy of PPD was firmly established through the first large-scale clinical trials conducted by Huebner and colleagues, which solidified PPD's status as the gold standard for detecting latent tuberculosis infection (Huebner et al. 1993). Despite its widespread use, the tuberculin skin test is not without limitations (Thompson et al. 1979; Rose et al. 1995; Mazana 2009; Kestler and Tyler 2022). The validity of the two-step tuberculin skin test was further supported by Menzies et al. (1994), who demonstrated its effectiveness in distinguishing true latent tuberculosis infection from boosted immune responses caused by prior *Bacillus Calmette-Guérin* (BCG) vaccination or past environmental exposure to Mtb. This distinction is crucial for accurate diagnosis and appropriate clinical management. The two-step testing approach is especially valuable in tuberculosis surveillance programs targeting healthcare workers and elderly populations, whose immune responses may diminish over time (Hallak et al. 1999). International studies have confirmed that this protocol improves the reliability of tuberculin skin testing in diverse populations with varying BCG vaccination statuses and environmental mycobacterial exposures (Kraut et al. 2004).

As protein purification methods progressed beyond simple precipitation, dialysis emerged as an essential technique for refining protein samples by removing small contaminants such as salts and buffer components. Among the earliest and most widely used purification methods, dialysis remains a fundamental tool in biochemical and molecular biology research (Gottschalk and Fellner 1997). This technique utilizes semi-permeable membranes, commonly made from materials like cellulose, that selectively allow small molecules, including salts, metabolites, and other low molecular weight impurities, to pass through while retaining larger macromolecules such as proteins inside the dialysis bag. Dialysis is particularly valuable for buffer exchange and the removal of small contaminants during protein purification workflows, thereby enhancing sample purity and preparing proteins for subsequent analyses or processing steps (Nagai et al. 1991; Phillips 1995; Andrew et al. 2001, 2002). In practical applications involving Mtb, lysed cultures or protein extracts are enclosed within dialysis bags made of cellulose or similar semi-permeable membranes with defined molecular weight cut-offs. These membranes allow only small molecules to diffuse out while retaining larger proteins. The dialysis bags are immersed in a large volume of buffered water or saline solution, which facilitates the diffusion-driven removal of unwanted small molecules and concentrates the retained protein solution inside the bag. Historically, antigens were prepared not only from culture filtrates but also by extracting dried organisms using organic and aqueous solvents. Among these methods, aqueous extracts demonstrated the highest antigenic activity. This activity could be further enhanced through processes such as dialysis, adsorption with serum globulin, and alcohol extraction (Wadsworth and Maltaner 1921). Following dialysis, the proteins retained within the membrane can be collected and concentrated if necessary. These purified proteins are then suitable for downstream analyses, including immunoassays, electrophoresis, or mass spectrometry (MS), enabling detailed characterization and application in tuberculosis research (Wilkins et al. 1999; Link et al. 1999; Scopes 1993). Automated dialysis systems have revolutionized protein and polymer purification workflows by enabling high-throughput processing with minimal manual intervention. These systems can simultaneously process up to 96 protein samples in milligram quantities, making them highly suitable for structural and functional analyses (Aoki et al. 2009; Kim et al. 2011). Compared to manual methods, automated dialysis is faster, more efficient, and uses less solvent, a benefit demonstrated in polymer purification and increasingly applied to Mtb protein purification (Schuett et al. 2020). By maintaining controlled conditions and continuous buffer exchange, these systems ensure reproducibility and consistency while effectively removing inhibitory small molecules and by-products (Dell'Aquila et al. 2007; Sharma et al. 2024). Automated dialysis is particularly valuable for overcoming challenges associated with expressing membrane or toxic proteins in cell-based systems, as it reduces cytotoxic effects and experimental variability. This enables rapid, reproducible, and parallel production of diverse proteins, which is critical for proteomic research and high-throughput screening of targets from challenging organisms such as Mtb. Moreover, advanced dialysis methods serve a crucial auxiliary role by efficiently removing small-molecule impurities and optimizing buffer conditions. This function is essential for enhancing the sensitivity and reliability of diagnostic assays and improving the efficacy of vaccine formulations (Dell'Aquila et al. 2007; Schuett et al. 2020; Sharma et al. 2024). In tuberculosis research, particular focus has been placed on the isolation and purification of key antigenic proteins such as Early Secretory Antigenic Target-6 (ESAT-6) and Culture Filtrate Protein-10 (CFP-10). These secreted proteins are highly immunogenic and produced by Mtb during early infection, eliciting strong host immune responses essential for early detection and immune targeting (Ganguly et al. 2008; Mojgani et al. 2020; Khosrobeygi et al. 2021; Suameitria Dewi et al. 2023). Both ESAT-6 and CFP-10 are encoded within the region of difference 1 (RD1) genomic locus of Mtb, which is notably absent in the *Bacillus Calmette-Guérin* (BCG) vaccine strain and most nontuberculous mycobacteria. This genomic distinction underscores their specificity and utility as diagnostic markers and vaccine candidates (Pym et al. 2002; Andersen and Doherty 2005; Brosch et al. 2007). Although modern dialysis membranes and automated systems have greatly enhanced buffer exchange and the removal of contaminants, the purification of key Mtb antigens such as ESAT-6 and CFP-10 primarily depends on chromatographic techniques, which provide superior resolution and specificity. Nevertheless, advanced dialysis methods remain an important complementary step, effectively eliminating small-molecule impurities and optimizing buffer conditions. This auxiliary role is critical for improving the sensitivity and reliability of diagnostic assays, as well as the effectiveness of vaccine formulations (Dell'Aquila et al. 2007; Schuett et al. 2020; Sharma et al. 2024).

Apart from dialysis, high-resolution electroelution has been employed to purify novel culture filtrate proteins from Mtb, aiding their immunological characterization (Weldingh et al. 2000). Co-expression and co-purification strategies have also proven effective for obtaining recombinant multiprotein complexes in quantities suitable for structural studies (Thakur et al. 2010). Additionally, advances in heterologous expression and purification of Mtb membrane proteins have been critical for understanding infection biology and developing new therapeutics. Together, these developments in automated and specialized purification techniques have transformed the study of Mtb proteins, supporting both fundamental research and the identification of potential drug targets. The integration of advanced dialysis technologies into purification pipelines complements chromatographic methods, improving protein purity and recovery. These advancements contribute substantially to global efforts in tuberculosis control by enabling the development of accurate diagnostics and effective vaccines.

Complementing dialysis, gel filtration chromatography (GFC) is a pivotal technique for separating proteins and other biomolecules based primarily on their size. Also known by various terms such as gel permeation, gel exclusion, size exclusion, and molecular sieve chromatography, GFC operates on the principle of partitioning molecules according to their ability to penetrate a porous stationary phase (Porath and Flodin 1959; Andrews 1964; Granath and Kvist 1967; Ó'Fágáin et al. 2017; Kummari and Bose 2022). The stationary phase typically consists of a column packed with porous beads made from cross-linked polymers such as agarose or dextran. Within the column, two distinct liquid volumes exist: the void volume (V_0), which is the liquid surrounding the beads, and the internal volume, the liquid contained within the bead pores (Yun et al. 1995; Svec and Huber 2006). The total column volume (V_t) is the sum of these two volumes (Ó'Fágáin et al. 2017). The separation mechanism hinges on the molecular size relative to the pore size of the beads. Molecules smaller than the pores enter and diffuse through the beads, resulting in longer retention times, whereas larger molecules are excluded from the pores and elute faster. This size-based partitioning effectively separates molecules according to their hydrodynamic volume, which depends on both molecular size and shape (Yun et al. 1995; Svec and Huber 2006; Janson 2011). During the 1990s, GFC was extensively utilized to separate secretory proteins of Mtb, shedding light on their roles in cell wall biosynthesis and host immune responses. Since then, GFC has become an indispensable tool for purifying Mtb proteins in their native, biologically active forms (Hoffman et al. 1990; Dobos et al. 1995; Hagel 2001). This technique has been successfully applied to isolate key secreted antigens such as the Antigen 85 complex (Ag85), the 38-kDa antigen, and various low molecular weight proteins ranging from 10 to 41 kDa. These proteins are critical for immune recognition and are integral to the development of diagnostic assays and vaccine candidates (Bentley-Hibbert et al. 1999; Senthil Kumar et al. 2002; Lee et al. 2006; Huygen 2014; Khosrobeygi et al. 2021). The ability of GFC to preserve the native conformation of these proteins facilitates detailed biochemical and immunological analyses, essential for understanding Mtb pathogenesis and host-pathogen interactions.

In addition to chromatographic methods, optimized protein extraction protocols that combine chemical solubilization with mechanical disruption techniques, such as sonication and bead beating, have substantially enhanced the yield and quality of Mtb proteins for downstream purification and proteomic analyses (Målen et al. 2010; Sajid 2017). These approaches are especially critical due to the complex, lipid-rich, and waxy cell wall of Mtb, which presents a significant barrier to efficient protein release. Furthermore, affinity purification techniques, often coupled with mass spectrometry, have been successfully applied to study protein-protein interactions and multiprotein complexes within Mtb (Dunham et al. 2012). These advanced methodologies provide valuable insights into the functional protein networks and molecular mechanisms that govern bacterial survival, virulence, and immune evasion (Bach et al. 2022).

Ion exchange chromatography (IEC) is a powerful and widely utilized technique for separating proteins based on their net electrical charge, a principle first established in the mid-twentieth century (Sober and Peterson 1958; Wang 1990; Jungbauer and Hahn 2009; Cummins et al. 2017a; Shaik 2020). The method exploits the reversible electrostatic interactions between charged analytes, such as proteins, peptides, or nucleic acids, and oppositely charged ionic functional groups covalently attached to the stationary phase of an ion-exchange resin. The stationary phase resin can be classified as either a cation exchanger, which contains negatively charged groups that bind positively charged ions (cations), or an anion exchanger, which contains positively charged groups that bind negatively charged ions (anions) (Patil and Okada 1999; Nesterenko and Haddad 2000). The binding affinity of proteins to the resin depends on several factors, including the protein's net charge, which is influenced by the solution's pH relative to the protein's isoelectric point (pI), as well as the ionic strength and composition of the buffer (Bahadir 2013; Cummins et al. 2017b; Shaik 2020). By gradually increasing the ionic strength of the mobile phase—typically through a salt gradient, or by adjusting the pH—proteins are selectively eluted based on their strength of interaction with the resin, with weakly bound proteins eluting first and strongly bound proteins eluting later. IEC is extensively employed in both research and industrial settings due to its high resolution, capacity, and versatility in purifying complex biomolecular mixtures. Early applications of immunoelectrophoresis (IEP) in tuberculosis studies demonstrated its utility in identifying changes in serum protein profiles during disease progression and recovery; for example, Horner and Elliott (1965) observed a reduction in beta-2 macroglobulin bands and an increase in gamma-1 globulin arcs in recovering tuberculosis patients, indicating an evolving antibody response during convalescence. Collazos et al. (1999) reported a significant decrease in serum beta-2 microglobulin levels during tuberculosis treatment, suggesting its potential use in monitoring therapy response. Reyes et al. (2013) described the purification of an IgA monoclonal antibody specific for the Acr protein using immunoaffinity chromatography with recombinant Acr as the ligand. The purification of key recombinant antigens from Mtb, such as ESAT-6 and Acr, typically involves a combination of ion exchange chromatography (IEC), affinity chromatography, and size exclusion chromatography (SEC).

The choice of resin in IEC critically influences protein separation efficiency and selectivity. Different resins possess distinct binding characteristics and capacities tailored for various proteins (Beck et al. 2023). Commonly used matrices include agarose-based resins like Sepharose Fast Flow (FF), which are popular due to their high flow rates and mechanical stability. However, these agarose-based resins may exhibit trace adsorption of positively charged proteins because of residual anionic sulfate groups present on the beads (Yu and Sun 2012). Strong anion-exchange resins such as Q Sepharose FF, Poros QE/M, and TSK-Gel Q-5PW differ in their pH operating ranges, binding efficiencies, and dynamic capacities for various proteins (Staby et al. 2001). For example, Q Sepharose XL, a strong anion exchanger, demonstrates high binding capacity and excellent salt tolerance, making it suitable for challenging separations (Staby et al. 2001). In contrast, weak anion exchangers like DEAE Sepharose FF show the expected pH-dependent retention behavior, with binding strength varying predictably with pH changes (Staby et al. 2007). Both strong and weak anion exchangers exhibit protein-specific binding affinities and elution profiles, underscoring the importance of empirical testing for each target protein (Staby et al. 2007). Notably, the phenomenon of trace adsorption of positively charged proteins on agarose-based anion exchangers persists even at low NaCl concentrations typical of IEC buffers, likely due to residual sulfate groups on the resin surface (Yu and Sun 2012). These studies provide valuable guidance for resin selection during process development, emphasizing that optimal resin choice depends on the specific protein, buffer conditions, and separation goals. Therefore, systematic evaluation of resins under relevant conditions is essential to maximize purity, yield, and process robustness (Staby et al. 2001, 2005; Staby et al. 2007).

Polymer-based chromatographic methods, including hydrophobic interaction chromatography (HIC) and size exclusion chromatography (SEC), have been essential tools for protein purification for over a decade. HIC separates proteins based on differences in surface hydrophobicity by exploiting reversible interactions between hydrophobic regions on protein surfaces and hydrophobic ligands attached to the chromatography resin. This technique is especially effective for purifying bacterial proteins, including those from Mtb, and is valued for preserving protein stability due to its mild, non-denaturing conditions (Fausnaugh et al. 1984; Kennedy 2001; O'Connor and Cummins 2017;

Murphy et al. 2011; Fekete et al. 2016; Ewonde et al. 2022). In Lussow et al. 1991, Lussow et al. demonstrated that HSP60 can serve as a carrier molecule for peptide antigens without the need for conventional adjuvants. De Bruyn et al. (2000) identified multiple forms of mycobacterial HSP60 distinguished by hydrophobicity and showed that lipids are non-covalently bound to HSP60-2 and homologous proteins such as *Escherichia coli* GroEL. Cao et al. (2010) developed a purification protocol combining HIC with IEC to produce a clinical-grade HSP65-MUC1 fusion protein. IEC itself has proven highly effective for purifying mycobacterial proteins. HIC was instrumental in purifying ESAT-6, a low molecular weight T cell antigen secreted by Mtb (Sorensen et al. 1995). Proteins such as the 38 kDa, 30 kDa (antigen 85B), and 16 kDa were isolated using various chromatographic methods and exhibited distinct immunological properties (Devi et al. 2002). The antigen 85 complex (30/31 kDa) and 38 kDa antigens were purified from culture filtrates using anion exchange chromatography and evaluated for their diagnostic potential in tuberculosis (Senthil Kumar et al. 2002). Monoclonal antibodies against $H_{37}Rv$ antigens were also produced and purified, with IEC yielding the highest purity (Lu et al. 2000). IEC has been used to isolate extracellular and secretory protein antigens from the $H_{37}Rv$ strain, including immunodominant proteins of 6, 27, 30, 38, and 64 kDa, which were characterized for their immunological properties (Tiwari et al. 2014). These purified antigens and antibodies show promise for developing more sensitive and specific tuberculosis diagnostics, especially when used in combination. Size exclusion chromatography (SEC) has proven particularly useful for enriching Mtb extracellular vesicles. Compared to traditional ultracentrifugation, SEC offers gentler separation that better preserves vesicle integrity (Ryan et al. 2022). Both SEC and HIC complement IEC by separating proteins based on size, shape, and hydrophobicity, whereas IEC separates based on charge. Overall, IEC's scalability, compatibility with automation, and high resolving power make it an ideal technique for purifying biologically active mycobacterial proteins (Cummins et al. 2017c).

Electrophoresis is a widely used, accessible technique for separating and detecting proteins based on their size, charge, and shape. The method involves applying an electric field to a protein sample, causing the proteins to migrate through a medium such as agarose gel, polyacrylamide gel, or capillary columns. Electrophoresis plays a crucial role in diagnostic applications and the characterization of Mtb proteins. Among electrophoretic methods, Sodium Dodecyl Sulfate Polyacrylamide Gel Electrophoresis (SDS-PAGE) is routinely employed in tuberculosis laboratories to separate proteins according to their molecular weight (Brown 1988; Guttman and Nolan 1994; Schuhmacher et al. 1996; Gallagher 2006; Khan et al. 2016; Kielkopf et al. 2021). SDS-PAGE involves treating proteins with the anionic detergent SDS, which denatures the proteins and imparts a uniform negative charge, thereby allowing separation solely based on size when subjected to an electric field. After electrophoresis, proteins are visualized using stains such as Coomassie Brilliant Blue or silver stain, which bind specifically to proteins (Al-Tubuly 2000). For higher-resolution analysis of Mtb proteins, two-dimensional gel electrophoresis (2DGE) is commonly employed. This technique separates proteins in two steps: first by their isoelectric point (pI) along the horizontal axis, and then by molecular weight along the vertical axis, providing much greater resolution than one-dimensional methods. Weldingh et al. (1998) applied 2DGE to analyze and purify novel culture filtrate proteins from Mtb, identifying several proteins that elicit strong immune responses. Thereafter, Wong et al. (1999) combined 2DGE with mass spectrometry and database searching to precisely identify expressed proteins and investigate their physiological roles, including the study of iron regulation in Mtb, which led to the identification of iron-regulated proteins such as Fur and aconitase. Rosenkrands et al. (2000) further demonstrated the power of 2D-PAGE by separating hundreds of proteins from various Mtb subcellular fractions, including culture filtrates, cell walls, and cytosolic extracts, achieving superior resolution compared to one-dimensional methods. Although 2DGE tends to favor the resolution of low molecular weight proteins, it complements other proteomic techniques like isotope-coded affinity tag (ICAT) technology, which is more effective for analyzing high molecular weight proteins. Together, these methods provide a more comprehensive and detailed understanding of the Mtb proteome (Schmidt et al. 2004). Recent methodological advancements have further enhanced 2DGE. Differential In-Gel Electrophoresis (DIGE) improves sensitivity and reproducibility by labeling different protein samples with distinct fluorescent cyanine (Cy) dyes, allowing simultaneous separation of up to three samples on a single gel. This multiplexing reduces gel-to-gel variability and enables precise comparative analyses (Viswanathan et al. 2006; Falvo et al. 2011; Minden 2012; Truong et al. 2014; König 2020).

Schmidt et al. (2004) compared two-dimensional difference gel electrophoresis (2D-DIGE) with isotope-coded affinity tag (ICAT) technology, uncovering complementary insights into protein quantification. While ICAT favored high molecular weight proteins, 2D-DIGE showed a preference for low molecular weight and cysteine-free proteins (Schmidt et al. 2004). Subsequent studies confirmed the complementary nature of these methods, with each technique identifying unique redox-sensitive proteins (Fu et al. 2008). More recently, Kaushik et al. (2022) identified 18 abundant Mtb proteins and 11 host proteins in urine, suggesting promising biomarkers for tuberculosis diagnosis. Collectively, these studies demonstrate the power of proteomic approaches in advancing our understanding of Mtb biology and in identifying potential therapeutic targets. Målen et al. (2010) enhanced the detection of membrane-associated and low-abundance proteins in Mtb by combining subcellular fractionation and protein enrichment prior to two-dimensional electrophoresis (2-DE) analysis. They utilized Triton X-114 detergent-phase separation together with high-resolution mass spectrometry, identifying 1417 proteins, including 395 proteins not previously described. This comprehensive approach revealed numerous membrane proteins critical to the pathogen's virulence. By profiling protein composition across different cellular compartments, their work has provided valuable insights that support the identification of novel targets for tuberculosis pathogenesis and drug development.

Additional electrophoretic techniques, such as Blue Native PAGE (BN-PAGE), enable the separation of membrane protein complexes under mild, non-denaturing conditions, making them particularly valuable for studying hydrophobic and complex membrane proteins (Reisinger and Eichacker 2006, 2008; Dresler et al. 2011). This method uses neutral detergents to solubilize proteins and Coomassie Brilliant Blue G-250 dye to impart a negative charge to protein complexes, allowing their separation based on molecular mass (Reisinger and Eichacker 2006, 2008). BN-PAGE is especially useful for investigating protein-protein interactions, which are critical for understanding bacterial physiology and pathogenesis, particularly in membrane proteins (Dresler et al. 2011). The technique is well regarded for its reproducibility and ease of use, although optimal solubilization conditions must be tailored for each specific membrane protein complex (Reisinger and Eichacker 2008). BN-PAGE has been successfully applied to various *Mycobacterium* species, including Mtb. For instance, Zheng et al. (2011) combined BN-PAGE with liquid chromatography-tandem mass spectrometry (LC-MS/MS) to identify 40 proteins in *Mycobacterium bovis BCG* membrane fractions, revealing significant protein-protein interactions involved in lipid transport and ATP synthesis. Houben et al. (2012) employed BN-PAGE to elucidate the composition of the ESX-5 membrane complex, a component of the type VII secretion system in Mtb, which is essential for secreting virulence factors. Esposito et al. (2012) investigated the dimerization of the heparin-binding hemagglutinin adhesin (HBHA) from Mtb using BN-PAGE, contributing to a better understanding of bacterial aggregation mechanisms. Leite et al. (2015) used BN-PAGE to characterize membrane protein complexes in *Mycobacterium avium* subsp. *paratuberculosis*, identifying antigenic complexes with potential diagnostic applications. Furthermore, 2-DE is widely used for analyzing posttranslational modifications (PTMs) like phosphorylation and glycosylation. Specialized stains and bioinformatics tools have been developed to detect and characterize these modifications directly on gels, providing detailed information on protein regulation and function.

Modern protein research extensively utilizes mass spectrometry (MS), a highly sensitive and precise analytical technique that measures the mass-to-charge ratio (m/z) of ionized molecules (Gygi & Aebersold, 2000; Trauger et al. 2002; Kowalczyk et al. 2020). The fundamental principle of MS involves ionizing chemical compounds to generate charged particles, which are then separated according to their m/z values using electric or magnetic fields. The resulting ions are detected to produce a mass spectrum, a graphical representation of ion intensity as a function of m/z, that enables accurate identification, quantification, and structural characterization of molecules in both purified samples and complex biological mixtures (Trauger et al. 2002). Mass spectrometry techniques are classified based on their ionization methods, mass analyzers, and analytical workflows. Among ionization techniques, electron ionization (EI) employs high-energy electrons to ionize gaseous molecules, often causing extensive fragmentation, and is primarily used for small-molecule analysis. Electrospray ionization (ESI) is a soft ionization method that produces multiply charged ions from liquid samples, making it ideal for analyzing proteins and large biomolecules (Garg and Zubair 2024). Matrix-assisted laser desorption/ionization (MALDI) uses laser pulses to ionize analytes embedded in a chemical matrix, facilitating the analysis of large biomolecules and enabling imaging mass spectrometry (Murray 2021; Udhayabanu et al. 2025). The other ionization methods include chemical ionization, atmospheric pressure chemical ionization (APCI), and photoionization, each suited to specific analytical contexts (Niu et al. 2020). In terms of workflows, tandem mass spectrometry (MS/MS) involves sequential mass analysis coupled with fragmentation steps that provide detailed molecular structural information, crucial for peptide sequencing and posttranslational modification mapping (Neagu et al. 2022). Affinity purification mass spectrometry (AP-MS) combines biochemical affinity purification of protein complexes with MS identification, enabling the study of protein-protein interactions and complex composition (Gnanasekaran and Pappu 2023). Quantitative MS approaches include label-free quantification, stable isotope labeling by amino acids in cell culture (SILAC), tandem mass tags (TMT), and targeted quantification methods such as selected reaction monitoring (SRM) or multiple reaction monitoring (MRM) (Mirza 2012; Calderón-Celis et al. 2017; van Bentum and Selbach 2021).

Mass spectrometry is indispensable in proteomics for comprehensive protein identification, isoform differentiation, characterization of posttranslational modifications, quantification of protein abundance and dynamics, mapping of protein-protein interactions, primarily through AP-MS, and investigation of protein structure and complex formation. A landmark study by Kelkar et al. (2011) conducted a proteogenomic analysis of $H_{37}Rv$ using high-resolution Fourier transform mass spectrometry (FT-MS), identifying 3176 proteins, which accounted for approximately 80% of the predicted proteome. This study also discovered around 250 novel peptides, leading to the annotation of 41 previously unknown protein-coding genes and the correction of 79 gene models (Kelkar et al. 2011). Potgieter et al. (2016) similarly utilized high-resolution mass spectrometry for *Mycobacterium smegmatis*, identifying 2887 open reading frames (ORFs) and validating translational start sites for 558 gene models, highlighting the importance of proteomic data in genome annotation (Potgieter et al. 2016). Beyond protein identification, investigating protein interactions is critical, particularly when uncharacterized proteins associate with proteins of known function. Such interaction data can provide valuable insights into the biological roles of novel proteins. Techniques integrating affinity purification with MS have been instrumental in elucidating these protein networks, advancing our understanding of cellular processes and pathogen biology (Dunham et al.

2012; Oeffinger 2012; Morris et al. 2014; Meyer and Selbach 2015; Liu et al. 2020; Lee and O'Reilly 2023; Serrano et al. 2025).

Beyond AP-MS, electrospray ionization tandem mass spectrometry (ESI-MS/MS) has been widely applied for rapid identification of Mtb proteins by analyzing unique mycolic acid profiles of the bacterial cell wall (Song et al. 2009; Szewczyk et al. 2013; López-Hernández et al. 2016; Magdalena et al. 2025). This technique enables fast and sensitive detection by targeting characteristic fatty acids specific to different mycobacterial species. Furthermore, ESI-MS/MS can differentiate drug-resistant strains based on variations in mycolic acid intensities and provides detailed structural insights through advanced multistage mass spectrometry (Magdalena et al. 2025). Currently, liquid chromatography coupled with tandem mass spectrometry (LC-MS/MS) stands as a gold standard analytical technique in proteomics, metabolomics, and drug development. This method combines the separation power of liquid chromatography with the highly sensitive and selective mass analysis capability of triple quadrupole or high-resolution mass spectrometers (Briki et al. 2024). One illustrative example of LC-MS/MS application in Mtb research involves the hypothetical protein Rv0341. Flyer et al. (2002) used LC-MS/MS to analyze peptides presented by major histocompatibility complex (MHC) class I molecules on macrophages infected with Mtb, identifying three peptides derived from Rv0341. This discovery advanced understanding of antigen presentation and immune recognition in tuberculosis. In addition, the application of LC-MS/MS to *Mycobacterium smegmatis* enabled the identification and relative quantification of 901 proteins, significantly surpassing previous methods and greatly enhancing gene annotation and validation efforts (Wang et al. 2005). More comprehensive proteomic analyses combining LC-MS/MS with subcellular fractionation and membrane enrichment techniques identified 1051 Mtb protein groups, including 183 transmembrane proteins (Bell et al. 2012). Proteomic studies using label-free LC-MS/MS have also revealed dynamic changes in Mtb extracellular proteins under nutrient starvation, identifying toxin-antitoxin system members implicated in dormancy (Albrethsen et al. 2013). Furthermore, isobaric tags for relative and absolute quantitation (iTRAQ) combined with 2D-LC-MS/MS has been employed to analyze serum from tuberculosis patients, identifying differentially expressed proteins with potential as biomarkers for treatment monitoring (Wang et al. 2015). Comparative proteomic analyses of different Mtb lineages have elucidated lineage-specific variations in growth and virulence proteins, informing targeted therapeutic strategies (Asare et al. 2020). Recent peptidomic studies have identified immunogenic peptides from secreted proteins, aiding vaccine candidate discovery and understanding of host-pathogen interactions (Mayer and Impens 2021). Moreover, LC-MS/MS analyses of macrophages infected with Mtb and treated with therapeutic compounds have provided insights into protein regulation relevant to treatment efficacy (Wu et al. 2024). In conclusion, AP-MS, ESI-MS/MS, and LC-MS/MS represent cornerstone technologies in Mtb proteomics. Their combined capabilities for sensitive, high-throughput, and detailed protein identification, quantification, and interaction analysis are indispensable for advancing tuberculosis research, facilitating biomarker discovery, and accelerating the development of diagnostics, therapeutics, and vaccines.

Cryo-Electron Microscopy (Cryo-EM) represents a next-generation method in protein detection and structural biology, having gained substantial prominence for its ability to elucidate protein structures at near-atomic resolution without the prerequisite of crystallization (Cheng et al. 2015; Frank 2017; Shen 2018). This technique involves rapidly freezing protein samples within a thin layer of vitreous ice, preserving their native conformation, followed by imaging with an electron microscope (Binshtein and Ohi 2015). Over the past decade, Cryo-EM has profoundly advanced our understanding of Mtb protein structures, offering critical insights into their biological functions and identifying novel therapeutic targets. A striking example of Cryo-EM's impact is the resolution of the cytochrome bd oxidase structure from Mtb at an exceptional 2.5 Å resolution (Safarian et al. 2021). This high-resolution structural analysis revealed a previously unidentified menaquinone-9 (MK-9) binding site within the enzyme complex, significantly enhancing our comprehension of the respiratory pathways Mtb employs for energy production and survival, especially under the hostile conditions encountered within the host. The identification of this novel binding site opens promising avenues for drug development, as molecules designed to inhibit MK-9 binding could disrupt bacterial respiration and thereby impair Mtb viability. This study exemplifies how Cryo-EM can uncover critical molecular details that inform the design of innovative therapeutics against persistent pathogens. Similarly, Saridakis et al. (2022) elucidated the Cryo-EM structure of the transcription termination factor Rho from Mtb at 3.3 Å resolution. Their work revealed a unique structural framework for Mtb Rho and provided mechanistic insights into its intrinsic resistance to the antibiotic bicyclomycin. Specifically, a leucine-to-methionine amino acid substitution in Mtb Rho was found to sterically hinder bicyclomycin binding, explaining why Mtb is naturally resistant to this drug. These findings deepen our understanding of bacterial transcription regulation and highlight potential strategies for overcoming antibiotic resistance. Concurrently, the Cryo-EM structure of the Mtb 50S ribosomal subunit complexed with clarithromycin was reported (Zhang et al. 2022). This study illuminated the dynamic interactions between the ribosome and macrolide antibiotics, providing valuable structural information that can guide the rational design of more effective antimicrobial agents targeting the ribosome. In 2023, a significant breakthrough was achieved with the determination of the Cryo-EM structures of the iron-loaded siderophore transporter IrtAB from Mtb (Sun et al. 2017). They resolved the structures of unliganded IrtAB and its complexes with ATP, ADP, and the ATP analogue adenylylimidodiphosphate (AMP-PNP) at resolutions ranging from 2.8 to 3.5 Å. These structures revealed ATP-driven conformational changes essential for iron acquisition, a vital process for Mtb survival and virulence, thus providing a molecular basis for targeting iron uptake pathways therapeutically (Sun and Sun 2021). More recently, Cryo-EM studies have elucidated the structure of the Rv1217c-Rv1218c transporter complex, shedding light on its role in rifampicin resistance (Wang et al. 2024). This discovery provides new molecular insights into mechanisms of antibiotic resistance in Mtb and identifies potential targets for overcoming drug resistance, a critical challenge in tuberculosis treatment. Collectively, these advances underscore the transformative impact of Cryo-EM in tuberculosis research. By revealing high-resolution structural details of key Mtb proteins and complexes, Cryo-EM has accelerated the development of targeted therapies, informed antibiotic design, and enhanced our understanding of bacterial physiology and pathogenicity.

Among the foremost advanced techniques in proteomics, protein microarrays have also emerged as a powerful and versatile tool for high-throughput analysis of protein interactions and functions (Stoevesandt et al. 2009). Protein microarrays enable the simultaneous screening of thousands of proteins against a wide array of ligands, including antibodies, small molecules, nucleic acids, and other proteins, thereby providing comprehensive insights into complex protein networks, signaling pathways, and cellular functional relationships (Wolf-Yadlin et al. 2009; Syu et al. 2020). This multiplexed approach facilitates rapid identification of binding partners, detection of posttranslational modifications, and discovery of biomarkers, making protein microarrays invaluable for elucidating disease mechanisms, validating drug targets, and advancing vaccine development (Montoya et al. 2025). Their scalability, high sensitivity, and adaptability render them essential for accelerating systems-level studies in tuberculosis and other infectious diseases. Microarray technologies have been extensively applied to Mtb research at the genomic and transcriptomic levels, yielding critical insights into bacterial gene regulation and adaptation. DNA microarrays became a widely adopted tool for analyzing global gene expression patterns under diverse physiological and stress conditions, including infection, hypoxia, nutrient deprivation, and exposure to antitubercular drugs (Deng et al. 2014; Li et al. 2020; Sun and Sun 2021). For example, DNA microarray studies characterizing the Mtb transcriptome during hypoxic conditions revealed gene expression changes associated with dormancy and persistence, shedding light on mechanisms underlying latent tuberculosis (Kendall et al. 2004; Murphy and Brown 2007; Balázsi et al. 2008; Hudock et al. 2017).

Complementing DNA microarrays, peptide microarrays, and focused protein arrays have been developed to interrogate specific antigenic regions and protein interactions with greater precision (Gaseitsiwe et al. 2010).

Kunnath-Velayudhan et al. (2010) pioneered the use of protein microarray technology to comprehensively analyze antibody responses against the entire Mtb proteome. Their study revealed that sera from tuberculosis suspects recognized approximately 10% of the bacterial proteome, with antibody responses predominantly targeting extracellular proteins during active disease and correlating with bacillary load. Building on this work, subsequent studies investigated the dynamic nature of antibody responses in both macaques and humans, linking these profiles to infection outcomes (Kunnath-Velayudhan et al. 2012). These proteome-wide approaches have provided an unbiased and global view of the Mtb immunoproteome, identifying potential biomarkers helpful in detecting acute infection and preclinical tuberculosis (Kunnath-Velayudhan and Porcelli 2013). More recently, advancements such as the High-Density Nucleic Acid Programmable Protein Array have refined the identification of antibody targets, uncovering both known and novel immunoreactive proteins with promising diagnostic potential (Song et al. 2017).

A functional Mtb proteome microarray, covering the majority of the proteome along with an open reading frame (ORF) library, has been developed to further enhance the utility of microarray technology (Deng et al. 2014). This platform demonstrated broad applicability in investigating global protein-protein interactions, small-molecule-protein binding, and serum biomarker discovery. Using this approach, researchers identified 59 PknG-interacting proteins, 30 bis-(3′-5′)-cyclic dimeric guanosine monophosphate (c-di-GMP) binding proteins, and 14 Mtb proteins capable of distinguishing patients with active tuberculosis from those who had recovered (Deng et al. 2014). Additionally, a two-way proteome microarray approach identified 84 potential human-Mtb interactors, providing insights into Mtb pathogenesis (Cao et al. 2019). Serum biomarker screening using proteome microarrays revealed 152 Mtb-antigen-specific IgGs with significantly higher levels in active TB patients compared to latent TB infection, with Rv2031c, Rv1408, and Rv2421c showing promising diagnostic potential (Cao et al. 2019). Furthermore, a comprehensive STPK protein interaction map was constructed, identifying 492 binding proteins and 1027 interactions, highlighting diverse functional roles including cell wall integrity regulation by PknG (Wu et al. 2017). Overall, microarray platforms have proven indispensable for advancing our understanding of Mtb biology, pathogenesis, and host-pathogen interactions. By enabling large-scale, multiplexed analyses, these technologies facilitate the identification of novel biomarkers and therapeutic targets, thereby accelerating research efforts aimed at combating tuberculosis.

The clustered regularly interspaced short palindromic repeat interference (CRISPRi) system has emerged as a particularly powerful and versatile genetic tool for precisely and efficiently repressing target gene expression in mycobacteria, including the formidable pathogen Mtb (Choudhary et al. 2015; Bosch et al. 2021; Wong and Rock 2021). This innovative technology offers significant advantages over traditional gene knockout approaches by enabling reversible and tunable gene silencing without introducing permanent genetic alterations. Unlike conventional gene editing methods that rely on generating double-strand breaks in the DNA, CRISPRi employs a catalytically inactive Cas9 (dCas9) protein. This dCas9 variant retains its DNA-binding capability but lacks the nuclease activity to cleave DNA (Choudhary et al. 2015). The specificity of dCas9 targeting is directed by a customizable single-guide RNA (sgRNA), which is designed to be complementary to a specific sequence within the promoter region or coding sequence of the target gene. Upon binding, the dCas9-sgRNA complex physically obstructs the transcriptional machinery, sterically hindering

RNA polymerase binding or elongation. This interference effectively silences the expression of the targeted gene at the transcriptional level (Rock et al. 2017; Singpanomchai and Ratthawongjirakul 2024). The key benefits of the CRISPRi system in mycobacterial research are manifold (Fleck and Grundner 2021). Firstly, its high specificity allows for precise targeting of individual genes, minimizing off-target effects that can confound experimental results. Secondly, the ability to achieve tunable and reversible gene repression provides researchers with a powerful means to study gene function dynamically. This is particularly valuable for investigating essential genes in Mtb, where traditional knockout approaches often lead to lethality, making their study challenging. By titrating the level of gene repression, researchers can explore the dose-dependent effects of gene dosage on bacterial physiology, metabolism, virulence, and drug susceptibility. Furthermore, CRISPRi facilitates high-throughput screening for drug targets and genetic determinants of drug resistance (Yimer et al. 2020; Yan et al. 2022). By systematically repressing genes across the Mtb genome, researchers can identify genes whose silencing leads to increased susceptibility to existing antibiotics or reveals novel drug vulnerabilities. This capability is especially critical given the increasing global burden of multidrug-resistant and extensively drug-resistant tuberculosis. In summary, the CRISPRi system represents a transformative advancement in mycobacterial genetics. Its ability to provide precise, targeted, and reversible gene silencing offers unprecedented opportunities to dissect complex biological pathways in Mtb, accelerating the discovery of novel therapeutic strategies and contributing significantly to the global fight against tuberculosis.

2.3 Analyzing Proteins in *Mycobacterium tuberculosis*

Analyzing proteins in *Mycobacterium tuberculosis* (Mtb) is crucial for uncovering the complex molecular mechanisms that drive its pathogenesis, identifying reliable drug targets, and advancing vaccine development. In recent years, computational and bioinformatics approaches have transformed this field by enabling high-throughput, cost-effective, and detailed analyses that integrate diverse data types. These methods provide a comprehensive characterization of Mtb proteins, allowing for in-depth examination of active site residues and their interactions, key factors in understanding protein activity, specificity, and function (Saikat 2023; Devlin et al. 2025).

A wide range of computational techniques has been effectively employed to investigate Mtb proteins at the atomic and molecular levels. Homology or comparative modeling predicts three-dimensional (3D) protein structures by using known templates, addressing the frequent lack of experimentally determined structures for many Mtb proteins. Molecular dynamics simulations complement this by providing dynamic views of protein flexibility and conformational changes over time. Additionally, protein-ligand docking and structure-based ligand optimization facilitate the analysis of binding affinities and interaction modes between proteins and small molecules, shedding light on how specific mutations influence ligand binding and alter protein function (Eswar et al. 2006; Vyas et al. 2012; Fakhar et al. 2016; Śledź and Caflisch 2018; Bertoline et al. 2023). These computational approaches have been pivotal in understanding the molecular mechanisms behind drug resistance in tuberculosis, where mutations often reduce drug binding affinity, leading to decreased therapeutic efficacy. The integration of advanced methods such as machine learning (ML), artificial intelligence (AI), bioinformatics, and cheminformatics databases has further strengthened the predictive capabilities of these analyses. By combining these tools, researchers can build sophisticated models that forecast the impact of mutations on protein structure and function, as well as their role in conferring resistance to drugs (Xie et al. 2023; Chen et al. 2024). Such predictive frameworks accelerate the discovery of resistance-associated mutations and support the rational design of new therapeutics.

Comprehensive bioinformatics analyses enable the modeling of 3D structures of Mtb proteins, the identification of novel structural motifs, the investigation of specific biochemical pathways, and the classification of proteins into functional categories. This holistic approach not only clarifies evolutionary relationships and biological lineages but also facilitates the precise assignment of functions to previously uncharacterized proteins. These structural and functional insights are invaluable for drug discovery, providing a rational foundation for pinpointing pharmacological targets and designing effective therapeutic agents against tuberculosis and other infectious diseases (Mészáros et al. 2011a, 2011b). Overall, the integration of computational and bioinformatics tools has revolutionized Mtb protein research by enabling detailed molecular characterization, unraveling drug resistance mechanisms, and guiding informed drug design. These advances are crucial for accelerating the development of novel interventions to combat tuberculosis, a significant global health threat.

The rapid expansion of genome sequencing data has enabled extensive analyses aimed at identifying clinically significant polymorphisms, thereby deepening our knowledge of drug resistance evolution in Mtb (Cohen et al. 2019; Koleske et al. 2023; Zein-Eddine et al. 2025). To organize and facilitate access to mutation data, several specialized databases have been developed, including the Genome-wide *Mycobacterium tuberculosis* Variation Database (GMTVD) (Chernyaeva et al. 2014), the Tuberculosis Drug Resistance Mutation Database (TBDReaMDB) (Sandgren et al. 2009), and MUBII-TB-DB (a database of mutations associated with antibiotic resistance in *Mycobacterium tuberculosis*) (Flandrois et al. 2014). While these resources are invaluable, they often require continual updating to incorporate the latest mutation records, which is critical for a comprehensive understanding of anti-TB drug resistance dynamics. Key genes implicated in drug resistance, such as *rpo*B, *inh*A, *kat*G, *pnc*A, *gyr*A, and *gyr*B, harbor mutations that disrupt protein function and impair drug binding, leading to antibiotic resistance (Zhang and Yew 2015; Phelan et al. 2016; Singh et al. 2021). Among genetic variations, single-nucleotide polymorphisms (SNPs) represent the most prevalent contributors to altered protein function and resistance phenotypes in Mtb. While many SNPs are silent or benign, non-synonymous mutations often exert deleterious effects by modifying amino acid residues critical for protein activity. Identifying these mutations in drug-resistant isolates is therefore essential for elucidating resistance mechanisms.

Alongside traditional phenotypic assays, in silico methods have become essential for prioritizing mutations by predicting their structural and functional effects, particularly in cases where quantitative phenotypic data are unavailable (Pires et al. 2016; Albanaz et al. 2017). By combining homology modeling with molecular dynamics simulations and protein-ligand docking, it is possible to assess the structural and biophysical impacts of mutations on drug targets (Kumar and Jena 2014; Mugumbate et al. 2021). These computational approaches facilitate the prediction of resistance-conferring mutations and support the rational design of new therapeutics. Platforms like the *Mycobacterium tuberculosis* Resistance Associated Polymorphisms Database (MycoTRAP-DB) enhance these efforts by linking genetic mutations to their predicted structural and functional consequences, thereby accelerating tuberculosis research and improving drug susceptibility testing.

Beyond well-characterized proteins, computational analyses of hypothetical Mtb proteins have uncovered critical enzymatic functions previously unrecognized. For instance, several hypothetical proteins have been identified as disulfide oxidoreductases, enzymes essential for maintaining redox homeostasis and proper protein folding, processes vital for bacterial survival and pathogenicity (Chim et al. 2011; Bocian-Ostrzycka et al. 2017; Carranza et al. 2021; Gómez-González et al. 2022). These findings, derived from in silico structural modeling, sequence analysis, and functional prediction, highlight the potential of hypothetical proteins as sources of novel biochemical activities contributing to Mtb's complex biology. Understanding these enzymatic roles expands knowledge of the pathogen's physiology and reveals new therapeutic targets, especially considering the crucial role of redox regulation in bacterial viability and virulence (Araújo et al. 2020; Rahman et al. 2022; Chakma et al. 2023).

Advances in computational biology have profoundly transformed tuberculosis (TB) vaccine research. The integration of genomics, structural biology, and bioinformatics has accelerated the identification and optimization of potential vaccine antigens (Liljeroos et al. 2015). This "Translational Systems Biology" approach, which combines experimental data with mathematical modeling, provides a deeper understanding of the complex biological processes underlying TB infection and immunity (Day et al. 2010). Structural and computational methods such as epitope mapping and prediction have yielded critical insights into immunogenic determinants, enabling more rational vaccine design (Ptak et al. 2023). Immunoinformatics and peptide discovery techniques have been instrumental in identifying and optimizing T cell epitopes from key Mtb antigens, supporting the development of multistage vaccines that elicit robust cellular immune responses (Ortega-Tirado et al. 2020). Building on this foundation, bio-immunoinformatics, a multidisciplinary field combining bioinformatics and immunology, offers promising strategies for designing more effective TB vaccines through precise prediction of both B cell and T cell epitopes (Ahmad et al. 2025). Recently, Nguyen and Kim (2024) demonstrated the power of integrating structural modeling, epitope prediction, and molecular conjugation techniques to enhance vaccine stability and immunogenicity. Their work highlights the translational potential of computational biology, providing a robust framework for rational vaccine development.

By merging sequence analysis, structural modeling, functional prediction, and immunoinformatics, these computational approaches create a comprehensive platform to elucidate the roles of hypothetical proteins in Mtb. This integrative strategy accelerates the development of targeted therapies and vaccines, deepening our understanding of Mtb biology and opening new avenues for combating tuberculosis through rational drug and vaccine design (Saikat 2023; Nguyen and Kim 2024). Overall, computational methodologies have significantly accelerated TB vaccine research, offering the potential for breakthroughs in vaccine efficacy, stability, and immunogenicity. The synergistic application of structural modeling, epitope prediction, and molecular conjugation exemplifies the impactful role of computational biology in advancing TB vaccine development (Ahmad et al. 2025).

Protein-ligand docking and structure-based virtual screening are essential computational techniques for analyzing Mtb proteins and identifying potential drug targets (Bishi et al. 2019; de Azevedo 2011). These methods simulate interactions between small molecules and specific binding sites on Mtb proteins, offering detailed molecular insights into protein function and ligand affinity (de Azevedo 2011). Molecular dynamics simulations further enhance understanding by revealing ligand-binding characteristics and aiding in the selection of compounds with higher affinity (de Azevedo 2011; Kingdon and Alderwick 2021). With the growing availability of Mtb protein structures, novel targets are now being explored using advanced in silico techniques such as ensemble docking and free energy calculations (Kingdon and Alderwick 2021). These approaches have successfully identified potential inhibitors for key proteins like PknB, a serine/threonine protein kinase involved in cell wall synthesis and metabolism (Vieira et al. 2021). Computational methods thus play a critical role in addressing drug resistance and guiding the development of new antituberculosis drugs (Bishi et al. 2019; Kingdon and Alderwick 2021). Recent studies have focused on designing novel inhibitors targeting DNA gyrase, a vital enzyme in Mtb DNA replication and a well-established drug target. Adeniji et al. (2020) utilized quantitative structure-activity relationship (QSAR) modeling to predict inhibitory activity and identify molecular features influencing anti-tubercular potency, demonstrating the value of computational methods in drug discovery. Building on this, Carpio Arévalo and Amorim (2022) applied virtual screening and molecular docking to discover promising DNA gyrase inhibitors, including a pyrrolo[1,2-a]quinazoline derivative with favorable binding and pharmacokinetic profiles. Sajid et al. (2024) further expanded the pool of candidate inhibitors by characterizing 2-(2-(aryl)-4-oxo-4,5-dihydrothiazol-5-yl)acetohydrazides, which showed strong binding affinity and desirable drug-like properties through docking analyses. Parallel efforts by Lee et al. (2024) employed machine learning-assisted drug repurposing to identify compounds DB11753 and DB14703 as dual inhibitors targeting both DNA gyrase A and PknB. Their study combined molecular dynamics simulations to rigorously evaluate the stability and binding affinities of these candidates, illustrating the power of integrating AI-driven approaches with traditional computational chemistry. More recently, Nyijime et al. (2025) conducted a comprehensive in silico investigation on DNA gyrase inhibitors, designing novel compounds and thoroughly assessing their pharmacokinetics and safety profiles. Molecular docking revealed key interactions within the enzyme's active site, while molecular dynamics simulations provided insights into the stability and flexibility of protein-ligand complexes over time. Together with prior research by Shallangwa and Adeniji (2021), Lee et al. (2024), and Elsaman et al. (2025), these findings firmly establish protein-ligand docking and structure-based optimization as critical and indispensable tools in the rational design of next-generation antituberculosis therapeutics.

Advanced computational approaches, including machine learning (ML) and artificial intelligence (AI), combined with bioinformatics and cheminformatics databases, are increasingly employed to predict the impact of genetic mutations on drug resistance in Mtb. These models utilize extensive genomic and phenotypic datasets to identify key resistance markers, predict resistance profiles for specific anti-TB drugs, and forecast co-occurring resistance patterns (Ekins et al. 2017; Rusic et al. 2024; Zhang et al. 2024). In this context, Deelder et al. (2019) applied classification-tree and gradient-boosted-tree models to a dataset of 16,688 Mtb isolates, achieving high predictive accuracy across most anti-TB drugs and establishing a foundation for ML-based resistance prediction. Jamal et al. (2020) expanded on this by developing a computational framework using various ML algorithms to predict resistance in specific Mtb genes, reaching an average accuracy of 85%. More recently, Sirri et al. (2021) employed advanced boosting algorithms and deep neural networks on large isolated datasets, attaining exceptional area under the curve (AUC) values between 0.97 and 0.99 across nine anti-TB drugs. The latest work by Paredes-Gutierrez et al. (2025) applied four distinct ML algorithms to predict resistance to both first- and second-line drugs, reporting high accuracy and reinforcing the clinical potential of ML techniques.

Complementing these advances, Chen et al. (2025) highlighted the importance of integrating the updated World Health Organization (WHO) mutation catalog with computational tools. Their study demonstrated that combining the latest WHO-curated mutation database with ML models significantly improves drug resistance prediction accuracy in Mtb. The updated catalog offers a comprehensive, validated framework of resistance-associated mutations, enhancing the reliability of computational predictions. This synergy between curated mutation data and AI-driven analytics represents a powerful strategy to accelerate tuberculosis diagnostics, personalize treatment regimens, and inform public health policies. Collectively, these studies illustrate the transformative potential of combining mutation catalogs with sophisticated AI and ML models, paving the way for rapid, cost-effective, and scalable drug susceptibility testing, critical steps toward controlling and ultimately eradicating tuberculosis.

In summary, the isolation, identification, and in-depth analysis of Mtb proteins are fundamental to advancing our understanding of the bacterium's biology, pathogenesis, and drug resistance mechanisms. Proteins act as key effectors of cellular processes, and their detailed study provides crucial insights into bacterial survival strategies and host-pathogen interactions. Integrating proteomic data with genomic and transcriptomic information offers a comprehensive view of bacterial physiology, facilitating the discovery of vulnerabilities that can be targeted for clinical intervention. Protein-focused research thus remains indispensable for improving tuberculosis control, enhancing patient outcomes, and addressing this persistent global health challenge.

Conflict of Interest The authors declare that they have no conflict of interest related to the content of this chapter.

Ethical Consideration This chapter is a review based entirely on data and information obtained from previously published and publicly available sources, all of which have been appropriately cited. The authors conducted no new experimental work involving human participants or animals. Therefore, no additional ethical approval was required. The authors have ensured proper attribution and compliance with ethical standards in the use of all referenced materials.

References

Abbasnia S, Hashem Asnaashari AM, Sharebiani H, Soleimanpour S, Mosavat A, Rezaee SA (2024) *Mycobacterium tuberculosis* and host interactions in the manifestation of tuberculosis. J Clin Tubercul Other Mycobacter Dis 36:100458

Adeniji SE, Adalumo OB, Ekoja FO (2020) Anti-tubercular modelling, molecular docking simulation and insight toward computational design of novel compounds as potent antagonist against DNA gyrase receptor. Med Microecol 5:100020

Ahmad A, Rukmana A, Khairinisa MA, Pitaloka DAE, Agus R, Ladju RB et al (2025) Comprehensive review on the application of bio-immunoinformatics in the development of highly effective new candidate vaccines against tuberculosis. Karbala Int J Mod Sci 11(2):2. https://doi.org/10.33640/2405-609X.3399

Albanaz ATS, Rodrigues CHM, Pires DEV, Ascher DB (2017) Combating mutations in genetic disease and drug resistance: understanding molecular mechanisms to guide drug design. Expert Opin Drug Discov 12(6):553–563

Albrethsen J, Agner J, Piersma SR, Højrup P, Pham TV, Weldingh K, Jimenez CR, Andersen P, Rosenkrands I (2013) Proteomic profiling of *Mycobacterium tuberculosis* identifies nutrient-starvation-responsive toxin-antitoxin systems. Mol Cell Proteomics 12(5):1180–1191

Al-Tubuly AA (2000) SDS-PAGE and Western blotting. Methods Mol Med 40:391–405

Amalia F, Syamsunarno MRAA, Triatin RD, Fatimah SN, Chaidir L, Achmad TH (2022) The role of amino acids in tuberculosis infection: a literature review. Meta 12(10):933

An Y, Ni R, Zhuang L, Yang L, Ye Z, Li L, Parkkila S, Aspatwar A, Gong W (2025) Tuberculosis vaccines and therapeutic drug: challenges and future directions. Mol Biomed 6:4

Andersen ÅB, Brennan P (1994) Proteins and antigens of *Mycobacterium tuberculosis*. In: Bloom BR (ed) Tuberculosis: pathogenesis, protection, and control. ASM Press, pp 307–332

Andersen P, Doherty TM (2005) The success and failure of BCG—implications for a novel tuberculosis vaccine. Nat Rev Microbiol 3(8):656–662

Andrews P (1964) Estimation of the molecular weights of proteins by Sephadex gel-filtration. Biochem J 91(2):222–233

Andrew SM, Titus JA, Zumstein L (2001) Dialysis and concentration of protein solutions. Curr Protoc Cell Biol. Appendix 3: Appendix 3C. https://doi.org/10.1002/0471143030.cba03cs04. PMID:18228285

Andrew SM, Titus JA, Zumstein L (2002) Dialysis and concentration of protein solutions. Curr Protoc Toxicol. Appendix 3:A.3H.1-5. https://doi.org/10.1002/0471140856.txa03hs10. PMID:20976673

Aoki M, Matsuda T, Tomo Y, Miyata Y, Inoue M, Kigawa T, Yokoyama S (2009) Automated system for high-throughput protein production using the dialysis cell-free method. Protein Expr Purif 68(1):128–136

Araújo CL, Blanco I, Souza L, Tiwari S, Pereira LC, Ghosh P, Azevedo V, Silva A, Folador A (2020) In silico functional prediction of hypothetical proteins from the core genome of Corynebacterium pseudotuberculosis biovar ovis. PeerJ 8:e9643

Arend SM, Andersen P, van Meijgaarden KE, Skjøt RLV, Subronto YW, van Dissel JT, Ottenhoff THM (2000) Detection of active tuberculosis infection by T cell responses to early-secreted antigenic target 6-kDa protein and culture filtrate protein 10. J Infect Dis 181(5):1850–1854

Arévalo JMC, Amorim JC (2022) Virtual screening, optimization and molecular dynamics analyses highlighting a pyrrolo[1,2-a]quinazoline derivative as a potential inhibitor of DNA gyrase B of *Mycobacterium tuberculosis*. Sci Rep 12:4742

Asare P, Otchere ID, Bedeley E, Brites D, Loiseau C, Baddoo NA et al (2020) Whole genome sequencing and spatial analysis identifies recent tuberculosis transmission hotspots in Ghana. Front Med (Lausanne) 7:161. https://doi.org/10.3389/fmed.2020.00161. PMID:32509791; PMCID:PMC7248928

Ayalew S, Wegayehu T, Wondale B, Tarekegn A, Tessema B, Admasu F, Piantadosi A, Sahi M, Gebresilase TT, Fredolini C, Mihret A (2024) Candidate serum protein biomarkers for active pulmonary tuberculosis diagnosis in tuberculosis endemic settings. BMC Infect Dis 24(1):1329

Bach F, Wejse C, Storgaard M, Patsche CB (2022) Is body height a prognostic marker for outcome of tuberculosis treatment? Infect Dis 54(7):538–541

Bahadir O (2013) Ion-exchange chromatography and its applications. In: Martin MA (ed) Column chromatography. InTech

Balázsi G, Heath AP, Shi L, Gennaro ML (2008) The temporal response of the *Mycobacterium tuberculosis* gene regulatory network during growth arrest. Mol Syst Biol 4:225

Barazani O, Erdos T, Chowdhury R, Kaur G, Venketaraman V (2025) New advances in the development and design of *Mycobacterium tuberculosis* vaccines: construction and validation of multi-epitope vaccines for tuberculosis prevention. Biology 14(4):417

Bass JA, Serio RA (1981) The use of repeat skin tests to eliminate the booster phenomenon in serial tuberculin testing. Am Rev Respir Dis 123(4 Pt 1):394–396

Beck J, Hochdaninger G, Carta G, Hahn R (2023) Resin structure impacts two-component protein adsorption and separation in anion exchange chromatography. J Chromatogr A 1705:464208

Bell C, Smith GT, Sweredoski MJ, Hess S (2012) Characterization of the *Mycobacterium tuberculosis* proteome by liquid chromatography mass spectrometry-based proteomics techniques: A comprehensive resource for tuberculosis research. J Proteome Res 11(1):119–130

Bentley-Hibbert SI, Quan X, Newman T, Huygen K, Godfrey HP (1999) Pathophysiology of antigen 85 in patients with active tuberculosis: antigen 85 circulates as complexes with fibronectin and immunoglobulin G. Infect Immun 67(2):581–588

Bertoline LMF, Lima AN, Krieger JE, Teixeira SK (2023) Before and after AlphaFold2: An overview of protein structure prediction. Frontiers in Bioinformatics 3:1120370

Binshtein E, Ohi MD (2015) Cryo-electron microscopy and the amazing race to atomic resolution. Biochemistry 54:3133

Bishi LY, Vedithi SC, Blundell TL, Mugumbate GC (2019) Computational deorphaning of *Mycobacterium tuberculosis* targets. In IntechOpen

Bocian-Ostrzycka KM, Grzeszczuk MJ, Banaś AM, Jagusztyn-Krynicka EK (2017) Bacterial thiol oxidoreductases—from basic research to new antibacterial strategies. Appl Microbiol Biotechnol 101(10):3977–3989

Bosch B, DeJesus MA, Poulton NC, Zhang W, Engelhart CA, Zaveri A, Lavalette S, Ruecker N, Trujillo C, Wallach JB, Li S, Ehrt S, Chait BT, Schnappinger D, Rock JM (2021) Genome-wide gene expression tuning reveals diverse vulnerabilities of *Mycobacterium tuberculosis*. Cell 184(17):4579–4592.e24

Briki M, Murisier A, Guidi M, Seydoux C, Buclin T, Marzolini C, Girardin FR, Thoma Y, Carrara S, Choong E, Decosterd LA (2024) Liquid chromatography coupled to tandem mass spectrometry (LC-MS/MS) methods for the therapeutic drug monitoring of cytotoxic anticancer drugs: An update. J Chromatogr B 1236:124039

Brosch R, Gordon SV, Garnier T, Eiglmeier K, Frigui W, Valenti P, Dos Santos S, Duthoy S, Lacroix C, Garcia-Pelayo C, Inwald JK, Golby P, Garcia JN, Hewinson RG, Behr MA, Quail MA, Churcher C, Barrell BG, Parkhill J, Cole ST (2007) Genome plasticity of BCG and impact on vaccine efficacy. Proc Natl Acad Sci 104(13):5596–5601

Brown EG (1988) Mixed anionic detergent/aliphatic alcohol-polyacrylamide gel electrophoresis alters the separation of proteins relative to conventional sodium dodecyl sulfate-polyacrylamide gel electrophoresis. Anal Biochem 174(1):337–348

Calderón-Celis F, Cid-Barrio L, Encinar JR, Sanz-Medel A, Calvete JJ (2017) Absolute venomics: absolute quantification of intact venom proteins through elemental mass spectrometry. J Proteome 164:33–42

Cao Z, Feng Y, Wei H, Fang M, Hu X, Yu Y, Wang L, Wan M (2010) Purification of clinical-grade recombinant HSP65-MUC1 fusion protein. Biotechnol Appl Biochem 57(1):9–15

Cao T, Lyu L, Jia H, Wang J, Du F, Pan L, Li Z, Xing A, Xiao J, Ma Y, Zhang Z (2019) A two-way proteome microarray strategy to identify novel Mycobacterium tuberculosis-human interactors. Front Cell Infect Microbiol 9:65. https://doi.org/10.3389/fcimb.2019.00065. PMID:30984625; PMCID:PMC6448480

Carranza C, Herrera MT, Guzmán-Beltrán S, Salgado-Cantú MG, Salido-Guadarrama I, Santiago E, Chávez-Galán L, Gutiérrez-González LH, González Y (2021) A dual marker for monitoring MDR-TB treatment: host-derived miRNAs and *M. tuberculosis*-derived RNA sequences in serum. Front Immunol 12:760468

Chai Q, Zhang Y, Liu CH (2018) *Mycobacterium tuberculosis*: an adaptable pathogen associated with multiple human diseases. Front Cell Infect Microbiol 8:158

Chai Q, Wang L, Liu CH, Ge B (2020) New insights into the evasion of host innate immunity by *Mycobacterium tuberculosis*. Cell Mol Immunol 17:901–913

Chakma V, Barman DN, Das SC, Hossain A, Momin MB, Tasneem M, Gupta SD (2023) In silico analysis of a novel hypothetical protein (YP_498675.1) from Staphylococcus aureus unravels the protein of tryptophan synthase beta superfamily (try-synth-beta_II). J Genet Eng Biotechnol 21(1):135

Chen J, Lin A, Luo P (2024) Advancing pharmaceutical research: A comprehensive review of cutting-edge tools and technologies. Curr Pharm Anal 21(1):1–19

Chen Y, Zhang X, Liang J, Jiang Q, Peierdun M, Xu P, Takiff HE, Gao Q (2025) Advantages of updated WHO mutation catalog combined with existing whole-genome sequencing-based approaches for mycobacterium tuberculosis resistance prediction. Genome Med 17:31. https://doi.org/10.1186/s13073-025-01458-0

Cheng Y, Grigorieff N, Penczek PA, Walz T (2015) A primer to single-particle cryo-electron microscopy. Cell. 23;161(3):438-449 161:438

Chernyaeva EN, Shulgina MV, Rotkevich MS, Dobrynin PV, Simonov SA, Shitikov EA, Ischenko DS, Karpova IY, Kostryukova ES, Ilina EN, Govorun VM, Zhuravlev VY, Manicheva OA, Yablonsky PK, Isaeva YD, Nosova EY, Mokrousov IV, Vyazovaya AA, Narvskaya OV, Lapidus AL, O'Brien SJ (2014) Genome-wide *Mycobacterium tuberculosis* variation (GMTV) database: A new tool for integrating sequence variations and epidemiology. BMC Genomics 15:308

Chim N, Habel JE, Johnston JM, Krieger I, Miallau L, Sankaranarayanan R, Morse RP, Bruning J, Swanson S, Kim H, Kim CY, Li H, Bulloch EM, Payne RJ, Manos-Turvey A, Hung LW, Baker EN, Lott JS, James MN, Terwilliger TC, Eisenberg DS, Sacchettini JC, Goulding CW (2011) The TB structural genomics consortium: A decade of progress. Tuberculosis 91(2):155–172

Choudhary E, Thakur P, Pareek M, Agarwal N (2015) Gene silencing by CRISPR interference in mycobacteria. Nat Commun 6:6267

Cohen KA, Manson AL, Desjardins CA, Abeel T, Earl AM (2019) Deciphering drug resistance in *Mycobacterium tuberculosis* using whole-genome sequencing: Progress, promise, and challenges. Genome Med 11:45

Collazos J, Martínez E, Mayo J (1999) Evolution of serum beta2-microglobulin concentrations during treatment of tuberculosis patients. Scand J Infect Dis 31(3):265–267. https://doi.org/10.1080/00365549950163554. PMID:10482055

Colover J, Kremer M, Lovelace RE, Silver JR (1962) Effect of intrathecal purified protein derivative of tuberculin in multiple sclerosis. J Neurol Neurosurg Psychiatry 25(1):79–85

Cummins PM, O'Connor BF, Ó'Fágáin C (2017a) Gel-filtration chromatography. Methods Mol Biol 1485:15–25

Cummins PM, Rochfort KD, O'Connor BF (2017b) Ion-exchange chromatography: basic principles and application. Methods Mol Biol 1485:209–223

Cummins PM, Rochfort KD, O'Connor BF (2017c) Hydrophobic interaction chromatography. Methods Mol Biol 1485:355–363

Day J, Schlesinger L, Friedman A (2010) Tuberculosis research: going forward with a powerful "translational systems biology" approach. Tuberculosis 90(1):1–8

de Azevedo WF Jr (2011) Protein targets for development of drugs against mycobacterium tuberculosis. Curr Med Chem 18(9):1255–1257. https://doi.org/10.2174/092986711795029564. PMID:21366537

De Bruyn J, Soetaert K, Buyssens P, Calonne I, De Coene JL, Gallet X, Brasseur R, Wattiez R, Falmagne P, Montrozier H, Lanéelle MA, Daffé M (2000) Evidence for specific and non-covalent binding of lipids to natural and recombinant Mycobacterium bovis BCG hsp60 proteins, and to the Escherichia coli homologue GroEL. Microbiology 146(Pt 7):1513–1524

Deelder W, Christakoudi S, Phelan J, Benavente ED, Campino S, McNerney R, Palla L, Clark TG (2019) Machine learning predicts accurately *Mycobacterium tuberculosis* drug resistance from whole genome sequencing data. Front Genet 10:922

Dell'Aquila R, Rodighiero MP, Ronco C (2007) Advances in the technology of automated, tidal, and continuous flow peritoneal dialysis. Perit Dial Int 27(Suppl 2):S2–S6

Deng J, Bi L, Zhou L, Guo SJ, Fleming J, Jiang HW, Zhou Y, Gu J, Zhong Q, Wang ZX, Liu Z, Deng RP, Gao J, Chen T, Li W, Wang JF, Wang X, Li H, Ge F, Zhu G, Zhang HN, Gu J, Wu FL, Zhang Z, Wang D, Hang H, Li Y, Cheng L, He X, Tao SC, Zhang XE (2014) *Mycobacterium tuberculosis* proteome microarray for global studies of protein function and immunogenicity. Cell Rep 9(6):2317–2329

Devi KR, Kumar KS, Ramalingam B, Alamelu R (2002) Purification and characterization of three immunodominant proteins (38, 30, and 16 kDa) of *Mycobacterium tuberculosis*. Protein Expr Purif 24(2):188–195

Devlin KL, Leach DT, Stratton KG, Lamichhane G, Lin VS, Beatty KE (2025) Proteomic characterization of mycobacterium tuberculosis subjected to carbon starvation. mSystems 10(5). e0153024, https://doi.org/10.1128/msystems.01530-24. PMID:40231840; PMCID:PMC12090744

Dobbelaer R, Knops J, Meulepas E, Uydebrouck M (1977) Standardization of a PPD preparation in Guinea-pigs and in man. J Biol Stand 5(3):187–195

Dobos KM, Swiderek K, Khoo KH, Brennan PJ, Belisle JT (1995) Evidence for glycosylation sites on the 45-kilodalton glycoprotein of *Mycobacterium tuberculosis*. Infect Immun 63(8):2846–2853

Dresler J, Klimentova J, Stulik J (2011) Bacterial protein complexes investigation using blue native PAGE. Microbiol Res 166(1):47–62

Dunham WH, Mullin M, Gingras AC (2012) Affinity-purification coupled to mass spectrometry: basic principles and strategies. Proteomics 12(10):1576–1590

Duong-Ly KC, Gabelli SB (2014) Salting out of proteins using ammonium sulfate precipitation. Methods Enzymol 541:85–94

Ekins S, Godbole AA, Kéri G, Orfi L, Pato J, Bhat RS, Verma R, Bradley EK, Nagaraja V (2017) Machine learning and docking models for *Mycobacterium tuberculosis* topoisomerase I. Tuberculosis 103:52–60

Elsaman T, Mohamed MA, Mohamed MS, Eltayib EM, Abdalla AE (2025) Microbial-based natural products as potential inhibitors targeting DNA gyrase B of mycobacterium tuberculosis: an in silico study. Front Chem 13:1524607. https://doi.org/10.3389/fchem.2025.1524607. PMID:39917046; PMCID:PMC11798933

Esposito C, Cantisani M, D'Auria G, Falcigno L, Pedone E, Galdiero S, Berisio R (2012) Mapping key interactions in the dimerization process of HBHA from mycobacterium tuberculosis: insights into bacterial agglutination. FEBS Lett 586(6):659–667. https://doi.org/10.1016/j.febslet.2012.01.047. PMID:22306117

Eswar N, Webb B, Marti-Renom MA, Madhusudhan MS, Eramian D, Shen MY, Pieper U, Sali A (2006) Comparative protein structure modeling using Modeller. Curr Protoc Bioinform. Chapter 5:unit 5.6

Ewonde RE, Lingg N, Eßer D, Eeltink S (2022) A protocol for setting-up robust hydrophobic interaction chromatography targeting the analysis of intact proteins and monoclonal antibodies. Analytical Science Advances 3(11–12):304–312

Fakhar Z, Naiker S, Alves CN, Govender T, Maguire GE, Lameira J, Lamichhane G, Kruger HG, Honarparvar B (2016) A comparative modeling and molecular docking study on *Mycobacterium tuberculosis* targets involved in peptidoglycan biosynthesis. J Biomol Struct Dyn 34(11):2399–2417

Falvo S, Di Carli M, Desiderio A, Benvenuto E, Moglia A, America T, Lanteri S, Acquadro A (2011) 2-D DIGE analysis of UV-C radiation-responsive proteins in globe artichoke leaves. Proteomics 11(24):4296–4309

Fausnaugh JL, Kennedy LA, Regnier FE (1984) Comparison of hydrophobic-interaction and reversed-phase chromatography of proteins. J Chromatogr A 317:141–155

Fekete S, Veuthey JL, Beck A, Guillarme D (2016) Hydrophobic interaction chromatography for the characterization of monoclonal antibodies and related products. J Pharm Biomed Anal 130:3–18

Flandrois JP, Lina G, Dumitrescu O (2014) MUBII-TB-DB: A database of mutations associated with antibiotic resistance in *Mycobacterium tuberculosis*. BMC Bioinform 15:107

Fleck N, Grundner C (2021) A Cas12a-based CRISPR interference system for multigene regulation in mycobacteria. J Biol Chem 297(2):100990

Flyer DC, Ramakrishna V, Miller C, Myers H, McDaniel M, Root K, Flournoy C, Engelhard VH, Canaday DH, Marto JA, Ross MM (2002) Identification by mass spectrometry of CD8+-T-cell *Mycobacterium tuberculosis* epitopes within the Rv0341 gene product. Infect Immun 70(6):2926–2932

Frank J (2017) Advances in the field of single-particle cryo-electron microscopy over the last decade. Nat Protoc 12:209–212

Fu C, Hu J, Liu T, Ago T, Sadoshima J, Li H (2008) Quantitative analysis of redox-sensitive proteome with DIGE and ICAT. J Proteome Res 7(9):3789–3802

Gallagher SR (2006) One-dimensional SDS gel electrophoresis of proteins. Curr Protoc Immunol 75(1):8.4.1–8.4.37

Ganguly N, Giang PH, Gupta C, Basu SK, Siddiqui I, Salunke DM, Sharma P (2008) *Mycobacterium tuberculosis* secretory proteins CFP-10, ESAT-6 and the CFP10:ESAT6 complex inhibit lipopolysaccharide-induced NF-κB transactivation by downregulation of reactive oxidative species (ROS) production. Immunol Cell Biol 86(1):98–106

Garg E, Zubair M (2024) Mass spectrometer. StatPearls Publishing, In StatPearls. Retrieved from https://www.ncbi.nlm.nih.gov/books/NBK574559/

Gaseitsiwe S, Valentini D, Mahdavifar S, Reilly M, Ehrnst A, Maeurer M (2010) Peptide microarray-based identification of *Mycobacterium tuberculosis* epitope binding to HLA-DRB10101, DRB11501, and DRB10401. Clin Vaccine Immunol 17(1):168–175

Gengenbacher M, Kaufmann SH (2012) *Mycobacterium tuberculosis*: success through dormancy. FEMS Microbiol Rev 36(3):514–532

Gideon HP, Flynn JL (2011) Latent tuberculosis: what the host "sees"? Immunol Res 50(2–3):202–212

Gnanasekaran P, Pappu HR (2023) Affinity purification-mass spectrometry (AP-MS) and co-immunoprecipitation (co-IP) technique to study protein-protein interactions. Methods Mol Biol 2690:81–85. https://doi.org/10.1007/978-1-0716-3327-4_7. PMID:37450138

Goletti D, Petruccioli E, Joosten SA, Ottenhoff TH (2016) Tuberculosis biomarkers: from diagnosis to protection. Infect Dis Reports 8(2):6568

Gómez-González PJ, Campino S, Phelan JE, Clark TG (2022) Portable sequencing of *Mycobacterium tuberculosis* for clinical and epidemiological applications. Brief Bioinform 23(5):bbac256

Gottschalk CW, Fellner SK (1997) (1997). History of the science of dialysis. Am J Nephrol 17(3–4):289–298

Granath KA, Kvist BE (1967) Molecular weight distribution analysis by gel chromatography on Sephadex. J Chromatogr A 28(1):69–81

Guld J, Bentzon MW, Bleiker MA, Griep WA, Magnusson M, Waaler H (1958) Standardization of a new batch of purified tuberculin (PPD) intended for international use. Bull World Health Organ 19(5):845–951

Guttman A, Nolan J (1994) Comparison of the separation of proteins by sodium dodecyl sulfate-slab gel electrophoresis and capillary sodium dodecyl sulfate-gel electrophoresis. Anal Biochem 221(2):285–289

Gygi SP, Aebersold R (2000) Mass spectrometry and proteomics. Curr Opin Chem Biol 4(5):489–494. https://doi.org/10.1016/s1367-5931(00)00121-6. PMID:11006534

Hagel L (2001) Gel-filtration chromatography. Current protocols in protein science, chapter 8, unit 8.3

Hallak KM, Schenk M, Neale AV (1999) Evaluation of the two-step tuberculin skin test in health care workers at an inner-city medical center. J Occup Environ Med 41(5):393–396

Hoffman D, Venters R, Franks S, Spicer L (1990) Use of gel filtration in the preparation of biological fluids for magnetic resonance spectroscopy. Magn Reson Med 13(3):507–513

Horner G, Elliott GB (1965) An immuno-electrophoretic apparatus: studies in tuberculosis. Can Med Assoc J 92(9):461–464

Houben EN, Bestebroer J, Ummels R, Wilson L, Piersma SR, Jiménez CR, Ottenhoff TH, Luirink J, Bitter W (2012) Composition of the type VII secretion system membrane complex. Mol Microbiol 86(2):472–484

Houben EN, Korotkov KV, Bitter W (2014) Take five—type VII secretion systems of Mycobacteria. Biochimica et Biophysica Acta (BBA) – molecular. Cell Res 1843(8):1707–1716

Hudock TA, Foreman TW, Bandyopadhyay N, Gautam US, Veatch AV, DN LB et al (2017) Hypoxia sensing and persistence genes are expressed during the intragranulomatous survival of mycobacterium tuberculosis. Am J Respir Cell Mol Biol 56(5):637–647. https://doi.org/10.1165/rcmb.2016-0239OC. PMID:28135421; PMCID:PMC5449490

Huebner RE, Schein MF, Bass JB Jr (1993) The tuberculin skin test. Clin Infect Dis 17(6):968–975

Huygen K (2014) The immunodominant T-cell epitopes of the mycolyl-transferases of the antigen 85 complex of *Mycobacterium tuberculosis*. Front Immunol 5:321

Jamal S, Khubaib M, Gangwar R, Grover S, Grover A, Hasnain SE (2020) Artificial intelligence and machine learning based prediction of resistant and susceptible mutations in *Mycobacterium tuberculosis*. Sci Rep 10:5487

Janson JC (2011) Protein purification: principles, high resolution methods, and applications, 3rd edn. Wiley

Jiang L, He L, Fountoulakis M (2004) Comparison of protein precipitation methods for sample preparation prior to proteomic analysis. J Chromatogr A 1023(2):317–320

Jungbauer A, Hahn R (2009) Ion-exchange chromatography. Methods Enzymol 463:349–371

Kaushik A, Bandyopadhyay S, Porwal C, Srinivasan A, Rukmangadachar LA, Hariprasad G, Kola S, Kataria J, Singh UB (2022) 2D-DIGE based urinary proteomics and functional enrichment studies to reveal novel *Mycobacterium tuberculosis* and human protein biomarker candidates for pulmonary tuberculosis. Biochem Biophys Res Commun 619:15–21

Kelkar DS, Kumar D, Kumar P, Balakrishnan L, Muthusamy B, Yadav AK, Shrivastava P, Marimuthu A, Anand S, Sundaram H, Kingsbury R, Harsha HC, Nair B, Prasad TS, Chauhan DS, Katoch K, Katoch VM, Kumar P, Chaerkady R, Ramachandran S, Dash D, Pandey A (2011) Proteogenomic analysis of *Mycobacterium tuberculosis* by high resolution mass spectrometry. Mol Cell Proteomics 10(12):M111.011627

Kendall SL, Rison SCG, Movahedzadeh F, Frita R, Stoker NG (2004) What do microarrays really tell us about *Mycobacterium tuberculosis*? Trends Microbiol 12(12):537–544

Kennedy RM (2001) Hydrophobic-interaction chromatography. Current protocols in protein science, Chapter 8, unit 8.4

Kestler B, Tyler SK (2022) Latent tuberculosis testing through the ages: the search for a sleeping killer. Am J Phys Lung Cell Mol Phys 322(3):L412–L419

Khan N, Mironov G, Berezovski MV (2016) Direct detection of endogenous microRNAs and their post-transcriptional modifications in cancer serum by capillary electrophoresis-mass spectrometry. Anal Bioanal Chem 408:2891–2899

Khosrobeygi M, Mosavari N, Salehi M, Mojgani N, Akbari M (2021) Isolation and purification of low molecular weight proteins from culture filtrate of *Mycobacterium tuberculosis* strain C. Iranian J Microbiol 13(2):273–282

Kielkopf CL, Bauer W, Urbatsch IL (2021) Sodium dodecyl sulfate–polyacrylamide gel electrophoresis of proteins. Cold Spring Harb Protoc 2021(12):pdb.rot102228

Kim HK, Magnusson M, Bentzon MW (1964) Tuberculin production: 4. Specificity of tuberculin of *Mycobacterium tuberculosis* and *M. bovis* (BCG strain). Acta Pathol Microbiol Scand 60(2):241–251

Kim Y, Babnigg G, Jedrzejczak R, Eschenfeldt WH, Li H, Maltseva N, Hatzos-Skintges C, Gu M, Makowska-Grzyska M, Wu R, An H, Chhor G, Joachimiak A (2011) High-throughput protein purification and quality assessment for crystallization. Methods 55(1):12–28

Kingdon ADH, Alderwick LJ (2021) Structure-based in silico approaches for drug discovery against *Mycobacterium tuberculosis*. Comput Struct Biotechnol J 19:3708–3719

Koch R (1882) Die Aetiologie der Tuberkulose. Berl Klin Wochenschr 19:221–230

Koleske BN, Jacobs WR Jr, Bishai WR (2023) The *Mycobacterium tuberculosis* genome at 25 years: lessons and lingering questions. J Clin Invest 133(19):e173156

König S (2020) Differential vs. comparative gel electrophoresis: new technology drives standardisation and quantification in protein two-dimensional gel electrophoresis. TrAC Trends Anal Chem 122:115731

Kowalczyk T, Ciborowski M, Kisluk J, Kretowski A, Barbas C (2020) Mass spectrometry based proteomics and metabolomics in personalized oncology. Biochim Biophys Acta (BBA) Mol Basis Dis 1866(5):165690

Kraut A, Coodin M, Plessis R, McLean D (2004) Predictors of positive tuberculin skin test (TST) results after 2-step TST among health care workers in Manitoba, Canada. Clin Infect Dis 1(11):e113

Kumar S, Jena L (2014) Understanding rifampicin resistance in tuberculosis through a computational approach. Genom Inform 12(4):276–282

Kummari R, Bose K (2022) Gel filtration chromatography. In: Textbook on cloning, expression and purification of recombinant proteins. Springer Nature, pp 199–219

Kunnath-Velayudhan S, Porcelli SA (2013) Recent advances in defining the immunoproteome of *Mycobacterium tuberculosis*. Front Immunol 11(4):335

Kunnath-Velayudhan S, Salamon H, Wang HY, Davidow AL, Molina DM, Huynh VT, Cirillo DM, Michel G, Talbot EA, Perkins MD, Felgner PL, Liang X, Gennaro ML (2010) Dynamic antibody responses to the *Mycobacterium tuberculosis* proteome. Proc Natl Acad Sci 107(33):14703–14708

Kunnath-Velayudhan S, Davidow AL, Wang HY, Molina DM, Huynh VT, Salamon H, Pine R, Michel

G, Perkins MD, Xiaowu L, Felgner PL, Flynn JL, Catanzaro A, Gennaro ML (2012) Proteome-scale antibody responses and outcome of *Mycobacterium tuberculosis* infection in nonhuman primates and in tuberculosis patients. J Infect Dis 206(5):697–705

Lai R, Gong DN, Williams T, Ogunsola AF, Cavallo K, Lindestam Arlehamn CS, Acolatse S, Beamer GL, Ferris MT, Sassetti CM, Lauffenburger DA, Behar SM (2023) Host genetic background is a barrier to broadly effective vaccine-mediated protection against tuberculosis. J Clin Invest 133(13):e167762

Lalvani A, Pareek M (2010) A 100 year update on diagnosis of tuberculosis infection. Br Med Bull 93(1):69–84

Landi S, Held HR, Gupta KC (1975) The multi-facets of tuberculin standardization. Dev Biol Stand 29:393–411

Lee K, O'Reilly FJ (2023) Cross-linking mass spectrometry for mapping protein complex topologies in situ. Essays Biochem 67(2):215–228

Lee J, Paik TH, Yoo YC, Lee J, Shin A, Song CH, Jo EK, Park JK (2006) Purification of native Ag85 complex, 38-kDa and MTB12 protein antigens from the culture filtrate of *Mycobacterium tuberculosis*. J Bacteriol Virol 36(4):211–218

Lee D, Islam MA, Natarajan S, Dudekula DB, Chung H, Park J, Oh B (2024) Identification of anti-tuberculosis drugs targeting DNA gyrase A and serine/threonine protein kinase PknB: A machine learning-assisted drug-repurposing approach. Trop Med Infect Dis 9(12):288

Lei Z, Wang J, Liu CH (2025) Developing next-generation tuberculosis vaccines based on pathogen–host interactions: towards a holistic perspective. hLife 3(4):164–171

Leite FL, Reinhardt TA, Bannantine JP, Stabel JR (2015) Envelope protein complexes of Mycobacterium avium subsp. paratuberculosis and their antigenicity. Vet Microbiol 175(2–4):275–285

Li KK, Qu DH, Zhang HN, Chen FY, Xu L, Wang MY et al (2020) Global discovery of the PstP interactions using M. tuberculosis proteome microarray and revealing novel connections with EthR. J Proteome 215:103650. https://doi.org/10.1016/j.jprot.2020.103650

Liljeroos L, Malito E, Ferlenghi I, Bottomley MJ (2015) Structural and computational biology in the Design of Immunogenic Vaccine Antigens. J Immunol Res 2015:156241

Link AJ, Eng J, Schieltz DM, Carmack E, Mize GJ, Morris DR, Garvik BM, Yates JR (1999) Direct analysis of protein complexes using mass spectrometry. Nat Biotechnol 17(7):676–682

Liu X, Salokas K, Weldatsadik RG, Gawriyski L, Varjosalo M (2020) Combined proximity labeling and affinity purificationmass spectrometry workflow for mapping and visualizing protein interaction networks. Nat Protoc 15(10):3182–3211. https://doi.org/10.1038/s41596-020-0365-x.

López-Hernández Y, Patiño-Rodríguez O, García-Orta ST, Pinos-Rodríguez JM (2016) Mass spectrometry applied to the identification of *Mycobacterium tuberculosis* and biomarker discovery. J Appl Microbiol 121(6):1485–1497

Lu X, Zhang X, Zhang Y (2000) Production, characterization and purification of monoclonal antibodies against antigens of *Mycobacterium tuberculosis* H37Rv strain. Zhonghua Jie He He Hu Xi Za Zhi 23(11):672–675

Lucas M, Ryan JM, Watkins J, Early K, Kruh-Garcia NA, Mehaffy C, Dobos KM (2021) Extraction and separation of Mycobacterial proteins. Methods Mol Biol 2314:77–107

Lussow AR, Barrios C, van Embden J, Van der Zee R, Verdini AS, Pessi A, Louis JA, Lambert PH, Del Giudice G (1991) Mycobacterial heat-shock proteins as carrier molecules. Eur J Immunol 21(10):2297–2302

Magdalena D, Karol M, Bartłomiej S, Konrad K (2025) Detection and accurate identification of Mycobacterium species by flow injection tandem mass spectrometry (FIA-MS/MS) analysis of mycolic acids. Sci Rep 15(1):13118

Målen H, Berven FS, Fladmark KE, Wiker HG (2007) Comprehensive analysis of exported proteins from *Mycobacterium tuberculosis* H37Rv. Proteomics 7(10):1702–1718

Målen H, Pathak S, Søfteland T, de Souza GA, Wiker HG (2010) Definition of novel cell envelope associated proteins in triton X-114 extracts of Mycobacterium tuberculosis H37Rv. BMC Microbiol 29(10):132

Målen H, De Souza GA, Pathak S, Søfteland T, Wiker HG (2011) Comparison of membrane proteins of *Mycobacterium tuberculosis* H37Rv and H37Ra strains. BMC Microbiol 11:1

Martini M, Besozzi G, Barberis I (2018) The never-ending story of the fight against tuberculosis: from Koch's bacillus to global control programs. J Prev Med Hyg 59(3):E241–E247

Mayer RL, Impens F (2021) Immunopeptidomics for next-generation bacterial vaccine development. Trends Microbiol 29(11):1034–1045

Mazana JS (2009) La tuberculosis y sus epónimos: Charles Mantoux (1877–1947). Revista Española de Sanidad Penitenciaria 11(1):17–23

Menzies R, Vissandjee B, Rocher I, Germain Y (1994) The booster effect in two-step tuberculin testing among young adults in Montreal. Ann Intern Med 120(3):190–198

Mészáros B, Simon I, Dosztányi Z (2011a) The expanding view of protein-protein interactions: complexes involving intrinsically disordered proteins. Phys Biol 8(3):035003

Mészáros B, Tóth J, Vértessy BG, Dosztányi Z, Simon I (2011b) Proteins with complex architecture as potential targets for drug design: a case study of *Mycobacterium tuberculosis*. PLoS Comput Biol 7(7):e1002118

Meyer K, Selbach M (2015) Quantitative affinity purification mass spectrometry: A versatile technology to study protein-protein interactions. Front Genet 6:237

Minden JS (2012) Two-dimensional difference gel electrophoresis. Methods Mol Biol 869:287–304

Mirza SP (2012) Quantitative mass spectrometry-based approaches in cardiovascular research. Circ Cardiovasc Genet 5(4):477

Mohammadnabi N, Shamseddin J, Emadi M, Bodaghi AB, Varseh M, Shariati A, Rezaei M, Dastranj M, Farahani A (2024) *Mycobacterium tuberculosis*: the mechanism of pathogenicity, immune responses, and diagnostic challenges. J Clin Lab Anal 38(23):e25122

Mojgani N, Babaie M, Shakibamehr N, Mohammad Taheri M, Mosavari N, Ghaempanah A, Soleymani BK (2020) Purification and biological analysis of specific antigens (ESAT6/CFP10) from *Mycobacterium tuberculosis*. Iran J Vet Sci Technol 12(2):59–67

Montoya Y, Deckerman P, Guler MO (2025) Protein recognition methods for diagnostics and therapy. BBA Adv 7:100149

Morris JH, Knudsen GM, Verschueren E, Johnson JR, Cimermancic P, Greninger AL, Pico AR (2014) Affinity purification-mass spectrometry and network analysis to understand protein-protein interactions. Nat Protoc 9(11):2539–2554

Mosavari N, Karimi A, Tadayon K, Shahhosseini G, Zavaran Hosseini A, Babaie M (2021) Evaluation of heating and irradiation methods for production of purified protein derivative (PPD) of *Mycobacterium tuberculosis*. Arch Razi Institute 75(4):439–449

Mugumbate G, Nyathi B, Zindoga A, Munyuki G (2021) Application of computational methods in understanding mutations in *Mycobacterium tuberculosis* drug resistance. Front Mol Biosci 8:643849

Murphy DJ, Brown JR (2007) Identification of gene targets against dormant phase *Mycobacterium tuberculosis* infections. BMC Infect Dis 26(7):84

Murphy PJ, Stone OJ, Anderson ME (2011) Automated hydrophobic interaction chromatography column selection for use in protein purification. J Vis Exp 55:3060

Murray KK (2021) Lasers for matrix-assisted laser desorption ionization. J Mass Spectrom 56(6):e4664

Nagai S, Wiker HG, Harboe M, Kinomoto M (1991) Isolation and partial characterization of major protein antigens in the culture fluid of *Mycobacterium tuberculosis*. Infect Immun 59(1):372–382

Nahid P, Pai M, Hopewell PC (2006) Advances in the diagnosis and treatment of tuberculosis. Proc Am Thorac Soc 3(1):103–110. https://doi.org/10.1513/pats.200511-119JH. PMID:16493157; PMCID:PMC2658675

Nayak S, Acharjya B (2012) Mantoux test and its interpretation. Indian Dermatol Online J 3(1):2–6

Neagu AN, Jayathirtha M, Baxter E, Donnelly M, Petre BA, Darie CC (2022) Applications of tandem mass spectrometry (MS/MS) in protein analysis for biomedical research. Molecules 27(8):2411

Nesterenko PN, Haddad PR (2000) Zwitterionic ion-exchangers in liquid chromatography. Anal Sci 16:565–574. https://doi.org/10.2116/analsci.16.565

Nguyen TL, Kim H (2024) Integrating immunoinformatics and computational epitope prediction for a vaccine candidate against respiratory syncytial virus. Infect Dis Model 9(3):763–774

Niu Y, Liu J, Yang R, Zhang J, Shao B (2020) Atmospheric pressure chemical ionization source as an advantageous technique for gas chromatography-tandem mass spectrometry. TrAC Trends Anal Chem 132:116053

Nyijime TA, Shallangwa GA, Uzairu A, Umar AB, Ibrahim MT, Madkhali HA, Hamadi A, Barnawi J, Abdalla M, Ahmed AU (2025) Computational design, pharmacokinetics, molecular docking and molecular dynamic simulation of novel anti-tubercular inhibitors. In Silico Res Biomed 1:100012

O'Connor BF, Cummins PM (2017) Hydrophobic interaction chromatography. Methods Mol Biol 1485:355–363. https://doi.org/10.1007/978-1-4939-6412-3_18. PMID:27730562

Oeffinger M (2012) Two steps forward—one step back: advances in affinity purification mass spectrometry of macromolecular complexes. Proteomics 12(10):1591–1608

Ó'Fágáin C, Cummins PM, O'Connor BF (2017) Gel-filtration chromatography. Methods Mol Biol 1485:15–25. https://doi.org/10.1007/978-1-4939-6412-3_2. PMID:27730546; PMCID:PMC7121854

Ortega-Tirado D, Arvizu-Flores AA, Velazquez C, Garibay-Escobar A (2020) The role of immunoinformatics in the development of T-cell peptide-based vaccines against *Mycobacterium tuberculosis*. Expert Rev Vaccines 19(9):831–841

Pal R, Bisht MK, Mukhopadhyay S (2022) Secretory proteins of *Mycobacterium tuberculosis* and their roles in modulation of host immune responses: focus on therapeutic targets. FEBS J 289(14):4146–4171

Paredes-Gutierrez G, Perea-Jacobo R, Acosta-Mesa HG, Mezura-Montes E, Morales Reyes JL, Zenteno-Cuevas R, Guerrero-Chevannier MÁ, Muñiz-Salazar R, Flores DL (2025) Predicting drug resistance in *Mycobacterium tuberculosis*: A machine learning approach to genomic mutation analysis. Diagnostics 15(3):279

Parish T, Stoker NG (eds) (2008) *Mycobacterium tuberculosis* protocols. Springer

Patil JM, Okada T (1999) Anion-modulated switching of retention properties of a zwitterionic stationary phase. Anal Commun 36:9–11

Phelan J, Coll F, McNerney R, Ascher DB, Pires DE, Furnham N, Coeck N, Hill-Cawthorne GA, Nair MB, Mallard K, Ramsay A, Campino S, Hibberd ML, Pain A, Rigouts L, Clark TG (2016) *Mycobacterium tuberculosis* whole genome sequencing and protein structure modelling provides insights into anti-tuberculosis drug resistance. BMC Med 14:31

Phillips AT (1995) Desalting, concentration, and buffer exchange by dialysis and ultrafiltration. In: Current protocols in protein science. First published December 1995

Pires DEV, Chen J, Blundell TL, Ascher DB (2016) In silico functional dissection of saturation mutagenesis: interpreting the relationship between phenotypes and changes in protein stability, interactions and activity. Sci Rep 6:19848

Pollock NR, MacIntyre AT, Blauwkamp TA, Blair L, Ho C, Calderon R, Franke MF (2021) Detection of *Mycobacterium tuberculosis* cell-free DNA to diagnose TB in pediatric and adult patients. Int J Tuberc Lung Dis 25(5):403

Porath J, Flodin P (1959) Gel filtration: A method for desalting and group separation. Nature 183(4676):1657–1659

Potgieter MG, Nakedi KC, Ambler JM, Nel AJ, Garnett S, Soares NC, Mulder N, Blackburn JM (2016) Proteogenomic analysis of *Mycobacterium smegmatis* using high resolution mass spectrometry. Front Microbiol 7:427

Ptak CP, Hsieh CL, Garnett JA, Rappuoli R, Lin YP (2023) Editorial: intersection of biophysical and structural approaches in vaccine design. Front Mol Biosci 10:1142656

Pym AS, Brodin P, Brosch R, Huerre M, Cole ST (2002) Loss of RD1 contributed to the attenuation of the live tuberculosis vaccines Mycobacterium bovis BCG and Mycobacterium microti. Mol Microbiol 46(3):709–717

Rahman MA, Heme UH, Parvez MAK (2022) In silico functional annotation of hypothetical proteins from the bacillus paralicheniformis strain Bac84 reveals proteins with biotechnological potentials and adaptational functions to extreme environments. PLoS One 17(10):e0276085

Reisinger V, Eichacker LA (2006) Analysis of membrane protein complexes by blue native PAGE. Proteomics 6(Suppl 2):6–15

Reisinger V, Eichacker LA (2008) Isolation of membrane protein complexes by blue native electrophoresis. Methods Mol Biol 424:423–431

Reyes F, Otero O, Camacho F, Sarmiento ME, Acosta A, Guillén G (2013) Purification of an IgA monoclonal antibody specific for the Acr protein of *Mycobacterium tuberculosis* by immunoaffinity chromatography. Malaysian J Med Sci 20(5):16–22

Rock JM, Hopkins FF, Chavez A, Diallo M, Chase MR, Gerrick ER, Pritchard JR, Church GM, Rubin EJ, Sassetti CM, Schnappinger D, Fortune SM (2017) Programmable transcriptional repression in mycobacteria using an orthogonal CRISPR interference platform. Nat Microbiol 2:16274

Rose DN, Schechter CB, Adler JJ (1995) Interpretation of the tuberculin skin test. J Gen Intern Med 10(11):635–642

Rosenkrands I, Weldingh K, Jacobsen S, Hansen CV, Florio W, Gianetri I, Andersen P (2000) Mapping and identification of *Mycobacterium tuberculosis* proteins by two-dimensional gel electrophoresis, microsequencing and immunodetection. Electrophoresis 21(5):935–948

Rusic D, Kumric M, Seselja Perisin A, Leskur D, Bukic J, Modun D et al (2024) Tackling the antimicrobial resistance “pandemic” with machine learning tools: a summary of available evidence. Microorganisms 12(5):842

Ryan JM, Dobos KM, Kruh-Garcia NA (2022) *Mycobacterium tuberculosis* extracellular vesicle enrichment through size exclusion chromatography. J Vis Exp 183

Safarian S, Opel-Reading HK, Wu D, Mehdipour AR, Hards K, Harold LK, Radloff M, Stewart I, Welsch S, Hummer G, Cook GM, Krause KL, Michel H (2021) The cryo-EM structure of the bd oxidase from M. Tuberculosis reveals a unique structural framework and enables rational drug design to combat TB. Nature Commun 12:5236

Saikat AS (2023) Computational approaches for molecular characterization and structure-based functional elucidation of a hypothetical protein from *Mycobacterium tuberculosis*. Genom Informat 21(2):e25

Sajid M (2017) Porous membrane protected micro-solid-phase extraction: A review of features, advancements and applications. Anal Chim Acta 965:36–53

Sajid Z, Akhtar T, Ahmad K, Haroon M (2024) Molecular docking simulation and ADMET/pharmacokinetic screening of newly designed 2-(2-(aryl)-4-oxo-4,5-dihydrothiazol-5-yl)acetohydrazides as potential anti-tubercular agents. Chemist Select 9(10):e202400000

Sandgren A, Strong M, Muthukrishnan P, Weiner BK, Church GM, Murray MB (2009) Tuberculosis drug resistance mutation database. PLoS Med 6(2):e2

Saraswat M, Musante L, Ravidá A, Shortt B, Byrne B, Holthofer H (2013) Preparative purification of recombinant proteins: current status and future trends. Biomed Res Int 2013:312709

Saridakis E, Vishwakarma R, Lai-Kee-Him J, Martin K, Simon I, Cohen-Gonsaud M, Coste F, Bron P, Margeat E, Boudvillain M (2022) Cryo-EM structure of transcription termination factor Rho from *Mycobacterium tuberculosis* reveals bicyclomycin resistance mechanism. Communicat Biol 5:120

Schmidt F, Donahoe S, Hagens K, Mattow J, Schaible UE, Kaufmann SH, Aebersold R, Jungblut PR (2004) Complementary analysis of the *Mycobacterium tuberculosis* proteome by two-dimensional electrophoresis and isotope-coded affinity tag technology. Mol Cell Proteomics 3(1):24–42

Schuett T, Kimmig J, Zechel S, Schubert US (2020) Automated polymer purification using dialysis. Polymers 12(9):2095

Schuhmacher M, Glocker MO, Wunderlin M, Przybylski M (1996) Direct isolation of proteins from sodium dodecyl sulfatepolyacrylamide gel electrophoresis and analysis by electrospray-ionization mass spectrometry. Electrophoresis 17(5):848–854

Scopes RK (1993) Protein purification: principles and practice, 3rd edn. Springer Science & Business Media

Senthil Kumar KS, Uma Devi KR, Alamelu R (2002) Isolation and evaluation of diagnostic value of two major secreted proteins of *Mycobacterium tuberculosis*. Indian J Chest Dis Allied Sci 44(4):225–232

Serrano LR, Pelin A, Arrey TN, Damoc NE, Richards AL, Zhou Y, Lancaster NM, Peters-Clarke TM, Pashkova A, Jang GM, Eckhardt M, Quarmby ST, Zeller M, Hermanson D, Stewart H, Hock C, Makarov A, Zabrouskov V, Krogan NJ, Coon JJ, Swaney DL (2025) Affinity purification mass spectrometry on the Orbitrap–Astral mass spectrometer enables high-throughput protein–protein interaction mapping. J Proteome Res 24(4):2006–2016

Shaik A (2020) Ion exchange chromatography, 1st edn. Springer, Singapore

Shallangwa GA, Adeniji SE (2021) Binding profile of protein–ligand inhibitor complex and structure based design of new potent compounds via computer-aided virtual screening. J Clin Tubercul Other Mycobacter Dis 24:100256

Sharma S, Shamy OE, Wilmington A, Keller B, Aslam N, Tran H, Rutherford P, Uribarri J (2024) Performance evaluation of an automated peritoneal dialysis solution generation system in patients using automated peritoneal dialysis. Kidney Intl Reports 9(6):1752–1757

Shen PS (2018) The 2017 Nobel prize in chemistry: Cryo-EM comes of age. Anal Bioanal Chem 410(8):2053–2057

Singh P, Jamal S, Ahmed F, Saqib N, Mehra S, Ali W, Roy D, Ehtesham NZ, Hasnain SE (2021) Computational modeling and bioinformatic analyses of functional mutations in drug target genes in *Mycobacterium tuberculosis*. Comput Struct Biotechnol J 19:2423–2446

Singpanomchai N, Ratthawongjirakul P (2024) The CRISPR-dCas9 interference system suppresses inhA gene expression in *Mycobacterium smegmatis*. Sci Rep 14(1):26116

Sirri N, Guyeux C, Sola C (2021) Advanced machine learning for predicting drug resistance in clinical isolates of *Mycobacterium tuberculosis* complex World Conference on Complex Systems, 2021

Śledź P, Caflisch A (2018) Protein structure-based drug design: from docking to molecular dynamics. Curr Opin Struct Biol 48:93–102. https://doi.org/10.1016/j.sbi.2017.10.010. PMID:29149726

Sober HA, Peterson EA (1958) Protein chromatography on ion exchange cellulose. Fed Proc 17(4):1116–1126

Song SH, Park KU, Lee JH, Kim EC, Kim JQ, Song J (2009) Electrospray ionization-tandem mass spectrometry analysis of the mycolic acid profiles for the identification of common clinical isolates of mycobacterial species. J Microbiol Methods 77(2):165–177

Song L, Wallstrom G, Yu X, Hopper M, Van Duine J, Steel J, Park J, Wiktor P, Kahn P, Brunner A, Wilson D, Jenny-Avital ER, Qiu J, Labaer J, Magee DM, Achkar JM (2017) Identification of antibody targets for tuberculosis serology using high-density nucleic acid programmable protein arrays. Mol Cell Proteomics 16(4 Suppl 1):S277–S289

Sørensen AL, Nagai S, Houen G, Andersen P, Andersen AB (1995) Purification and characterization of a low-molecular-mass T-cell antigen secreted by Mycobacterium tuberculosis. Infect Immun 63(5):1710–1717

Sreelatha S, Nagarajan U, Natarajan S (2023) Protein targets in *Mycobacterium tuberculosis* and their inhibitors for therapeutic implications: A narrative review. Int J Biol Macromol 243:125022

Staby A, Jensen RH, Bensch M, Hubbuch J, Dünweber DL, Krarup J, Nielsen J, Lund M, Kidal S, Hansen TB, Jensen IH (2001) Comparison of chromatographic ion-exchange resins: II. More strong anion-exchange resins. J Chromatogr A 908(1–2):149–161

Staby A, Jensen IH, Sand MB, Hansen RG, Jacobsen JH, Andersen LA, Gerstenberg M, Bruus UK, Krarup J, Nielsen J, Lund M, Kidal S, Hansen TB (2005) Comparison of chromatographic ion-exchange resins: IV. Strong and weak cation-exchange resins and heparin resins. J Chromatogr A 1069(1):65–77

Staby A, Jensen IH, Sand MB, Hansen RG, Jacobsen JH, Andersen LA, Gerstenberg M, Bruus UK, Krarup J, Nielsen J, Lund M, Kidal S, Hansen TB (2007) Comparison of chromatographic ion-exchange resins: VI. Weak anion-exchange resins. J Chromatogr A 1164(1–2):82–94

Stamm CE, Pasko BL, Chaisavaneeyakorn S, Franco LH, Nair VR, Weigele BA, Alto NM, Shiloh MU (2019) Screening *Mycobacterium tuberculosis* secreted proteins identifies Mpt64 as a eukaryotic membrane-binding bacterial effector. mSphere 4(3):e00354–e00319

Stasyk T, Huber LA (2004) Zooming in: fractionation strategies in proteomics. Proteomics 4(12):3704–3716

Stoevesandt O, Taussig MJ, He M (2009) Protein microarrays: high-throughput tools for proteomics. Expert Rev Proteomics 6(2):145–157. https://doi.org/10.1586/epr.09.2. PMID:19385942; PMCID:PMC7105755

Suameitria Dewi DNS, Mertaniasih NM, Soedarsono, Hagino K, Yamazaki T, Ozeki Y, Artama WT, Kobayashi H, Inouchi E, Yoshida Y, Ishikawa S, Shaban AK, Tateishi Y, Nishiyama A, Ato M, Matsumoto S (2023) Antibodies against native proteins of *Mycobacterium tuberculosis* can detect pulmonary tuberculosis patients. Sci Rep 13:12685

Sun B, Sun Y (2021) Diagnostic performance of DNA microarray for detecting rifampicin and isoniazid resistance in *Mycobacterium tuberculosis*. J Thorac Dis 13(7):4448–4454

Sun S, Li L, Yang F, Wang X, Fan F, Yang M, Chen C, Li X, Wang HW, Sui SF (2017) Cryo-EM structures of the ATP-bound Vps4^{E233Q} hexamer and its complex with Vta1 at near-atomic resolution. Nat Commun 8:16064

Sun MR, Xing JY, Li XT, Fang R, Zhang Y, Li ZL, Song NN (2025) Recent advances in research on *Mycobacterium tuberculosis* virulence factors and their role in pathogenesis. J Microbiol Immunol Infect

Svec F, Huber CG (2006 Apr 1) (2006). Monolithic materials: promises, challenges, achievements. Anal Chem 78(7):2101–2107

Syu GD, Dunn J, Zhu H (2020) Developments and applications of functional protein microarrays. Mol Cell Proteomics 19(6):916–927

Szewczyk R, Kowalski K, Janiszewska-Drobinska B, Druszczyńska M (2013) Rapid method for mycobacterium tuberculosis identification using electrospray ionization tandem mass spectrometry analysis of mycolic acids. Diagn Microbiol Infect Dis 76(3):298–305. https://doi.org/10.1016/j.diagmicrobio.2013.03.025

Temur N, Eren EE, Celik I, Yıldız IE, Nisari M, Unal IS, Celik C, Ildiz N, Ocsoy I (2025) The recent development in the diagnosis of *Mycobacterium tuberculosis*. Smart Med 4(2):e70007

Thakur KG, Praveena T, Gopal B (2010) Structural and biochemical bases for the redox sensitivity of Mycobacterium tuberculosis RslA. J Mol Biol 397(5):1199-208. https://doi.org/10.1016/j.jmb.2010.02.026. Epub 2010 Feb 22. PMID: 20184899; PMCID: PMC2877774

Thompson NJ, Glassroth JL, Snider DE Jr, Farer LS (1979) The booster phenomenon in serial tuberculin testing. Am Rev Respir Dis 119(4):587–597

Tiwari D, Tiwari RP, Chandra R, Bisen PS, Haque S (2014) Efficient ELISA for diagnosis of active tuberculosis employing a cocktail of secretory proteins of *Mycobacterium tuberculosis*. Folia Biol 60(1):10–20

Trauger SA, Webb W, Siuzdak G (2002) Peptide and protein analysis with mass spectrometry. J Spectrosc 16:15–28

Truong PQ, Hammer E, Salazar MG, Ha DT, Huong NL, Hieu DM, Hoa NT, Thuy PT, Volker U (2014) 2D DIGE proteomic analysis of multidrug resistant and susceptible clinical *Mycobacterium tuberculosis* isolates. J Integrat Omics 4(1):28–36

Udhayabanu L, Sridharan KS, Antony T (2025) Pitfalls in using matrix-assisted laser desorption ionization-time of flight mass spectrometry for the identification of Mycobacterium. Biomed Biotechnol Res J (BBRJ) 9(1):113–117

van Bentum M, Selbach M (2021) An introduction to advanced targeted acquisition methods. Mol Cell Proteomics 20:100165

Velayati AA, Mitaria S, Farnia P, Farnia P, Ghanavi J, Maleknia M, Murase Y, Hoffner S (2024) Exploring the genetic landscape of *Mycobacterium tuberculosis*: unlocking the differences between latent and active tuberculosis. Intl J Mycobacteriol 13(4):410–419

Vieira TF, Martins FG, Moreira JP, Barbosa T, Sousa SF (2021) In silico identification of possible inhibitors for protein kinase B (PknB) of mycobacterium tuberculosis. Molecules 26(20):6162

Viswanathan S, Unlü M, Minden JS (2006) Two-dimensional difference gel electrophoresis. Nat Protoc 1(3):1351–1358

Vyas VK, Ukawala RD, Ghate M, Chintha C (2012) Homology modeling a fast tool for drug discovery: current perspectives. Indian J Pharm Sci 74(1):1–17

Wadsworth AB, Maltaner F (1921 Jan 1) (1921). Purification and concentration of antigens for complement fixation by methods of dialysis, adsorption, and extraction. J Exp Med 33(1):119–123

Wang NW (1990) Ion exchange in purification. Bioprocess Technol 9:359–400

Wang R, Prince JT, Marcotte EM (2005) Mass spectrometry of the M. Smegmatis proteome: protein expression levels correlate with function, operons, and codon bias. Genome Res 15(8):1118–1126

Wang C, Wei LL, Shi LY, Pan ZF, Yu XM, Li TY, Liu CM, Ping ZP, Jiang TT, Chen ZL, Mao LG, Li ZJ, Li JC (2015) Screening and identification of five serum proteins as novel potential biomarkers for cured pulmonary tuberculosis. Sci Rep 5:15615

Wang Y, Gao S, Wu F, Gong Y, Mu N, Wei C, Wu C, Wang J, Yan N, Yang H, Zhang Y, Liu J, Wang Z, Yang X, Lam SM, Shui G, Li S, Da L, Guddat LW, Rao Z, Zhang L (2024) Cryo-EM structures of a mycobacterial ABC transporter that mediates rifampicin resistance. Proc Natl Acad Sci 121(37):e2403421121

Weldingh K, Rosenkrands I, Jacobsen S, Rasmussen PB, Elhay MJ, Andersen P (1998) Two-dimensional electrophoresis for analysis of *Mycobacterium tuberculosis* culture filtrate and purification and characterization of six novel proteins. Infect Immun 66(8):3492–3500

Weldingh K, Hansen A, Jacobsen S, Andersen P (2000) High resolution electroelution of polyacrylamide gels for the purification of single proteins from mycobacterium tuberculosis culture filtrate. Scand J Immunol 51(1):79–86

Wilkins MR, Gasteiger E, Bairoch A, Sanchez JC, Williams KL, Appel RD, Hochstrasser DF (1999) Protein identification and analysis tools in the ExPASy server. Methods Mol Biol 112:531–552

Wingfield P (2001) Protein precipitation using ammonium sulfate. Curr Protoc Protein Sci 13:A.3F.1–A.3F.8

Wolf-Yadlin A, Sevecka M, MacBeath G (2009) Dissecting protein function and signaling using protein microarrays. Curr Opin Chem Biol 13(4):398–405. https://doi.org/10.1016/j.cbpa.2009.06.027. PMID:19660979; PMCID:PMC2811369

Wong AI, Rock JM (2021) CRISPR interference (CRISPRi) for targeted gene silencing in Mycobacteria. Methods Mol Biol 2314:343–364

Wong DK, Lee BY, Horwitz MA, Gibson BW (1999) Identification of fur, aconitase, and other proteins expressed by *Mycobacterium tuberculosis* under conditions of low and high concentrations of iron by combined two-dimensional gel electrophoresis and mass spectrometry. Infect Immun 67(1):327–336

Wu FL, Liu Y, Jiang HW, Luan YZ, Zhang HN, He X, Xu ZW, Hou JL, Ji LY, Xie Z, Czajkowsky DM, Yan W, Deng JY, Bi LJ, Zhang XE, Tao SC (2017) The Ser/Thr protein kinase protein-protein interaction map of M. Tuberculosis. Mol Cell Proteomics 16(8):1491–1506

Wu YL, Li YT, Liu GB, Wu JL, Liu XR, Gao XX, Huang QD, Liang J, Ouyang JY, Ding YR, Wu JY, Lu YB, Gao YC, Cai XZ, Zhang JA (2024) LC–MS-based quantitation of proteomic changes induced by Norcantharidin in MTB-treated macrophages. Proteome Sci 22:13

Xie X, Yu T, Li X, Zhang N, Foster LJ, Peng C, Huang W, He G (2023) Recent advances in targeting the "undruggable" proteins: from drug discovery to clinical trials. Signal Transduct Target Ther 8:335

Yan MY, Zheng D, Li SS, Ding XY, Wang CL, Guo XP, Zhan L, Jin Q, Yang J, Sun YC (2022) Application of combined CRISPR screening for genetic and chemical-genetic interaction profiling in *Mycobacterium tuberculosis*. Sci Adv 8(47):eadd5907

Yang H, Tang H, Chen XX, Zhang CJ, Zhu PP, Ding H, Chen W, Lin H (2016) Identification of secretory proteins in *Mycobacterium tuberculosis* using pseudo amino acid composition. Biomed Res Int 2016:5413903

Yang J, Zhang L, Qiao W, Luo Y (2023) *Mycobacterium tuberculosis*: pathogenesis and therapeutic targets. MedComm 4(5):e353

Yang Y, Chen YZ, Xia T (2024) Optimizing antigen selection for the development of tuberculosis vaccines. Cell Insight 3(3):100163

Yazdi MK, Beheshtnejad AH, Hedayatpour A, Sharifi-Yazdi S, Mehrani F (2018) Comparative study of *Mycobacterium tuberculosis* and Mycobacterium bovis protein profiles. J Surgery Med 2(3):223–227

Yimer SA, Kalayou S, Homberset H, Birhanu AG, Riaz T, Zegeye ED, Lutter T, Abebe M, Holm-Hansen C, Aseffa A, Tønjum T (2020) Lineage-specific proteomic signatures in the *Mycobacterium tuberculosis* complex reveal differential abundance of proteins involved in virulence, DNA repair, CRISPR-Cas, bioenergetics and lipid metabolism. Front Microbiol 11:550760

Yu LL, Sun Y (2012) Trace adsorption of positively charged proteins onto Sepharose FF and Sepharose FF-based anion exchangers. J Chromatogr A 1253:105–109

Yun KS, Zhu C, Parcher JF (1995) (1995). Theoretical relationships between the void volume, mobile phase volume, retention volume, adsorption, and Gibbs free energy in chromatographic processes. Anal Chem 67(3):613–619

Zein-Eddine R, Le Meur A, Skouloubris S, Jelsbak L, Refrégier G, Myllykallio H (2025) Genome wide analyses reveal the role of mutator phenotypes in *Mycobacterium tuberculosis* drug resistance emergence. NPJ Antimicrobial Resist 3:35

Zhang Y, Yew WW (2015) Mechanisms of drug resistance in *Mycobacterium tuberculosis*: update 2015. Int J Tuberc Lung Dis 19(11):1276–1289

Zhang W, Li Z, Sun Y, Cui P, Liang J, Xing Q et al (2022) Cryo-EM structure of mycobacterium tuberculosis 50S ribosomal subunit bound with clarithromycin reveals dynamic and specific interactions with macrolides. Emerg Microbes Infect 11(1):293–305. https://doi.org/10.1080/22221751.2021.2022439. PMID:34935599; PMCID:PMC8786254

Zhang X, Zhang D, Zhang X, Zhang X (2024) Artificial intelligence applications in the diagnosis and treatment of bacterial infections. Front Microbiol 15:1449844

Zheng J et al (2011) Combing blue native polyacrylamide gel electrophoresis with liquid chromatography tandem mass spectrometry as an effective strategy for analyzing potential membrane protein complexes of Mycobacterium bovis bacillus Calmette-Guérin. BMC Genomics 12:40

Zhuang L, Yang L, Li L, Ye Z, Gong W (2024) *Mycobacterium tuberculosis*: immune response, biomarkers, and therapeutic intervention. MedComm 5(1):e419

3 Protein Architecture and Composition in *Mycobacterium tuberculosis*

Abstract

The protein architecture of *Mycobacterium tuberculosis* (Mtb) demonstrates remarkable complexity and adaptability, emblematic of its evolutionary refinement as a highly successful pathogen. The Mtb proteome can be broadly classified into four principal categories: core, accessory, transcriptionally plastic, and uncharacterized proteins. Core proteins are highly conserved across all Mtb strains and essential for fundamental cellular functions and bacterial viability; they form the structural and metabolic foundation required for critical processes such as DNA replication, transcription, and cell wall biosynthesis. Conversely, accessory proteins exhibit considerable variability among strains, endowing Mtb with strain-specific traits including virulence, environmental adaptation, and antibiotic resistance. These proteins are vital for enabling the pathogen to thrive in diverse ecological niches and to overcome selective pressures imposed by environmental factors and antimicrobial agents. Distinguished not by strain distribution but by regulatory dynamics, transcriptionally plastic proteins exhibit differential expression in response to environmental changes and host-derived cues, allowing Mtb to modulate its physiological state during infection rapidly. This regulatory flexibility supports the pathogen's ability to enter dormancy, mount stress responses, and transition between metabolic states. A substantial portion of the Mtb proteome remains uncharacterized or annotated as hypothetical, with functions yet to be elucidated. Nevertheless, recent advances in integrative bioinformatics and experimental proteomics have begun to clarify the roles of many such proteins, revealing novel contributions to bacterial survival, pathogenicity, and immune evasion. The complex interplay among these protein categories illustrates a highly sophisticated regulatory network that governs Mtb's growth, dormancy, stress adaptation, and persistence. This dynamic and adaptable protein architecture is fundamental to the bacterium's capacity to endure hostile host environments, evade immune surveillance, and establish chronic infections. Consequently, a comprehensive understanding of the composition, regulation, and functional plasticity of these protein classes is imperative. Such knowledge will drive the development of innovative diagnostics, next-generation vaccines, and targeted therapeutics, ultimately advancing more effective strategies for tuberculosis control and eradication.

Keywords

Mycobacterium tuberculosis · Cell wall · Molecular chaperones · Core functional proteins · Transcriptional plasticity proteins · Uncharacterized proteins

P. Farnia et al., *Proteins in Mycobacterium Tuberculosis*, Advances in Experimental Medicine and Biology 1484, https://doi.org/10.1007/978-3-031-96883-9_3

3.1 Introduction

A breakthrough in the study of Mtb proteins occurred with the publication of the complete genome sequence of the reference laboratory strain $H_{37}Rv$ by Cole et al. (1998). This seminal work offered an extensive genetic blueprint that has since served as a foundational resource, driving advances in understanding the pathogen's biology and accelerating the development of new therapeutic strategies against tuberculosis. The genome of $H_{37}Rv$ comprises approximately 4,411,529 base pairs with a notably high guanine plus cytosine (G + C) content of 65.6%, ranking it as the second-largest bacterial genome sequenced at the time, surpassed only by *Escherichia coli.* Within this genome, 3924 open reading frames (ORFs) were identified, accounting for roughly 91% of the bacterium's coding potential. Functional annotation revealed that about 40% of the proteins could be assigned precise functions based on database comparisons, and 44% shared similarity with known proteins. In comparison, approximately 16% lacked recognizable homologues, suggesting the presence of novel, Mtb-specific proteins with potentially unique roles (Cole et al. 1998). Correspondence analyses of amino acid composition in Mtb and other high-GC bacteria have demonstrated a statistically significant preference for amino acids encoded by G + C-rich codons, such as alanine, glycine, proline, and arginine. In contrast, amino acids encoded predominantly by A + T-rich codons, including asparagine, isoleucine, lysine, phenylalanine, and tyrosine, are comparatively underrepresented. This compositional bias, closely tied to the high genomic GC content of MTB, is thought to influence not only codon usage but also the resulting protein structure and functional evolution within the organism (Andersson and Sharp 1996; Kunst et al. 1997; Cole et al. 1998). The availability of the complete genome sequence enabled researchers to delve deeper into Mtb biology, facilitating the identification of novel drug targets and vaccine candidates, thereby significantly advancing tuberculosis research and clinical practice (Pankhurst et al. 2016; Satta et al. 2018; Cohen et al. 2019; Koleske et al. 2023; Razzak et al. 2023; Smith et al. 2023; Otchere et al. 2024; Sweeney et al. 2025). The advent of whole-genome sequencing (WGS) technologies, including platforms such as Illumina and Oxford Nanopore Technology (ONT), has revolutionized the ability to obtain high-quality, comprehensive DNA sequence data for the entire Mtb genome. This advancement has made genome sequencing increasingly accessible and efficient, enabling detailed insights into the genetic diversity, evolution, and transmission dynamics of Mtb worldwide (Ford et al. 2012; Witney et al. 2016a, 2016b; Votintseva et al. 2017; Wollenberg et al. 2017; Bainomugisa et al. 2024; Hunt et al. 2019; Phelan et al. 2019; Smith et al. 2020; Gómez-González et al. 2022; Hasan et al. 2023; Thorpe et al. 2024; Bejaoui et al. 2025). Genomic analyses of Mtb have elucidated key genetic differences between the attenuated H37Ra strain and its virulent counterpart, $H_{37}Rv$. Early investigations revealed significant genomic rearrangements mediated by the IS6110 insertion sequence, including the deletion of the RvD2 region unique to $H_{37}Ra$ (Brosch et al. 1999). Expanding on this foundation, Zheng et al. (2008) characterized 198 single-nucleotide variants (SNVs), 53 insertions, and 21 deletions, distinguishing the two strains, with the majority of these polymorphisms localized within repetitive regions such as IS6110 and the PE/PPE gene families. Subsequent studies further demonstrated notable genomic variability even among laboratory-maintained $H_{37}Rv$ strains; Ioerger et al. (2010) identified 73 shared polymorphisms across six laboratory stocks, most of which were subsequently attributed to sequencing and assembly errors in the original $H_{37}Rv$ reference genome. Kohli et al. (2012) documented pronounced differences within the PE/PPE gene family between $H_{37}Ra$ and $H_{37}Rv$, affecting critical physico-chemical properties, phosphorylation sites, and protein-protein interaction domains. Comprehensive analyses of $H_{37}Ra$'s reduced virulence further identified three single-nucleotide polymorphisms with potential functional consequences: S219L in PhoP, A219E in MazG, and I228M in EspK (Jia et al. 2017). Sanoussi et al. (2021) extended these insights through lineage-specific investigations focusing on Lineage 5 (L5) isolates. Their work revealed unique gene clusters and differential expression of virulence-associated genes, including novel ESX-5 paralogs and regulators PhoP and MazG, highlighting lineage-specific adaptations. Notably, variable gene content among L5 sub-lineages (e.g., L5.3.1 versus L5.3.2) challenges the traditional view of a static Mtb genome and suggests ongoing adaptive evolution driven by host or environmental pressures. Most recently, Chitale et al. (2022) refined the $H_{37}Rv$ reference genome using advanced assembly techniques (Bact-Builder), incorporating approximately 6400 base pairs absent from the original sequence. This updated assembly identified ten novel genomic regions, including insertions within PE/PPE genes and novel paralogs of *esxN* and *esxJ* within the ESX-5 locus. Their analysis also revealed 109 SNVs and 35 small insertions/deletions relative to the prior reference, resolving longstanding inconsistencies and confirming that widely used $H_{37}Rv$ strains (e.g., NR123, TMC102) harbor these structural variations. Collectively, these findings underscore the significant role of genomic plasticity, particularly within repetitive and regulatory regions, in shaping the divergent virulence profiles of $H_{37}Ra$ and $H_{37}Rv$. This growing body of knowledge not only advances our understanding of Mtb pathogenesis but also highlights novel genetic targets for therapeutic intervention.

In fact, WGS has become an indispensable tool in tuberculosis (TB) control by providing detailed insights into genetic variation, drug resistance mutations, and transmission dynamics. This comprehensive genomic information enhances TB prevention, diagnosis, treatment monitoring, and epidemiological surveillance efforts worldwide. WGS enables rapid and accurate detection of drug resistance, allowing for tailored treatment regimens and improved patient outcomes (Witney et al. 2016a, b; Dohál et al. 2022; Hazra et al. 2023). The integration of multi-platform sequencing strategies further enhances data quality and resolution, overcoming limitations inherent to individual technologies. Consequently, these advances are transforming TB control efforts by enabling timely outbreak detection, guiding personalized treatment regimens, and informing public health interventions aimed at reducing the global burden of tuberculosis (Walker et al. 2017;Zhang et al. 2024a, b). To complement genomic data, proteomics, the large-scale study of proteins, has been employed to predict and validate gene products. Proteomic approaches commonly utilize mass spectrometry (MS) and two-dimensional gel electrophoresis (2-DE), which allow for protein identification, quantification, and metabolic pathway analysis (Kim et al. 2004; Encarnación et al. 2005; López 2007; Aslam et al. 2017; Chandramouli and Qian 2009; Birhanu 2023; Palande et al. 2024). Moreover, proteomics investigates protein-protein interactions and enables comparative analyses across different organisms or experimental conditions to detect changes in protein expression (Al-Amrani et al. 2021; Hollander et al. 2021). These tools facilitate the prediction of protein structures, identification of posttranslational modifications (PTMs), and annotation of protein functions based on sequence homology and experimental evidence (Ertelt et al. 2024). An open reading frame (ORF) is defined as a continuous stretch of codons beginning with a start codon (e.g., ATG) and ending with a stop codon (e.g., TAA, TAG, TGA). Identifying ORFs is fundamental to predicting gene locations and potential functions within a genome. Annotation of ORFs involves comparing sequences to known genes in other organisms to infer function (Orr et al. 2020; Varabyou et al. 2023). This process is critical for understanding the genetic and functional landscape of Mtb and has broad implications in biological research. Recent advances in machine learning, such as Random Forest classifiers, have further improved gene prediction accuracy by integrating diverse data types, including sequence features, gene expression profiles (e.g., RNA-Seq), and ribosome binding data (e.g., Ribo-Seq) (Díaz-Uriarte et al. 2006; Sinha et al. 2024). These sophisticated models accelerate the identification and annotation of protein-coding genes, thereby advancing functional genomics studies. Posttranslational modifications (PTMs) in Mtb, such as phosphorylation, glycosylation, ubiquitination, acetylation, methylation, and lipidation, regulate various aspects of bacterial physiology and pathogenicity (van Els et al. 2014; Arora et al. 2021). Therefore, to comprehensively understand the protein composition of Mtb, whole-genome sequencing is typically followed by proteomic identification and expression analysis. Proteomic research has significantly accelerated tuberculosis research and therapeutic development. The protein architecture of Mtb reflects its complex biology and remarkable adaptability, which underpin its ability to survive hostile environments and establish persistent infections. Central to this architecture is the core proteome, which comprises approximately 82.5% of the total Mtb proteome and is conserved across Mycobacterium species (Cole et al. 1998; Yuan et al. 2020; Mészáros et al. 2011; Devlin et al. 2025). This core set of proteins can be broadly categorized into about major functional classes, each representing essential biological processes critical for bacterial viability, adaptation, and pathogenicity (Cook et al. 2009). Among these categories, enzymatic proteins form a substantial portion, driving the metabolic pathways necessary for energy production, biosynthesis, and cellular maintenance (Warner 2014). Structural proteins contribute significantly by maintaining the integrity and unique composition of the Mtb cell wall and membrane, which are critical for protection against environmental stress and host immune defenses (Jacobo-Delgado et al. 2023). Regulatory proteins orchestrate gene expression and cellular responses, enabling Mtb to modulate its physiology in response to changing conditions (Tucker et al. 2007). Transport proteins facilitate the controlled uptake of nutrients and the efflux of toxic substances, maintaining intracellular homeostasis. Additionally, virulence factors and immunogenic proteins play pivotal roles in mediating interactions with the host, promoting immune evasion, and establishing infection (Chandra et al. 2022; Abbasnia et al. 2024). Chaperones and heat shock proteins assist in proper protein folding and repair, especially under stress conditions encountered during infection. Cofactor-binding proteins support enzymatic activity by ensuring the correct association of essential cofactors, while signal transduction proteins enable Mtb to sense and respond to environmental cues, facilitating adaptive responses (Stetler et al. 2010; Wan et al. 2020; Gu et al. 2025).

Based on recent advances in proteomics and the extensive data accumulated from numerous Mtb samples, the proteins of Mtb can be broadly classified into four main categories: core, accessory, transcriptionally plastic, and uncharacterized proteins. Core proteins are highly conserved across all Mtb strains and are indispensable for the bacterium's fundamental survival, growth, and cellular functions (Marmiesse et al. 2004). These proteins are involved in essential biological processes, including DNA replication, transcription, translation, cell division, and the maintenance of cell shape and structural integrity. Owing to their critical roles, core proteins are subject to strong purifying selection, resulting in minimal genetic variability. Notable examples include FtsZ, a tubulin-like protein that initiates cell division; MreB, an actin-like protein responsible for preserving the rod-shaped morphology of the bacterium; and Wag31, which regulates polar growth and is essential for cell elongation and compartmentalization (Marmiesse et al. 2004; Hett and Rubin 2008; Velayati and Farnia 2012; Robinson et al. 2012). Other core proteins include ribosomal components, DNA and RNA polymerases, enzymes involved in central metabolism, and structural proteins that maintain the unique cell wall of Mtb. In contrast, accessory proteins are not universally essential and exhibit considerable variation across different Mtb strains. These proteins often facilitate the bacterium's adaptation to specific environmental niches, contribute to immune evasion, enhance virulence, or confer antibiotic resistance. Unlike core proteins, accessory proteins are subject to more dynamic genetic processes, including gene gain and loss, horizontal gene transfer, and diversifying selection, which collectively drive their increased variability (Ligon et al. 2012; Supply et al. 2013). Prominent examples include members of the PE/PPE protein family, which play critical roles in antigenic variation and modulation of host immune responses (Brennan 2017;

Mukhopadhyay and Balaji 2011; Ates 2020; Koleske et al. 2025). Additionally, components of specialized secretion systems, such as the ESX (Type VII) secretion systems, are implicated in virulence, while various accessory proteins participate in mechanisms underlying drug resistance (Bitter et al. 2009; Roy et al. 2020). Accessory proteins also encompass lineage- or strain-specific factors that enable Mtb to persist and proliferate within particular hosts or geographic regions.

The third category, transcriptionally plastic proteins(TP), exhibit dynamic regulation of expression in response to environmental cues and host-induced stresses, enabling Mtb to adapt effectively during infection, latency, and immune system challenges (Bei et al. 2023). Unlike the relatively stable expression patterns of core or accessory proteins, these proteins are characterized by their regulatory flexibility, allowing the bacterium to precisely modulate essential processes such as metabolism and virulence. Lastly, uncharacterized proteins comprise a substantial fraction, approximately 29%, of the Mtb proteome, with their functions remaining largely unexplored. Given the increasing challenge posed by multidrug-resistant tuberculosis, these proteins represent promising avenues for novel drug targets and vaccine development. A deeper understanding of both transcriptionally plastic and uncharacterized proteins is therefore critical for advancing tuberculosis research and improving therapeutic strategies. While classifying proteins into these four functional groups facilitates a clearer understanding of Mtb biology, it is important to acknowledge that protein functions often overlap across categories, underscoring the intricate complexity of bacterial adaptation and survival mechanisms.

3.2 Characterization of Core Functional Proteins of *Mycobacterium tuberculosis*

Core proteins in Mtb constitute a highly conserved group of molecules essential for the bacterium's survival, growth, and pathogenicity. Comparative genomic analyses across mycobacterial species have played a pivotal role in identifying core genes that represent promising targets for therapeutic intervention against tuberculosis and leprosy. Early investigations identified 219 core genes encoding proteins unique to mycobacteria, which are highly conserved across multiple strains and critical for bacterial viability and pathogenic potential (Marmiesse et al. 2004). Employing transposon mutagenesis, genes indispensable for optimal growth in Mtb were uncovered, many of which lack identifiable orthologs in other bacterial species (Sassetti et al. 2003). Further comprehensive analyses identified 1080 core orthologous clusters enriched in essential cellular functions, while also noting that pathogenic mycobacterial species have acquired additional genes that facilitate survival within host cells (Malhotra et al. 2017). Building upon these insights, subsequent refinement of the core genome led to the identification of eight putative broad-spectrum drug targets within a conserved set of 1251 proteins (Dar et al. 2020). Collectively, these findings underscore the critical value of comparative genomics in identifying viable drug targets and enhancing our understanding of mycobacterial pathogen evolution. Core proteins, encoded by core genes, are fundamental to essential biological processes that sustain cellular life.

Proteins involved in fundamental cellular processes such as DNA replication, transcription, translation, cell division, and the maintenance of cell shape are critical for the survival and propagation of Mtb. Among these, the protein FtsZ, a tubulin homologue, plays a pivotal role in bacterial cytokinesis. FtsZ polymerizes in a GTP-dependent manner to assemble a contractile ring, termed the Z-ring, at the prospective site of cell division. This structure serves as a scaffold, recruiting other division machinery components, and generates the necessary force for septum constriction, ultimately leading to daughter cell separation (Balas et al. 1998; Figge et al. 2004; Hoffmann et al. 2010; van Teeffelen et al. 2011; Harnagel et al. 2021; McQuillen and Xiao 2020).

The *ftsZ* gene encodes this essential cytoskeletal protein, whose expression and intracellular levels are tightly regulated to ensure precise coordination of cell division. Notably, FtsZ levels decline during the stationary phase of growth, underscoring stringent regulatory mechanisms controlling its availability in Mtb (Dziadek et al. 2003; Roy and Ajitkumar 2005). Structural insights into Mtb FtsZ began with crystallographic studies by Leung et al. (2004), who resolved the protein in complex with guanosine diphosphate (GDP) and an inhibitor, revealing fold similarities to small GTPases. The protein features a central core domain containing a nucleotide-binding pocket accommodating GTP or GDP and conserved motifs such as the phosphate-binding loop (P-loop) and two switch regions that undergo conformational changes upon nucleotide binding and hydrolysis. These molecular "switches" modulate the polymerization-depolymerization cycles of FtsZ, essential for dynamic Z-ring assembly during division (Niemann et al. 2001; Leung et al. 2004; Scheffers and Driessen 2001). Comparatively, Mtb FtsZ exhibits slower polymerization rates and reduced intrinsic GTPase activity relative to homologues from other bacteria, potentially reflecting unique structural flexibility and adaptation to its complex lifestyle (Leung et al. 2004). Chen et al. (2007) detailed the kinetic properties of Mtb FtsZ, reporting subunit turnover half-times around 42 seconds in vitro and 25 seconds in vivo, significantly slower than *Escherichia coli* FtsZ dynamics. This decelerated turnover aligns with the advanced regulatory mechanisms modulating ftsZ expression and protein activity in Mtb. Consequently, exploring regulatory elements governing ftsZ, including promoter architecture and posttranslational modifications, may elucidate crucial control points within the cell division pathway (Chen et al. 2007). FtsZ's functionality and midcell localization are also regulated via accessory proteins that respond to cellular stress. For instance, in the related nonpathogenic *Mycobacterium smegmatis*, the accessory protein FipA is essential for proper FtsZ localization during oxidative stress. Disruption of the *fipA* gene leads to aberrant FtsZ distribution when cells are exposed to hydrogen peroxide, a defect reversible by complementation with the Mtb *fip*A gene, highlighting FipA's vital role in stress-responsive division control (Sureka et al. 2010).

Further structural studies have illuminated FtsZ's conformational plasticity. Li et al. (2013) demonstrated that GDP-bound Mtb FtsZ protofilaments adopt curved conformations facilitating a hinge-like opening, hypothesized to generate constriction forces during cytokinesis. Lazo et al. (2019) identified a distinctive arrangement of the T9 loop consistent with oligomeric assembly states that may underlie polymerization dynamics. More recently, Alnami et al. (2021) discovered two cryptic allosteric binding sites in nucleotide-free FtsZ modulated by the flexible H7 helix; these sites bind coumarin derivatives, presenting promising targets for therapeutic inhibition. Altogether, the extensive biochemical and structural characterization of Mtb FtsZ underscores its unique kinetic and conformational properties combined with tightly regulated expression and modulation by accessory factors. As the central organizer of bacterial cytokinesis, FtsZ represents an attractive target for novel antitubercular agents aimed at disrupting cell division mechanisms vital for Mtb survival and infection propagation (Leung et al. 2004; Hong et al. 2013; McQuillen and Xiao 2020; Chen et al. 2007; Sureka et al. 2010; Li et al. 2013; Lazo et al. 2019; Alnami et al. 2021).

Utilizing atomic force microscopy (AFM), we effectively demonstrated the localization of FtsZ at the mid-cell or subpolar regions during the exponential growth phase of Mtb (Figs. 3.1 and 3.2). Notably, both the number and spatial distribution of FtsZ rings exhibited variability between drug-susceptible and drug-resistant Mtb (Farnia et al. 2010; Velayati and Farnia 2012, 2016). These observations highlight the pivotal role of accessory proteins, such as FipA, in stabilizing and accurately positioning the Z-ring under diverse environmental stresses, a process essential for the successful completion of bacterial cell division. Due to its essential function in the cytokinesis of bacilli, FtsZ has attracted significant interest as a highly promising target for the development of novel antitubercular therapies. Inactivation of FtsZ disrupts Mtb cell division, resulting in impaired proliferation and viability (Kumar et al. 2010; Tovar-Nieto et al. 2024), thereby underscoring its potential as a critical

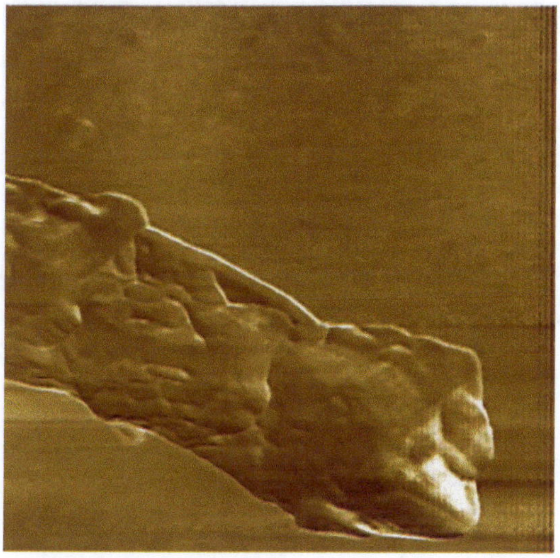

Fig. 3.1 Visualizing FtsZ ring formation in TB bacteria using atomic force microscopy. This image presents an atomic force microscopy (AFM) scan of a single Mtb bacillus at the exponential phase of life. A prominent circumferential band, also referred to as a proto-dividing ring, is discernible along the cell body. This band corresponds to the localization of the FtsZ protein, a highly conserved tubulin-like protein that polymerizes to form the so-called FtsZ ring at the future site of cell division. The FtsZ ring serves as the primary scaffold for the bacterial division machinery, orchestrating the recruitment of additional proteins required for cytokinesis. Its precise assembly and spatial positioning are essential for septum formation and subsequent daughter cell separation. The visualization of this structure through AFM provides compelling evidence of active division processes within the bacillus and underscores the dynamic cytoskeletal organization of Mtb during its cell cycle. Such high-resolution imaging advances our understanding of bacterial morphogenesis and the spatial regulation of critical division proteins, highlighting the efficacy of AFM as a tool for elucidating key aspects of mycobacterial cell biology. *Experimental Work by Prof. Parissa -Farnia and Prof. Gennady Konstantinovich Zhavnerko; Mycobacteriology Research Center, National Research Institute of Tuberculosis and Lung Disease, Shahid Beheshti University of Medical Sciences, Tehran, Iran & the Republican Research and Practical Centre for Epidemiology & and Microbiology, Filimonova 23, Minsk, Belarus

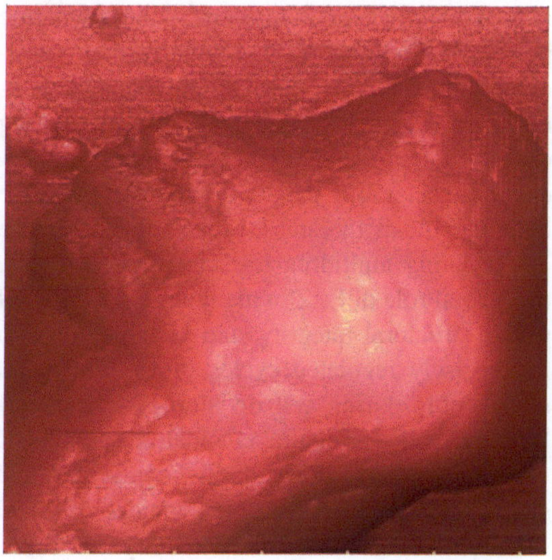

Fig. 3.2 FtsZ ring in the polar region of *Mycobacterium tuberculosis*

Atomic force microscopy (AFM) image of a single MTB bacilli, revealing the detailed topographical features of the bacterial surface at nanometer resolution. The prominent ring-like band observed near one pole of the cell represents the localized assembly of the FtsZ protein, known as the FtsZ ring. This structure is a critical component of the bacterial division machinery, designating the site of septum formation and cytokinesis. In Mtb, the FtsZ ring may localize at mid-cell or subpolarly, including near the head or tail regions, reflecting the organism's capacity for asymmetric cell division and its ability to adapt division processes in response to environmental and physiological cues. This high-resolution visualization underscores the dynamic spatial regulation of cell division proteins in pathogenic mycobacteria.

*Experimental work by Prof. Parissa -Farnia and Prof. Gennady Konstantinovich Zhavnerko; Mycobacteriology Research Center, National Research Institute of Tuberculosis and Lung Disease, Shahid Beheshti University of Medical Sciences, Tehran, Iran & The Republican Research and Practical Centre for Epidemiology & and Microbiology, Filimonova 23, Minsk, Belarus

point of intervention in antimicrobial drug discovery. A variety of compounds have demonstrated efficacy against Mtb by specifically targeting FtsZ and inhibiting its GTPase activity. Notably, SRI-3072, totarol, zantrin, berberine, and various berberine derivatives have shown inhibitory effects on FtsZ function, leading to attenuated bacilli growth (Anderson et al. 2012; Zhang et al. 2023). Similarly, small molecules such as TB-E12 and 297-F have been reported to suppress FtsZ polymerization and activity in Mtb (Lin et al. 2018). More recently, a high-throughput screening of FDA-approved drugs conducted by Tovar-Nieto et al. (2024) identified multiple repurposed compounds capable of disrupting FtsZ function. These drugs exhibited the ability to bind directly to FtsZ, interfere with its dynamic polymerization and GTPase activity, and consequently impair Z-ring formation, which is essential for proper cell division. The use of existing drugs to target FtsZ presents a compelling strategy for accelerating the development of effective antituberculosis therapies, especially in the face of escalating multidrug resistance. Collectively, the pivotal role of FtsZ in maintaining bacilli cell shape, structural integrity, and viability through its precisely regulated cycles of assembly and disassembly during cell division is well established. Detailed structural elucidation of the FtsZ GTPase domain, alongside its conformational switching mechanisms, has yielded critical molecular insights into its regulatory functions. This expanding knowledge base, when integrated with advancements in drug repurposing strategies and the visualization of FtsZ dynamics under various stress conditions, paves the way for innovative therapeutic interventions. Therefore, targeting the intrinsic vulnerability of this essential cytoskeletal protein in Mtb presents a compelling opportunity for the development of next-generation antitubercular agents.

Among the core proteins integral to cell division and maintenance of cellular morphology in Mtb, Wag31 emerges as a multifaceted regulator predominantly involved in cell wall synthesis and spatial organization (Kapoor et al. 2025). Wag31 is the mycobacterial ortholog of DivIVA, a well-characterized bacterial protein known for its role in subcellular localization and morphogenesis. Typically, Wag31 localizes to the cell poles and septum, where it functions as a scaffold protein that recruits and regulates a suite of factors critical for cell division, chromosome segregation, and cell wall biosynthesis (Kang et al. 2008; Jani et al. 2010; Melzer et al. 2022; Habibi Arejan et al. 2023). Wag31 localizes distinctly at the cell poles and septal region immediately preceding cytokinesis, thereby demarcating sites of polar growth and division (Fig. 3.3). Beyond its involvement in cell wall construction, Wag31 plays a crucial role in chromosomal segregation by mediating the recruitment of the segregation protein ParA to cell poles (Pióro et al. 2018). Although Wag31 itself is not directly involved in

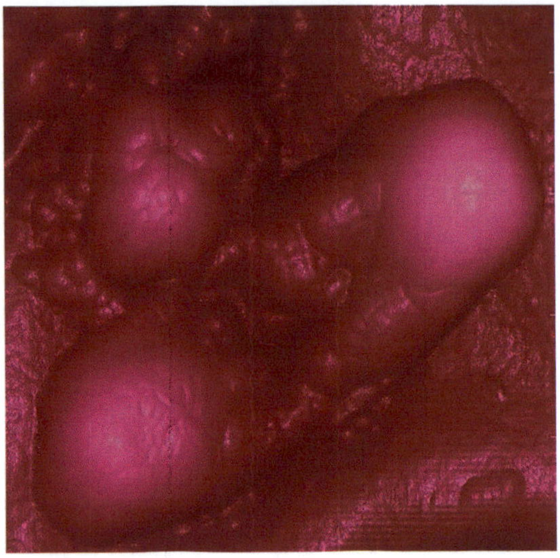

Fig. 3.3 High-resolution AFM imaging reveals distinct polar and septal features in *Mycobacterium tuberculosis*. Atomic force microscopy (AFM) image of a single *Mycobacterium tuberculosis* bacilli, illustrating the surface topography and morphology at nanometer resolution. During cell division, the essential cell division protein Wag31 (DivIVA homologue) localizes distinctly at the cell poles and the septal region immediately preceding cytokinesis. This spatial distribution of Wag31 demarcates the sites of polar growth and septation, coordinating cell wall synthesis at the poles and the division septum
*Experimental work by Prof. Parissa -Farnia and Prof. Gennady Konstantinovich Zhavnerko; Mycobacteriology Research Center, National Research Institute of Tuberculosis and Lung Disease, Shahid Beheshti University of Medical Sciences, Tehran, Iran & The Republican Research and Practical Centre for Epidemiology & and Microbiology, Filimonova 23, Minsk, Belarus

the enzymatic activities of cell wall metabolism, it is essential for directing cell wall synthetic processes to the bacterial poles, a spatial regulatory mechanism that remains to be fully elucidated at the molecular level (Habibi Arejan et al. 2023). Structurally, Wag31, similar to other DivIVA homologues, is characterized by extensive coiled-coil α-helical domains that facilitate oligomerization and membrane association (McAlinden et al. 2003). While high-resolution atomic structures of Mtb Wag31 remain to be determined, structural studies of the DivIVA family reveal elongated coiled-coil motifs that confer scaffold functionality by mediating oligomeric assemblies at the cell poles (Oliva et al. 2003). This architecture enables Wag31 to interact with key protein partners and coordinate the assembly and activity of the cell wall synthesis machinery. Functionally, Wag31 plays a critical protective role by stabilizing penicillin-binding protein 3 (PBP3) under oxidative stress conditions, thereby preserving cell wall integrity, a function that highlights its importance in stress resilience (Mukherjee et al. 2009; Plocinski et al. 2012). As a DivIVA homologue, Wag31 is indispensable for maintaining the characteristic rod-shaped morphology of the bacillus and for directing polar peptidoglycan elongation, crucial to cell growth. Regulatory control of Wag31 is mediated, in part, by phosphorylation through the serine/threonine protein kinase PknA, with phosphorylation-dependent modulation of Wag31 activity representing a potential therapeutic target for tuberculosis (Lee et al. 2014). The expression of Wag31 is also positively influenced by the stringent response, a central stress adaptation mechanism that contributes to Mtb virulence and survival under hostile host conditions (Kang et al. 2005, 2008). Phenotypically, depletion or functional disruption of Wag31 leads to pronounced morphological defects, including aberrant polar peptidoglycan synthesis and impaired cell elongation, underscoring its fundamental role in structural maintenance and growth (Kang et al. 2008). Beyond its structural and enzymatic roles, Wag31 has been implicated in modulating host immune responses; notably, its C-terminal domain suppresses the production of immunomodulatory cytokines IL-10 and IL-17 by activated human T cells, without affecting interferon-gamma (IFN-γ) secretion. This immunomodulatory effect suggests a role for Wag31 in facilitating immune evasion and contributing to pathogenesis (Samten et al. 2016). Therefore, it can be concluded that Wag31 functions as a vital multifunctional protein in Mtb, orchestrating structural integrity, regulatory pathways, and immunomodulatory roles that are indispensable for cell division, cell shape maintenance, stress adaptation, and host immune system interactions. A comprehensive understanding of the complex mechanisms underlying Wag31's function and regulation holds significant potential to inform the development of innovative therapeutic strategies aimed at targeting and disrupting these critical processes in tuberculosis.

Another group of essential structural core proteins of Mtb are those involved in cell wall synthesis and maintenance, which play a critical role in preserving cellular integrity and viability. A prime example includes proteins associated with mycolic acid biosynthesis, such as the enzyme encoded by the Rv2524c gene, located at approximately 2,840,123 base pairs on the genome. Mycolic acids constitute a major lipid component of the complex mycobacterial cell wall, conferring its characteristic impermeability and resistance to environmental stresses and antibiotics (Gu et al. 2003; Takayama et al. 2005; Bhatt et al. 2007; PaweŁczyk and Kremer 2014; Batt et al. 2020). Equally important are proteins involved in the synthesis of arabinogalactan. This key polysaccharide interlinks the peptidoglycan layer and mycolic acids, thereby constituting an essential covalent bridge within the cell wall matrix. For instance, the protein encoded by Rv3806c, located near the 4,268,925 base pair positions in the genome, contributes to the biosynthesis of arabinogalactan (Gao et al. 2024). This protein, along with others within its pathway, facilitates the assembly of this complex polysaccharide network crucial for maintaining cell wall architecture and integrity (Qin et al. 2024).

These cell wall components encompass a variety of functional categories, including transporters, receptors, lipoproteins, and surface-exposed proteins, each playing a vital role in mediating the dynamic interface between Mtb and its environment. Collectively, they influence nutrient acquisition, host-pathogen interactions, and immune evasion mechanisms. Among the extensively studied transport proteins are the ATP-binding cassette (ABC) transporters, which constitute approximately 2.5% of the Mtb genome. A notable example is the ABC transporter encoded by the Rv0986 gene (located at 1,101,803 base pairs), which facilitates the translocation of nutrients and ions across the bacterial membrane (Braibant et al. 2000; Choudhuri et al. 2002; Abu Saim Saikat et al. 2020; Cassio Barreto de Oliveira and Balan 2020). Additionally, the inner membrane ABC transporter encoded by Rv0194 (located at 226,878 base pairs) plays a key role in drug efflux, thereby contributing to antibiotic resistance (Braibant et al. 2000; Danilchanka et al. 2008). Mtb harbors 38 identified ABC transporters responsible for transporting a diverse range of substrates, including ions, sugars, amino acids, and drugs (Braibant et al. 2000; Xu et al. 2014). These transporters are critical for nutrient uptake, pathogenicity, and drug resistance (Soni et al. 2020). The ABC transporter family in Mtb consists of roughly equal numbers of importers and exporters. Importers are essential for survival in nutrient-limited environments and represent promising targets for drug and vaccine development (Soni et al. 2020). Exporters contribute to cellular detoxification, nutrient recycling, and the efflux of antibiotics, thereby facilitating bacterial survival and promoting the emergence of drug-resistant strains (de Oliveira and Balan 2020). The fact that ABC transporter genes constitute about 2.5% of the Mtb genome underscores their fundamental role in the bacterium's physiology and pathogenesis (Braibant et al. 2000).

Among the structural receptor proteins present in Mtb, MspA-like proteins act as channel-forming receptors that enable the transport of hydrophobic molecules across the bacterial cell envelope. A particularly significant receptor in this category is the necrosis-inducing toxin CpnT, encoded by Rv3903c (located at approximately 4,387,892 base pairs). CpnT fulfills a dual role by participating in nutrient uptake and simultaneously inducing necrosis in host cells (Niederweis et al. 2010; Danilchanka et al. 2015). Structurally, CpnT comprises two distinct domains: an N-terminal channel domain responsible for nutrient translocation, and a secreted C-terminal domain (CTD) that exhibits cytotoxic activity by triggering necrotic cell death in eukaryotic host cells (Danilchanka et al. 2015). The CTD exerts NAD^+-glycohydrolase enzymatic function, depleting intracellular NAD^+ levels and thereby promoting necrosis (Pajuelo et al. 2018). Besides its role in virulence, CpnT serves as an outer membrane porin in *Mycobacterium bovis BCG*, facilitating the penetration of diverse antibiotics and reactive nitrogen species, such as nitric oxide, and thus influencing the bacterium's susceptibility to these compounds (Danilchanka et al. 2015). Importantly, genomic investigations have revealed a higher prevalence of nonsynonymous mutations within the cpnT gene among drug-resistant strains of Mtb, suggesting that this gene is under selective pressure during infection and antibiotic treatment (Danilchanka et al. 2015; Farnia et al. 2018). Structural studies of MspA, a homologous mycobacterial porin, have demonstrated a unique homooctameric structure that forms a singular central channel. These insights provide a detailed understanding of the architecture and function of porin-like proteins within the mycobacterial outer membrane (Faller et al. 2004).

Lipoproteins constitute a distinct class of structural proteins integral to host-pathogen interactions. For instance, LppM, encoded by Rv2171, has been shown to subvert host immune defenses by impairing phagosomal acidification following bacterial internalization, thereby facilitating the intracellular persistence of Mtb within host macrophages (Barthe et al. 2016; Deboosère et al. 2017). Beyond modulating phagosomal maturation, LppM influences host immune responses by altering the secretion profiles of pro-inflammatory chemokines, effectively reshaping the host environment to favor bacterial survival (Deboosère et al. 2017). While LppM is not traditionally classified as a core protein, its pivotal role in Mtb pathogenesis and intracellular persistence underscores its functional significance. Among the well-characterized core structural proteins implicated in cell wall biosynthesis and maintenance are members of the LytR-CpsA-Psr (LCP) family. The Mtb genome encodes six LCP proteins, categorized according to domain architecture: three proteins harboring both LCP and LytR_C domains (Rv3267, Rv3484, and Rv0822), two containing only the LytR_C domain (VirR and Rv2700), and one possessing solely the LCP domain (Rv3840) (Kawai et al. 2011; Grzegorzewicz et al. 2016; Köster et al. 2017; Malm et al. 2018; Salgueiro-Toledo et al. 2025). Notably, Rv3267 (located at 3,647,885 base pairs) functions as the principal ligase responsible for the covalent attachment of arabinogalactan to the peptidoglycan layer, a critical process essential for preserving cell wall rigidity and overall functionality. These LCP genes are constituents of the mycobacterial core genome and are highly conserved across diverse mycobacterial species, including multiple members of the Mtb complex (Marmiesse et al. 2004). The essentiality and evolutionary conservation of LCP proteins highlight their fundamental role in maintaining cell wall integrity, positioning them as promising targets for the development of novel antituberculosis agents aimed at disrupting the mycobacterial cell envelope (Grzegorzewicz et al. 2016; Xu et al. 2022).

Ag85 family Proteins (RV3804c; location 4,265,642 bp) include Ag85A, Ag85B, and Ag85C and represent a group of highly conserved and essential proteins, classifying them as core protein components of the bacterium (Kremer et al. 2002; Ronning et al. 2000; Huygen 2014). These proteins possess mycolyltransferase activity, catalyzing the formation of trehalose dimycolate (TDM) and cell wall arabinogalactan-linked mycolic acids (Warrier et al. 2012; Touchette et al. 2017). Ag85 proteins are conserved across mycobacterial species and induce strong T cell responses in infected hosts (Huygen 2014). Inhibition of Ag85C specifically disrupts TDM synthesis and Mtb envelope integrity, highlighting its potential as a drug target (Warrier et al. 2012). The Ag85 complex is abundantly secreted and plays a key role in tuberculosis pathogenesis by triggering host immune responses (Tang et al. 2012). Understanding Ag85 biology and regulation is crucial for developing novel vaccines and diagnostics against Mtb infection (Tang et al. 2012).

Several enzymatic proteins in Mtb are classified as core components due to their indispensable roles in fundamental metabolic processes and their remarkable conservation across all strains within the Mtb complex. These enzymes constitute the backbone of essential metabolic pathways, primarily glycolysis and the tricarboxylic acid (TCA) cycle, which are crucial for energy production, biosynthesis, and maintaining cellular homeostasis. Glycolysis represents the primary pathway for glucose catabolism and energy generation in Mtb. Among the key enzymes involved are glucokinases, which catalyze the phosphorylation of glucose to glucose-6-phosphate, the critical first step in glycolysis (Marrero et al. 2013). Notably, Mtb expresses two main forms of glucokinase: polyphosphate glucokinase (PpgK) and GlkA (Hsieh et al. 1996; Ganapathy et al. 2015; Chugh et al. 2024). The coexistence of these two isoforms provides metabolic redundancy, enabling the bacterium to sustain robust glucose metabolism under varying environmental and nutritional conditions, an adaptability that supports its survival and growth in the host (Marrero et al. 2013).

The TCA cycle plays a pivotal role not only in energy production but also in supplying precursors for numerous biosynthetic pathways (Tian et al. 2005). Several key core enzymes sustain this central metabolic pathway. Citrate synthase catalyzes the condensation of oxaloacetate and acetyl-CoA to form citrate, the first committed step of the cycle (Ferraris et al. 2015). Intriguingly, Mtb possesses two isoforms of citrate synthase; one of these isoforms is particularly active under hypoxic conditions, contributing to metabolic reprogramming that facilitates bacterial persistence during oxygen limitation (Murphy and Brown 2007). Moreover, the activity of citrate synthase is finely regulated by metabolites such as pyruvate and cysteine, allowing Mtb to adjust its metabolic flux dynamically in response to intracellular and environmental cues (Pathirage et al. 2023).

Another crucial enzyme is isocitrate dehydrogenase, which converts isocitrate to α-ketoglutarate while simultaneously generating reducing equivalents in the form of NADH or NADPH. These cofactors are essential for maintaining intracellular redox balance and driving anabolic reactions. Mtb harbors two conserved and immunogenic isoforms of isocitrate dehydrogenase, underscoring its fundamental metabolic importance (Banerjee et al. 2004). Importantly, Mtb exhibits a noncanonical variant of the TCA cycle; it lacks the typical α-ketoglutarate dehydrogenase (Tian et al. 2005). Instead, it utilizes alternative enzymes such as α-ketoglutarate decarboxylase and succinic semialdehyde dehydrogenases that bypass this step. This unique metabolic adaptation supports Mtb's survival and persistence during nutrient stress and hostile host conditions. Other enzymes integral to the completion of the TCA cycle include aconitase, fumarase, malate dehydrogenase, and succinate dehydrogenase. These enzymes orchestrate sequential transformations of metabolic intermediates, facilitating energy extraction from carbon substrates and linking the cycle to the respiratory chain for efficient ATP generation (Hartman et al. 2014). The reliability and efficiency of these enzymes are vital for sustaining Mtb's energy metabolism and providing key intermediates for biosynthetic needs. Beyond glycolysis and the TCA cycle, core proteins involved in oxidative phosphorylation are critical for the generation of adenosine triphosphate (ATP), the cell's energy currency (Zhang et al. 2024a, b). These proteins maintain cellular energy homeostasis and are conserved throughout the Mtb complex, which further justifies their classification as core metabolic enzymes. Amino acid biosynthesis is another essential aspect of Mtb metabolism supported by core enzymes such as aspartate aminotransferase and glutamate dehydrogenase. Aspartate aminotransferase catalyzes the reversible transamination between aspartate and α-ketoglutarate, playing a central role in nitrogen metabolism (Jansen et al. 2020). Concurrently, glutamate dehydrogenase facilitates the formation of glutamate, a key amino acid involved in nitrogen assimilation and broader metabolic regulation. Lipid biosynthesis enzymes represent an indispensable subset of core metabolic proteins. Acetyl-CoA carboxylase (Rv3285) catalyzes the ATP-dependent carboxylation of acetyl-CoA to malonyl-CoA, marking the committed and regulated entry point into fatty acid synthesis. KasA (Rv2524c), another vital enzyme in this pathway, is responsible for the elongation steps required for mycolic acid biosynthesis (Tong 2005; Zu et al. 2013; Salie and Thelen 2016). Mycolic acids are unique, long-chain fatty acids that constitute a significant portion of the mycobacterial cell wall. Their presence imparts impermeability and resistance to antimicrobial agents, thereby contributing to Mtb's virulence and pathogenic resilience.

The wide array of conserved enzymatic core proteins underscores the remarkable metabolic versatility of Mtb. This versatility enables the pathogen to adapt to myriad host environments and fluctuating nutrient availability, which is critical during different stages of infection and latency (Chang and Guan 2021). Lipid metabolism, in particular, plays a dominant role during infection since fatty acids serve as a major carbon and energy source. Consequently, enzymes involved in fatty acid and mycolic acid biosynthesis are central to Mtb's survival within the host. In summary, these core enzymes collectively empower Mtb with exceptional metabolic flexibility, facilitating its persistence and pathogenicity across diverse environments, especially within the human host (Ehrt et al. 2018). The essential nature and high degree of conservation of these proteins make them promising targets for the development of novel antitubercular therapeutics. A deeper understanding of their structure, function, and regulation is crucial for advancing strategies aimed at combating tuberculosis by disrupting the bacterium's core metabolic machinery.

Sigma factors are indispensable regulatory proteins that orchestrate transcription initiation by associating with RNA polymerase and directing its binding to specific promoter sequences. Mtb encodes 13 sigma factors, including the essential primary sigma factor σ^A and at least 6 additional sigma factors implicated in virulence and pathogenesis (Manganelli 2014). These interchangeable RNA polymerase subunits confer promoter specificity, enabling Mtb to rapidly adjust gene expression in response to fluctuating environmental conditions encountered during infection (Rodrigue et al. 2006; Lee et al. 2008; Manganelli 2014; Grosse-Siestrup et al. 2021). The sigma factors of Mtb are broadly categorized into three principal classes: SigA, SigB, and SigC.

SigA functions as the primary housekeeping sigma factor, regulating constitutive expression of essential genes required for basal cellular processes, with stable expression maintained irrespective of growth phase or environmental stressors (Hu and Coates 1999; Paget and Helmann 2003; Chauhan et al. 2016). In contrast, SigB, an alternative sigma factor, is upregulated during the stationary phase and under diverse stress conditions, fulfilling functions analogous to the σ^S factor described in other bacteria and facilitating adaptation to hostile environments (Hurst-Hess et al. 2019). SigC, classified within the extracytoplasmic function (ECF) sigma factor family, is notably critical for Mtb's virulence and lethality in murine models, although it is not essential for bacterial viability under standard conditions (Sun et al. 2004; Chang et al. 2012; Grosse-Siestrup et al. 2021). Collectively, SigA and SigB are regarded as core regulatory sigma factors fundamental to transcription initiation and stress adaptation. Conversely, SigC, while integrated within the broader sigma factor regulatory network, exhibits traits characteristic of accessory sigma factors due to its specialized roles in stress response modulation and virulence-associated mechanisms (Manganelli 2014; Chauhan et al. 2016).

RNA regulatory proteins play a pivotal role in modulating gene expression, particularly through posttranscriptional regulatory mechanisms. These proteins interact with RNA molecules to influence their stability, translation efficiency, and degradation, processes that are critical for Mtb to adapt to diverse host environments and withstand various stress conditions. Key members of this regulatory repertoire include NusA, RNase J, and noncoding RNA-associated proteins (Beisel and Storz 2010; Schwenk and Arnvig 2018; Fathizadeh et al. 2020; Ostrik et al. 2021). NusA (encoded by Rv2841c, located at 3,148,385 base pairs) is a central regulator of transcriptional elongation, pausing, termination, and antitermination in Mtb. Crystallographic analyses reveal that Mtb NusA comprises four domains, including an N-terminal domain structurally analogous to ABC-ATPases and a C-terminal module endowed with RNA-binding capacity (Gopal et al. 2001). Genome-wide mapping studies demonstrate that NusA associates extensively with RNA polymerase throughout the Mtb chromosome, with consistent distribution patterns observed in both exponential and stationary growth phases (Uplekar et al. 2013). NusA enhances regulatory processes such as transcriptional pausing and termination, exhibiting notably high-affinity binding to the ribosomal RNA (rrn) nut site, particularly involving the conserved boxC region downstream of boxA (Arnvig et al. 2004; Beuth et al. 2005). The ubiquitous presence of NusA along gene bodies correlates positively with gene expression levels, underscoring its role as a facilitator of transcriptional elongation (Uplekar et al. 2013). Furthermore, the structural features and surface properties of NusA provide mechanistic insights into its interactions with RNA, RNA polymerase, and its discriminative recognition of transcriptional pause sites (Gopal et al. 2001). RNase J plays an indispensable role in RNA metabolism by mediating RNA processing, degradation, and overall maintenance of RNA homeostasis. Its activity is essential for Mtb growth and survival, underscoring its critical function within the bacterial transcriptome regulatory network (Martini et al. 2022; Bao et al. 2023). Noncoding RNA (ncRNA) proteins in Mtb can be broadly categorized into small RNAs (sRNAs), long noncoding RNAs (lncRNAs), and circular RNAs (circRNAs). sRNAs are instrumental in coordinating stress responses and facilitating adaptation to the intracellular host environment. lncRNAs, in turn, participate in intricate regulatory networks that modulate messenger RNA (mRNA) expression during infection, frequently acting as competing endogenous RNAs (ceRNAs) that regulate microRNA activity, thereby influencing posttranscriptional gene regulation (Niu et al. 2020; Vijayaraghavan et al. 2025).

Chaperones and heat shock proteins (HSPs) constitute approximately 2–5% of the total protein content in Mtb and are indispensable for maintaining protein homeostasis by ensuring proper folding and preventing aggregation under various stress conditions (Hu et al. 2020). Although functionally positioned between core and accessory proteins, chaperones are generally classified as core proteins due to their conditionally essential roles in bacterial survival, particularly under hostile host-induced stresses such as exposure to reactive nitrogen species (Parbhoo et al. 2022). Prominent examples of these molecular chaperones include DnaK (HSP70; Rv0350, located at 419,835 base pairs), which assists in protein folding and mitigates aggregation during stress (Nachappa et al. 2022); GroEL1 (Rv3417c, located at 3,835,272 base pairs), which is specialized for functions in biofilm formation and mycolic acid biosynthesis; and GroEL2, both crucial for protein folding and adaptation to stress (Ojha et al. 2005; Liu et al. 2024). ClpB (Rv0384c, located at 459,456 base pairs) acts as a vital protein disaggregase, reactivating aggregated proteins that would otherwise be toxic, thereby enhancing cellular stress tolerance (Lupoli et al. 2018). Additionally, HrpG (HSP90; Rv0351, located at 421,709 base pairs), a small heat shock protein highly expressed during the stationary phase, confers protection against nitric oxide-induced stress (Deng et al. 2021).

Collectively, secretory and export proteins, cell wall-associated proteins, heat shock proteins, and cytoplasmic chaperones constitute fundamental components of the Mtb core proteome. Each group performs unique and essential functions that underpin bacterial physiology, adaptation, and interaction with the host environment, thereby facilitating Mtb's survival and pathogenicity (Pal et al. 2022). The identification and characterization of core proteins in clinical Mtb isolates are of paramount importance, as these highly conserved proteins support critical cellular processes and represent universal targets for the development of diagnostics, vaccines, and therapeutics. Such targets are particularly valuable because they transcend strain-specific variability and circumvent issues related to drug resistance. For instance, a comprehensive pangenome analysis of multiple Mtb strains identified 1251 core proteins, among which eight were prioritized as potential drug targets based on their essentiality and functional relevance (Dar et al. 2020). Similarly, screening of a phage secretome library revealed ten proteins indispensable for Mtb growth and survival, including notable factors such as Nrp, PssA, and MmpL5, highlighting additional candidates for therapeutic intervention (Chiliza et al. 2017). The proteasome core complex, encoded by the prcBA gene cluster, has been demonstrated to be essential for Mtb persistence during chronic infection in murine models, underscoring its critical role in bacterial endurance under host immune pressure (Gandotra et al. 2007). In addition, the unfoldase ClpC1, a component of the Clp proteolytic system, is pivotal for maintaining proteostasis by regulating the degradation of proteins containing intrinsically disordered termini. ClpC1 is indispensable for Mtb's extracellular growth and survival within macrophages, reinforcing its significance in bacterial pathogenesis and stress adaptation (Lunge et al. 2020). These findings collectively emphasize the indispensable roles that core proteins play in Mtb biology and virulence, and they illuminate promising avenues for the design of novel drugs and diagnostic tools aimed at advancing tuberculosis control and treatment efforts.

3.3 Characterization of Accessory Proteins of *Mycobacterium tuberculosis*

Accessory proteins in Mtb, many of which are well-established virulence factors, play a critical role in the bacterium's survival, persistence, and pathogenicity within the human host. Although these proteins are often dispensable for bacterial growth under standard in vitro conditions, they become indispensable during infection, particularly within the hostile intracellular environment of host macrophages. The accessory proteome of Mtb encompasses a broad and diverse group including secreted effectors, regulatory enzymes, stress response mediators, and immunomodulatory proteins (Rahlwes et al. 2023). Prominent among these accessory proteins are substrates of the ESX secretion systems, members of the PE/PPE protein families, and various kinases and phosphatases, all of which are classified as virulence proteins due to their direct involvement in pathogenesis, immune evasion, and adaptation to the host immune milieu (Ramon-Luing et al.

2023). Virulence proteins collectively constitute approximately 5–10% of the total Mtb proteome and significantly contribute to the pathogen's ability to circumvent host defenses, establish infection, and cause disease (Chandra et al. 2022). A comprehensive understanding of these virulence determinants is essential for advancing therapeutic and vaccine development efforts.

Functionally, virulence proteins can be categorized into distinct groups such as secreted effectors translocated via the ESX (type VII secretion) systems, PE and PPE family proteins, phosphatases, kinases, and other virulence-associated factors (Singh et al. 2006; Bhuwan et al. 2016; Echeverria-Valencia et al. 2017; Ramon-Luing et al. 2023; Rahlwes et al. 2023). The ESX systems, numbering ESX-1 through ESX-5, represent specialized secretion apparatuses that facilitate substrate transport across Mtb's complex cell envelope, thereby mediating interactions with host cells and subverting immune responses. ESX-1, encoded within the region of difference 1 (RD1) locus, is the most extensively characterized system and secretes key virulence factors such as ESAT-6 (Rv3875) and CFP-10 (Rv3874). These proteins form a heterodimeric complex that disrupts phagosomal membranes, enabling Mtb to escape into the host cell cytosol and evade intracellular killing (McLaughlin et al. 2007; Abdallah et al. 2009; Frigui et al. 2008; Samten et al. 2009; Bottai et al. 2012; Houben et al. 2014; Anil Kumar et al. 2016; Damen et al. 2020).ESX-3 is primarily involved in metal ion homeostasis, notably facilitating iron acquisition by secreting substrates such as EsxG and EsxH, which assist in siderophore-mediated iron uptake. ESX-5 oversees secretion of the PE and PPE proteins, which contribute to antigenic variation and modulate host-pathogen interactions (Bitter et al. 2009; Tufariello et al. 2016; Daher et al. 2025). Among the major virulence-associated protein families unique to mycobacteria are the PE and PPE proteins, named after their conserved proline (P) and glutamic acid (E) residues in their N-terminal domains. These proteins account for approximately 10% of the Mtb genome and are predominantly localized to the cell wall, where they contribute to virulence, immune modulation, nutrient transport, and host-pathogen interplay. The PE/PPE proteins are classified into subgroups, including PE_PGRS, PPE-PPW, PPE-SVP, and PPE-MPTR, each characterized by distinctive sequence motifs and functional attributes (Sampson 2011). Notably, PE/PPE proteins influence Mtb survival within host cells through interactions with Toll-like receptors (TLRs), triggering signaling cascades that regulate inflammatory cytokine secretion (Guo et al. 2023; Priyanka et al. 2024). For example, PE11 is highly upregulated during macrophage infection and in patients with active tuberculosis, where it enhances resistance to environmental stressors and antibiotics, underscoring its role in pathogenic persistence (Zhang et al. 2025).

Phosphatases and kinases are pivotal enzymes that regulate cellular processes through post-translational modifications, primarily phosphorylation and dephosphorylation (Ardito et al. 2017). Although these enzymes are not essential for Mtb survival under in vitro growth conditions, they assume critical importance during infection and persistence within the host. Consequently, they are functionally designated as accessory proteins. These enzymes facilitate Mtb's adaptation to environmental fluctuations, modulate host-pathogen interactions, and sustain pathogenicity. Notable among the phosphatases and kinases are SapM, a secreted acid phosphatase that inhibits phagosome-lysosome fusion by dephosphorylating phosphatidylinositol 3-phosphate (PI3P); PtpA, a tyrosine phosphatase that suppresses host immune responses via dephosphorylation of signaling proteins; and serine/threonine kinases PknE and PknG, which modulate host cellular pathways to promote intracellular survival (Kang et al. 2005; Poirier and Av-Gay 2012; Kumar and Narayanan 2012; Puri et al. 2013; Wang et al. 2017; Margenat et al. 2023). Additional virulence factors of significance include INT (Tuberculosis Necrotizing Toxin), which induces necroptosis through intracellular NAD^+ degradation, and RipA, a peptidoglycan hydrolase secreted via the twin-arginine translocation (TAT) pathway, implicated in bacterial virulence and potentially influencing β-lactam antibiotic susceptibility

(Tak et al. 2019). Together, these virulence proteins contribute substantively to immune evasion by modulating host immune responses, thereby preventing effective bacterial clearance. They facilitate bacterial survival within host cells by inhibiting phagosome maturation and manipulating cell death pathways, and participate in granuloma formation and associated tissue pathology. Immunogenic proteins of Mtb are paramount in eliciting host immune responses and represent promising candidates for vaccine development (Sethi et al. 2024; Barazani et al. 2025). Among the key immunogenic determinants are ESAT-6 and ESAT-6-like proteins, which are integral to virulence and immune modulation. ESAT-6 (Rv3875, located at 4,352,609 base pairs) serves as a principal T cell antigen and is indispensable in vaccine design.

The PE and PPE protein families, which contribute to antigenic variation and complex host-pathogen interactions, are also recognized by the host immune system and induce both humoral and cellular immune responses. Representative proteins include PE18 (Rv1788, located at 2,026,477 base pairs), PE31 (Rv3477, located at 3,894,093 base pairs), PE26 (Rv1789, located at 2,026,790 base pairs), PPE55 (Rv3347c, located at 3,743,711 base pairs), and Rv2536 (located at 2,860,452 base pairs) (D'Souza et al. 2023; Zhang et al. 2025).

Cofactor-binding proteins in Mtb are indispensable mediators of a wide array of metabolic and enzymatic functions. These proteins interact with specific cofactors—nonprotein chemical entities essential for the catalytic activity of numerous enzymes, thus facilitating critical cellular processes such as redox reactions and biosynthetic pathways (Graciano and Liu 2025). Functionally, cofactor-binding proteins are broadly categorized into core and accessory classes based on their essentiality and contextual roles. Core cofactor-binding proteins underpin fundamental cellular processes, including basal metabolism and survival, whereas accessory proteins often contribute to stress adaptation, virulence, and host-pathogen interactions.These proteins can be further delineated into three primary groups according to their cofactor specificity and functional roles. First, proteins involved in the biosynthesis of the molybdenum cofactor (Moco), exemplified by MoaC2 (Rv0864), are crucial for the transformation of guanosine triphosphate (GTP) into the active Moco form. Structural analyses have demonstrated that MoaC2 interacts with intermediates such as pteridinebenzomonophosphate during this biosynthetic process, highlighting its mechanistic role in Moco assembly (Srivastava et al. 2013). Second, proteins that bind transition metals such as iron and manganese, exemplified by Rv0233, an R2 homologue, incorporate a heterodinuclear manganese/iron-carboxylate cofactor critical for catalyzing complex redox reactions and generating radical species. Notably, Rv0233 exhibits a remodeled R2 protein scaffold that enhances ligand affinity and oxidase activity, reflecting evolutionary adaptation to its specific biochemical functions (Andersson and Högbom 2009; Powell et al. 2023). Third, cofactor-binding proteins involved in the biosynthesis and processing of mycofactocin—an emerging redox cofactor—include MftB and MftC. MftB functions as a molecular chaperone that binds the mycofactocin precursor MftA with high specificity, facilitating its proper maturation. MftC, a radical S-adenosylmethionine (rSAM) enzyme, catalyzes the oxidative decarboxylation of the C-terminal tyrosine residue of MftA, a reaction that requires the presence of MftB and is critical for generating the active mycofactocin cofactor (Mawuenyega et al. 2005; Ayikpoe et al. 2018; Mendauletova and Latham 2022).

Signal transduction proteins in Mtb are indispensable for sensing and responding to environmental stimuli and play a critical role in modulating host cell signaling pathways to facilitate bacterial infection and survival (Johnson and McDonough 2018). Among these, specific proteins are classified as core components due to their essentiality for bacterial growth and viability, exemplified by the serine/threonine protein kinase PknB. Conversely, other proteins such as PknA, which plays a significant role in cell division, are not strictly required for in vitro survival and are therefore deemed accessory proteins (Turapov et al. 2018). Similarly, two-component regulatory systems, including MtrA-MtrB, are vital for adaptation during infection but are not necessarily essential for growth under standard laboratory conditions (Zahrt and Deretic 2000). Calmodulin-like proteins, such as CAMLP, typically exhibit nonessential roles during in vitro culture but contribute substantially to bacterial survival and regulatory functions within the host environment.The principal groups of signal transduction proteins in Mtb are as follows: (i) serine/threonine protein kinases (STPKs), which phosphorylate proteins on serine and threonine residues to modulate diverse cellular processes including cell division, metabolism, and stress responses, for instance, PknA (Rv0014c; location 15,590 bp), required for cell division and survival, and PknB (Rv0020c), involved in regulating cell wall biosynthesis; (ii) two-component systems, composed of a sensor kinase and corresponding response regulator, which are essential for adapting to environmental fluctuations and host-pathogen interactions, exemplified by MtrA-MtrB (Rv3246c-Rv3245c) and PrrA-PrrB (Rv0903c-Rv0902c); and (iii) calmodulin-like proteins such as CAMLP (Rv1211; location 1,354,243 bp), which participate in calcium-dependent signaling pathways and may regulate transcriptional regulators (Av-Gay et al. 1999; Rosenkrands et al. 2000; Boitel et al. 2003; Rengarajan et al. 2005; Nowak et al. 2006; Haydel et al. 2012; Advani et al. 2014; DeJesus et al. 2017; Li et al. 2020; Choo et al. 2022).

Mtb harbors a family of seven cutinase-like proteins (CULPs) that exhibit diverse enzymatic activities. While these proteins demonstrate varying degrees of esterase and lipase functions, they notably lack classical cutinase activity (West et al. 2009). Among these, two secreted cutinase-like proteins, Rv1984c and Rv3452, demonstrate distinct substrate specificities; notably, Rv3452 possesses phospholipase A2 activity, which has been shown to induce macrophage lysis, thereby enhancing Mycobacterium tuberculosis pathogenicity (Schué et al. 2010). Additionally, Culp6, a cell wall-associated protein, elicits robust T lymphocyte responses in Mtb-infected mice and tuberculosis patients, highlighting its potential as a candidate antigen for subunit vaccine development (Strong and West 2018). Mtb also possesses an itaconate dissimilation pathway, wherein the bifunctional enzyme Rv2498c participates in both itaconate dissimilation and leucine catabolism. This enzyme plays a key role in Mtb's intrinsic resistance to itaconate, a host-derived antimicrobial metabolite, and has been implicated in virulence within murine infection models (Wang et al. 2019). Collectively, cutinase-like proteins are generally classified as accessory proteins given their specialized, nonessential roles under laboratory conditions, although some may exhibit conservation across strains. Similarly, the itaconate dissimilation enzyme (Rv2498c; located at 2,812,355 bp) is also regarded as an accessory protein.

Regulatory proteins constitute a substantial component of the Mtb proteome, accounting for approximately 5 to 10% of total proteins, and are integral to the pathogen's ability to adapt to environmental fluctuations and host-derived stresses (Raman et al. 2005; West et al. 2009; Lin et al. 2024). These regulators orchestrate critical processes including gene expression, metabolic regulation, and stress response modulation. They can be broadly categorized into transcription factors, transcriptional regulators of metal ion homeostasis, sigma factors, and RNA-binding regulatory proteins. Examples of key transcription factors include RbpA, a master regulator of gene expression, and ResR/McdR (Rv1830), which modulates protein translation and contributes to antibiotic tolerance. Among transcriptional regulators of metal ion homeostasis are SmtB (Rv2358; located at 2,641,246 bp) and Zur (Rv2359; located at 2,641,650 bp), which regulate zinc uptake and release (Smith et al. 2009; Pal et al. 2023; Wang et al. 2024). Additionally, RicR (Rv0190; located at 221,871 bp) serves as a copper-responsive transcriptional repressor that controls genes involved in copper detoxification (Shi et al. 2014; Wang et al. 2024).

In summary, virulence proteins, often regarded as a subset of accessory proteins, together with regulatory proteins, constitute a highly interconnected and complex network that underlies Mtb's virulence, immune evasion, stress adaptation, and drug resistance mechanisms. Their secretion via specialized export systems and involvement in diverse cellular processes underscore their pivotal role in pathogenesis. Consequently, these multifunctional proteins represent promising targets for the development of novel antituberculosis therapeutics and vaccines, especially in light of the mounting challenges posed by drug-resistant Mtb strains.

3.4 Characterization of Transcriptional Plasticity Proteins in *Mycobacterium tuberculosis*

Transcriptional plasticity proteins (TPs) in Mtb are defined as proteins encoded by genes whose expression levels dynamically fluctuate in response to diverse environmental stresses and conditions. This transcriptional plasticity underpins the bacterium's remarkable ability to modulate gene expression genome-wide, thereby facilitating adaptation, survival, and pathogenicity under the varying pressures imposed by host and environmental contexts (Silva et al. 2021). Typically conserved across Mtb strains, TP-encoding genes enable the pathogen to reprogram transcriptional activity rapidly and precisely in response to external stimuli. Although these proteins are often nonessential for growth under standard in vitro conditions, they are critical for survival within hostile microenvironments characterized by oxidative stress, nutrient limitation, and immune-mediated challenges. Comprehensive genome-wide analyses involving 894 RNA sequencing datasets spanning 73 distinct environmental conditions have demonstrated pronounced variability in the transcriptional plasticity of Mtb genes, which correlates strongly with gene function and essentiality (Orgeur et al. 2024). Moreover, intrinsic genetic features such as gene length, GC content, and operon size exert independent influences on transcriptional plasticity, beyond canonical trans-regulatory mechanisms. Importantly, this dynamic transcriptional landscape is conserved across related mycobacterial species, including *Mycobacterium smegmatis* and *Mycobacterium abscessus*, underscoring its evolutionary significance (Bei et al. 2024). Genes exhibiting high transcriptional plasticity predominantly encode proteins implicated in critical adaptive functions, including iron metabolism, stress response pathways, and host-pathogen interactions. In contrast, essential genes generally display low transcriptional plasticity, maintaining stable expression to preserve core cellular processes. This regulatory balance enables Mtb to fine-tune gene expression in response to diverse environmental challenges such as hypoxia, oxidative stress, nutrient deprivation, and antibiotic exposure (Namouchi et al. 2016; Bei et al. 2024; Wang et al. 2024). A central component of Mtb's transcriptional response to hypoxia and dormancy is the transcription factor DosR (Rv3133c), which orchestrates the expression of nearly all hypoxia-inducible genes by binding to specific promoter motifs upstream of its target loci (Park et al. 2003; Boon and Dick 2002). DosR is indispensable for mycobacterial survival under oxygen-limited conditions, with its disruption resulting in bacterial death during hypoxia (Boon and Dick 2002). The DosR regulon is governed by a two-component signal transduction system comprising the sensor kinases DosS and DosT along with the response regulator DosR itself (Chen et al. 2013). Structural analyses of DosR reveal distinctive features, including an atypical (βα)_4 fold in the N-terminal domain and a unique conformation of the linker region that forms a four-helix bundle critical for dimerization (Wisedchaisri et al. 2008). DosS and DosT act as gas sensors, becoming activated under hypoxic conditions or upon binding to nitric oxide (NO) or carbon monoxide (CO), with DosS additionally functioning as a potential redox sensor (Sivaramakrishnan and de Montellano 2013). Intriguingly, overexpression of DosR during aerobic growth induces its regulon without adversely affecting bacterial replication either in vitro or within macrophages, underscoring its regulatory flexibility (Flores-Valdez et al. 2015). Recent multi-omics studies in *Mycobacterium bovis BCG* have further revealed that DosR exerts broad regulatory influence beyond hypoxia-responsive genes, impacting transcriptional and protein expression profiles related to amino acid biosynthesis and general metabolism (Cui et al. 2023). Collectively, these insights illustrate the functional plasticity of DosR and underscore its potential as a pivotal regulatory target for therapeutic interventions against mycobacterial infections.

SigH (Rv3223c), an extracytoplasmic function (ECF) sigma factor, exemplifies functional plasticity in Mtb by orchestrating the bacterial response to oxidative and redox stress. Initially characterized by Raman et al. (2001), SigH regulates an extensive transcriptional network that transiently suppresses energy-intensive processes such as lipid biosynthesis immediately following oxidative insult, only to reactivate these pathways during recovery phases (Rohde et al. 2007). This temporal modulation allows Mtb to conserve resources and effectively repair cellular damage. Furthermore, SigH induces a suite of transcriptional regulators that remodel global gene expression to repress lipid and virulence factor synthesis until stress conditions subside, while simultaneously upregulating genes involved in detoxification and protein quality control (Dutta et al. 2012; Sharp et al. 2016). Functional analyses have demonstrated that deletion of SigH compromises Mtb's capacity to maintain redox homeostasis and endure oxidative stress, underscoring its critical role in both bacterial physiology and pathogenicity (Singh et al. 2025). The SigH regulon thus comprises genes pivotal for adaptation to hostile host environments, contributing directly to Mtb persistence and virulence.

WhiB3 represents another transcription factor exhibiting remarkable functional adaptability. This DNA-binding protein senses fluctuations in environmental gases such as nitric oxide (NO) and oxygen gradients, as well as intracellular redox states reflected by levels of reducing equivalents like NADH, generated during respiration, and NADPH, which accumulates during β-oxidation of host fatty acids (Saini et al. 2012). WhiB3 harbors a redox-sensitive 4Fe-4S cluster that dynamically detects host-derived signals crucial during infection (Kumar et al. 2011; Mehta and Singh 2019; Wan et al. 2023). Functionally, WhiB3 maintains redox equilibrium by facilitating recovery from oxidative and nitrosative stress; Mtb strains deficient in WhiB3 exhibit heightened susceptibility to reactive oxygen and nitrogen species (Saini et al. 2012; Mehta and Singh 2019). It integrates environmental redox cues with central metabolism, regulating metabolic shifts necessary for adaptation to fluctuating host conditions (Singh et al. 2007; Mahatha et al. 2020). While not essential for bacterial replication in vivo, WhiB3 influences host-pathogen dynamics by modulating immune responses, highlighting its role in pathogenesis (Steyn et al. 2002). Moreover, WhiB3 interacts with RpoV, the primary sigma factor in virulent Mtb strains, underscoring its regulatory versatility (Steyn et al. 2002). Experimental evidence supports WhiB3 as an intracellular redox sensor coordinating early granuloma formation, thereby linking redox homeostasis directly to host-pathogen interactions (Pacl et al. 2018; Mehta and Singh 2019). The conformational plasticity of its 4Fe-4S cluster upon exposure to reactive species exemplifies its functional flexibility, facilitating precise modulation of gene expression crucial for survival under hostile conditions (Saini et al. 2012). Additional regulatory proteins underscore Mtb's capacity for adaptive plasticity. MprA (Rv0981), a component of a two-component regulatory system, governs stress adaptation during environmental shifts and host invasion (He et al. 2006). It directly regulates the espA operon, essential for the functionality of the ESX-1 secretion system (Pang et al. 2013). Together with DosR, MprA positively regulates the expression of Rv0081, a gene implicated in a putative formate hydrogenlyase complex (He et al. 2011). Modulation of ESX-1 activity by the MprAB system influences the secretion of key virulence factors such as ESAT-6 and EspB, thereby affecting host cytokine responses (Pang et al. 2013). Although not directly linked to MprA, the study of Rv1698 has revealed a novel class of channel-forming outer membrane proteins in mycobacteria that may facilitate transport processes across the outer membrane (Siroy et al. 2008). These findings highlight the intricate regulatory networks and membrane structures fundamental to Mtb survival and pathogenesis. AhpC (Rv2428), an alkyl hydroperoxide reductase, exemplifies a stress-response protein whose role becomes critical under oxidative stress conditions. Although nonessential under normal growth, AhpC can compensate for KatG deficiency in isoniazid-resistant strains (Sherman et al. 1996; Heym et al. 1997; Chauhan and Mande 2001). Notably, while AhpC overexpression correlates with isoniazid resistance, it does not confer direct resistance to the drug; rather, it serves as a compensatory mechanism protecting KatG-deficient strains from organic peroxide toxicity (Sherman et al. 1996; Heym et al. 1997). Structurally, Mtb AhpC is a unique decameric enzyme stabilized primarily through ionic interactions that are essential for its catalytic activity (Chauhan and Mande 2001). Although AhpC overexpression does not restore virulence in KatG-deficient strains, it likely contributes to maintaining peroxide homeostasis when KatG activity is diminished or absent (Heym et al. 1997). Collectively, these findings illuminate the intricate interplay between AhpC, KatG, and isoniazid resistance mechanisms in Mtb. Overall, the multifaceted roles and extraordinary functional plasticity of these regulatory proteins exemplify Mtb's sophisticated adaptive strategies, enabling survival and persistence within hostile host environments.

3.5 Uncharacterized Proteins in *Mycobacterium tuberculosis*

A substantial proportion of the Mtb proteome remains uncharacterized, comprising proteins whose functions are either unknown or predicted solely through computational approaches. These proteins, often designated as hypothetical or conserved hypothetical, present both significant challenges and opportunities for tuberculosis research. Elucidating their roles is critical, as they may participate in essential bacterial processes including pathogenicity, stress survival, drug resistance, and immune evasion, thereby representing promising novel therapeutic targets (Qin et al. 2013; Sharma and Bisht 2017; Ijaq et al. 2022; Yang et al. 2019; Jaiswal et al. 2024).

Comprehensive proteomic analyses of Mtb $H_{37}Rv$ culture filtrates have identified 257 exported proteins, of which 159 are predicted to utilize the general secretory (Sec) pathway (Målen et al. 2007). Notably, approximately 30% of these secreted proteins are hypothetical or uncharacterized, underscoring a significant fraction of the secretome whose functional contributions remain to be defined. This highlights an urgent need for a focused investigation into the roles these proteins play in bacterial survival and pathogenesis. Increasing evidence supports the premise that hypothetical proteins may be integral to drug resistance mechanisms and thus represent attractive candidates for novel therapeutic interventions (Sharma and Bisht 2017). Bioinformatics-driven analyses have the potential to reveal previously unknown resistance pathways and identify new drug targets in multidrug-resistant tuberculosis strains (Ramakrishnan et al. 2015). Label-free quantitative proteomics has detected 68 proteins exhibiting significant differential expression between multidrug-resistant and susceptible strains, including uncharacterized oxidoreductases and membrane proteins (Phong et al. 2015). Likewise, two-dimensional differential gel electrophoresis identified 41 protein spots with altered regulation in rifampicin- and isoniazid-resistant isolates, among which several were uncharacterized (Sharma et al. 2015). Comparative proteomic profiling of aminoglycoside-resistant strains revealed increased abundance of 12 proteins, three of which remain functionally unelucidated. These findings support the notion that uncharacterized proteins may contribute to novel resistance mechanisms and could serve as potential diagnostic markers or therapeutic targets (Sharma and Bisht 2017). Moreover, proteins implicated in iron regulation, stress response, and protein turnover have been linked to drug resistance, reflecting complex adaptive networks within Mtb. An illustrative case of a protein transitioning from hypothetical to functionally inferred is Rv0216 (located at 258,913 bp). Initially annotated as uncharacterized due to limited experimental evidence, Rv0216 has been demonstrated to be essential for Mtb survival during infection. Structural modeling predicts a double hotdog fold—a motif typical of hydratase enzymes involved in lipid metabolism (Castell et al. 2005). However, despite structural homology, Rv0216 lacks catalytic activity against established substrates, positioning it as an intermediate entity: no longer strictly hypothetical, yet incompletely characterized given the absence of direct biochemical validation. Its essentiality in vivo underscores its potential as a high-priority therapeutic target. In silico analyses have expanded the catalog of uncharacterized Mtb proteins. For example, Rv0986 is predicted to function as an ATP-binding cassette (ABC) transporter implicated in lipoprotein and macrolide export, potentially linking it to drug resistance (de Oliveira and Balan 2020). Rv1708 is associated with cell division, whereas Rv2714 (located at 3,027,065 bp) remains hypothetical but shares conserved features with known protein families. Another notable protein, Rv3651 (located at 4,091,841 bp), possesses a triple sensor-domain architecture but lacks a defined effector domain, suggesting a role in environmental signal integration without a currently understood downstream function. The PE_PGRS family protein PE_PGRS20 (Rv1068c, located at 1,190,757 bp) has been linked to virulence and may contribute to drug resistance, exemplifying the functional relevance of uncharacterized proteins in pathogenicity (De Maio et al. 2020). Recent advances have leveraged comprehensive computational pipelines to predict the structures and functions of hypothetical proteins in the $H_{37}Rv$ strain (Gazi et al. 2016). Among identified candidates are proteins harboring GTPase domains and putative virulence factors, which represent promising targets for drug development or vaccine design. Conserved GTP-binding proteins such as Era, Obg, LepA, and EngA exhibit critical roles in cellular signaling and pathogenesis through their GTPase activities (Morimoto et al. 2002). Notably, four hypothetical proteins—Rv0679c, Rv0906, Rv3627c, and Rv3811—contain GTPase domains and have been prioritized as potential drug targets (Beg et al. 2021). These insights enrich our understanding of Mtb pathogenesis and open new avenues for therapeutic innovation. In conclusion, uncharacterized proteins constitute a largely untapped frontier in tuberculosis research. Continued integration of bioinformatic prediction with targeted experimental validation promises to deepen our understanding of Mtb's complex biology and pathogenic mechanisms. Deciphering the functions and interactions of these elusive proteins holds considerable potential to identify novel drug targets and vaccine candidates, thereby transforming strategies for tuberculosis treatment and prevention. This endeavor is especially critical amid the escalating threat posed by multidrug-resistant and extensively drug-resistant tuberculosis strains, which challenge global public health and necessitate innovative therapeutic approaches. Ultimately, elucidating the roles of these uncharacterized proteins may prove pivotal to overcoming current obstacles and achieving more effective control and eradication of tuberculosis worldwide.

Conflict of Interest The authors declare that they have no conflict of interest related to the content of this chapter.

Ethical Consideration This chapter is based on conclusions drawn from the authors' research as well as data and information available in the literature, all of which have been appropriately cited. The figures and images presented originate from the authors' studies, which did not involve experiments on humans, animals, or cell cultures. All referenced experimental work was conducted using laboratory and wild-type strains of Mycobacterium tuberculosis (Mtb), following established standard protocols. The research project received approval from the Scientific Committee of the NRITLD, Tehran, Iran. The findings and interpretations presented are grounded in a rigorous analysis of existing research combined with the authors' experimental work.

References

Abbasnia S, Hashem Asnaashari AM, Sharebiani H, Soleimanpour S, Mosavat A, Rezaee SA (2024) Mycobacterium tuberculosis and host interactions in the manifestation of tuberculosis. J Clin Tuberc Other Mycobact Dis 36:100458

Abdallah AM, Verboom T, Weerdenburg EM, Gey van Pittius NC, Mahasha PW, Jiménez C, Parra M, Cadieux N, Brennan MJ, Appelmelk BJ, Bitter W (2009) PPE and PE_PGRS proteins of Mycobacterium marinum are transported via the type VII secretion system ESX-5. Mol Microbiol 73(3):329–340

Advani M, Rajagopalan M, Reddy P (2014) Calmodulin-like protein from Mycobacterium tuberculosis H37Rv is required during infection. Sci Rep 4:6861

Al-Amrani S, Al-Jabri Z, Al-Zaabi A, Alshekaili J, Al-Khabori M (2021) Proteomics: concepts and applications in human medicine. World J Biol Chem 12(5):57–69

Alnami A, Norton RS, Pena HP, Haider S, Kozielski F (2021) Conformational flexibility of a highly conserved helix controls cryptic pocket formation in FtsZ. J Mol Biol 433(15):167061

Anderson DE, Kim MB, Moore JT, O'Brien TE, Sorto NA, Grove CI et al (2012) Comparison of small molecule inhibitors of the bacterial cell division protein FtsZ and identification of a reliable cross-species inhibitor. ACS Chem Biol 7(11):1918–1928

Andersson CS, Högbom M (2009) A Mycobacterium tuberculosis ligand-binding Mn/Fe protein reveals a new cofactor in a remodeled R2-protein scaffold. Proc Natl Acad Sci USA 106(14):5633–5638

Andersson SGE, Sharp PM (1996 Apr) Codon usage in the Mycobacterium tuberculosis complex. Microbiology (Reading). 142(Pt 4):915–925

Anil Kumar V, Goyal R, Bansal R, Singh N, Sevalkar RR, Kumar A, Sarkar D (2016) EspR-dependent ESAT-6 protein secretion of Mycobacterium tuberculosis requires the presence of virulence regulator PhoP. J Biol Chem 291(36):19018–19030

Ardito F, Giuliani M, Perrone D, Troiano G, Lo ML (2017) The crucial role of protein phosphorylation in cell signaling and its use as targeted therapy (review). Int J Mol Med 40(2):271–280

Arnvig KB, Pennell S, Gopal B, Colston MJ (2004) A high-affinity interaction between NusA and the rrn nut site in Mycobacterium tuberculosis. Proc Natl Acad Sci USA 101(22):8325–8330

Arora G, Bothra A, Prosser G, Arora K, Sajid A (2021) Role of post-translational modifications in the acquisition of drug resistance in mycobacterium tuberculosis. FEBS J 288(11):3375–3393. https://doi.org/10.1111/febs.15699

Aslam B, Basit M, Nisar MA, Khurshid M, Rasool MH (2017) Proteomics: technologies and their applications. J Chromatogr Sci 55(2):182–196

Ates LS (2020 Jan) New insights into the mycobacterial PE and PPE proteins provide a framework for future research. Mol Microbiol 113(1):4–21

Ayikpoe R, Salazar J, Majestic B, Latham JA (2018) Mycofactocin biosynthesis proceeds through 3-Amino-5-[(p-hydroxyphenyl)methyl]-4,4-dimethyl-2-pyrrolidinone (AHDP); direct observation of MftE specificity toward MftA. Biochemistry 57(37):5379–5383

Bainomugisa A, Pandey S, O'Connor B, Syrmis M, Whiley D, Sintchenko V, Coin LJM, Marais BJ, Coulter C (2024) Sustained transmission over two decades of a previously unrecognised MPT64 negative Mycobacterium tuberculosis strain in Queensland, Australia: a whole genome sequencing study. Lancet Reg Health West Pac 47:101105

Balas D, Fernández-Moreira E, De La Campa AG (1998 Jun) Molecular characterization of the gene encoding the DNA gyrase a subunit of Streptococcus pneumoniae. J Bacteriol 180(11):2854–2861

Banerjee S, Nandyala A, Podili R, Katoch VM, Murthy KJ, Hasnain SE (2004) Mycobacterium tuberculosis (Mtb) isocitrate dehydrogenases show strong B cell response and distinguish vaccinated controls from TB patients. Proc Natl Acad Sci USA 101(34):12652–12657

Bao L, Hu J, Zhan B, Chi M, Li Z, Wang S et al (2023) Structural insights into RNase J that plays an essential role in mycobacterium tuberculosis RNA metabolism. Nat Commun 14(1):2280. https://doi.org/10.1038/s41467-023-38023-6

Barazani O, Erdos T, Chowdhury R, Kaur G, Venketaraman V (2025) New advances in the development and Design of Mycobacterium tuberculosis vaccines: construction and validation of multi-epitope vaccines for tuberculosis prevention. Biology (Basel). 14(4):417

Barthe P, Veyron-Churlet R, de Visch A, Gilleron M, Saliou JM, Tomavo S et al (2016) Mycobacterium tuberculosis LppM displays an original structure and domain composition linked to a dual localization. Structure 24(10):1788–1794

Batt SM, Minnikin DE, Besra GS (2020) The thick waxy coat of mycobacteria, a protective layer against antibiotics and the host's immune system. Biochem J 477(10):1983–2006

Beg M, Hejazi I, Thakur S, Athar F (2021) Domain wise differentiation of Mycobacterium tuberculosis H37Rv hypothetical proteins: a roadmap to discover bacterial survival potentials. Biotechnol Appl Biochem 67:296

Bei C, Zhu J, Culviner PH, Rubin EJ, Fortune SM, Gao Q, Liu Q (2023) Genetically encoded transcriptional plasticity underlies stress adaptation in Mycobacterium tuberculosis. bioRxiv [Preprint] 2023.08.20.553992

Bei C, Zhu J, Culviner PH, Gan M, Rubin EJ, Fortune SM, Gao Q, Liu Q (2024) Genetically encoded transcriptional plasticity underlies stress adaptation in mycobacterium tuberculosis. Nature Commun 15(1):3088

Beisel CL, Storz G (2010) Base pairing small RNAs and their roles in global regulatory networks. FEMS Microbiol Rev 34:866–882

Bejaoui S, Fehri NE, Ferchichi MA, Jemmali B (2025) Overview of bovine tuberculosis in Tunisia: risk factors, diagnostic approaches, and current situation: bovine tuberculosis: risk factors, diagnostic approaches, and current situation. Afr J Manage Eng Technol 3(1)

Beuth B, Pennell S, Arnvig KB, Martin SR, Taylor IA (2005) Structure of a Mycobacterium tuberculosis NusA–RNA complex. EMBO J 24(20):3576–3587

Bhatt A, Fujiwara N, Bhatt K, Gurcha SS, Kremer L, Chen B, Chan J, Porcelli SA, Kobayashi K, Besra GS, Jacobs WR Jr (2007) Deletion of kasB in Mycobacterium tuberculosis causes loss of acid-fastness and subclinical latent tuberculosis in immunocompetent mice. Proc Natl Acad Sci USA 104(12):5157–5162

Bhuwan M, Arora N, Sharma A, Khubaib M, Pandey S, Chaudhuri TK, Hasnain SE, Ehtesham NZ (2016) Interaction of Mycobacterium tuberculosis virulence factor RipA with chaperone MoxR1 is required for transport through the TAT secretion system. mBio 7(2):e02259

Birhanu AG (2023) Mass spectrometry-based proteomics as an emerging tool in clinical laboratories. Clin Proteomics 20(1):32

Bitter W, Houben ENG, Bottai D, Brodin P, Brown EJ, Cox JS et al (2009) Systematic genetic nomenclature for type VII secretion systems. PLoS Pathog 5(10):e1000507

Boitel B, Ortiz-Lombardía M, Durán R, Pompeo F, Cole ST, Cerveñansky C, Alzari PM (2003 Sep) PknB kinase activity is regulated by phosphorylation in two Thr residues and dephosphorylation by PstP, the cognate phospho-Ser/Thr phosphatase. Mycobacterium tuberculosis Mol Microbiol 49(6):1493–1508

Boon C, Dick T (2002 Dec) Mycobacterium bovis BCG response regulator essential for hypoxic dormancy. J Bacteriol 184(24):6760–6767

Bottai D, Di Luca M, Majlessi L, Frigui W, Simeone R, Sayes F, Bitter W, Brennan MJ, Leclerc C, Batoni G, Campa M, Brosch R, Esin S. Disruption of the ESX-5 system of Mycobacterium tuberculosis causes loss of PPE protein secretion, reduction of cell wall integrity and strong attenuation. Mol Microbiol 2012 Mar;83(6):1195–209. –

Braibant M, Gilot P, Content J (2000 Oct) The ATP binding cassette (ABC) transport systems of Mycobacterium tuberculosis. FEMS Microbiol Rev 24(4):449–467

Brennan MJ (2017) The enigmatic PE/PPE multigene family of Mycobacteria and tuberculosis vaccination. Infect Immun 85(6):e00969–e00916

Brosch R, Philipp WJ, Stavropoulos E, Colston MJ, Cole ST, Gordon SV (1999 Nov) Genomic analysis reveals variation between Mycobacterium tuberculosis H37Rv and the attenuated M. Tuberculosis H37Ra strain. Infect Immun 67(11):5768–5774

Cassio Barreto de Oliveira M, Balan A (2020) The ATP-binding cassette (ABC) transport systems in Mycobacterium tuberculosis: structure, function, and possible targets for therapeutics. Biology (Basel). 9(12):443

Castell A, Johansson P, Unge T, Jones TA, Bäckbro K (2005 Jul) Rv0216, a conserved hypothetical protein from Mycobacterium tuberculosis that is essential for bacterial survival during infection, has a double hotdog fold. Protein Sci 14(7):1850–1862

Chandra P, Grigsby SJ, Philips JA (2022 Dec) Immune evasion and provocation by Mycobacterium tuberculosis. Nat Rev Microbiol 20(12):750–766

Chandramouli K, Qian PY (2009 Dec) Proteomics: challenges, techniques and possibilities to overcome biological sample complexity. Hum Genomics Proteomics 8(2009):239204

Chang DPS, Guan XL (2021) Metabolic versatility of Mycobacterium tuberculosis during infection and dormancy. Meta 11(2):88

Chang A, Smollett KL, Gopaul KK, Chan BH, Davis EO (2012) Mycobacterium tuberculosis H37Rv sigC is expressed from two promoters but is not autoregulatory. Tuberculosis 92(1):48–55

Chauhan R, Mande SC (2001) Characterization of the Mycobacterium tuberculosis H37Rv alkyl hydroperoxidase AhpC points to the importance of ionic interactions in oligomerization and activity. Biochem J 354(Pt 1):209–215. https://doi.org/10.1042/0264-6021:3540209. PMID: 11171096; PMCID: PMC1221645

Chauhan R, Ravi J, Datta P, Chen T, Schnappinger D, Bassler KE, Balazsi G, Gennaro ML (2016) Reconstruction and topological characterization of the sigma factor regulatory network of Mycobacterium tuberculosis. Nat Commun 7:11062

Chen Y, Anderson DE, Rajagopalan M, Erickson HP (2007) Assembly dynamics of Mycobacterium tuberculosis FtsZ. J Biol Chem 282(38):27736–27743

Chen T, He L, Deng W, Xie J (2013 Jan) The Mycobacterium DosR regulon structure and diversity revealed by comparative genomic analysis. J Cell Biochem 114(1):1–6

Chiliza TE, Pillay M, Pillay B (2017) Identification of unique essential proteins from a Mycobacterium tuberculosis F15/LAM4/KZN phage secretome library. Pathog Dis. 75(1)

Chitale P, Lemenze AD, Fogarty EC, Shah A, Grady C, Odom-Mabey AR, Johnson WE, Yang JH, Eren AM, Brosch R, Kumar P (2022) A comprehensive update to the Mycobacterium tuberculosis H37Rv reference genome. Nat Commun 13(1):7068

Choo M, Oh S, Jo S, Jin X, Song Y, Wen H, Park S, Kang S (2022) Highly conserved protein Rv1211 in Mycobacterium tuberculosis is a natively unfolded protein that binds to a calmodulin antagonist, trifluoperazine. Biochem Biophys Res Commun 610:182–187

Choudhuri BS, Bhakta S, Barik R, Basu J, Kundu M, Chakrabarti P (2002) Overexpression and functional characterization of an ABC (ATP-binding cassette) transporter encoded by the genes drrA and drrB of Mycobacterium tuberculosis. Biochem J 367(1):279–285

Chugh S, Tiwari P, Suri C, Gupta SK, Singh P, Bouzeyen R, Kidwai S, Srivastava M, Rameshwaram NR, Kumar Y, Asthana S, Singh R (2024) Polyphosphate kinase-1 regulates bacterial and host metabolic pathways involved in pathogenesis of Mycobacterium tuberculosis. Proc Natl Acad Sci USA 121(2):e2309664121

Cohen KA, Manson AL, Desjardins CA, Abeel T, Earl AM (2019) Deciphering drug resistance in Mycobacterium tuberculosis using whole-genome sequencing: progress, promise, and challenges. Genome Med 11(1):45

Cole ST, Brosch R, Parkhill J, Garnier T, Churcher C, Harris D, Gordon SV et al (1998) Deciphering the biology of Mycobacterium tuberculosis from the complete genome sequence. Nature 393(6685):537–544

Cook GM, Berney M, Gebhard S, Heinemann M, Cox RA, Danilchanka O, Niederweis M (2009) Physiology of mycobacteria. Adv Microb Physiol 55(81–182):318–319

Cui Y, Dang G, Wang H, Tang Y, Lv M, Liu S, Song N (2023 Nov) DosR's multifaceted role on Mycobacterium bovis BCG revealed through multi-omics. Front Cell Infect Microbiol 21(13):1292864

D'Souza C, Kishore U, Tsolaki AG (2023) The PE-PPE family of Mycobacterium tuberculosis: proteins in disguise. Immunobiology 228(2):152321

Daher W, Le Moigne V, Tasrini Y, Parmar S, Sexton DL, Aguilera-Correa JJ, Berdal V, Tocheva EI, Herrmann JL, Kremer L (2025) Deletion of ESX-3 and ESX-4 secretion systems in Mycobacterium abscessus results in highly impaired pathogenicity. Commun Biol 8(1):166

Damen MPM, Phan TH, Ummels R, Rubio-Canalejas A, Bitter W, Houben ENG (2020) Modification of a PE/PPE substrate pair reroutes an Esx substrate pair from the mycobacterial ESX-1 type VII secretion system to the ESX-5 system. J Biol Chem 295(18):5960–5969

Danilchanka O, Mailaender C, Niederweis M (2008 Jul) Identification of a novel multidrug efflux pump of Mycobacterium tuberculosis. Antimicrob Agents Chemother 52(7):2503–2511

Danilchanka O, Pires D, Anes E, Niederweis M (2015 Apr) The Mycobacterium tuberculosis outer membrane channel protein CpnT confers susceptibility to toxic molecules. Antimicrob Agents Chemother 59(4):2328–2336

Dar HA, Zaheer T, Ullah N, Bakhtiar SM, Zhang T, Yasir M, Azhar EI, Ali A (2020) Pangenome analysis of Mycobacterium tuberculosis reveals Core-drug targets and screening of promising Lead compounds for drug discovery. Antibiotics (Basel) 9(11):819

De Maio F, Berisio R, Manganelli R, Delogu G (2020 Dec) PE_PGRS proteins of Mycobacterium tuberculosis: a specialized molecular task force at the forefront of host-pathogen interaction. Virulence 11(1):898–915

de Oliveira CB, Balan A (2020) The ATP-binding cassette (ABC) transport systems in Mycobacterium tuberculosis: structure, function, and possible targets for therapeutics. Biology (Basel) 9(12):443

Deboosère N, Iantomasi R, Queval CJ, Song OR, Veyron-Churlet R, Jouny S et al (2017 Jan) LppM impact on the colonization of macrophages by Mycobacterium tuberculosis. Cell Microbiol 19(1):e12619

DeJesus MA, Gerrick ER, Xu W, Park SW, Long JE, Boutte CC, Rubin EJ, Schnappinger D, Ehrt S, Fortune SM, Sassetti CM, Ioerger TR (2017) Comprehensive essentiality analysis of the Mycobacterium tuberculosis genome via saturating transposon mutagenesis. MBio 8(1):e02133–e02116

Deng Y, Li F, Hu Z, Yue C, Tang YZ (2021) Expression patterns of the heat shock protein 90 (Hsp90) gene suggest its possible involvement in maintaining the dormancy of dinoflagellate resting cysts. Int J Mol Sci 22(20):11054

Devlin KL, Leach DT, Stratton KG, Lamichhane G, Lin VS, Beatty KE (2025) Proteomic characterization of Mycobacterium tuberculosis subjected to carbon starvation. mSystems 10(5):e0153024

Díaz-Uriarte R, Alvarez de Andrés S (2006) Gene selection and classification of microarray data using random forest. BMC Bioinformatics 7:3. https://doi.org/10.1186/1471-2105-7-3. PMID: 16398926; PMCID: PMC1363357

Dohál M, Dvořáková V, Šperková M, Pinková M, Spitaleri A, Norman A, Cabibbe AM, Rasmussen EM, Porvazník I, Škereňová M, Solovič I (2022) Whole genome sequencing of multidrug-resistant Mycobacterium tuberculosis isolates collected in The Czech Republic, 2005–2020. Sci Rep 12(1):7149

Dutta NK, Mehra S, Martinez AN, Alvarez X, Renner NA, Morici LA, Pahar B, MacLean AG, Lackner AA, Kaushal D (2012) The stress-response factor SigH modulates the interaction between Mycobacterium tuberculosis and host phagocytes. PLoS One 7(1):e28958

Dziadek J, Rutherford SA, Madiraju MV, Atkinson MAL, Rajagopalan M (2003 Jun) Conditional expression of Mycobacterium smegmatis ftsZ, an essential cell division gene. Microbiology (Reading). 149(Pt 6):1593–1603

Echeverria-Valencia G, Flores-Villalva S, Espitia CI (2017) Virulence factors and pathogenicity of Mycobacterium. Mycobacterium Res Develop. IntechOpen

Ehrt S, Schnappinger D, Rhee KY (2018 Aug) Metabolic principles of persistence and pathogenicity in Mycobacterium tuberculosis. Nat Rev Microbiol 16(8):496–507

Encarnación S, Hernández M, Martínez-Batallar G, Contreras S, Vargas Mdel C, Mora J (2005) Comparative proteomics using 2-D gel electrophoresis and mass spectrometry as tools to dissect stimulons and regulons in bacteria with sequenced or partially sequenced genomes. Biol Proced Online 7:117–135

Ertelt M, Mulligan VK, Maguire JB, Lyskov S, Moretti R, Schiffner T, Meiler J, Schoeder CT (2024) Combining machine learning with structure-based protein design to predict and engineer post-translational modifications of proteins. PLoS Comput Biol 20(3):e1011939

Faller M, Niederweis M, Schulz GE (2004) The structure of a mycobacterial outer-membrane channel. Science 303(5661):1189–1192

Farnia P, Mohammad RM, Merza MA, Tabarsi P, Zhavnerko GK, Ibrahim TA, Kuan HO, Ghanavei J, Farnia P, Ranjbar R, Poleschuyk NN, Titov LP, Owlia P, Kazampour M, Setareh M, Sheikolslami M, Migliori GB, Velayati AA (2010) Growth and cell-division in extensive (XDR) and extremely drug resistant (XXDR) tuberculosis strains: transmission and atomic force observation. Int J Clin Exp Med 3(4):308–314. PMID: 21072265; PMCID: PMC2971545

Farnia P, Farhadi T, Farnia P, Ghanavi J, Velayati AA (2018) A review on the C-terminal domain of channel protein with necrosis-inducing toxin as a novel necrotizing toxin of Mycobacterium tuberculosis. Biomed Biotechnol Res J 2(2):100–105

Fathizadeh H, Hayat SM, Dao S, Ganbarov K, Tanomand A, Asgharzadeh M, Kafil HS (2020) Long non-coding RNA molecules in tuberculosis. Int J Biol Macromol 156:340–346

Ferraris DM, Spallek R, Oehlmann W, Singh M, Rizzi M (2015 Feb) Structures of citrate synthase and malate dehydrogenase of Mycobacterium tuberculosis. Proteins 83(2):389–394

Figge RM, Divakaruni AV, Gober JW (2004 Mar) MreB, the cell shape-determining bacterial actin homologue, co-ordinates cell wall morphogenesis in Caulobacter crescentus. Mol Microbiol 51(5):1321–1332

Flores-Valdez MA, Freches D, Bruffaerts N, Romano M, Schoolnik G, Dolganov G, Huygen K (2015) Overexpression of DosR in Mycobacterium tuberculosis does not affect aerobic replication in vitro or in murine macrophages. Ann Microbiol 65:713–720

Ford C, Yusim K, Ioerger T, Feng S, Chase M, Greene M, Korber B, Fortune S. Mycobacterium tuberculosis--heterogeneity revealed through whole genome sequencing. Tuberculosis (Edinb) 2012 May;92(3):194–201

Frigui W, Bottai D, Majlessi L, Monot M, Josselin E, Brodin P, Garnier T, Gicquel B, Martin C, Leclerc C, Cole ST, Brosch R (2008) Control of Mycobacterium tuberculosis ESAT-6 secretion and specific T cell recognition by PhoP. PLoS Pathog 4(2):e33

Ganapathy U, Marrero J, Calhoun S, Eoh H, de Carvalho LPS, Rhee K et al (2015 Aug) Two enzymes with redundant fructose bisphosphatase activity sustain gluconeogenesis and virulence in Mycobacterium tuberculosis. Nat Commun 10(6):7912

Gandotra S, Schnappinger D, Monteleone M, Hillen W, Ehrt S (2007 Dec) In vivo gene silencing identifies the Mycobacterium tuberculosis proteasome as essential for the bacteria to persist in mice. Nat Med 13(12):1515–1520

Gao S, Wu F, Gurcha SS, Batt SM, Besra GS, Rao Z, Zhang L (2024 Mar) Structural analysis of phosphoribosyltransferase-mediated cell wall precursor synthesis in Mycobacterium tuberculosis. Nat Microbiol 15(9):976–987

Gazi MA, Kibria MG, Mahfuz M, Islam MR, Ghosh P, Afsar MN, Khan MA, Ahmed T (2016) Functional, structural and epitopic prediction of hypothetical proteins of Mycobacterium tuberculosis H37Rv: an in silico approach for prioritizing the targets. Gene 591(2):442–455

Gómez-González PJ, Campino S, Phelan JE, Clark TG (2022) Portable sequencing of Mycobacterium tuberculosis for clinical and epidemiological applications. Briefings Bioinform 23(5):bbac256

Gopal B, Haire LF, Gamblin SJ, Dodson EJ, Lane AN, Papavinasasundaram KG, Colston MJ, Dodson G (2001) Crystal structure of the transcription elongation/anti-termination factor NusA from Mycobacterium tuberculosis at 1.7 a resolution. J Mol Biol 314(5):1087–1095

Graciano A, Liu A (2025) Protein-derived cofactors: chemical innovations expanding enzyme catalysis. Chem Soc Rev 54:4502–4530

Grosse-Siestrup BT, Gupta T, Helms S, Tucker SL, Voskuil MI, Quinn FD, Karls RK (2021) A role for Mycobacterium tuberculosis sigma factor C in copper nutritional immunity. Int J Mol Sci 22(4):2118

Grzegorzewicz AE, de Sousa-d'Auria C, McNeil MR, Huc-Claustre E, Jones V, Petit C, Angala SK, Zemanová J, Wang Q, Belardinelli JM, Gao Q, Ishizaki Y, Mikušová K, Brennan PJ, Ronning DR, Chami M, Houssin C, Jackson M (2016) Assembling of the Mycobacterium tuberculosis Cell Wall Core. J Biol Chem 291(36):18867–18879

Gu S, Chen J, Dobos KM, Bradbury EM, Belisle JT, Chen X (2003 Dec) Comprehensive proteomic profiling of the membrane constituents of a Mycobacterium tuberculosis strain. Mol Cell Proteomics 2(12):1284–1296

Gu J, He Y, He C, Zhang Q, Huang Q, Bai S, Wang R, You Q, Wang L (2025) Advances in the structures, mechanisms and targeting of molecular chaperones. Signal Transduct Target Ther 10:84

Guo F, Wei J, Song Y, Li B, Qian Z, Wang X, Wang H, Xu T (2023 Sep) Immunological effects of the PE/PPE family proteins of Mycobacterium tuberculosis and related vaccines. Front Immunol 27(14):1255920

Habibi Arejan N, Ensinck D, Diacovich L, Patel PB, Quintanilla SY, Emami Saleh A, Gramajo H, Boutte CC (2023 Jan) Polar protein Wag31 both activates and inhibits cell wall metabolism at the poles and septum. Front Microbiol 12(13):1085918

Harnagel A, Lopez Quezada L, Park SW, Baranowski C, Kieser K, Jiang X, Roberts J, Vaubourgeix J, Yang A, Nelson B, Fay A, Rubin E, Ehrt S, Nathan C, Lupoli TJ (2021 Feb) Nonredundant functions of Mycobacterium tuberculosis chaperones promote survival under stress. Mol Microbiol 115(2):272–289

Hartman T, Weinrick B, Vilchèze C, Berney M, Tufariello J, Cook GM et al (2014) Succinate dehydrogenase is the regulator of respiration in Mycobacterium tuberculosis. PLoS Pathog 10(11):e1004510

Hasan Z, Razzak SA, Kanji A, Shakoor S, Hasan R (2023) Whole-genome sequencing reveals genotypic resistance in phenotypically susceptible mycobacterium tuberculosis clinical isolates. Int J Mycobacteriol 12(2):179–183. https://doi.org/10.4103/ijmy.ijmy_101_23

Haydel SE, Malhotra V, Cornelison GL, Clark-Curtiss JE (2012) The prrAB two-component system is essential for Mycobacterium tuberculosis viability and is induced under nitrogen-limiting conditions. J Bacteriol 194(2):354–361

Hazra D, Lam C, Chawla K, Sintchenko V, Dhyani VS, Venkatesh BT (2023) Impact of whole-genome sequencing of Mycobacterium tuberculosis on treatment outcomes for MDR-TB/XDR-TB: a systematic review. Pharmaceutics 15(12):2782

He H, Hovey R, Kane J, Singh V, Zahrt TC (2006 Mar) MprAB is a stress-responsive two-component system that directly regulates expression of sigma factors SigB and SigE in Mycobacterium tuberculosis. J Bacteriol 188(6):2134–2143

He H, Bretl DJ, Penoske RM, Anderson DM, Zahrt TC (2011) Components of the Rv0081-Rv0088 locus, which encodes a predicted formate hydrogenlyase complex, are coregulated by Rv0081, MprA, and DosR in mycobacterium tuberculosis. J Bacteriol 193(19):5105–5118. https://doi.org/10.1128/JB.05433-11

Hett EC, Rubin EJ (2008) Bacterial growth and cell division: a mycobacterial perspective. Microbiol Mol Biol Rev 72(1):126–156

Heym B, Stavropoulos E, Honoré N, Domenech P, Saint-Joanis B, Wilson TM, Collins DM, Colston MJ, Cole ST (1997 Apr) Effects of overexpression of the alkyl hydroperoxide reductase AhpC on the virulence and isoniazid resistance of Mycobacterium tuberculosis. Infect Immun 65(4):1395–1401

Hoffmann A, Bukau B, Kramer G (2010 Jun) Structure and function of the molecular chaperone trigger factor. Biochim Biophys Acta 1803(6):650–661

Hollander M, Do T, Will T, Helms V (2021 Sep) Detecting rewiring events in protein-protein interaction networks based on transcriptomic data. Front Bioinform 8(1):724297

Hong W, Deng W, Xie J (2013) The structure, function, and regulation of mycobacterium FtsZ. Cell Biochem Biophys 65(2):97–105. https://doi.org/10.1007/s12013-012-9410-3

Houben EN, Korotkov KV, Bitter W (2014 Aug) Take five - type VII secretion systems of Mycobacteria. Biochim Biophys Acta 1843(8):1707–1716

Hsieh PC, Shenoy BC, Samols D, Phillips NF (1996) Cloning, expression, and characterization of polyphosphate glucokinase from Mycobacterium tuberculosis. J Biol Chem 271(9):4909–4915

Hu Y, Coates AR (1999 Jan) Transcription of two sigma 70 homologue genes, sigA and sigB, in stationary-phase Mycobacterium tuberculosis. J Bacteriol 181(2):469–476

Hu C, Yang J, Qi Z, Wu H, Wang B, Zou F, Mei H, Liu J, Wang W, Liu Q (2020) Heat shock proteins: biological functions, pathological roles, and therapeutic opportunities. MedComm 3(3):e161

Hunt M, Bradley P, Lapierre SG, Heys S, Thomsit M, Hall MB, Malone KM, Wintringer P, Walker TM, Cirillo DM, Comas I (2019 Dec) Antibiotic resistance prediction for Mycobacterium tuberculosis from genome sequence data with Mykrobe. Wellcome open research 2(4):191

Hurst-Hess K, Biswas R, Yang Y, Rudra P, Lasek-Nesselquist E, Ghosh P (2019) Mycobacterial SigA and SigB Cotranscribe essential housekeeping genes during exponential growth. MBio 10(3):e00273–e00219

Huygen K (2014) The immunodominant T-cell epitopes of the mycolyl-transferases of the antigen 85 complex of Mycobacterium tuberculosis. Front Immunol 5:321

Ijaq J, Chandra D, Ray MK, Jagannadham MV (2022) Investigating the functional role of hypothetical proteins from an Antarctic Bacterium Pseudomonas sp. Lz4W: emphasis on identifying proteins involved in Cold adaptation. Front Genet 13:825269

Ioerger TR, Feng Y, Ganesula K, Chen X, Dobos KM, Fortune S, Jacobs WR Jr, Mizrahi V, Parish T, Rubin E, Sassetti C (2010) Variation among genome sequences of H37Rv strains of Mycobacterium tuberculosis from multiple laboratories. J Bacteriol 192(14):3645–3653

Jacobo-Delgado YM, Rodríguez-Carlos A, Serrano CJ, Rivas-Santiago B (2023 May) Mycobacterium tuberculosis cell-wall and antimicrobial peptides: a mission impossible? Front Immunol 17(14):1194923

Jaiswal LK, Singh RK, Nayak T, Kakkar A, Kandwal G, Singh VS, Gupta A (2024) A comparative analysis of mycobacterial ribonucleases: towards a therapeutic novel drug target. Infect Genet Evol 123:105645

Jani C, Eoh H, Lee JJ, Hamasha K, Sahana MB, Han J-S et al (2010 Dec) Regulation of polar peptidoglycan biosynthesis by Wag31 phosphorylation in mycobacteria. BMC Microbiol 29(10):327

Jansen RS, Mandyoli L, Hughes R, Wakabayashi S, Pinkham JT, Selbach B, Guinn KM, Rubin EJ, Sacchettini JC, Rhee KY (2020) Aspartate aminotransferase Rv3722c governs aspartate-dependent nitrogen metabolism in Mycobacterium tuberculosis. Nat Commun 11:1960

Jia X, Yang L, Dong M, Chen S, Lv L, Cao D, Fu J, Yang T, Zhang J, Zhang X, Shang Y, Wang G, Sheng Y, Huang H, Chen F (2017 Mar) The bioinformatics analysis of comparative genomics of Mycobacterium tuberculosis complex (MTBC) provides insight into dissimilarities between intraspecific groups differing in host association, virulence, and epitope diversity. Front Cell Infect Microbiol 21(7):88

Johnson RM, McDonough KA (2018) Cyclic nucleotide signaling in Mycobacterium tuberculosis: an expanding repertoire. Pathog Dis. 76(5):fty048

Kang CM, Abbott DW, Park ST, Dascher CC, Cantley LC, Husson RN (2005) The Mycobacterium tuberculosis serine/threonine kinases PknA and PknB: substrate identification and regulation of cell shape. Genes Dev 19(14):1692–1704

Kang CM, Nyayapathy S, Lee JY, Suh JW, Husson RN (2008 Mar) Wag31, a homologue of the cell division protein DivIVA, regulates growth, morphology and polar cell wall synthesis in mycobacteria. Microbiology (Reading). 154(Pt 3):725–735

Kapoor Y, Khurana H, Dutta D, Chakraborty A, Priya A, Singh A, Kamat SS, Dhar N, Pucadyil TJ, Nandicoori VK (2025 May) Wag31, a membrane tether, is crucial for lipid homeostasis in mycobacteria. elife 22(14):RP104268

Kawai Y, Marles-Wright J, Cleverley RM, Emmins R, Ishikawa S, Kuwano M, Heinz N, Bui NK, Hoyland CN, Ogasawara N, Lewis RJ, Vollmer W, Daniel RA, Errington J (2011) A widespread family of bacterial cell wall assembly proteins. EMBO J 30(24):4931–4941

Kim H, Page GP, Barnes S (2004 Jan) Proteomics and mass spectrometry in nutrition research. Nutrition 20(1):155–165

Kohli S, Singh Y, Sharma K, Mittal A, Ehtesham NZ, Hasnain SE (2012 Aug) Comparative genomic and proteomic analyses of PE/PPE multigene family of Mycobacterium tuberculosis $H_{37}Rv$ and $H_{37}Ra$ reveal novel and interesting differences with implications in virulence. Nucleic Acids Res 40(15):7113–7122

Koleske BN, Jacobs WR Jr, Bishai WR (2023) The Mycobacterium tuberculosis genome at 25 years: lessons and lingering questions. J Clin Invest 133(19):e173156

Koleske B, Shen J, Gupta M, Bishai WR (2025 Jun) In vivo profiling of the PE/PPE proteins of Mycobacterium tuberculosis reveals diverse contributions to virulence. Front Microbiol 18(16):1634229

Köster S, Upadhyay S, Chandra P, Papavinasasundaram K, Yang G, Hassan A et al (2017) Mycobacterium tuberculosis is protected from NADPH oxidase and LC3-associated phagocytosis by the LCP protein CpsA. Proc Natl Acad Sci U S A. 114(41):E8711–E8720. https://doi.org/10.1073/pnas.1707792114. PMID: 28973896; PMCID: PMC5642705. Erratum in: Proc Natl Acad Sci USA 2017 Nov 7;114(45):E9752

Kremer L, Guérardel Y, Gurcha SS, Locht C, Besra GS (2002 Oct) Temperature-induced changes in the cell-wall components of Mycobacterium thermoresistibile. Microbiology (Reading). 148(Pt 10):3145–3154

Kumar D, Narayanan S (2012 Jun) pknE, a serine/threonine kinase of Mycobacterium tuberculosis modulates multiple apoptotic paradigms. Infect Genet Evol 12(4):737–747

Kumar K, Awasthi D, Berger WT, Tonge PJ, Slayden RA, Ojima I (2010 Aug) Discovery of anti-TB agents that target the cell-division protein FtsZ. Future Med Chem 2(8):1305–1323

Kumar A, Farhana A, Guidry L, Saini V, Hondalus M, Steyn AJ (2011 Dec) Redox homeostasis in mycobacteria: the key to tuberculosis control? Expert Rev Mol Med 16(13):e39

Kunst F, Ogasawara N, Moszer I, Albertini AM, Alloni G, Azevedo V, Bertero MG et al (1997) The complete genome sequence of the gram-positive bacterium Bacillus subtilis. Nature 390(6657):249–256

Lazo EO, Jakoncic J, RoyChowdhury S, Awasthi D, Ojima I (2019) Novel T9 loop conformation of filamenting temperature-sensitive mutant Z from Mycobacterium tuberculosis. Acta Crystallogr F Struct Biol Commun 75(Pt 5):359–367

Lee JH, Karakousis PC, Bishai WR (2008 Jan) Roles of SigB and SigF in the Mycobacterium tuberculosis sigma factor network. J Bacteriol 190(2):699–707

Lee JJ, Kan CM, Lee JH, Park KS, Jeon JH, Lee SH (2014) Phosphorylation-dependent interaction between a serine/threonine kinase PknA and a putative cell division protein Wag31 in Mycobacterium tuberculosis. New Microbiol 37(4):525–533

Leung AK, Lucile White E, Ross LJ, Reynolds RC, DeVito JA, Borhani DW (2004) Structure of Mycobacterium tuberculosis FtsZ reveals unexpected, G protein-like conformational switches. J Mol Biol 342(3):953–970. https://doi.org/10.1016/j.jmb.2004.07.061. PMID: 15342249

Li Y, Hsin J, Zhao L, Cheng Y, Shang W, Huang KC, Wang HW, Ye S (2013) FtsZ protofilaments use a hinge-opening mechanism for constrictive force generation. Science 341(6144):392–395

Li D, Gao C, Zhao L, Zhang Y (2020) Inflammatory response is modulated by lincRNACox2 via the NF−κB pathway in macrophages infected by Mycobacterium tuberculosis. Mol Med Rep 21(6):2513–2521

Ligon LS, Hayden JD, Braunstein M (2012 Mar) The ins and outs of Mycobacterium tuberculosis protein export. Tuberculosis (Edinb) 92(2):121–132

Lin Y, Zhang H, Zhu N, Wang X, Han Y, Chen M, Jiang J, Si S (2018 May) Identification of TB-E12 as a novel FtsZ inhibitor with anti-tuberculosis activity. Tuberculosis (Edinb) 110:79–85

Lin H, Xing J, Wang H, Wang S, Fang R, Li X, Li Z, Song N (2024) Roles of lipolytic enzymes in Mycobacterium tuberculosis pathogenesis. Front Microbiol 15:1329715

Liu X, Hu J, Wang W, Yang H, Tao E, Ma Y, Sha S (2024) Mycobacterial biofilm: mechanisms, clinical problems, and treatments. Int J Mol Sci 25(14):7771

Lopez JL (2007) Two-dimensional electrophoresis in proteome expression analysis. J Chromatogr B 849(1–2):190–202

Lupoli TJ, Vaubourgeix J, Burns-Huang K, Gold B (2018) Targeting the Proteostasis network for Mycobacterial drug discovery. ACS Infect Dis. 4(4):478–498

Mahatha AC, Mal S, Majumder D, Saha S, Ghosh A, Basu J, Kundu M (2020) RegX3 activates whiB3 under acid stress and subverts lysosomal trafficking of Mycobacterium tuberculosis in a WhiB3-dependent manner. Front Microbiol 11:572433

Målen H, Berven FS, Fladmark KE, Wiker HG (2007 May) Comprehensive analysis of exported proteins from Mycobacterium tuberculosis H37Rv. Proteomics 7(10):1702–1718

Malhotra S, Vedithi SC, Blundell TL (2017) Decoding the similarities and differences among mycobacterial species. PLoS Negl Trop Dis 11(8):e0005883

Malm S, Maaß S, Schaible UE, Ehlers S, Niemann S (2018) In vivo virulence of Mycobacterium tuberculosis depends on a single homologue of the LytR-CpsA-Psr proteins. Sci Rep 8(1):3936

Manganelli R (2014) Sigma factors: key molecules in Mycobacterium tuberculosis physiology and virulence. Microbiol Spectr 2(1):MGM2-0007-2013. https://doi.org/10.1128/microbiolspec.MGM2-0007-2013. PMID: 26082107

Margenat M, Betancour G, Irving V, Costábile A, García-Cedrés T, Portela MM, Carrión F, Herrera FE, Villarino

A (2023 Jun) Characteristics of Mycobacterium tuberculosis PtpA interaction and activity on the alpha subunit of human mitochondrial trifunctional protein, a key enzyme of lipid metabolism. Front Cell Infect Microbiol 22(13):1095060

Marmiesse M, Brodin P, Buchrieser C, Gutierrez C, Simoes N, Vincent V, Glaser P, Cole ST, Brosch R (2004 Feb) Macro-array and bioinformatic analyses reveal mycobacterial 'core' genes, variation in the ESAT-6 gene family and new phylogenetic markers for the Mycobacterium tuberculosis complex. Microbiology (Reading). 150(Pt 2):483–496

Marrero J, Trujillo C, Rhee KY, Ehrt S (2013 Jan) Glucose phosphorylation is required for Mycobacterium tuberculosis persistence in mice. PLoS Pathog 9(1):e1003116

Martini MC, Hicks ND, Xiao J, Alonso MN, Barbier T, Sixsmith J, Fortune SM, Shell SS (2022) Loss of RNase J leads to multi-drug tolerance and accumulation of highly structured mRNA fragments in Mycobacterium tuberculosis. PLoS Pathog 18(7):e1010705

Mawuenyega KG, Forst CV, Dobos KM, Belisle JT, Chen J, Bradbury EM, Bradbury AR, Chen X (2005 Jan) Mycobacterium tuberculosis functional network analysis by global subcellular protein profiling. Mol Biol Cell 16(1):396–404

McAlinden A, Smith TA, Sandell LJ, Ficheux D, Parry DA, Hulmes DJ (2003) Alpha-helical coiled-coil oligomerization domains are almost ubiquitous in the collagen superfamily. J Biol Chem 278(43):42200–42207

McLaughlin B, Chon JS, MacGurn JA, Carlsson F, Cheng TL, Cox JS, Brown EJ (2007) A mycobacterium ESX-1-secreted virulence factor with unique requirements for export. PLoS Pathog 3(8):e105

McQuillen R, Xiao J (2020 May) Insights into the structure, function, and dynamics of the bacterial Cytokinetic FtsZ-ring. Annu Rev Biophys 6(49):309–341

Mehta M, Singh A (2019) Mycobacterium tuberculosis WhiB3 maintains redox homeostasis and survival in response to reactive oxygen and nitrogen species. Free Radic Biol Med 131:50–58. https://doi.org/10.1016/j.freeradbiomed.2018.11.015

Melzer ES, Kado T, García-Heredia A, Gupta KR, Meniche X, Morita YS et al (2022) Cell wall damage reveals spatial flexibility in peptidoglycan synthesis and a nonredundant role for RodA in mycobacteria. J Bacteriol 204(6):e0054021. https://doi.org/10.1128/jb.00540-21

Mendauletova A, Latham JA (2022) Biosynthesis of the redox cofactor mycofactocin is controlled by the transcriptional regulator MftR and induced by long-chain acyl-CoA species. J Biol Chem 298(1):101474

Mészáros B, Tóth J, Vértessy BG, Dosztányi Z, Simon I (2011) Proteins with complex architecture as potential targets for drug design: a case study of mycobacterium tuberculosis. PLoS Comput Biol 7(7):e1002118. https://doi.org/10.1371/journal.pcbi.1002118. PMID: 21814507; PMCID: PMC3140968

Morimoto T, Loh PC, Hirai T, Asai K, Kobayashi K, Moriya S, Ogasawara N (2002 Nov) Six GTP-binding proteins of the era/Obg family are essential for cell growth in Bacillus subtilis. Microbiology (Reading) 148(Pt 11):3539–3552

Mukherjee P, Sureka K, Datta P, Hossain T, Barik S, Das KP et al (2009 Jul) Novel role of Wag31 in protection of mycobacteria under oxidative stress. Mol Microbiol 73(1):103–119

Mukhopadhyay S, Balaji KN (2011) The PE and PPE proteins of Mycobacterium tuberculosis. Tuberculosis (Edinb) 91(5):441–447

Murphy DJ, Brown JR (2007) Identification of gene targets against dormant phase Mycobacterium tuberculosis infections. BMC Infect Dis 7:84

Nachappa OA, Neelambike SM, Ramachandra NB (2022) Differential expression of the Mycobacterium tuberculosis heat shock protein genes in response to drug-induced stress. Tuberculosis (Edinb) 134:102201

Namouchi A, Gómez-Muñoz M, Frye SA, Moen LV, Rognes T, Tønjum T, Balasingham SV (2016) The Mycobacterium tuberculosis transcriptional landscape under genotoxic stress. BMC Genomics 17:791

Niederweis M, Danilchanka O, Huff J, Hoffmann C, Engelhardt H (2010) Mycobacterial outer membranes: in search of proteins. Trends Microbiol 18(3):109–116

Niemann HH, Knetsch ML, Scherer A, Manstein DJ, Kull FJ (2001) Crystal structure of a dynamin GTPase domain in both nucleotide-free and GDP-bound forms. EMBO J 20(21):5813–5821

Niu ZS, Wang WH, Dong XN, Tian LM (2020) Role of long noncoding RNA-mediated competing endogenous RNA regulatory network in hepatocellular carcinoma. World J Gastroenterol 26(29):4240–4260

Nowak E, Panjikar S, Morth JP, Jordanova R, Svergun DI, Tucker PA (2006 Feb) Structural and functional aspects of the sensor histidine kinase PrrB from Mycobacterium tuberculosis. Structure 14(2):275–285

Ojha A, Anand M, Bhatt A, Kremer L, Jacobs WR Jr, Hatfull GF (2005) GroEL1: a dedicated chaperone involved in mycolic acid biosynthesis during biofilm formation in mycobacteria. Cell 123(5):861–873

Orgeur M, Sous C, Madacki J, Brosch R (2024) Evolution and emergence of Mycobacterium tuberculosis. FEMS Microbiol Rev 48(2):fuae006

Orr MW, Mao Y, Storz G, Qian SB (2020) Alternative ORFs and small ORFs: shedding light on the dark proteome. Nucleic Acids Res 48(3):1029–1042

Ostrik AA, Azhikina TL, Salina EG (2021 Jan) Small noncoding RNAs and their role in the pathogenesis of Mycobacterium tuberculosis infection. Biochemistry (Mosc) 86(Suppl 1):S109–S119

Otchere ID, Asante-Poku A, Akpadja KF, Diallo AB, Sanou A, Asare P, Osei-Wusu S, Onyejepu N, Diarra B, Dagnra YA, Kehinde A (2024 May) Opinion review of drug resistant tuberculosis in West Africa: tackling the challenges for effective control. Front Public Health 16(12):1374703

Pacl HT, Reddy VP, Saini V, Chinta KC, Steyn AJC (2018) Host-pathogen redox dynamics modulate Mycobacterium tuberculosis pathogenesis. Pathog Dis 76(5):fty036

Paget MS, Helmann JD (2003) The sigma70 family of sigma factors. Genome Biol 4:203

Pajuelo D, Gonzalez-Juarbe N, Tak U, Sun J, Orihuela CJ, Niederweis M (2018) NAD+ depletion triggers macrophage necroptosis, a cell death pathway exploited by Mycobacterium tuberculosis. Cell Rep 24(2):429–440

Pal R, Bisht MK, Mukhopadhyay S (2022 Jul) Secretory proteins of Mycobacterium tuberculosis and their roles in modulation of host immune responses: focus on therapeutic targets. FEBS J 289(14):4146–4171

Pal P, Khan MY, Sharma S, Kumar Y, Mangla N, Kaushal PS, Agarwal N (2023) ResR/McdR-regulated protein translation machinery contributes to drug resilience in Mycobacterium tuberculosis. Commun Biol 6(1):708

Palande A, Patil S, Veeram A, Sahoo SS, Lodhiya T, Maurya P, Muralikrishnan B, Chugh J, Mukherjee R (2024) Proteomic analysis of the Mycobacterium tuberculosis outer membrane for potential implications in uptake of small molecules. ACS Infect Dis 10(3):890–906

Pang X, Samten B, Cao G, Wang X, Tvinnereim AR, Chen X, Howard ST (2013) MprAB regulates the espA operon in Mycobacterium tuberculosis and modulates ESX-1 function and host cytokine response. J Bacteriol 195(1):66–75

Pankhurst LJ, del Ojo EC, Votintseva AA, Walker TM, Cole K, Davies J, Fermont JM, Gascoyne-Binzi DM, Kohl TA, Kong C, Lemaitre N (2016) Rapid, comprehensive, and affordable mycobacterial diagnosis with whole-genome sequencing: a prospective study. Lancet Respir Med 4(1):49–58

Parbhoo T, Mouton JM, Sampson SL (2022 Sep) Phenotypic adaptation of Mycobacterium tuberculosis to host-associated stressors that induce persister formation. Front Cell Infect Microbiol 27(12):956607

Park HD, Guinn KM, Harrell MI, Liao R, Voskuil MI, Tompa M, Schoolnik GK, Sherman DR (2003 May) Rv3133c/dosR is a transcription factor that mediates the hypoxic response of Mycobacterium tuberculosis. Mol Microbiol 48(3):833–843

Pathirage R, Favrot L, Petit C, Yamsek M, Singh S, Mallareddy JR, Rana S, Natarajan A, Ronning DR (2023) Mycobacterium tuberculosis CitA activity is modulated by cysteine oxidation and pyruvate binding. RSC Med Chem 14(5):921–933

PaweŁczyk J, Kremer L (2014) The molecular genetics of mycolic acid biosynthesis. Microbiol Spectr. 2(4):MGM2-0003-2013

Phelan JE, Lim DR, Mitarai S, de Sessions PF, Tujan MA, Reyes LT, Medado IA, Palparan AG, Naim AN, Jie S, Segubre-Mercado E (2019) Mycobacterium tuberculosis whole genome sequencing provides insights into the Manila strain and drug-resistance mutations in The Philippines. Sci Rep 9(1):9305

Phong TQ, Ha do TT, Volker U, Hammer E (2015) Using a label free quantitative proteomics approach to identify changes in protein abundance in multidrug-resistant Mycobacterium tuberculosis. Indian. J Microbiol 55(2):219–230

Pióro M, Małecki T, Portas M, Magierowska I, Trojanowski D, Sherratt D et al (2018) Competition between DivIVA and the nucleoid for ParA binding promotes segrosome separation and modulates mycobacterial cell elongation. Mol Microbiol 111:204–220

Plocinski P, Arora N, Sarva K, Blaszczyk E, Qin H, Das N, Plocinska R, Ziolkiewicz M, Dziadek J, Kiran M, Gorla P, Cross TA, Madiraju M, Rajagopalan M (2012 Dec) Mycobacterium tuberculosis CwsA interacts with CrgA and Wag31, and the CrgA-CwsA complex is involved in peptidoglycan synthesis and cell shape determination. J Bacteriol 194(23):6398–6409

Poirier V, Av-Gay Y (2012) Mycobacterium tuberculosis modulators of the macrophage's cellular events. Microbes Infect 14(13):1211–1219

Powell MM, Rao G, Britt RD, Rittle J (2023) Enzymatic hydroxylation of aliphatic C–H bonds by a Mn/Fe cofactor. J Am Chem Soc 145(30):16526–16537

Priyanka T, Sharma S, Sharma M (2024) Role of PE/PPE proteins of Mycobacterium tuberculosis in triad of host mitochondria, oxidative stress and cell death. Microb Pathog 193:106757

Puri RV, Reddy PV, Tyagi AK (2013) Secreted acid phosphatase (SapM) of Mycobacterium tuberculosis is indispensable for arresting phagosomal maturation and growth of the pathogen in Guinea pig tissues. PLoS One 8(7):e70514

Qin L, Gao S, Wang J, Zheng R, Lu J, Hu Z (2013) The conservation and application of three hypothetical protein coding genes for direct detection of Mycobacterium tuberculosis in sputum specimens. PLoS One 8(9):e73955

Qin L, Xu J, Chen J, Wang S, Zheng R, Cui Z, Liu Z, Wu X, Wang J, Huang X, Wang Z (2024) Cell-autonomous targeting of arabinogalactan by host immune factors inhibits mycobacterial growth. elife 13:RP92737

Rahlwes KC, Dias BRS, Campos PC, Alvarez-Arguedas S, Shiloh MU (2023 Dec) Pathogenicity and virulence of Mycobacterium tuberculosis. Virulence 14(1):2150449

Ramakrishnan G, Ochoa-Montano B, Raghavender US, Mudgal R, Joshi AG, Chandra NR, Sowdhamini R, Blundell TL, Srinivasan N (2015) Enriching the annotation of Mycobacterium tuberculosis H37Rv proteome using remote homology detection approaches: insights into structure and function. Tuberculosis 95(1):14–25

Raman S, Song T, Puyang X, Bardarov S, Jacobs WR Jr, Husson RN (2001 Oct) The alternative sigma factor SigH regulates major components of oxidative and heat stress responses in Mycobacterium tuberculosis. J Bacteriol 183(20):6119–6125

Raman R, Sasisekharan V, Sasisekharan R (2005) Structural insights into biological roles of protein-glycosaminoglycan interactions. Chem Biol 12(3):267–277

Ramon-Luing LA, Palacios Y, Ruiz A, Téllez-Navarrete NA, Chavez-Galan L (2023) Virulence factors of Mycobacterium tuberculosis as modulators of cell death mechanisms. Pathogens 12(6):839

Razzak SA, Hasan Z, Azim MK, Kanji A, Shakoor S, Hasan R (2023 Jul) A versatile tool for precise variant calling in Mycobacterium tuberculosis genetic polymorphisms. BioRxiv 26:2023

Rengarajan J, Bloom BR, Rubin EJ (2005) Genome-wide requirements for Mycobacterium tuberculosis adaptation and survival in macrophages. Proc Natl Acad Sci USA 102(23):8327–8332

Robinson A, Causer RJ, Dixon NE (2012 Mar) Architecture and conservation of the bacterial DNA replication machinery, an underexploited drug target. Curr Drug Targets 13(3):352–372

Rodrigue S, Provvedi R, Jacques PE, Gaudreau L, Manganelli R (2006 Nov) The sigma factors of Mycobacterium tuberculosis. FEMS Microbiol Rev 30(6):926–941

Rohde K, Yates RM, Purdy GE, Russell DG (2007 Oct) Mycobacterium tuberculosis and the environment within the phagosome. Immunol Rev 219(1):37–54

Ronning DR, Klabunde T, Besra GS, Vissa VD, Belisle JT, Sacchettini JC (2000 Feb) Crystal structure of the secreted form of antigen 85C reveals potential targets for mycobacterial drugs and vaccines. Nat Struct Biol 7(2):141–146

Rosenkrands I, Weldingh K, Jacobsen S, Hansen CV, Florio W, Gianetri I, Andersen P (2000) Mapping and identification of Mycobacterium tuberculosis proteins by two-dimensional gel electrophoresis, microsequencing and immunodetection. Electrophoresis 21(5):935–948

Roy S, Ajitkumar P (2005 Apr) Transcriptional analysis of the principal cell division gene, ftsZ, of Mycobacterium tuberculosis. J Bacteriol 187(7):2540–2550

Roy S, Ghatak D, Das P, BoseDasgupta S (2020) ESX secretion system: the gatekeepers of mycobacterial survivability and pathogenesis. Eur J Microbiol Immunol (Bp) 10(4):202–209

Saikat AS, Kabir ML, Khalipha AB (2020) An in silico approach for structural and functional annotation of uncharacterized protein Rv0986 present in Mycobacterium tuberculosis. Eur J Med Health Sci 2(3):61–67

Saini V, Farhana A, Steyn AJ (2012) Mycobacterium tuberculosis WhiB3: a novel iron-sulfur cluster protein that regulates redox homeostasis and virulence. Antioxid Redox Signal 16(7):687–697

Salgueiro-Toledo VC, Bertol J, Gutierrez C, Serrano-Mestre JL, Ferrer-Luzon N, Vázquez-Iniesta L et al (2025) Maintenance of cell wall remodeling and vesicle production are connected in Mycobacterium tuberculosis. elife 13:RP94982

Salie MJ, Thelen JJ (2016) Regulation and structure of the heteromeric acetyl-CoA carboxylase. Biochim Biophys Acta 1861(9 Pt B):1207-1213

Sampson SL (2011) Mycobacterial PE/PPE proteins at the host-pathogen interface. Clin Dev Immunol 2011:497203

Samten B, Wang X, Barnes PF (2009 Dec) Mycobacterium tuberculosis ESX-1 system-secreted protein ESAT-6 but not CFP10 inhibits human T-cell immune responses. Tuberculosis (Edinb) 89(Suppl 1):S74–S76

Samten B, Fannin S, Sarva K, Yi N, Madiraju M, Rajagopalan M (2016) Modulation of human T cell cytokines by the Mycobacterium tuberculosis-secreted protein Wag31. Tuberculosis (Edinb) 101(Suppl):S99–S104

Sanoussi CN, Coscolla M, Ofori-Anyinam B, Otchere ID, Antonio M, Niemann S, Parkhill J, Harris S, Yeboah-Manu D, Gagneux S, Rigouts L (2021) Mycobacterium tuberculosis complex lineage 5 exhibits high levels of within-lineage genomic diversity and differing gene content compared to the type strain H37Rv. Microbial genomics 7(7):000437

Sassetti CM, Boyd DH, Rubin EJ (2003 Apr) Genes required for mycobacterial growth defined by high density mutagenesis. Mol Microbiol 48(1):77–84

Satta G, Lipman M, Smith GP, Arnold C, Kon OM, McHugh TD (2018) Mycobacterium tuberculosis and whole-genome sequencing: how close are we to unleashing its full potential? Clin Microbiol Infect 24(6):604–609

Scheffers D-J, Driessen AJM (2001) The polymerization mechanism of the bacterial cell division protein FtsZ. FEBS Lett 506(1):6–10

Schué M, Maurin D, Dhouib R, Bakala N'Goma JC, Delorme V, Lambeau G, Carrière F, Canaan S (2010 Jun) Two cutinase-like proteins secreted by Mycobacterium tuberculosis show very different lipolytic activities reflecting their physiological function. FASEB J 24(6):1893–1903

Schwenk S, Arnvig KB (2018) Regulatory RNA in Mycobacterium tuberculosis: back to basics. Pathog Dis. 76(4):fty035

Sethi G, Varghese RP, Lakra AK, Nayak SS, Krishna R, Hwang JH (2024) Immunoinformatics and structural aided approach to develop multi-epitope based subunit vaccine against Mycobacterium tuberculosis. Sci Rep 14(1):15923

Sharma D, Bisht DM (2017 Mar) Tuberculosis hypothetical proteins and proteins of unknown function: hope for exploring novel resistance mechanisms as well as future target of drug resistance. Front Microbiol 21(8):465

Sharma D, Kumar B, Lata M, Joshi B, Venkatesan K, Shukla S, Bisht D (2015) Comparative proteomic analysis of aminoglycosides resistant and susceptible Mycobacterium tuberculosis clinical isolates for exploring potential drug targets. PLoS One 10(10):e0139414

Sharp JD, Singh AK, Park ST, Lyubetskaya A, Peterson MW, Gomes AL, Potluri LP, Raman S, Galagan JE, Husson RN (2016) Comprehensive definition of the SigH regulon of Mycobacterium tuberculosis reveals transcriptional control of diverse stress responses. PLoS One 11(3):e0152145

Sherman DR, Mdluli K, Hickey MJ, Arain TM, Morris SL, Barry CE IIIrd, Stover CK (1996) Compensatory ahpC gene expression in isoniazid-resistant mycobacterium tuberculosis. Science 272(5268):1641–1643. https://doi.org/10.1126/science.272.5268.1641. PMID: 8658136

Shi X, Festa RA, Ioerger TR, Butler-Wu S, Sacchettini JC, Darwin KH, Samanovic MI (2014) The copper-responsive RicR regulon contributes to Mycobacterium tuberculosis virulence. MBio 5(1):e00876–e00813

Silva WTAF, Otto SP, Immler S (2021) Evolution of plasticity in production and transgenerational inheritance of small RNAs under dynamic environmental conditions. PLoS Genet 17(5):e1009581

Singh A, Mai D, Kumar A, Steyn AJ (2006) Dissecting virulence pathways of Mycobacterium tuberculosis through protein-protein association. Proc Natl Acad Sci USA 103(30):11346–11351

Singh A, Guidry L, Narasimhulu KV, Mai D, Trombley J, Redding KE, Giles GI, Lancaster JR Jr, Steyn AJ (2007) Mycobacterium tuberculosis WhiB3 responds to O2 and nitric oxide via its [4Fe-4S] cluster and is essential for nutrient starvation survival. Proc Natl Acad Sci USA 104(28):11562–11567

Singh DK, Ahmed M, Akter S, Shivanna V, Bucşan AN, Mishra A, Golden NA, Didier PJ, Doyle LA, Hall-Ursone S, Roy CJ, Arora G, Dick EJ Jr, Jagannath C, Mehra S, Khader SA, Kaushal D (2025) Prevention of tuberculosis in cynomolgus macaques by an attenuated Mycobacterium tuberculosis vaccine candidate. Nat Commun 16:1957

Sinha PR, Balasubramanian R, Hegde SR (2024 May) Integrated sequence and -omic features reveal novel small proteome of Mycobacterium tuberculosis. Front Microbiol 15(15):1335310

Siroy A, Mailaender C, Harder D, Koerber S, Wolschendorf F, Danilchanka O, Wang Y, Heinz C, Niederweis M (2008) Rv1698 of Mycobacterium tuberculosis represents a new class of channel-forming outer membrane proteins. J Biol Chem 283(26):17827–17837

Sivaramakrishnan S, de Montellano PR (2013) The DosS-DosT/DosR mycobacterial sensor system. Biosensors (Basel) 3(3):259–282

Smith KF, Bibb LA, Schmitt MP, Oram DM (2009 Mar) Regulation and activity of a zinc uptake regulator, Zur. Corynebacterium diphtheriae J Bacteriol 191(5):1595–1603

Smith C, Halse TA, Shea J, Modestil H, Fowler RC, Musser KA, Escuyer V, Lapierre P (2020) Assessing nanopore sequencing for clinical diagnostics: a comparison of next-generation sequencing (NGS) methods for Mycobacterium tuberculosis. J Clin Microbiol 59(1):10–128

Smith JP, Cohen T, Dowdy D, Shrestha S, Gandhi NR, Hill AN (2023 Jan) Quantifying Mycobacterium tuberculosis transmission dynamics across global settings: a systematic analysis. Am J Epidemiol 192(1):133–145

Soni DK, Dubey SK, Bhatnagar R (2020) ATP-binding cassette (ABC) import systems of Mycobacterium tuberculosis: target for drug and vaccine development. Emerg Microbes Infect 9(1):207–220

Srivastava VK, Srivastava S, Arora A, Pratap JV (2013) Structural insights into putative molybdenum cofactor biosynthesis protein C (MoaC2) from Mycobacterium tuberculosis H37Rv. PLoS One 8(3):e58333

Stetler RA, Gan Y, Zhang W, Liou AK, Gao Y, Cao G, Chen J (2010 Oct) Heat shock proteins: cellular and molecular mechanisms in the central nervous system. Prog Neurobiol 92(2):184–211

Steyn AJ, Collins DM, Hondalus MK, Jacobs WR Jr, Kawakami RP, Bloom BR (2002) Mycobacterium tuberculosis WhiB3 interacts with RpoV to affect host survival but is dispensable for in vivo growth. Proc Natl Acad Sci USA 99(5):3147–3152

Strong EJ, West NP (2018) Use of soluble extracellular regions of MmpL (SERoM) as vaccines for tuberculosis. Sci Rep 8:5604

Sun R, Converse PJ, Ko C, Tyagi S, Morrison NE, Bishai WR (2004 Apr) Mycobacterium tuberculosis ECF sigma factor sigC is required for lethality in mice and for the conditional expression of a defined gene set. Mol Microbiol 52(1):25–38

Supply P, Marceau M, Mangenot S, Roche D, Rouanet C, Khanna V et al (2013) Genomic analysis of smooth tubercle bacilli provides insights into ancestry and pathoadaptation of Mycobacterium tuberculosis. Nat Genet 45:172–179

Sureka K, Hossain T, Mukherjee P, Chatterjee P, Datta P, Kundu M, Basu J (2010) Novel role of phosphorylation-dependent interaction between FtsZ and FipA in mycobacterial cell division. PLoS One 5(1):e8590

Sweeney MI, Carranza CE, Tobin DM (2025) Understanding Mycobacterium tuberculosis through its genomic diversity and evolution. PLoS Pathog 21(2):e1012956

Tak U, Vlach J, Garza-Garcia A, William D, Danilchanka O, de Carvalho LPS, Saad JS, Niederweis M (2019) The tuberculosis necrotizing toxin is an NAD+ and NADP+ glycohydrolase with distinct enzymatic properties. J Biol Chem 294(9):3024–3036

Takayama K, Wang C, Besra GS (2005 Jan) Pathway to synthesis and processing of mycolic acids in Mycobacterium tuberculosis. Clin Microbiol Rev 18(1):81–101

Tang X, Deng W, Xie J (2012) Novel insights into Mycobacterium antigen Ag85 biology and implications in countermeasures for *M. tuberculosis*. Crit Rev Eukaryot Gene Expr 22(3):179–187

Thorpe J, Sawaengdee W, Ward D, Campos M, Wichukchinda N, Chaiyasirinroje B, Thanraka A, Chumpol J, Phelan JE, Campino S, Mahasirimongkol S (2024) Multi-platform whole genome sequencing for tuberculosis clinical and surveillance applications. Sci Rep 14(1):5201

Tian J, Bryk R, Shi S, Erdjument-Bromage H, Tempst P, Nathan C (2005 Aug) Mycobacterium tuberculosis appears to lack α-ketoglutarate dehydrogenase and encodes pyruvate dehydrogenase in widely separated genes. Mol Microbiol 57(3):859–868

Tong L (2005 Aug) Acetyl-coenzyme A carboxylase: crucial metabolic enzyme and attractive target for drug discovery. Cell Mol Life Sci 62(16):1784–1803

Touchette MH, Van Vlack ER, Bai L, Kim J, Cognetta AB 3rd, Previti ML, Backus KM, Martin DW, Cravatt BF, Seeliger JC (2017) A screen for protein-protein interactions in live Mycobacteria reveals a functional link between the virulence-associated lipid transporter LprG and the Mycolyltransferase antigen 85A. ACS Infect Dis. 3(5):336–348

Tovar-Nieto AM, Flores-Padilla LE, Rivas-Santiago B, Trujillo-Paez JV, Lara-Ramirez EE, Jacobo-Delgado YM, López-Ramos JE, Rodríguez-Carlos A (2024) The repurposing of FDA-approved drugs as FtsZ inhibitors against Mycobacterium tuberculosis: an in Silico and in vitro study. Microorganisms 12(8):1505

Tucker PA, Nowak E, Morth JP (2007) Two-component systems of *Mycobacterium tuberculosis*—structure-based approaches. Methods Enzymol 423:479–501. Academic

Tufariello JM, Chapman JR, Kerantzas CA, Wong KW, Vilchèze C, Jones CM, Cole LE, Tinaztepe E, Thompson V, Fenyö D, Niederweis M (2016) Separable roles for Mycobacterium tuberculosis ESX-3 effectors in iron acquisition and virulence. Proc Natl Acad Sci 113(3):E348–E357

Turapov O, Forti F, Kadhim B, Ghisotti D, Sassine J, Straatman-Iwanowska A, Bottrill AR, Moynihan PJ, Wallis R, Barthe P, Cohen-Gonsaud M, Ajuh P, Vollmer W, Mukamolova GV (2018) Two faces of CwlM, an essential PknB substrate in Mycobacterium tuberculosis. Cell Rep 25(1):57–67.e5

Uplekar S, Rougemont J, Cole ST, Sala C (2013 Jan) High-resolution transcriptome and genome-wide dynamics of RNA polymerase and NusA in Mycobacterium tuberculosis. Nucleic Acids Res 41(2):961–977

van Els CACM, Corbière V, Smits K, van Gaans-van den Brink JAM, MCM P, Mascart F, Meiring HD, Locht C (2014) Toward understanding the essence of post-translational modifications for the mycobacterium tuberculosis immunoproteome. Front Immunol 5:361. https://doi.org/10.3389/fimmu.2014.00361

van Teeffelen S, Wang S, Furchtgott L, Huang KC, Wingreen NS, Shaevitz JW, Gitai Z (2011) The bacterial actin MreB rotates, and rotation depends on cell-wall assembly. Proc Natl Acad Sci USA 108(38):15822–15827. https://doi.org/10.1073/pnas.1108999108. PMID: 21903929; PMCID: PMC3179079

Varabyou A, Erdogdu B, Salzberg SL, Pertea M (2023) Investigating open reading frames in known and novel transcripts using ORFanage. bioRxiv [Preprint] 2023.03.23.533704

Velayati AA, Farnia P (2012 Sep) Division-cycle in Mycobacterium tuberculosis. Int J Mycobacteriol 1(3):111–117

Velayati AA, Farnia P (2016) Atlas of *Mycobacterium tuberculosis*, 1st edn. Academic Press, London. Hardback ISBN: 9780128038086. eBook ISBN: 9780128038802

Vijayaraghavan M, Gadad SS, Dhandayuthapani S (2025) Long non-coding RNA transcripts in Mycobacterium tuberculosis–host interactions. Non-Coding RNA Res 11:281–293

Votintseva AA, Bradley P, Pankhurst L, del Ojo EC, Loose M, Nilgiriwala K, Chatterjee A, Smith EG, Sanderson N, Walker TM, Morgan MR (2017 May) Same-day diagnostic and surveillance data for tuberculosis via whole-genome sequencing of direct respiratory samples. J Clin Microbiol 55(5):1285–1298

Walker TM, Merker M, Kohl TA, Crook DW, Niemann S, Peto TE (2017) Whole genome sequencing for M/XDR tuberculosis surveillance and for resistance testing. Clin Microbiol Infect 23(3):161–166

Wan Q, Song D, Li H, He M (2020) Stress proteins: the biological functions in virus infection, present and challenges for target-based antiviral drug development. Signal Transduct Target Ther 5:125

Wan T, Horová M, Khetrapal V, Li S, Jones C, Schacht A, Sun X, Zhang L (2023) Structural basis of DNA binding by the WhiB-like transcription factor WhiB3 in mycobacterium tuberculosis. J Biol Chem 299(6):104777. https://doi.org/10.1016/j.jbc.2023.104777 . Epub 2023 May 2. PMID: 37142222; PMCID: PMC10245118

Wang J, Ge P, Qiang L, Tian F, Zhao D, Chai Q, Zhu M, Zhou R, Meng G, Iwakura Y, Gao GF, Liu CH (2017) The mycobacterial phosphatase PtpA regulates the expression of host genes and promotes cell proliferation. Nat Commun 8:244

Wang H, Fedorov AA, Fedorov EV, Hunt DM, Rodgers A, Douglas HL, Garza-Garcia A, Bonanno JB, Almo SC, de Carvalho LPS (2019) An essential bifunctional enzyme in Mycobacterium tuberculosis for itaconate dissimilation and leucine catabolism. Proc Natl Acad Sci USA 116(32):15907–15913

Wang S, Fang R, Wang H, Li X, Xing J, Li Z, Song N (2024) The role of transcriptional regulators in metal ion homeostasis of mycobacterium tuberculosis. Front Cell Infect Microbiol 14:1360880. https://doi.org/10.3389/fcimb.2024.1360880

Warner DF (2014) Mycobacterium tuberculosis metabolism. Cold Spring Harb Perspect Med 5(4):a021121

Warrier T, Tropis M, Werngren J, Diehl A, Gengenbacher M, Schlegel B, Schade M, Oschkinat H, Daffe M, Hoffner S, Eddine AN, Kaufmann SH (2012 Apr) Antigen 85C inhibition restricts Mycobacterium tuberculosis growth through disruption of cord factor biosynthesis. Antimicrob Agents Chemother 56(4):1735–1743

West NP, Chow FM, Randall EJ, Wu J, Chen J, Ribeiro JM, Britton WJ (2009 Jun) Cutinase-like proteins of Mycobacterium tuberculosis: characterization of their variable enzymatic functions and active site identification. FASEB J 23(6):1694–1704

Wisedchaisri G, Wu M, Sherman DR, Hol WGJ (2008) Crystal structures of the response regulator DosR from Mycobacterium tuberculosis suggest a helix rearrangement mechanism for phosphorylation activation. J Mol Biol 378(1):227–242

Witney AA, Cosgrove CA, Arnold A, Hinds J, Stoker NG, Butcher PD (2016a Mar) Clinical use of whole genome sequencing for Mycobacterium tuberculosis. BMC Med 23(14):46

Witney AA, Cosgrove CA, Arnold A, Hinds J, Stoker NG, Butcher PD (2016b) Clinical use of whole genome sequencing for Mycobacterium tuberculosis. BMC Med 14(1):46

Wollenberg KR, Desjardins CA, Zalutskaya A, Slodovnikova V, Oler AJ, Quiñones M, Abeel T, Chapman SB, Tartakovsky M, Gabrielian A, Hoffner S (2017 Feb) Whole-genome sequencing of Mycobacterium tuberculosis provides insight into the evolution and genetic composition of drug-resistant tuberculosis in Belarus. J Clin Microbiol 55(2):457–469

Xu Z, Zhou A, Yao Y (2014) ATP-binding cassette transporters and transmembrane transport in Mycobacterium tuberculosis: a review. Wei Sheng Wu Xue Bao 54(6):608–615. Chinese

Xu X, Dong B, Peng L, Gao C, He Z, Wang C, Zeng J (2022 Dec) Anti-tuberculosis drug development via targeting the cell envelope of Mycobacterium tuberculosis. Front Microbiol 21(13):1056608

Yang Z, Zeng X, Tsui SK (2019) Investigating function roles of hypothetical proteins encoded by the Mycobacterium tuberculosis H37Rv genome. BMC Genom 21;20(1):394

Yuan P, He L, Chen D, Sun Y, Ge Z, Shen D, Lu Y (2020 Feb) Proteomic characterization of Mycobacterium tuberculosis reveals potential targets of bostrycin. J Proteome 10(212):103576

Zahrt TC, Deretic V (2000 Jul) An essential two-component signal transduction system in Mycobacterium tuberculosis. J Bacteriol 182(13):3832–3838

Zhang H, Chen Y, Zhang Y, Qiao L, Chi X, Han Y et al (2023) Identification of anti-Mycobacterium tuberculosis agents targeting the interaction of bacterial division proteins FtsZ and SepFe. Acta Pharm Sin B 13(5):2056–2070

Zhang X, Zhao R, Qi Y, Yan X, Qi G, Peng Q (2024a) The progress of Mycobacterium tuberculosis drug targets. Front Med 11:1455715

Zhang H, Tang M, Li D, Xu M, Ao Y, Lin L (2024b Jun) Applications and advances in molecular diagnostics: revolutionizing non-tuberculous mycobacteria species and subspecies identification. Front Public Health 19(12):1410672

Zhang Z, Dong L, Li X, Deng T, Wang Q (2025 Jun) The PE/PPE family proteins of Mycobacterium tuberculosis: evolution, function, and prospects for tuberculosis control. Front Immunol 17(16):1606311

Zheng H, Lu L, Wang B, Pu S, Zhang X, Zhu G, Shi W, Zhang L, Wang H, Wang S, Zhao G (2008) Genetic basis of virulence attenuation revealed by comparative genomic analysis of Mycobacterium tuberculosis strain H37Ra versus H37Rv. PLoS One 3(6):e2375

Zu X, Zhong J, Luo D, Tan J, Zhang Q, Wu Y, Liu J, Cao R, Wen G, Cao D (2013) Chemical genetics of acetyl-CoA carboxylases. Molecules 18(2):1704–1719

4 Contribution of Proteins to Membrane and Cell Wall Structures in *Mycobacterium tuberculosis*

Abstract

The cell wall of *Mycobacterium tuberculosis* (Mtb) is an exceptionally complex, multilayered structure that is fundamental to the bacterium's survival, pathogenicity, and resistance to both host immune defenses and antibiotics. This cell envelope consists of several distinct components: a surface capsule, an outer membrane, and a peptidoglycan layer that is covalently attached to arabinogalactan. The arabinogalactan is further esterified with mycolic acids, forming the characteristic mycolyl-arabinogalactan-peptidoglycan (mAGP) complex, which is anchored to the underlying plasma membrane. The assembly and maintenance of this intricate architecture depend on a suite of specialized proteins that regulate lipid transport, enzymatic modifications, and structural integrity. Among these, the Antigen 85 Complex (Ag85) is particularly important, as it catalyzes the transfer of mycolic acids to arabinogalactan, facilitating the synthesis of trehalose dimycolate (TDM), a major virulence factor. The inner membrane transporter MmpL3 is critical for translocating trehalose monomycolate (TMM) across the plasma membrane, a prerequisite for TDM synthesis and cell wall stability. Additionally, the MmpS4/MmpL4 and MmpS5/MmpL5 complexes mediate the secretion of siderophores, which are essential for iron acquisition under nutrient-limited conditions and thereby support bacterial growth. Enzymes such as cyclopropane synthases (e.g., CmaA2) and methyltransferases (e.g., MmaA3) chemically modify mycolic acids, enhancing the pathogen's ability to evade host immune responses. Outer membrane proteins (OMPs), including Rv1698 and Rv1973, contribute to the structural integrity and selective permeability of the cell envelope, although their precise biological functions are still being elucidated. Collectively, these membrane- and cell wall-associated proteins orchestrate the biosynthesis, modification, and transport of essential cell wall components, underpinning the unique composition of the Mtb cell envelope that confers intrinsic resistance to antibiotics and host defenses. A comprehensive understanding of these proteins is crucial for the development of novel antituberculosis therapeutics and vaccines, as they represent promising targets for disrupting the protective barrier of Mtb.

Keywords

Mycobacterium tuberculosis · Peptidoglycan (PG) biosynthesis · Outer membrane-associated proteins · Mycolic acids · Glycolipid biosynthesis proteins · Arabinogalactan (AG) biosynthesis

P. Farnia et al., *Proteins in Mycobacterium Tuberculosis*, Advances in Experimental Medicine and Biology 1484, https://doi.org/10.1007/978-3-031-96883-9_4

4.1 Introduction

The contribution of proteins to the membrane and cell wall structures in *Mycobacterium tuberculosis* (Mtb) is fundamental to the bacterium's survival, pathogenicity, and ability to resist hostile environments. Mtb possesses a uniquely complex and highly specialized cell wall architecture, which includes an elaborate, highly hydrophobic capsule as its outermost layer. Directly beneath this capsule lies the mycomembrane (MM), a distinctive lipid bilayer that plays a critical role in maintaining barrier function. At the core of the cell wall is a robust peptidoglycan (PG) layer covalently linked to arabinogalactan (AG), which is further esterified with mycolic acids, together forming the mycolyl-arabinogalactan-peptidoglycan (mAGP) complex. This mAGP complex is essential for maintaining the structural integrity, mechanical strength, and characteristic hydrophobicity of the cell wall (Minnikin et al. 1982; Draper 1982; Brennan and Nikaido 1995; Brennan 1994; Daffé and Draper 1997; Rastogi et al. 2001; Draper and Daffé 2004; Kremer et al. 2005; Alsteens et al. 2008; Alderwick et al. 2015; Daffé and Marrakchi 2019; Brown et al. 2023).

The plasma membrane, which constitutes the innermost layer of the cell envelope, functions as a selective barrier separating the cytoplasm from the complex cell wall components. Unlike Gram-negative bacteria, Mtb lacks a conventional periplasmic space; instead, the mAGP complex physically links the plasma membrane to the outer mycomembrane, creating a continuous and highly impermeable envelope (Minnikin et al. 1982; Brennan and Nikaido 1995; Rastogi et al. 2001). The capsule of Mtb has been visualized as an electron-transparent zone (ETZ) surrounding the cell envelope in conventional electron microscopy preparations (Chapman et al. 1959; Frehel et al. 1986; Paul and Beveridge 1994; Daffé and Etienne 1999). Chemically, the capsule is composed predominantly of polysaccharides, with smaller proportions of proteins and lipids contributing to its structure and function (Ortalo-Magné et al. 1995; Lemassu et al. 1996; Daffé and Etienne 1999). The presence and structural integrity of the capsule are highly dependent on culturing conditions. In laboratory settings, detergents and mechanical agitation are commonly employed to prevent bacterial clumping, but these practices can disrupt or remove the capsule (Wright et al. 1998; Schwebach et al. 2001, 2002). Because the capsule is only loosely attached to the cell wall, it can be readily shed into the surrounding environment. A breakthrough in confirming the existence of the Mtb capsule came in 2010 when Sani et al. utilized cryo-electron microscopy (cryo-EM) to visualize a distinct, loosely organized, electron-transparent outermost layer external to the cell wall and mycomembrane. This study also explained why the capsule had been overlooked in earlier work, attributing its frequent absence to routine culturing methods involving detergents such as Tween-80 that strip away this fragile layer (Sani et al. 2010). Subsequent studies, including those by Velayati and Farnia (2016), employed atomic force microscopy (AFM) to validate the capsule's presence further and characterize its physical properties. The capsule's dominant polysaccharide component is a glycogen-like α-glucan, which forms a neutral polysaccharide layer. This layer is complemented by smaller amounts of arabinomannan and mannan, which collectively contribute to the capsule's biochemical complexity and functional versatility (Stokes and Waddell 2009; Kalscheuer et al. 2019). Functionally, the capsule acts as an antiphagocytic barrier, playing a critical role in immune evasion by preventing recognition and uptake by host phagocytes. Additionally, it restricts the penetration of large macromolecules, thereby enhancing the bacterium's virulence and contributing to its intrinsic drug resistance (Stokes and Waddell 2009; Zhai et al. 2019). Kalscheuer et al. (2019) specifically demonstrated that the capsule contributes structurally to Mtb's drug resistance phenotype, underscoring its importance beyond a mere physical barrier. Expanding on this foundation, Sharma et al. (2024) performed a comparative proteomic analysis of capsule-associated proteins derived from aminoglycoside-resistant and sensitive Mtb isolates. This study employed two-dimensional gel electrophoresis, mass spectrometry, and bioinformatics tools to identify proteins differentially expressed in resistant strains, shedding light on the molecular mechanisms underlying drug resistance and capsule function. Using two-dimensional gel electrophoresis, mass spectrometry, and bioinformatics, Sharma et al. (2024) identified eight proteins that were significantly upregulated in aminoglycoside-resistant Mtb strains. These included Rv3029c and Rv2110c, which are involved in metabolism and respiration; Rv2462c, associated with cell wall processes; Rv3804c, implicated in lipid metabolism; Rv2416c and Rv2623, linked to virulence and detoxification; Rv0020c, a regulatory protein; and Rv0639, related to information pathways (Rosenkrands et al. 2000a, b; Gu et al. 2003; DeJesus et al. 2017; Minato et al. 2019; Sharma et al. 2024). Interactome analysis using STRING 12.0 revealed that these proteins form interconnected networks, suggesting coordinated roles in mediating aminoglycoside resistance (Sharma et al. 2024). Furthermore, molecular docking studies demonstrated favorable binding affinities between aminoglycoside antibiotics such as amikacin and kanamycin and capsular proteins Rv2462c, Rv3804c, and Rv2623, supporting the hypothesis that these proteins influence drug efficacy (Palande et al. 2024; Sharma et al. 2024). Collectively, these findings underscore the multifaceted nature of aminoglycoside resistance in Mtb, highlighting that capsular proteins contribute not only to maintaining structural integrity and facilitating host interactions but also to modulating bacterial susceptibility to antibiotics. The observed upregulation of proteins involved in metabolism, cell wall maintenance, and virulence within the capsule suggests that it functions as a dynamic interface critical for bacterial survival and antibiotic resistance (Sharma et al. 2024).

The mycomembrane, a concept first proposed by Minnikin in 1982, is a lipid bilayer characterized by an inner leaflet of mycolic acids covalently bound to arabinogalactan within the mAGP complex and an outer leaflet composed of free lipids (Minnikin et al. 1982; Brennan and Nikaido 1995). The inner leaflet contains extremely long-chain fatty acids (C60–90), including alpha-mycolates, methoxymycolates, and ketomycolates, each bearing distinctive structural features such as cyclopropane rings and oxygenated groups (Yuan and Barry 1996; Kolattukudy et al. 1997; Indrigo et al. 2002; Takayama et al. 2005; Dover et al. 2004). These mycolic acids confer the cell wall with its hallmark hydrophobicity and impermeability, serving as a formidable barrier against desiccation, antibiotics, and host immune defenses. The outer leaflet consists of non-covalently associated lipids and glycolipids, including trehalose-6′-dimycolate (TDM), which imparts the characteristic waxy and acid-fast properties of Mtb's outer membrane. Other key components of the outer leaflet include trehalose monomycolate (TMM), a precursor in TDM biosynthesis; phthiocerol dimycocerosates (PDIMs), which enhance hydrophobicity and aid immune evasion; sulfolipids (SLs), which regulate cell wall permeability and modulate host immune responses; and phosphatidylinositol mannosides (PIMs), which are essential for the biosynthesis of lipoglycans such as lipoarabinomannan (LAM) (Noll et al. 1956; Indrigo et al. 2002; Hunter 2006; Ishikawa et al. 2009; Werninghaus et al. 2009; Bowdish et al. 2009; Marrakchi et al. 2014; Holzheimer et al. 2021; Brown et al. 2023). TDM exhibits unique immunostimulatory properties, including the induction of granuloma formation and the enhancement of cell-mediated immune responses, acting as an adjuvant (Ishikawa et al. 2009; Schoenen et al. 2010; Lang et al. 2013; Braganza et al. 2018; Dubé et al. 2022). Other important cell envelope components such as PIMs, lipomannan (LM), and lipoarabinomannan (LAM) span the cell envelope, anchoring to the plasma membrane and extending outward to interact with host immune cells, thereby modulating immune responses (Daffé and Draper 1997; Morita et al. 2005; Hoffmann et al. 2008; Bansal-Mutalik and Nikaido 2014; Minnikin et al. 2015; Chiaradia et al. 2017; Daffé and Marrakchi 2019; Batt et al. 2020; Brown et al. 2023).

Early studies proposed that macrophage receptors such as Marco and Toll-like receptors (TLRs) might recognize TDM (Bowdish et al. 2009), while other investigations suggested involvement of FcRγ-coupled receptors (Werninghaus et al. 2009). However, Ishikawa et al. (2009) clarified that Mincle (macrophage inducible C-type lectin) is the essential receptor for TDM, directly binding to TDM and signaling through the FcRγ-CARD9 pathway. Notably, macrophages deficient in Mincle fail to produce cytokine responses upon TDM stimulation, confirming its critical role in immune recognition (Ishikawa et al. 2009; Schoenen et al. 2010; Lang et al. 2013; Hunter 2016; Braganza et al. 2018; Dubé et al. 2022).

Phosphatidyl-myo-inositol mannosides (PIMs) are essential glycolipids within the mycobacterial cell envelope, serving as critical precursors for the biosynthesis of lipomannan (LM) and lipoarabinomannan (LAM) (Daffé and Draper 1997; Morita et al. 2005; Hoffmann et al. 2008; Guerin et al. 2010; Bansal-Mutalik and Nikaido 2014; Minnikin et al. 2015; Chiaradia et al. 2017; Daffé and Marrakchi 2019; Batt et al. 2020; Brown et al. 2023). These glycolipids are anchored to the plasma membrane and extend outward, facilitating interactions with host cells (Ilangumaran et al. 1995). The biosynthesis of PIMs involves key enzymatic activities, including those of phosphatidyl-myo-inositol mannosyltransferases and the acyltransferase PatA, which are crucial for their proper assembly and function (Sancho-Vaello et al. 2020). PIMs engage with host immune receptors such as C-type lectins, including the mannose receptor (MR) and DC-SIGN, thereby modulating immune responses (Torrelles et al. 2006). Notably, higher-order PIMs, such as PIM5 and PIM6, exhibit preferential binding to MR, whereas lower-order PIMs like PIM2 show weaker association. The degree of acylation further influences receptor affinity; triacylated PIMs are recognized more effectively than their tetra-acylated counterparts (Torrelles et al. 2006). Regardless of acylation status, PIMs activate macrophages to secrete tumor necrosis factor-alpha (TNF-α) through Toll-like receptor 2 (TLR2) signaling pathways (Gilleron et al. 2003). LAM molecules can integrate into host cell membranes via their glycosylphosphatidylinositol (GPI) anchors, localizing within plasma membrane domains enriched in GPI-anchored proteins. This integration is facilitated by both the acyl chains and the mannoside core glycan structure of the LAM-GPI anchor, underscoring their role in host-pathogen interactions (Ilangumaran et al. 1995). Advances in understanding the structure and biosynthesis of these glycolipids have been instrumental in elucidating their biological functions, with synthetic fragments of PIMs and LAMs serving as valuable tools for functional studies (Cao and Williams 2010).

In addition to glycolipids, phospholipids such as phosphatidylethanolamine (PE), phosphatidylinositol (PI), phosphatidylglycerol, and cardiolipin (CL) are integral components of the mycobacterial plasma membrane and participate in cell wall biosynthesis. The mycobacterial membrane is particularly rich in PE and PI, whereas phosphatidylglycerol is present at lower levels. Phthiocerol dimycocerosates (PDIMs), a distinct class of lipids, contribute significantly to the hydrophobicity and impermeability of the cell wall, enhancing the bacterium's defense against environmental stresses (Niederweis et al. 2010; Chiaradia et al. 2017; Daffé and Marrakchi 2019; Rens et al. 2021). Proteins associated with the mycomembrane play vital roles in maintaining bacterial functionality and survival. Outer membrane proteins (OMPs) of Mtb are essential for nutrient uptake, secretion, and interactions with the host (Stephan et al. 2004; Hoffmann et al. 2008; Song et al. 2008; Niederweis et al. 2010; Danilchanka et al. 2014; Merrick et al. 2023; Jaisinghani et al. 2024; Ravindran Nair et al. 2025). Among these, porins are β-barrel proteins that form hydrophilic channels facilitating the transport of small molecules across the otherwise impermeable outer membrane. At least two types of porins have been characterized in Mtb: a cation-selective channel with a conductance of approximately 0.7 nS and a larger 3-nS channel (Kartmann et al. 1999). OmpATb, a member of the OmpA family, forms pores ranging from 1.4 to 1.8 nm in diameter and is critical for Mtb's adaptation to acidic environments and survival within macrophages (Senaratne et al. 1998; Niederweis 2003). Introducing the MspA porin from *Mycobacterium smegmatis* into

Mycobacterium tuberculosis and *Mycobacterium bovis BCG* results in increased outer membrane permeability, faster bacterial growth, and heightened sensitivity to multiple antibiotics, including beta-lactams, isoniazid, ethambutol, and streptomycin (Mailaender et al. 2004). These findings underscore the pivotal role of porins in nutrient uptake, growth regulation, and antibiotic susceptibility, positioning them as promising targets for drug development (Jones et al. 2014). Supporting this, Velayati and Farnia (2016) identified pore-like structures in Mtb resembling porins found in other bacteria, further emphasizing their physiological importance. Nevertheless, additional research is necessary to fully identify and characterize the specific porins that contribute to Mtb's survival and pathogenicity.

Other outer membrane-associated proteins include Rv1698 (also known as MctB) and Rv1973, both predicted to be β-barrel outer membrane proteins. Rv1698/MctB forms channels essential for copper homeostasis, a critical factor for bacterial virulence and survival. Loss of MctB results in increased intracellular copper accumulation and reduced bacterial burden in animal infection models, highlighting its functional significance (Siroy et al. 2008; Wolschendorf et al. 2011; Song et al. 2008). These proteins exhibit structural features characteristic of Gram-negative bacterial outer membrane proteins, including β-barrel folds and signal peptides, reinforcing their role in maintaining outer membrane integrity and function (Niederweis et al. 2010). Their identification and characterization are essential for advancing our understanding of Mtb physiology and pathogenicity and represent promising targets for novel antituberculosis therapies (Niederweis et al. 2010).

Among important channel proteins, CpnT (Rv3903c) plays a dual role in nutrient uptake and induction of host cell death. CpnT is critical for bacterial survival within macrophages, mediating necrotic cell death through its C-terminal toxic domain (Danilchanka et al. 2014). Structurally, CpnT is an NAD^+ glycohydrolase composed of two distinct domains: the C-terminal tuberculosis necrotizing toxin (TNT) domain, which functions as a toxin, and the N-terminal domain, which is essential for secretion. The N-terminal domain forms a channel that facilitates nutrient transport across the outer membrane, while the C-terminal TNT domain induces necrotic cell death in host cells, thereby promoting bacterial escape and dissemination (Song et al. 2008; Niederweis et al. 2010; Danilchanka et al. 2014; Izquierdo Lafuente et al. 2021; Pajuelo et al. 2021; Famelis et al. 2023). CpnT is secreted via the ESX-5 type VII secretion system, underscoring its multifunctional role as both a nutrient transporter and a virulence factor critical to Mtb pathogenesis (Pajuelo et al. 2021).

Glycolipid biosynthesis proteins are central to the synthesis and modification of glycolipids integral to the mycomembrane. A key component of this process is the Antigen 85 (Ag85) Complex, which comprises three mycolyltransferases: FbpA (Rv3804c), FbpB (Rv1886c), and FbpC2 (Rv0129c) (Lim et al. 2000; Daffé 2000; Kremer et al. 2002; Puech et al. 2002; Takayama et al. 2005; Backus et al. 2014; Goins et al. 2018). The Ag85 enzymes catalyze the transfer of mycolic acids between trehalose molecules, facilitating the formation of trehalose dimycolate (TDM) and other essential cell wall components (Goins et al. 2018). Their enzymatic mechanism follows a two-step ping-pong process: initially, the enzyme binds trehalose monomycolate (TMM) at the active site, where a conserved serine residue attacks the ester bond of the mycolate moiety, forming a covalent acyl-enzyme intermediate and releasing free trehalose. Subsequently, a second trehalose molecule binds at a distinct site within the active-site cleft, serving as the acceptor. The mycolate group is then transferred from the acyl-enzyme intermediate to the acceptor trehalose, producing TDM. This interfacial mechanism minimizes exposure of the hydrophobic mycolic acid chains to the aqueous environment, ensuring efficient lipid transfer. The three Ag85 isoforms, Ag85A, Ag85B, and Ag85C, exhibit distinct substrate specificities and catalytic efficiencies, which are modulated by a secondary carbohydrate-binding site influencing substrate recognition and binding dynamics (Backus et al. 2014). Ag85A preferentially catalyzes TDM formation, whereas

Ag85C demonstrates higher activity toward producing TMM, reflecting functional specialization among the isoforms (Backus et al. 2014). Additionally, environmental factors such as glucose availability can shift the balance between TDM and glucose monomycolate synthesis, illustrating the dynamic remodeling of cell wall glycolipids in response to nutrient conditions and bacterial growth states (Matsunaga et al. 2008). This adaptability enables Mtb to fine-tune its cell wall composition, optimizing structural integrity and interactions with the host environment (Goins et al. 2018; Backus et al. 2014; Matsunaga et al. 2008). Polyketide synthases (PKSs) play a crucial role in the biosynthesis of complex lipids within the Mtb cell wall, producing diverse lipid molecules essential for maintaining the bacterium's protective barrier and pathogenicity (Sankaranarayanan et al. 2004; Quadri 2014). The Mtb genome encodes multiple PKS genes, including three type III PKSs (Constant et al. 2002; Trivedi et al. 2005; Yu et al. 2012; Nivina et al. 2019; Ramos et al. 2020; Bon et al. 2022). Among these, PKS11 and PKS18 are novel type III PKSs involved in synthesizing long-chain α-pyrones and exhibit specificity for long-chain acyl-CoA substrates. Crystal structure analysis of PKS11 revealed a deep hydrophobic tunnel accommodating substrates and confirmed its role in producing methyl-branched alkylpyrones (Gokulan et al. 2013). Fatty acyl-AMP ligases (FAALs) activate fatty acids and transfer them to PKSs, facilitating the production of unusual acyl chains characteristic of mycobacterial lipids (Mohanty et al. 2011). Given their central roles in critical lipid biosynthetic pathways, PKSs and FAALs represent attractive targets for antituberculosis drug development (Mohanty et al. 2011; Quadri 2014). Other protein groups significantly influencing lipid composition and modulating immune responses include PPE51 (Rv3136), which functions as a trehalose transporter across the mycomembrane, thereby impacting the synthesis of trehalose dimycolate (TDM) (Babu Sait et al. 2022). PPE51 plays a critical role in Mtb pathogenesis and immune modulation. It is co-transcribed with PPE50, forming a functional unit, and the two proteins interact both in vitro and in vivo, suggesting coordinated activity (Mukku et al. 2024). Importantly, PPE51 inhibits autophagy by suppressing Toll-like receptor 2 (TLR2)-dependent signaling pathways, which enhances bacterial survival within host cells (Strong et al. 2022). Additionally, PPE51 acts as a solute-specific channel in the outer membrane, facilitating nutrient transport while maintaining the impermeability of the mycomembrane (Wang et al. 2020). The PE/PPE protein family, to which PPE51 belongs, is characterized by N-terminal motifs, either PE (Pro-Glu) or PPE (Pro-Pro-Glu), and includes over 170 proteins implicated in virulence, immune modulation, and host-pathogen interactions that promote bacterial survival (Daleke et al. 2011; Ekiert and Cox 2014; Guo et al. 2023). This family contains both tuberculosis-specific and conserved T cell epitopes; notably, an immunodominant epitope spanning residues 118–135 is highly conserved among PPE proteins, underscoring their immunological significance (Okkels et al. 2003). Many PE/PPE proteins are surface-exposed or secreted via type VII secretion systems, where they play critical roles in immune evasion and pathogenesis (Ly and Liu 2020; Bunduc et al. 2020; Pal et al. 2022; Famelis et al. 2023). The multifaceted involvement of PPE51 in nutrient uptake, immune modulation, and drug susceptibility highlights its importance in Mtb physiology and identifies it as a promising target for therapeutic intervention and vaccine development. Another notable surface-associated protein is heparin-binding hemagglutinin adhesin (HbhA, Rv0475), which mediates adhesion to epithelial cells and macrophages by binding to sulfated glycoconjugates such as heparin and chondroitin sulfate (Delogu and Brennan 1999; Verbelen et al. 2008; Esposito et al. 2011; Raze et al. 2018; Dessenne et al. 2025). The C-terminal lysine-rich domain of HbhA facilitates interactions with host cells, and methylation of lysine residues modulates immune recognition, inducing protective T cell responses, particularly in latent tuberculosis. Structurally, HbhA forms elongated dimers and dynamically localizes between the mycobacterial surface and intracytosolic lipid inclusions, reflecting its versatile role in host-pathogen interactions. In contrast to Gram-negative bacteria, Mtb lacks a conventional periplasmic space. Instead, it possesses a complex cell envelope architecture comprising the inner membrane, peptidoglycan layer, arabinogalactan layer, and mycolic acid layer, which together form the mycolyl-arabinogalactan-peptidoglycan (mAGP) complex (Draper 1982; Brennan 1994; Ortalo-Magne et al. 1996; Daffé and Draper 1997; Draper and Daffé 2004; Kremer et al. 2005; Alsteens et al. 2008; Alderwick et al. 2015; Vincent et al. 2018; Daffé and Marrakchi 2019). The space between the inner membrane and the mAGP complex is a critical compartment for nutrient processing, cell wall assembly, and immune modulation. Within this region, several key classes of proteins contribute to these essential functions, including transport and trafficking proteins such as members of the mycobacterial membrane protein Large (MmpL) family, siderophore secretion proteins, lipid trafficking proteins, enzymes involved in cell wall synthesis, and other proteins indispensable for bacterial survival and virulence and trafficking proteins are integral to the biogenesis of the mycobacterial cell envelope, with MmpL3 (Rv0206c) serving Transport as a pivotal inner membrane transporter (Cole et al. 1998; Grzegorzewicz et al. 2012; Varela et al. 2012; Chalut 2016; Degiacomi et al. 2017; Viljoen et al. 2017; Melly and Purdy 2019; Ma et al. 2020). MmpL3 is essential for flipping trehalose monomycolate (TMM), a precursor of trehalose dimycolate (TDM) and mycolic acids linked to arabinogalactan—across the plasma membrane into the cell envelope (Degiacomi et al. 2017; Su et al. 2021). Structural analyses, including crystal structures and molecular dynamics simulations, reveal that MmpL3 facilitates lipid transport through coordinated movements of its periplasmic domain and transmembrane helices, powered by the proton motive force (Su et al. 2021). This transport is critical for synthesizing the hydrophobic mycomembrane, which underpins bacterial survival, virulence, and antibiotic resistance. Once exported, TMM is further processed by enzymes such as the Antigen 85 Complex, which converts it into TDM or incorporates it into the mycomembrane. MmpL3 specifically binds TMM and phosphatidylcholine (PC), underscoring its role in shuttling diverse lipids to fortify the cell wall. Disruption of MmpL3 results in the accumulation of TMM precursors, defective cell division, and rapid bacterial death, highlighting its essential function and validating it as a promising therapeutic target (Belardinelli et al. 2022; Fay et al. 2019; Su et al. 2019; Batt et al. 2020; Alsayed and Gunosewoyo 2023). Complementing MmpL3, MmpS4 (Rv0450c) and MmpS5 (Rv0451c) are small membrane proteins that interact with their respective partners, MmpL4 (Rv0451c) and MmpL5 (Rv0676c), forming complexes essential for siderophore secretion in Mtb. These accessory proteins function analogously to periplasmic adapter proteins in Gram-negative tripartite efflux pumps, with MmpS4 binding the periplasmic loop of MmpL4 and MmpS5 interacting specifically with MmpL5 (Tahlan et al. 2012; Wells et al. 2013; Earp et al. 2025). These complexes mediate the export of siderophores such as mycobactins and carboxymycobactins, which are vital for iron acquisition under iron-limiting conditions (Cole et al. 1998; Deshayes et al. 2013; Zhang et al. 2020). The protein Rv0455c, involved in siderophore secretion, interacts with both MmpS4/MmpL4 and MmpS5/MmpL5 complexes. Although Rv0455c can be detected extracellularly, its primary role is within the cell envelope, ensuring efficient siderophore transport and maintaining membrane integrity. Disruption of these systems impairs siderophore secretion, leading to iron starvation and attenuated virulence in animal models, underscoring their importance in bacterial survival and pathogenicity (Earp et al. 2025). Additionally, lipid trafficking proteins such as PgfA (Rv2613c) cooperate with MmpL3 to coordinate lipid transport and assembly of the mycomembrane, further emphasizing the complex orchestration required for cell envelope biogenesis (Gupta et al. 2022).MmpL proteins play a crucial role in transporting various cell wall lipids, including trehalose monomycolate (TMM), phthiocerol dimycocerosate (PDIM), and sulfolipid-1 (SL-1), positioning them as promising targets for antituberculosis drug development (Bailo et al. 2015).

Mtb preferentially associates with lipid-rich foamy macrophages and exploits host-derived fatty acids and cholesterol as nutrient sources (Lovewell et al. 2016). To ensure its intracellular survival, Mtb interferes with Rab GTPase-controlled membrane trafficking pathways, thereby arresting phagosome maturation within macrophages (Vergne et al. 2004). Central to cell wall biogenesis and pathogenesis is the Antigen 85 Complex, composed of FbpA (Rv3804c), FbpB (Rv1886c), and FbpC (Rv0129c), which function as mycolyltransferases (De Wit et al. 1994; Harth et al. 1996; Rosenkrands et al. 2000b; Kremer et al. 2002; Takayama et al. 2005; Backus et al. 2014; Fay et al. 2019; Vinod et al. 2020; Melkie et al. 2022; García-Ruiz et al. 2022). These enzymes catalyze the transfer of mycolic acids to the arabinogalactan layer and convert TMM into trehalose dimycolate (TDM). Despite the absence of a classical periplasmic space in Mtb, the Ag85 complex localizes to the cell envelope, integrating lipid trafficking with mycomembrane assembly. Other significant cell envelope components include mannose-containing glycolipids such as lipomannans (LM) and lipoarabinomannans (LAM), which are vital for host cell interaction and immune modulation (Nigou et al. 2003; Kaur et al. 2008; Pitarque et al. 2008; Fukuda et al. 2013; Dulberger et al. 2020; Sparks et al. 2024). LM and LAM biosynthesis begins with the formation of phosphatidyl-myo-inositol mannosides (PIMs), initiated by mannosyltransferases PimA (Rv2610c) and PimB (Rv2188c). PimA transfers mannose to phosphatidylinositol at the 2-position, while PimB transfers mannose to the 6-position (Guerin et al. 2009). PimA is essential for Mtb growth both in vitro and in vivo, with its depletion resulting in bacterial clearance in mouse models (Boldrin et al. 2014). Although PimB was initially implicated in LAM biosynthesis (Schaeffer et al. 1999), subsequent studies suggest gene redundancy or alternative pathways, as PimB disruption reduces LAM and LM levels without affecting higher-order PIMs (Torrelles and Schlesinger 2010). Notably, PimB mutants exhibit increased macrophage death during infection (Torrelles and Schlesinger 2010). PIMs serve as precursors to LM and LAM, which are transported to the bacterial surface where they play critical roles in modulating host-pathogen interactions. This complex interplay of proteins and glycolipids within the cell envelope underscores its indispensable role in Mtb biology and pathogenicity (Berg et al. 2007; Lea-Smith et al. 2007; Guerin et al. 2009; Mishra et al. 2011; Rainczuk et al. 2012; Boldrin et al. 2021; Correia-Neves et al. 2022). Surface-exposed LM and LAM contribute significantly to immune evasion by inhibiting pro-inflammatory cytokine production and promoting anti-inflammatory cytokines such as interleukin-10 (IL-10), thereby facilitating bacterial survival within host cells (Józefowski et al. 2008; Källenius et al. 2016). Structural alterations in LM and LAM affect cell wall integrity, acid-fast staining properties, antibiotic sensitivity, and virulence (Fukuda et al. 2013). Their immunomodulatory functions are mediated through interactions with immune receptors including Toll-like receptor 2 (TLR2), DC-SIGN, and Dectin-2 (Correia-Neves et al. 2022). Prolonged exposure to these glycolipids can dampen immune responses in myeloid cells and T lymphocytes, potentially benefiting the host by limiting active tuberculosis progression (Correia-Neves et al. 2022). The cell wall core, known as the mycolyl-arabinogalactan-peptidoglycan (mAGP) complex, consists of peptidoglycan (PG) covalently linked to arabinogalactan (AG), which is further esterified with mycolic acids. This complex provides structural integrity and shape to the bacterial cell, with PG conferring rigidity and AG serving as the scaffold for mycolic acid attachment (Draper 1982; Brennan 1994; Ortalo-Magné et al. 1996; Daffé and Draper 1997; Draper and Daffé 2004; Kremer et al. 2005; Payne et al. 2007; Alsteens et al. 2008; Koul et al. 2011; Alderwick et al. 2015; Velayati and Farnia 2016; Daffé and Marrakchi 2019). Advances in analytical techniques and chemical methodologies have elucidated intricate details of the mycobacterial cell wall and its lipid constituents, deepening our understanding of

Mtb's unique envelope architecture (Draper 1982; Brennan 1994; Ortalo-Magné et al. 1996; Daffé and Draper 1997; Draper and Daffé 2004; Kremer et al. 2005; Alsteens et al. 2008).

The peptidoglycan (PG) layer in Mtb is located external to the inner membrane and serves a critical function in maintaining the bacterial cell's rigidity, structural integrity, and shape. This essential component underpins bacterial viability and is a major target for numerous antibiotics (Paul and Beveridge 1994; Daffé and Draper 1997; Hoffmann et al. 2008; Maitra et al. 2019; Vilchèze 2020; Holzheimer et al. 2021; Gong et al. 2023a, 2023b). PG is a polysaccharide composed of alternating residues of N-acetylglucosamine (GlcNAc) and muramic acid (MurNAc), linked by β(1 → 4) glycosidic bonds. Notably, the muramic acid residues can be either N-acetylated or N-glycosylated, a modification unique to mycobacteria that influences cell wall properties (Lederer et al. 1975; Mahapatra et al. 2000; Mahapatra et al. 2005; Raymond et al. 2005; Maitra et al. 2019; Vilchèze 2020; Holzheimer et al. 2021; Gong et al. 2023a, b). These polysaccharide strands are acylated at the muramic acid residues by a pentapeptide stem composed of L-alanyl-D-isoglutaminyl-meso--diaminopimelyl-D-alanyl-D-alanine. The synthesis of this pentapeptide is catalyzed by a series of Mur ligases (MurC, MurD, MurE, and MurF), with the D-Ala:D-Ala ligase DdlA facilitating the formation of the terminal D-Ala-D-Ala dipeptide. Cross-linking of these pentapeptides forms the three-dimensional PG meshwork. In Mtb, two primary types of cross-links are observed: (i) 4 → 3 linkages formed by the D,D-transpeptidase activity of penicillin-binding proteins PonA1 and PonA2, which connect meso-diaminopimelic acid (meso-DAP) to D-alanine; and (ii) 3 → 3 linkages catalyzed by L,D-transpeptidases (LdtMt1 to LdtMt5), which link two meso-DAP residues (Lavollay et al. 2008; Cordillot et al. 2013; Maitra et al. 2019; Vilchèze 2020). The 3 → 3 cross-links provide localized tensile strength in regions where 4 → 3 cross-links alone are insufficient to maintain the wild-type cell morphology (Kuk et al. 2022; Maitra et al. 2019). PG biosynthesis is a complex, multistage process involving a coordinated network of specialized enzymes that mediate polymerization, cross-linking, and integration with other cell wall components. This process occurs in three main stages: cytoplasmic, membrane-associated, and extracellular. During the cytoplasmic stage, precursor molecules required for PG assembly are synthesized through enzyme-catalyzed reactions. ATP-dependent Mur ligases (Mur synthetases) catalyze the sequential addition of amino acids to the stem peptide, utilizing ATP hydrolysis for energy (Maitra et al. 2019; Vilchèze 2020). The Mur enzyme family—MurA, MurB, MurC, MurD, MurE, MurF, and MurG—plays essential roles in this phase. MurA (Rv2155c) catalyzes the transfer of an enolpyruvyl group from phosphoenolpyruvate (PEP) to UDP-GlcNAc, generating UDP-GlcNAc-enolpyruvate. MurB (Rv2156c) subsequently reduces this intermediate to UDP-MurNAc. MurC (Rv2157c) initiates the pentapeptide side chain by adding L-alanine to UDP-MurNAc. MurD (Rv2158c) incorporates D-glutamate, MurE (Rv2159c) adds meso-DAP, a residue critical for cross-linking, and MurF (Rv2160c) completes the pentapeptide by attaching D-alanine-D-alanine (Mahapatra et al. 2000; Barreteau et al. 2008; Basavannacharya et al. 2009; Typas et al. 2012; Munshi et al. 2013; Jankute et al. 2015; Kieser et al. 2015a, b; Raghavendra et al. 2018; Catalão et al. 2019; Oluwole and Olatunji 2022).

In the membrane-associated stage, MraY (Rv2154c) transfers the phospho-MurNAc-pentapeptide moiety to undecaprenyl phosphate (C55-P), producing lipid I, which marks the initiation of membrane-bound PG biosynthesis. MurG then catalyzes the addition of GlcNAc to lipid I, forming lipid II (MurNAc-GlcNAc-pentapeptide-PP-undecaprenol) (Mahapatra et al. 2000; Barreteau et al. 2008; Basavannacharya et al. 2009; Typas et al. 2012; Munshi et al. 2013; Jankute et al. 2015; Kieser et al. 2015a, b; Raghavendra et al. 2018; Saxena et al. 2018; Catalão et al. 2019; Gaur and Bera 2022; Oluwole and Olatunji 2022). Lipid II is then translocated across the cytoplasmic membrane to the cell wall by MurJ, a flippase responsible for transporting

PG precursors. However, the detailed mechanism of this translocation remains incompletely understood (Ruiz 2008). Once in the cell wall matrix, lipid II undergoes polymerization. Transglycosylases, including penicillin-binding proteins (PBPs) and shape, elongation, division, and sporulation (SEDS) proteins, catalyze the polymerization of MurNAc-GlcNAc subunits into glycan strands (Kumar et al. 2021). Transpeptidases, particularly high molecular weight PBPs, mediate the formation of 4 → 3 cross-links between meso-DAP and D-Ala residues of adjacent pentapeptide chains. In Mtb, remodeling of 4 → 3 cross-links into 3 → 3 cross-links is facilitated by the coordinated action of PG endopeptidases and L,D-transpeptidases (Maitra et al. 2019; Kumar et al. 2021).Overall, PG biosynthesis in Mtb is a highly orchestrated process involving multiple enzymes and stages, organized into complex networks with varying susceptibilities to antibiotics. This complexity underscores the importance of further research to elucidate PG assembly mechanisms fully and to develop strategies to combat drug-resistant Mtb strains.

The Mur enzymes, MurA, MurB, MurC, MurD, MurE, MurF, and MurG, are particularly critical during the cytoplasmic stage of PG biosynthesis and represent potential targets for novel antitubercular agents (Shinde et al. 2021; Moraes et al. 2015). MurA (Rv2155c; location 2,414,934 bp) initiates peptidoglycan (PG) biosynthesis by transferring an enolpyruvyl group from phosphoenolpyruvate (PEP) to UDP-N-acetylglucosamine (UDP-GlcNAc), forming UDP-GlcNAc-enolpyruvate. Subsequently, MurB (Rv0482; location 570,539 bp) reduces this intermediate to UDP-N-acetylmuramic acid (UDP-MurNAc) (Mahapatra et al. 2000; Barreteau et al. 2008; Basavannacharya et al. 2009; Typas et al. 2012; Munshi et al. 2013; Jankute et al. 2015; Kieser et al. 2015a, b; Raghavendra et al. 2018; Saxena et al. 2018; Catalão et al. 2019; Gaur and Bera 2022; Oluwole and Olatunji 2022). During the membrane-associated phase, MraY (Rv2154c; location 2,413,348 bp) catalyzes the transfer of the phospho-MurNAc-pentapeptide moiety to undecaprenyl phosphate (C55-P), generating lipid I, a critical step initiating membrane-bound PG synthesis (Bugg et al. 2006). MraY is a validated drug target, with inhibitors such as capuramycin analogs demonstrating potent activity against Mtb, including nonreplicating forms under hypoxic conditions (Pandey et al. 2022). Following this, extracellular enzymes including PonA1, PonA2, LdtB, and Lcp1 contribute to PG maturation and cell wall assembly (Kieser et al. 2015a, b; Filippova et al. 2016). PonA1 (Rv0050; location 53,663 bp) and PonA2 (Rv3682; location 4,121,916 bp) are bifunctional penicillin-binding proteins (PBPs) possessing both transglycosylase and D,D-transpeptidase activities. They polymerize glycan strands and catalyze the formation of classical 4 → 3 peptide cross-links between D-alanine and meso-diaminopimelic acid (mDAP) residues (Gupta et al. 2010; Kieser et al. 2015a, b; Pidgeon et al. 2019; Kado et al. 2023; Chiang and West 2024). LdtB (Rv2518c; location 2,834,109 bp) is responsible for generating the nonclassical 3 → 3 cross-links that predominate in mycobacterial PG, which are essential for maintaining the rod-shaped morphology and structural integrity of the cell wall (Kieser et al. 2015a, b; Pidgeon et al. 2019; Kado et al. 2023). LdtA (Rv0116c; location 140,267 bp) acts redundantly with LdtB, compensating for its loss and contributing to pathogenesis by reinforcing PG during host adaptation (Hett et al. 2008; Machowski et al. 2014; Schoonmaker et al. 2014; Angala et al. 2014).

Lcp1 has been identified as an essential phosphotransferase that covalently links arabinogalactan (AG) to the PG layer, a critical step for cell wall integrity (Harrison et al. 2016). Recent studies emphasize the transcriptional plasticity of Mtb, which allows dynamic regulation of PG biosynthesis genes in response to environmental stresses, facilitating survival under antibiotic pressure and hostile host conditions (Liu et al. 2025a, b). Structural and biochemical analyses reveal that PG synthesis enzymes exhibit distinct substrate specificities and catalytic mechanisms, enabling Mtb to maintain cell wall plasticity and adapt to diverse stressors (Hett et al. 2008;

Machowski et al. 2014; Schoonmaker et al. 2014; Angala et al. 2014; Küssau et al. 2020). Functional redundancies among L,D-transpeptidases and compensatory interactions with other hydrolases underscore their collective importance in bacterial viability. Chemically, arabinogalactan (AG) consists of three arabinan chains, each composed of approximately 27 D-arabinofuranosyl (Araf) residues attached to a homogalactan core of about 32 D-galactofuranosyl (Galf) residues. The galactan backbone is a linear polymer of alternating β-D-galactofuranosyl units linked by β(1 → 5) and β(1 → 6) glycosidic bonds, providing a flexible scaffold for arabinan attachment. The arabinan chains primarily contain linear α-D-arabinofuranosyl residues, with branching introduced by 3,5-linked α-D-arabinofuranosyl units further substituted at both positions by additional α-D-arabinofuranosyl residues. The nonreducing termini of these arabinan chains feature pentaarabinosyl units with 2-linked arabinofuranosyl residues that serve as key seroactive epitopes of AG (Draper 1982; Brennan 1994; Ortalo-Magné et al. 1996; Daffé and Draper 1997; Kremer et al. 2005; Alderwick et al. 2015; Brown et al. 2023).

Arabinogalactan (AG) biosynthesis in Mtb begins with the assembly of the galactan backbone, a process initiated by two key bifunctional galactofuranosyltransferases: GlfT1 (Rv3782; location 4,228,347 bp) and GlfT2 (Rv3808c; location 4,270,366 bp). GlfT1 catalyzes the initial step by transferring the first two galactofuranose (Galf) residues onto the lipid-linked acceptor decaprenyl-P-P-GlcNAc-Rha (GL2), forming Galf-β(1 → 4)-Rhap and Galf-β(1 → 5)-Galf glycosidic bonds. This priming step generates the substrate for GlfT2, which extends the galactan chain by adding alternating β(1 → 5) and β(1 → 6)-linked Galf units, producing a lipid-linked galactan polymer of approximately 30 Galf residues. GlfT2 contains a single bifunctional active site that catalyzes both linkage types. These enzymes are encoded within the AG biosynthetic gene cluster, reflecting their coordinated role in galactan assembly (Alderwick et al. 2006, 2018; Jankute et al. 2015; Poulin and Lowary 2016; Wu et al. 2017; Gong et al. 2023a, b). The arabinan domain is assembled by transferring arabinofuranose (Araf) residues to the galactan backbone, mediated by two critical proteins: arabinosyltransferase AftA (Rv3792; location 4,237,932 bp) and the GtrA family protein GtrA (Rv3789; location 4,235,374 bp). AftA, a membrane-bound GT-C superfamily glycosyltransferase, catalyzes the priming step by transferring the first arabinofuranose residue from decaprenyl-monophosphoryl-arabinose (DPA) to the galactan chain, initiating arabinan assembly. This step is essential for Mtb survival and growth and represents a promising drug target due to AftA's substrate specificity. GtrA is involved in the DPA biosynthesis pathway and facilitates recruitment of AftA to the membrane, effectively bridging arabinan assembly to the galactan backbone (Alderwick et al. 2006; Jankute et al. 2015; Wu et al. 2017; Gong et al. 2023a, b).The galactan backbone, composed of about 30 alternating β-(1 → 5) and β-(1 → 6)-linked Galf residues, supports up to two highly branched arabinan chains, each containing approximately 31 Araf residues. Subsequent elongation and branching of arabinan chains are mediated by specialized arabinofuranosyltransferases (AraTs), including AftB (Rv3805c; location 4,266,953 bp), AftC (Rv2673; location 2,989,291 bp), EmbA (Rv3794; location 4,243,233 bp), and EmbB (Rv3795; location 4,246,514 bp). AftB, a GT-C glycosyltransferase, uses DPA as a donor to add terminal β(1 → 2)-linked arabinofuranose residues to the nonreducing ends of arabinan chains in AG and lipoarabinomannan (LAM), forming the terminal hexa-arabinofuranoside motif ($Araf_6$) required for mycolic acid attachment. AftC introduces α(1 → 3)-linked branches into arabinan chains, increasing structural complexity and enabling proper mycolic acid attachment. Deletion of AftC reduces AG-bound mycolates and increases trehalose dimycolate (TDM) production; notably, AftC is absent in nonpathogenic mycobacteria, underscoring its role in pathogenicity (Jankute et al. 2017). EmbA and EmbB are membrane-bound GT-C enzymes containing conserved DDX motifs that elongate arabinan chains by synthesizing linear α(1 → 5)-linked arabinofuranose residues, contributing to terminal motifs necessary for mycolic acid attachment. These enzymes are inhibited by ethambutol, which disrupts arabinan elongation and the AG-mycolic acid linkage (Lamichhane et al. 2003; Berg et al. 2007; Birch et al. 2008, 2010; Alderwick et al. 2015).

Following arabinogalactan synthesis, its covalent linkage to peptidoglycan (PG) is facilitated by a phosphodiester bond involving a unique linker disaccharide, forming the mycolyl-arabinogalactan-peptidoglycan (mAGP) complex. This complex constitutes the structural core of the mycobacterial cell wall, providing rigidity, impermeability, and resistance to environmental stresses, and is essential for Mtb viability and pathogenicity (McNeil et al. 1990; Becq et al. 2007; Kaur et al. 2008; Jankute et al. 2012, 2015). Key proteins involved in this process include the integral membrane phosphotransferase Lcp1 (mtbc0_003475; location 3,670,028 bp), which links AG to PG through an α-L-rhamnose-(1 → 3)-α-D-GlcNAc-phosphodiester bond, ensuring structural cohesion vital for bacterial survival (Baumgart et al. 2016; Abrahams and Besra 2018). Additionally, mycolyltransferases attach long-chain mycolic acids to the arabinan chains of AG, completing the assembly of the mAGP complex. This attachment underlies the acid-fast staining properties and low permeability of the cell wall, contributing to intrinsic antibiotic resistance. The mAGP complex also supports the outer mycomembrane, a lipid-rich barrier protecting Mtb from hostile environments and immune defenses. Both Lcp1 and mycolyltransferases are indispensable for integrating AG into the cell wall structure, highlighting their importance in maintaining bacterial viability and pathogenicity (Baumgart et al. 2016; Abrahams and Besra 2018). Furthermore, the RmlA-RmlD enzyme cascade is essential for synthesizing dTDP-L-rhamnose, a critical precursor for the linker disaccharide that connects AG to PG in Mtb (Li et al. 2022; Guérin et al. 2022; Kampff et al. 2023; Aguilera-Correa et al. 2024).The synthesis of dTDP-L-rhamnose, a critical sugar for the mycobacterial cell wall, involves a sequential enzymatic pathway catalyzed by RmlA (Rv3464; glucose-1-phosphate thymidylyltransferase), RmlB (Rv3465; dTDP-D-glucose 4,6-dehydratase; location 3,882,834 bp), RmlC (Rv3465c; dTDP-6-deoxy-D-xylo-4-hexulose 3,5-epimerase; location 3,882,834 bp), and RmlD (Rv3266; dTDP-6-deoxy-L-lyxo-4-hexulose reductase; location 3,646,895 bp). The pathway begins with RmlA catalyzing the conversion of glucose-1-phosphate and dTTP into dTDP-glucose, a rate-limiting and regulatory step. RmlB then dehydrates dTDP-glucose to dTDP-4-keto-6-deoxy-D-glucose, which RmlC epimerizes to dTDP-6-deoxy-L-lyxo-4-hexulose. Finally, RmlD reduces this intermediate to produce dTDP-L-rhamnose. This sugar is incorporated into the unique linker disaccharide that bridges arabinogalactan (AG) and peptidoglycan (PG), ensuring structural integrity and contributing to Mtb virulence and resistance to host defenses (Giraud and Naismith 2000; Giraud et al. 2000; Li et al. 2006; Qu et al. 2007). Disruption of this pathway impairs cell wall integrity, leading to growth defects and reduced survival within host macrophages. Given that these enzymes are absent in humans, the RmlA-RmlD pathway represents a promising target for novel antituberculosis therapies (Kantardjieff et al. 2004; Harathi et al. 2016; Brown et al. 2024; Qu et al. 2021).

Several proteins play indirect but crucial roles in AG biosynthesis by supplying essential precursors or modifying related structures. Central to this is the production of decaprenylphosphoryl-D-arabinose (DPA), a vital precursor for the mycobacterial cell wall. Key enzymes such as DprE1, UbiA, and PrsA participate in this pathway, and their absence leads to lethal effects and altered cell morphology (Kolly et al. 2014). The pathway initiates with the conversion of sedoheptulose 7-phosphate into D-ribose 5-phosphate, catalyzed by the transketolase Rv1449c, providing the ribose phosphate backbone for subsequent steps (Wolucka 2008). The putative phospholipid phosphatase Rv3807c is believed to dephosphorylate decaprenylphosphoryl-β-D-5-phosphoribose to yield decaprenylphosphoryl-ribose (DPR), an intermediate in DPA biosynthesis (Cai et al. 2014). The final critical step is the epimerization of DPR to DPA by the DprE1/DprE2 enzyme complex, essential for producing the activated arabinose donor required exclusively for AG assembly. This step is indispensable for bacterial viability and represents a promising drug target (Kolly et al. 2014). Additionally, the transmembrane protein Rv3789, although not directly involved in DPA synthesis, plays a pivotal role by recruiting AftA, the first arabinosyltransferase initiating arabinan chain formation on the galactan backbone. Loss of Rv3789 results in reduced arabinan content and accumulation of DPA, highlighting its regulatory function in arabinan biosynthesis rather than DPA production itself (Kolly et al. 2015). Another key enzyme, Ppm1 (polyprenol monophosphomannose synthase; Rv2051c; location 2,308,131 bp), primarily functions in lipoarabinomannan (LAM) biosynthesis but also contributes indirectly to early arabinan synthesis stages. Ppm1 generates mannosylated phosphatidylinositol mannoside (PIM) scaffolds, essential substrates for adding arabinofuranose residues during LAM assembly, facilitating early biosynthetic steps of arabinan-containing molecules (Berg et al. 2007; Kaur et al. 2008; Mishra et al. 2011; Rana et al. 2024; Gurcha et al. 2002).

GtrA (Rv3789; location 4,235,374 bp) plays a vital role in arabinan biosynthesis in Mtb. Rv3789, a short transmembrane protein of the GtrA family, is cotranscribed with the essential gene dprE1, which encodes a key enzyme in the decaprenyl-phospho-arabinose (DPA) biosynthetic pathway. Rv3789 directly interacts with AftA, the priming arabinosyltransferase responsible for adding the initial arabinofuranosyl residue to the galactan backbone of arabinogalactan (AG), thereby playing a crucial role in arabinogalactan biosynthesis (Kolly et al. 2015; Alderwick et al. 2006). This priming step catalyzed by AftA is indispensable for mycobacterial viability. Specifically, AftA transfers a single arabinose unit to the C-5 position of a precisely positioned internal 6-linked galactofuranosyl residue within the galactan chain (Gong et al. 2023a, b). Unlike the Emb proteins, which also participate in arabinan synthesis, AftA is not inhibited by the frontline antitubercular drug ethambutol (Alderwick et al. 2006). Deletion of Rv3789 results in reduced arabinan content and accumulation of decaprenylphosphoryl-arabinose (DPA), confirming its critical function in arabinan biosynthesis (Kolly et al. 2015; Minato et al. 2019). Structural and interaction studies indicate that Rv3789 does not function as a decaprenylphosphoryl-arabinose (DPA) flippase; instead, it acts as an anchoring protein that recruits the arabinosyltransferase AftA to the site of arabinogalactan (AG) assembly (Kolly et al. 2015). Together, these proteins form a complex and coordinated network essential for sustaining AG biosynthesis, which is critical for maintaining the structural integrity and functionality of the mycobacterial cell wall.

Electron microscopy and atomic force microscopy investigations have revealed that the Mtb cell wall is composed of three covalently linked layers (Watanabe et al. 2001; Hoffmann et al. 2008). These layers include (1) the basal peptidoglycan (PG) layer; (2) an intermediate electron-transparent layer (ETL) consisting of the arabinogalactan-mycolate complex; and (3) an outer dense layer characterized by heavy metal deposits that surround the lipid-rich ETL. Studies by Velayati and Farnia have demonstrated variations in cell wall thickness among clinical isolates. Drug-susceptible strains exhibit an average cell wall thickness of 15.6 ± 1.3 nm, whereas multidrug-resistant (MDR) and extensively drug-resistant (XDR) strains show significantly increased thicknesses of 17.1 ± 1.03 nm and 20.2 ± 1.5 nm, respectively (Velayati et al. 2009; Velayati and Farnia 2016).

Mycolic acids are essential long-chain fatty acids covalently attached to arabinogalactan within the Mtb cell wall, forming a critical component of the mycolyl-arabinogalactan-peptidoglycan (mAGP) complex (Draper 1982; Brennan 1994; Daffé and Draper 1997; Kremer et al. 2005; Daffé and Marrakchi 2019). These molecules are fundamental to maintaining the structural integrity, low permeability, and pathogenicity of the Mtb cell envelope. Constituting up to 60% of the cell envelope's dry weight, mycolic acids are classified into three major types: alpha-, methoxy-, and keto-mycolic acids, each imparting distinct physicochemical and immunological properties (Daffé 2008; Marrakchi et al. 2014;

Ndlandla et al. 2016). The structural diversity of mycolic acids produces unique chromatographic fingerprints that can differentiate *Mycobacterium* species (Verschoor et al. 2012; Arora et al. 2018). While the majority of mycolic acids are esterified to cell wall components, some free mycolic acids are secreted during in vitro growth (Wang et al. 2023). Different classes of mycolic acids vary in their capacity to induce immune responses and modulate host lipid metabolism, thereby influencing the progression of tuberculosis (Wang et al. 2023). Notably, antibodies targeting mycolic acids show promise as diagnostic serum biomarkers for active tuberculosis, with combinations of methoxy- and alpha-mycolic acids demonstrating higher antigenicity compared to keto- and alpha-mycolic acid combinations (Ndlandla et al. 2016).

The biosynthesis of mycolic acids in Mtb is a complex, multistep process involving four main classes of enzymes: fatty acid synthases, modifying enzymes, condensation/reduction enzymes, and transport/integration proteins. Fatty acid synthases include two systems: FAS-I, a multifunctional polypeptide enzyme responsible for de novo synthesis of C16:0 and C26:0 fatty acids (Zimhony et al. 2000; Schweizer and Hofmann 2004; Takayama et al. 2005), and FAS-II, which consists of monofunctional enzymes and acyl carrier proteins that elongate FAS-I products to form long-chain mycolic acid precursors (Schaeffer et al. 2001; Singh et al. 2011). Key components of FAS-II include β-ketoacyl-ACP synthases KasA and KasB, which exhibit high specificity for long-chain acyl-ACPs and are inhibited by compounds such as cerulenin and thiolactomycin (Schaeffer et al. 2001; Sacco et al. 2007; Bhatt et al. 2007). The FAS-II system is organized into specialized, interconnected complexes, with dehydratases HadAB and HadBC and methyltransferases playing crucial roles in determining substrate specificity and enzymatic activity (Cantaloube et al. 2011; Singh et al. 2011).

Modification enzymes introduce specific functional groups that define the structural subtypes of mycolic acids. Cyclopropane synthases, such as cyclopropane mycolic acid synthase A2 (CmaA2), catalyze the addition of trans-cyclopropane rings to oxygenated mycolates, while prokaryotic cyclopropane synthase A (PcaA) and mycolic acid methyltransferase A2 (MmaA2) catalyze cis-cyclopropanation of both alpha- and oxygenated mycolates (Banerjee et al. 2011). Notably, CmaA2 exhibits dual functionality, acting on oxygenated mycolates as both a cis- and trans-cyclopropane synthase, specifically mediating trans-cyclopropanation of methoxy- and keto-mycolates (Glickman et al. 2000; Barkan et al. 2009). MmaA2 is essential for distal cyclopropanation of α-mycolates and also contributes to cis-cyclopropanation of methoxymycolates (Glickman 2003). Functional redundancy between CmaA2 and MmaA2 exists in catalyzing cis-cyclopropanation of oxygenated mycolates (Barkan et al. 2009). The stereochemistry of these cyclopropane modifications profoundly influences Mtb virulence and host inflammatory responses; for example, trans-cyclopropanation of trehalose dimycolate (TDM) by CmaA2 suppresses inflammation and reduces virulence, as evidenced by hypervirulence in cmaA2 null mutants and increased macrophage-stimulating activity of TDM lacking trans-cyclopropane rings (Rao et al. 2006). Methyltransferases such as MmaA4 (also known as Hma) catalyze the introduction of a methyl branch and adjacent hydroxyl group, precursors for keto- and methoxy-mycolates (Boissier et al. 2006). Subsequently, MmaA3 O-methylates the secondary alcohol generated by MmaA4 to form methoxy-mycolates (Yuan and Barry 1996). Beyond their biosynthetic roles, these enzymes impact drug susceptibility; for instance, MmaA4 is required for sensitivity to the antitubercular drug thiacetazone, with its absence conferring resistance (Alahari et al. 2009).

Desaturases, specifically fatty acid desaturase A1 (DesA1) and fatty acid desaturase A2 (DesA2), introduce double bonds into fatty acid chains and function in coordination with hydroxyacyl-acyl carrier protein dehydratase subunits HadAB and HadBC. Their expression is tightly regulated by the mycolic acid desaturase regulator (MadR), a transcriptional repressor that balances saturated and unsaturated mycolate levels to maintain cell wall integrity (Hashimoto et al. 2006; Jamet et al. 2015; Cooper et al. 2022). Condensation and reduction enzymes catalyze the final assembly of mycolic acids. Polyketide synthase 13 (Pks13) performs the critical condensation step, joining the meromycolate chain and α-alkyl branch (Portevin et al. 2004; Marrakchi et al. 2014). Cyclopropane mycolic acid reductase A (CmrA) reduces the β-keto group formed during condensation, stabilizing the mycolic acid molecule (Lea-Smith et al. 2007). Transport and integration proteins such as MmpL3 mediate the translocation of trehalose monomycolate (TMM) across the plasma membrane. Inhibition of MmpL3 disrupts this transport, causing toxic intracellular accumulation of TMM and bacterial death (Melly and Purdy 2019). The Antigen 85 Complex (FbpA, FbpB, FbpC) functions as mycolyltransferases, transferring mycolic acids from TMM to arabinogalactan, or converting TMM into trehalose dimycolate (TDM), a key virulence factor (Takayama et al. 2005; Grzegorzewicz et al. 2012).

Polyketide-associated protein A5 (PapA5) is an acyltransferase involved in the biosynthesis of virulence-associated lipids, specifically phthiocerol dimycocerosate esters (PDIMs) and phenolic glycolipids (PGLs) (Buglino et al. 2004; Onwueme et al. 2005; Chavadi et al. 2012). These lipids significantly influence bacterial virulence and contribute to cell wall structural integrity (Forrellad et al. 2013; Richards et al. 2019). The crystal structure of PapA5 reveals a two-domain architecture with conserved histidine and aspartic acid residues critical for its enzymatic activity (Buglino et al. 2004). PapA5 catalyzes the diesterification of phthiocerol and phthiodiolone with mycocerosate in PDIM synthesis, demonstrating chain length specificity and dual esterification activity in the final steps of dimycocerosate lipid biosynthesis (Onwueme et al. 2005; Touchette et al. 2015). Furthermore, PapA5 is essential for PGL production, utilizing diol-containing polyketide compounds as acyl-acceptor substrates (Chavadi et al. 2012). PapA5, an acyltransferase involved in the biosynthesis of virulence-associated lipids such as phthiocerol dimycocerosate esters (PDIMs) and phenolic glycolipids (PGLs), can be phosphorylated by *Mycobacterium tuberculosis* Ser/Thr kinases PknB and PknE, suggesting a potential regulatory mechanism for DIM biosynthesis (Touchette et al. 2015). Regulatory proteins play a critical role in tightly controlling mycolic acid synthesis, enabling Mtb to adapt to varying growth conditions and environmental signals. Two key transcriptional regulators, mycolic acid biosynthesis regulator (MabR) and fatty acid degradation regulator (FadR), modulate the fatty acid synthase II (FAS-II) system, with MabR acting as a transcriptional activator. MabR's DNA-binding affinity is influenced by long-chain acyl-CoA molecules, allowing fine-tuned gene expression regulation in response to cellular lipid levels (Tsai et al. 2017). Additionally, serine/threonine protein kinases phosphorylate hydroxyacyl-acyl carrier protein dehydratase subunits HadAB and HadBC, negatively regulating their enzymatic activity in a growth phase-dependent manner (Slama et al. 2011). Alternative sigma factors and regulatory proteins such as RelA/SpoT Homolog (Rel(Mtb)) and cyclic AMP receptor protein (CRP) activate genes involved in complex lipid biosynthesis, facilitating adaptation to nutrient deprivation (Dover et al. 2007). Furthermore, two-component regulatory systems and eukaryotic-like serine/threonine protein kinases coordinate essential processes including cell division, morphogenesis, and regulation of bioactive lipid profiles (Dover et al. 2007). Collectively, these sophisticated regulatory networks enable Mtb to adjust its cell wall composition dynamically, optimizing survival and pathogenicity under diverse environmental conditions (PaweŁczyk and Kremer 2014). Given the critical importance of mycolic acid biosynthesis for Mtb survival, enzymes such as InhA, Polyketide Synthase 13 (Pks13), mycobacterial membrane protein large 3 (MmpL3), and acetyl-CoA carboxylase subunit D6 (AccD6) represent promising targets for antituberculosis drug development (Lavania et al. 2022). Inhibiting these proteins disrupts cell wall integrity, leading to bacterial death and offering valuable avenues for novel therapeutic strategies, particularly against multidrug-resistant (MDR) and extensively drug-resistant (XDR) strains.

The plasma membrane (cytoplasmic membrane) of Mtb is a vital component of the cell envelope, providing structural integrity, regulating solute transport, and facilitating essential metabolic processes. Approximately 7 nm thick, it exhibits a characteristic triple-layered profile of biological membranes (Brodie et al. 1979). The membrane is asymmetric, with the outer layer (facing the cell wall) thicker and denser than the inner layer (facing the cytoplasm). The outer layer primarily consists of mannose-containing phospholipids such as phosphatidylinositol mannosides (PIMs), approximately 3 nm thick and reactive to cytochemical saccharide stains. The inner layer measures about 2 nm in thickness. The membrane also contains lipopolysaccharides common to Actinomycetales (Yagi et al. 2005). It provides osmotic protection, regulates solute traffic between the cytoplasm and the environment, and serves as the platform for synthesizing key cell wall components, including peptidoglycan, arabinogalactan, and mycolic acids. Proteomic analyses of the Mtb $H_{37}Rv$ strain have identified numerous membrane-associated and integral membrane proteins involved in cell envelope functions and energy metabolism (Gu et al. 2003; Xiong et al. 2005). Detergent extraction methods, particularly with Triton X-114, have proven effective in detecting putative membrane proteins (Sinha et al. 2002). Recent advances in cryo-electron microscopy have enabled structural elucidation of vital Mtb membrane proteins, providing mechanistic insights and facilitating structure-based drug design and vaccine discovery (Bendre et al. 2021). Among proteins critical for plasma membrane function is PimA (Rv2610c; location 2,937,865 bp), an essential mannosyltransferase that catalyzes the transfer of a mannose residue from GDP-mannose to phosphatidylinositol at the 2-position of the myo-inositol ring, initiating phosphatidylinositol mannoside (PIM) biosynthesis (Korduláková et al. 2002; Guerin et al. 2009). PimB' subsequently adds a second mannose at the 6-position, followed by acylation (Guerin et al. 2009). PimA is an amphitropic enzyme that binds membranes via its N-terminal domain, which is necessary for its activation (Guerin et al. 2007). The essentiality of PimA for mycobacterial growth has been demonstrated both in vitro and in vivo, making it a promising drug target (Boldrin et al. 2014; Korduláková et al. 2002). Inhibition of PimA reduces PIM levels and leads to bacterial clearance in mouse infection models (Boldrin et al. 2014). Further downstream, Ppm1 (Rv2051c; location 2,308,131 bp) synthesizes polyprenol phosphate mannose (PPM), a critical mannose donor for lipomannan and lipoarabinomannan production (Gurcha et al. 2002). Ppm1 is essential for mycobacterial survival and lipoglycan synthesis, as demonstrated in *Mycobacterium smegmatis* (Rana et al. 2024). Notably, Mtb Ppm1 exhibits a unique two-domain structure, whereas in *Mycobacterium smegmatis*, these domains are encoded by two separate genes, Msppm1 and Msppm2. Msppm2 interacts with Msppm1, stabilizing it within the membrane, analogous to the interaction between mammalian Dpm1 and Dpm3 proteins (Baulard et al. 2003). Lipomannan and lipoarabinomannan are vital glycolipids maintaining cell wall integrity and modulating host-pathogen interactions and immune evasion (Korduláková et al. 2002; Morita et al. 2005; Umesiri et al. 2010; Rana et al. 2024; Boldrin et al. 2014; Batt et al. 2020). Transport proteins are indispensable for Mtb's structural integrity by facilitating lipid trafficking and cell wall biosynthesis. Among these, MmpL3 (Rv0206c; location 244,484 bp), a member of the resistance-nodulation-cell division (RND) permease superfamily, functions as a homotrimer with 12 transmembrane segments and two large periplasmic domains, structurally resembling Gram-negative efflux pumps (Sandhu and Akhter 2015; Belardinelli et al. 2022).MmpL3 is a critical transporter in Mtb responsible for shuttling trehalose monomycolate (TMM), a key precursor for essential cell wall components, including trehalose dimycolate (TDM) and mycolic acids bound to arabinogalactan (Belardinelli et al. 2022; Shao et al. 2020). Additionally, MmpL3 participates in heme uptake, with the secreted protein Rv0203 transferring heme to MmpL3's extracellular domain (Owens et al. 2013). Inhibition of MmpL3 leads to intracellular accumulation of TMM precursors, disrupting cell wall synthesis and impairing bacterial survival (Belardinelli et al. 2022; Shao et al. 2020). This transporter is essential for Mtb replication both in vitro and during infection of human macrophages, making it a promising drug target. Inhibitors, including bedaquiline-like compounds, act either by directly blocking MmpL3's transport function or by disrupting the proton motive force necessary for its activity (Chalut 2016; Degiacomi et al. 2017; Melly and Purdy 2019; Su et al. 2021). The MmpL family in Mtb comprises 13 proteins that mediate lipid trafficking across the plasma membrane and contribute to cell envelope biogenesis and virulence. Functionally redundant complexes formed by MmpL4/MmpS4 and MmpL5/MmpS5 are essential for siderophore export, with their inactivation increasing susceptibility to certain antibiotics (Wells et al. 2013; Maharjan et al. 2024). MmpS5 facilitates MmpL5 assembly into a homotrimer on the inner membrane, enhancing its drug efflux capability (Yamamoto et al. 2021). Other MmpL proteins include MmpL7 (Rv0194), which transports virulence-associated lipids like phthiocerol dimycocerosates (PDIM), and MmpL10 (Rv1183), which facilitates trehalose-containing glycolipid transport (Jain and Cox 2005; Moolla et al. 2021; Biegas et al. 2021; Luna-Herrera et al. 2025).

The plasma membrane of Mtb also harbors key proteins involved in energy production via the electron transport chain (ETC) and ATP synthesis. Complex I NADH dehydrogenases (NuoA-NuoN; Rv3145-Rv3157) oxidize NADH to NAD^+ while transferring electrons to menaquinone, generating the proton motive force (PMF) (Pohl et al. 2007; Sharma et al. 2009; Cook et al. 2014; Iqbal et al. 2018; Foo et al. 2020; Bajeli et al. 2020). The non-proton-pumping NADH dehydrogenase NDH-2 (Rv1854c) also transfers electrons to menaquinone but does not contribute to PMF. Complex II (SdhA-SdhD; Rv0247c-Rv0250c) functions in the tricarboxylic acid cycle by oxidizing succinate to fumarate and transferring electrons to menaquinone (Bandara et al. 2021; Xu et al. 2023; Esteban-Amo et al. 2024). ATP synthesis is mediated by the F_1F_0-ATP synthase complex, which harnesses the PMF generated by the ETC. The F_0 membrane-embedded portion includes subunits a (AtpB), b, b′, and c (AtpE), while the cytoplasmic F_1 catalytic domain contains α (AtpA), β, γ, δ, and ε subunits (Zhang et al. 2020; De Jonge et al. 2007). Proton flow through the F_0 c-ring drives conformational changes in F_1 that catalyze ATP synthesis from ADP and inorganic phosphate. The diarylquinoline drug bedaquiline (R207910) targets this complex by binding subunits c and a, disrupting proton translocation and ATP synthesis, leading to bacterial death. Mutations in atpE, encoding subunit c, confer bedaquiline resistance (De Jonge et al. 2007; Petrella et al. 2006). Unique features of Mtb ATP synthase, such as the fused bδ-subunit and specialized C-terminal domains of α and ε subunits, contribute to its regulation and efficiency. Notably, in slow-growing mycobacteria like Mtb, ATP synthase is active in ATP synthesis but naturally inhibited from ATP hydrolysis, preventing wasteful energy consumption and maintaining energy homeostasis under hypoxic or nonreplicating conditions (Bogdanović et al. 2018; Pohl et al. 2007; Haagsma et al. 2010; Rao et al. 2008). This makes ATP synthase indispensable for both replicating and dormant Mtb, underscoring its importance as a drug target (Kamariah et al. 2019; Cook et al. 2014). Additional plasma membrane-associated systems include arabinosyltransferases critical for cell wall biosynthesis, notably the Emb proteins, EmbA, EmbB, and EmbC. These membrane-bound enzymes catalyze polymerization of arabinose residues, essential for forming the arabinan domains of arabinogalactan (AG) and lipoarabinomannan (LAM) (Amin et al. 2008; Zhang et al. 2020). EmbA and EmbB primarily cooperate to synthesize the terminal hexaarabinoside motif in AG, crucial for cell wall integrity and mycolylation. Both possess distinct N-terminal substrate recognition domains and C-terminal catalytic domains for arabinose transfer (Escuyer et al. 2001; Zhang et al. 2020). Disruption of embA or embB in M.

smegmatis impairs AG synthesis and increases cell wall permeability, while embA is essential for Mtb viability (Amin et al. 2008; Escuyer et al. 2001). EmbC specializes in arabinan biosynthesis for LAM, containing multiple transmembrane helices and an extracytoplasmic domain extending arabinan chains during LAM assembly. Unlike M. smegmatis, where embC mutants lack LAM but remain viable, embC is essential in Mtb, highlighting LAM's critical role in pathogenicity (Amin et al. 2008; Escuyer et al. 2001).

AftB catalyzes the terminal step of arabinan biosynthesis in the Mtb cell wall and is insensitive to ethambutol, a frontline antitubercular drug that inhibits Emb-mediated arabinan polymerization (Seidel et al. 2007). Collectively, these plasma membrane-associated arabinosyltransferases orchestrate the biosynthesis of arabinan-containing cell wall components, vital for maintaining cell wall structure, permeability, and modulating host immune interactions.

Enzymes involved in lipid biosynthesis, including the fatty acid synthase systems FAS-I and FAS-II, as well as Acetyl-CoA Carboxylase Subunit D6 (AccD6), are closely associated with the plasma membrane of *Mycobacterium tuberculosis* (Mtb). The plasma membrane serves as a specialized platform where these enzymes localize and function, facilitating the synthesis of lipid precursors essential for mycolic acid production. Studies have demonstrated that lipid biosynthetic enzymes are compartmentalized within distinct membrane subdomains, enabling efficient coordination of lipid assembly and modification processes. This strategic membrane association supports substrate channeling and the integration of newly synthesized lipids into the cell envelope, which is critical for maintaining the structural integrity and impermeability of the mycobacterial cell wall. Thus, the plasma membrane acts not only as a physical barrier but also as an active site orchestrating the biosynthesis of complex lipids vital to Mtb's survival and pathogenicity (Alderwick et al. 2015; Abrahams et al. 2016; Shaku et al. 2020; Puffal et al. 2022; Chiang et al. 2024).

Type VII secretion systems (T7SS) are specialized protein transport complexes embedded in the plasma membrane of pathogenic mycobacteria, such as Mtb, and play a critical role in virulence by exporting key effector proteins. The ESX-5 T7SS membrane complex is a large assembly of approximately 1500 kDa, composed of core components EccB5, EccC5, EccD5, and EccE5, which together form a stable secretion apparatus (Houben et al. 2012). Another important system, the ESX-1 T7SS, is particularly vital for Mtb pathogenesis and is believed to contribute to the formation of a translocon that spans the mycomembrane, facilitating protein secretion across the complex cell envelope (Das et al. 2011). In M. smegmatis, the ESX-3 T7SS core complex consists of EccB3, EccC3, EccD3, and EccE3 subunits arranged in a 1:1:2:1 stoichiometry (Famelis et al. 2019). The EccC3 protein acts as a coupling ATPase with four ATPase domains connected to the membrane via a stalk domain. Adjacent to this stalk is a domain of unknown function (DUF) identified as an essential ATPase domain for secretion activity. Structural studies suggest that conformational changes in the stalk domain, triggered by substrate binding and ATP hydrolysis, drive the translocation of substrates through the secretion channel to the periplasm (Famelis et al. 2019).

Overall, the T7SS machinery forms a highly complex, multi-protein membrane-embedded system that enables Mtb to secrete virulence factors and other proteins essential for infection and survival within the host. Its unique architecture and mechanism distinguish it from other bacterial secretion systems and make it a promising target for novel antimicrobial therapies.

In conclusion, the cell wall of Mtb is a complex, multilayered structure essential for bacterial survival, immune evasion, and pathogenicity. It consists of the mycomembrane, a hydrophobic barrier rich in mycolic acids and glycolipids, the peptidoglycan-arabinogalactan complex that provides mechanical strength, and the plasma membrane, which supports vital metabolic functions. Key proteins such as MmpL3 and the Antigen 85 Complex are indispensable for transporting and integrating mycolic acids into the cell wall. MmpL3 functions as a critical transporter of trehalose monomycolate. At the same time, the Antigen 85 Complex catalyzes the transfer of mycolic acids to arabinogalactan and converts trehalose monomycolate into trehalose dimycolate, a major virulence factor. The plasma membrane plays a central role in lipid biosynthesis and energy production, housing enzymes like PimA and PimB′ that participate in phosphatidylinositol mannoside synthesis, as well as respiratory complexes responsible for oxidative phosphorylation. These processes are fundamental for maintaining cellular homeostasis and supporting the growth and survival of Mtb. Targeting the essential proteins and pathways involved in cell wall biosynthesis and maintenance offers promising therapeutic avenues for tuberculosis treatment (Chauhan et al. 2024). Inhibitors directed against key enzymes such as MmpL3 and DprE, which play critical roles in the assembly and integrity of the mycobacterial cell wall, can effectively impair bacterial replication and viability. This approach highlights the potential of developing novel antituberculosis drugs that exploit specific vulnerabilities within the mycobacterial cell envelope. Such targeted therapies hold significant promise for improving treatment outcomes, particularly in combating multidrug-resistant and extensively drug-resistant strains of tuberculosis.

Conflict of Interest The authors declare that they have no conflict of interest related to the content of this chapter.

Ethical Consideration This chapter is a review based entirely on data and information obtained from previously published and publicly available sources, all of which have been appropriately cited. The authors conducted no new experimental work involving human participants or animals. Therefore, no additional ethical approval was required. The authors have ensured proper attribution and compliance with ethical standards in the use of all referenced materials.

References

Abrahams KA, Besra GS (2018) Mycobacterial cell wall biosynthesis: a multifaceted antibiotic target. Parasitology 145(2):116–133. https://doi.org/10.1017/S0031182016002377

Abrahams KA, Chung C-w, Ghidelli-Disse S, Rullas J, Rebollo-López MJ, Gurcha SS et al (2016) Identification of KasA as the cellular target of an anti-tubercular scaffold. Nat Commun 7(1):12581

Aguilera-Correa JJ, Wei F, Leclercq LD, Tasrini Y, Mullapudi E, Daher W et al (2024) A dTDP-L-rhamnose 4-epimerase required for glycopeptidolipid biosynthesis in Mycobacterium abscessus. J Biol Chem 300(11):107852. https://doi.org/10.1016/j.jbc.2024.107852

Alahari A, Alibaud L, Trivelli X, Gupta R, Lamichhane G, Reynolds RC et al (2009) Mycolic acid methyltransferase, MmaA4, is necessary for thiacetazone susceptibility in Mycobacterium tuberculosis. Mol Microbiol 71(6):1263–1277. https://doi.org/10.1111/j.1365-2958.2009.06604.x

Alderwick LJ, Seidel M, Sahm H, Besra GS, Eggeling L (2006) Identification of a novel arabinofuranosyltransferase (AftA) involved in cell wall arabinan biosynthesis in Mycobacterium tuberculosis. J Biol Chem 281(23):15653–15661. https://doi.org/10.1074/jbc.M600045200

Alderwick LJ, Harrison J, Lloyd GS, Birch HL (2015) The mycobacterial Cell Wall—peptidoglycan and arabinogalactan. Cold Spring Harb Perspect Med 5(8):a021113. https://doi.org/10.1101/cshperspect.a021113

Alderwick LJ, Birch HL, Krumbach K, Bott M, Eggeling L, Besra GS (2018) AftD functions as an α1 → 5 Arabinofuranosyltransferase involved in the biosynthesis of the mycobacterial Cell Wall Core. The Cell Surface 1:2–14. https://doi.org/10.1016/j.tcsw.2017.10.001

Alsayed SSR, Gunosewoyo H (2023) Tuberculosis: pathogenesis, current treatment regimens and new drug targets. Int J Mol Sci 24(6):5202. https://doi.org/10.3390/ijms24065202

Alsteens D, Verbelen C, Dague E, Raze D, Baulard AR, Dufrêne YF (2008) Organization of the mycobacterial cell wall: a nanoscale view. Pflugers Arch - Eur J Physiol 456(1):117–125. https://doi.org/10.1007/s00424-007-0386-0

Amin AG, Goude R, Shi L, Zhang J, Chatterjee D, Parish T (2008) EmbA is an essential arabinosyltransferase in mycobacterium tuberculosis. Microbiology 154(1):240–248

Angala SK, Belardinelli JM, Huc-Claustre E, Wheat WH, Jackson M (2014) The cell envelope glycoconjugates of mycobacterium tuberculosis. Crit Rev Biochem Mol Biol 49(5):361–399

Arora A, Kaul B, Singh A (2018) Multiple peroneal nerve abscesses: the first presentation of borderline Tuberculoid leprosy. Biomedical and Biotechnology Research Journal (BBRJ) 2(2):159–160

Backus KM, Dolan MA, Barry CS, Joe M, McPhie P, Boshoff HIM et al (2014) The three Mycobacterium tuberculosis antigen 85 isoforms have unique substrates and activities determined by non-active site regions. J Biol Chem 289(36):25041–25053. https://doi.org/10.1074/jbc.M114.584581

Bailo R, Bhatt A, Ainsa JA (2015) Lipid transport in Mycobacterium tuberculosis and its implications in virulence and drug development. Biochem Pharmacol 96(3):159–167. https://doi.org/10.1016/j.bcp.2015.05.001

Bajeli S, Baid N, Kaur M, Pawar GP, Chaudhari VD, Kumar A (2020) Terminal respiratory oxidases: a targetables vulnerability of mycobacterial bioenergetics? Front Cell Infect Microbiol 10:589318

Bandara RA, Chen ZR, Hu J (2021) Potential of helper-dependent adenoviral vectors in CRISPR-cas9-mediated lung gene therapy. Cell Biosci 11(1):145

Banerjee R, Vats P, Dahale S, Kasibhatla SM, Joshi R (2011) Comparative genomics of cell envelope components in mycobacteria. PLoS One 6(5):e19280. https://doi.org/10.1371/journal.pone.0019280

Bansal-Mutalik R, Nikaido H (2014) Mycobacterial outer membrane is a lipid bilayer and the inner membrane is unusually rich in diacyl phosphatidylinositol dimannosides. Proc Natl Acad Sci 111(13):4958–4963. https://doi.org/10.1073/pnas.1403078111

Barkan D, Liu Z, Sacchettini JC, Glickman MS (2009) Mycolic acid cyclopropanation is essential for viability, drug resistance, and cell wall integrity of Mycobacterium tuberculosis. Chem Biol 16(5):499–509. https://doi.org/10.1016/j.chembiol.2009.04.001

Barreteau H, Kovac A, Boniface A, Sova M, Gobec S, Blanot D (2008) Cytoplasmic steps of peptidoglycan biosynthesis. FEMS Microbiol Rev 32(2):168–207. https://doi.org/10.1111/j.1574-6976.2008.00104.x

Basavannacharya C, Robertson G, Munshi T, Keep NH, Bhakta S (2009) ATP-dependent MurE ligase in Mycobacterium tuberculosis: biochemical and structural characterisation. Tuberculosis 90(1):16–24. https://doi.org/10.1016/j.tube.2009.10.007

Batt SM, Minnikin DE, Besra GS (2020) The thick waxy coat of mycobacteria, a protective layer against antibiotics and the host's immune system. Biochem J 477(10):1983–2006. https://doi.org/10.1042/BCJ20200194

Baulard AR, Gurcha SS, Engohang-Ndong J, Gouffi K, Locht C, Besra GS (2003) In vivo interaction between the polyprenol phosphate mannose synthase Ppm1 and the integral membrane protein Ppm2 from mycobacterium smegmatis revealed by a bacterial two-hybrid system. J Biol Chem 278(4):2242–2248

Baumgart M, Schubert K, Bramkamp M, Frunzke J (2016) Impact of LytR-CpsA-Psr proteins on cell wall biosynthesis in Corynebacterium glutamicum. J Bacteriol 198(22):3045–3059. https://doi.org/10.1128/JB.00406-16

Becq J, Gutierrez MC, Rosas-Magallanes V, Rauzier J, Gicquel B, Neyrolles O, Deschavanne P (2007) Contribution of horizontally acquired genomic islands to the evolution of the tubercle bacilli. Mol Biol Evol 24(8):1861–1871. https://doi.org/10.1093/molbev/msm111

Belardinelli JM et al (2022) Therapeutic efficacy of antimalarial drugs targeting DosRS signaling in Mycobacterium abscessus. Sci Transl Med 14(651):eabj3860. https://doi.org/10.1126/scitranslmed.abj3860

Bendre AD, Peters PJ, Kumar J (2021) Recent insights into the structure and function of mycobacterial membrane proteins facilitated by Cryo-EM. J Membr Biol 254(3):321–341

Berg S, Kaur D, Jackson M, Brennan PJ (2007) The glycosyltransferases of Mycobacterium tuberculosis—roles in the synthesis of arabinogalactan, lipoarabinomannan, and other glycoconjugates. Glycobiology 17(6):35R–56R. https://doi.org/10.1093/glycob/cwm010

Bhatt A, Fujiwara N, Bhatt K, Gurcha SS, Kremer L, Chen B et al (2007) Deletion of kasB in Mycobacterium tuberculosis causes loss of acid-fastness and subclinical latent tuberculosis in immunocompetent mice. Proc Natl Acad Sci 104(12):5157–5162. https://doi.org/10.1073/pnas.0608654104

Biegas KJ, Swarts BM (2021) Chemical probes for tagging mycobacterial lipids. Curr Opin Chem Biol 65:57–65

Birch HL, Alderwick LJ, Bhatt A, Rittmann D, Krumbach K, Singh A et al (2008) Biosynthesis of mycobacterial arabinogalactan: identification of a novel α (1→ 3) arabinofuranosyltransferase. Mol Microbiol 69(5):1191–1206. https://doi.org/10.1111/j.1365-2958.2008.06354.x

Birch HL, Alderwick LJ, Appelmelk BJ, Maaskant J, Bhatt A, Singh A et al (2010) A truncated lipoglycan from mycobacteria with altered immunological properties. Proc Natl Acad Sci 107(6):2634–2639. https://doi.org/10.1073/pnas.0915201107

Bogdanović N, Sundararaman L, Kamariah N, Tyagi A, Bhushan S, Ragunathan P et al (2018) Structure and function of mycobacterium-specific components of F-ATP synthase subunits α and ε. J Struct Biol 204(3):420–434

Boissier F, Bardou F, Guillet V, Uttenweiler-Joseph S, Daffé M, Quémard A, Mourey L (2006) Further insight into S-adenosylmethionine-dependent methyltransferases: structural characterization of Hma, an enzyme essential for the biosynthesis of oxygenated mycolic acids in Mycobacterium tuberculosis. J Biol Chem 281(7):4434–4445. https://doi.org/10.1074/jbc.M510250200

Boldrin F, Ventura M, Degiacomi G, Ravishankar S, Sala C, Svetlikova Z et al (2014) The phosphatidyl-myo-inositol mannosyltransferase PimA is essential for mycobacterium tuberculosis growth in vitro and in vivo. J Bacteriol 196(19):3441–3451

Boldrin F, Anso I, Alebouyeh S, Sevilla IA, Geijo M, Garrido JM et al (2021) The phosphatidylmyo-inositol dimannoside acyltransferase PatA is essential for Mycobacterium tuberculosis growth in vitro and in vivo. J Bacteriol 203(1):e00439–e00420. https://doi.org/10.1128/JB.00439-20

Bon C, Cabantous S, Julien S, Guillet V, Chalut C, Rima J et al (2022) Solution structure of the type I polyketide synthase Pks13 from Mycobacterium tuberculosis. BMC Biol 20(1):147. https://doi.org/10.1186/s12915-022-01351-y

Bowdish DM et al (2009) MARCO, TLR2, and CD14 are required for macrophage cytokine responses to mycobacterial trehalose dimycolate and Mycobacterium tuberculosis. PLoS Pathog 5(6):e1000474. https://doi.org/10.1371/journal.ppat.1000474

Braganza CD, Teunissen T, Timmer MSM, Stocker BL (2018) Identification and biological activity of synthetic macrophage inducible C-type lectin ligands. Front Immunol 8:1940. https://doi.org/10.3389/fimmu.2017.01940

Brennan PJ (1994) The microbiology of mycobacterium leprae, part II. Reflections on major developments and those responsible for them. Int J Lepr Other Mycobact Dis 62(4):594–598

Brennan PJ, Nikaido H (1995) The envelope of mycobacteria. Annu Rev Biochem 64:29–63

Brodie AF, Kalra VK, Lee S-h, Cohen NS (1979) Properties of energy-transducing systems in different types of membrane preparations from mycobacterium phlei-preparation, resolution, and reconstitution. In: Methods in enzymology, vol 55. Academic Press, pp 175–200

Brown LK, Van Schalkwyk C, De Villiers AK, Marx FM (2023) Impact of interventions for tuberculosis prevention and care in South Africa - a systematic review of mathematical modelling studies. S Afr Med J 113(3):125–134. https://doi.org/10.7196/SAMJ.2023.v113i3.16812

Brown TS et al (2024) Genotype-phenotype characterization of serial Mycobacterium tuberculosis isolates in Bedaquiline-resistant tuberculosis. Clin Infect Dis 78(2):269–276. https://doi.org/10.1093/cid/ciad596

Bugg TD, Lloyd AJ, Roper DI (2006) Phospho-MurNAc-pentapeptide translocase (MraY) as a target for antibacterial agents and antibacterial proteins. Infect Disord Drug Targets 6(2):85–106. https://doi.org/10.2174/187152606784112128

Buglino J, Onwueme KC, Ferreras JA, Quadri LE, Lima CD (2004) Crystal structure of PapA5, a phthiocerol dimycocerosyl transferase from mycobacterium tuberculosis. J Biol Chem 279(29):30634–30642

Bunduc CM, Bitter W, Houben ENG (2020) Structure and function of the mycobacterial type VII secretion systems. Ann Rev Microbiol 74:315–335. https://doi.org/10.1146/annurev-micro-012420-081657

Cai L, Zhao X, Jiang T, Qiu J, Owusu L, Ma Y et al (2014) Prokaryotic expression, identification and bioinformatics analysis of the Mycobacterium tuberculosis Rv3807c gene encoding the putative enzyme committed to Decaprenylphosphoryl-d-arabinose synthesis. Indian J Microbiol 54(1):46–51. https://doi.org/10.1007/s12088-013-0418-8

Cantaloube S, Veyron-Churlet R, Haddache N, Daffé M, Zerbib D (2011) The Mycobacterium tuberculosis FAS-II dehydratases and methyltransferases define the specificity of the mycolic acid elongation complexes. PLoS One 6(12):e29564. https://doi.org/10.1371/journal.pone.0029564

Cao B, Williams SJ (2010) Chemical approaches for the study of the mycobacterial glycolipids phosphatidylinositol mannosides, lipomannan and lipoarabinomannan. Nat Prod Rep 27(6):919–947. https://doi.org/10.1039/b919446j

Catalão MJ, Filipe SR, Pimentel M (2019) Revisiting anti-tuberculosis therapeutic strategies that target the peptidoglycan structure and synthesis. Front Microbiol 10:190. https://doi.org/10.3389/fmicb.2019.00190

Chalut C (2016) MmpL transporter-mediated export of cell-wall associated lipids and siderophores in mycobacteria. Tuberculosis 100:32–45. https://doi.org/10.1016/j.tube.2016.06.004

Chapman LF, Goodell H, Wolff HG (1959) Changes in tissue vulnerability induced during hypnotic suggestion. J Psychosom Res 4(2):99–105. https://doi.org/10.1016/0022-3999(59)90023-6

Chauhan M, Barot R, Yadav R, Joshi K, Mirza S, Chikhale R et al (2024) The Mycobacterium tuberculosis cell wall: an alluring drug target for developing newer anti-TB drugs—a perspective. Chem Biol Drug Des. https://doi.org/10.1111/cbdd.14612

Chavadi SS, Onwueme KC, Edupuganti UR, Jerome J, Chatterjee D, Soll CE, Quadri LEN (2012) The mycobacterial acyltransferase PapA5 is required for biosynthesis of cell wall-associated phenolic glycolipids. Microbiol (Reading) 158(Pt 5):1379–1387

Chiang CY, West NP (2024) The fall of the mycobacterial cell wall: interrogating peptidoglycan synthesis for novel anti-TB agents. PeerJ 12:e18404. https://doi.org/10.7717/peerj.18404

Chiang Y, Pan S-W, Feng J-Y, Su W-J, Chen Y-M (2024) Genotype, drug susceptibility profiles and associated clinical features of mycobacterium Abscessus (Mabs). In: Isolates in patients with Mabs pulmonary and Extrapulmonary disease. B26 unlocking NTM/BRONCHIECTASIS mysteries. American Thoracic Society, pp A3180–A318A

Chiaradia L, Lefebvre C, Parra J, Marcoux J, Burlet-Schiltz O, Etienne G, Tropis M, Daffé M (2017) Dissecting the mycobacterial cell envelope and defining the composition of the native mycomembrane. Sci Rep 7:12807. https://doi.org/10.1038/s41598-017-12718-4

Cole ST, Brosch R, Parkhill J, Garnier T, Churcher C, Harris D et al (1998) Deciphering the biology of Mycobacterium tuberculosis from the complete genome sequence. Nature 393(6685):537–544. https://doi.org/10.1038/31159

Constant P, Perez E, Malaga W, Lanéelle MA, Saurel O, Daffé M, Guilhot C (2002) Role of the pks15/1 gene in the biosynthesis of phenolglycolipids in the Mycobacterium tuberculosis complex. Evidence that all strains synthesize glycosylated p-hydroxybenzoic methyl esters and that strains devoid of phenolglycolipids harbor a frameshift mutation in the pks15/1 gene. J Biol Chem 277(41):38148–38158. https://doi.org/10.1074/jbc.M206538200

Cook GM, Hards K, Vilchèze C, Hartman T, Berney M (2014) Energetics of respiration and oxidative phosphorylation in mycobacteria. Microbiol Spectr 2(3)

Cooper C, Peterson EJR, Bailo R, Pan M, Singh A, Moynihan P et al (2022) MadR mediates acyl CoA-dependent regulation of mycolic acid desaturation in mycobacteria. Proc Natl Acad Sci 119(8):e2111059119. https://doi.org/10.1073/pnas.2111059119

Cordillot, M., Dubée, V., Triboulet, S., Dubost, L., Marie, A., Hugonnet, J. E., ... & Mainardi, J. L. (2013). In vitro cross-linking of Mycobacterium tuberculosis peptidoglycan by L,D-transpeptidases and inactivation of these enzymes by carbapenems. Antimicrob Agents Chemother, 57(12), 5940–5945. doi:https://doi.org/10.1128/AAC.01663-13

Correia-Neves M, Nigou J, Mousavian Z, Sundling C, Källenius G (2022) Immunological hyporesponsiveness in tuberculosis: the role of mycobacterial glycolipids. Front Immunol 13:1035122. https://doi.org/10.3389/fimmu.2022.1035122

Daffé M (2000) The mycobacterial antigens 85 complex–from structure to function and beyond. Trends Microbiol 8(10):438–440

Daffé M (2008) The global architecture of the mycobacterial cell envelope

Daffé M, Draper P (1997) The envelope layers of mycobacteria with reference to their pathogenicity. Adv Microb Physiol 39:131–203. https://doi.org/10.1016/S0065-2911(08)60016-8

Daffé M, Etienne G (1999) The capsule of Mycobacterium tuberculosis and its implications for pathogenicity. Tuber Lung Dis 79(3):153–169. https://doi.org/10.1054/tuld.1998.0200

Daffé M, Marrakchi H (2019) Unraveling the structure of the mycobacterial envelope. Microbiology. Spectrum 7(4). https://doi.org/10.1128/microbiolspec.GPP3-0027-2018

Daleke MH, Cascioferro A, de Punder K, Ummels R, Abdallah AM, van der Wel N et al (2011) Conserved pro-Glu (PE) and pro-pro-Glu (PPE) protein domains target LipY lipases of pathogenic mycobacteria to the cell surface via the ESX-5 pathway. J Biol Chem 286(21):19024–19034. https://doi.org/10.1074/jbc.M110.204966

Danilchanka O, Sun J, Pavlenok M, Maueroder C, Speer A, Siroy A et al (2014) An outer membrane channel protein of Mycobacterium tuberculosis with exotoxin activity. Proc Natl Acad Sci 111(18):6750–6755. https://doi.org/10.1073/pnas.1400136111

Das C, Ghosh TS, Mande SS (2011) Computational analysis of the ESX-1 region of mycobacterium tuberculosis: insights into the mechanism of type VII secretion system. PLoS One 6(11):e27980

De Jonge MI, Pehau-Arnaudet G, Fretz MM, Romain F, Bottai D, Brodin P et al (2007) ESAT-6 from mycobacterium tuberculosis dissociates from its putative chaperone CFP-10 under acidic conditions and exhibits membrane-lysing activity. J Bacteriol 189(16):6028–6034

De Wit L, Palou M, Content J (1994) Nucleotide sequence of the 85B-protein gene of Mycobacterium bovis BCG and Mycobacterium tuberculosis. DNA Seq 4(5):267–270. https://doi.org/10.3109/10425179409020848

Degiacomi G, Benjak A, Madacki J, Boldrin F, Provvedi R, Palù G et al (2017) Essentiality of mmpL3 and impact of its silencing on Mycobacterium tuberculosis gene expression. Sci Rep 7:43495. https://doi.org/10.1038/srep43495

DeJesus MA, Gerrick ER, Xu W, Park SW, Long JE, Boutte CC, Rubin EJ, Schnappinger D, Ehrt S, Fortune SM, Sassetti CM, Ioerger TR (2017) Comprehensive essentiality analysis of the mycobacterium tuberculosis genome via saturating transposon mutagenesis. mBio 8(1):e02133–e02116. https://doi.org/10.1128/mBio.02133-16

Delogu G, Brennan MJ (1999) Functional domains present in the mycobacterial hemagglutinin. HBHA Journal of Bacteriology 181(24):7464–7469. https://doi.org/10.1128/JB.181.24.7464-7469.1999

Deshayes C, Angala SK, Marion E, Brandli I, Babonneau J, Preisser L et al (2013) Regulation of Mycolactone, the Mycobacterium ulcerans toxin, depends on nutrient source. PLoS Negl Trop Dis 7(11):e2502. https://doi.org/10.1371/journal.pntd.0002502

Dessenne C, Mariller C, Vidal O, Huvent I, Guerardel Y, Elass-Rochard E, Rossez Y (2025) Glycan-mediated adhesion mechanisms in antibiotic-resistant bacteria. BBA Advances 7:100156. https://doi.org/10.1016/j.bbadva.2025.100156

Dover LG, Corsino PE, Daniels IR, Cocklin SL, Tatituri RV, Besra GS, Fütterer K (2004) Crystal structure of the TetR/CamR family repressor Mycobacterium tuberculosis EthR implicated in ethionamide resistance. J Mol Biol 340(5):1095–1105. https://doi.org/10.1016/j.jmb.2004.06.003

Dover LG, Alahari A, Gratraud P, Gomes JM, Bhowruth V, Reynolds RC et al (2007) EthA, a common activator of thiocarbamide-containing drugs acting on different mycobacterial targets. Antimicrob Agents Chemother 51(3):1055–1063

Draper P (1982) The anatomy of mycobacteria. In: Ratledge C, Stanford J (eds) The biology of the mycobacteria (Vol. 1, pp. 9–52). Academic Press

Draper P, Daffé M (2004) The cell envelope of Mycobacterium tuberculosis with special reference to the capsule and outer permeability barrier. In: Cole ST, Eisenach KD, McMurray DN, Jacobs WR Jr (eds) Tuberculosis and the tubercle bacillus. ASM Press, pp 261–273. https://doi.org/10.1128/9781555817657.ch17

Dubé JY, McIntosh F, Behr MA (2022) Mice dually disrupted for Nod2 and Mincle manifest early bacteriological control but late susceptibility during Mycobacterium tuberculosis infection. Front

Immunol 13:862992. https://doi.org/10.3389/fimmu.2022.862992
Dulberger CL, Rubin EJ, Boutte CC (2020) The mycobacterial cell envelope – a moving target. Nat Rev Microbiol 18(1):47–59. https://doi.org/10.1038/s41579-019-0273-7
Earp JC, Garaeva AA, Meikle V, Niederweis M, Seeger NA (2025) Structural basis of siderophore export and drug efflux by Mycobacterium tuberculosis. Nat Commun 16:1934. https://doi.org/10.1038/s41467-025-56888-6
Ekiert DC, Cox JS (2014) Structure of a PE-PPE-EspG complex from Mycobacterium tuberculosis reveals molecular specificity of ESX protein secretion. Proc Natl Acad Sci 111(41):14758–14763. https://doi.org/10.1073/pnas.1409345111
Escuyer VE, Lety M-A, Torrelles JB, Khoo K-H, Tang J-B, Rithner CD et al (2001) The role of the embA and embB gene products in the biosynthesis of the terminal hexaarabinofuranosyl motif of mycobacterium smegmatisarabinogalactan. J Biol Chem 276(52):48854–48862
Esposito C, Marasco D, Delogu G, Pedone E, Berisio R (2011) Heparin-binding hemagglutinin HBHA from Mycobacterium tuberculosis affects actin polymerisation. Biochem Biophys Res Commun 410(2):339–344. https://doi.org/10.1016/j.bbrc.2011.05.159
Esteban-Amo MJ, Jiménez-Cuadrado P, Serrano-Lorenzo P, de la Fuente M, Simarro M (2024) Succinate dehydrogenase and human disease: novel insights into a well-known enzyme. Biomedicine 12(9)
Famelis N, Rivera-Calzada A, Degliesposti G, Wingender M, Mietrach N, Skehel JM et al (2019) Architecture of the mycobacterial type VII secretion system. Nature 576(7786):321–325
Famelis N, Geibel S, van Tol D (2023) Mycobacterial type VII secretion systems. Biol Chem 404(7):691–702. https://doi.org/10.1515/hsz-2022-0350
Fay A, Czudnochowski N, Rock JM, Johnson JR, Krogan NJ, Rosenberg O, Glickman MS (2019) Two accessory proteins govern MmpL3 mycolic acid transport in mycobacteria. mBio 10(3):e00850–e00819. https://doi.org/10.1128/mBio.00850-19
Filippova EV, Kieser KJ, Luan CH, Wawrzak Z, Kiryukhina O, Rubin EJ, Anderson WF (2016) Crystal structures of the transpeptidase domain of the Mycobacterium tuberculosis penicillin binding protein PonA1 reveal potential mechanisms of antibiotic resistance. FEBS J 283(24):4524–4538. https://doi.org/10.1111/febs.13738
Foo CS-Y, Pethe K, Lupien A (2020) Oxidative phosphorylation—an update on a new, essential target space for drug discovery in mycobacterium tuberculosis. Appl Sci 10(7):2339
Forrellad MA, Klepp LI, Gioffré A, Sabio y García J, Morbidoni HR, de la Paz Santangelo M et al (2013) Virulence factors of the mycobacterium tuberculosis complex. Virulence 4(1):3–66
Frehel C, De Chastellier C, Lang T, Rastogi N (1986) Evidence for inhibition of fusion of lysosomal and Prelysosomal compartments with phagosomes in macrophages infected with pathogenic Mycobacterium avium. Infect Immun 52(1):252–262. https://doi.org/10.1128/iai.52.1.252-262.1986
Fukuda, T., Matsumura, T., Ato, M., Hamasaki, M., Nishiuchi, Y., Murakami, Y., … & Morita, Y. S. (2013). Critical roles for lipomannan and lipoarabinomannan in cell wall integrity of mycobacteria and pathogenesis of tuberculosis. mBio, 4(1), e00472–e00412. doi:https://doi.org/10.1128/mBio.00472-12
García-Ruiz V, Orduña P, Castillo-Rodal AI, Flores-Rodríguez TJ, López-Vidal Y (2022) Proteome profile changes induced by heterologous overexpression of Mycobacterium tuberculosis-derived antigens PstS-1 (Rv0934) and Ag85B (Rv1886c) in Mycobacterium microti. Biomolecules 12(12):1836. https://doi.org/10.3390/biom12121836
Gaur V, Bera S (2022) Recent developments on UDP-N-acetylmuramoyl-L-alanine-D-gutamate ligase (Mur D) enzyme for antimicrobial drug development: an emphasis on in-silico approaches. Current Res Pharmacol Drug Discov 3:100137
Gilleron M, Quesniaux VFJ, Puzo G (2003) Acylation state of the phosphatidylinositol Hexamannosides and its implication in toll-like receptor response. J Biol Chem 278(32):29880–29889. https://doi.org/10.1074/jbc.M303446200
Giraud MF, Naismith JH (2000) The rhamnose pathway. Curr Opin Struct Biol 10(6):687–696. https://doi.org/10.1016/S0959-440X(00)00145-7
Giraud MF, Leonard GA, Field RA, Berlind C, Naismith JH (2000) RmlC, the third enzyme of dTDP-L-rhamnose pathway, is a new class of epimerase. Nat Struct Biol 7(5):398–402. https://doi.org/10.1038/75169
Glickman MS (2003) The mmaA2 gene of Mycobacterium tuberculosis encodes the distal cyclopropane synthase of the alpha-mycolic acid. J Biol Chem 278(10):7844–7849. https://doi.org/10.1074/jbc.M212458200
Glickman MS, Cox JS, Jacobs WR Jr (2000) A novel mycolic acid cyclopropane synthetase is required for cording, persistence, and virulence of Mycobacterium tuberculosis. Mol Cell 5(4):717–727. https://doi.org/10.1016/S1097-2765(00)80250-6
Goins CM, Schreidah CM, Dajnowicz S, Ronning DR (2018) Structural basis for lipid binding and mechanism of the Mycobacterium tuberculosis Rv3802 phospholipase. J Biol Chem 293(4):1363–1372. https://doi.org/10.1074/jbc.RA117.000132
Gokulan K, O'Leary SE, Russell WK, Russell DH, Lalgondar M, Begley TP et al (2013) Crystal structure of Mycobacterium tuberculosis polyketide synthase 11 (PKS11) reveals intermediates in the synthesis of methylbranched alkylpyrones. J Biol Chem 288(23):16484–16494. https://doi.org/10.1074/jbc.M113.473132
Gong Y, Wang J, Li F, Zhu B (2023a) Polysaccharides and glycolipids of Mycobacterium tuberculosis and their induced immune responses. Scand J Immunol 97(3):e13261. https://doi.org/10.1111/sji.13261
Gong Y, Wei C, Wang J, Mu N, Lu Q, Wu C et al (2023b) Structure of the priming arabinosyltransferase AftA required for AG biosynthesis of Mycobacterium tuber-

culosis. Proc Natl Acad Sci 120(23):e2302858120. https://doi.org/10.1073/pnas.2302858120

Grzegorzewicz AE, Pham H, Gundi VAKB, Scherman MS, North EJ, Hess T et al (2012) Inhibition of mycolic acid transport across the Mycobacterium tuberculosis plasma membrane. Nat Chem Biol 8(4):334–341. https://doi.org/10.1038/nchembio.794

Gu S et al (2003) Comprehensive proteomic profiling of the membrane constituents of a Mycobacterium tuberculosis strain, vol 2

Guerin ME, Kordulakova J, Schaeffer F, Svetlikova Z, Buschiazzo A, Giganti D et al (2007) Molecular recognition and interfacial catalysis by the essential phosphatidylinositol mannosyltransferase PimA from mycobacteria. J Biol Chem 282(28):20705–20714

Guerin ME, Kaur D, Somashekar B, Gibbs S, Gest P, Chatterjee D et al (2009) New insights into the early steps of phosphatidylinositol mannoside biosynthesis in mycobacteria: PimB′ is an essential enzyme of mycobacterium smegmatis. J Biol Chem 284(38):25687–25696

Guerin ME, Korduláková J, Alzari PM, Brennan PJ, Jackson M (2010) Molecular basis of phosphatidyl-myo-inositol mannoside biosynthesis and regulation in mycobacteria. J Biol Chem 285(44):33577–33583

Guérin H, Kulakauskas S, Chapot-Chartier MP (2022) Structural variations and roles of rhamnose-rich cell wall polysaccharides in gram-positive bacteria. J Biol Chem 298(10):102488

Guo F, Wei J, Song Y, Li B, Qian Z, Wang X et al (2023) Immunological effects of the PE/PPE family proteins of Mycobacterium tuberculosis and related vaccines. Front Immunol 14:1255920. https://doi.org/10.3389/fimmu.2023.1255920

Gupta R, Lavollay M, Mainardi JL, Arthur M, Bishai WR, Lamichhane G (2010) The Mycobacterium tuberculosis protein LdtMt2 is a nonclassical transpeptidase required for virulence and resistance to amoxicillin. Nat Med 16(4):466–469. https://doi.org/10.1038/nm.2120

Gupta KR, Gwin CM, Rahlwes KC, Biegas KJ, Wang C, Park JH et al (2022) An essential periplasmic protein coordinates lipid trafficking and is required for asymmetric polar growth in mycobacteria. eLife 11:e80395. https://doi.org/10.7554/eLife.80395

Gurcha SS, Baulard AR, Kremer L, Locht C, Moody DB, Muhlecker W et al (2002) Ppm1, a novel polyprenol monophosphomannose synthase from Mycobacterium tuberculosis. Biochem J 365(2):441–450. https://doi.org/10.1042/bj20020107

Haagsma AC, Driessen NN, Hahn MM, Lill H, Bald D (2010) ATP synthase in slow- and fast-growing mycobacteria is active in ATP synthesis and blocked in ATP hydrolysis direction. FEMS Microbiol Lett 313(1):68–74

Harathi N, Pulaganti M, Anuradha CM, Kumar Chitta S (2016) Inhibition of mycobacterium-RmlA by molecular modeling. Dynamics simulation, and docking. Adv Bioinform 2016:9841250. https://doi.org/10.1155/2016/9841250

Harrison J, Lloyd G, Joe M, Lowary TL, Reynolds E, Walters-Morgan H et al (2016) Lcp1 is a phosphotransferase responsible for ligating arabinogalactan to peptidoglycan in Mycobacterium tuberculosis. mBio 7(4):e00972–e00916. https://doi.org/10.1128/mBio.00972-16

Harth G, Lee BY, Wang J, Clemens DL, Horwitz MA (1996) Novel insights into the genetics, biochemistry, and immunocytochemistry of the 30-kilodalton major extracellular protein of Mycobacterium tuberculosis. Infect Immun 64(8):3038–3047

Hashimoto K, Kawasaki H, Akazawa K, Nakamura J, Asakura Y, Kudo T et al (2006) Changes in composition and content of mycolic acids in glutamate-overproducing Corynebacterium glutamicum. Biosci Biotechnol Biochem 70(1):22–30. https://doi.org/10.1271/bbb.70.22

Hett EC, Chao MC, Deng LL, Rubin EJ (2008) A mycobacterial enzyme essential for cell division synergizes with resuscitation-promoting factor. PLoS Pathog 4(2):e1000001. https://doi.org/10.1371/journal.ppat.1000001

Hoffmann C, Leis A, Niederweis M, Plitzko JM, Engelhardt H (2008) Disclosure of the mycobacterial outer membrane: Cryo-electron tomography and vitreous sections reveal the lipid bilayer structure. Proc Natl Acad Sci 105(10):3963–3967

Holzheimer M, Buter J, Minnaard AJ (2021) Chemical synthesis of Cell Wall constituents of Mycobacterium tuberculosis. Chem Rev 121(15):9554–9643. https://doi.org/10.1021/acs.chemrev.1c00043

Houben D, Demangel C, Van Ingen J, Perez J, Baldeón L, Abdallah AM et al (2012) ESX-1-mediated translocation to the cytosol controls virulence of mycobacteria. Cell Microbiol 14(8):1287–1298

Hunter JD (2006) Ventilator associated pneumonia. Postgrad Med J 82(965):172–178

Hunter RL (2016) Tuberculosis as a three-act play: a new paradigm for the pathogenesis of pulmonary tuberculosis. Tuberculosis 97:8–17. https://doi.org/10.1016/j.tube.2015.11.010

Ilangumaran S et al (1995) Integration of mycobacterial lipoarabinomannans into glycosylphosphatidylinositol-rich domains of lymphomonocytic cell plasma membranes. J Immunol 155(3):1334–1342

Indrigo J, Hunter RL Jr, Actor JK (2002) Influence of trehalose 6,6′-dimycolate (TDM) during mycobacterial infection of bone marrow macrophages. Microbiology 148(7):1991–1998. https://doi.org/10.1099/00221287-148-7-1991

Ishikawa E et al (2009) Direct recognition of the mycobacterial glycolipid, trehalose dimycolate, by C-type lectin Mincle. J Exp Med 206(13):2879–2888. https://doi.org/10.1084/jem.20091750

Iqbal IK, Bajeli S, Akela AK, Kumar A (2018) Bioenergetics of mycobacterium: an emerging landscape for drug discovery. Pathogens 7(1):24

Izquierdo Lafuente B, Ummels R, Kuijl C, Bitter W, Speer A (2021) Mycobacterium tuberculosis toxin CpnT is an ESX-5 substrate and requires three type

VII secretion Systems for Intracellular Secretion. mBio 12(2):e02983–e02920. https://doi.org/10.1128/mBio.02983-20

Jain M, Cox JS (2005) Interaction between polyketide synthase and transporter suggests coupled synthesis and export of virulence lipid in M. tuberculosis. PLoS Pathog 1(1):e2

Jaisinghani N et al (2024) Proteomics from compartment-specific APEX2 labeling in Mycobacterium tuberculosis reveals type VII secretion substrates in the cell wall. Cell Chemical Biology 31(3):523–533.e4. https://doi.org/10.1016/j.chembiol.2023.10.013

Jamet S, Quentin Y, Coudray C et al (2015) Evolution of mycolic acid biosynthesis genes and their regulation during starvation in Mycobacterium tuberculosis. J Bacteriol 197(24):3797–3811. https://doi.org/10.1128/JB.00433-15

Jankute M, Grover S, Rana AK, Besra GS (2012) Arabinogalactan and lipoarabinomannan biosynthesis: structure, biogenesis and their potential as drug targets. Future Microbiol 7(1):129–147. https://doi.org/10.2217/fmb.11.123

Jankute M, Cox JA, Harrison J, Besra GS (2015) Assembly of the mycobacterial cell wall. Ann Rev Microbiol 69:405–423. https://doi.org/10.1146/annurev-micro-091014-104121

Jankute M, Alderwick LJ, Noack S, Veerapen N, Nigou J, Besra GS (2017) Disruption of mycobacterial AftB results in complete loss of terminal β(1 → 2) Arabinofuranose residues of Lipoarabinomannan. ACS Chem Biol 12(1):183–190. https://doi.org/10.1021/acschembio.6b00898

Jones CM, Wells RM, Madduri AVR, Renfrow MB, Ratledge C, Moody DB, Niederweis M (2014) Self-poisoning of Mycobacterium tuberculosis by interrupting siderophore recycling. Proc Natl Acad Sci 111(5):1945–1950. https://doi.org/10.1073/pnas.1311402111

Józefowski S, Sobota A, Kwiatkowska K (2008) How Mycobacterium tuberculosis subverts host immune responses. BioEssays 30(10):943–954. https://doi.org/10.1002/bies.20815

Kado T, Akbary Z, Motooka D, Sparks IL, Melzer ES, Nakamura S et al (2023) A cell wall synthase accelerates plasma membrane partitioning in mycobacteria. eLife 12:e81924. https://doi.org/10.7554/eLife.81924

Källenius G, Correia-Neves M, Buteme H, Hamasur B, Svenson SB (2016) Lipoarabinomannan, and its related glycolipids, induce divergent and opposing immune responses to Mycobacterium tuberculosis depending on structural diversity and experimental variations. Tuberculosis 96:120–130. https://doi.org/10.1016/j.tube.2015.09.005

Kalscheuer R, Palacios A, Anso I, Cifuente J, Anguita J, Jacobs WR Jr, Guerin ME, Prados-Rosales R (2019) The Mycobacterium tuberculosis capsule: a cell structure with key implications in pathogenesis. Biochem J 476(14):1995–2016. https://doi.org/10.1042/BCJ20190324

Kampff Z, van Sinderen D, Mahony J (2023) Cell wall polysaccharides of streptococci: a genetic and structural perspective. Biotechnol Adv 69:108279. https://doi.org/10.1016/j.biotechadv.2023.108279

Kamariah N, Huber RG, Nartey W, Bhushan S, Bond PJ, Grüber G (2019) Structure and subunit arrangement of mycobacterial F1FO ATP synthase and novel features of the unique mycobacterial subunit δ. J Struct Biol 207(2):199–208

Kantardjieff KA, Kim CY, Naranjo C, Waldo GS, Lekin T, Segelke BW et al (2004) Mycobacterium tuberculosis RmlC epimerase (Rv3465): a promising drug-target structure in the rhamnose pathway. Acta Crystallogr D Biol Crystallogr 60(5):895–902. https://doi.org/10.1107/S0907444904004924

Kartmann B, Stenger S, Niederweis M (1999) Porins in the cell wall of Mycobacterium tuberculosis. J Bacteriol 181(20):6543–6546

Kaur D, Obregón-Henao A, Pham H, Chatterjee D, Brennan PJ, Jackson M (2008) Lipoarabinomannan of mycobacterium: mannose capping by a multifunctional terminal mannosyltransferase. Proc Natl Acad Sci 105(39):17973–17977. https://doi.org/10.1073/pnas.0807761105

Kieser KJ, Baranowski C, Chao MC, Long JE, Sassetti CM, Waldor MK et al (2015a) Peptidoglycan synthesis in Mycobacterium tuberculosis is organized into networks with varying drug susceptibility. Proc Natl Acad Sci 112(42):13087–13092. https://doi.org/10.1073/pnas.1514135112

Kieser KJ, Boutte CC, Kester JC, Baer CE, Barczak AK, Meniche X et al (2015b) Phosphorylation of the peptidoglycan synthase PonA1 governs the rate of polar elongation in mycobacteria. PLoS Pathog 11(6):e1005010. https://doi.org/10.1371/journal.ppat.1005010

Kolattukudy PE, Fernandes ND, Azad AK, Fitzmaurice AM, Sirakov TD (1997) Biochemistry and molecular genetics of cell-wall lipid biosynthesis in mycobacteria. Mol Microbiol 24(2):263–270

Kolly GS, Boldrin F, Sala C, Dhar N, Hartkoorn RC, Ventura M et al (2014) Assessing the essentiality of the decaprenyl-phospho-d-arabinose pathway in Mycobacterium tuberculosis using conditional mutants. Mol Microbiol 92(1):194–211. https://doi.org/10.1111/mmi.12546

Kolly GS, Mukherjee R, Kilacsková E, Abriata LA, Raccaud M, Blaško J et al (2015) GtrA protein Rv3789 is required for Arabinosylation of arabinogalactan in mycobacterium tuberculosis. J Bacteriol 197(23):3686–3697

Kordulákóvá J, Gilleron M, Mikusova K, Puzo G, Brennan PJ, Gicquel B, Jackson M (2002) Definition of the first mannosylation step in phosphatidylinositol mannoside synthesis. PimA is essential for growth of mycobacteria. J Biol Chem 277(35):31335–31344

Koul A, Arnoult E, Lounis N, Guillemont J, Andries K (2011) The challenge of new drug discovery for tuberculosis. Nature 469(7331):483–490

Kremer L, Maughan WN, Wilson RA, Dover LG, Besra GS (2002) The M. Tuberculosis antigen 85 complex and mycolyltransferase activity. Lett Appl Microbiol 34(4):233–237

Kremer K, Arnold C, Cataldi A, Gutierrez MC, Haas WH, Panaiotov S, Skuce RA, Supply P, van der Zanden AG, van Soolingen D (2005) Discriminatory power and reproducibility of novel DNA typing methods for Mycobacterium tuberculosis complex strains. J Clin Microbiol 43(11):5628–5638. https://doi.org/10.1128/JCM.43.11.5628-5638.2005

Kumar SU, Saleem A, Kumar DT, Preethi VA, Younes S, Zayed H et al (2021) A systemic approach to explore the mechanisms of drug resistance and altered signaling cascades in extensively drug-resistant tuberculosis. In: Donev R, Karabencheva-Christova T (eds) Advances in protein chemistry and structural biology, vol 127. Academic Press, pp 343–364. https://doi.org/10.1016/bs.apcsb.2021.02.006

Kuk ACY, Hao A, Lee SY (2022) Structure and mechanism of the lipid Flippase MurJ. Annu Rev Biochem 91:705–729

Küssau T, Van Wyk N, Johansen MD, Alsarraf HMAB, Neyret A, Hamela C et al (2020) Functional characterization of the N-Acetylmuramyl-l-alanine amidase, Ami1, from Mycobacterium abscessus. Cells 9(11):2410. https://doi.org/10.3390/cells9112410

Lamichhane G, Zignol M, Blades NJ, Geiman DE, Dougherty A, Grosset J et al (2003) A postgenomic method for predicting essential genes at subsaturation levels of mutagenesis: application to Mycobacterium tuberculosis. Proc Natl Acad Sci 100(12):7213–7218. https://doi.org/10.1073/pnas.1231432100

Lang JM, Darling AE, Eisen JA (2013) Phylogeny of bacterial and archaeal genomes using conserved genes: Supertrees and Supermatrices. PLoS One 8(4):e62510

Lavania M, Datta G, Gupta UD (2022) Potential approaches for the diagnosis and treatment of drug-resistant leprosy. Journal of Preventive, Diagnostic and Treatment Strategies in Medicine 1(1):21–25

Lavollay M, Arthur M, Fourgeaud M, Dubost L, Marie A, Veziris N et al (2008) The peptidoglycan of stationary-phase Mycobacterium tuberculosis predominantly contains cross-links generated by L, D-transpeptidation. J Bacteriol 190(12):4360–4366. https://doi.org/10.1128/JB.00239-08

Lea-Smith DJ, Pyke JS, Tull D, McConville MJ, Coppel RL, Crellin PK (2007) The reductase that catalyzes mycolic motif synthesis is required for efficient attachment of mycolic acids to arabinogalactan. J Biol Chem 282(15):11000–11008

Lea-Smith DJ, Martin KL, Pyke JS, Tull D, McConville MJ, Coppel RL, Crellin PK (2008) Analysis of a new mannosyltransferase required for the synthesis of phosphatidylinositol mannosides and lipoarbinomannan reveals two lipomannan pools in corynebacterineae. J Biol Chem 283(11):6773–6782. https://doi.org/10.1074/jbc.M707139200

Lederer E, Adam A, Ciorbaru R, Petit JF, Wietzerbin J (1975) Cell walls of mycobacteria and related organisms; chemistry and immunostimulant properties. Mol Cell Biochem 7(2):87–104. https://doi.org/10.1007/BF01731567

Lemassu A, Ortalo-Magné A, Bardou F, Silve G, Lanéelle MA, Daffé M (1996) Extracellular and surface-exposed polysaccharides of non-tuberculous mycobacteria. Microbiol (Reading) 142(Pt 6):1513–1520

Li W, Xin Y, McNeil MR, Ma Y (2006) rmlB and rmlC genes are essential for growth of mycobacteria. Biochem Biophys Res Commun 342(1):170–178. https://doi.org/10.1016/j.bbrc.2006.01.130

Li B, Zhang Y, Guo Q, He S, Fan J, Xu L et al (2022) Antibacterial peptide RP557 increases the antibiotic sensitivity of mycobacterium abscessus by inhibiting biofilm formation. Sci Total Environ 807:151855

Lim H, Lay R et al (2000) Cloning and expression of immunoreactive antigens from Mycobacterium tuberculosis. Clinical Diagnostic Laboratory Immunology 7(4):600–606

Liu Y, Brown CM, Erramilli S, Guu SY, Tseng PS, Wang YJ et al (2025a) Structural insights into terminal arabinosylation of mycobacterial cell wall arabinan. Nat Commun 16:3973. https://doi.org/10.1038/s41467-025-58196-5

Liu S, Xiao G, Li P, Xu Y, Fan X, Tian L et al (2025b) Plasma-based ultrasensitive detection of Mycobacterium tuberculosis ESAT6/CFP10 fusion antigen using a CRISPR-driven aptamer fluorescence testing (CRAFT). Biosens Bioelectron 284:117566. https://doi.org/10.1016/j.bios.2025.117566

Lovewell RR, Sassetti CM, VanderVen BC (2016) Chewing the fat: lipid metabolism and homeostasis during M. Tuberculosis infection. Curr Opin Microbiol 29:30–36. https://doi.org/10.1016/j.mib.2015.10.002

Luna-Herrera J, Corneho-Baez AA, Delgadillo-Gutierrez K (2025) Relevance of efflux pumps in the development of drug resistance in mycobacterial infections. In: Wang L, Bi L, Hu X, Zhang Y (eds) Diagnosis and treatment of mycobacterial infections. IntechOpen, London

Ly A, Liu J (2020) Mycobacterial virulence factors: surface-exposed lipids and secreted proteins. Int J Mol Sci 21(11):3985. https://doi.org/10.3390/ijms21113985

Ma S, Huang Y, Xie F, Gong Z, Zhang Y, Stojkoska A, Xie J (2020) Transport mechanism of Mycobacterium tuberculosis MmpL/S family proteins and implications in pharmaceutical targeting. Biol Chem 401(3):331–348. https://doi.org/10.1515/hsz-2019-0320

Machowski EE, Senzani S, Ealand C, Kana BD (2014) Comparative genomics for mycobacterial peptidoglycan remodelling enzymes reveals extensive genetic multiplicity. BMC Microbiol 14:75. https://doi.org/10.1186/1471-2180-14-75

Mahapatra S, Crick DC, Brennan PJ (2000) Comparison of the UDP-N-acetylmuramate:l-alanine ligase enzymes from Mycobacterium tuberculosis and Mycobacterium leprae. J Bacteriol 182(24):6827–6830. https://doi.org/10.1128/JB.182.24.6827-6830.2000

Mahapatra S, Scherman H, Brennan PJ, Crick DC (2005) N-Glycosylation of the nucleotide precursors of peptidoglycan biosynthesis of Mycobacterium spp. is altered by drug treatment. J Bacteriol 187(7):2341–2347. https://doi.org/10.1128/JB.187.7.2341-2347.2005

Maharjan R, Zhang Z, Klenotic PA, Gregor WD, Purdy GE, Yu EW (2024) Cryo-EM structure of the mycobacterium smegmatis MmpL5-AcpM complex. MBio 15(12):e03035–e03024

Mailaender C, Reiling N, Engelhardt H, Bossmann S, Ehlers S, Niederweis M (2004) The MspA porin promotes growth and increases antibiotic susceptibility of both Mycobacterium bovis BCG and Mycobacterium tuberculosis. Microbiology 150(4):853–864. https://doi.org/10.1099/mic.0.26902-0

Maitra A, Munshi T, Healy J, Martin LT, Vollmer W, Keep NH, Bhakta S (2019) Cell wall peptidoglycan in Mycobacterium tuberculosis: an Achilles' heel for the TB-causing pathogen. FEMS Microbiol Rev 43(5):548–575. https://doi.org/10.1093/femsre/fuz016

Marrakchi H, Lanéelle MA, Daffé M (2014) Mycolic acids: structures, biosynthesis, and beyond. Chem Biol 21(1):67–85. https://doi.org/10.1016/j.chembiol.2013.11.011

Matsunaga I, Naka T, Talekar RS, McConnell MJ, Katoh K, Nakao H et al (2008) Mycolyltransferase-mediated glycolipid exchange in mycobacteria. J Biol Chem 283(43):28835–28841. https://doi.org/10.1074/jbc.M805724200

McNeil M, Daffe M, Brennan PJ (1990) Evidence for the nature of the link between the arabinogalactan and peptidoglycan of mycobacterial cell walls. J Biol Chem 265(30):18200–18206

Melkie ST, Arias L, Farroni C, Makek MJ, Goletti D, Vilaplana C (2022) The role of antibodies in tuberculosis diagnosis, prophylaxis and therapy: a review from the ESGMYC study group. Eur Respir Rev 31(165):210218. https://doi.org/10.1183/16000617.0218-2021

Melly G, Purdy GE (2019) MmpL proteins in physiology and pathogenesis of M. tuberculosis. Microorganisms 7(3):70. https://doi.org/10.3390/microorganisms7030070

Merrick B, Sergaki C, Edwards L, Moyes DL, Kertanegara M, Prossomariti D et al (2023) Modulation of the gut microbiota to control antimicrobial resistance (AMR)-a narrative review with a focus on Faecal microbiota transplantation (FMT). Infect Dis Rep 15(3):238–254

Minato Y, Gohl DM, Thiede JM, Chacón JM, Harcombe WR, Maruyama F, Baughn AD (2019) Genomewide assessment of Mycobacterium tuberculosis conditionally essential metabolic pathways. mSystems 4(4):e00070–e00019. https://doi.org/10.1128/mSystems.00070-19

Minnikin DE, Minnikin SM, Goodfellow M, Stanford JL (1982) The mycolic acids of Mycobacterium chelonei. J Gen Microbiol 128(4):817–822. https://doi.org/10.1099/00221287-128-4-817

Minnikin DE et al (2015) Ancient mycobacterial lipids: key reference biomarkers in charting the evolution of tuberculosis. Tuberculosis 95(Suppl 1):S133–S139. https://doi.org/10.1016/j.tube.2015.02.009

Mishra AK, Driessen NN, Appelmelk BJ, Besra GS (2011) Lipoarabinomannan and related glycoconjugates: structure, biogenesis and role in Mycobacterium tuberculosis physiology and host-pathogen interaction. FEMS Microbiol Rev 35(6):1126–1157. https://doi.org/10.1111/j.1574-6976.2011.00276.x

Mohanty D, Sankaranarayanan R, Gokhale RS (2011) Fatty acyl-AMP ligases and polyketide synthases are unique enzymes of lipid biosynthetic machinery in Mycobacterium tuberculosis. Tuberculosis 91(5):448–455. https://doi.org/10.1016/j.tube.2011.04.006

Moolla N, Bailo R, Marshall R, Bavro VN, Bhatt A (2021) Structure-function analysis of MmpL7-mediated lipid transport in mycobacteria. Cell Surface 7:100062

Moraes GL, Gomes GC, Monteiro de Sousa PR, Alves CN, Govender T, Kruger HG et al (2015) Structural and functional features of enzymes of mycobacterium tuberculosis peptidoglycan biosynthesis as targets for drug development. Tuberculosis (Edinb) 95(2):95–111

Morita YS et al (2005) Compartmentalization of lipid biosynthesis in mycobacteria. J Biol Chem 280(22):21645–21652

Mukku RP, Poornima K, Yadav S, Raghunand TR (2024) Delineating the functional role of the PPE50 (Rv3135) – PPE51 (Rv3136) gene cluster in the pathophysiology of Mycobacterium tuberculosis. Microbes Infect 26(3):105248. https://doi.org/10.1016/j.micinf.2023.105248

Munshi T, Gupta A, Evangelopoulos D, Guzman JD, Gibbons S, Keep NH, Bhakta S (2013) Characterisation of ATP-dependent Mur ligases involved in the biogenesis of cell wall peptidoglycan in Mycobacterium tuberculosis. PLoS One 8(3):e60143. https://doi.org/10.1371/journal.pone.0060143

Nair RR, Meikle V, Dubey S, Pavlenok M, Niederweis M (2025) Master control of protein secretion by Mycobacterium tuberculosis. bioRxiv. https://doi.org/10.1101/2025.03.13.643117

Ndlandla FL, Ejoh V, Stoltz AC, Naicker B, Cromarty AD, van Wyngaardt S et al (2016) Standardization of natural mycolic acid antigen composition and production for use in biomarker antibody detection to diagnose active tuberculosis. J Immunol Methods 435:50–59. https://doi.org/10.1016/j.jim.2016.05.010

Niederweis M (2003) Mycobacterial porins--new channel proteins in unique outer membranes. Mol Microbiol 49(5):1167–1177. https://doi.org/10.1046/j.1365-2958.2003.03662.x

Niederweis M, Danilchanka O, Huff J, Hoffmann C, Engelhardt H (2010) Mycobacterial outer membranes: in search of proteins. Trends Microbiol 18(3):109–116. https://doi.org/10.1016/j.tim.2009.12.005

Nigou J, Gilleron M, Puzo G (2003) Lipoarabinomannans: from structure to biosynthesis. Biochimie 85(1–2):153–166. https://doi.org/10.1016/S0300-9084(03)00048-8

Nivina A, Yuet KP, Hsu J, Khosla C (2019) Evolution and diversity of assembly-line polyketide synthases. Chem Rev 119(24):12524–12547. https://doi.org/10.1021/acs.chemrev.9b00525

Noll H, Bloch H, Asselineau J, Lederer E (1956) The chemical structure of the cord factor of Mycobacterium tuberculosis. Biochim Biophys Acta 20:299–309

Okkels LM, Brock I, Follmann F, Agger EM, Arend SM, Ottenhoff TH et al (2003) PPE protein (Rv3873) from DNA segment RD1 of Mycobacterium tuberculosis: strong recognition of both specific T-cell epitopes and epitopes conserved within the PPE family. Infect Immun 71(11):6116–6123. https://doi.org/10.1128/IAI.71.11.6116-6123.2003

Oluwole AO, Olatunji OS (2022) Photocatalytic degradation of tetracycline in aqueous systems under visible light irradiation using needle-like SnO2 nanoparticles anchored on exfoliated g-C3N4. Environ Sci Eur 34:5. https://doi.org/10.1186/s12302-021-00588-7

Onwueme KC, Vos CJ, Zurita J, Soll CE, Quadri LEN (2005) Identification of phthiodiolone ketoreductase, an enzyme required for production of mycobacterial diacyl phthiocerol virulence factors. J Bacteriol 187(14):4760–4766. https://doi.org/10.1128/JB.187.14.4760-4766.2005

Ortalo-Magné A, Dupont MA, Lemassu A, Andersen AB, Gounon P, Daffé M (1995) Molecular composition of the outermost capsular material of the tubercle bacillus. Microbiology 141(7):1609–1620. https://doi.org/10.1099/13500872-141-7-1609

Ortalo-Magné A, Lemassu A, Lanéelle MA, Bardou F, Silve G, Gounon P, Marchal G, Daffé M (1996) Identification of the surface-exposed lipids on the cell envelopes of Mycobacterium tuberculosis and other mycobacterial species. J Bacteriol 178(2):456–461. https://doi.org/10.1128/jb.178.2.456-461.1996

Owens CP, Chim N, Graves AB, Harmston CA, Iniguez A, Contreras H et al (2013) The mycobacterium tuberculosis secreted protein Rv0203 transfers heme to membrane proteins MmpL3 and MmpL11. J Biol Chem 288(30):21714–21728

Pajuelo D, Tak U, Zhang L, Danilchanka O, Tischler AD, Niederweis M (2021) Toxin secretion and trafficking by Mycobacterium tuberculosis. Nat Commun 12:6592. https://doi.org/10.1038/s41467-021-26925-1

Pal R, Bisht MK, Mukhopadhyay S (2022) Secretory proteins of Mycobacterium tuberculosis and their roles in modulation of host immune responses: focus on therapeutic targets. FEBS J 289(14):4146–4171. https://doi.org/10.1111/febs.16420

Palande A, Patil S, Veeram A, Sahoo SS et al (2024) Proteomic analysis of the Mycobacterium tuberculosis outer membrane for potential implications in uptake of small molecules, vol 10. ACS Infectious Diseases

Pandey P, Chatterjee S, Berida T, Doerksen RJ, Roy S (2022) Identification of potential non-nucleoside MraY inhibitors for tuberculosis chemotherapy using structure-based virtual screening. J Biomol Struct Dyn 40(11):4832–4849. https://doi.org/10.1080/07391102.2020.1862705

Paul TR, Beveridge TJ (1994) Preservation of surface lipids and determination of ultrastructure of Mycobacterium kansasii by freeze-substitution. Infect Immun 62(4):1542–1550

PaweŁczyk J, Kremer L (2014) The molecular genetics of mycolic acid biosynthesis. Microbiol Spectr 2(4):Mgm2-0003-2013

Payne DJ, Gwynn MN, Holmes DJ, Pompliano DL (2007) Drugs for bad bugs: confronting the challenges of antibacterial discovery. Nat Rev Drug Discov 6(1):29–40

Peltier P, Beláňová M, Dianišková P, Zhou R, Zheng RB, Pearcey JA et al (2010) Synthetic UDP-furanoses as potent inhibitors of mycobacterial galactan biogenesis. Chem Biol 17(12):1356–1366. https://doi.org/10.1016/j.chembiol.2010.10.013

Petrella S, Cambau E, Chauffour A, Andries K, Jarlier V, Sougakoff W (2006) Genetic basis for natural and acquired resistance to the diarylquinoline R207910 in mycobacteria. Antimicrob Agents Chemother 50(8):2853–2856

Pidgeon SE, Apostolos AJ, Nelson JM, Shaku M, Rimal B, Islam MN et al (2019) L,D-transpeptidase specific probe reveals spatial activity of peptidoglycan cross-linking. ACS Chem Biol 14(10):2185–2196. https://doi.org/10.1021/acschembio.9b00427

Pitarque S, Larrouy-Maumus G, Payré B, Jackson M, Puzo G, Nigou J (2008) The immunomodulatory lipoglycans, lipoarabinomannan and lipomannan, are exposed at the mycobacterial cell surface. Tuberculosis 88(6):560–565. https://doi.org/10.1016/j.tube.2008.04.002

Pohl T, Bauer T, Dörner K, Stolpe S, Sell P, Zocher G, Friedrich T (2007) Iron–sulfur cluster N7 of the NADH: ubiquinone oxidoreductase (complex I) is essential for stability but not involved in electron transfer. Biochemistry 46(22):6588–6596

Portevin D, de Sousa-D'Auria C, Houssin C, Grimaldi C, Chami M, Guilhot C (2004) A polyketide synthase catalyzes the last condensation step of mycolic acid biosynthesis in mycobacteria and related organisms. Proc Natl Acad Sci 101(1):314–319. https://doi.org/10.1073/pnas.0305439101

Poulin MB, Lowary TL (2016) Chemical insight into the mechanism and specificity of GlfT2, a bifunctional galactofuranosyltransferase from mycobacteria. J Org Chem 81(18):8123–8130. https://doi.org/10.1021/acs.joc.6b01501

Puech V, Guilhot C, Perez E, Tropis M, Armitige LY, Gicquel B, Daffé M (2002) Evidence for a partial redundancy of the fibronectin-binding proteins for the transfer of mycoloyl residues onto the cell wall arabinogalactan termini of Mycobacterium tuberculosis. Mol Microbiol 44(4):1109–1122. https://doi.org/10.1046/j.1365-2958.2002.02953.x

Puffal J, Sparks IL, Brenner JR, Li X, Leszyk JD, Hayashi JM et al (2022) Compartmentalized cell envelope biosynthesis in mycobacterium tuberculosis. bioRxiv:2022.01. 07.475471

Qu H, Xin Y, Dong X, Ma Y (2007) An rmlA gene encoding d-glucose-1-phosphate thymidylyltransferase is essential for mycobacterial growth. FEMS Microbiol Lett 275(2):237–243. https://doi.org/10.1111/j.1574-6968.2007.00890.x

Qu M, Zhou X, Li H (2021) BCG vaccination strategies against tuberculosis: updates and perspectives. Hum Vaccin Immunother 17(12):5284–5295

Quadri LE (2014) Biosynthesis of mycobacterial lipids by polyketide synthases and beyond. Crit Rev Biochem Mol Biol 49(3):179–211. https://doi.org/10.3109/10409238.2014.896859

Raghavendra T, Patil S, Mukherjee R (2018) Peptidoglycan in Mycobacteria: chemistry, biology and intervention. Glycoconj J 35(5):421–432. https://doi.org/10.1007/s10719-018-9836-5

Rainczuk A, Yamaryo-Botte Y, Brammananth R, Stinear TP, Seemann T, Coppel RL et al (2012) The lipoprotein LpqW is essential for the Mannosylation of periplasmic glycolipids in Corynebacteria. J Biol Chem 287(52):42726–42738. https://doi.org/10.1074/jbc.M112.373415

Ramos B, Gordon SV, Cunha MV (2020) Revisiting the expression signature of pks15/1 unveils regulatory patterns controlling phenolphtiocerol and phenolglycolipid production in pathogenic mycobacteria. PLoS One 15(5):e0229700. https://doi.org/10.1371/journal.pone.0229700

Rana N, Grover P, Singh H (2024) Recent developments and future perspectives of purine derivatives as a promising scaffold in drug discovery. Curr Top Med Chem 24(6):541–579. https://doi.org/10.2174/0115680266290152240110074034

Rao V, Gao F, Chen B, Jacobs WR Jr, Glickman MS (2006) Trans-cyclopropanation of mycolic acids on trehalose dimycolate suppresses Mycobacterium tuberculosis–induced inflammation and virulence. J Clin Invest 116(6):1660–1667. https://doi.org/10.1172/JCI27335

Rao SP, Alonso S, Rand L, Dick T, Pethe K (2008) The protonmotive force is required for maintaining ATP homeostasis and viability of hypoxic, nonreplicating mycobacterium tuberculosis. Proc Natl Acad Sci 105(33):11945–11950

Rastogi N, Legrand E, Sola C (2001) The mycobacteria: an introduction to nomenclature and pathogenesis. Revue Scientifique et Technique (International Office of Epizootics) 20(1):21–54. https://doi.org/10.20506/rst.20.1.1265

Raymond JB, Mahapatra S, Crick DC, Pavelka MS Jr (2005) Identification of the namH gene, encoding the hydroxylase responsible for the N-glycosylation of the mycobacterial peptidoglycan. J Biol Chem 280(1):326–333. https://doi.org/10.1074/jbc.M411006200

Raze D, Verwaerde C, Deloison G, Werkmeister E, Coupin B, Loyens M et al (2018) Heparin-binding hemagglutinin Adhesin (HBHA) is involved in Intracytosolic lipid inclusions formation in mycobacteria. Front Microbiol 9:2258. https://doi.org/10.3389/fmicb.2018.02258

Rens C et al (2021) Roles for phthiocerol dimycocerosate lipids in Mycobacterium tuberculosis pathogenesis. Microbiology 167(3):001042. https://doi.org/10.1099/mic.0.001042

Richards JP, Cai W, Zill NA, Zhang W, Ojha AK (2019) Adaptation of mycobacterium tuberculosis to biofilm growth is genetically linked to drug tolerance. Antimicrob Agents Chemother 63(11)

Rosenkrands I, Weldingh K, Jacobsen S, Hansen CV, Florio W, Gianetri I, Andersen P (2000a) Mapping and identification of Mycobacterium tuberculosis proteins by two-dimensional gel electrophoresis, microsequencing and immunodetection. Electrophoresis 21(5):935–948. https://doi.org/10.1002/(SICI)1522-2683(20000301)21:5<935::AID--ELPS935>3.0.CO;2-P

Rosenkrands I et al (2000b) Towards the proteome of Mycobacterium tuberculosis, vol 21

Ruiz N (2008) Bioinformatics identification of MurJ (MviN) as the peptidoglycan lipid II flippase in Escherichia coli. Proc Natl Acad Sci USA 105(40):15553–15557

Sacco E, Covarrubias AS, O'Hare HM, Carroll P, Eynard N, Jones TA et al (2007) The missing piece of the type II fatty acid synthase system from Mycobacterium tuberculosis. Proc Natl Acad Sci 104(36):14628–14633. https://doi.org/10.1073/pnas.0704132104

Sait B, Rizwan M, Koliwer-Brandl H, Stewart J, Swarts B, Jacobsen M et al (2022) PPE51 mediates uptake of trehalose across the mycomembrane of Mycobacterium tuberculosis. Sci Rep 12:2097. https://doi.org/10.1038/s41598-022-06109-7

Sancho-Vaello E et al (2020) The structure of the antimicrobial human cathelicidin LL-37 shows oligomerization and channel formation in the presence of membrane mimics. Sci Rep 10:17356. https://doi.org/10.1038/s41598-020-74401-5

Sandhu P, Akhter Y (2015) The internal gene duplication and interrupted coding sequences in the MmpL genes of mycobacterium tuberculosis: towards understanding the multidrug transport in an evolutionary perspective. Int J Med Microbiol 305(3):413–423

Sani M, Houben EN, Geurtsen J, Pierson J, de Punder K, van Zon M, Wever B, Piersma SR, Jiménez CR, Daffé M, Appelmelk BJ, Bitter W, van der Wel N, Peters PJ (2010) Direct visualization by cryo-EM of the mycobacterial capsular layer: a labile structure containing ESX-1-secreted proteins. PLoS Pathog 6(3):e1000794. https://doi.org/10.1371/journal.ppat.1000794

Sankaranarayanan R, Saxena P, Marathe UB, Gokhale RS, Shanmugam VM, Rukmini R (2004) A novel tunnel in mycobacterial type III polyketide synthase reveals the structural basis for generating diverse metabolites. Nat Struct Mol Biol 11(9):894–900. https://doi.org/10.1038/nsmb811

Saxena A, Srivastava R, Srivastava BS (2018) Differential expression of proteins of Mycobacterium bovis BCG during adaptation to anaerobic non replicating persistence. Int J Pulmonol Infect Dis 1:1–5

Schaeffer ML, Khoo K-H, Besra GS, Chatterjee D, Brennan PJ, Belisle JT, Inamine JM (1999) The pimB gene of mycobacterium tuberculosis encodes a mannosyltransferase involved in lipoarabinomannan biosynthesis. J Biol Chem 274(44):31625–31631

Schaeffer ML, Agnihotri G, Kallender H, Brennan PJ, Lonsdale JT (2001) Expression, purification, and characterization of the Mycobacterium tuberculosis acyl carrier protein, AcpM. Biochimica et Biophysica Acta (BBA) – molecular and cell biology of. Lipids 1532(1–2):67–78. https://doi.org/10.1016/S1388-1981(01)00116-0

Schoenen H et al (2010) Cutting edge: Mincle is essential for recognition and adjuvanticity of the mycobacterial cord factor and its synthetic analog trehalose-dibehenate. J Immunol 184(5):2756–2760. https://doi.org/10.4049/jimmunol.0904013

Schoonmaker MK, Bishai WR, Lamichhane G (2014) Nonclassical transpeptidases of Mycobacterium tuberculosis alter cell size, morphology, the cytosolic matrix, protein localization, virulence, and resistance to β-lactams. J Bacteriol 196(7):1394–1402. https://doi.org/10.1128/JB.01396-13

Schwebach JR, Casadevall A, Schneerson R, Dai Z, Wang X, Robbins JB, Glatman-Freedman A (2001) Expression of a Mycobacterium tuberculosis Arabinomannan antigen in vitro and in vivo. Infect Immun 69(9):5671–5678. https://doi.org/10.1128/iai.69.9.5671-5678.2001

Schwebach JR, Glatman-Freedman A, Gunther-Cummins L, Dai Z, Robbins JB, Schneerson R, Casadevall A (2002) Glucan is a component of the Mycobacterium tuberculosis surface that is expressed in vitro and in vivo. Infect Immun 70(5):2566–2575. https://doi.org/10.1128/iai.70.5.2566-2575.2002

Schweizer E, Hofmann J (2004) Microbial type I fatty acid synthases (FAS): major players in a network of cellular FAS systems. Microbiol Mol Biol Rev 68(3):501–517. https://doi.org/10.1128/MMBR.68.3.501-517.2004

Seidel M, Alderwick LJ, Sahm H, Besra GS, Eggeling L (2007) Topology and mutational analysis of the single Emb arabinofuranosyltransferase of Corynebacterium glutamicum as a model of Emb proteins of mycobacterium tuberculosis. Glycobiology 17(2):210–219

Senaratne RH et al (1998) Expression of a gene for a porin-like protein of the OmpA family from Mycobacterium tuberculosis H37Rv. J Bacteriol 180(14):3541–3547

Shaku M, Ealand C, Kana BD (2020) Cell surface biosynthesis and remodeling pathways in mycobacteria reveal new drug targets. Front Cell Infect Microbiol 10:603382

Shao L, Qiu C, Zheng L, Yang Y, Yang X, Liang Q et al (2020) Comparison of diagnostic accuracy of the GeneXpert ultra and cell-free nucleic acid assay for tuberculous meningitis: a multicentre prospective study. Int J Infect Dis 98:441–446

Sharma S, Sharma M, Bose M (2009) Mycobacterium tuberculosis infection of human monocyte-derived macrophages leads to apoptosis of T cells. Immunol Cell Biol 87(3):226–234

Sharma D, Gautam S, Srivastava N, Khan AM, Bisht D (2024) Comparative proteomic analysis of capsule proteins in aminoglycoside-resistant and sensitive Mycobacterium tuberculosis clinical isolates: unraveling potential drug targets. International Journal of Mycobacteriology 13(2):197–205. https://doi.org/10.4103/ijmy.ijmy_47_24

Shinde Y, Ahmad I, Surana S, Patel H (2021) The mur enzymes chink in the Armour of mycobacterium tuberculosis cell wall. Eur J Med Chem 222:113568

Singh V, Mani I, Chaudhary DK, Somvanshi P (2011) The β-ketoacyl-ACP synthase from Mycobacterium tuberculosis as potential drug targets. Curr Med Chem 18(9):1318–1324. https://doi.org/10.2174/092986711795029573

Sinha S, Arora S, Kosalai K, Namane A, Pym AS, Cole ST (2002) Proteome analysis of the plasma membrane of mycobacterium tuberculosis. Comp Func Geno 3(6):470–483

Siroy A, Mailaender C, Harder D, Koerber S, Wolschendorf F, Danilchanka O et al (2008) Rv1698 of Mycobacterium tuberculosis represents a new class of channel-forming outer membrane proteins. J Biol Chem 283(26):17827–17837. https://doi.org/10.1074/jbc.M800866200

Slama N, Leiba J, Eynard N, Daffé M, Kremer L, Quémard A, Molle V (2011) Negative regulation by Ser/Thr phosphorylation of HadAB and HadBC dehydratases from mycobacterium tuberculosis type II fatty acid synthase system. Biochem Biophys Res Commun 412(3):401–406

Song H, Sandie R, Wang Y, Andrade-Navarro MA, Niederweis M (2008) Identification of outer membrane proteins of Mycobacterium tuberculosis. Tuberculosis 88(6):526–544. https://doi.org/10.1016/j.tube.2008.05.005

Sparks IL, Kado T, Prithviraj M, Nijjer J, Yan J, Morita YS (2024) Lipoarabinomannan mediates localized cell wall integrity during division in mycobacteria. Nat Commun 15:2191. https://doi.org/10.1038/s41467-024-46565-5

Stephan J, Mailaender C, Etienne G, Daffé M, Niederweis M (2004) Multidrug resistance of a porin deletion mutant of mycobacterium smegmatis. Antimicrob Agents Chemother 48(11):4163–4170

Stokes RW, Waddell SJ (2009) Adjusting to a new home: Mycobacterium tuberculosis gene expression in response to an intracellular lifestyle. Future Microbiol 4(10):1317–1335. https://doi.org/10.2217/fmb.09.94

Strong EJ, Wang J, Ng TW, Porcelli SA, Lee S (2022) Mycobacterium tuberculosis PPE51 inhibits autophagy by suppressing toll-like receptor 2-dependent signaling. MBio 13(3):e02974–e02921. https://doi.org/10.1128/mbio.02974-21

Su CC, Klenotic PA, Cui M, Lyu M, Morgan CE, Yu EW (2021) Structures of the mycobacterial membrane protein MmpL3 reveal its mechanism of lipid transport. PLoS Biol 19(8):e3001370. https://doi.org/10.1371/journal.pbio.3001370

Tahlan K, Wilson R, Kastrinsky DB, Arora K, Nair V, Fischer E et al (2012) SQ109 targets MmpL3, a membrane transporter of trehalose monomycolate involved in mycolic acid donation to the cell wall core of Mycobacterium tuberculosis. Antimicrob Agents Chemother 56(4):1797–1809. https://doi.org/10.1128/AAC.05708-11

Takayama K, Wang C, Besra GS (2005) Pathway to synthesis and processing of mycolic acids in Mycobacterium tuberculosis. Clin Microbiol Rev 18(1):81–101. https://doi.org/10.1128/CMR.18.1.81-101.2005

Torrelles JB, Schlesinger LS (2010) Diversity in Mycobacterium tuberculosis mannosylated cell wall determinants impacts adaptation to the host. Tuberculosis 90(2):84–93. https://doi.org/10.1016/j.tube.2010.02.003

Torrelles JB, Azad AK, Schlesinger LS (2006) Fine discrimination in the recognition of individual species of Phosphatidyl-myo-inositol Mannosides from Mycobacterium tuberculosis by C-type lectin pattern recognition receptors. J Immunol 177(3):1805–1816. https://doi.org/10.4049/jimmunol.177.3.1805

Touchette MH, Holsclaw CM, Previti ML, Solomon VC, Leary JA, Bertozzi CR, Seeliger JC (2015) The rv1184c locus encodes Chp2, an acyltransferase in mycobacterium tuberculosis polyacyltrehalose lipid biosynthesis. J Bacteriol 197(1):201–210

Trivedi OA, Arora P, Vats A, Ansari MZ, Tickoo R, Sridharan V et al (2005) Dissecting the mechanism and assembly of a complex virulence mycobacterial lipid. Mol Cell 17(5):631–643. https://doi.org/10.1016/j.molcel.2005.02.009

Tsai YT, Salzman V, Cabruja M, Gago G, Gramajo H (2017) Role of long-chain acyl-CoAs in the regulation of mycolic acid biosynthesis in mycobacteria. Open Biol 7(7)

Typas A, Banzhaf M, Gross CA, Vollmer W (2012) From the regulation of peptidoglycan synthesis to bacterial growth and morphology. Nat Rev Microbiol 10(2):123–136. https://doi.org/10.1038/nrmicro2677

Umesiri FE, Sanki AK, Boucau J, Ronning DR, Sucheck SJ (2010) Recent advances toward the inhibition of mAG and LAM synthesis in mycobacterium tuberculosis. Med Res Rev 30(2):290–326

Varela C, Rittmann D, Singh A, Krumbach K, Bhatt K, Eggeling L et al (2012) MmpL genes are associated with mycolic acid metabolism in mycobacteria and corynebacteria. Chem Biol 19(4):498–506. https://doi.org/10.1016/j.chembiol.2012.03.006

Velayati AA, Farnia P (2016) Atlas of Mycobacterium tuberculosis. Academic

Velayati AA, Farnia P, Ibrahim TA, Haroun RZ, Kuan HO, Ghanavi J et al (2009) Differences in cell wall thickness between resistant and nonresistant strains of Mycobacterium tuberculosis: using transmission electron microscopy. Chemotherapy 55(5):303–307. https://doi.org/10.1159/000226425

Verbelen C, Dupres V, Raze D, Bompard C, Locht C, Dufrêne YF (2008) Interaction of the mycobacterial heparin-binding hemagglutinin with actin, as evidenced by single-molecule force spectroscopy. J Bacteriol 190(23):7614–7620. https://doi.org/10.1128/JB.00974-08

Vergne I, Chua J, Singh SB, Deretic V (2004) Cell biology of mycobacterium tuberculosis phagosome. Annu Rev Cell Dev Biol 20:367–394

Verschoor JA, Baird MS, Grooten J (2012) Towards understanding the functional diversity of cell wall mycolic acids of Mycobacterium tuberculosis. Prog Lipid Res 51(4):325–339. https://doi.org/10.1016/j.plipres.2012.05.002

Vilchèze C (2020) Mycobacterial cell wall: a source of successful targets for old and new drugs. Appl Sci 10(7):2278. https://doi.org/10.3390/app10072278

Viljoen A, Dubois V, Girard-Misguich F, Blaise M, Herrmann JL, Kremer L (2017) The diverse family of MmpL transporters in mycobacteria: from regulation to antimicrobial developments. Mol Microbiol 104(6):889–904. https://doi.org/10.1111/mmi.13675

Vincent AT, Nyongesa S, Morneau I, Reed MB, Tocheva EI, Veyrier FJ (2018) The mycobacterial cell envelope: a relict from the past or the result of recent evolution? Front Microbiol 9:2341. https://doi.org/10.3389/fmicb.2018.02341

Vinod V, Vijayrajratnam S, Vasudevan AK, Biswas R (2020) The cell surface adhesins of Mycobacterium tuberculosis. Microbiol Res 232:126392. https://doi.org/10.1016/j.micres.2019.126392

Wang Q, Boshoff HIM, Harrison JR, Ray PC, Green SR, Wyatt PG, Barry CE (2020) PE/PPE proteins mediate nutrient transport across the outer membrane of Mycobacterium tuberculosis. Science 367(6482):1147–1151. https://doi.org/10.1126/science.aav5912

Wang H, Liu D, Zhou X (2023) Effect of mycolic acids on host immunity and lipid metabolism. Int J Mol Sci 25(1):396. https://doi.org/10.3390/ijms25010396

Watanabe M, Aoyagi Y, Ridell M, Minnikin DE (2001) Separation and characterization of individual mycolic acids in representative mycobacteria. Microbiol (Reading). 147(Pt 7):1825–1837

Wells RM, Jones CM, Xi Z, Speer A, Danilchanka O, Doornbos KS et al (2013) Discovery of a siderophore export system essential for virulence of Mycobacterium tuberculosis. PLoS Pathog 9(1):e1003120. https://doi.org/10.1371/journal.ppat.1003120

Werninghaus K et al (2009) Adjuvanticity of a synthetic cord factor analogue for subunit Mycobacterium tuberculosis vaccination requires FcRγ-Syk-Card9-dependent innate immune activation. J Exp Med 206(1):89–97. https://doi.org/10.1084/jem.20081445

Wesener DA, Levengood MR, Kiessling LL (2017) Comparing galactan biosynthesis in Mycobacterium tuberculosis and Corynebacterium diphtheriae. J Biol Chem 292(7):2944–2955. https://doi.org/10.1074/jbc.M116.759340

Wolschendorf F, Ackart D, Shrestha TB, Hascall-Dove L, Nolan S, Lamichhane G et al (2011) Copper resistance is essential for virulence of Mycobacterium tuberculosis. Proc Natl Acad Sci 108(4):1621–1626. https://doi.org/10.1073/pnas.1009261108

Wolucka BA (2008) Biosynthesis of D-arabinose in mycobacteria—a novel bacterial pathway

with implications for antimycobacterial therapy. FEBS J 275(11):2691–2711. https://doi.org/10.1111/j.1742-4658.2008.06395.x
Wright PW, Wallace RJ Jr, Wright NW, Brown BA, Griffith DE (1998) Sensitivity of Fluorochrome microscopy for detection of Mycobacterium tuberculosis versus nontuberculous mycobacteria. J Clin Microbiol 36(4):1046–1049
Wu Y, Xiong DC, Chen SC, Wang YS, Ye XS (2017) Total synthesis of mycobacterial arabinogalactan containing 92 monosaccharide units. Nat Commun 8:14851. https://doi.org/10.1038/ncomms14851
Xiong Y, Wu X, Mahmud T (2005) A homologue of the mycobacterium tuberculosis PapA5 protein, rif-orf20, is an acetyltransferase involved in the biosynthesis of antitubercular drug rifamycin B by Amycolatopsis mediterranei S699. Chembiochem 6(5):834–837
Xu C, Yue Y, Xiong S (2023) Mycobacterium tuberculosis Rv0928 protein facilitates macrophage control of mycobacterium infection by promoting mitochondrial intrinsic apoptosis and ROS-mediated inflammation. Front Microbiol 14:1291358
Yagi T, Wachino J, Kurokawa H, Suzuki S, Yamane K, Doi Y et al (2005) Practical methods using boronic acid compounds for identification of class C beta-lactamase-producing Klebsiella pneumoniae and Escherichia coli. J Clin Microbiol 43(6):2551–2558
Yamamoto K, Nakata N, Mukai T, Kawagishi I, Ato M (2021) Coexpression of MmpS5 and MmpL5 contributes to both efflux transporter MmpL5 trimerization and drug resistance in mycobacterium tuberculosis. MSphere 6(1):00518-20. https://doi.org/10.1128/msphere
Yu L, Cao N, Wang L, Xiao C, Guo M, Chu J et al (2012) Oxytetracycline biosynthesis improvement in Streptomyces rimosus following duplication of minimal PKS genes. Enzym Microb Technol 50(6–7):318–324. https://doi.org/10.1016/j.enzmictec.2012.03.001
Yuan Y, Barry CE (1996) A common mechanism for the biosynthesis of methoxy and cyclopropyl mycolic acids in Mycobacterium tuberculosis. Proc Natl Acad Sci 93(23):12828–12833. https://doi.org/10.1073/pnas.93.23.12828
Zhai W, Wu F, Zhang Y, Fu Y, Liu Z (2019) The immune escape mechanisms of Mycobacterium tuberculosis. Int J Mol Sci 20(2):340. https://doi.org/10.3390/ijms20020340
Zhang L, Hendrickson RC, Meikle V, Lefkowitz EJ, Ioerger TR, Niederweis M (2020) Comprehensive analysis of iron utilization by Mycobacterium tuberculosis. PLoS Pathog 16(2):e1008337. https://doi.org/10.1371/journal.ppat.1008337
Zimhony O, Cox JS, Welch JT, Vilchèze C, Jacobs WR Jr (2000) Pyrazinamide inhibits the eukaryotic-like fatty acid synthetase I (FASI) of Mycobacterium tuberculosis. Nat Med 6(9):1043–1047. https://doi.org/10.1038/79558

5 Protein-Mediated Virulence in *Mycobacterium tuberculosis*

Abstract

The pathogenic potential of *Mycobacterium tuberculosis* (Mtb), the etiological agent of tuberculosis, is attributable primarily to its complex and highly specialized repertoire of molecular determinants, comprising both proteins and lipids. These virulence factors operate synergistically to facilitate evasion of host immune defenses, establish persistent infection, and sustain prolonged survival within the host environment, often for months or even years. These factors can be systematically classified according to their functional roles, molecular composition, and spatial localization within or in proximity to the bacterial cell, collectively underscoring Mtb's extraordinary capacity to adapt to, manipulate, and thrive within hostile host environments. Central to Mtb's defense and survival strategy is its robust lipid and fatty acid metabolism, which orchestrates the biosynthesis of a complex, lipid-rich, waxy cell wall. This structure serves as a formidable physical and biochemical barrier against immune-mediated clearance and antimicrobial agents. Enzymes such as KasA and AccD6 are indispensable in the biosynthetic pathway of mycolic acids, long-chain fatty acids that confer the cell envelope with structural rigidity, impermeability, and resistance to chemical and enzymatic attack. Complementing this, proteins such as Rv2952 contribute to the synthesis of phenolic glycolipids (PGLs), which effectively "cloak" the bacterial surface by masking antigenic epitopes, thereby facilitating immune evasion. Among the most prominent lipid virulence factors is trehalose dimycolate (TDM), commonly referred to as the "cord factor." TDM plays a critical role in granuloma formation. These dense aggregates of immune cells characterize tuberculosis pathology and modulate host cell death pathways in a manner that promotes bacterial persistence and dissemination. Supporting this protective lipid barrier are lipoproteins such as LprG and LpqH, which modulate host immune signaling pathways to attenuate inflammatory responses, thereby enhancing bacterial survival. Concurrently, proteins such as Rv2700 contribute to maintaining cell wall stability, particularly under environmental and immunological stress. Collectively, these factors significantly impede adequate immune clearance, enabling Mtb to persist despite sustained host immune pressure. A particularly potent virulence mechanism employed by Mtb involves specialized secretion systems, notably the ESX-1 Type VII secretion system (T7SS). This molecular apparatus functions analogously to a syringe, translocating effector proteins directly into host cells. Key secreted proteins such as ESAT-6 and CFP-10 disrupt host cell membrane integrity and modulate immune

P. Farnia et al., *Proteins in Mycobacterium Tuberculosis*, Advances in Experimental Medicine and Biology 1484, https://doi.org/10.1007/978-3-031-96883-9_5

responses, while accessory Esp proteins facilitate their secretion and delivery. The PE/PPE protein families further enhance immune evasion by promoting antigenic variation, thereby enabling Mtb to circumvent adaptive immune recognition. A defining feature of Mtb's survival strategy is its sophisticated regulation of host cell death pathways. The bacterium inhibits apoptosis, the programmed cell death pathway, via proteins such as NuoG, which interfere with mitochondrial stress signaling cascades. Conversely, Mtb can selectively induce apoptosis through ESAT-6 when such cell death is advantageous for bacterial dissemination. Moreover, Mtb actively induces necrosis, a pro-inflammatory and lytic form of cell death, through toxins such as tuberculosis necrotizing toxin (TNT) and lipid virulence factors including phthiocerol dimycocerosates (PDIMs). These effectors compromise host cell membrane integrity and deplete critical metabolites such as NAD^+, facilitating bacterial escape and spread. Additionally, Mtb exploits ferroptosis, an iron-dependent regulated cell death pathway associated with oxidative stress, by elevating intracellular free iron within immune cells, thereby promoting damaging oxidative reactions. In contrast, Mtb suppresses pyroptosis, a highly inflammatory programmed cell death, via the Zmp1 protein, which inhibits inflammasome activation, thus dampening inflammation and enabling silent intracellular replication. Within macrophages, the primary immune effector cells responsible for bacterial clearance, Mtb deploys a diverse array of proteins to neutralize host antimicrobial defenses. For example, AhpC detoxifies reactive oxygen species, while KatG mitigates nitrosative stress and is responsible for activating the frontline antibiotic isoniazid. Serine/threonine protein kinases such as PknG regulate intracellular homeostasis and facilitate bacterial survival under immune pressure. Proteases including ZmpA and ZmpB degrade host proteins and modulate immune responses, occasionally contributing to host tissue pathology. To sustain metabolism in nutrient-limited environments, Mtb employs specialized metal transporters to acquire and maintain homeostasis of essential micronutrients such as iron and zinc, effectively circumventing host-imposed nutritional immunity. In summary, Mtb possesses a remarkably sophisticated and multifaceted arsenal of virulence factors that collectively enable it to withstand intense immune pressures. By constructing an almost impenetrable cell wall, deploying specialized secretion systems, modulating host immune pathways, and adapting to the hostile intracellular milieu, Mtb has optimized its capacity for survival and propagation. A comprehensive understanding of the interplay among these diverse systems not only enriches our insight into tuberculosis pathogenesis but also informs the rational design of improved vaccines and therapeutic interventions aimed at ultimately controlling this ancient and devastating disease.

Keywords

Mycobacterium tuberculosis · Virulence proteins · Phthiocerol dimycocerosates (PDIM) · Phenolic glycolipids (PGL) · Type VII secretion systems (ESX) · PE-PGRS subfamily · Tuberculosis necrotizing toxin

5.1 Introduction

Mycobacterium tuberculosis (Mtb) represents an intracellular pathogen predominantly disseminated via airborne droplets, commonly characterized as droplet nuclei, which possess dimensions ranging from 1 to 5 μm. These droplets are expelled into the atmosphere when individuals afflicted with smear-positive tuberculosis (TB) engage in coughing, sneezing, or even verbal communication. Each droplet may harbor two to three bacilli of Mtb. Due to their diminutive size, these droplets have the ability to circumvent the nasopharyngeal and tracheobronchial regions, ultimately reaching the distal airways. Conversely, they may become entrapped within the upper airway or oropharyngeal cavity, potentially resulting in extrapulmonary manifestations of tuberculosis, such as oropharyngeal or cervical lymph node tuberculosis (Laal 2012; Barrios-Payán et al. 2012; Fennelly and Jones-López 2015). Macrophages and dendritic cells predominantly internalize Mtb within the lower respiratory tract. Nevertheless, investigations of sputum specimens from individuals with TB have revealed that neutrophils also serve as significant phagocytic cells that become infected with Mtb. Furthermore, there is corroborative evidence indicating that Mtb can be detected in non-myelocytic cells in patients diagnosed with TB (Eum et al. 2010; Lerner et al. 2015; Randall et al. 2015; Ganbat et al. 2016). Bacilli that successfully evade the upper airways ultimately access the alveolar spaces, where they encounter alveolar macrophages, dendritic cells, and neutrophils (Schluger and Rom 1998; Ulrichs and Kaufmann 2006; Eum et al. 2010; Bussi and Gutierrez 2019). Mtb is internalized by phagocytic cells through the process of phagocytosis, which is orchestrated by an assortment of receptors contingent upon the specific type of immune cell involved. In macrophages, this process is facilitated by a diverse array of receptors, including dectin-1, complement receptor 3, Toll-like receptors, mannose receptor, dendritic cell-specific intercellular adhesion molecule (ICAM)-3-grabbing nonintegrin (DC-SIGN), Fc receptors, scavenger receptors, and CD14 (Schluger and Rom 1998; Ulrichs and Kaufmann 2006; Guirado et al. 2013; Randall et al. 2015; Ganbat et al. 2016). The interplay between Mtb and these immune receptors is crucial in shaping the host immune response and bears significant implications for the formulation of novel therapeutic strategies. A comprehensive understanding of the mechanisms underlying phagocytosis and receptor interactions is paramount for the development of targeted treatments and the enhancement of vaccine efficacy against tuberculosis. Once internalized by macrophages, Mtb inhabits a specialized compartment referred to as the phagosome. In typical physiological conditions, phagosomes undergo maturation and subsequently fuse with lysosomes to form phagolysosomes, which create acidic environments specifically tailored for the degradation of pathogens. However, Mtb effectively impedes this maturation process by obstructing acidification at a pH level of approximately 6.4 and by inhibiting the fusion of phagosomes with lysosomes (Armstrong and Hart 1971; Clemens and Horwitz 1995; Deretic and Fratti 1999; Ramachandra et al. 2005; Deretic 2008; Lee et al. 2020a, b; Liu et al. 2024a, b). The capacity of Mtb to endure and persist within the host is predominantly ascribed to a diverse array of virulence factors (Lyu et al . 2024). These factors are generally classified according to their functional roles in the pathogenic process, which encompass the preservation of cell envelope integrity, the operation of specialized secretion systems, the modulation of immune responses, the acquisition of essential metal ions, the regulation of gene expression, and the execution of enzymatic functions that promote bacterial survival. I) Proteins that are integral to the structure of the cell envelope and lipid metabolism are critical for the preservation of the bacterial cell wall and the modulation of host immune responses (Gago and Gramajo 2018; Stokas et al. 2020; Sun et al. 2025). Notable among these are enzymes such as AccD6 and KasA, which are indispensable for the biosynthesis of mycolic acids, a fundamental component that contributes to the impermeability of the cell wall and provides resistance against immune-mediated assaults (Slayden and Barry 2002; Kurth et al. 2009; Luckner et al. 2009; Tsai et al. 2006). Lipids, including phthiocerol dimycocerosates (PDIM) and phenolic glycolipids (PGL), facilitate Mtb's evasion of the immune system by concealing pathogen-associated molecular patterns (PAMPs) (Cambier et al. 2017; Layre 2020). II) Specialized secretion systems, particularly the Type VII secretion systems (ESX), play a pivotal role in Mtb's ability to thrive within host cells. For instance, the ESX-1 system is responsible for the secretion of CFP-10 and ESAT-6, which disrupt the integrity of phagosomal membranes, thereby assisting Mtb in evading degradation within lysosomal compartments (Poirier and Av-Gay 2012; Sachdeva et al. 2020; Pal et al. 2022; Passos et al. 2024). III) Proteins that are rich in glycine also contribute to immune evasion and the establishment of chronic infection. The PE-PGRS subfamily, for example, enhances adhesion to host cells while simultaneously facilitating evasion from the immune system (Hill et al. 2025; Borhani et al. 2024). IV) Certain enzymatic proteins are adept at manipulating host cell apoptosis to the advantage of Mtb. For example, the tuberculosis necrotizing toxin (TNT) induces necroptosis by depleting NAD+, which results in mitochondrial dysfunction and subsequent cell death (Maueröder et al. 2016; Pajuelo et al. 2018; Kim et al. 2024). The catalase-peroxidase enzyme KatG is instrumental in Mtb's resistance to oxidative stress while concurrently influencing apoptosis within infected macrophages (Chandra et al. 2022; Fan et al. 2023; Vu et al. 2024). Additional proteins assist Mtb in counteracting host antimicrobial mechanisms. For instance, SodA serves to neutralize reactive oxygen species (ROS), whereas NirBD protects against nitrosative stress (Bose and Farnia 1995; Piddington et al. 2001; Wang et al. 2018; Farnese et al. 2017). VI) The acquisition and regulation of metals are of paramount importance for the survival of *Mycobacterium tuberculosis* (Mtb). This bacterium necessitates essential metals such as iron, zinc, and copper, while also requiring stringent control over their concentrations to avert toxic consequences. Siderophores, including mycobactin, which are translocated by the MmpL4/MmpL5 transport proteins, facilitate the sequestration of iron from the host organism (Sritharan 2016; Sandhu and Akhter 2017; Chao et al. 2019; Earp et al. 2025). VII) Regulatory proteins, encompassing two-component systems such as PhoP/PhoR and sigma factors including SigH, are essential in assisting Mtb to acclimate to environmental pressures encountered during infection. These regulatory elements have a significant impact on lipid metabolism and the modulation of stress responses, including oxidative stress and hypoxic conditions within granulomatous structures (Groisman 2001; Gonzalo Asensio et al. 2006; Manganelli 2014; Dutta et al. 2012). VIII) Proteolytic enzymes serve dual functions in the degradation of host proteins and the activation of Mtb factors that are critical for pathogenicity. MycP1 and MycP3 are essential for the optimal operation of Type VII secretion systems, whereas metalloproteases facilitate the breakdown of the host's extracellular matrix, thus exacerbating tissue destruction during the advanced stages of tuberculosis (Ohol et al. 2010; Forrellad et al. 2013; van Winden 2021). IX) Proteins that exert direct influences on the host immune response are pivotal in undermining the host's defensive mechanisms. For example, the CFP-10/ESAT-6 complex disrupts the integrity of phagosome membranes, thereby enabling Mtb to translocate into the cytosol and evade immune surveillance (De Jonge et al. 2007; Bates et al. 2024; Koiri et al. 2025). Furthermore, proteins such as SapM and PknG inhibit the normal maturation of phagosomes, consequently obstructing autophagic processes and permitting Mtb to persist within host cells (Pieters 2008; Zulauf et al. 2018; Zhang et al. 2024). Some proteins of Mtb, particularly those belonging to the PE/PPE family, have been empirically associated with virulence, although their exact functional roles remain under investigation. As research progresses concerning these proteins, they may emerge as promising targets for novel therapeutic strategies. Collectively, these diverse classes of immune-modulating proteins exemplify the intricate strategies employed by Mtb to elude detection and establish protracted infections. Their roles in immune evasion render them significant candidates for prospective drug and vaccine development. A comprehensive understanding of these molecular mechanisms will be crucial for the formulation of more efficacious tools aimed at combating this enduring global health challenge.

5.2 Cell Envelope Proteins and Lipid Metabolism as Virulence Factors in *Mycobacterium tuberculosis*

The cell wall of Mtb is a critical factor in its virulence and survival within the host. It consists of a complex combination of lipids, carbohydrates, and proteins (Jankute et al. 2015; Bah et al. 2020). Prominent components such as lipoarabinomannan (LAM), trehalose dimycolate (TDM, or cord factor), and mycolic acids contribute significantly to Mtb's ability to evade host immune defenses and persist under harsh conditions (Rajni et al. 2011; Ghazaei 2018). These lipid elements modulate host immunity by inhibiting phagosome maturation and reducing neutrophil migration (Cox et al. 1999; Rajni et al. 2011). Proteomic analyses have revealed over 500 proteins associated with the cell wall, many of which are involved in small-molecule metabolism and macromolecular biosynthesis (Wolfe et al. 2010). Among these, AccD6, encoded by the accD6 gene (Rv2247, located at 2520743 bp), plays a pivotal role in mycolic acid synthesis. It catalyzes the carboxylation of acetyl-CoA to malonyl-CoA, a key step in the fatty acid synthase II (FAS-II) pathway, providing precursors for the elongation of meromycolic acids (Daniel 2006; Pawelczyk et al. 2011, 2017; Odriozola et al. 1977; Sacco et al. 2007; Lefebvre et al. 2018). The crystal structure of AccD6 has been elucidated both in its apo form and in complex with haloxyfop-R, an herbicide that inhibits its activity (Reddy et al. 2014). Structural studies revealed two inhibitor-binding sites per subunit: one overlapping the active site and another located in an adjacent pocket, highlighting its potential as a drug target (Zhang et al. 2004; Reddy et al. 2014; Qin et al. 2024). The enzyme's substrate-binding pocket interacts with acetyl-CoA and biotin, enabling the carboxylation reaction. Crucial residues such as Lys401, Lys403, and Lys404 participate in electrostatic interactions with the phosphate groups of acetyl-CoA (Reddy et al. 2014). Functionally, AccD6 forms a complex with AccA3 and AccE to constitute the functional acetyl-CoA carboxylase (ACC) enzyme, which preferentially carboxylates acetyl-CoA to produce malonyl-CoA, a substrate for fatty acid synthases (Daniel 2006; Kurth et al. 2009; Gago et al. 2011). This process is vital for maintaining cell envelope integrity and virulence in Mtb. AccD6 is regulated at both transcriptional and posttranslational levels, with its expression peaking during exponential growth phases (Liu et al. 2018). The proper assembly of these lipids ensures cell envelope integrity, which is crucial for bacterial virulence. Moreover, mycolic acid covers pathogen-associated molecular patterns (PAMPs), hindering their recognition by the host immune system. Disruption of the accD6 gene impairs mycolic acid synthesis, alters cell morphology, and reduces virulence in vivo. Notably, accD6 is essential for Mtb but not for nonpathogenic strains such as *Mycobacterium smegmatis*. KasA and KasB are β-ketoacyl synthases that play essential roles in mycolic acid biosynthesis in Mtb, contributing critically to cell wall formation and bacterial virulence (Slayden and Barry 3rd 2000). KasA catalyzes the elongation of palmitoyl-CoA into 40-carbon monounsaturated fatty acids, while KasB further extends these fatty acids to 54-carbon multiunsaturated hydrocarbons (Slayden and Barry 3rd 2000). Both enzymes exhibit high specificity for long-chain acyl-acyl carrier proteins (acyl-ACPs) and are susceptible to inhibition by compounds such as cerulenin and thiolactomycin (Schaeffer et al. 2001). Mutations in KasA have been linked to isoniazid resistance in Mtb (Slayden and Barry 3rd 2000). Additionally, phosphorylation of KasA and KasB by serine-threonine protein kinases downregulates their enzymatic activity, indicating a regulatory mechanism that may help Mtb adapt to changing environmental conditions (Alsayed and Gunosewoyo 2023). The enzyme KasA is crucial for synthesizing mycolic acids, fundamental molecules that impart structural integrity and environmental resistance to the Mtb cell wall (Pawelczyk et al. 2011; Li et al. 2022). As a member of the β-ketoacyl synthase family, KasA adopts a thiolase fold typical of fatty acid biosynthesis enzymes (Singh et al. 2011; Rudraraju et al. 2022). Its active site contains key residues such as Cys171 and His345, which catalyze the Claisen condensation reaction essential for fatty acyl chain elongation (Luckner et al. 2009; Lefebvre et al. 2020; Dover et al. 2021). Crystal structures of KasA, both in its apo form and complexed with inhibitors like thiolactomycin (TLM), have provided detailed insights into its substrate-binding channel and catalytic mechanism. TLM binds within the malonyl-binding site and the phosphopantetheine tunnel, which accommodate the long-chain fatty acid substrates (Luckner et al. 2009; Rudraraju et al. 2022). By maintaining the lipid-rich cell envelope, KasA enables Mtb to resist oxidative and nitrosative stresses encountered inside macrophages, thereby promoting bacterial survival. Genetic studies further confirm that kasA is essential for Mtb viability both in vitro and in vivo (Rudraraju et al. 2022; J. Yang et al. 2023). Virulence-associated proteins in Mtb arise from various genes and mechanisms. In a signature-tagged transposon mutagenesis screen, 16 mutants with reduced virulence were identified, with numerous affected genes involved in lipid metabolism or membrane transport (Camacho et al. 1999). The Rv2952 gene, located at position 3,304,441 base pairs, encodes a methyltransferase enzyme that plays a crucial role in the biosynthesis of phenolic glycolipids (PGLs), which are important virulence factors in Mtb.

PGLs help Mtb evade immune detection, disrupt phagosome maturation, and suppress host immune responses, thereby enhancing the pathogen's ability to persist and cause disease (Pérez et al. 2004; Huet et al. 2009; Guzmán et al. 2024; Guzmán-Téllez et al. 2024). The Rv2952 enzyme methylates lipid components of phthiocerol and glycosylated phenolphthiotriol dimycocerosates, key intermediates in the biosynthesis of PGLs and phthiocerol dimycocerosates (DIMs). These methylation modifications are essential for maintaining the structure and function of the bacterial cell envelope. PGLs mask pathogen-associated molecular patterns (PAMPs) and interfere with cytokine signaling pathways, reducing immune activation (Guzmán et al. 2024). The Rv2952 gene is part of a larger gene cluster involved in lipid biosynthesis, including pathways for DIM and p-HBAD production (Constant et al. 2002), indicating coordinated regulation of these virulence factors.

Polyketide synthases (PKSs) in Mtb play crucial roles in synthesizing complex cell wall lipids and virulence factors. PpsE, a type I PKS, is involved in phthiocerol dimycocerosate (PDIM) synthesis and interacts with the MmpL7 transporter, suggesting coupled synthesis and export of virulence lipids (Jain and Cox 2005; Onwueme et al. 2005). PKS11, a type III PKS, synthesizes methyl-branched alkylpyrones using long-chain acyl-CoA substrates and both malonyl-CoA and methylmalonyl-CoA as cofactors (Gokulan et al. 2013; Saxena et al. 2003). Fatty acyl-AMP ligases (FAALs) activate fatty acids and transfer them to PKSs, contributing to the production of unusual acyl chains in mycobacterial lipids (Mohanty et al. 2011). The availability of methyl malonyl CoA (MMCoA) controls the size and abundance of PDIM and SL-1, with increased MMCoA levels leading to enhanced virulence lipid synthesis (Jain et al. 2007). This metabolic coupling is particularly relevant during infection, as Mtb catabolizes host lipids, potentially increasing MMCoA flux through lipid biosynthetic pathways and promoting virulence lipid production (Cannon et al. 2011; Ouellet et al. 2013). LprG, a lipoprotein found in the cell envelope of Mtb and encoded by the Rv1411c gene (located at position 1,587,772 base pairs), plays a crucial role in the bacterium's ability to survive and cause disease. It helps present lipoarabinomannan (LAM), a major glycolipid, on the bacterial surface. This surface exposure of LAM is essential for Mtb's entry into macrophages, its ability to block the fusion of phagosomes with lysosomes, and its survival inside host cells (Drage et al. 2010; Gaur et al. 2014; Martinot et al. 2016; Ramon-Luing et al. 2023). LprG binds triacylated glycolipids like LAM using a specialized hydrophobic pocket. This not only enhances their detection by Toll-like receptor 2 (TLR2) on immune cells but may also contribute to the proper assembly of the bacterial cell wall (Drage et al. 2010). By controlling where LAM is positioned in the cell envelope, LprG significantly influences how the immune system perceives the pathogen (Shukla et al. 2014). What is particularly intriguing is that LprG also seems to have anti-inflammatory properties. It can suppress inflammation triggered by lipopolysaccharides (LPS) in macrophages by interfering with key signaling pathways, such as NF-κB, AP-1, and MAPK (Abekura et al. 2019). When LprG is missing or dysfunctional, Mtb shows reduced levels of LAM on its surface, struggles to enter macrophages, and becomes significantly less virulent

(Shukla et al. 2014). These insights underscore the multifaceted and critical role LprG plays in Mtb pathogenesis.

LpqH, a 19-kilodalton (kDa) lipoprotein produced by Mtb, plays a central role in host-pathogen interactions and immune modulation. Structurally, LpqH is rich in β-sheets and random coils, which support the formation of multiple B cell and cytotoxic T lymphocyte (CTL) epitopes, making it highly immunogenic (Liu et al. 2021). Functionally, LpqH acts as an agonist for Toll-like receptor 2 (TLR2), triggering the production of pro-inflammatory cytokines such as interleukin-1 beta (IL-1β). It also influences cellular pathways like autophagy and programmed cell death (apoptosis). It activates the nucleotide-binding domain, leucine-rich repeat-containing receptor protein 3 (NLRP3) inflammasome, a key component of the host's innate immune defense (Rastogi and Briken 2022; Chen et al. 2024). Interestingly, prolonged exposure to LpqH has been shown to inhibit major histocompatibility complex class II (MHC-II)-mediated antigen presentation. This suppression impairs the activation of cluster of differentiation 4-positive ($CD4^+$) T cells, enabling MTB to persist intracellularly and evade the host's adaptive immune response (Mir and Sharma 2013; O'Neil et al. 2021; Witt 2023). High-resolution structural studies have revealed that LpqH possesses a unique three-dimensional fold with a conserved surface cleft, which may be essential for its interaction with host immune components (Chatterjee et al. 2022). While LpqH contributes significantly to Mtb virulence, other lipoproteins, such as lipoprotein Q subunit S (LpqS), also support bacterial survival inside host macrophages (Sakthi and Narayanan 2013). Not all lipoproteins are directly involved in immune modulation. For example, the LpqY-SugA-SugB-SugC transporter system plays a specialized role in trehalose recycling, a vital process for maintaining the integrity of the mycolic acid-rich cell envelope and supporting overall Mtb virulence (Kalscheuer et al. 2010). Importantly, LpqH expression is specific to slow-growing mycobacteria such as Mtb and M. bovis, and it is notably absent in fast-growing, non-pathogenic mycobacterial species (Liu et al. 2021). This selective expression further emphasizes LpqH's role in MTB pathogenesis. Rv2700(Location: 3015203 bp), recently proposed to be named cei (cell envelope integrity factor), plays a crucial role in maintaining the structural integrity of the cell envelope and supporting the virulence of Mtb (Ballister et al. 2019). When this gene is deleted, the resulting mutant bacteria show increased sensitivity to antibiotics, a more permeable cell envelope, and significantly reduced virulence in mouse infection models. Structurally, Rv2700 is composed of a transmembrane helix and an extracellular LytR_C domain, which is homologous to Virulence Regulator Rv0431 (VirR). These features are similar to those found in a group of enzymes called LytR-CpsA-Psr (LCP) enzymes, which are known for anchoring arabinogalactan to peptidoglycan, both key components of the bacterial cell wall. Interestingly, mutations in LCP enzymes produce effects that closely resemble those seen in strains lacking cei, supporting the idea that Rv2700 carries out a related function in MTB. Moreover, Rv2700 has been identified in genetic screens as a suppressor of nitric oxide sensitivity, suggesting it may help the bacterium resist nitrosative stress within the host. While the exact mechanism behind this protective role is still unclear, these findings further highlight Rv2700's importance in MTB pathogenesis and stress resilience. Trehalose dimycolate (TDM), commonly known as cord factor, is one of the most important glycolipid virulence factors produced by Mtb (Ryll et al. 2001; Kan-Sutton et al. 2009; Ishikawa et al. 2009; Queiroz and Riley 2017; Holzheimer et al. 2021). Structurally, TDM is made up of a trehalose sugar core bound to two mycolic acid chains, forming a molecule that plays a multifaceted role in MTB's interaction with the host. TDM helps the bacterium manipulate the host immune system. It binds to immune receptors such as macrophage-inducible C-type lectin (Mincle), triggering the release of pro-inflammatory cytokines like tumor necrosis factor-alpha (TNF-α) and interleukin-6 (IL-6). This immune activation is critical for the formation of granulomas, organized immune cell structures that contain the infection. TDM also interferes with the normal maturation of phagosomes. These compartments engulf and attempt to destroy bacteria, possibly by engaging Mincle in coordination with Fc gamma receptors (FcγRs) to block phagosome-lysosome fusion (Hunter 2016; Holzheimer et al. 2021; Parbhoo et al. 2022). In addition, TDM stimulates the production of matrix metalloproteinase-9 (MMP-9), an enzyme involved in remodeling granulomas, which may influence disease progression. Remarkably, TDM forms stable monolayers that are resistant to drying out, a feature that could help MTB survive harsh environmental conditions outside the host (Keating et al. 2021).The biosynthesis of TDM is orchestrated by enzymes such as polyketide synthase 13 (Pks13) and the antigen 85 complex (Ag85 complex), which convert trehalose monomycolate (TMM) into the final TDM structure.

5.3 Type VII Secretion System (T7SS) as Virulence Factors in *Mycobacterium tuberculosis*

The Type VII Secretion System (T7SS) in Mtb is a highly specialized and complex protein transport machinery that plays a pivotal role in the bacterium's ability to cause disease and survive within the hostile environment of the host. Unlike many other bacteria, Mtb possesses a unique and exceptionally impermeable cell envelope composed of a thick, lipid-rich outer membrane and an inner membrane, forming a formidable double-membrane barrier. This complex structure restricts the passage of molecules and proteins, rendering conventional secretion systems, such as the Sec and Tat pathways, insufficient for transporting many essential virulence factors. T7SS has explicitly evolved to overcome this challenge by enabling the translocation of a distinct set of effector proteins across the mycobacterial cell envelope (Houben et al. 2014; Bunduc et al. 2020; Famelis and van Tol 2023). These secreted proteins are crucial for manipulating host immune responses, facilitating intracellular survival, and promoting bacterial dissemination. By effectively breaching this impermeable barrier, T7SS allows Mtb to deliver virulence factors directly into the host cell environment, where they can disrupt normal cellular processes, evade immune detection, and establish persistent infections. The critical role of T7SS in Mtb pathogenicity is underscored by its involvement in processes such as phagosomal membrane disruption, modulation of host signaling pathways, and interference with antigen presentation. Without this secretion system, Mtb's ability to survive within macrophages and cause active disease is severely compromised. Therefore, T7SS represents not only a fundamental biological adaptation to the unique mycobacterial cell envelope but also a promising target for novel therapeutic strategies aimed at combating tuberculosis. The Type VII Secretion System (T7SS) in Mtb is a highly specialized and complex protein transport machinery that plays a pivotal role in the bacterium's ability to cause disease and survive within the hostile environment of the host. Unlike many other bacteria, Mtb possesses a unique and exceptionally impermeable cell envelope composed of a thick, lipid-rich outer membrane and an inner membrane, forming a formidable double-membrane barrier. This complex structure restricts the passage of molecules and proteins, rendering conventional secretion systems, such as the Sec and Tat pathways, insufficient for transporting many essential virulence factors. T7SS has evolved specifically to overcome this challenge by enabling the translocation of a distinct set of effector proteins across the mycobacterial cell envelope. These secreted proteins are crucial for manipulating host immune responses, facilitating intracellular survival, and promoting bacterial dissemination. By effectively breaching this impermeable barrier, T7SS allows Mtb to deliver virulence factors directly into the host cell environment, where they can disrupt normal cellular processes, evade immune detection, and establish persistent infections. The critical role of T7SS in Mtb pathogenicity is underscored by its involvement in processes such as phagosomal membrane disruption, modulation of host signaling pathways, and interference with antigen presentation. Without this secretion system, Mtb's ability to survive within macrophages and cause active disease is severely compromised. Therefore, T7SS represents not only a fundamental biological adaptation to the unique mycobacterial cell envelope but also a promising target for novel therapeutic strategies aimed at combating tuberculosis (Ates et al. 2017). Mtb encodes five homologous but functionally distinct T7SS subtypes, designated ESX-1 to ESX-5 (Bottai et al. 2011). Each subtype is responsible for secreting specific classes of substrates, including Esx proteins (WXG100 family), Esp proteins, and PE/PPE proteins, which are critical for the pathogen's ability to interact with its host and establish infection (Brodin et al. 2006 &2011; Shah and Briken 2016; Ates et al. 2017; Bunduc et al. 2020, 2021; Damen et al. 2020; Garrett and Palmer 2024). These systems play a central role in transporting effector proteins across the unique and impermeable mycobacterial cell envelope, which is vital for nutrient acquisition, immune evasion, and pathogenesis. The WXG100 protein family is a group of small proteins characterized by their conserved WXG motif. Members of this protein family are approximately 100 amino acids in length. They are predominantly found in Gram-positive bacteria, including Mtb, where they play critical roles in bacterial survival and virulence. The WXG motif, consisting of tryptophan (W), any amino acid (X), and glycine (G), is centrally located within the protein sequence and serves as a signature for this superfamily. The WXG100 proteins are secreted as full-length molecules and adopt a predominantly α-helical structure. Their tertiary structure often forms a four-helix bundle, which can exist either as homodimers or heterodimers (Callahan et al. 2010; Poulsen et al. 2014; Wang et al. 2020). The most studied members of this family are ESAT-6 (EsxA) and CFP-10 (EsxB) from Mtb, which form a heterodimeric complex. This complex is stabilized by hydrophobic interactions, salt bridges, and hydrogen bonds, resulting in a highly stable structure. The flexible N- and C-terminal regions of these proteins contribute to their ability to interact with host cell membranes and other bacterial factors (Arbing et al. 2010; Prest et al. 2024). The WXG100 protein family is divided into three subfamilies based on sequence similarity: ESAT-6-like proteins, CFP-10-like proteins, and sagEsxA-like proteins (similar to EsxA from *Streptococcus agalactiae*). These subfamilies share less than 15% sequence identity, making them highly diverse despite their conserved structural features (Poulsen et al. 2014). Bioinformatic analyses have revealed that genes encoding WXG100 proteins are often organized in bi-cistronic operons, where one gene encodes a CFP-10-like protein and the adjacent gene encodes an ESAT-6-like protein (Poulsen et al. 2014). This genetic arrangement ensures co-expression and facilitates the formation of heterodimeric complexes.

ESAT-6 (Early Secreted Antigenic Target 6 kDa), encoded by the Rv3875 (Location: 4352609 bp) gene within the RD1 locus, is a highly immunogenic protein secreted via the ESX-1 secretion system (Type VII secretion system). Chemically, ESAT-6 is a small protein with a molecular weight of approximately 6 kDa, characterized by a conserved WXG motif essential for its structural stability and functional activity (Brodin et al. 2004; Sreejit et al. 2014). It ESAT-6 consists of 95 amino acids and adopts an α-helical structure with two long helices connected by a flexible loop, forming a helix-turn-helix hairpin configuration. This amphipathic structure allows ESAT-6 to interact with host cell membranes and other proteins. The flexible N- and C-terminal regions of ESAT-6 enhance its ability to bind host receptors and disrupt phagosomal membranes (Renshaw et al. 2005; Bates and Crowell 2023). Under acidic conditions typical of phagosomes, ESAT-6 undergoes self-association into homodimers or tetramers, which contribute to membrane disruption by destabilizing lipid bilayers or potentially forming pore-like structures (Bates and Crowell 2023). ESAT-6, as a major virulence factor, plays a critical role in phagosomal membrane destabilization, which is essential for bacterial survival and replication within host cells. This process begins with ESAT-6 binding shallowly to the lipid bilayer of the phagosomal membrane, a mechanism that has been implicated in mycobacterial cytosolic translocation and phagosomal rupture (Anes et al.

2023). Unlike classical pore-forming toxins, ESAT-6 does not create transmembrane pores but instead induces structural changes in the lipid bilayer. These changes facilitate the disruption of phagosomal membranes, allowing TB bacilli to escape into the cytosol (Danilchanka et al. 2014; Peng and Sun 2016; Koiri et al. 2025). Additionally, ESAT-6 decreases the compressibility modulus of the membrane, making it more deformable and susceptible to structural changes. The binding of ESAT-6 triggers tubular and bud-like deformations on the surface of the phagosomal membrane (De Leon et al. 2012; Peng and Sun 2016; Koiri et al. 2025). These deformations occur in stages and are concentration-dependent. Initially, small tubules form as ESAT-6 oligomers bind to the membrane and reduce its tension. Over time, these tubules transition into spherical buds due to localized curvature changes and lipid phase separation. The phase separation involves lipid domains enriched in sphingomyelin or cholesterol segregating from other regions, further destabilizing the membrane (Roux et al. 2005; De Leon et al. 2012; Peng and Sun 2016; Koiri et al. 2025). As these structural changes progress, ESAT-6 polymerizes into fibrils on the membrane surface. These fibrils exert mechanical forces on the membrane, generating radial forces that push outward and tangential forces that stretch the membrane laterally. Such forces lead to vesiculation and eventual rupture of the phagosomal membrane. ESAT-6-mediated fibril formation follows a nucleation-elongation mechanism, with a lag phase preceding rapid elongation (Peng and Sun 2016; Bates et al. 2024; Koiri et al. 2025). During this process, growing fibrils apply increasing strain on the membrane, promoting lipid phase separation and weakening its integrity. The cumulative effects result in vesiculation and bud fission, where spherical buds detach from the membrane as small vesicles. Experimental observations have shown that vesicle sizes decrease significantly during this process, indicating extensive remodeling of the phagosomal structure. The rupture of the phagosomal membrane has profound consequences for both bacterial survival and host immune responses. The disruption allows Mtb to escape into the cytosol, where it can replicate and evade degradation by lysosomal enzymes. ESAT-6 can also act as a pathogen-associated molecular pattern (PAMP), triggering innate immune responses through its interaction with host immune sensors such as NLRP3 inflammasomes (Jamilloux and Henry 2013; Rastogi and Briken 2022; Prakash et al. 2023; Passos et al. 2024). First, the ESAT-6 is recognized as a potent immunogenic protein secreted by Mtb via the ESX-1 secretion system and interacts with Toll-like receptors (TLRs), leading to downstream signaling cascades that promote pro-inflammatory cytokine production and immune cell recruitment. During infection, ESAT-6 contributes to the activation of the NLRP3 inflammasome, a key innate immune sensor (Mishra et al. 2010; Basu et al. 2018; Kelley et al. 2019). This activation is driven by cytosolic Mtb DNA and other PAMPs released during phagosomal membrane rupture mediated by ESAT-6. The NLRP3 inflammasome activation leads to caspase-1 processing of pro-IL-1β into mature IL-1β, an inflammatory cytokine that exacerbates tissue damage and promotes disease progression (Leemans et al. 2011; Kelley et al. 2019). Also, ESAT-6 and the ESX-1 secretion system are associated with the production of type I interferons (IFN-α and IFN-β), which are triggered by PAMP recognition. These interferons play dual roles in infection, contributing to both immune defense and increased susceptibility to pathogen-mediated cell death (Anes et al. 2024; Passos et al. 2024). CFP-10 (Culture Filtrate Protein 10 kDa), also known as EsxB, is a secretory protein encoded by the Rv3874 gene within the RD1 locus of Mtb. It belongs to the WXG100 protein family and is secreted via the ESX-1 Type VII secretion system.

CFP-10 plays an essential role in tuberculosis pathogenesis, primarily functioning as part of a heterodimeric complex with ESAT-6 (EsxA) (Kurenuma et al. 2009; Forrellad et al. 2013; Sreejit et al. 2014; Bosserman et al. 2018; Malakar et al. 2023). Structurally, CFP-10 forms a tight 1:1 complex with ESAT-6, stabilized by hydrophobic interactions and intermolecular salt bridges. The complex adopts a four-helix bundle configuration, with CFP-10 contributing a long, flexible C-terminal arm. This arm is critical for binding to host cell surfaces, particularly macrophages and monocytes, as demonstrated through fluorescence microscopy studies (Renshaw et al. 2005; Bates and Crowell 2023). The interaction facilitates modulation of host cell behavior to favor bacterial survival and replication. The secretion of the CFP-10/ESAT-6 complex is mediated by the ESX-1 machinery, where Rv3871, an AAA ATPase, recognizes the C-terminal signal sequence of CFP-10 (Brodin et al. 2005; Champion et al. 2014; Das et al. 2013). This recognition allows the complex to be transported through the ESX-1 system and delivered to host cells. The CFP-10/ESAT-6 complex has been implicated in membrane destabilization, inflammasome activation, and immune evasion during infection, highlighting its importance in Mtb's virulence. CFP-10 selectively activates human neutrophils through G-Protein-coupled receptors (GPCRs), triggering intracellular calcium (Ca^{2+}) signaling and reactive oxygen species (ROS) production via NADPH oxidase activation. This response is pertussis toxin-sensitive, confirming GPCR-dependent pathways. Neutrophils uniquely respond to CFP-10, while monocytes and lymphocytes show no Ca^{2+} flux, highlighting its specific role in neutrophil recruitment during infection. This activity promotes inflammation and bacterial dissemination. (Renshaw et al. 2005; De Jonge et al. 2007; Welin et al. 2015). CFP-10 can also modulate immune responses by downregulating NF-κB signaling and suppressing reactive oxygen species production in macrophages, aiding bacterial survival within host cells.CFP-10 also influences autophagy suppression and antigen presentation (Mohanty et al. 2015). The CFP-10/ESAT-6 complex inhibits autophagic pathways that would otherwise degrade intracellular pathogens. Furthermore, this complex reduces MHC class II expression on infected cells by interfering with antigen presentation mechanisms, impairing T-cell-mediated immunity (Mir and Sharma 2013). CFP-10's phosphorylation status critically modulates Mtb virulence. PKnA Kinase phosphorylates CFP-10 at residues such as Thr^{34}, and phosphomimetic mutations (e.g., $Thr^{34}Glu$) reduce secretion of ESAT-6. This dynamic phosphorylation regulates bacterial survival in macrophages and mice, demonstrating that CFP-10 phosphorylation directly impacts virulence factor delivery and pathogen persistence. (Sreejit et al. 2014; Malakar et al. 2023) . CFP-10 also interacts with components of Mtb's secretion pathways, including Rv0686 (a signal-recognition GTPase) and the SecYEG translocon. These interactions suggest CFP-10 aids in coordinating the secretion of virulence factors through overlapping pathways, such as the SecA/SRP systems, even though it lacks a classical signal peptide. This role is independent of its partnership with ESAT-6. Therefore, CFP-10 is considered a multifunctional virulence factor that contributes to tuberculosis pathogenesis by orchestrating neutrophil-driven inflammation, regulating ESAT-6 secretion via phosphorylation, and interacting with core secretion machinery. Its genomic position (Rv3874) within the RD1 locus highlights its essential role in Mtb virulence, as RD1 is absent in nonpathogenic strains like *Mycobacterium bovis BCG*.

Esp proteins, encoded by genes such as espACD (Rv3616c-Rv3614c) within the ESX-1 locus, are crucial for the virulence of Mtb (Raghavan et al. 2008; Garces et al. 2010; Chen et al. 2024; Kumar et al. 2016; Guo et al. 2021). These proteins are essential components of the ESX-1 Type VII secretion system (T7SS), which exports major virulence factors like ESAT-6 (EsxA) and CFP-10 (EsxB). The Esp protein family includes EspA, EspB, EspC, EspD, and EspK, each contributing to pathogenesis by modulating immune responses, enabling intracellular survival, and facilitating secretion of additional virulence proteins. Their distinct structures and functions underline their importance in Mtb virulence (Raghavan et al. 2008; Garces et al. 2010; Chen et al. 2024; Kumar et al. 2016; Guo et al. 2021). EspA, an alanine- and glycine-rich protein encoded within the espACD operon, forms filamentous appendages on the bacterial surface that assist in ESAT-6 and CFP-10 translocation into host cells. These filaments stabilize the ESX-1 secretion apparatus and are essential for maintaining its integrity (Chen et al. 2024). Beyond this role, EspA also contributes independently to virulence. Mutations that prevent EspA disulfide bond formation attenuate Mtb virulence while leaving ESX-1-dependent immune responses unaffected (Garces et al. 2010). EspB is a multifunctional protein with PE (Pro-Glu) and PPE (Pro-Pro-Glu)-like domains, encoded by Rv3881c and secreted via ESX-1 (Ekiert and Cox 2014; Korotkova et al. 2015; Sengupta et al. 2023). Structurally, it forms elongated helical molecules with a disordered C-terminal domain cleaved by MycP1 protease. The mature form oligomerizes into heptameric, donut-shaped structures under acidic conditions (Ekiert and Cox 2014; Korotkova et al. 2015; Sengupta et al. 2023). Functionally, EspB disrupts phagosomal membranes and induces cytotoxicity, facilitating cytoplasmic escape. It remains secretion-competent through interaction with EspK, which prevents premature oligomerization (Solomonson et al. 2015; Phan 2023). EspC, another espACD-encoded substrate, plays a structural role in ESX-1 apparatus stabilization and ESAT-6/CFP-10 secretion (Arnedo 2018). It forms filamentous structures that may serve as needles for translocation. EspC also modulates host immunity by suppressing macrophage pro-inflammatory responses. Disruption of espC impairs immune modulation and reduces Mtb virulence (Chirakos et al. 2020; Freire et al. 2018; Roy et al. 2020). EspG functions as a chaperone-like protein for PE/PPE substrates, stabilizing them in secretion-competent conformations before export via T7SS. It interacts with PE/PPE proteins through helical domains, preventing intracellular aggregation and ensuring efficient secretion (Tran et al. 2021; Damen et al. 2020; Garrett and Palmer 2024; Yan et al. 2024). By aiding in EspB secretion, EspG contributes to immune modulation and Mtb intracellular survival. Collectively, Esp proteins support the ESX-1 machinery, promote host immune evasion, and ensure the secretion of virulence factors like ESAT-6 and CFP-10. The genes encoding these proteins, espACD (EspA/C), Rv3881c (EspB), and espG (EspG), are conserved in pathogenic mycobacteria but absent in nonpathogenic strains like *Mycobacterium bovis BCG*, emphasizing their roles as virulence determinants and potential drug targets. The PE (Proline-Glutamic acid) and PPE (Proline-Proline-Glutamic acid) protein families are unique to Mtb, comprising approximately 10% of its genome with about 107 PE and 69 PPE genes (Cole et al. 1998; Gey van Pittius et al. 2006; Akhter et al. 2012; McGuire et al. 2012; Gómez-González et al. 2023). These proteins are markedly expanded in pathogenic mycobacteria compared to non-pathogens like *Mycobacterium smegmatis*, suggesting roles in virulence, immune evasion, and host interaction. PE proteins have conserved N-terminal domains forming two-helix bundles that heterodimerize with PPE proteins. PPE proteins also contain conserved domains with additional helices, forming a four-helix bundle in complex with PE partners. Their variable C-terminal domains often harbor disordered regions or repeat arrays implicated in host interactions. Some possess enzymatic domains or eukaryotic-like motifs that enhance functionality. Functionally, PE/PPE proteins localize to the bacterial surface or are secreted, mainly via the ESX-5 system, and engage host immunity. For instance, PE_PGRS33 enhances Mtb uptake by macrophages (Sampson 2011; Palucci et al. 2016), and PPE2 inhibits NADPH oxidase activity by blocking subunits p67phox and p47phox, thus reducing reactive oxygen species (ROS) production (Srivastava et al. 2019; Bisht et al. 2023).

PE_PGRS29 (Rv1468) contains a ubiquitin-associated domain that binds polyubiquitin chains, recruiting autophagy receptors like p62, and promoting LC3-associated autophagy (Chai et al. 2019; De Maio et al. 2020; Strong et al. 2020). In contrast, PE_PGRS47 impedes autophagy by binding Rab1A, a protein necessary for autophagy initiation (Saini et al. 2016; Strong and Lee 2021). PE/PPE genes are often organized in bicistronic operons, facilitating heterodimer formation and secretion. For example, PE25 and PPE41 form stable complexes secreted via ESX-5, crucial for intracellular survival and immune modulation (Korotkova et al. 2014; Fishbein et al. 2015; Ates 2020; Bunduc et al. 2021). Their co-localization with ESX system components underscores their interdependence. Several PE/

PPE proteins are notable virulence factors. PPE34 (Rv1917c) interacts with Toll-like receptor 2 (TLR2), inducing IL-10 and suppressing pro-inflammatory cytokines, thus promoting bacterial persistence (Sampson et al. 2001; Voskuil et al. 2004; Menéndez et al. 2007; D'Souza et al. 2023; Bohada-Lizarazo et al. 2024).

PE-PGRS11 (Rv0754) enhances oxidative stress resistance and upregulates anti-apoptotic genes like Bcl2 and COX-2 via TLR2 signaling (De Maio et al. 2020). PE-PGRS17 (Rv0978c) promotes TNF-α secretion in macrophages via ERK signaling, inducing host cell death and immune niche remodeling (Bansal et al. 2010; Chen et al. 2024; Guo et al. 2023). PE-PGRS33 (Rv1818c), a well-studied member, also triggers TNF-α production and enhances macrophage entry through TLR2, balancing immune activation and evasion (Palucci et al. 2016; Kramarska et al. 2021; Guo et al. 2023). PPE26 (Rv1789) interferes with phagolysosomal fusion by suppressing actin cytoskeletal rearrangement and modulating cytokine production, increasing IL-10 and reducing IL-12. It activates the TLR2-p38-MAPK pathway to support immune evasion (Mukku et al. 2023; Ehtram et al. 2025). PPE60 (Rv3478), by contrast, enhances pro-inflammatory cytokines like TNF-α, IL-1β, and IL-6 via TLR2, promoting Th1/Th17 responses and dendritic cell maturation (Su et al. 2018; Sharma et al. 2022). Collectively, PE/PPE proteins act as key virulence factors through mechanisms such as TLR2 signaling modulation (e.g., PPE34, PE-PGRS11), oxidative stress resistance (e.g., PE-PGRS11), autophagy manipulation (e.g., PE-PGRS29, PE-PGRS47), and immune evasion (e.g., PE-PGRS33, PPE26, PPE57). They are secreted primarily via T7SS, with ESX-5 specialized for their export. Their genomic organization near secretion components like EccC ATPases and MycP proteases enables efficient translocation (Ateş et al. 2019; Gómez-González et al. 2023; Chen et al. 2024). The ability of PE/PPE proteins to modulate immune responses is central to Mtb pathogenesis. Their absence in nonpathogenic species such as M. smegmatis highlights their evolutionary adaptation for virulence. The interplay between PE/PPE proteins and other ESX substrates, such as Esx and Esp proteins, reflects a coordinated strategy to subvert host defenses. The ESX-5 system (Rv1782-Rv1798) mediates PE/PPE secretion, while ESX-1 facilitates ESAT-6/CFP-10 delivery to induce necrosis. In conclusion, the T7SS in Mtb is a highly specialized system that exports Esx, Esp, and PE/PPE proteins to evade host immunity and sustain intracellular survival by exporting various virulence factors, including surface-exposed lipids and secreted proteins, which are essential for the pathogen's ability to establish infection and persist within the host.

5.4 Modulators of Host Cell Death Proteins in *Mycobacterium tuberculosis*

Mtb employs a sophisticated arsenal of virulence factors, both protein and nonprotein components, to manipulate host cell death pathways, ensuring its survival, replication, and eventual dissemination (Ramon-Luing et al. 2023). Early in infection, Mtb blocks programmed cell death by inhibiting apoptosis, then later promotes necrosis to escape from dying host cells (Srinivasan et al. 2016). In doing so, it interferes with multiple death pathways, apoptosis, necrosis, pyroptosis, necroptosis, and autophagy, shifting the balance between protective and pathological outcomes to favor its persistence (Mohareer et al. 2018; Nisa et al. 2022; Ramon-Luing et al. 2023; Vu et al. 2024; Ruan et al. 2024). Apoptosis normally restricts bacterial growth by packaging pathogens into apoptotic bodies that phagocytes clear. Mtb subverts this defense by three coordinated strategies: modulating anti-apoptotic proteins, disrupting mitochondrial signaling, and manipulating cytokine responses (Zhai et al. 2019; Nisa et al. 2022; Rodríguez-González and Gutiérrez-Kobeh 2024).

First, Mtb skews the balance toward cell survival by upregulating Bcl-2 and Mcl-1 while downregulating pro-apoptotic Bax and Bad, thereby preventing mitochondrial outer membrane permeabilization (MOMP), cytochrome c release, and caspase-9 activation. Key players in this process are NuoG (Rv3151; located at 3,516,746 bp) and PknE (Rv1743; located at 1,969,004 bp). NuoG, a core subunit of the type I NADH dehydrogenase (NDH-1) complex, contains iron-sulfur centers essential for electron transfer from NADH to ubiquinone, generating NAD^+ and ubiquinol (Velmurugan et al. 2007; Miller et al. 2010; Vilchèze et al. 2018; Grivennikova et al. 2024). The anti-apoptotic activity is mediated through the neutralization of phagosomal reactive oxygen species (ROS) generated by host NOX2, which suppresses TNF-α-mediated apoptosis (Miller et al. 2010). The inhibition of apoptosis extends to neutrophils, delaying bacterial acquisition by dendritic cells and subsequent CD4 T cell priming (Blomgran et al. 2012). The NDH-1 complex's activity is further modulated by the orphan response regulator Rv3143, which interacts with the NuoD subunit. Deletion of Rv3143 affects bacterial growth under various stress conditions and alters gene expression related to nitrate respiration, suggesting a role in adapting to host environments (Płocińska et al. 2022). SecA2 is an accessory ATPase within the Sec secretion system that plays a specialized role in exporting a distinct subset of virulence factors essential for the pathogenesis of Mtb. In contrast to the essential housekeeping ATPase SecA1, which facilitates general protein secretion, SecA2 focuses specifically on exporting proteins that enable the bacterium to evade host immune defenses—particularly by countering oxidative stress and blocking phagosome maturation (Braunstein et al. 2001, 2003; Kurtz et al. 2006; Zulauf et al. 2018). Among the proteins secreted via the SecA2 pathway are superoxide dismutase A (SodA) and catalase-peroxidase (KatG), both of which are critical for neutralizing reactive oxygen intermediates produced by host macrophages. This detoxifying function helps protect the bacterium from oxidative damage. Interestingly, SodA lacks a classical signal sequence, which suggests that SecA2 is capable of recognizing and exporting proteins outside the scope of the canonical SecA1 system (Braunstein et al. 2003; Kurtz et al. 2006; Echeverria-Valencia et al. 2018; Wengenack et al. 1999; Miller and Braunstein 2017). Functionally, SecA2 relies on ATP binding and hydrolysis to power the translocation of its substrates. It also physically interacts with both the canonical SecA1 and the SecYEG translocon, indicating a potential cooperative mechanism in protein export. Structurally, while SecA2 shares considerable similarity with SecA1, it lacks certain regions, such as the helical wing domain, which may contribute to its unique substrate specificity (Bensing et al. 2014; Dong et al. 2023). Cellular localization studies have shown that SecA2 is mainly cytosolic, in contrast to SecA1, which is found both in the cytosol and associated with the cell envelope. This distribution reflects their distinct, yet complementary, roles in protein secretion. Beyond its role in countering oxidative stress, SecA2 is also involved in secreting other virulence factors, including SapM and PknG. These proteins interfere with the fusion of phagosomes and lysosomes, a process that is crucial for bacterial degradation within host cells. By disrupting this fusion, SecA2 contributes to immune evasion and helps the bacterium suppress the production of pro-inflammatory cytokines such as TNF-α and IL-6 (Sullivan et al. 2012). The importance of SecA2 is underscored by studies of Mtb mutants lacking this protein. These mutants show increased phagosome acidification, elevated levels of reactive oxygen and nitrogen intermediates, higher rates of macrophage apoptosis, and significantly reduced bacterial growth in both macrophages and animal infection models. These findings highlight the essential role of SecA2 in the virulence of Mtb (Braunstein et al. 2003; Zulauf et al. 2018). Therefore, SecA2 acts as a specialized accessory ATPase that exports a targeted set of virulence-associated proteins. Through this activity, it enables Mtb to neutralize host oxidative defenses, block phagosome maturation, and modulate immune responses—ultimately supporting bacterial survival and enhancing pathogenicity within the host.

PknE is a serine/threonine protein kinase encoded by the pknE gene (Rv1743; Location 1,969,004 bp) in Mtb that plays a crucial role in helping the bacterium survive within the hostile environment of the host. This kinase suppresses multiple apoptotic pathways, reduces the production of pro-inflammatory cytokines, and modulates key host cell signaling processes, including the phosphorylation of Akt, a central regulator of cell survival (Kumar and Narayanan 2012; Jayakumar et al. 2008). Structurally, PknE belongs to the family of serine/threonine kinases, enzymes that phosphorylate specific serine or threonine residues on target proteins, thereby altering their activity and enabling PknE to influence host cell signaling during infection (Av-Gay and Everett 2000; Grundner et al. 2005).One of PknE's most significant functions is its ability to suppress apoptosis, the programmed cell death mechanism that macrophages use to eliminate infected cells. Under stress conditions such as hypoxia or exposure to nitric oxide (NO), a potent antimicrobial molecule produced by immune cells, PknE protects infected macrophages from premature death by phosphorylating Akt kinase. This phosphorylation activates survival pathways that inhibit apoptotic signaling, allowing *M. tuberculosis* to maintain its intracellular niche and replicate within host cells. The pknE promoter is responsive to NO stress, highlighting PknE's role in sensing and adapting to the host's hostile environment (Jayakumar et al. 2008; Parandhaman and Narayanan 2014). Beyond apoptosis suppression, PknE also modulates immune responses by interfering with the production of pro-inflammatory cytokines such as TNF-α and IL-6, which are critical for effective immune defense. Deletion of pknE leads to increased apoptosis and reduced necrosis in infected macrophages, demonstrating how PknE finely balances host cell death pathways to favor bacterial survival. Moreover, the absence of PknE results in a marked decline in these cytokines during infection, further dampening innate immune responses and promoting bacterial persistence (Kumar et al. 2009; Sullivan et al. 2012; Jayakumar et al. 2008; Parandhaman and Narayanan 2014). In addition to its effects on apoptosis and cytokine production, PknE influences intracellular signaling pathways such as the Mitogen-Activated Protein Kinase (MAPK) cascade, which is essential for cellular responses to infection. Studies using pknE deletion mutants revealed defective phosphorylation kinetics of MAPKs and transcription factors, which delay and reduce inflammatory responses, thereby creating a more favorable environment for bacterial survival (Parandhaman and Narayanan 2014). This modulation of host signaling also impacts co-receptor expression relevant to HIV coinfection, illustrating the broader implications of PknE activity. Hence, PknE acts as a sophisticated molecular tool that enables Mtb to sense nitric oxide stress and manipulate host cell survival pathways through Akt phosphorylation. By inhibiting apoptosis and modulating immune signaling, PknE helps the bacterium preserve its intracellular environment, evade immune clearance, and establish persistent infection. This underscores PknE's critical role as a virulence factor in tuberculosis pathogenesis and highlights its potential as a target for therapeutic intervention (Kumar and Narayanan 2012; Jayakumar et al. 2008; Parandhaman and Narayanan 2014).

Mtb has evolved a remarkable ability to manipulate host cell death pathways to support its survival and promote dissemination. While apoptosis, a form of programmed, immunologically silent cell death, helps protect the host by limiting infection, necrosis benefits the pathogen by causing cell rupture and inflammation, aiding in its spread (Srinivasan and Briken 2014). Mtb induces necrotic cell death through several mechanisms, including necroptosis, pyroptosis, and ferroptosis (Vu et al. 2024). In contrast to apoptosis, necrosis leads to the lysis of infected macrophages, releasing their contents and triggering strong inflammatory responses that facilitate Mtb's transmission. A major contributor to Mtb-induced necrosis is the Tuberculosis Necrotizing Toxin (TNT), which corresponds to the C-terminal domain of the CpnT protein encoded by the Rv3903c gene. TNT acts as an NAD^+ glycohydrolase, breaking down intracellular NAD^+ a molecule crucial for maintaining cellular energy metabolism (Sun et al. 2015; Tak et al. 2019; Pajuelo and Niederweis 2020). By depleting NAD^+, TNT disrupts energy homeostasis and undermines the cell's membrane repair systems, ultimately inducing necrotic death in infected macrophages. This form of necrosis is mechanistically distinct from apoptosis and involves a regulated necroptotic process, primarily driven by the RIPK1/RIPK3-MLKL signaling axis (Pajuelo et al. 2021). Through necroptosis, infected macrophages undergo membrane rupture, releasing live bacteria into the extracellular environment, which not only enhances bacterial dissemination but also helps Mtb escape immune elimination. To reach the host cell cytosol and deliver TNT, Mtb relies on the Type VII secretion systems (including ESX-1, ESX-2, and ESX-4) that compromise the integrity of the phagosomal membrane (Izquierdo Lafuente et al. 2021). Interestingly, Mtb also produces a protective protein known as the Immunity Factor for TNT (IFT), which binds and neutralizes TNT to avoid harming the bacterium itself (Pajuelo et al. 2021). Collectively, these mechanisms show how Mtb strategically deploys TNT to subvert host cell death programs in ways that enhance its persistence and pathogenic potential. Beyond TNT-mediated necrosis, Mtb also exerts control over host lipid mediators, especially prostaglandin E_2 (PGE_2), an eicosanoid synthesized from arachidonic acid that plays a multifaceted role in immune regulation during infection (Kaul et al. 2012; Sun et al. 2015; Pellegrini et al. 2020; Rangel Moreno et al. 2002). In the early stages of infection, low levels of PGE_2 help maintain mitochondrial stability and promote plasma membrane repair, supporting cell survival. However, as the infection progresses, elevated PGE_2 levels suppress key pro-inflammatory cytokines like TNF-α and IL-6, and impair antigen presentation. This shift contributes to immunosuppression and creates a favorable niche for the bacterium to persist (Pellegrini et al. 2021). Overall, Mtb-induced necrosis, driven by TNT-mediated NAD^+ depletion and regulated necroptosis, compromises host energy balance and membrane integrity while enabling bacterial escape and spread. Concurrently, Mtb's manipulation of PGE_2 modulates immune responses, suppressing host defenses and sustaining long-term infection. Together, these complex and coordinated strategies illustrate how Mtb expertly hijacks host cell death and signaling pathways to avoid immune clearance and thrive within the host.

Pyroptosis and ferroptosis are two distinct, regulated forms of cell death that Mtb skillfully manipulates to enhance its survival, persistence, and spread within the host. Although both processes involve complex molecular signaling with important consequences for immune response, Mtb has evolved sophisticated strategies to subvert them to its advantage. Pyroptosis is an inflammatory type of programmed cell death triggered by the activation of inflammasomes, intracellular multiprotein complexes that detect danger signals. Once activated, these inflammasomes initiate the cleavage of caspase-1, which then processes the precursors of pro-inflammatory cytokines, interleukin-1β (IL-1β) and interleukin-18 (IL-18), into their mature, active forms. These cytokines are crucial for host defense, as they promote inflammation, attract immune cells, and help eliminate infected cells (Bergsbaken and Cookson 2009; Liu et al. 2024a, b). However, Mtb has evolved effective mechanisms to suppress pyroptosis, thereby evading immune surveillance and persisting within macrophages. At the heart of this suppression is Zmp1, a zinc-dependent metalloprotease encoded by the *zmp1* gene (Rv0198c) (Ferraris et al. 2011). Zmp1 prevents the activation of the NLRP3 inflammasome, blocking caspase-1 activity and the subsequent maturation of IL-1β (Master et al. 2008; Ma et al. 2021). By halting this inflammasome-mediated pathway, Zmp1 effectively prevents pyroptosis, allowing Mtb to remain inside host cells without triggering the potent immune responses that would otherwise lead to its elimination. Zmp1's role extends beyond pyroptosis inhibition. It also interferes with phagosome maturation, the process in which phagosomes fuse with lysosomes to form phagolysosomes, where engulfed pathogens are usually degraded. Zmp1 blocks this fusion, protecting Mtb from lysosomal destruction (Zhang et al. 2023). When *zmp1* is experimentally deleted, inflammasome activation is restored, leading to heightened caspase-1 activity, increased IL-1β production, and complete phagosome maturation. This results in more efficient bacterial clearance by macrophages. Additionally, studies show that administering exogenous IL-1β can compensate for Zmp1 absence by promoting phagosome maturation and enhancing the host's ability to clear the infection (Zhang et al. 2023). On a structural level, Zmp1 is part of the M13 family of zinc-dependent metalloproteases, sharing significant homology with human enzymes such as neprilysin and endothelin-converting enzyme-1 (Ferraris et al. 2011). However, Zmp1 features unique active-site elements that distinguish it from these human counterparts. These structural differences make it an appealing target for drug development aimed at reactivating inflammasome function and enhancing bacterial clearance. Zmp1 serves as a critical virulence factor that helps *Mtb* evade host defenses by blocking pyroptosis and phagosome maturation (Ferraris et al. 2018; Ramon-Luing et al. 2023). Its unique structural and functional features as a zinc-dependent metalloprotease not only underpin its role in pathogenesis but also underscore its promise as a potential therapeutic target for restoring immune function and combating tuberculosis.

Ferroptosis is a newly characterized, regulated form of cell death marked by the intracellular accumulation of iron, uncontrolled lipid peroxidation, and oxidative injury to cellular membranes (Tang et al. 2021; Ruan et al. 2024). This iron-dependent pathway plays a nuanced, dual role in the context of tuberculosis (TB). While ferroptosis can act as a host defense mechanism by promoting macrophage death and limiting intracellular iron availability, thus restricting Mtb replication, Mtb can also exploit it to promote its survival and dissemination through manipulation of host iron metabolism and antioxidant defenses (Amaral et al. 2019; Wang et al. 2024; Dawi et al. 2025). Mtb subverts the host's iron regulatory systems to elevate free iron levels within infected macrophages. This is accomplished by increasing expression of transferrin receptors, which enhance iron uptake, while downregulating ferroportin, the only known iron exporter, effectively trapping iron inside the cell. The accumulation of intracellular free iron accelerates Fenton chemistry, generating reactive oxygen species (ROS) that drive lipid peroxidation, a defining feature of ferroptosis. This oxidative damage compromises membrane integrity, leading to ferroptotic cell death and necrotic-like outcomes. The resulting rupture of host membranes releases viable bacteria into the extracellular environment, thereby facilitating continued infection and spread (Gan 2023). Beyond disrupting iron balance, Mtb also targets the host's antioxidant defense system to promote ferroptosis. Two key regulators—glutathione (GSH) and glutathione peroxidase 4 (GPX4)—ordinarily act to suppress lipid peroxidation. GSH scavenges lipid peroxides, while GPX4 enzymatically reduces lipid hydroperoxides to nontoxic alcohols (Amaral et al. 2022; Fei et al. 2024; Dawi et al. 2025). During Mtb infection, however, both GSH levels and GPX4 activity are diminished in macrophages, intensifying oxidative stress and driving ferroptotic damage. Experimental studies show that reduced GPX4 expression correlates with heightened lipid peroxidation and necrosis in infected cells and lung tissue. Significantly, pharmacological blockade of ferroptosis using Ferrostatin-1 (Fer-1) or iron chelators mitigates necrotic cell death, lowers bacterial loads, and improves survival in TB models (Amaral et al. 2019; Dawi et al. 2025). In addition to promoting macrophage death, ferroptosis affects host lipid metabolism and inflammatory signaling during TB infection.

Mtb enhances the synthesis of polyunsaturated fatty acids (PUFAs) in infected macrophages, which serve as prime substrates for lipid peroxidation. Moreover, the pathogen modulates host **lipoxygenase** activity, amplifying oxidative membrane damage. These alterations not only fuel ferroptotic death but also disrupt granuloma architecture—the structured immune microenvironment crucial for bacterial containment, thus enabling Mtb to escape immune control and spread to surrounding tissues (Dawi et al. 2025). By integrating these mechanisms, Mtb turns ferroptosis into a strategic advantage. It reshapes the intracellular environment to support its persistence by manipulating iron handling, impairing antioxidant responses, and driving membrane damage through lipid peroxidation. This process works synergistically with other immune evasion strategies, such as inhibition of pyroptosis via the zinc-dependent metalloprotease Zmp1 (Rv0198c), allowing the bacterium to evade destruction while promoting host tissue damage and bacterial dissemination. Thus, ferroptosis represents a key axis of TB pathogenesis that *Mtb* expertly manipulates to bypass immune defenses and enhance its survival. Therapeutic strategies targeting ferroptosis, such as the use of Ferrostatin-1 or iron chelators, offer promising avenues for limiting disease progression while preserving host tissue. Still, further investigation is needed to fully unravel the molecular interplay between ferroptosis and host immunity during Mtb infection. In summary, Mtb employs a sophisticated array of virulence factors to manipulate host cell death pathways in ways that favor its survival, replication, and dissemination. By selectively modulating these pathways, Mtb not only avoids immune clearance but also creates a permissive environment for chronic infection.

To evade host defenses, Mtb actively inhibits apoptosis, a noninflammatory, programmed form of cell death that typically aids in pathogen clearance. One key player in this process is the serine/threonine kinase PknE, which interferes with intrinsic apoptotic signaling and dampens pro-inflammatory cytokine production, thereby promoting bacterial persistence within host macrophages. Conversely, Mtb induces necroptosis, a lytic and inflammatory form of cell death that benefits the pathogen by causing plasma membrane rupture and releasing viable bacteria into surrounding tissues. This process is mediated by virulence factors such as ESAT-6, a Type VII secretion system effector that disrupts phagosomal membranes, and the tuberculosis necrotizing toxin (TNT), which depletes intracellular NAD^{+} and impairs membrane repair mechanisms. Mtb also suppresses pyroptosis, an inflammatory form of cell death driven by inflammasome activation. Through the action of Zmp1, a zinc-dependent metalloprotease, Mtb prevents the assembly of the NLRP3 inflammasome, blocking caspase-1 activation and subsequent release of pro-inflammatory cytokines like IL-1β. This allows Mtb to persist within macrophages while avoiding the robust immune responses that pyroptosis would normally trigger. In addition, Mtb exploits ferroptosis, an iron-dependent form of regulated necrosis characterized by lipid peroxidation and oxidative membrane damage. By manipulating host iron metabolism, upregulating iron import via transferrin receptors and downregulating iron export via ferroportin, Mtb increases intracellular iron availability. Simultaneously, it suppresses antioxidant defenses, including glutathione (GSH) and glutathione peroxidase 4 (GPX4), thereby amplifying oxidative stress and triggering ferroptotic cell death. The resulting tissue damage and necrosis facilitate bacterial escape and dissemination. Together, these interconnected strategies illustrate Mtb's remarkable ability to subvert host cell death machinery to its advantage. By selectively suppressing protective cell death pathways (like apoptosis and pyroptosis) and inducing destructive ones (like necroptosis and ferroptosis), Mtb creates an intracellular niche that supports long-term survival and enables the establishment of chronic, often latent infection. Understanding these mechanisms provides critical insights into TB pathogenesis and reveals promising targets for host-directed therapies aimed at restoring immune function and enhancing bacterial clearance.

5.5 The Lipid Virulence Factors in *Mycobacterium tuberculosis*

Mtb relies heavily on a variety of lipid molecules embedded in its complex cell wall to manipulate the host's immune system and ensure its survival. Among these, phthiocerol dimycocerosates (PDIMs), phenolic glycolipids (PGLs), trehalose dimycolate (TDM, also known as cord factor), and lipoarabinomannan (LAM) stand out for their unique structures and influential roles in infection. Each of these lipids contributes in its way to helping the bacterium evade immune detection, influence host cell death, and establish a persistent infection (Jackson 2014; Gago et al. 2017; Queiroz and Riley 2017; Holzheimer et al. 2021; Mashabela et al. 2019). PDIMs and PGLs are closely related lipids that act like a molecular cloak, masking the bacterial surface to hide pathogen-associated molecular patterns (PAMPs) from the host's pattern recognition receptors, such as Toll-like receptors. This clever disguise prevents the immune system from recognizing the invader and launching an effective attack (Greenfield et al. 2007; Quigley et al. 2017; Cambier et al. 2020; Holzheimer et al. 2021; Rens et al. 2021). PDIMs are complex lipids consisting of long-chain phthiocerols esterified to multi-methyl-branched mycocerosic acids, while phenolic glycolipids (PGLs) are glycosylated derivatives of phenolphthiocerols. Both lipid types are highly hydrophobic, contributing significantly to the overall hydrophobicity of the Mtb cell surface, which reduces its accessibility to host immune defenses. The genes responsible for PDIM biosynthesis are clustered within the Mtb genome and include enzymes such as polyketide synthases and transporters that orchestrate lipid assembly and export (Camacho et al. 2001; Jain and Cox 2005; Trivedi et al. 2005). PGLs utilize much of this same biosynthetic machinery but require additional genes like pks15/1, which encodes a polyketide synthase essential for synthesizing the phenolphthiocerol backbone unique to PGLs (Constant et al. 2002; Yu et al. 2012; Guzmán et al. 2024). Importantly, some Mtb strains carry mutations in pks15/1 that abolish PGL production, while others maintain this capacity, leading to strain-specific differences in lipid profiles and pathogenicity (Ramos and Cunha 2020). PDIMs are established as key virulence factors of Mtb due to their diverse roles in manipulating host-pathogen interactions and promoting bacterial survival. During infection, PDIMs are transferred from the bacterial envelope to host cell membranes, where they alter membrane properties by decreasing polarity and adopting a conical shape that disrupts lipid organization (Augenstreich et al. 2019, 2025; Cambier et al. 2014; Barczak et al. 2017). This modification enhances the function of host receptors such as complement receptor 3, facilitating increased phagocytosis by macrophages, while simultaneously boosting the activity of bacterial membrane-permeabilizing effectors like the virulence factor EsxA (ESAT-6) (Augenstreich et al. 2019, 2025). Functionally, PDIMs modulate host immune responses by interacting with host membranes and supporting other virulence determinants such as the ESX-1 secretion system. This cooperation promotes phagosomal membrane disruption, enabling Mtb to escape into the cytosol and evade immune clearance (Bussi and Gutierrez 2019; Augenstreich et al. 2019). PDIMs also shape the host cellular environment to favor bacterial persistence by preventing phagosomal acidification and modulating immune signaling pathways (Jamwal et al. 2016; Quigley et al. 2017). Complementing these effects, PGLs suppress the production of critical pro-inflammatory cytokines, including tumor necrosis factor-alpha (TNF-α) and interleukin-12 (IL-12), thereby dampening the host immune response and facilitating chronic infection (Domingo-Gonzalez et al. 2016; Chen et al. 2024; Chandra et al. 2022).

Trehalose-6,6′-dimycolate (TDM), commonly known as cord factor, is a critical lipid virulence factor of Mtb. Structurally, TDM is a glycolipid composed of two mycolic acid molecules esterified to a trehalose sugar, making it a significant and biologically active component of the mycobacterial cell wall. Mycolic acids are unique, long-chain fatty acids that confer distinctive hydrophobicity and impermeability to the Mtb envelope, which are essential for bacterial survival and pathogenicity. TDM functions as a prototypical pathogen-associated molecular pattern (PAMP), signaling the presence of mycobacteria to innate immune cells. Its ability to induce inflammatory and granulomatous responses comparable to whole mycobacteria in animal models, combined with its adjuvant properties that enhance cellular immune responses, supports its classification as a PAMP (Ryll et al. 2001; Indrigo et al. 2002; Kan-Sutton et al. 2009; Lang 2013; Chandra et al. 2022). By inhibiting phagosome maturation and modulating host immune signaling, TDM enhances Mtb's ability to evade immune clearance and establish persistent infection (Kan-Sutton et al. 2009; Lerner et al. 2018). Early studies investigated Toll-like receptors (TLRs) as potential pattern recognition receptors

(PRRs) for TDM. Geisel et al. (2005) demonstrated that TDM-coated beads stimulated TNF and IL-6 production in macrophages in vitro and induced leukocyte infiltration in mice. While TLR2 and TLR4 knockout mice showed no defect in leukocyte recruitment, Myd88-deficient mice exhibited a marked reduction, suggesting involvement of other Myd88-dependent receptors such as those for IL-1 or IL-18. Bowdish et al. (2009) later reported diminished macrophage responses to TDM in TLR2- and TLR4-deficient cells and implicated the scavenger receptor MARCO in TDM binding. Since scavenger receptors typically mediate endocytosis rather than activation, MARCO was proposed to facilitate TDM delivery to TLRs. Additionally, Ozeki et al. (2006) found that scavenger receptor A (SR-A) binds cord factor and modulates inflammatory responses. More recent research revealed that TDM activates macrophages and dendritic cells via a distinct pathway involving the C-type lectin receptor Mincle and downstream Syk-Card9 signaling. Werninghaus et al. (2009) showed that macrophage activation by TDM and its synthetic analog TDB is independent of TLRs and Myd88 but requires Syk-Card9-Bcl10-Malt1 signaling. The adapter protein FcRγ, but not Dap12, was necessary, implicating Mincle (Clec4e), which is inducibly expressed in macrophages upon glycolipid stimulation. Initially identified as a sensor of necrotic cells binding SAP130 (Yamasaki et al. 2008), Mincle was later shown to recognize fungal carbohydrates (Yamasaki et al. 2009; Wells et al. 2008). Ishikawa et al. (2009) identified TDM as a direct ligand of Mincle, and Mincle-deficient mice failed to mount granulomatous responses to TDM. Parallel studies confirmed that TDM and TDB fail to activate macrophages lacking Mincle and that Mincle is essential for their Th1/Th17 adjuvant effects in vivo (Schoenen et al. 2010). These findings firmly establish Mincle as the critical receptor mediating TDM recognition in mice, though its role in human macrophages remains to be confirmed.

Discrepancies regarding TLR involvement in TDM recognition may arise from differences in macrophage types, culture conditions, and TDM presentation. Generally, TLR and C-type lectin receptor (CLR) pathways synergize to enhance macrophage activation, as exemplified by Dectin-1 and TLR2 responses to fungal components (Dennehy et al. 2008, 2009). In the case of TDM, TLR ligand stimulation induces Mincle expression, priming macrophages for heightened sensitivity to TDM (Matsumoto et al. 1999). Bowdish et al.'s (2009) MARCO-TLR2-Myd88 pathway may thus license macrophages for TDM responsiveness via Mincle upregulation. Supporting this, Lee et al. (2012) showed that Myd88-deficient neutrophils exhibit impaired synergistic TNF production and adhesion responses to combined TLR2 ligand and TDM stimulation, correlating with deficient Mincle induction. Biosynthesis of TDM involves multiple steps, including the transfer of mycolic acids from glucose monomycolate (GMM) to trehalose by antigen 85 proteins (Matsunaga et al. 2008; Takayama et al. 2005). The production of TDM is dynamically regulated in response to environmental conditions, with glucose availability triggering a switch from TDM to GMM synthesis (Matsunaga et al. 2008). The cording morphology, often associated with TDM, is a reliable predictor of mycobacterial virulence and involves multiple genes (Glickman 2008). The pathway for mycolic acid synthesis and processing in Mtb involves fatty acid synthetases, polyketide synthases, and various transferases, with the final steps including the transport of trehalose monomycolate outside the cell and subsequent formation of TDM (Takayama et al. 2005). Collectively, TDM's unique structure and multifaceted immunomodulatory functions underscore its critical role as a lipid virulence factor of Mtb, significantly contributing to the bacterium's pathogenicity and representing a promising target for therapeutic intervention (Lerner et al. 2018).

Lipoarabinomannan (LAM) is another significant key lipid virulence factor of Mtb that plays a vital role in enabling the bacterium to persist within macrophages (Chan et al. 1991). Structurally, LAM consists of a mannosylated polysaccharide core capped with mannose residues, classifying it as mannose-capped LAM (ManLAM). This unique molecular architecture allows LAM to interact intricately with host immune cells, modulating immune responses to favor bacterial survival (Källenius et al. 2016; Zhou et al. 2019). One of LAM's primary mechanisms is the inhibition of phagosomal maturation. LAM integrates into macrophage membrane lipid rafts through its glycosylphosphatidylinositol anchor, which is essential for blocking phagosome-lysosome fusion (Welin et al. 2008). It interferes with calcium signaling by blocking the increase of cytosolic Ca^{2+} and disrupting a Ca^{2+}/calmodulin-dependent phosphatidylinositol 3-kinase (PI3K) hvps34 cascade, which is critical for generating phosphatidylinositol 3-phosphate (PI3P) on phagosomes (Vergne et al. 2003). This inhibition prevents the delivery of lysosomal cargo and syntaxin 6 from the trans-Golgi network, further impairing phagosomal maturation (Fratti et al. 2003). By these mechanisms, LAM effectively halts phagosome-lysosome fusion, allowing Mtb to survive and replicate within macrophages. Beyond phagosomal arrest, LAM exerts multiple immunomodulatory effects. It scavenges reactive oxygen species, protecting Mtb from oxidative stress, and inhibits protein kinase C activity and transcriptional activation of interferon-inducible genes (Chan et al. 1991; Correia-Neves et al. 2019; Liu et al. 2022). LAM also suppresses macrophage apoptosis by interfering with tumor necrosis factor-alpha (TNF-α) signaling, preventing programmed cell death that would otherwise limit bacterial replication (Correia-Neves et al. 2019; Liu et al. 2022). Additionally, LAM reduces expression of galectin-9 and components of the TNF pathway in macrophages, further impairing their ability to control intracellular Mtb growth (Chávez-Galán et al. 2017). LAM also dampens the host immune response by suppressing the production of key pro-inflammatory cytokines, thereby promoting chronic infection (Sasindran and Torrelles 2011; Vergne et al. 2015; Palčeková et al. 2023). Through these combined actions, LAM manipulates host signaling pathways—including calcium signaling and cytokine networks—to create a permissive environment that facilitates Mtb persistence. Together with other lipid virulence factors such as phthiocerol dimycocerosates (PDIMs), phenolic glycolipids (PGLs), and trehalose dimycolate (TDM), LAM forms part of a sophisticated lipid arsenal that Mtb employs to evade immune detection, manipulate host cell death pathways, and establish persistent infections. PDIMs and PGLs mask the bacterial surface and modulate host membranes to avoid immune recognition and promote survival. TDM influences immune cell organization and delays bacterial destruction, while LAM protects the bacteria by inhibiting apoptosis and oxidative damage. This intricate interplay of lipid virulence factors is central to Mtb's ability to persist within the host. Understanding these mechanisms offers promising avenues for developing novel therapeutic strategies aimed at disrupting lipid-mediated immune evasion and improving tuberculosis control.

In conclusion, Mtb deploys a broad and sophisticated repertoire of proteins that work together to establish infection, evade immune defenses, and ensure bacterial survival within the host. These virulence factors can be broadly categorized into several functional classes. Proteins involved in lipid metabolism strengthen the bacterial cell envelope, enhancing resistance to host antimicrobial mechanisms. Others participate in modifying the cell surface to mask pathogen-associated molecular patterns, thereby modulating immune recognition and response. Structural proteins maintain cell envelope integrity, especially under stress conditions, preserving the bacterium's protective barrier. Secreted effectors disrupt host cell membranes and manipulate immune signaling pathways to favor bacterial persistence. Additionally, antigenic variation is facilitated by families of surface proteins that help the bacterium evade adaptive immunity. Protective enzymes defend against oxidative and nitrosative stresses encountered within host cells, while regulatory proteins such as kinases fine-tune intracellular signaling to adapt to the hostile host environment. Together, these diverse protein classes form an integrated network that allows Mtb to survive, replicate, and persist within the host, making them crucial targets for the development of new and more effective tuberculosis therapies.

Conflict of Interest The authors declare that they have no conflict of interest related to the content of this chapter.

Ethical Consideration This chapter is a review based entirely on data and information obtained from previously published and publicly available sources, all of which have been appropriately cited. The authors conducted no new experimental work involving human participants or animals. Therefore, no additional ethical approval was required. The authors have ensured proper attribution and compliance with ethical standards in the use of all referenced materials.

References

Abekura F, Park J, Kwak CH, Ha SH, Cho SH, Chang YC et al (2019) Esculentoside B inhibits inflammatory response through JNK and downstream NF-κB signaling pathway in LPS-triggered murine macrophage RAW 264.7 cells. Int Immunopharmacol 68:156–163

Akhter Y, Ehebauer MT, Mukhopadhyay S, Hasnain SE (2012) The PE/PPE multigene family codes for virulence factors and is a possible source of mycobacterial antigenic variation: perhaps more? Biochimie 94(1):110–116

Alsayed SSR, Gunosewoyo H (2023) Tuberculosis: pathogenesis, current treatment regimens and new drug targets. Int J Mol Sci 24(6):5202

Amaral EP, Costa DL, Namasivayam S, Riteau N, Kamenyeva O, Mittereder L, Mayer-Barber KD, Andrade BB, Sher A (2019) A major role for ferroptosis in Mycobacterium tuberculosis-induced cell death and tissue necrosis. J Exp Med 216(3):556–570

Amaral EP, Foreman TW, Namasivayam S, Hilligan KL, Kauffman KD, Bomfim CCB, Costa DL, Barreto-Duarte B, Gurgel-Rocha C, Santana MF, Cordeiro-Santos M, Du Bruyn E, Riou C, Aberman K, Wilkinson RJ, Barber DL, Mayer-Barber KD, Andrade BB, Sher A (2022) GPX4 regulates cellular necrosis and host resistance in Mycobacterium tuberculosis infection. J Exp Med 219(11):e20220504

Anes E, Pires D, Mandal M, Azevedo-Pereira JM (2023) ESAT-6 a major virulence factor of Mycobacterium tuberculosis. Biomolecules 13(6):968

Anes E, Azevedo-Pereira JM, Pires D (2024) Role of type I interferons during mycobacterium tuberculosis and HIV infections. Biomolecules. 14(7):848

Arbing MA, Kaufmann M, Phan T, Chan S, Cascio D, Eisenberg D (2010) The crystal structure of the Mycobacterium tuberculosis Rv3019c-Rv3020c ESX complex reveals a domain-swapped heterotetramer. Protein Sci 19(9):1692–1702

Arnedo PJS (2018) Functional characterization of the virulence determinant ESX-1 from Mycobacterium tuberculosis: Ecole Polytechnique Fédérale de Lausanne

Armstrong JA, Hart PD (1971) Response of cultured macrophages to Mycobacterium tuberculosis, with observations on fusion of lysosomes with phagosomes. J Exp Med 134(3 Pt 1):713–740

Ates LS (2020) New insights into the mycobacterial PE and PPE proteins provide a framework for future research. Mol Microbiol 113(1):4–21

Ates I, Katipoglu B, Copur B, Yilmaz N (2017) A rare cause of hypophysitis: tuberculosis. Endocr Regul 51(4):213–215

Ateş M, Çiftçi M, Oruç E, Celebi A, Akcakavak G. (2019) Mycobacterium bovis tuberculosis in a neonatal Holstein calf

Augenstreich JHE, Ferré G, Czaplicki G, Jolibois F, Destainville N, Guilhot C, Milon A, Astarie-Dequeker C, Chavent M (2019) The conical shape of DIM lipids promotes mycobacterium tuberculosis infection of macrophages. Proc Natl Acad Sci USA 116(51):25649–25658. https://doi.org/10.1073/pnas.1910368116. Epub 2019 Nov 22. PMID: 31757855; PMCID: PMC6926010

Augenstreich JPA, Allen CNS, Poddar A, Chen H, Srinivasan L, Briken V (2025) Dynamic interplay of autophagy and membrane repair during mycobacterium tuberculosis infection. PLoS Pathog 21(1):e1012830. https://doi.org/10.1371/journal.ppat.1012830

Av-Gay Y, Everett M (2000) The eukaryotic-like Ser/Thr protein kinases of mycobacterium tuberculosis. Trends Microbiol 8(5):238–244

Bah A, Sanicas M, Nigou J, Guilhot C, Astarie-Dequeker C, Vergne I (2020) The lipid virulence factors of Mycobacterium tuberculosis exert multilayered control over autophagy-related pathways in infected human macrophages. Cells 9(3):666

Ballister ER, Samanovic MI, Darwin KH (2019) Mycobacterium tuberculosis Rv2700 contributes to cell envelope integrity and virulence. J Bacteriol 201(19):e00228-19

Bansal K, Sinha AY, Ghorpade DS, Togarsimalemath SK, Patil SA, Kaveri SV et al (2010) Src homology 3-interacting domain of Rv1917c of mycobacterium tuberculosis induces selective maturation of human dendritic cells by regulating PI3K-MAPK-NF-κB signaling and drives Th2 immune responses. J Biol Chem 285(47):36511–36522

Barczak AK, Avraham R, Singh S, Luo SS, Zhang WR, Bray M-A et al (2017) Systematic, multiparametric analysis of mycobacterium tuberculosis intracellular infection offers insight into coordinated virulence. PLoS Pathog 13(5):e1006363

Barrios-Payán J, Saqui-Salces M, Jeyanathan M, Alcántara-Vazquez A, Castañon-Arreola M, Rook G, Hernandez-Pando R (2012) Extrapulmonary locations of Mycobacterium tuberculosis DNA during latent infection. J Infect Dis 206(8):1194–1205

Basu S, Fowler BJ, Kerur N, Arnvig KB, Rao NA (2018) NLRP3 inflammasome activation by mycobacterial ESAT-6 and dsRNA in intraocular tuberculosis. Microb Pathog 114:219–224

Bates B, Crowell EL (2023) Ophthalmic manifestations of tuberculosis. Curr Opin Ophthalmol 34(6):529–534

Bates TA, Trank-Greene M, Nguyenla X, Anastas A, Gurmessa SK, Merutka IR, Dixon SD, Shumate A, Groncki AR, Parson MAH, Ingram JR, Barklis E, Burke JE, Shinde U, Ploegh HL, Tafesse FG (2024) ESAT-6 undergoes self-association at phagosomal pH and an ESAT-6-specific nanobody restricts M. tuberculosis growth in macrophages. eLife 12:RP91930

Bensing BASR, Yen YT, Sullam PM (2014) Selective transport by SecA2: an expanding family of customized motor proteins. Biochim Biophys Acta, Mol Cell Res 1843(8):1674–1686. https://doi.org/10.1016/j.bbamcr.2013.10.019

Bergsbaken TFS, Cookson BT (2009) Pyroptosis: host cell death and inflammation. Nat Rev Microbiol 7(2):99–109. https://doi.org/10.1038/nrmicro2070. PMID: 19148178; PMCID: PMC2910423

Bisht MK, Pal R, Dahiya P, Naz S, Sanyal P, Nandicoori VK et al (2023) The PPE2 protein of mycobacterium tuberculosis is secreted during infection and facilitates mycobacterial survival inside the host. Tuberculosis 143:102421

Blomgran R, Desvignes L, Briken V, Ernst JD (2012) Mycobacterium tuberculosis inhibits neutrophil apoptosis, leading to delayed activation of naive CD4 T cells. Cell Host Microbe 11(1):81–90

Bohada-Lizarazo DP, Bravo-Sanabria KD, Cárdenas-Malpica P, Rodríguez R (2024) Comparative genomic analysis of mycobacterium tuberculosis isolates circulating in North Santander, Colombia. Trop Med Infect Disease 9(9):197

Borhani ST, Vakili K, Jafari Y, Ghomi Z, Zandi B, Roozbahani F, Rostami SF, Malekzadegan Y, Feyzi K, Sameni F (2024) Immune system responses against Mycobacterium tuberculosis and its mechanisms to escape from the immune system. J Med Bacteriol 12(3):44–61

Bose M, Farnia P (1995) Proinflammatory cytokines can significantly induce human mononuclear phagocytes to produce nitric oxide by a cell maturation-dependent process. Immunol Lett 48(1):59–64

Bosserman RE, Thompson CR, Nicholson KR, Champion PA (2018) Esx paralogs are functionally equivalent to ESX-1 proteins but are dispensable for virulence in Mycobacterium marinum. J Bacteriol 200(10):e00726–e00717

Bottai D, Majlessi L, Simeone R, Frigui W, Laurent C, Lenormand P, Chen J, Rosenkrands I, Huerre M, Leclerc C, Cole ST, Brosch R (2011) ESAT-6 secretion-independent impact of ESX-1 genes espF and espG1 on virulence of Mycobacterium tuberculosis. J Infect Dis 203(8):1155–1164

Bowdish DM, Sakamoto K, Kim MJ, Kroos M, Mukhopadhyay S, Leifer CA et al (2009) MARCO, TLR2, and CD14 are required for macrophage cytokine responses to Mycobacterial trehalose dimycolate and Mycobacterium tuberculosis. PLoS Pathog 5(6):e1000474

Braunstein MBA, Kurtz S, Jacobs WR Jr (2001) Two nonredundant SecA homologues function in mycobacteria. J Bacteriol 183(24):6979–6990. https://doi.org/10.1128/JB.183.24.6979-6990.2001. PMID: 11717254; PMCID: PMC95544

Braunstein M, Espinosa BJ, Chan J, Belisle JT, Jacobs WR Jr (2003) SecA2 functions in the secretion of superoxide dismutase A and in the virulence of Mycobacterium tuberculosis. Mol Microbiol 48(2):453-464. https://doi.org/10.1046/j.1365-2958.2003.03438.x. PMID: 12675804

Brodin P, Rosenkrands I, Andersen P, Cole ST, Brosch R (2004) ESAT-6 proteins: protective antigens and virulence factors? Trends Microbiol 12(11):500–508

Brodin P, de Jonge MI, Majlessi L, Leclerc C, Nilges M, Cole ST, Brosch R (2005) Functional analysis of early secreted antigenic target-6, the dominant T-cell antigen of Mycobacterium tuberculosis, reveals key residues involved in secretion, complex formation, virulence, and immunogenicity. J Biol Chem 280(42):33953–33959

Brodin P, Majlessi L, Marsollier L, De Jonge MI, Bottai D, Demangel C et al (2006) Dissection of ESAT-6 system 1 of mycobacterium tuberculosis and impact on immunogenicity and virulence. Infect Immun 74(1):88–98

Bunduc CM, Bitter W, Houben ENG (2020) Structure and function of the mycobacterial type VII secretion systems. Ann Rev Microbiol 74:315–335

Bunduc CM, Fahrenkamp D, Wald J, Ummels R, Bitter W, Houben ENG, Marlovits TC (2021) Structure and dynamics of a mycobacterial type VII secretion system. Nature 593(7859):445–448

Bussi C, Gutierrez MG (2019) Mycobacterium tuberculosis infection of host cells in space and time. FEMS Microbiol Rev 43(4):341–361

Callahan B, Nguyen K, Collins A, Valdes K, Caplow M, Crossman DK, Steyn AJ, Eisele L, Derbyshire KM (2010) Conservation of structure and protein-protein interactions mediated by the secreted mycobacterial proteins EsxA, EsxB, and EspA. J Bacteriol 192(1):326–335

Camacho LR, Ensergueix D, Perez E, Gicquel B, Guilhot C (1999) Identification of a virulence gene cluster of Mycobacterium tuberculosis by signature-tagged transposon mutagenesis. Mol Microbiol 34(2):257–267

Camacho LR, Constant P, Raynaud C, Lanéelle MA, Triccas JA, Gicquel B, Daffé M, Guilhot C (2001) Analysis of the phthiocerol dimycocerosate locus of Mycobacterium tuberculosis: evidence that this lipid is involved in the cell wall permeability barrier. J Biol Chem 276(23):19845–19854

Cambier C, Falkow S, Ramakrishnan L (2014) Host evasion and exploitation schemes of mycobacterium tuberculosis. Cell 159(7):1497–1509

Cambier CJ, O'Leary SM, O'Sullivan MP, Keane J, Ramakrishnan L (2017) Phenolic glycolipid facilitates mycobacterial escape from microbicidal tissue-resident macrophages. Immunity 47(3):552–565

Cambier CJ, Banik SM, Buonomo JA, Bertozzi CR (2020) Spreading of a mycobacterial cell-surface lipid into host epithelial membranes promotes infectivity. elife 9:e60648

Cannon J, O'Driscoll M, Litman G (2011) Specific lipid recognition is a general feature of CD300 and TREM molecules. Immunogenetics 64(1):39–47

Chai Q, Wang X, Qiang L, Zhang Y, Ge P, Lu Z et al (2019) A mycobacterium tuberculosis surface protein recruits ubiquitin to trigger host xenophagy. Nat Commun 10(1):1973

Chirakos AE, Balaram A, Conrad W, Champion PA (2020) Modeling tubercular ESX-1 secretion using Mycobacterium marinum. Microbiol Mol Biol Rev 84(4). https://doi.org/10.1128/mmbr.00082-19

Champion MM, Williams EA, Pinapati RS, Champion PA (2014) Correlation of phenotypic profiles using targeted proteomics identifies mycobacterial esx-1 substrates. J Proteome Res 13(11):5151–5164

Chan J, Fan X, Hunter SW, Brennan PJ, Bloom BR (1991) Lipoarabinomannan, a possible virulence factor involved in persistence of mycobacterium tuberculosis within macrophages. Infect Immun 59(5):1755–1761

Chandra P, Grigsby SJ, Philips JA (2022) Immune evasion and provocation by Mycobacterium tuberculosis. Nat Rev Microbiol 20(12):750–766

Chao A, Sieminski PJ, Owens CP, Goulding CW (2019) Iron Acquisition in Mycobacterium tuberculosis. Chem Rev 119(2):1193–1220

Chatterjee S, Kundapura SV, Basak AJ, Mukherjee D, Dash S, Ganguli N, Das AK, Mukherjee G, Samanta D, Ramagopal UA (2022) High-resolution crystal structure of LpqH, an immunomodulatory surface lipoprotein of Mycobacterium tuberculosis reveals a distinct fold and a conserved cleft on its surface. Int J Biol Macromol 210:494–503

Chávez-Galán L, Ramon-Luing L, Carranza C, Garcia I, Sada-Ovalle I (2017) Lipoarabinomannan decreases Galectin-9 expression and tumor necrosis factor pathway in macrophages favoring Mycobacterium tuberculosis intracellular growth. Front Immunol 8:1659

Chen G, Shen L, Hu H, Feng Y, Wen D, Liu Y, Zhai H, Sun W, Wang M, Lei X et al (2024) Sulforaphane inhibits oxidative stress and may exert anti-pyroptotic effects by modulating NRF2/NLRP3 signaling pathway in Mycobacterium tuberculosis-infected macrophages. Microorganisms 12(6):1191

Clemens DL, Horwitz MA (1995) Characterization of the Mycobacterium tuberculosis phagosome and evidence that phagosomal maturation is inhibited. J Exp Med 181(1):257–270

Cole S, Brosch R, Parkhill J, Garnier T, Churcher C, Harris D et al (1998) Deciphering the biology of mycobacterium tuberculosis from the complete genome sequence. Nature 396(6707):190

Constant P, Perez E, Malaga W, Lanéelle MA, Saurel O, Daffé M, Guilhot C (2002) Role of the pks15/1 gene in the biosynthesis of phenolglycolipids in the Mycobacterium tuberculosis complex: evidence that all strains synthesize glycosylated p-hydroxybenzoic methyl esters and that strains devoid of phenolglycolipids harbor a frameshift mutation in the pks15/1 gene. J Biol Chem 277(41):38148–38158

Correia-Neves M, Fröberg G, Korshun L, Viegas S, Vaz P, Ramanlal N, Bruchfeld J, Hamasur B, Brennan P, Källenius G (2019) Biomarkers for tuberculosis: the case for lipoarabinomannan. ERJ Open Res 5(1):00115–02018

Cox JS, Chen B, McNeil M, Jacobs WR Jr (1999) Complex lipid determines tissue-specific replication of Mycobacterium tuberculosis in mice. Nature 402(6757):79–83

Damen MPM, Phan TH, Ummels R, Rubio-Canalejas A, Bitter W, Houben ENG (2020) Modification of a PE/PPE substrate pair reroutes an Esx substrate pair from the mycobacterial ESX-1 type VII secretion system to the ESX-5 system. J Biol Chem 295(18):5960–5969

Daniel TM (2006) The history of tuberculosis. Respir Med 100(11):1862–1870

Danilchanka O, Sun J, Pavlenok M, Niederweis M (2014) An outer membrane channel protein of Mycobacterium tuberculosis with exotoxin activity. Proc Natl Acad Sci USA 111(18):6750–6755

Das K, Saikolappan S, Dhandayuthapani S (2013) Differential expression of miRNAs by macrophages infected with virulent and avirulent Mycobacterium tuberculosis. Tuberculosis 93:S47–S50

Dawi J, Affa S, Kafaja K, Misakyan Y, Kades S, Dayal S, Fardeheb S, Narasimhan A, Tumanyan K, Venketaraman V (2025) The role of ferroptosis and cuproptosis in tuberculosis pathogenesis: implications for therapeutic strategies. Curr Issues Mol Biol 47(2):99

De Jonge MI, Pehau-Arnaudet G, Fretz MM, Romain F, Bottai D, Brodin P et al (2007) ESAT-6 from mycobacterium tuberculosis dissociates from its putative chaperone CFP-10 under acidic conditions and exhibits membrane-lysing activity. J Bacteriol 189(16):6028–6034

De Leon J, Jiang G, Ma Y, Rubin E, Fortune S, Sun J (2012) Mycobacterium tuberculosis ESAT-6 exhibits a unique membrane-interacting activity that is not found in its ortholog from non-pathogenic Mycobacterium smegmatis. J Biol Chem 287(53):44184–44191

De Maio F, Berisio R, Manganelli R, Delogu G (2020) PE_PGRS proteins of mycobacterium tuberculosis: a specialized molecular task force at the forefront of host–pathogen interaction. Virulence 11(1):898–915

Dennehy KM, Ferwerda G, Faro-Trindade I, Pyz E, Willment JA, Taylor PR et al (2008) Syk kinase is required for collaborative cytokine production induced through Dectin-1 and Toll-like receptors. Eur J Immunol 38(2):500–506

Dennehy KM, Willment JA, Williams DL, Brown GD (2009) Reciprocal regulation of IL-23 and IL-12 following co-activation of Dectin-1 and TLR signaling pathways. Eur J Immunol 39(5):1379–1386

Deretic V (2008) Autophagy, an immunologic magic bullet: Mycobacterium tuberculosis phagosome maturation block and how to bypass it. Future Microbiol 3(5):517–524

Deretic V, Fratti RA (1999) Mycobacterium tuberculosis phagosome. Mol Microbiol 31(6):1603–1609

Domingo-Gonzalez R, Prince O, Cooper A, Khader SA (2016) Cytokines and chemokines in Mycobacterium tuberculosis infection. Microbiol Spectrum 4(5):TBTB2-0018-2016

Dong W, Wang G, Bai Y, Li Y, Zhao L, Lu W et al (2023) Repurposing an antioxidant to kill mycobacterium tuberculosis by targeting the 50S subunit of the ribosome. Biomolecul 13(12):1793

Dover LG, Thompson AR, Sutcliffe IC, Sangal V (2021) Phylogenomic reappraisal of fatty acid biosynthesis, mycolic acid biosynthesis and clinical relevance among members of the genus Corynebacterium. Front Microbiol 12:802532

Drage MG, Tsai HC, Pecora ND, Cheng TY, Arida AR, Shukla S, Rojas RE, Seshadri C, Moody DB, Boom WH, Sacchettini JC, Harding CV (2010) Mycobacterium tuberculosis lipoprotein LprG (Rv1411c) binds triacylated glycolipid agonists of toll-like receptor 2. Nat Struct Mol Biol 17(9):1088–1095

Dutta NK, Mehra S, Martinez AN, Alvarez X, Renner NA, Morici LA, Pahar B, Maclean AG, Lackner AA, Kaushal D (2012) The stress-response factor SigH modulates the interaction between Mycobacterium tuberculosis and host phagocytes. PLoS One 7(1):e28958

D'Souza C, Kishore U, Tsolaki AG (2023) The PE-PPE family of Mycobacterium tuberculosis: proteins in disguise. Immunobiology 228(2):152321

Earp JC, Garaeva AA, Meikle V, Niederweis M, Seeger MA (2025) Structural basis of siderophore export and drug efflux by Mycobacterium tuberculosis. Nat Commun 16(1):1934

Echeverria-Valencia G, Flores-Villalva S, Espitia C (2018) Virulence factors and pathogenicity of mycobacterium. In: Mycobacterium – research and development. IntechOpen

Ehtram A, Shariq M, Quadir N, Jamal S, Pichipalli M, Zarin S et al (2025) Deciphering the functional roles of PE18 and PPE26 proteins in modulating mycobacterium tuberculosis pathogenesis and immune response. Front Immunol 16:1517822

Ekiert DC, Cox JS (2014) Structure of a PE–PPE–EspG complex from mycobacterium tuberculosis reveals molecular specificity of ESX protein secretion. Proc Natl Acad Sci 111(41):14758–14763

Eum SY, Kong JH, Hong MS, Lee YJ, Kim JH, Hwang SH, Cho SN, Via LE, Barry CE (2010) Neutrophils are the predominant infected phagocytic cells in the airways of patients with active pulmonary TB. Chest 137(1):122–128

Famelis NGS, van Tol D (2023) Mycobacterial type VII secretion systems. Biol Chem 404(7):691–702. https://doi.org/10.1515/hsz-2022-0350

Fan X, Zhao X, Wang R, Li M, Luan X, Wang R et al (2023) A novel multistage antigens ERA005f confer protection against mycobacterium tuberculosis by driving Th-1 and Th-17 type T cell immune responses. Front Immunol 14:1276887

Farnese FS, Oliveira JA, Paiva EA, Menezes-Silva PE, da Silva AA, Campos FV, Ribeiro C (2017) The involvement of nitric oxide in integration of plant physiological and ultrastructural adjustments in response to arsenic. Front Plant Sci 8:516

Fennelly KP, Jones-López EC (2015) Quantity and quality of inhaled dose predicts immunopathology in tuberculosis. Front Immunol 6:313

Fei Q, Zhang J, Chen L, Shi M, Wang Q, Xu F, Shi J, Qin Y (2024) Modulating ferroptosis and mycobactericidal activity in lung epithelial cells via YY1/iNOS pathway. Life Sci 358:123131

Ferraris DMSD, Petrera A, Marini S, Amstutz B, Coletta M, Sander P, Rizzi M (2011) Crystal structure of mycobacterium tuberculosis zinc-dependent metalloprotease-1 (Zmp1), a metalloprotease involved in pathogenicity. J Biol Chem 286(37):32475–32482. https://doi.org/10.1074/jbc.M111.271809. Epub 2011 Aug 3. PMID: 21813647; PMCID:PMC3173161

Fishbein S, Van Wyk N, Warren R, Sampson S (2015) Phylogeny to function: PE/PPE protein evolution and impact on M ycobacterium tuberculosis pathogenicity. Mol Microbiol 96(5):901–916

Forrellad MA, Klepp LI, Gioffré A, Sabio y García J, Morbidoni HR, de la Paz Santangelo M et al (2013) Virulence factors of the mycobacterium tuberculosis complex. Virulence 4(1):3–66

Freire ILS, dos Santos FR, de Menezes LCC, de Medeiros AB, Pinto JTJM, da Silva BCO (2018) Aspects of tuberculosis in the elderly. Acta Scientiarum Health Sci:40

Fratti RA, Chua J, Vergne I, Deretic V (2003) Mycobacterium tuberculosis glycosylated phosphatidylinositol causes phagosome maturation arrest. Proc Natl Acad Sci USA 100(9):5437–5442

Ferraris DMMR, Rossi F, Rizzi M (2018) Mycobacterium tuberculosis molecular determinants of infection, survival strategies, and vulnerable targets. Pathogens. 7(1):17. https://doi.org/10.3390/pathogens7010017

Gago GDL, Gramajo H (2018) Lipid metabolism and its implication in mycobacteria-host interaction. Curr Opin Microbiol 41:36–42. https://doi.org/10.1016/j.mib.2017.11.020. Epub 2017 Dec 19. PMID: 29190491; PMCID: PMC5862736

Gago G, Diacovich L, Arabolaza A, Tsai SC, Gramajo H (2011) Fatty acid biosynthesis in actinomycetes. FEMS Microbiol Rev 35(3):475–497

Gan B (2023) Ferroptosis hijacking by Mycobacterium tuberculosis. Nat Commun 14:1431

Ganbat D, Seehase S, Richter E, Vollmer E, Reiling N, Fellenberg K, Gaede KI, Kugler C, Goldmann T (2016) Mycobacteria infect different cell types in the human lung and cause species dependent cellular changes in infected cells. BMC Pulm Med 16:1–16

Garces A, Atmakuri K, Chase MR, Woodworth JS, Krastins B, Rothchild AC, Ramsdell TL, Lopez MF, Behar SM, Sarracino D et al (2010) EspA acts as a critical mediator of ESX1-dependent virulence in Mycobacterium tuberculosis by affecting bacterial cell wall integrity. PLoS Pathog 6(6):e1000957

Garrett SRHA, Palmer T (2024) Multiple variants of the type VII secretion system in gram-positive bacteria. Microlife 5:uqae013. https://doi.org/10.1093/femsml/uqae013. PMID: 38957458; PMCID: PMC11217815

Gaur RL, Ren K, Blumenthal A, Bhamidi S, Gibbs S, Jackson M et al (2014) LprG-mediated surface expression of lipoarabinomannan is essential for virulence of Mycobacterium tuberculosis. PLoS Pathog 10(9):e1004376

Geisel RE, Sakamoto K, Russell DG, Rhoades ER (2005) In vivo activity of released cell wall lipids of Mycobacterium bovis bacillus Calmette-Guerin is due principally to trehalose mycolates. J Immunol 174(8):5007–5015

Gey van Pittius NC, Sampson SL, Lee H, Kim Y, Van Helden PD, Warren RM (2006) Evolution and expansion of the mycobacterium tuberculosis PE and PPE multigene families and their association with the duplication of the ESAT-6 (esx) gene cluster regions. BMC Evol Biol 6(1):95

Ghazaei C (2018) Mycobacterium tuberculosis and lipids: insights into molecular mechanisms from persistence to virulence. J Res Med Sci 23:63

Glickman MS (2008) Cording, cord factors, and trehalose dimycolate. In: The mycobacterial cell envelope. ASM Press, pp 63–73

Gokulan K, O'Leary SE, Russell WK, Russell DH, Lalgondar M, Begley TP et al (2013) Crystal structure of mycobacterium tuberculosis polyketide synthase 11 (PKS11) reveals intermediates in the synthesis of methylbranched alkylpyrones. J Biol Chem 288(23):16484–16494

Gómez-González PJ, Grabowska AD, Tientcheu LD, Tsolaki AG, Hibberd ML, Campino S et al (2023) Functional genetic variation in pe/ppe genes contributes to diversity in mycobacterium tuberculosis lineages and potential interactions with the human host. Front Microbiol 14:1244319

Gonzalo Asensio J, Maia C, Ferrer NL, Barilone N, Laval F, Soto CY, Winter N, Daffé M, Gicquel B, Martín C, Jackson M (2006) The virulence-associated two-component PhoP-PhoR system controls the biosynthesis of polyketide-derived lipids in Mycobacterium tuberculosis. J Biol Chem 281(3):1313–1316

Greenfield EM, Beidelschies MA, Tatro JM, Goldberg VM, Hise AG (2007) Bacterial pathogen-associated molecular patterns stimulate biological activity of orthopaedic wear particles by activating cognate Toll-like receptors. J Biol Chem 282(25):18278–18284

Grivennikova VG, Gladyshev GV, Zharova TV, Borisov VB (2024) Proton-translocating NADH–ubiquinone oxidoreductase: interaction with artificial electron acceptors, inhibitors, and potential medicines. Int J Mol Sci 25(24):13421

Groisman EA (2001) The pleiotropic two-component regulatory system PhoP-PhoQ. J Bacteriol 183(6):1835–1842

Grundner C, Gay LM, Alber T (2005) Mycobacterium tuberculosis serine/threonine kinases PknB, PknD, PknE, and PknF phosphorylate multiple FHA domains. Protein Sci 14(7):1918–1921

Guirado E, Schlesinger LS, Kaplan G (2013) Macrophages in tuberculosis: friend or foe. Semin Immunopathol 35(5):563–583

Guo Q, Bi J, Wang H, Zhang X (2021) Mycobacterium tuberculosis ESX-1-secreted substrate protein EspC promotes mycobacterial survival through endoplasmic reticulum stress-mediated apoptosis. Emerg Microb Infect 10(1):19-36

Guo F, Wei J, Song Y, Li B, Qian Z, Wang X et al (2023) Immunological effects of the PE/PPE family proteins of mycobacterium tuberculosis and related vaccines. Front Immunol 14:1255920

Guzmán LE, Cambier CJ, Cheng TY, Naqvi KF, Shiloh MU, Moody DB, Bertozzi CR (2024) Bioorthogonal metabolic labeling of the virulence factor phenolic glycolipid in mycobacteria. ACS Chem Biol 19(3):642–651

Guzmán-Téllez P, Rivero-Silva MA, Moreno-Altamirano MMB, Sánchez-García FJ (2024) Macrophages, metabolism, mitochondria, circadian rhythms and the pathogen: the multidimensional nature of tuberculosis. In: Monocytes and macrophages in development, regeneration, and disease. Academic Press, pp 341–364

Hill K, Owori R, Naisanga M, Owarwo N, Mills S, Stagg HR, Mpagama S, Sekaggya Wiltshire C, Sloan D (2025) Multimorbidity in tuberculosis (TB) and its impact on patient care (MITICare): a cross sectional study nested within a prospective cohort study protocol. Wellcome Open Res 10:282

Holzheimer M, Buter J, Minnaard AJ (2021) Chemical synthesis of Cell Wall constituents of Mycobacterium tuberculosis. Chem Rev 121(15):9554–9643

Houben ENG, Korotkov KV, Bitter W (2014) Take five – type VII secretion systems of mycobacteria. Biochimica et Biophysica Acta (BBA) Mol Cell Res 1843(8):1707–1716

Huet G, Constant P, Malaga W, Lanéelle MA, Kremer K, van Soolingen D, Daffé M, Guilhot C (2009) A lipid profile typifies the Beijing strains of Mycobacterium tuberculosis: identification of a mutation responsible for a modification of the structures of phthiocerol dimycocerosates and phenolic glycolipids. J Biol Chem 284(40):27101–27113

Hunter RL (2016) Tuberculosis as a three-act play: A new paradigm for the pathogenesis of pulmonary tuberculosis. Tuberculosis 97:8–17

Indrigo J, Hunter RL Jr, Actor JK (2002) Influence of trehalose 6,6′-dimycolate (TDM) during mycobacterial infection of bone marrow macrophages. Microbiology 148(7):1991–1998

Ishikawa E, Ishikawa T, Morita YS, Toyonaga K, Yamada H, Takeuchi O, Kinoshita T, Akira S, Yoshikai Y, Yamasaki S (2009) Direct recognition of the mycobacterial glycolipid, trehalose dimycolate, by C-type lectin Mincle. J Exp Med 206(13):2879–2888

Izquierdo Lafuente BUR, Kuijl C, Bitter W, Speer A (2021) Mycobacterium tuberculosis toxin CpnT is an ESX-5 substrate and requires three type VII secretion Systems for Intracellular Secretion. MBio 12(2):e02983–e02920. https://doi.org/10.1128/mBio.02983-20. PMID: 33653883; PMCID: PMC8092274

Jackson M (2014) The mycobacterial cell envelope-lipids. Cold Spring Harb Perspect Med 4(10):a021105

Jain M, Cox JS (2005) Interaction between polyketide synthase and transporter suggests coupled synthesis and export of virulence lipid in M. tuberculosis. PLoS Pathog 1(1):e2

Jain M, Petzold CJ, Schelle MW, Leavell MD, Mougous JD, Bertozzi CR, Leary JA, Cox JS (2007) Lipidomics reveals control of Mycobacterium tuberculosis virulence lipids via metabolic coupling. Proc Natl Acad Sci USA 104(12):5133–5138

Jamilloux Y, Henry T (2013) Inflammasomes, innate immunity platforms. Med Sci 29(11):975–984

Jamwal SV, Mehrotra P, Singh A, Siddiqui Z, Basu A, Rao KVS (2016) Mycobacterial escape from macrophage phagosomes to the cytoplasm represents an alternate adaptation mechanism. Sci Rep 6:23089

Jankute M, Cox JA, Harrison J, Besra GS (2015) Assembly of the mycobacterial Cell Wall. Ann Rev Microbiol 69:405–423

Jayakumar D, Jacobs WR Jr, Narayanan S (2008) Protein kinase E of Mycobacterium tuberculosis has a role in the nitric oxide stress response and apoptosis in a human macrophage model of infection. Cell Microbiol 10(10):2594–2605

Källenius G, Correia-Neves M, Buteme H, Hamasur B, Svenson SB (2016) Lipoarabinomannan and its related glycolipids induce divergent and opposing immune responses to Mycobacterium tuberculosis depending on structural diversity and experimental variations. Tuberculosis 97:120–130

Kalscheuer R, Weinrick B, Veeraraghavan U, Besra GS, Jacobs WR Jr (2010) Trehalose-recycling ABC transporter LpqY-SugA-SugB-SugC is essential for virulence of Mycobacterium tuberculosis. Proc Natl Acad Sci USA 107(50):21761–21766

Kan-Sutton CJC, Hunter RL Jr (2009) Trehalose 6,6′-dimycolate on the surface of mycobacterium tuberculosis modulates surface marker expression for antigen presentation and costimulation in murine macrophages. Microbes Infect 11(1):40–48. https://doi.org/10.1016/j.micinf.2008.10.006. Epub 2008 Nov 1.PMID: 19007905; PMCID: PMC2680729

Kaul VBD, Singh Y, Van Kaer L, Peters-Golden M, Bishai WR, Das G (2012) An important role of prostanoid receptor EP2 in host resistance to mycobacterium tuberculosis infection in mice. J Infect Dis 206(12):1816–1825. https://doi.org/10.1093/infdis/jis609

Keating T, Lethbridge S, Allnutt JC, Hendon-Dunn CL, Thomas SR, Alderwick LJ, Taylor SC, Bacon J (2021) Mycobacterium tuberculosis modifies cell wall carbohydrates during biofilm growth with a concomitant reduction in complement activation. Cell Surface 7:100065

Kelley N, Jeltema D, Duan Y, He Y (2019) The NLRP3 Inflammasome: an overview of mechanisms of activation and regulation. Int J Mol Sci 20(13):3328

Kim SY, Choi JA, Choi S, Kim KK, Song CH, Kim EM (2024) Advances in an in vitro tuberculosis infection model using human lung organoids for host-directed therapies. PLoS Pathog 20(7):e1012295

Koiri D, Nandi M, Hameem PMA, Aher JB, Kumar A, Behura A, Meher G, Choudhary V, Choubey S, Saleem M (2025) Real-time visualization reveals Mycobacterium tuberculosis ESAT-6 disrupts phagosome-like compartment via fibril-mediated vesiculation. Cell Rep 44(3):115328

Korotkova N, Freire D, Phan TH, Ummels R, Creekmore CC, Evans TJ et al (2014) Structure of the mycobacterium tuberculosis type VII secretion system chaperone ESPG. Chaperones Substrate Spec Type VII Secret Sys Reconn 94(2):45

Korotkova N, Piton J, Wagner JM, Boy-Röttger S, Japaridze A, Evans TJ et al (2015) Structure of EspB, a secreted substrate of the ESX-1 secretion system of mycobacterium tuberculosis. J Struct Biol 191(2):236–244

Kramarska E, Squeglia F, De Maio F, Delogu G, Berisio R (2021) PE_PGRS33, an important virulence factor of mycobacterium tuberculosis and potential target of host humoral immune response. Cells 10(1):161

Kumar D, Narayanan S (2012) pknE, a serine/threonine kinase of mycobacterium tuberculosis modulates multiple apoptotic paradigms. Infect Genet Evol 12(4):737–747. https://doi.org/10.1016/j.meegid.2011.09.008. Epub 2011 Sep 16

Kumar P, Kumar D, Parikh A, Rananaware D, Gupta M, Singh Y, Nandicoori VK (2009) The mycobacterium tuberculosis protein kinase K modulates activation of transcription from the promoter of mycobacterial monooxygenase operon through phosphorylation of the transcriptional regulator VirS. J Biol Chem 284(17):11090–11099

Kumar VA, Goyal R, Bansal R, Singh N, Sevalkar RR, Kumar A, Sarkar D (2016) EspR-dependent ESAT-6 protein secretion of mycobacterium tuberculosis requires the presence of virulence regulator PhoP. J Biol Chem 291(36):19018-19030. https://doi.org/10.1074/jbc.M116.746289. Epub 2016 Jul 21. PMID: 27445330; PMCID: PMC5009273

Kurenuma T, Kawamura I, Hara H, Uchiyama R, Daim S, Dewamitta SR, Sakai S, Tsuchiya K, Nomura T, Mitsuyama M (2009) The RD1 locus in the

Mycobacterium tuberculosis genome contributes to activation of caspase-1 via induction of potassium ion efflux in infected macrophages. Infect Immun 77(9):3992–4001

Kurth DG, Gago GM, de la Iglesia A, Bazet Lyonnet B, Lin TW, Morbidoni HR, Tsai SC, Gramajo H (2009) ACCase 6 is the essential acetyl-CoA carboxylase involved in fatty acid and mycolic acid biosynthesis in mycobacteria. Microbiology 155(8):2664–2675

Kurtz S, McKinnon KP, Runge MS, Ting JP, Braunstein M (2006) The SecA2 secretion factor of mycobacterium tuberculosis promotes growth in macrophages and inhibits the host immune response. Infect Immun 74(12):6855-6864. https://doi.org/10.1128/IAI.01022-06. Epub 2006 Oct 9. PMID: 17030572; PMCID: PMC1698048

Laal S (2012) How does Mycobacterium tuberculosis establish infection? J Infect Dis 206(8):1157–1159

Lang R (2013) Recognition of the mycobacterial cord factor by Mincle: relevance for granuloma formation and resistance to tuberculosis. Front Immunol 4:5

Layre E (2020) Trafficking of Mycobacterium tuberculosis envelope components and release within extracellular vesicles: host-pathogen interactions beyond the wall. Front Immunol 11:1230

Lee WB, Kang JS, Yan JJ, Lee MS, Jeon BY, Cho SN et al (2012) Neutrophils promote mycobacterial trehalose dimycolate-induced lung inflammation via the Mincle pathway. PLoS Pathog 8(1):e1002614

Lee RS, Proulx JF, McIntosh F, Behr MA, Hanage WP (2020a) Previously undetected super-spreading of Mycobacterium tuberculosis revealed by deep sequencing. elife 9:e53245

Lee HJ, Woo Y, Hahn TW, Jung YM, Jung YJ (2020b) Formation and maturation of the phagosome: A key mechanism in innate immunity against intracellular bacterial infection. Microorganisms 8(9):1298

Leemans JC, Cassel SL, Sutterwala FS (2011) Sensing damage by the NLRP3 inflammasome. Immunol Rev 243(1):152–162

Lefebvre C, Boulon R, Ducoux M, Gavalda S, Laval F, Jamet S, Eynard N et al (2018) HadD, a novel fatty acid synthase type II protein, is essential for alpha-and epoxy-mycolic acid biosynthesis and mycobacterial fitness. Sci Rep 8(1):6034

Lefebvre C, Frigui W, Slama N, Lauzeral-Vizcaino F, Constant P, Lemassu A, Parish T et al (2020) Discovery of a novel dehydratase of the fatty acid synthase type II critical for ketomycolic acid biosynthesis and virulence of Mycobacterium tuberculosis. Sci Rep 10(1):2112

Lerner TR, Borel S, Gutierrez MG (2015) The innate immune response in human tuberculosis. Cell Microbiol 17(9):1277–1285

Lerner TR, Queval CJ, Fearns A, Repnik U, Griffiths G, Gutierrez MG (2018) Phthiocerol dimycocerosates promote access to the cytosol and intracellular burden of mycobacterium tuberculosis in lymphatic endothelial cells. BMC Biol 16(1):1

Li S, Poulton NC, Chang JS, Azadian ZA, DeJesus MA, Ruecker N et al (2022) CRISPRi chemical genetics and comparative genomics identify genes mediating drug potency in Mycobacterium tuberculosis. Nat Microbiol 7:766–779

Liu XX, Shen MJ, Liu WB, Ye BC (2018) Transcriptional and post-translational regulation of AccD6 in Mycobacterium smegmatis. FEMS Microbiol Lett 365(9):fny074

Liu L, Zhai K, Chen Y, Chen X, Wang G, Wu L (2021) Effect and mechanism of Mycobacterium tuberculosis lipoprotein LpqH in NLRP3 inflammasome activation in mouse Ana-1 macrophage. Biomed Res Int 2021:8239135

Liu D, Gu L, Zhang R, Liu L, Shen Y, Shao Y et al (2022) Utility of urine lipoarabinomannan (LAM) in diagnosing mycobacteria infection among hospitalized HIV-positive patients. Int J Infect Dis 118:65–70

Liu Y, Pan R, Ouyang Y, Gu W, Xiao T, Yang H, Tang L, Wang H, Xiang B, Chen P (2024a) Pyroptosis in health and disease: mechanisms, regulation and clinical perspective. Signal Transduct Target Ther 9(1):245

Liu Y, Zhang L, Chen Y, Ruan G, Liu Y, Chen S, Xie L, Wu F, Shi X, Liu X (2024b) Incidence and risk factors of active tuberculosis among hospitalized patients with latent tuberculosis infection in China: A cohort study. Infect Drug Resist 17:953–960

Luckner SR, Machutta CA, Tonge PJ, Kisker C (2009) Crystal structures of Mycobacterium tuberculosis KasA show mode of action within cell wall biosynthesis and its inhibition by thiolactomycin. Structure 17(7):1004–1013

Lyu J, Narum DE, Baldwin SL, Larsen SE, Bai X, Griffith DE, Dartois V, Naidoo T, Steyn AJC, Coler RN, Chan ED (2024) Understanding the development of tuberculous granulomas: insights into host protection and pathogenesis, a review in humans and animals. Front Immunol 15:1427559

Ma JZS, Gao X, Wang R, Liu J, Zhou X, Zhou Y (2021) The roles of Inflammasomes in host defense against mycobacterium tuberculosis. Pathogens. 10(2):120. https://doi.org/10.3390/pathogens10020120

Malakar B, Chauhan K, Sanyal P, Naz S, Kalam H, Vivek-Ananth RP, Singh LV, Samal A, Kumar D, Nandicoori VK (2023) Phosphorylation of CFP10 modulates Mycobacterium tuberculosis virulence. MBio 14:e01232–e01223

Manganelli R (2014) Sigma factors: key molecules in Mycobacterium tuberculosis physiology and virulence. *Microbiology*. Spectrum 2(1):MGM2-0007-2013

Martinot AJ, Farrow M, Bai L, Layre E, Cheng TY, Tsai JH et al (2016) Mycobacterial metabolic syndrome: LprG and Rv1410 regulate Triacylglyceride levels, growth rate and virulence in Mycobacterium tuberculosis. PLoS Pathog 12(1):e1005351

Mashabela GT, de Wet TJ, Warner DF (2019) Mycobacterium tuberculosis metabolism. *Microbiology*. Spectrum 7(4):GPP3-0067-2019

Master S, Rampini SK, Davis AS, Keller C, Ehlers S, Springer B, Timmins GS, Sander P, Deretic V (2008)

Mycobacterium tuberculosis prevents inflammasome activation. Cell Host Microbe 3(4):224–232

Matsumoto S, Yukitake H, Furugen M, Matsuo T, Mineta T, Yamada T (1999) Identification of a novel DNA-binding protein from Mycobacterium bovis bacillus Calmette-Guérin. Microbiol Immunol 43(11):1027–1036

Matsunaga I, Naka T, Talekar RS, McConnell MJ, Katoh K, Nakao H, Otsuka A, Behar SM, Yano I, Moody DB, Sugita M (2008) Mycolyltransferase-mediated glycolipid exchange in mycobacteria. J Biol Chem 283(43):28835–28841

Maueröder C, Chaurio R, Dumych TM, Podolska MD, Lootsik S, Culemann S et al (2016) A blast without power—cell death induced by the tuberculosis-necrotizing toxin fails to elicit adequate immune responses. Cell Death Differ 23(6):1016–1025

McGuire AM, Weiner B, Park ST, Wapinski I, Raman S, Dolganov G et al (2012) Comparative analysis of mycobacterium and related Actinomycetes yields insight into the evolution of mycobacterium tuberculosis pathogenesis. BMC Genomics 13(1):120

Menéndez MC, Buxton RS, Evans JT, Gascoyne-Binzi D, Barlow RE, Hinds J et al (2007) Genome analysis shows a common evolutionary origin for the dominant strains of mycobacterium tuberculosis in a UK south Asian community. Tuberculosis 87(5):426–436

Miller JL, Velmurugan K, Cowan MJ, Briken V (2010) The type I NADH dehydrogenase of mycobacterium tuberculosis counters phagosomal NOX2 activity to inhibit TNF-α-mediated host cell apoptosis. PLoS Pathog 6(4):e1000864

Miller BKZK, Braunstein M (2017) The sec pathways and Exportomes of mycobacterium tuberculosis. Microbiol Spectr 5(2). https://doi.org/10.1128/microbiolspec.TBTB2-0013-2016

Mir SA, Sharma S (2013) Role of MHC class Ib molecule, H2-M3 in host immunity against tuberculosis. Vaccine 31(37):3818–3825. https://doi.org/10.1016/j.vaccine.2013.04.005. Epub 2013 Apr 28. PMID: 23628242

Mishra BB, Moura-Alves P, Sonawane A, Hacohen N, Griffiths G, Moita LF, Anes E (2010) Mycobacterium tuberculosis protein ESAT-6 is a potent activator of the NLRP3/ASC inflammasome. Cell Microbiol 12(8):1046–1063

Mohanty D, Sankaranarayanan R, Gokhale RS (2011) Fatty acyl-AMP ligases and polyketide synthases are unique enzymes of lipid biosynthetic machinery in Mycobacterium tuberculosis. Tuberculosis 91(5):448–455

Mohanty P, Naaz F, Singh P, Chauhan D, Gupta U (2015) Molecular characterization of mycobacterial ribonucleotide reductase (RNR) and its implication as a novel drug target. Bioenergetics 4(129):2

Mohareer K, Asalla S, Banerjee S (2018) Cell death at the cross roads of host-pathogen interaction in mycobacterium tuberculosis infection. Tuberculosis 113:99–121

Mukku RP, Poornima K, Chakraborty K, Raghunand TR (2023) The PPE25 (Rv1787)-PE18 (Rv1788)-PPE26 (Rv1789) gene cluster is involved in immune evasion by mycobacterium tuberculosis. bioRxiv:2023.04.07.535983

Nisa A, Kipper FC, Panigrahy D, Tiwari S, Kupz A, Subbian S (2022) Different modalities of host cell death and their impact on mycobacterium tuberculosis infection. Am J Phys Cell Phys 323(5):C1444–C1C74

Odriozola JM, Ramos JA, Bloch K (1977) Fatty acid synthetase activity in Mycobacterium smegmatis: characterization of the acyl carrier protein-dependent elongating system. Biochimica et Biophysica Acta (BBA) Lipids Lipid Metabol 488(2):207–217

Ohol YM, Goetz DH, Chan K, Shiloh MU, Craik CS, Cox JS (2010) Mycobacterium tuberculosis MycP1 protease plays a dual role in regulation of ESX-1 secretion and virulence. Cell Host Microbe 7(3):210–220

O'Neil TR, Hu K, Truong NR, Arshad S, Shacklett BL, Cunningham AL, Nasr N (2021) The role of tissue resident memory CD4 T cells in herpes simplex viral and HIV infection. Viruses 13(3):359

Onwueme KC, Vos CJ, Zurita J, Ferreras JA, Quadri LE (2005) The dimycocerosate ester polyketide virulence factors of mycobacteria. Prog Lipid Res 44(5):259–30

Ouellet H, Chow E, Guan S, Cox J, Burlingame A, Ortiz de Montellano P (2013) Genetic and mass spectrometric tools for elucidating the physiological function(s) of cytochrome P450 enzymes from Mycobacterium tuberculosis. Methods Mol Biol 987:79–94

Ozeki Y, Tsutsui H, Kawada N, Suzuki H, Kataoka M, Kodama T et al (2006) Macrophage scavenger receptor down-regulates mycobacterial cord factor-induced proinflammatory cytokine production by alveolar and hepatic macrophages. Microb Pathog 40(4):171–176

Pajuelo D, Gonzalez-Juarbe N, Tak U, Sun J, Orihuela CJ, Niederweis M (2018) NAD+ depletion triggers macrophage necroptosis, a cell death pathway exploited by Mycobacterium tuberculosis. Cell Rep 24(2):429–440

Pajuelo DG-JN, Niederweis M (2020) NAD hydrolysis by the tuberculosis necrotizing toxin induces lethal oxidative stress in macrophages. Cell Microbiol 22(1):e13115. https://doi.org/10.1111/cmi.13115. Epub 2019 Oct 23. PMID:31509891; PMCID: PMC6925628

Pajuelo D, Tak U, Zhang L, Danilchanka O, Tischler AD, Niederweis M (2021) Toxin secretion and trafficking by Mycobacterium tuberculosis. Nature Commun 12(1):6592

Pal R, Bisht MK, Mukhopadhyay S (2022) Secretory proteins of Mycobacterium tuberculosis and their roles in modulation of host immune responses: focus on therapeutic targets. FEBS J 289(13):3891–3914

Palčeková Z, Obregón-Henao A, De K, Walz A, Lam H, Philp J, Angala SK, Patterson J, Pearce C, Zuberogoitia S, Avanzi C, Nigou J, McNeil M, Muñoz Gutiérrez JF, Gilleron M, Wheat WH, Gonzalez-Juarrero M, Jackson M (2023) Role of succinyl substituents in the mannose-capping of lipoarabinomannan and control of inflammation in Mycobacterium tuberculosis infection. PLoS Pathog 19(9):e1011636

Palucci I, Camassa S, Cascioferro A, Sali M, Anoosheh S, Zumbo A et al (2016) PE_PGRS33 contributes to mycobacterium tuberculosis entry in macrophages through interaction with TLR2. PLoS One 11(3):e0150800

Parandhaman DK, Narayanan S (2014) Cell death paradigms in the pathogenesis of Mycobacterium tuberculosis infection. Front Cell Infect Microbiol 4:31

Parbhoo T, Schurz H, Mouton JM, Sampson SL (2022) Persistence of Mycobacterium tuberculosis in response to infection burden and host-induced stressors. Front Cell Infect Microbiol 12:981827

Passos BBS, Araújo-Pereira M, Vinhaes CL, Amaral EP, Andrade BB (2024) The role of ESAT-6 in tuberculosis immunopathology. Front Immunol 15:1383098

Pawelczyk J, Brzostek A, Kremer L, Dziadek B, Rumijowska-Galewicz A, Fiolka M, Dziadek J (2011) AccD6, a key carboxyltransferase essential for mycolic acid synthesis in Mycobacterium tuberculosis, is dispensable in a nonpathogenic strain. J Bacteriol 193(24):6960–6972

Pawelczyk J, Viljoen A, Kremer L, Dziadek J (2017) The influence of AccD5 on AccD6 carboxyltransferase essentiality in pathogenic and non-pathogenic Mycobacterium. Sci Rep 7:42692

Pérez E, Constant P, Lemassu A, Laval F, Daffé M, Guilhot C (2004) Characterization of three glycosyltransferases involved in the biosynthesis of the phenolic glycolipid antigens from the mycobacterium tuberculosis complex. J Biol Chem 279(41):42574–42583

Pellegrini JMMC, Morelli MP, Schander JA, Tateosian NL, Amiano NO, Rolandelli A, Palmero DJ, Levi A, Ciallella L, Colombo MI, García VE (2021) PGE2 displays immunosuppressive effects during human active tuberculosis. Sci Rep 11(1):13559. https://doi.org/10.1038/s41598-021-92667-1. Erratum in: Sci Rep. 2021;11(1):18138. https://doi.org/10.1038/s41598-021-97720-7. PMID: 34193890; PMCID: PMC8245456

Pellegrini JM, Martin C, Morelli MP, Schander JA, Tateosian NL, Amiano NO, Rolandelli A, Palmero DJ, Levi A, Ciallella L, Colombo MI, Garcia VE (2020) PGE2 displays immunosuppressive effects during human active tuberculosis. Sci Rep 10(1):13559

Peng X, Sun J (2016) Mechanism of ESAT-6 membrane interaction and its roles in pathogenesis of Mycobacterium tuberculosis. Toxicon 116:29–34

Phan TH (2019) Chaperones and substrate specificity, type VII secretion systems reconnected [doctoral thesis]. Amsterdam: Vrije Universiteit Amsterdam

Piddington DL, Fang FC, Laessig T, Cooper AM, Orme IM, Buchmeier NA (2001) Cu, Zn superoxide dismutase of Mycobacterium tuberculosis contributes to survival in activated macrophages that are generating an oxidative burst. Infect Immun 69(8):4980–4987

Pieters J (2008) Mycobacterium tuberculosis and the macrophage: maintaining a balance. Cell Host Microbe 3(6):399–407

Poirier V, Av-Gay Y (2012) Mycobacterium tuberculosis modulators of the macrophage's cellular events. Microbes Infect 14(13):1211–1219

Poulsen CPS, Holton SJ, Wilmanns M, Song YH (2014) WXG100 protein superfamily consists of three subfamilies and exhibits an α-helical C-terminal conserved residue pattern. PLoS One 9(2):e89313. https://doi.org/10.1371/journal.pone.0089313

Płocińska R, Wasik K, Płociński P, Lechowicz E, Antczak M, Błaszczyk E et al (2022) The orphan response regulator Rv3143 modulates the activity of the NADH dehydrogenase complex (Nuo) in mycobacterium tuberculosis via protein–protein interactions. Front Cell Infect Microbiol 12:909507

Prakash AVPI, Park JW, Bae JP, Lee GS, Kang TJ (2023) NLRP3 Inflammasome as therapeutic targets in inflammatory diseases. Biomol Ther (Seoul) 31(4):395–401. https://doi.org/10.4062/biomolther.2023.099. PMID:37376952; PMCID: PMC10315342

Prest RJ, Korotkov KV, Champion PA (2024) The regulatory functions of ESX-1 substrates, EspE and EspF, are separable from secretion. J Bacteriol 206(9):e0027124

Qin Y, Chen Z, Peng Y, Xiao Y, Zhong T, Yu X (2024) Deep learning methods for protein structure prediction. MedComm Adv Online Publ 3

Queiroz A, Riley LW (2017) Bacterial immunostat: Mycobacterium tuberculosis lipids and their role in the host immune response. Rev Soc Bras Med Trop 50(1):9–18

Quigley J, Hughitt VK, Velikovsky CA, Mariuzza RA, El-Sayed NM, Briken V (2017) The cell wall lipid PDIM contributes to Phagosomal escape and host cell exit of Mycobacterium tuberculosis. MBio 8(2):e00148–e00117

Raghavan S, Manzanillo P, Chan K, Dovey C, Cox JS (2008) Secreted transcription factor controls mycobacterium tuberculosis virulence. Nature 454(7205):717–721

Rajni, Rao N, Meena L (2011) Biosynthesis and virulent behavior of lipids produced by Mycobacterium tuberculosis: LAM and cord factor: an overview. Biotechnol Res Int 2011:274693

Ramachandra L, Smialek JL, Shank SS, Convery M, Boom WH, Harding CV (2005) Phagosomal processing of mycobacterium tuberculosis antigen 85B is modulated independently of mycobacterial viability and phagosome maturation. Infect Immun 73(2):1097–1105

Ramon-Luing LA, Palacios Y, Ruiz A, Téllez-Navarrete NA, Chavez-Galan L (2023) Virulence factors of mycobacterium tuberculosis as modulators of cell death mechanisms. Pathogens 12(6)

Ramos BGS, Cunha MV (2020) Revisiting the expression signature of pks15/1 unveils regulatory patterns controlling phenolphtiocerol and phenolglycolipid production in pathogenic mycobacteria. PLoS One 15(5):e0229700. https://doi.org/10.1371/journal.pone.0229700. PMID: 32379829; PMCID: PMC7205293

Randall PJ, Hsu NJ, Quesniaux V, Ryffel B, Jacobs M (2015) Mycobacterium tuberculosis infection of the 'non-classical immune cell'. Immunol Cell Biol 93(9):789–795

Rangel Moreno JEGI, La Luz D, García Hernández M, Aguilar Leon D, Marquez R, Hernández PR (2002) The role of prostaglandin E2 in the immunopathogenesis of experimental pulmonary tuberculosis. Immunology 106(2):257–266. https://doi.org/10.1046/j.1365-2567.2002.01403.x. PMID: 12047755; PMCID: PMC1782721

Rastogi S, Briken V (2022) Interaction of mycobacteria with host cell inflammasomes. Front Immunol 13:791136

Reddy MC, Breda A, Bruning JB, Sherekar M, Valluru S, Thurman C, Ehrenfeld H, Sacchettini JC (2014) Structure, activity, and inhibition of the Carboxyltransferase β-subunit of acetyl coenzyme A carboxylase (AccD6) from Mycobacterium tuberculosis. Antimicrob Agents Chemother 58(10):6122–6132

Renshaw PS, Lightbody KL, Veverka V, Muskett FW, Kelly G, Frenkiel TA, Gordon SV, Hewinson RG, Burke B, Norman J, Williamson RA, Carr MD (2005) Structure and function of the complex formed by the tuberculosis virulence factors CFP-10 and ESAT-6. EMBO J 24(14):2491–2498

Rens CCJ, Sexton DL et al (2021) Roles for phthiocerol dimycocerosate lipids in mycobacterium tuberculosis pathogenesis. Microbiology (Reading) 167. https://doi.org/10.1099/mic.0.001042

Rodríguez-González J, Gutiérrez-Kobeh L (2024) Apoptosis and its pathways as targets for intracellular pathogens to persist in cells. Parasitol Res 123(1):60

Roux A, Cuvelier D, Nassoy P, Prost J, Bassereau P, Goud B (2005) Role of curvature and phase transition in lipid sorting and fission of membrane tubules. EMBO J 24(8):1537–1545

Roy M, Bose M, Bankoti K, Kundu A, Dhara S, Das AK (2020) Biochemical characterization of VapC46 toxin from mycobacterium tuberculosis. Mol Biotechnol 62(6):335–343

Ruan HLM, Lai H, Niu L, Zhao Z, Liu T, Lei S, Ying B (2024) Host-pathogen dialogues in different cell death modes during mycobacterium tuberculosis infection. Interdiscipl Med 2(2):e20230044

Rudraraju RS, Daher SS, Gallardo-Macias R, Wang X, Neiditch MB, Freundlich JS (2022) Mycobacterium tuberculosis KasA as a drug target: structure-based inhibitor design. Front Cell Infect Microbiol 12:1008213

Ryll R, Kumazawa Y, Yano I (2001) Immunological properties of trehalose dimycolate (cord factor) and other mycolic acid-containing glycolipids—a review. Microbiol Immunol 45(12):801–811

Sacco E, Covarrubias AS, O'Hare HM, Carroll P, Eynard N, Jones TA, Parish T, Daffé M, Bäckbro K, Quémard A (2007) The missing piece of the type II fatty acid synthase system from Mycobacterium tuberculosis. Proc Natl Acad Sci USA 104(37):14628–14633

Sachdeva K, Goel M, Sudhakar M, Mehta M, Raju R, Raman K, Singh A, Sundaramurthy V (2020) Mycobacterium tuberculosis (Mtb) lipid mediated lysosomal rewiring in infected macrophages modulates intracellular Mtb trafficking and survival. J Biol Chem 295(27):9192–9210

Saini V, Cumming BM, Guidry L, Lamprecht DA, Adamson JH, Reddy VP et al (2016) Ergothioneine maintains redox and bioenergetic homeostasis essential for drug susceptibility and virulence of mycobacterium tuberculosis. Cell Rep 14(3):572–585

Sakthi S, Narayanan S (2013) The lpqS knockout mutant of Mycobacterium tuberculosis is attenuated in macrophages. Microbiol Res 168(7):407–414

Sampson SL (2011) Mycobacterial PE/PPE proteins at the host-pathogen interface. Clin Dev Immunol 2011:497203. https://doi.org/10.1155/2011/497203. Epub 2011 Jan 26. PMID: 21318182; PMCID: PMC3034920

Sampson S, Lukey P, Warren R, Van Helden P, Richardson M, Everett M (2001) Expression, characterization and subcellular localization of the mycobacterium tuberculosis PPE gene Rv1917c. Tuberculosis 81(5–6):305–317

Sampson SL, Mansfield KG, Carville A, Magee DM, Quitugua T, Howerth EW et al (2011) Extended safety and efficacy studies of a live attenuated double leucine and pantothenate auxotroph of mycobacterium tuberculosis as a vaccine candidate. Vaccine 29(29–30):4839–4847

Sandhu P, Akhter Y (2017) Siderophore transport by MmpL5-MmpS5 protein complex in Mycobacterium tuberculosis. J Inorg Biochem 170:75–84

Santos JL, Yamasaki PR, Chin CM, Takashi CH, Pavan FR, Leite CQ (2009) Synthesis and in vitro anti mycobacterium tuberculosis activity of a series of phthalimide derivatives. Bioorg Med Chem 17(11):3795–3799

Sasindran SJ, Torrelles JB (2011) Mycobacterium tuberculosis infection and inflammation: what is beneficial for the host and for the bacterium? Front Microbiol 2:2

Saxena P, Yadav G, Mohanty D, Gokhale RS (2003) A new family of type III polyketide synthases in Mycobacterium tuberculosis. J Biol Chem 278(45):44780–44790

Schaeffer ML, Agnihotri G, Volker C, Kallender H, Brennan PJ, Lonsdale JT (2001) Purification and biochemical characterization of the Mycobacterium tuberculosis β-ketoacyl-acyl carrier protein synthases KasA and KasB. J Biol Chem 276(50):47029–47037

Schluger NW, Rom WN (1998) The host immune response to tuberculosis. Am J Respir Crit Care Med 157(3 Pt 1):679–691

Schoenen H, Bodendorfer B, Hitchens K, Manzanero S, Werninghaus K, Nimmerjahn F, Aggerbeck H et al (2010) Cutting edge: Mincle is essential for recognition and adjuvanticity of the mycobacterial cord factor and its synthetic analog trehalose-dibehenate. J Immunol 184(6):2756–2760

Sengupta N, Padmanaban S, Dutta S (2023) Cryo-EM reveals the membrane-binding phenomenon of EspB, a virulence factor of the mycobacterial type VII secretion system. J Biol Chem 299(4):104589. https://doi.org/10.1016/j.jbc.2023.104589. PMID: 36889587; PMCID: PMC10140165

Shah S, Briken V (2016) Modular organization of the ESX-5 secretion system in Mycobacterium tuberculosis. Front Cell Infect Microbiol 6:49

Sharma A, Vashistt J, Shrivastava R (2022) Mycobacterium fortuitum fabG4 knockdown studies: implication as pellicle and biofilm specific drug target. J Basic Microbiol 62(12):1504–1513

Shukla S, Richardson ET, Athman JJ, Shi L, Wearsch PA, McDonald D, Banaei N, Boom WH, Jackson M, Harding CV (2014) Mycobacterium tuberculosis lipoprotein LprG binds lipoarabinomannan and determines its cell envelope localization to control phagolysosomal fusion. PLoS Pathog 10(10):e1004471

Singh V, Mani I, Chaudhary DK, Somvanshi P (2011) The β-ketoacyl-ACP synthase from Mycobacterium tuberculosis as potential drug targets. Curr Med Chem 18(9):1318–1324

Slayden RA, Barry CE 3rd (2000) The genetics and biochemistry of isoniazid resistance in mycobacterium tuberculosis. Microbes Infect 2(6):659-669. https://doi.org/10.1016/s1286-4579(00)00359-2. PMID: 10884617

Slayden, R. A., & Barry, C. E., (2002). The role of KasA and KasB in the biosynthesis of meromycolic acids and isoniazid resistance in Mycobacterium tuberculosis. Tuberculosis, 82(4–5)-, 149–160

Solomonson M, Setiaputra D, Makepeace KA, Lameignere E, Petrotchenko EV, Conrady DG et al (2015) Structure of EspB from the ESX-1 type VII secretion system and insights into its export mechanism. Structure 23(3):571–583

Sreejit G, Ahmed A, Parveen N, Jha V, Valluri VL, Ghosh S, Mukhopadhyay S (2014) The ESAT-6 protein of Mycobacterium tuberculosis interacts with beta-2-microglobulin (β2M) affecting antigen presentation function of macrophage. PLoS Pathog 10(10):e1004446

Sritharan M (2016) Iron homeostasis in Mycobacterium tuberculosis: mechanistic insights into Siderophore-mediated iron uptake. J Bacteriol 198(18):2399–2409

Srinivasan L, Gurses SA, Hurley BE, Miller JL, Karakousis PC, Briken V (2016) Identification of a transcription factor that regulates host cell exit and virulence of mycobacterium tuberculosis. PLoS Pathog 12(5):e1005652

Srinivasan LAS, Briken V (2014) Interaction of mycobacterium tuberculosis with host cell death pathways. Cold Spring Harb Perspect Med 4(8):a022459

Srivastava S, Battu MB, Khan MZ, Nandicoori VK, Mukhopadhyay S (2019) Mycobacterium tuberculosis PPE2 protein interacts with p67phox and inhibits reactive oxygen species production. J Immunol 203(5):1218–1229

Stokas H, Rhodes H, Purdy G (2020) Modulation of the Mycobacterium tuberculosis cell envelope between replicating and non-replicating persistent bacteria. Tuberculosis 125:102007

Strong EJ, Jurcic Smith KL, Saini NK, Ng TW, Porcelli SA, Lee S (2020) Identification of autophagy-inhibiting factors of mycobacterium tuberculosis by high-throughput loss-of-function screening. Infect Immun 88(12). https://doi.org/10.1128/iai.00269-20

Strong EJ, Lee S (2021) Targeting autophagy as a strategy for developing new vaccines and host-directed therapeutics against mycobacteria. Front Microbiol 11:614313

Sullivan JT, Young EF, McCann JR, Braunstein M (2012) The Mycobacterium tuberculosis SecA2 system subverts phagosome maturation to promote growth in macrophages. Infect Immun 80(3):996–1006

Sun J, Siroy A, Lokareddy RK, Speer A, Doornbos KS, Cingolani G, Niederweis M (2015) The tuberculosis necrotizing toxin kills macrophages by hydrolyzing NAD. Nat Struct Mol Biol 22(9):672–678

Sun MR, Xing JY, Li XT, Fang R, Zhang Y, Li ZL, Song NN (2025) Recent advances in research on Mycobacterium tuberculosis virulence factors and their role in pathogenesis. J Microbiol Immunol Infect. advance online publication

Su H, Zhang Z, Liu Z, Peng B, Kong C, Wang H et al (2018) Mycobacterium tuberculosis PPE60 antigen drives Th1/Th17 responses via toll-like receptor 2–dependent maturation of dendritic cells. J Biol Chem 293(26):10287–10302

Takayama K, Wang C, Besra GS (2005) Pathway to synthesis and processing of mycolic acids in Mycobacterium tuberculosis. Clin Microbiol Rev 18(1):81–101

Tak UVJ, Garza-Garcia A, de Carvalho LP, Saad JS, Niederweis M et al (2019) The tuberculosis necrotizing toxin is an NAD+ and NADP+ glycohydrolase with distinct enzymatic properties. J Biol Chem 294(9):3024–3036. https://doi.org/10.1074/jbc.RA118.006733

Tang D, Chen X, Kang R, Kroemer G (2021) Ferroptosis: molecular mechanisms and health implications. Cell Res 31(2):107–125

Tran W, Kusay AS, Hawkins PM, Cheung C-Y, Nagalingam G, Pujari V et al (2021) Synthetic sansanmycin analogues as potent mycobacterium tuberculosis translocase I inhibitors. J Med Chem 64(23):17326–17345

Trivedi OAAP, Vats A, Ansari MZ, Tickoo R, Sridharan V, Mohanty D, Gokhale RS (2005) Dissecting the mechanism and assembly of a complex virulence mycobacterial lipid. Mol Cell 17(5):631–643. https://doi.org/10.1016/j.molcel.2005.02.009

Tsai MC, Chakravarty S, Zhu G, Xu J, Tanaka K, Koch C, Tufariello J, Flynn J, Chan J (2006) Characterization of the tuberculous granuloma in murine and human lungs: cellular composition and relative tissue oxygen tension. Cell Microbiol 8(2):218–232

Ulrichs T, Kaufmann SH (2006) New insights into the function of granulomas in human tuberculosis. J Pathol 208(2):261–269

van Winden VJC (2021) The mycosins of pathogenic mycobacteria: digesting the role of conserved proteases in type VII secretion. Doctoral dissertation, Vrije Universiteit Amsterdam

Velmurugan K, Chen B, Miller JL, Azogue S, Gurses S, Hsu T et al (2007) Mycobacterium tuberculosis nuoG is a virulence gene that inhibits apoptosis of infected host cells. PLoS Pathog 3(7):e110

Vergne I, Chua J, Deretic V (2003) Tuberculosis toxin blocking phagosome maturation inhibits a novel Ca2+/calmodulin-PI3K hVPS34 cascade. J Exp Med 198(4):653–659

Vergne I, Gilleron M, Nigou J (2015) Manipulation of the endocytic pathway and phagocyte functions by mycobacterium tuberculosis lipoarabinomannan. Front Cell Infect Microbiol 4:187

Vilchèze C, Weinrick B, Leung LW, Jacobs WR Jr (2018) Plasticity of mycobacterium tuberculosis NADH dehydrogenases and their role in virulence. Proc Natl Acad Sci 115(7):1599–1604

Voskuil M, Visconti K, Schoolnik G (2004) Mycobacterium tuberculosis gene expression during adaptation to stationary phase and low-oxygen dormancy. Tuberculosis 84(3–4):218–227

Vu A, Glassman I, Campbell G, Yeganyan S, Nguyen J, Shin A, Venketaraman V (2024) Host cell death and modulation of immune response against Mycobacterium tuberculosis infection. Int J Mol Sci 25(11):6255

Wang Y, Branicky R, Noë A, Hekimi S (2018) Superoxide dismutases: dual roles in controlling ROS damage and regulating ROS signaling. J Cell Biol 217(6):1915–1928

Wang S, Zhou K, Yang X, Zhang B, Zhao Y, Xiao Y et al (2020) Structural insights into substrate recognition by the type VII secretion system. Protein Cell 11:124–137

Wang SGQ, Zhou L, Xia X (2024) Ferroptosis: a double-edged sword. Cell Death Discov 10(1):265. https://doi.org/10.1038/s41420-024-02037-9. PMID: 38816377; PMCID: PMC11139933

Werninghaus K, Babiak A, Groß O, Hölscher C, Dietrich H, Agger EM et al (2009) Adjuvanti city of a synthetic cord factor analogue for subunit mycobacterium tuberculosis vaccination requires FcRγ–Syk–Card9–dependent innate immune activation. J Exp Med 206(1):89–97

Welin A, Winberg ME, Abdalla H, Särndahl E, Rasmusson B, Stendahl O, Lerm M (2008) Incorporation of Mycobacterium tuberculosis lipoarabinomannan into macrophage membrane rafts is a prerequisite for the phagosomal maturation block. Infect Immun 76(7):2882–2887

Welin A, Björnsdottir H, Winther M, Christenson K, Oprea T, Karlsson A, Forsman H, Dahlgren C, Bylund J (2015) CFP-10 from Mycobacterium tuberculosis selectively activates human neutrophils through a pertussis toxin-sensitive chemotactic receptor. Infect Immun 83(1):205–213

Wells CA, Salvage-Jones JA, Li X, Hitchens K, Butcher S, Murray RZ et al (2008) The macrophage-inducible C-type lectin, mincle, is an essential component of the innate immune response to Candida albicans. J Immunol 180(11):7404–7413

Wengenack NL, Hoard HM, Rusnak F (1999) Isoniazid oxidation by Mycobacterium tuberculosis KatG: A role for superoxide which correlates with isoniazid susceptibility. J Am Chem Soc 121(41):9748–9749

Witt KD (2023) Role of MHC class I pathways in Mycobacterium tuberculosis antigen presentation. Front Cell Infect Microbiol 13:1107884

Wolfe LM, Mahaffey SB, Kruh NA, Dobos KM (2010) Proteomic definition of the cell wall of Mycobacterium tuberculosis. J Proteome Res 9(11): 5816–5826

Yamasaki S, Ishikawa E, Sakuma M, Hara H, Ogata K, Saito T (2008) Mincle is an ITAM-coupled activating receptor that senses damaged cells. Nat Immunol 9(10):1179–1188. https://doi.org/10.1038/ni.1651. Epub 2008 Sep 7. PMID: 18776906

Yan MY, Li H, Qu YM, Li SS, Zheng D, Guo XP et al (2024) CRISPR screening and comparative LC-MS analysis identify genes mediating efficacy of Delamanid and Pretomanid against mycobacterium tuberculosis. Adv Sc 11(39):2400176

Yang J, Zhang L, Qiao W, Luo Y (2023) Mycobacterium tuberculosis: pathogenesis and therapeutic targets. MedComm Adv Online Publ

Yu J, Tran V, Li M, Huang X, Niu C, Wang D, Zhu J, Wang J, Gao Q, Liu J (2012) Both phthiocerol dimycocerosates and phenolic glycolipids are required for virulence of Mycobacterium marinum. Infect Immun 80(4):1381–1389

Zhai W, Wu F, Zhang Y, Fu Y, Liu Z (2019) The immune escape mechanisms of mycobacterium tuberculosis. Int J Mol Sci 20(2):340

Zhang Y, Vilchèze C, Jacobs WR Jr (2004) Mechanisms of drug resistance in mycobacterium tuberculosis. In: Tuberculosis and the tubercle bacillus, pp 115–140

Zhang QAMS, Li P, Xie J (2023) The dynamics of mycobacterium tuberculosis phagosome and the fate of infection. Cell Signal 108:110715

Zhang X, Zhao R, Qi Y, Yan X, Qi G, Peng Q (2024) The progress of Mycobacterium tuberculosis drug targets. Front Med 11:1455715

Zhou KL, Li X, Zhang XL, Pan Q (2019) Mycobacterial mannose-capped lipoarabinomannan: a modulator bridging innate and adaptive immunity. Emerg Microb Infect 8(1):1168–1177

Zulauf KE, Sullivan JT, Braunstein M (2018) The SecA2 pathway of Mycobacterium tuberculosis exports effectors that work in concert to arrest phagosome and autophagosome maturation. PLoS Pathog 14(4):e1007011

A Key Metabolic Protein in Active *Mycobacterium tuberculosis*: Insights into Carbon, Nitrogen, and Sulfur Metabolism

6

Abstract

Mycobacterium tuberculosis (Mtb) employs a highly adaptable network of metabolic pathways that are pivotal for its survival and pathogenesis within the host during both exponential growth and persistent infection phases. Central carbon metabolism in Mtb exhibits remarkable flexibility, enabling the bacterium to utilize diverse carbon sources efficiently. In the absence of glucose, Mtb preferentially metabolizes fatty acids as primary carbon substrates. This metabolic shift is supported by the glyoxylate shunt and methyl citrate cycle, which replenish tricarboxylic acid (TCA) cycle intermediates essential for energy production and biosynthesis. Key enzymes such as isocitrate lyase *(icl)* and methylcitrate lyase (*mcl*) facilitate the catabolism of fatty acids and maintain TCA cycle functionality, thereby sustaining bacterial growth under nutrient-limited conditions. Further enhancing metabolic adaptability, Mtb modulates central carbon metabolism through lysine acetylation, a posttranslational modification that regulates enzyme activity, particularly within fatty acid metabolic pathways. This regulatory mechanism allows Mtb to fine-tune its metabolic responses and optimize carbon utilization in response to fluctuating environmental nutrient availability.

Nitrogen metabolism in Mtb is equally versatile, characterized by the capacity to utilize a variety of nitrogen sources. Amino acids such as glutamine, glutamate, aspartate, and asparagine serve as superior nitrogen donors compared to inorganic ammonium (NH_4^+), reflecting Mtb's adaptation to the nutrient milieu of the host, where these amino acids are abundant. Alanine dehydrogenase (*ald*) exemplifies the complexity of nitrogen metabolism by functioning dually in alanine utilization and ammonium assimilation. Mtb exhibits limited homeostatic control over certain intracellular amino acid pools and demonstrates the ability to co-metabolize multiple nitrogen sources simultaneously, underscoring the dynamic nature of its nitrogen metabolic network.

Sulfur metabolism plays a critical role in maintaining redox balance and supporting Mtb virulence. The sulfate assimilation pathway is central to the biosynthesis of sulfur-containing metabolites such as cysteine (Cys), which serves as a precursor for low molecular weight thiols, including mycothiol (MSH) and ergothioneine (EGT). These thiols are essential antioxidants that protect Mtb from oxidative stress encountered within host macrophages. The trans-sulfuration pathway, which converts methionine (*Met*) to cysteine, links methylation processes to antioxidant

P. Farnia et al., *Proteins in Mycobacterium Tuberculosis*, Advances in Experimental Medicine and Biology 1484, https://doi.org/10.1007/978-3-031-96883-9_6

metabolism, further contributing to sulfur homeostasis. Sulfotransferases (*Stfs*) utilize 3′-phosphoadenosine 5′-phosphosulfate (PAPS) as a sulfate donor to catalyze the sulfation of lipids and other molecules, thereby influencing Mtb's virulence and survival mechanisms. Overall, Mtb's carbon, nitrogen, and sulfur metabolic pathways are intricately interconnected, with each influencing the others to create a robust and flexible metabolic network. This metabolic integration is fundamental to Mtb's ability to thrive within the hostile environment of the host. A comprehensive understanding of these pathways is critical for identifying novel therapeutic targets aimed at disrupting Mtb's metabolic adaptability and pathogenicity. Despite the challenges posed by Mtb's metabolic resilience, ongoing research into its metabolic mechanisms continues to provide valuable insights that will inform the development of innovative antituberculosis therapies.

Keywords

Mycobacterium tuberculosis · Metabolic proteins · Central carbon metabolism (CCM) · Gluconeogenesis · Nitrogen metabolism · Sulfur metabolism

6.1 Introduction

Mycobacterium tuberculosis (Mtb), the pathogen responsible for tuberculosis, demonstrates exceptional adaptability to a wide range of challenging environments throughout its lifecycle. Primarily, Mtb inhabits and replicates within the phagosomal compartment of host macrophages, a highly dynamic intracellular niche characterized by variable nutrient availability (Warner 2014; Huang et al. 2018). Although macrophages serve as the primary site for Mtb replication, accumulating evidence indicates that the pathogen is not restricted to this environment. Mtb is capable of escaping from the intracellular milieu, surviving extracellularly, persisting within alternative cell types, and adapting to the heterogeneous and evolving microenvironments found within granulomatous lesions (Grosset 2003; Bussi and Gutierrez 2019; Olaniyan et al. 2023). This extraordinary adaptability of Mtb is underpinned by its capacity to endure and thrive across a broad spectrum of microenvironments, each characterized by distinct nutrient profiles and immune pressures. Central to this adaptability are metabolic proteins, comprising a diverse array of enzymes and transporters, that orchestrate the biochemical pathways necessary for the pathogen's survival and proliferation under variable conditions (Bussi and Gutierrez 2019). These metabolic proteins fulfill dual roles: they support essential metabolic functions during periods of active growth and replication in nutrient-rich environments, and they mediate specialized adaptations that are critical for persistence under nutrient limitation and immune stress within host cells. Throughout its complex lifecycle, Mtb relies on its metabolic proteins to regulate key processes such as central carbon metabolism, energy production, biosynthesis of vital macromolecules (including lipids and nucleotides), and the maintenance of redox homeostasis (Beste and McFadden 2012). These interconnected metabolic processes are indispensable for the bacterium's survival, enabling it to transition between environments ranging from the nutrient-abundant conditions of laboratory culture to the nutrient-deprived, immune-challenged tissues of the host. A defining feature of Mtb's metabolic network is the high degree of integration within its central carbon metabolism. This network encompasses glycolysis and gluconeogenesis, the tricarboxylic acid (TCA) cycle, the glyoxylate shunt, lipid and cholesterol metabolism, co-metabolic nodes, and a wide variety of transporters that facilitate the uptake and utilization of diverse substrates (Borah et al. 2021; Chen et al. 2024). Such metabolic versatility allows Mtb to exploit a broad range of carbon sources, including carbohydrates (e.g., glucose), lipids (e.g., fatty acids and cholesterol), and amino acids.

During periods of active replication, Mtb preferentially utilizes carbohydrates as its primary carbon source. The glycolytic pathway is initiated by enzymes such as glucokinase (*glkA*), which catalyzes the phosphorylation of glucose to glucose-6-phosphate, thereby committing glucose to cellular metabolism (Cumming and Steyn 2015; Mackenzie et al. 2020; Chang and Guan 2021). Subsequent steps in glycolysis are driven by key enzymes, including phosphofructokinase (*pfkA*) and pyruvate kinase (*pykA*), which collectively generate adenosine triphosphate (ATP) and pyruvate. These reactions not only provide energy for cellular processes but also yield metabolic intermediates that serve as precursors for biosynthetic pathways (Phong et al. 2013; Noy et al. 2016; Snášel and Pichová 2019). When carbohydrate sources become depleted, particularly during the chronic stages of infection, Mtb undergoes a critical metabolic shift toward lipid metabolism to sustain its energy and biosynthetic demands. This shift is facilitated by specialized fatty acid transporters, including fatty acyl-CoA synthetase D5 (fadD5) and the mammalian cell entry protein 4D (mce4D), which import host-derived lipids into the bacterial cell. Once internalized, these lipids are catabolized through β-oxidation, a process that breaks down fatty acids into acetyl-coenzyme A (acetyl-CoA). The acetyl-CoA generated serves as a key metabolic intermediate, entering either the tricarboxylic acid (TCA) cycle or the glyoxylate shunt, depending on the environmental conditions and metabolic requirements of the bacterium. The TCA cycle, a central hub of energy production and biosynthesis, is catalyzed by enzymes such as citrate synthase (*gltA*) and isocitrate dehydrogenase (*icd1*). This cycle oxidizes acetyl-CoA to produce reducing equivalents in the form of nicotinamide adenine dinucleotide (NADH) and flavin adenine dinucleotide ($FADH_2$). These reducing agents subsequently fuel oxidative phosphorylation, generating ATP to meet the bacterium's energy needs. In addition to energy production, the TCA cycle provides essential intermediates that serve as precursors for the biosynthesis of amino acids, lipids, and nucleotides, thereby supporting cellular growth and maintenance (Serafini et al. 2019). In contrast, the glyoxylate shunt offers an alternative metabolic pathway that bypasses the carbon dioxide (CO_2)-producing steps of the TCA cycle. This bypass is catalyzed by enzymes such as isocitrate lyase (*icl*) and malate synthase (*glcB*), which convert isocitrate into succinate and malate without the loss of carbon atoms as CO_2. By conserving carbon skeletons, the glyoxylate shunt is particularly important for Mtb's persistence within macrophages, where lipids are abundant but carbohydrates are scarce (Ahn et al. 2016). This pathway enables the bacterium to efficiently utilize fatty acids as carbon sources while minimizing carbon loss, thus supporting long-term survival under nutrient-limited intracellular conditions. Parallel to these pathways, the methylcitrate cycle (MCC) plays a crucial role in detoxifying propionyl-coenzyme A (propionyl-CoA), a potentially toxic intermediate derived from the catabolism of odd-chain fatty acids and cholesterol side chains. The MCC enzymes, including methylcitrate synthase (*prpC*), methylcitrate dehydratase (*prpD*), and 2-methylisocitrate lyase (*prpB*), convert propionyl-CoA into pyruvate and succinate. These products are then funneled back into central metabolic pathways, preventing the accumulation of toxic metabolites and ensuring metabolic homeostasis (Upton and McKinney 2007; Limenitakis et al. 2013). Cholesterol metabolism further exemplifies Mtb's metabolic versatility. This process is mediated by the vitamin B12-dependent enzyme methylmalonyl-CoA mutase (mcm), which converts cholesterol-derived intermediates into metabolites that not only provide energy but also serve as precursors for the biosynthesis of complex cell wall lipids such as phthiocerol dimycocerosate (PDIM) and sulfolipid-1. These lipids function as storage sinks for excess propionyl-CoA and contribute significantly to immune evasion and virulence, highlighting the intricate link between metabolism and pathogenicity. Mtb's metabolic adaptability is further enhanced by co-metabolism nodes, which represent strategic points of integration between distinct metabolic pathways. For example, trehalose recycling connects glycolysis with cell wall biosynthesis by converting glucose-6-phosphate into trehalose precursors, thereby linking energy metabolism with structural maintenance. Similarly, the pentose phosphate pathway (PPP) operates in conjunction with glycolysis to generate nicotinamide adenine dinucleotide phosphate (NADPH), which is essential for maintaining redox balance, and ribose-5-phosphate, a precursor for nucleotide synthesis. These integrative metabolic nodes ensure efficient substrate utilization and maintain metabolic equilibrium, particularly under environmental stresses encountered during infection (Takahashi-Iñiguez et al. 2012). Transport systems also play an indispensable role in supporting central carbon metabolism by facilitating nutrient uptake and waste removal. For instance, glycerol transporters enable Mtb to utilize glycerol as a carbon source under laboratory culture conditions, while fatty acid transporters are critical for lipid uptake during infection. Sulfate transporters encoded by the cysTWA operon mediate the uptake of extracellular sulfate, which feeds into sulfur metabolism pathways that are vital for the synthesis of sulfur-containing biomolecules and redox homeostasis. Sulfate assimilation supports cysteine biosynthesis through the coordinated actions of sulfite reductase (*cys*I) and serine acetyltransferase (*cysE*) (Hatzios and Bertozzi 2011; Chen et al. 2019). Cysteine is a critical amino acid that plays a pivotal role in maintaining redox homeostasis, particularly by contributing to antioxidant systems such as mycothiol, which protect Mtb against oxidative stress. Proteins involved in nitrogen metabolism enable Mtb to acquire nitrogen from host-derived sources, including ammonium and amino acids, which are essential for the synthesis of nucleotides and amino acids. For example, glutamine synthetase (*glnA1*) catalyzes the ATP-dependent conversion of ammonium into glutamine, a key nitrogen donor in multiple biosynthetic pathways (Xu et al. 2023). Aspartate kinase (*lysC*) initiates lysine biosynthesis from aspartate, while acetolactate synthase (*ilvB*) catalyzes the first committed step in the biosynthesis of branched-chain amino acids, underscoring the importance of nitrogen assimilation for essential amino acid production.

Proteins regulating carbon, nitrogen, and sulfur metabolism collectively enhance Mtb's ability to survive and proliferate within the nutrient-limited and hostile environments encountered during infection. By dynamically switching between metabolic pathways such as glycolysis, gluconeogenesis, lipid metabolism, and alternative cycles, including the methylcitrate cycle (MCC) and the glyoxylate shunt, depending on substrate availability, these proteins ensure efficient energy generation and resource conservation. Furthermore, specialized enzymes contribute to the bacterium's defense against oxidative stress imposed by host immune responses. They maintain redox balance through NADPH production and cysteine-dependent antioxidant systems, thereby protecting cellular components from damage. The expression of genes encoding these metabolic proteins is tightly regulated in response to environmental signals, enabling Mtb to optimize energy production and biosynthetic processes under stress conditions. Mutants lacking key enzymes such as isocitrate lyase (*icl*) or phosphoenolpyruvate carboxykinase (*pckA*) exhibit significantly reduced virulence, reflecting impaired carbon flux through the glyoxylate shunt and gluconeogenesis pathways, respectively (Rhee 2001). Disruption of these metabolic proteins or their corresponding genes markedly compromises the pathogen's ability to persist within host tissues. Metabolic proteins involved in carbon, nitrogen, and sulfur metabolism thus constitute the cornerstone of Mtb's survival strategies during infection. They facilitate metabolic adaptation in response to fluctuating nutrient availability and promote resistance to host immune pressures through mechanisms including lipid biosynthesis and oxidative stress defense. A comprehensive understanding of these metabolic adaptations provides valuable insights into potential therapeutic targets aimed at undermining Mtb's metabolic flexibility and persistence.

6.2 Central Carbon Metabolism (CCM) in *Mycobacterium tuberculosis*

Mycobacterium tuberculosis (Mtb) relies heavily on carbon metabolism proteins (CMPs) to thrive in diverse environments by utilizing a variety of carbon sources, including carbohydrates, lipids, and amino acids. These proteins are central to Mtb's metabolic adaptability and can be categorized according to their roles in key metabolic pathways and transport systems that enable the bacterium to harness these carbon sources for survival and virulence efficiently. CMPs are structurally specialized to perform distinct functions within the central carbon metabolism (CCM) network, participating in glycolysis, the tricarboxylic acid (TCA) cycle, the glyoxylate shunt, the methylcitrate cycle (MCC), and various transport mechanisms, each contributing to Mtb's metabolic flexibility (Rhee et al. 2011; Guasch Ferré et al. 2020; Borah et al. 2021). These interconnected pathways allow Mtb to metabolize a broad spectrum of carbon substrates, including carbohydrates such as glucose, lipids like fatty acids and cholesterol, and amino acids. The glycolysis/gluconeogenesis pathway plays a particularly pivotal role in Mtb metabolism by supporting both energy production and biosynthetic needs. Glycolysis breaks down glucose into pyruvate, generating adenosine triphosphate (ATP) and nicotinamide adenine dinucleotide (NADH), which are essential for the bacterium's energy requirements during its life cycle (Beste et al. 2011, 2012; Beste and McFadden 2012; Lofthouse 2013; Marrero et al. 2013; Basu et al. 2018; Borah et al. 2019; Fieweger et al. 2019). This pathway not only supports ATP generation but also facilitates the biosynthesis of amino acids and lipids by converting glucose into metabolic intermediates, ensuring Mtb's survival in nutrient-rich environments (Long 1954; Bloch and Segal 1956; Sassetti and Rubin 2003; Wheeler et al. 2005; Marrero et al. 2013; Phong et al. 2013; Baughn and Rhee 2014; Chaudhry and Varacallo 2025). Glycolysis in Mtb involves ten essential proteins: glucose kinase A (glkA), phosphoglucose isomerase (pgi), phosphofructokinase A (pfkA), fructose-bisphosphate aldolase (fba), triosephosphate isomerase (tpi), glyceraldehyde-3-phosphate dehydrogenase (gapdh), phosphoglycerate kinase (pgk), phosphoglycerate mutase (pgm), enolase (eno), and pyruvate kinase A (pykA). These enzymes catalyze sequential reactions that convert glucose into pyruvate, facilitating ATP production and providing intermediates for biosynthetic pathways. Mtb expresses two critical glucokinases, polyphosphate glucokinase (PPGK) and glucokinase A (GLKA), which initiate glucose metabolism by phosphorylating glucose to glucose-6-phosphate, integrating it into the central carbon metabolic network (Marrero et al. 2013). PPGK is indispensable for normal growth when glucose serves as the primary carbon source, whereas GLKA is nonessential under standard growth conditions. However, the combined deletion of both enzymes significantly impairs Mtb's ability to persist in mouse lungs during chronic infection, underscoring the vital importance of glucose phosphorylation for long-term bacterial survival (Marrero et al. 2013). PPGK is unique in its ability to utilize either inorganic polyphosphate or ATP as a phosphoryl donor, a capability enabled by specialized phosphate-binding motifs. In contrast, GLKA is a monomeric enzyme that relies solely on ATP and contains conserved binding sites for both glucose and ATP. These enzymes catalyze the first committed step in glycolysis, trapping glucose inside the cell by converting it into glucose-6-phosphate. This intermediate is then rearranged into fructose-6-phosphate by glucose-6-phosphate isomerase, preparing it for further downstream processing (Phong et al. 2013; Jiang et al. 2022). In related species such as *Mycobacterium smegmatis*, a homologous glucose kinase A (*glkA*) gene has been identified and shown to efficiently phosphorylate glucose using ATP, suggesting that this metabolic mechanism is evolutionarily conserved across mycobacteria (Pimentel-Schmitt et al. 2007). Beyond glucose metabolism, Mtb synthesizes a capsular α-glucan through enzymes encoded by glycogen synthesis genes (*glg*). This glucan contributes to bacterial persistence in animal models, potentially by facilitating immune evasion or serving as a carbon reservoir during nutrient limitation (Sambou et al. 2008).

Mtb also contains four glutamine synthetase A (*glnA1-4*) genes, although only glutamine synthetase A1 (*GlnA1*) is abundantly expressed and essential for maintaining bacterial homeostasis (Harth et al. 2005). GlnA1 plays a pivotal role in nitrogen metabolism by converting ammonium into glutamine. This crucial nitrogen donor supports biosynthetic reactions and redox homeostasis, both vital for adaptation to the hostile host environment (Harth et al. 2005). These interconnected metabolic strategies illustrate the complex and versatile means by which Mtb utilizes available nutrients to sustain infection. Following glucose phosphorylation, phosphofructokinase (*PFK*) catalyzes the conversion of fructose-6-phosphate into fructose-1,6-bisphosphate, a critical regulatory step that commits the substrate to the glycolytic pathway. ATP levels and the overall cellular energy status tightly regulate this reaction.

Mtb encodes two isoforms of PFK: phosphofructokinase A (*PfkA*; Rv3010c) and phosphofructokinase B (*PfkB*; Rv2029c), which differ in their functional properties and expression patterns (Snášel et al. 2021). PfkA serves as the primary isoenzyme essential for growth on glucose, whereas PfkB compensates under conditions where PfkA activity is inhibited (Phong et al. 2013; Snášel et al. 2021). Inhibition of glycolysis at the PFK step leads to the accumulation of toxic sugar-phosphates, resulting in loss of cellular viability (Phong et al. 2013). Under hypoxic conditions, the presence of glucose exacerbates toxicity in nonreplicating mycobacteria, thereby limiting their survival over time (Phong et al. 2013). Interestingly, Mtb can manipulate host glycolysis by upregulating microRNA-21, which targets the host phosphofructokinase muscle isoform (PFK-M). This downregulation reduces the production of the pro-inflammatory cytokine IL-1β, thereby promoting bacterial survival (Hackett et al. 2020). These findings highlight the finely tuned metabolic crosstalk between Mtb and its host during infection. Fructose-1,6-bisphosphate aldolase (MtFBA) is another essential glycolytic enzyme, critical for both actively replicating and dormant Mtb, making it a promising drug target (de la Paz Santangelo et al. 2011). MtFBA catalyzes the reversible cleavage of fructose-1,6-bisphosphate into glyceraldehyde 3-phosphate (G3P) and dihydroxyacetone phosphate (DHAP) through a zinc-dependent mechanism (Capodagli et al. 2014). Central to its catalytic function is the active site loop, where the glutamate residue E169 facilitates a water-mediated deprotonation and protonation process (Capodagli et al. 2014). Structural studies classify MtFBA as a tetrameric class IIa aldolase with unique reaction and substrate-binding characteristics (Capodagli et al. 2014). During catalysis, MtFBA forms a Schiff base intermediate with DHAP, stabilized by a conserved lysine-glutamate dyad, which ensures catalytic efficiency (Choi et al. 2006).

Triosephosphate isomerase (*tpi*), encoded at position 1,615,564 bp in the Mtb genome, plays a dual role in glycolysis and gluconeogenesis by interconverting DHAP and G3P. This step is crucial for maintaining metabolic efficiency by ensuring all three-carbon intermediates flow smoothly through glycolysis (Trujillo et al. 2014). Following this, glyceraldehyde-3-phosphate dehydrogenase (GapA) catalyzes the oxidation of G3P to 1,3-bisphosphoglycerate, concurrently reducing NAD^+ to NADH. GapA contains conserved NAD^+-binding domains essential for its redox activity, supplying NADH necessary for oxidative phosphorylation (Wolfson-Stofko et al. 2013). Biochemical characterization of Mtb triosephosphate isomerase reveals it functions as a homodimer with optimal enzymatic activity at pH 6–8 and 37 °C (Mathur et al. 2006). Although previously considered essential, a *tpi* knockout strain was found viable when cultured in media containing both glycolytic and gluconeogenic carbon sources, but not when grown on single carbon sources alone (Trujillo et al. 2014). The crystal structure of *tpi* provides valuable insights into its kinetic behavior and active site configuration (Connor et al. 2011). Importantly, *tpi* is required for Mtb viability both in vitro and in vivo, reinforcing its potential as a drug target (Hernández-Viveros et al. 2021).

Phosphoglycerate kinase (PGK) is a conserved enzyme that plays a central role in both glycolysis and gluconeogenesis by catalyzing the substrate-level phosphorylation of ADP using 1,3-bisphosphoglycerate to form ATP and 3-phosphoglycerate (Rojas-Pirela et al. 2020). In Mtb, glucose phosphorylation is not only critical for metabolism but also essential for persistence during chronic infection in mice (Marrero et al. 2013). Although Mtb predominantly metabolizes fatty acids to adapt to host environments, it retains a fully functional glycolytic pathway to maintain metabolic flexibility (Noy et al. 2016). Following PGK activity, phosphoglycerate mutase (*GpmI*) catalyzes the conversion of 3-phosphoglycerate into 2-phosphoglycerate, which is then dehydrated by enolase (Eno) to form phosphoenolpyruvate (PEP). Pyruvate kinase A (*PykA*) completes glycolysis by transferring a phosphate group from PEP to ADP, producing ATP and pyruvate. This final step provides immediate cellular energy and reducing equivalents via NADH. Pyruvate can either enter the tricarboxylic acid (TCA) cycle for further oxidation or be converted to lactate under anaerobic conditions. In Mtb, PykA is vital for growth on fermentable substrates and fatty acids and is allosterically activated by glucose-6-phosphate and AMP (Noy et al. 2016). Its activity is further modulated by intermediates from intersecting metabolic pathways, reflecting the tightly regulated nature of Mtb's central metabolism (Snášel and Pichová 2019). Understanding this metabolic control is essential, as it significantly contributes to Mtb's ability to persist and adapt within the host (Snášel and Pichová 2019).

The TCA cycle in Mtb serves as a central metabolic hub, playing crucial roles in energy production, biosynthesis of essential biomolecules, and adaptation to diverse and often hostile host environments (Tian et al. 2005; Ferraris et al. 2015). This cycle is driven by eight key enzymes: citrate synthase (*gltA*), aconitase, isocitrate dehydrogenase (icd1), α-ketoglutarate decarboxylase *(kgd)*, succinic semialdehyde dehydrogenases (gabD1 and gabD2), succinate dehydrogenase (*sdhA* and *sdhB*), fumarase (*fumC*), and malate dehydrogenase (mdh) (Tian et al. 2005). These enzymes enable Mtb to efficiently metabolize carbon substrates, generating ATP and key intermediates required for anabolic processes (Eoh et al. 2022). What distinguishes the Mtb TCA cycle from the classical Krebs or citric acid cycle in most aerobic organisms is its unique configuration, especially under nutrient-limited or stress conditions. The cycle begins with citrate synthase (*gltA*, Rv0896; location 999,472 bp), a homodimeric enzyme with conserved substrate-binding sites that catalyzes the condensation of acetyl-CoA and oxaloacetate to form citrate, marking the entry point of carbon into the cycle. Citrate is then converted into isocitrate through the intermediate cis-aconitate by aconitase, an iron-sulfur cluster-dependent enzyme critical for maintaining oxidative flux through the TCA cycle. Subsequently, isocitrate dehydrogenase *(icd1*, Rv0066c; location 72,274 bp) catalyzes the oxidative decarboxylation of isocitrate to α-ketoglutarate, concurrently generating nicotinamide adenine dinucleotide phosphate (NADPH). This step integrates the TCA cycle with biosynthetic and redox-balancing processes vital for Mtb's intracellular survival. Unlike many organisms, Mtb lacks the canonical α-ketoglutarate dehydrogenase (KDH) enzyme. Instead, it employs α-ketoglutarate decarboxylase (*kgd*, Rv1248c; location 1,389,357 bp) to convert α-ketoglutarate into succinic semialdehyde (SSA) (Tian et al. 2005; Beste et al. 2013; Maksymiuk et al. 2015; Ferraris et al. 2015; Tovilla-Coutiño et al. 2020). This adaptation bypasses the conventional KDH-dependent conversion of α-ketoglutarate to succinyl-CoA, providing Mtb with greater flexibility to regulate carbon flux under varying oxygen tensions and nutrient availabilities. The SSA produced is further oxidized to succinate by two isoforms of succinic semialdehyde dehydrogenase, *GabD1* (Rv0234c) and *GabD2* (Rv1731), completing an alternative pathway from α-ketoglutarate to succinate. Succinate is then oxidized to fumarate by succinate dehydrogenase, composed of subunits *SdhA* (Rv3318) and *SdhB* (Rv3319). This enzyme is particularly important as it links the TCA cycle to the electron transport chain (ETC) by transferring electrons to quinones, making it a key player in Mtb's aerobic energy metabolism, especially during active replication. Fumarase converts fumarate into malate, which is subsequently oxidized by malate dehydrogenase to regenerate oxaloacetate, closing the cycle. These transformations are essential not only for energy generation but also for maintaining pools of TCA intermediates that feed into biosynthetic pathways for amino acids, nucleotides, and lipids.

A distinctive feature of the Mtb TCA cycle is its branched or bifurcated architecture, consisting of oxidative and reductive arms. This design allows the bacterium to reroute metabolic flux in response to environmental conditions such as hypoxia or nutrient depletion, thereby optimizing the balance between energy production and carbon conservation. During conditions favoring lipid catabolism or limited carbohydrate availability, Mtb relies more heavily on the reductive arm of the TCA cycle to minimize carbon loss while maintaining redox balance. This metabolic flexibility, exemplified by the substitution of KDH with Kgd and the presence of bifurcated TCA pathways, enables Mtb to dynamically remodel its metabolism in response to the host's immune pressures and fluctuating nutrient supplies (Baughn et al. 2009; Howard and Khader 2020; Chang et al. 2021). By sustaining its central carbon metabolism under both aerobic and microaerophilic conditions, Mtb ensures its long-term persistence, pathogenesis, and ability to withstand antimicrobial stress.

Fumarase (*fumC*, Rv1098c; location 1,226,141 bp) plays a crucial role in the metabolism of Mtb and has emerged as a promising target for new tuberculosis therapies. This enzyme facilitates the hydration of fumarate to malate, a vital reaction in maintaining the flow of metabolites through the reductive arm of the tricarboxylic acid (TCA) cycle (Ruecker et al. 2017; Ahn and Boshoff 2017). Following this step, malate dehydrogenase (*mdh*, Rv1240; location 1,383,213 bp) catalyzes the oxidation of malate to oxaloacetate. This reaction not only replenishes TCA cycle intermediates but also generates nicotinamide adenine dinucleotide (NADH), which is essential for oxidative phosphorylation and cellular energy production. In parallel, the glyoxylate shunt, involving isocitrate lyase (*icl*) and malate synthase (*glcB*), functions alongside the TCA cycle, particularly under nutrient-limited conditions. This pathway helps conserve carbon during lipid metabolism, making it especially important when carbohydrates are scarce (Munoz-Elias and McKinney 2005; Marrero et al. 2010; Borah et al. 2021). Such metabolic adaptability allows Mtb to survive and replicate across a variety of nutrient environments within the host. Key metabolic enzymes such as succinate dehydrogenase (SDH) and those involved in the glyoxylate shunt represent attractive drug targets. SDH is a dual-function enzyme that bridges the TCA cycle with the respiratory chain, playing a central role in energy generation (Hards et al. 2020; Arnold and Finley 2023). Inhibiting SDH disrupts both metabolic flow and respiration in Mtb and has been shown to enhance the activity of other drugs that target bacterial energy production (Adolph et al. 2024). Loss of SDH function results in increased oxygen consumption and a failure to exit the stationary phase, highlighting its regulatory role in bacterial respiration (Hartman et al. 2014). Similarly, enzymes of the glyoxylate bypass are essential for Mtb's ability to persist during latent infection, emphasizing their therapeutic potential (Negi and Sharma 2024).

In addition to carbon metabolism, Mtb has evolved a highly specialized and flexible system for lipid and cholesterol utilization. The TB bacilli rely heavily on host-derived fatty acids and cholesterol as primary carbon sources, particularly during chronic stages of infection. Genetic studies have identified a network of approximately 83 proteins directly involved in lipid metabolism (Sturm et al. 2024; Parkin et al. 2025). These proteins span various functions, including lipid uptake, degradation, biosynthesis, and storage, and their diversity underscores the complexity of Mtb's lipid metabolic machinery. During active infection, Mtb uses mammalian cell entry (*Mce*) complexes to transport lipids across its thick, waxy cell envelope. Despite their name, Mce proteins are primarily involved in lipid transport rather than host cell entry. These complexes enable the bacterium to import essential host lipids, such as cholesterol and fatty acids, which fuel its intracellular survival (Zhang and Xie 2011). The Mtb genome encodes four major mammalian cell entry (*mce*) operons: mce1, mce2, mce3, and mce4, each comprising six substrate-binding proteins (SBPs) designated MceA through MceF (Casali and Riley 2007;

Pandey and Sassetti 2008; Perkowski et al. 2016; Fenn et al. 2020; Zaychikova and Danilenko 2020; Bashiri 2021; Klepp and y Garcia 2022). These operons are differentially regulated depending on the bacterial growth phase and the stage of infection. For example, mce1 is typically active during early culture growth, whereas mce4 predominates in the stationary phase. During in vivo infection, multiple operons, including mce1, mce3, and mce4, are expressed concurrently, reflecting their distinct roles throughout the infection cycle (Kumar et al. 2003). Mce proteins contribute not only to nutrient uptake but also to modulating host-pathogen interactions. For instance, Mce3C facilitates Mtb invasion by triggering β2 integrin-mediated signaling pathways, enhancing bacterial entry into host cells (Zhang et al. 2018). Mce2E suppresses macrophage immune responses by inhibiting mitogen-activated protein kinase (MAPK) signaling and promotes epithelial cell growth by preventing the polyubiquitination and degradation of eukaryotic elongation factor *1A1* (eEF1A1) (Qiang et al. 2019). Structurally, Mce SBPs assemble into hydrophobic channels that enable lipid transport across the complex mycobacterial cell envelope. These proteins form tunnel-like structures resembling an ATP-binding cassette (ABC)-type transporter system, similar to lipid transporters found in other bacteria. Their architecture includes specialized helical domains that provide stability and facilitate selective substrate transport (Chen et al. 2023a, b). Among the Mce complexes, Mce4 is particularly critical for cholesterol uptake. The process begins with cholesterol being converted to cholestenone by cholesterol oxidase (*ChoD*, Rv3409c; location 3,826,991 bp), which initiates the cholesterol degradation pathway. Subsequent enzymes, such as 3-ketosteroid-Δ1-dehydrogenase (*KstD*, Rv3537; location 3,975,369 bp), catalyze the removal of the cholestenone side chain to facilitate further breakdown. Fatty acids and cholesterol imported via Mce complexes undergo β-oxidation in the cytoplasm, producing acetyl-coenzyme A (acetyl-CoA) or propionyl-coenzyme A (propionyl-CoA). This process is initiated by enzymes like fatty acyl-CoA synthetase D5 (*FadD5*, Rv0166; location 194,993 bp), which activate fatty acids by conjugating them to coenzyme A. The resulting acyl-CoA molecules are subsequently processed by acyl-CoA dehydrogenases, such as members of the EchA family, and enoyl-CoA hydratases encoded by various fad genes. These enzymes contain conserved catalytic domains essential for substrate recognition and catalyzing redox reactions that ultimately drive ATP production. Acetyl-CoA generated from β-oxidation feeds into the tricarboxylic acid (TCA) cycle, where it contributes to the production of reducing equivalents, nicotinamide adenine dinucleotide (NADH) and flavin adenine dinucleotide ($FADH_2$), which fuel oxidative phosphorylation and cellular energy generation (Martínez-Reyes and Chandel 2020). Under carbon-limited conditions, acetyl-CoA may alternatively enter the glyoxylate shunt, enabling Mtb to conserve carbon for biosynthesis. The ability of Mtb to exploit host-derived lipids is central to its persistence during infection, especially under the nutrient-limiting pressures imposed by the host immune response (Lee et al. 2013). The Mce4 complex plays a critical role in this adaptation. Loss of mce4 function severely impairs bacterial survival within interferon-gamma (IFN-γ)-activated macrophages and during later stages of infection in mouse models, underscoring the essential role of lipid uptake systems, particularly Mce4, in Mtb's survival and virulence within hostile host environments (Pandey and Sassetti 2008; Wipperman et al. 2014; D'Arcy et al. 2011; Petrilli et al. 2020; Asthana et al. 2021; Chen et al. 2023a, b; Sturm et al. 2024). Once imported, cholesterol is catabolized to generate acetyl-CoA and propionyl-CoA, which serve as critical intermediates for energy production and biosynthetic processes. Cholesterol degradation initiates with the activity of cholesterol oxidase (*ChoD*), followed by enzymes such as 3-ketosteroid-Δ1-dehydrogenase (*KstD*), which remove the side chain of cholesterol-derived molecules, rendering them accessible for central metabolism. Propionyl-CoA, a potentially toxic byproduct of cholesterol and odd-chain fatty acid catabolism, is detoxified through the methylcitrate cycle (MCC). This cycle depends on three key enzymes:

methylcitrate synthase (PrpC, Rv1131; location 1,256,132 bp), methylcitrate dehydratase (*PrpD*, Rv1130; location 1,254,555 bp), and 2-methylisocitrate lyase (*PrpB*). Together, these enzymes convert propionyl-CoA into pyruvate and succinate, which are then funneled into the TCA cycle or redirected into biosynthetic pathways (Lovewell et al. 2016; Gago et al. 2018; Wilburn et al. 2018). This metabolic adaptability highlights Mtb's capacity to survive and thrive during the prolonged course of infection. In Mtb, lipids serve not only as substrates for degradation but also as vital precursors for synthesizing key virulence-associated molecules such as phthiocerol dimycocerosate (PDIM) and sulfolipid-1 (SL-1). The biosynthesis of PDIM is driven by polyketide synthases like Pks1 (Rv2946c; location 3,291,503 bp) and Mas, which depend on methylmalonyl-CoA, a metabolite derived from propionyl-CoA catabolism (Mulholland et al. 2024). This connection demonstrates how lipid degradation intermediates are seamlessly integrated into the assembly of complex, cell wall-associated molecules. Similarly, the production of SL-1 is mediated by Pks2 (Rv3825c; location 4,293,225 bp) and supported by transport proteins such as MmpL8 (Rv3823c; location 4,288,260 bp), which facilitate the export of these specialized lipids to the bacterial surface. Both PDIM and SL-1 play critical roles in Mtb pathogenesis by modulating host immune responses, aiding immune evasion, and promoting survival within macrophages (Shi et al. 2010; Chang et al. 2021; Samuels et al. 2022; Devlin et al. 2025). During its exponential growth phase, Mtb strategically balances lipid storage with the synthesis of structural lipids, prioritizing replication over the accumulation of energy reserves such as triacylglycerol (TAG). Although TAG storage becomes more prominent during dormancy, Mtb maintains a baseline level of TAG synthesis even during active growth. This process is primarily orchestrated by enzymes such as Tgs4 (Rv3088; location 3,454,340 bp), which catalyze the formation of ester bonds between glycerol-3-phosphate and fatty acyl-CoA derivatives. This low-level TAG production is complemented by lipase activity, particularly from α/β hydrolases, which recycle lipids stored in intracellular droplets. Such recycling is essential for preserving membrane integrity, regulating lipid turnover, and preventing the disruptive accumulation of lipid droplets that could compromise cellular homeostasis (Maurya et al. 2019; Patil and Jain 2021; Sarkar et al. 2025). A hallmark of Mtb lipid metabolism is its compartmentalized organization, which enables the bacterium to balance energy production and cell wall biosynthesis efficiently. Fatty acid β-oxidation, responsible for generating acetyl-CoA used in energy metabolism, occurs both in the cytoplasm and in membrane-associated regions. Concurrently, biosynthetic pathways channel β-ketoacyl-ACP intermediates through key enzymes such as KasA and KasB, driving the synthesis of mycolic acids—long-chain fatty acids that are essential components of the mycobacterial cell wall, contributing to its structural integrity and impermeability. This metabolic duality is tightly regulated by sophisticated control systems that direct more than 70% of the acetyl-CoA produced toward structural lipid biosynthesis rather than energy storage. Lipid-degrading enzymes play a crucial role in maintaining this balance by preventing excessive expansion of lipid droplets and ensuring the metabolic flexibility necessary for rapid bacterial growth (Takayama et al. 2005; Beites et al. 2021). Such fine-tuned regulation allows Mtb to swiftly adapt to environmental fluctuations while preserving the ability to enter dormancy when conditions become unfavorable. The differential expression of triacylglycerol synthase (*tgs*) genes and the precise routing of lipid substrates across distinct cellular compartments underpin this remarkable adaptability (Voskuil et al. 2004; Sirakova et al. 2006; Balázsi et al. 2011). Collectively, these mechanisms establish a meticulously regulated lipid metabolic profile during active growth, enabling Mtb to support replication efficiently while minimizing unnecessary energy storage. Approximately 83 proteins have been identified as direct contributors to Mtb lipid metabolism, encompassing diverse functions such as transport, degradation, biosynthesis, and structural assembly. Among these are transporters like LucA (mtbc0_003948;

location 4,192,657 bp), β-oxidation enzymes such as FadD5 (Rv0166; location 194,993 bp), cholesterol-degrading enzymes including *KstD* (Rv3537; location 3,975,369 bp), and key enzymes of the methylcitrate cycle (*MCC*) such as *PrpC* and *PrpB*. The network also includes biosynthetic enzymes like polyketide synthases *Pks1* and *Pks2* (Rv2946c-Rv3825c), storage-associated proteins such as *Tgs1* (Rv3130), and structural enzymes like KasA and KasB involved in mycolic acid synthesis. The structural diversity of these proteins, from tunnel-like substrate channels in Mce complexes to multi-domain polyketide synthases, reflects the high degree of specialization within Mtb's lipid metabolic machinery (Daniel et al. 2011; Mekonnen et al. 2021; Kim et al. 2023; Lin et al. 2024).

Lipid and cholesterol metabolism in Mtb form an intricate enzymatic network that enables the bacterium to exploit host-derived lipids during active growth effectively. This metabolic versatility is essential for rapid biomass expansion and plays a crucial preparatory role for eventual persistence within the host. Upon entering dormancy, especially under stress conditions, Mtb shifts its reliance toward host lipids as primary carbon sources, carefully modulating both lipid biosynthesis and degradation to ensure survival and prolonged infection. A central pathway within this network is the methylcitrate cycle (MCC), which detoxifies and utilizes propionyl-CoA, a potentially harmful byproduct generated during the degradation of cholesterol and odd-chain fatty acids (Muñoz-Elías et al. 2006; Savvi et al. 2008; Zhang et al. 2020; Chang et al. 2021; Qi et al. 2021; Samuels et al. 2022). The MCC is particularly vital during the exponential growth phase when lipid metabolism is at its peak. This pathway begins with methylcitrate synthase (encoded by *prpC*), which catalyzes the condensation of propionyl-CoA with oxaloacetate to form methylcitrate. Methylcitrate is then dehydrated to methylisocitrate by methylcitrate dehydratase (encoded by prpD), and finally cleaved into pyruvate and succinate by isocitrate lyases ICL1 and ICL2 (encoded by *icl1* and *icl2*). These isocitrate lyases also contribute to the glyoxylate shunt, underscoring their multifunctional role in Mtb's central metabolism (Upton and McKinney 2007; Dolan and Welch 2018; Huang et al. 2023). Unlike many bacteria, Mtb lacks a dedicated 2-methylisocitrate lyase, making ICL1 and ICL2 indispensable for both the MCC and the glyoxylate cycle (Muñoz-Elías and McKinney 2005; Muñoz-Elías et al. 2006; Griffin et al. 2012; Pham et al. 2017). Structurally, methylcitrate synthase functions as a homodimer with active sites tailored to bind CoA derivatives. Methylcitrate dehydratase belongs to the enolase-like family and contains conserved metal-binding sites critical for its enzymatic activity. The isocitrate lyases process both isocitrate and methylisocitrate, linking lipid catabolism directly to central metabolic pathways (Gerike et al. 1998; McKinney et al. 2000; Muñoz-Elías and McKinney 2005; Muñoz-Elías et al. 2006; Upton and McKinney 2007; Eoh and Rhee 2014). The MCC products, pyruvate and succinate, feed into glycolysis and the tricarboxylic acid (TCA) cycle, fueling energy production and biosynthesis during periods of rapid bacterial proliferation. This pathway is tightly regulated by transcriptional activators such as MtPrpR, which senses acyl-CoA molecules and iron availability through its [4Fe-4S] cluster, ensuring responsiveness to environmental cues and cellular needs (Tang et al. 2019). Through its dual role in detoxification and energy integration, the MCC exemplifies the metabolic flexibility that underpins Mtb's ability to thrive within the host environment, particularly during active growth (Chang et al. 2021). The lipid metabolic network in *Mycobacterium tuberculosis* (Mtb) is highly dynamic and tightly regulated, playing a crucial role in the bacterium's survival strategy. The intricate interplay between lipid degradation, biosynthesis, storage, and detoxification, coordinated by a diverse array of specialized enzymes and regulatory systems, enables Mtb to adapt rapidly to shifting environmental conditions, sustain rapid growth, and prepare for long-term persistence within the host. This central role of lipid metabolism forms a cornerstone of Mtb's pathogenesis and resilience.

Co-metabolism in Mtb represents a sophisticated process that integrates multiple host-derived carbon sources into central metabolic pathways, allowing the bacterium to survive and maintain virulence across diverse host environments (de Carvalho et al. 2010; Planck and Rhee 2021). Unlike classical central carbon pathways such as glycolysis, the tricarboxylic acid (TCA) cycle, the glyoxylate shunt, and the methylcitrate cycle (MCC), co-metabolism is distinguished by its unique capacity to simultaneously utilize multiple host-derived carbon substrates, including cholesterol, fatty acids, glycerol, and lactate, without segregating metabolism into distinct compartments (Serafini et al. 2019). This metabolic flexibility is critical for Mtb's adaptation to varying nutrient availabilities encountered during infection, thereby ensuring its survival and virulence within the host (de Carvalho et al. 2010; Chang and Guan 2021; Planck and Rhee 2021). During the exponential growth phase, co-metabolism is supported by transport proteins that facilitate nutrient uptake and efflux while preserving metabolic flexibility. Among these, the Mce4 complex, a member of the mammalian cell entry (Mce) protein family, is a key molecular system involved in lipid metabolism and host interaction. Encoded by the mce4 operon, the mce4A gene produces a 43 kDa protein that localizes to the bacterial cell wall during the stationary phase of growth. This strategic localization is critical for facilitating Mtb's entry into host cells and enhancing its ability to persist intracellularly (Saini et al. 2008; Khan et al. 2016). Functionally, the Mce4 complex operates as an ATP-dependent transporter specialized in the uptake of steroids, particularly cholesterol. This function is essential for Mtb survival when cholesterol or structurally related steroids serve as the primary carbon source (Mohn et al. 2008). The interaction between cholesterol and the Mce4A protein follows a defined four-step binding mechanism primarily driven by entropy, indicating that the binding process is energetically favorable and involves conformational changes that increase system disorder. This insight highlights the transporter's specificity and efficiency in capturing host-derived lipids (Khan et al. 2016). Expression of the mce4 operon is tightly regulated by a promoter region responsive to environmental cues such as surface stress, hypoxia, and the presence of cholesterol—conditions that mimic the intracellular environment within host macrophages. This regulation ensures operon upregulation precisely when its function is most required (Rathor et al. 2013). Notably, mce4 expression predominates during the stationary phase and throughout infection, reinforcing its critical role in supporting Mtb's long-term survival within the host (Saini et al. 2008; Rathor et al. 2013). Taken together, these features underscore the Mce4 complex as a vital component of Mtb's adaptive strategy. By enabling efficient cholesterol uptake, enhancing intracellular survival, and promoting persistence within the host, the Mce4 system plays a central role in the pathogen's ability to maintain infection under challenging environmental conditions (Pandey and Sassetti 2008; Miner et al. 2009; Klepp and Garcia 2022).

Bacterial ATP-binding cassette (ABC) systems represent one of the largest superfamilies of paralogous proteins, characterized by a highly conserved ATP-hydrolyzing domain known as the nucleotide-binding domain (NBD) (Davidson et al. 2008). Structurally, ABC transporters consist of two hydrophobic transmembrane domains (TMDs) that form the substrate translocation pathway, coupled with two hydrophilic NBDs typically located on the cytosolic side of the membrane. In bacterial importers, these domains are generally encoded as separate polypeptides that assemble into a functional complex, whereas most exporters feature a fusion of the N-terminal TMD and C-terminal NBD within a single polypeptide chain, although variations in domain organization exist (Davidson et al. 2008). In Mtb, ABC transporters are integral membrane proteins essential for a wide range of physiological functions, including nutrient uptake, drug efflux, and virulence (Cassio Barreto de Oliveira and Balan 2020; Akhtar and Turner 2022). Typically, these systems comprise three key components: NBDs, membrane-spanning domains (MSDs), and, in importers, substrate-binding proteins (SBPs) that confer substrate specificity (Braibant et al. 2000). The Mtb genome encodes at least 26 complete and 11 incomplete ABC transporter systems, with a relatively balanced distribution between importers and exporters (Braibant et al. 2000). Collectively, these transporters facilitate the uptake of peptides, carbohydrates, and iron, contribute to multidrug resistance, and enhance Mtb's survival in hostile host environments by modulating immune responses (Cassio Barreto de Oliveira and Balan 2020; Soni et al. 2020). Among these systems, the LpqY-SugABC transporter, encoded by the lpqY-sugABC operon (Rv1235-Rv1238), plays a pivotal role in recycling trehalose, a disaccharide critical for mycolic acid synthesis and bacterial virulence (Kalscheuer et al. 2010; Cassio Barreto de Oliveira and Balan 2020; Furze et al. 2021). Structural studies have revealed that LpqY binds trehalose through a unique lipoprotein rotation mechanism, enhancing substrate specificity (Liu and Ma 2020). This system exhibits high selectivity for trehalose, which is absent in mammalian hosts, indicating that its primary function is recycling trehalose released during cell wall turnover rather than scavenging nutrients from the host (Kalscheuer et al. 2010). This recycling process is vital for maintaining cell wall integrity and has been linked to the antimicrobial and anti-biofilm properties of trehalose analogs in mycobacteria (Wolber et al. 2017). Mtb encodes at least four ABC transporters specific for carbohydrates: LpqY/SugABC, UspABC, Rv2038c-41c, and UgpAEBC (de la Torre et al. 2021). Phylogenetic and structural bioinformatics analyses have shown that two of these transporter systems, Rv2038c-41c and UgpAEBC, are exclusive to pathogenic mycobacteria, suggesting specialized roles in virulence and adaptation to the intracellular environment. The components of these four ABC importers are phylogenetically distinct and exhibit structural differences in their regulatory regions, transmembrane domains, and ligand-binding pockets, which define substrate specificity and evolutionary divergence (de la Torre et al. 2021). Despite this diversity, the interface between transmembrane and nucleotide-binding domains remains highly conserved, indicating a shared mechanism of energy coupling across these transporters. The ABC transporter Rv1273c has been identified as a crucial virulence factor in Mtb, promoting intracellular survival within macrophages by modulating the host immune response (Gupta et al. 2020). Silencing of Rv1273c significantly reduces bacterial viability inside macrophages, underscoring its role in resisting host defenses. This transporter suppresses pro-inflammatory cytokines such as tumor necrosis factor-alpha (TNF-α) and interleukin-12 (IL-12), while enhancing the production of the anti-inflammatory cytokine interleukin-10 (IL-10), thereby facilitating immune evasion. Furthermore, Rv1273c inhibits macrophage apoptosis by downregulating caspase-3 activation, which prolongs the survival of infected cells and favors bacterial replication. Mechanistically, Rv1273c interacts with host pattern recognition receptors, including Toll-like receptor 2 (TLR2) and Toll-like receptor 4 (TLR4), leading to the downregulation of nuclear factor kappa B (NF-κB) and mitogen-activated protein kinase (MAPK) signaling pathways and reduced expression of antimicrobial peptides. In vivo studies confirm that loss of Rv1273c attenuates Mtb virulence and decreases bacterial loads, highlighting its potential as a therapeutic target (Gupta et al. 2020). Beyond Rv1273c, ABC importers in Mtb represent promising targets for drug delivery and vaccine development. Their ATP-dependent substrate uptake mechanisms can be exploited to design antimicrobials that mimic natural substrates, potentially overcoming the impermeable mycobacterial cell wall (Soni et al. 2020). The surface exposure and immunogenicity of many ABC transporter components further enhance their attractiveness as vaccine candidates. Importantly, since ABC importers are absent in mammalian hosts, targeting these systems offers high specificity with minimal off-target effects (Soni et al. 2020).

In addition to ABC transporters, the major facilitator superfamily (MFS) transporters contribute significantly to membrane transport in Mtb (Pao et al. 1998; Reddy et al. 2012; Yan 2013; Li et al. 2017). MFS transporters typically consist of 12 transmembrane α-helices arranged into two symmetrical bundles of six helices each, although some members possess 14 helices (Reddy et al. 2012; Pao et al. 1998). These proteins are believed to have evolved from a 2-transmembrane segment hairpin structure through triplication and duplication events (Reddy et al. 2012). Despite low sequence conservation, MFS members share structural similarities, including a deep central hydrophilic cavity that facilitates substrate translocation (Madej 2014). This superfamily encompasses transporters for a wide range of substrates, often utilizing symport mechanisms coupled to proton gradients (Madej and Kaback 2013). Evolutionary analyses suggest that MFS proteins arose through intragenic multiplication, with functional segments assembling in different orders, contributing to their sequence diversity (Madej et al. 2013). Unlike ABC transporters, MFS proteins rely on electrochemical gradients rather than ATP hydrolysis to mediate uniport, symport, or antiport functions, enabling efficient substrate transport across the membrane.

Resistance-nodulation-division (RND) family transporters, including MmpL4 and MmpL5, are intricately connected to carbon metabolism in Mtb through their roles in exporting complex lipids and siderophores. These exported molecules are essential for iron acquisition and maintaining the integrity of the mycobacterial cell envelope during the co-metabolism of host-derived carbon sources (Melly and Purdy 2019; Maharjan et al. 2024). Lipids serve dual functions as both carbon sources and structural components within central metabolic pathways, while siderophores facilitate iron uptake, which is critical for enzymatic reactions involved in carbon metabolism (Sritharan et al. 2016). In addition to their role in substrate export, RND transporters contribute to metabolic homeostasis by removing toxic metabolic intermediates and byproducts, thereby enabling efficient processing of carbon substrates. Their involvement in drug efflux further supports Mtb's persistence within the host by protecting vital metabolic processes from antibiotic-induced stress. Structurally, these transporters form proton gradient-driven complexes that span the bacterial membranes, facilitating the effective extrusion of metabolites and toxins. This function is essential for detoxification, virulence, and the sustained activity of carbon metabolism during infection (Melly and Purdy 2019; Maharjan et al. 2024; Chalut 2016). Central Carbon Metabolism (CCM), encompassing key pathways such as glycolysis, gluconeogenesis, the pentose phosphate pathway, the tricarboxylic acid (TCA) cycle, the glyoxylate shunt, and the methylcitrate cycle, plays a vital role in the survival, adaptability, and virulence of Mtb. The bacterium's ability to shift between and integrate these metabolic routes enables it to persist within the hostile environment of host macrophages, evade immune defenses, and develop resistance to antibiotics. A deeper understanding of the complex interplay among these pathways is critical not only for elucidating the mechanisms underlying Mtb pathogenesis but also for guiding the development of innovative therapeutic strategies aimed at improving tuberculosis treatment outcomes.

6.3 Nitrogen Metabolism in *Mycobacterium tuberculosis*

Nitrogen metabolism in Mtb encompasses the biochemical pathways that enable the bacterium to acquire, assimilate, and regulate nitrogen, an essential element required for synthesizing amino acids, nucleotides, and cell wall components. This metabolic adaptability is critical for Mtb's survival across diverse host environments and during both active replication and dormant phases of infection. Central to nitrogen assimilation is the conversion of inorganic ammonium (NH_4^+) into organic nitrogen compounds, primarily glutamine and glutamate, which serve as key nitrogen donors in numerous biosynthetic pathways. Ammonium uptake is facilitated by the ammonium transporter AMtb, a highly hydrophobic transmembrane protein belonging to the ammonium transporter family (TC 2.A.49). AMtb is characterized by 10 to 12 transmembrane domains that enable efficient import of NH_4^+ into the cytoplasm (Coutts et al. 2002; Gouzy et al. 2014; Bizior et al. 2023; Ensinck et al. 2024). The activity of AMtb is tightly regulated by the PII signal transduction protein GlnK, encoded by the *glnK* gene. GlnK functions as a homotrimer with a flexible T-loop structure that senses the cell's nitrogen status by binding allosteric effectors such as ADP, ATP, and 2-oxoglutarate (Atkinson and Ninfa 1998; Arcondéguy et al. 2001; Shetty et al. 2010; Huergo et al. 2013; Watzer et al. 2019). Under nitrogen-limiting conditions, GlnK binds ATP and 2-oxoglutarate and remains localized in the cytosol, permitting ammonium uptake via AMtb. Conversely, when nitrogen is abundant, GlnK binds ADP and translocates to the membrane, where it physically interacts with and inhibits AMtb, effectively halting further NH_4^+ influx. The regulatory function of GlnK is controlled by GlnD, a bifunctional uridylyltransferase encoded by glnD, which senses intracellular nitrogen levels and modifies GlnK through uridylylation or deuridylylation. During nitrogen starvation, GlnD catalyzes the addition of uridylyl groups to GlnK, promoting ammonium uptake. In contrast, under nitrogen-replete conditions, GlnD removes these groups, leading to AMtb inactivation and preventing excess ammonium accumulation (Müller et al. 2006; Leigh and Dodsworth 2007; Read et al. 2007; Jain et al. 2013). Once inside the cell, ammonium is incorporated into glutamine via the high-affinity glutamine synthetase-glutamate synthase (GS-GOGAT) pathway. The first step is catalyzed by glutamine synthetase (GlnA1), encoded by *glnA1* (Rv2220c), a hexameric enzyme of the ATP-grasp fold family with a monomeric mass of approximately 57.3 kDa (Hirschfield et al. 1990; Tullius et al. 2003; Harth et al. 2005; Read et al. 2007; Gao et al. 2023). GlnA1 catalyzes the ATP-dependent condensation of ammonium with glutamate to form glutamine, which serves as a crucial nitrogen donor for the biosynthesis of macromolecules. Glutamine is subsequently processed by glutamate synthase (GOGAT), a heterodimeric complex composed of GltB and GltD, encoded by *gltB* and *gltD* (Rv2221c and Rv2222c) (Krajewski et al. 2005; Read et al. 2007). This enzyme pair, members of the pyridoxal phosphate-dependent aminotransferase family, catalyzes the transfer of an amide group from glutamine to 2-oxoglutarate, producing two molecules of glutamate (Castillo et al. 2000; García-Gutiérrez et al. 2018; Rai et al. 2025). This reaction ensures efficient channeling of inorganic nitrogen into key biosynthetic intermediates, supporting Mtb's metabolic needs during infection (Agapova et al. 2019; Selim et al. 2021; Xu et al. 2023). Nitrogen metabolism in Mtb involves multiple interconnected pathways that enable the bacterium to acquire, assimilate, and regulate nitrogen, an essential element for synthesizing amino acids, nucleotides, and cell wall components. Beyond the primary glutamine synthetase-glutamate synthase (GS-GOGAT) pathway, glutamate dehydrogenase (GDH) plays a significant role by mediating the reversible oxidative deamination of glutamate into α-ketoglutarate and ammonium. Although GDH is not the main route for nitrogen assimilation, it is critical for nitrogen recycling and glutamate catabolism, particularly under fluctuating nutrient conditions (Plaitakis et al. 2017). Glutamine serves as a central molecule in nitrogen metabolism within Mtb, acting as the primary nitrogen donor inside infected macrophages and facilitating the biosynthesis of a wide array of amino acids and nitrogen-containing compounds (Borah et al. 2019). Isotope tracer studies have demonstrated extensive redistribution of nitrogen from glutamine to synthesize multiple amino acids, highlighting its pivotal role in nitrogen donation (Borah et al. 2019). The regulation of glutamine and glutamate metabolism is tightly controlled by the serine/threonine protein kinase PknG, which phosphorylates the glycogen accumulation regulator A (GarA). Phosphorylated GarA inhibits glutamate synthase (GOGAT) and activates glutamate dehydrogenase (GDH), thereby shifting metabolism toward glutaminolysis and modulating glutamate levels according to cellular demand. This regulatory mechanism is essential for maintaining nitrogen homeostasis and adapting to the intracellular environment during infection (Xu et al. 2023). Glutamine synthetase (*GlnA1*) contributes to Mtb virulence not only by synthesizing poly-L-glutamate/glutamine, a unique component of the mycobacterial cell wall, but also by being secreted extracellularly, where it modulates phagosomal pH and inhibits phagosome-lysosome fusion, promoting intracellular persistence (Harth et al. 1994).

A secondary ammonium assimilation pathway involves alanine dehydrogenase (*ald*), encoded by ald (Rv2780). Ald is an NAD^+-dependent oxidoreductase that catalyzes the reversible deamination of L-alanine to pyruvate and ammonium (Andersen et al. 1992; Feng et al. 2002; Giffin et al. 2012; Jeong et al. 2015). Structurally, Ald forms a hexamer with distinct NAD-binding and substrate-binding domains. Upon coenzyme binding, the substrate-binding domain undergoes an approximately 16° rotation toward the NAD domain, facilitating catalysis. Key residues such as His96 and Asp270 are essential for enzymatic activity (Ågren et al. 2008). The expression of ald is upregulated under hypoxia and nutrient starvation, conditions that mimic dormancy, suggesting a role in Mtb persistence (Ågren et al. 2008; Saxena et al. 2014). Due to its structural uniqueness and functional relevance in latent infection, Ald has emerged as a potential drug target. Structure-based virtual screening has identified inhibitors with IC_{50} values ranging from 35 to 80 μM (Saxena et al. 2014). Recently, a deletion of an adenine nucleotide at position 266 in the ald gene (ald:c.266delA) was identified, causing a frameshift mutation that results in the loss of glutamine at position 89 and introduces a premature stop codon. This mutation produces a truncated alanine dehydrogenase enzyme likely to be non-functional (Velayati et al. 2024). Genetic and structural analyses reveal that this mutation impairs Ald function, adversely affecting nitrogen metabolism in Mtb and compromising the bacterium's ability to survive and replicate, particularly under dormancy conditions. This mutation has been consistently observed across both clinical isolates and laboratory strains, suggesting a regulatory switch in nitrogen metabolism between active and dormant states. Ammonium assimilation in Mtb is a precisely regulated, multi-step process that begins with ammonium uptake mediated by the transporter AMtb, whose activity is controlled by the regulatory proteins GlnK and GlnD. Following uptake, ammonium is incorporated into glutamine-by-glutamine synthetase (*GlnA1*) and subsequently converted to glutamate via the glutamate synthase complex (GOGAT). These pathways ensure a reliable supply of nitrogen for essential biosynthetic processes. Auxiliary enzymes such as glutamate dehydrogenase (*GDH*) and alanine dehydrogenase (*ald*) enhance Mtb's metabolic flexibility, particularly under nutrient-limited or dormancy-inducing conditions. These mechanisms collectively underscore the critical role of nitrogen metabolism in Mtb's pathophysiology and highlight it as a promising target for therapeutic intervention. Nitrate and nitrite assimilation constitute a vital secondary nitrogen metabolism pathway in Mtb, particularly under hypoxic conditions. This pathway involves the enzymatic reduction of nitrate (NO_3^-) to nitrite (NO_2^-), followed by further reduction to ammonium (NH_4^+), which is then incorporated into essential biosynthetic processes to support bacterial survival. The initial step is catalyzed by the nitrate reductase complex NarGHJI, encoded by the *narGHJI* operon (Rv1161-Rv1164) (Sohaskey and Wayne 2003; Malm et al. 2009). NarGHJI reduces nitrate to nitrite, marking the first committed step in nitrate assimilation. While the *narGHJI* genes are constitutively expressed, the *nark2x* operon, responsible for nitrate and nitrite transport, is induced under hypoxic conditions (Sohaskey and Wayne 2003). NarGHJI extends its function beyond nitrate metabolism, contributing to nitrate respiration under anaerobic conditions and playing a role in virulence, as demonstrated in *Mycobacterium bovis BCG* during infection in immunodeficient mice (Weber et al. 2000). Structurally, NarGHJI is a molybdoenzyme composed of four subunits: *NarG* (Rv1161; Location 1,287,328 bp), *NarH* (Rv1162; Location 1,291,065 bp), *NarJ* (Rv1163; Location 1,292,798 bp), and *NarI* (Rv1164; Location 1,293,406 bp). NarG serves as the catalytic subunit, harboring a molybdenum cofactor essential for nitrate reduction (Malm et al. 2009; Giffin et al. 2012; Akhtar et al. 2013; Almeida et al. 2017; Simcox and Rohde 2024). NarH is an iron-sulfur protein that facilitates electron transfer within the complex by shuttling electrons from quinol oxidation at NarI to NarG, utilizing iron-sulfur clusters as electron carriers (Cole et al.

1998; Banerjee et al. 2007; Zhang et al. 2020; Marcela Rodriguez et al. 2022). NarI, a membrane-embedded cytochrome b subunit, anchors the complex and transfers electrons from the quinol pool to NarH. NarJ functions as a chaperone, assisting in complex assembly and ensuring proper insertion of cofactors such as molybdenum and iron-sulfur clusters into NarG and NarH during maturation. Together, these subunits coordinate nitrate reduction, enabling Mtb to utilize nitrate as an alternative nitrogen source under hypoxic or nutrient-limited conditions. The critical importance of this pathway is underscored by mutants lacking narG, nirB, or glnR, which fail to grow on nitrate (Malm et al. 2009). Following nitrate reduction, nitrite is further reduced to ammonium by the nitrite reductase NirBD, encoded by the *nirBD* operon. NirBD is a cytoplasmic NAD(P)H-dependent nitrite reductase composed of two subunits: NirB (MT18B_0318; Location 301,205 bp) and NirD (MT18B_0319; Location 303,792 bp). The crystal structure of NirD reveals a double β-sandwich fold related to Rieske proteins but notably lacks iron-sulfur clusters (Izumi et al. 2012). Expression of nirBD is significantly upregulated during hypoxia and plays a critical role in Mtb dormancy. Mutants lacking NirBD display reduced viability in hypoxic and macrophage-based dormancy models, highlighting its importance for bacterial persistence (Akhtar et al. 2013). While NirBD reduces nitrite, another enzyme, NirA, functions as a sulfite reductase. NirA contains a [Fe_4-S_4] cluster and a siroheme cofactor, with a unique covalent cysteine-tyrosine bond in its active site critical for enzymatic activity (Schnell et al. 2005). This enzymatic system ensures efficient conversion of nitrite to ammonium when nitrate serves as the nitrogen source. GlnR, a master nitrogen regulator in mycobacteria, controls the transcriptional regulation of nitrate and nitrite assimilation. Mutational studies demonstrate that ΔglnR mutants fail to grow on nitrate media but can be rescued by plasmid-based expression of nirBD, confirming GlnR's role in activating nitrite reductase gene transcription (Malm et al. 2009; Akhtar et al. 2013). GlnR also regulates central nitrogen metabolism and other pathways, including activation of *nirB* and *nirD*, which are essential for glutamine/glutamate biosynthesis and biofilm formation in Mycobacterium abscessus (Fan et al. 2023).

The nitrate transporter *NarK2* (Rv1737c; Location 1,964,183 bp) belongs to the major facilitator superfamily (MFS) and functions as a Type I H^+/nitrate symporter, importing nitrate in response to decreasing oxygen levels (Wang et al. 2015; Giffin et al. 2012). Structurally, NarK2 comprises 12 transmembrane helices that enable proton-coupled symport of nitrate or nitrite across the membrane. Although primarily active under hypoxia, NarK2 remains present during exponential growth and can function if nitrate is available (Giffin et al. 2012; Reddy et al. 2012; Goddard et al. 2017; Willson et al. 2019). Under aerobic conditions, NarK2 activity is inhibited by an oxygen-dependent mechanism likely linked to the cellular redox state rather than direct oxygen sensing or protein degradation, allowing reversible regulation of nitrate uptake according to environmental oxygen levels. Functional studies demonstrate NarK2's essential role in nitrogen assimilation under low oxygen, contributing to metabolic flexibility during infection. Mutational analyses have identified key regions and amino acids critical for NarK2 function (Giffin et al. 2012). Differences in NarK2 expression between Mtb and *Mycobacterium bovis* contribute to their distinct nitrate reduction capabilities (Sohaskey and Modesti 2009). Understanding NarK2's structure and function may facilitate the development of specific inhibitors for tuberculosis control (Wang et al. 2015). Nitrate and nitrite assimilation pathways, governed by intricate enzymatic systems and transcriptional regulation, allow Mtb to exploit alternative nitrogen sources effectively. This metabolic adaptability is crucial for the bacterium's survival and long-term persistence in diverse and changing host environments, especially under conditions of low oxygen and nutrient scarcity, while also enabling nitrate utilization during periods of active growth when environmental conditions are favorable. Amino acid metabolism is an essential process that enables mycobacteria to utilize amino acids as sources of nitrogen and carbon, thereby supporting their growth and adaptation within the host environment. This metabolic pathway is initiated by specific enzymes belonging to distinct protein families, each encoded by well-characterized genetic loci. One of the key enzymes involved in this process is alanine dehydrogenase, encoded by the gene ald.

Alanine dehydrogenase (*ald*) in Mtb is a hexameric enzyme that catalyzes the reversible conversion of pyruvate and ammonia to L-alanine (Ågren et al. 2008). Each subunit contains an NAD-binding domain and a substrate-binding domain, with conformational changes occurring upon coenzyme binding (Tripathi and Ramachandran 2008). Ald plays a crucial role in Mtb metabolism, particularly during nutrient starvation and anaerobic conditions (Feng et al. 2002). It is upregulated during the persistent phase and is considered a potential drug target (Tripathi and Ramachandran 2008). Ald is essential for alanine utilization and sustained anaerobic growth, as demonstrated by studies on *Mycobacterium smegmatis* ald mutants (Feng et al. 2002). During hypoxic nonreplicating persistence, ald is involved in metabolic remodeling and maintaining the optimal NADH/NAD$^+$ ratio, which is crucial for efficient reactivation and growth resumption upon reoxygenation (Giffin et al. 2016). Genetic studies have demonstrated that Mtb strains lacking the ald gene fail to grow when alanine is the sole nitrogen source but exhibit normal growth when ammonium or glutamine is available. This finding confirms ald's role as a secondary but vital enzyme in nitrogen assimilation. The expression of ald is regulated according to nitrogen availability: it is upregulated under nitrogen-rich conditions, such as in the presence of ammonium, asparagine, or aspartate, and downregulated during nitrogen starvation (Feng et al. 2002; Giffin et al. 2012, 2016; Yelamanchi and Surolia 2021). This regulation suggests that ald's activity is prioritized when nitrogen is abundant, allowing the bacterium to exploit available nitrogen sources efficiently. Beyond its role in nitrogen metabolism, ald also contributes to maintaining cellular redox balance by regenerating NAD$^+$ during the oxidation of alanine (Giffin et al. 2016). This function links nitrogen assimilation with energy homeostasis, supporting the bacterium's metabolic demands. Importantly, ald's significance extends beyond nitrogen assimilation; it aids Mtb survival in acidic environments, such as the phagosomes of host macrophages, by producing ammonium that helps neutralize intracellular pH (Giffin et al. 2016; Jeong and Oh 2019). Ald also supports bacterial persistence during hypoxic stress by maintaining metabolic flexibility, allowing the pathogen to adapt to hostile host conditions. Overall, *ald* exemplifies the metabolic adaptability of Mtb, functioning as a bifunctional enzyme that integrates nitrogen assimilation, redox homeostasis, and environmental stress resistance. This versatility enables the bacterium to thrive within diverse host niches through redundant and flexible metabolic pathways, underscoring the importance of amino acid metabolism in both bacterial survival and the outcome of tuberculosis infection.

Mtb preferentially utilizes host-derived amino acids such as glutamate, aspartate, glutamine, and asparagine, which are actively transported into the bacterial cytoplasm through specific transport systems. This aligns with your research interests in Mtb, particularly its resistance mechanisms and nitrogen pathways. Although specific transporters such as GltT (glutamate transporter) and AsnA (asparagine transporter) have been hypothesized in earlier studies, direct evidence for their roles in Mtb remains limited. Instead, amino acid uptake is likely facilitated by solute carrier (SLC) family proteins or ATP-binding cassette (ABC) transporters, which are characterized by transmembrane domains specialized for substrate binding and transport (Xu et al. 2014; de Oliveira and Balan 2020). Glutamate occupies a central position in Mtb metabolism, serving both as a nitrogen donor and a carbon source. However, a dedicated glutamate transporter has not been definitively identified in Mtb; homologues of the *glnH* and *glnQ* genes, known to mediate glutamine transport in other bacteria, suggest potential mechanisms for glutamate uptake (Krajewski et al. 2005). Glutamate acts as the substrate for glutamine synthetase (GS), which catalyzes its ATP-dependent conversion with ammonia into glutamine, thereby linking nitrogen assimilation to critical biosynthetic and pathogenic functions in Mtb. GS is a validated drug target due to its essential role in bacterial metabolism and virulence (Parveen et al. 2023). Early studies by Lyon et al. (1967) demonstrated the presence of an inducible glutamate transport system in Mtb, with uptake rates significantly higher in induced cells, indicating a close relationship between glutamate transport and its oxidation. Harper et al. (2008) explored nitrogen metabolism regulation in Mtb, drawing parallels with related actinobacteria such as *Corynebacterium glutamicum* and *Streptomyces coelicolor*, which provides valuable insights into nitrogen signaling pathways in mycobacteria. Complementing this, Amon et al. (2009) performed genomic analyses identifying various nitrogen transporters and enzymes in mycobacteria, noting a reduction in nitrogen metabolism-related genes in pathogenic species compared to environmental relatives. Moreira-Teixeira et al. (2017), using combined quantum mechanics/molecular mechanics (QM/MM) simulations, provided detailed insights into the catalytic mechanism of human glutamine synthetase. Their work identified key intermediate states and elucidated the critical roles of active site residues in facilitating the incorporation of ammonia into glutamate.

Glutamine synthetase, encoded by *glnA1*, plays a pivotal role in Mtb nitrogen metabolism and virulence. Among the four GS isoforms (GlnA1-GlnA4) present in Mtb, GlnA1 is the most abundantly expressed and essential for maintaining bacterial homeostasis (Harth et al. 2005). It catalyzes the synthesis of L-glutamine from glutamate and ammonia, a reaction critical for Mtb survival within host macrophages and animal models (Tullius et al. 2003; Harth et al. 2005). Structural studies of GlnA1 have elucidated its catalytic mechanism, revealing potential sites for drug targeting (Krajewski et al. 2005). Inhibition of GlnA1 by compounds such as L-methionine-S, R-sulfoximine (MSO) has been shown to impair Mtb growth both in vitro and in vivo, underscoring the promise of GS inhibitors as novel antituberculosis therapeutics (Harth et al. 2005).

Asparagine serves as a significant nitrogen source for Mtb, playing a vital role in its survival and virulence. Mtb assimilates nitrogen and resists acid stress by utilizing the asparagine transporter AnsP2 and the secreted asparaginase *AnsA*, which hydrolyzes asparagine to release ammonia (Gouzy et al. 2014). This ammonia production is essential for arresting phagosome acidification and promoting intracellular replication. Beyond asparagine, Mtb accesses multiple amino acids within macrophages, including glutamate, glutamine, aspartate, alanine, glycine, and valine, with glutamine identified as the predominant nitrogen donor (Borah et al. 2019). Aspartate also functions as a primary nitrogen source, imported via the AnsP1 transporter, which is crucial for host colonization (Gouzy et al. 2013). Understanding these nitrogen metabolic pathways is critical for developing novel antituberculosis therapies, especially in the context of emerging multidrug-resistant strains (Xu et al. 2023).

The serine/threonine protein kinase *PknG* plays a central regulatory role in Mtb nitrogen metabolism by modulating amino acid levels and virulence. PknG senses intracellular concentrations of glutamate and aspartate and, in response, phosphorylates its substrate GarA (also known as OdhI) (Rieck et al. 2014; Lisa et al. 2021). GarA regulates the tricarboxylic acid (TCA) cycle and glutamate catabolism, making this phosphorylation event essential for bacterial growth and survival, particularly under nutrient-limited conditions and during infection (Cowley et al. 2004). Disruption of PknG activity leads to glutamate accumulation and reduced glutamine synthesis, highlighting its critical role in maintaining amino acid homeostasis (Cowley et al. 2004). PknG-mediated phosphorylation of GarA inhibits α-ketoglutarate decarboxylase and modulates glutamate dehydrogenase activity, thereby impacting glutamate metabolism (O'Hare et al. 2008; Rieck et al. 2014). The loss of either pknG or garA impairs intracellular nitrogen balance and glutamate metabolism. However, with opposing effects: deletion of pknG causes defects in glutamate catabolism, whereas garA disruption results in intracellular glutamate depletion (Rieck et al. 2014). This PknG-GarA regulatory axis is critical for mycobacterial physiology, redox homeostasis, and persistence during latency-like conditions, contributing to stable granuloma formation and virulence (Khan et al. 2017).

Mtb demonstrates remarkable metabolic flexibility by synthesizing specific amino acids de novo when host-derived sources are limited, enabling survival under nutrient-restricted conditions encountered during infection. A key enzyme in this process is serine hydroxymethyltransferase (SHMT), which catalyzes the conversion of serine to glycine while generating one-carbon units essential for methylation reactions (Yelamanchi and Surolia 2021; Jiang and Shi 2021; Amalia et al. 2022; Guida 2024). Mtb expresses two SHMT isoforms, *SHM1* and *SHM2*, each with distinct structural and functional properties. Notably, SHM1 exhibits an unusual 1:1 ratio of pyridoxal 5′-phosphate to enzyme dimer, distinguishing it within the transferase family (Chaturvedi and Bhakuni 2003). SHMT contributes both to amino acid biosynthesis and redox balance, underscoring the importance of de novo amino acid synthesis in Mtb's nitrogen metabolism.

Another critical pathway supporting nitrogen metabolism is the proteasome-mediated regulatory system, which maintains protein homeostasis, facilitates metabolic adaptation, and supports virulence. This system is especially vital under stress conditions such as oxidative damage, reactive nitrogen intermediates, and nutrient limitation encountered within host macrophages, where selective degradation of damaged, misfolded, or regulatory proteins preserves cellular integrity (Elharar et al. 2014; Bode and Darwin 2014; Gengenbacher and Kaufmann 2012; Samanovic et al. 2013; Becker et al. 2019). At the core of this system lies the 20S proteasome particle, encoded by the *prcA* and *prcB* genes. The 20S proteasome is a cylindrical complex composed of four stacked heptameric rings: two outer rings formed by seven identical α-subunits (*PrcA*) and two inner rings formed by seven identical β-subunits (*PrcB*), creating an $\alpha_7\beta_7\beta_7\alpha_7$ architecture (Coux et al. 1996; Lin et al. 2006; Gandotra et al. 2010). The α-rings regulate substrate access through gated entryways, while the β-rings contain the proteolytic active sites responsible for protein degradation.

Proteins targeted for degradation are post-translationally tagged with the prokaryotic ubiquitin-like protein Pup in a process termed pupylation (Burns et al. 2010; Sutter et al. 2010; Von Rosen et al. 2021). This modification is catalyzed by PafA (proteasome accessory factor A), which ligates Pup to specific lysine residues on substrates in an ATP-dependent manner. Pupylated proteins are recognized by Mpa (mycobacterial proteasomal ATPase), a hexameric AAA+ ATPase that binds Pup via N-terminal coiled-coil domains. Mpa utilizes ATP hydrolysis to unfold substrates and translocate them through its central channel into the 20S proteasome core for degradation. This tightly coordinated system ensures selective and regulated protein turnover, sustaining cellular homeostasis and adaptation in Mtb (Burns et al. 2010; Sutter et al. 2010; Von Rosen et al. 2021). Beyond protein quality control, the proteasome system plays a pivotal role in regulating nitrogen metabolism by targeting the transcriptional repressor HrcA for degradation. This degradation is essential for derepressing and robustly expressing chaperonin genes (*groEL* and *groES*) (Aravindhan et al. 2009; Becker et al. 2019; Hakiem and Batra 2022). These chaperonins are critical for proper folding and activity of enzymes such as the nitrite reductase complex NirBD, which catalyzes the reduction of nitrate to ammonium. Mutants deficient in proteasome function fail to utilize nitrate due to impaired NirBD activity stemming from insufficient chaperonin expression (Akhtar et al. 2013). Furthermore, growth on nitrate specifically enhances HrcA degradation, indicating regulated adaptation of the Pup-proteasome system to metabolic demands. The transcriptional regulator GlnR plays a central role in nitrogen metabolism during Mtb's exponential growth phase, ensuring efficient utilization of available nitrogen sources. GlnR belongs to the OmpR/PhoB subfamily of response regulators and orchestrates the expression of genes involved in nitrogen assimilation, carbon metabolism, and stress adaptation (Shi et al. 2025). GlnR also represses its gene and those encoding glutamine synthetase (*GlnA1)* and glutamate synthase (GOGAT), forming a tightly controlled regulatory network.

Collectively, these findings emphasize that the Pup-proteasome system, comprising *Pup, PafA, Mpa, PrcA/PrcB*, and accessory factors, forms a coordinated network essential for protein homeostasis, metabolic flexibility, and immune evasion during active infection. This system enables Mtb to efficiently acquire, assimilate, and respond to available nitrogen sources within host environments, aligning with critical research interests in Mtb pathophysiology and therapeutic targeting.

Mtb relies on complex nitrogen metabolism pathways that are essential for its survival, virulence, and adaptability across diverse host environments. The bacterium efficiently exploits various nitrogen sources, including ammonium, nitrate, and amino acids, through specialized transporters and enzymes such as glutamine synthetase (*GlnA1*) and nitrate reductase (*NarGHJI*) (Amon et al. 2008; Xu et al. 2023). During dormancy, these pathways enable Mtb to persist under hypoxic and nutrient-limited conditions by prioritizing nitrogen assimilation and recycling (Khan and Sarkar 2012). Regulatory proteins like GlnA1 and the asparagine transporter AnsP2 not only govern nitrogen assimilation but also influence pathogenesis, directly linking metabolic flexibility to virulence (Xu et al. 2023).

Mtb exhibits remarkable metabolic versatility, demonstrating greater efficiency in utilizing amino acids compared to inorganic nitrogen sources, and possesses the ability to co-metabolize multiple nitrogen substrates simultaneously (Agapova et al. 2018). Alanine dehydrogenase exemplifies this adaptability by fulfilling dual roles in ammonium assimilation and alanine catabolism, thereby maintaining redox balance and nitrogen homeostasis (Agapova et al. 2018). Targeting these critical enzymes and pathways—particularly GlnA1 and nitrate reductase—offers promising avenues for developing novel therapeutic strategies aimed at disrupting Mtb's survival mechanisms and advancing tuberculosis treatment.

6.4 Sulfur Metabolism in *Mycobacterium tuberculosis*

Sulfur metabolism is a critical pathway in Mtb, underpinning its virulence, redox defense, and overall survival within the host (Schnell and Schneider 2010). Although active during the exponential growth phase, sulfur metabolism becomes particularly important and is upregulated during dormancy and various stress conditions. During active growth, Mtb assimilates sulfur primarily to synthesize essential sulfur-containing metabolites such as cysteine, methionine, and mycothiol, all of which are vital for maintaining normal cellular functions and supporting bacterial proliferation. Beyond these fundamental metabolites, sulfur metabolism also contributes to the biosynthesis of key virulence factors, including sulfolipid-1 (SL-1), which plays an important role in modulating host-pathogen interactions (Schelle and Bertozzi 2006; Hatzios and Bertozzi 2011). The pathway is tightly regulated by complex mechanisms that dynamically respond to environmental cues, ensuring metabolic flexibility and adaptation throughout the Mtb lifecycle (Schnell and Schneider 2010; Wheeler et al. 2005; Hatzios et al. 2011). The assimilation of sulfur in Mtb begins with the active uptake of inorganic sulfate (SO_4^{2-}) via a specialized sulfate transporter complex. This high-affinity ATP-binding cassette (ABC) transporter system likely comprises homologues of sulfur mobilization protein I (SufI), sulfate/thiosulfate permeases CysT and CysW, and the ATP-binding protein CysA. Together, these components facilitate efficient sulfate import, ensuring a steady intracellular supply critical for downstream sulfur metabolism and Mtb's pathogenicity (Schelle and Bertozzi 2006; Hatzios et al. 2011). Once inside the cytoplasm, sulfate undergoes activation through a coordinated two-step enzymatic process. Initially, ATP sulfurylase, encoded by cysD, catalyzes the formation of adenosine 5′-phosphosulfate (APS) from ATP and sulfate. Subsequently, APS kinase, encoded by cysN, phosphorylates APS to generate 3′-phosphoadenosine 5′-phosphosulfate (PAPS), which serves as the universal sulfate donor for a variety of sulfation reactions essential to Mtb's growth, defense mechanisms, and virulence during infection (Pinto et al. 2004; Schelle and Bertozzi 2006). Notably, in Mtb, the enzymatic activities of ATP sulfurylase and APS kinase are combined within a single bifunctional enzyme encoded by the cysNC gene. Structural analyses of the CysC domain reveal a conserved APS kinase fold with well-defined binding sites, presenting challenges for the rational design of selective inhibitors targeting this enzyme (Poyraz et al. 2015).The cysteine biosynthesis pathway is indispensable for Mtb survival, with several key enzymes orchestrating this process. One critical step involves serine acetyltransferase (CysE), which forms a functional complex with O-acetylserine sulfhydrylase A (CysK1). This complex catalyzes the conversion of O-acetylserine into cysteine, a reaction essential for bacterial persistence under stress and during infection (Joshi et al. 2019; Watanabe et al. 2021).

Sulfotransferases (STs) constitute a diverse family of enzymes that transfer sulfate groups from the universal sulfate donor PAPS to a wide range of substrates, thereby participating in numerous biological functions across species (Negishi et al. 2001; Mistry et al. 2024). In Mtb, sulfotransferase 3 (Stf3) mediates specific sulfation reactions through substrate recognition domains, contributing to the bacterium's metabolic versatility and adaptability (Klaassen and Boles 1997; Pinto et al. 2004; Hatzios et al. 2011; Paritala and Carroll 2013; Hirschler et al. 2021). Regulation of 3′-phosphoadenosine 5′-phosphosulfate (PAPS) and its byproduct adenosine 3′-phosphate 5′-phosphate (PAP) is tightly controlled by the enzyme 3′-phosphoadenosine-5′-phosphatase (CysQ), encoded by the Rv2131c gene. CysQ functions to prevent the accumulation of PAP, which otherwise acts as a potent inhibitor of ATP sulfurylase (CysD), thereby maintaining continuous sulfur flux and preventing feedback inhibition within the sulfur assimilation pathway (Hatzios et al. 2011; Erickson et al. 2015). Within this pathway, adenosine 5′-phosphosulfate (APS) serves as a critical branch point intermediate. The sulfate-activating complex, composed of ATP sulfurylase (CysD) and APS kinase (CysN), catalyzes the formation of APS, which can then enter the reductive branch of sulfur metabolism. Here, APS reductase, encoded by cysH, catalyzes the reduction of APS to sulfite through a mechanism involving a transient enzyme-thiosulfonate intermediate and requiring a conserved iron-sulfur cluster for catalytic activity (Carroll et al. 2005; Chartron et al. 2006; Williams et al. 2014). Alternatively, APS can be phosphorylated by the APS kinase domain (CysC) of the bifunctional CysN enzyme to form PAPS, which supplies activated sulfate for sulfolipid biosynthesis and other sulfation reactions (Williams et al. 2014; Poyraz et al. 2015). These pathways are essential for the biosynthesis of sulfur-containing amino acids such as cysteine and methionine.

Mtb exhibits exceptional metabolic flexibility in cysteine biosynthesis, utilizing multiple distinct routes to maintain sulfur homeostasis under varied environmental conditions. Three major cysteine biosynthesis pathways have been identified: (1) the sulfide-dependent pathway, (2) the cystathionine-based transsulfuration pathway, and (3) a novel pathway involving a protein-bound thiocarboxylate as the sulfur donor (Burns et al. 2005; O'Leary et al. 2008; Ågren et al. 2008; Steiner-Rebrova et al. 2014; Hatzios and Bertozzi 2011; Khan et al. 2024). The novel thiocarboxylate-dependent pathway is encoded by a gene cluster comprising a metalloprotease (Mec+, encoded by Rv1334), a sulfur carrier protein (CysO, encoded by Rv1335), and an O-phosphoserine sulfhydrylase (CysM, encoded by Rv1336). In this system, the sulfur carrier protein CysO is converted into its thiocarboxylate form by MoeZ, a molybdopterin biosynthesis-like protein and paralog of the ThiF family. CysM then uses O-phosphoserine as a substrate to transfer sulfur from thiocarboxylated CysO to O-phosphoserine, forming a CysO-cysteine adduct. This adduct is subsequently hydrolyzed by the JAMM (JAB1/MPN/Mov34) motif-containing metalloprotease Mec+, releasing free cysteine and regenerating CysO. This pathway is particularly important under oxidative stress because thiocarboxylates are more resistant to oxidation than free thiols, providing an advantage in the oxidative environment of host macrophages (Burns et al. 2005; O'Leary et al. 2008; Ågren et al. 2008; Hatzios and Bertozzi 2011; Burns-Huang and Mundhra 2019; Khan et al. 2024). In addition to the CysO-dependent system, Mtb expresses two other cysteine synthases: CysK1 and CysK2. CysK1 catalyzes the conventional reaction between O-acetylserine and sulfide, while CysK2, similar to CysM, uses O-phosphoserine and sulfide. The presence of these multiple, partially redundant enzymes ensures robust cysteine biosynthesis even if one pathway is compromised. Transcriptomic studies have shown that the mec + –cysO-cysM gene cluster is upregulated in response to oxidative stress, highlighting the adaptive significance of this pathway for survival within the host (Steiner et al. 2014; Khan et al. 2021). Cysteine produced through these pathways is essential for synthesizing low molecular weight thiols such as mycothiol (MSH) and ergothioneine (EGT), which are critical for maintaining redox balance and protecting Mtb from oxidative damage. The CysO-thiocarboxylate pathway provides a stable source of sulfide for cysteine biosynthesis, especially under the highly oxidizing conditions encountered within macrophages (Burns et al. 2005; Ågren et al. 2008; Jurgenson et al. 2008; Choudhary et al. 2022; Khan et al. 2021, 2024).

The trans-sulfuration pathway in Mtb serves as a vital link between methylation and antioxidant metabolism by converting methionine to cysteine (Xia et al. 2020; Bandyopadhyay et al. 2022). This pathway begins with cystathionine β-synthase (Cbs), which catalyzes the condensation of homocysteine and serine to form cystathionine. Notably, Mtb Cbs is a heme-less, pyridoxal-5′-phosphate (PLP)-dependent enzyme uniquely activated by S-adenosylmethionine (SAM). SAM binding enhances Cbs catalytic activity and protects it from proteasomal degradation, thereby supporting both methionine and redox metabolism. Following cystathionine formation, cystathionine γ-lyase (CysK1) catalyzes its conversion to cysteine and α-ketobutyrate (Conter et al. 2020). Genetic studies demonstrate that Cbs deficiency significantly impairs Mtb survival under homocysteine overload conditions, highlighting the essential role of the Cbs-SAM axis in coordinating sulfur metabolism and maintaining redox balance (Bandyopadhyay et al. 2022).

Sulfur is indispensable for iron-sulfur (Fe-S) cluster biogenesis in Mtb, a process critical for the function of essential enzymes involved in respiration, metabolism, and DNA repair. The sulfur mobilization (Suf) pathway serves as the principal system for Fe-S cluster assembly in Mtb, particularly under oxidative stress or iron-limiting conditions. This pathway comprises specialized proteins with distinct structural and functional roles (Garcia et al. 2019; Blahut et al. 2020; Esquilin-Lebron et al. 2021; Tripathi et al. 2022; Elchennawi et al. 2023). Cysteine desulfurase (SufS) initiates the process by catalyzing the removal of sulfur from cysteine, generating a persulfide intermediate on its active-site cysteine residue. As a pyridoxal 5′-phosphate (PLP)-dependent enzyme, SufS forms a homodimer with a conserved PLP-binding domain essential for its desulfurase activity. The persulfide sulfur is then transferred to the scaffold complex for Fe-S cluster assembly (Blahut et al. 2020; Esquilin-Lebron et al. 2021; Tripathi et al. 2022; Elchennawi et al. 2023). The sulfur acceptor and shuttle protein SufU transiently binds the persulfide sulfur generated by cysteine desulfurase (SufS) and facilitates its delivery to the scaffold proteins responsible for iron-sulfur (Fe-S) cluster assembly. SufU's conserved structural fold enables efficient and specific interactions with both SufS and downstream scaffold components, ensuring precise sulfur transfer while preventing the accumulation of toxic sulfur species within the cell (Albrecht et al. 2010; Braymer et al. 2021; Hudspeth et al. 2022).

The scaffold complex, composed of SufB, SufC, and SufD, serves as the critical assembly site for nascent Fe-S clusters. Within this complex, SufB acts as the primary scaffold, coordinating the incorporation of both iron and sulfur atoms essential for cluster formation. SufC, an ATP-binding cassette (ABC)-type ATPase, provides the energy required for cluster assembly and transfer through ATP hydrolysis. SufD contributes structural stability to the complex and may facilitate iron acquisition and delivery to the scaffold. Together, these proteins form a stable, multi-subunit complex wherein SufC adopts a canonical ABC-type ATPase fold, and SufB and SufD provide unique domains specialized for cluster coordination and protein-protein interactions (Nachin et al. 2003; Watanabe et al. 2005; Chahal et al. 2009; Krishnan et al. 2020). SufT, a recently identified Fe-S cluster transfer protein within the Mtb Suf pathway, likely facilitates the maturation and transfer of Fe-S clusters to target apoproteins. Although its precise structural characteristics remain less defined, SufT contains conserved cysteine-rich motifs hypothesized to mediate transient cluster binding and transfer, thereby supporting the efficient delivery of Fe-S clusters to diverse recipient proteins (Tripathi et al. 2022). Two key low molecular weight thiols, mycothiol (MSH) and ergothioneine (EGT), play significant roles in maintaining Mtb's redox homeostasis. Mycothiol functions as a major redox couple, protecting the bacterium against oxidative stress and the effects of antituberculosis drugs. Ergothioneine contributes to preserving bioenergetic homeostasis, shielding Mtb from oxidative damage, and enhancing virulence. The biosynthesis of ergothioneine involves a series of enzymes, including γ-glutamylcysteine synthetase (EgtA), sulfoxide synthase (EgtB), C-S lyase (EgtC), methyltransferase (EgtD), and β-lyase (EgtE) (Goncharenko et al. 2015; Sheridan et al. 2016; Wei et al. 2024).

Sulfotransferases (Stfs) catalyze the sulfation of specific metabolites, such as S881, which negatively regulates Mtb's virulence. Among these enzymes, Stf3 is a key sulfotransferase that utilizes 3′-phosphoadenosine 5′-phosphosulfate (PAPS) as a sulfate donor, branching from the main sulfate assimilation pathway to mediate these sulfation reactions (Klaassen and Boles 1997; Negishi et al. 2001; Hatzios and Bertozzi 2011). This intricate network of sulfur metabolism pathways underscores Mtb's remarkable metabolic flexibility and resilience within host environments. From sulfate uptake through specialized transporters to the biosynthesis of critical sulfur-containing compounds like mycothiol and ergothioneine, each step is precisely regulated to support the bacterium's survival, virulence, and adaptation to challenging host conditions. A comprehensive understanding of these pathways offers valuable insights for the development of novel therapeutic strategies targeting Mtb's sulfur metabolism.

Conflict of Interest The authors declare that they have no conflict of interest related to the content of this chapter.

Ethical Consideration This chapter is a review based entirely on data and information obtained from previously published and publicly available sources, all of which have been appropriately cited. The authors conducted no new experimental work involving human participants or animals. Therefore, no additional ethical approval was required. The authors have ensured proper attribution and compliance with ethical standards in the use of all referenced materials.

References

Adolph C, Cheung C-Y, McNeil MB, Jowsey WJ, Williams ZC, Hards K, Harold LK, Aboelela A, Bujaroski RS, Buckley BJ (2024) A dual-targeting succinate dehydrogenase and F1Fo-ATP synthase inhibitor rapidly sterilizes replicating and non-replicating mycobacterium tuberculosis. Cell Chem Biol 31(4):683–698.e687

Agapova A, Hunt DM, Petridis M, Garza-Garcia A, Sohaskey CD, d. Carvalho LPS (2018) Nitrogen utilisation by the metabolic generalist pathogen mycobacterium tuberculosis. bioRxiv:385278

Agapova A, Serafini A, Petridis M, Hunt DM, Garza-Garcia A, Sohaskey CD, de Carvalho LPS (2019) Flexible nitrogen utilisation by the metabolic generalist pathogen Mycobacterium tuberculosis. eLife 8:e41129

Ågren D, Stehr M, Berthold CL, Kapoor S, Oehlmann W, Singh M, Schneider G (2008) Threedimensional structures of apo-and holo-L-alanine dehydrogenase from mycobacterium tuberculosis reveal conformational changes upon coenzyme binding. J Mol Biol 377(4):1161–1173

Ågren D, Schnell R, Oehlmann W, Singh M, Schneider G (2008) Cysteine synthase (CysM) of Mycobacterium tuberculosis is an O-phosphoserine sulfhydrylase: evidence for an alternative cysteine biosynthesis pathway in mycobacteria. J Biol Chem 283:31567–31574

Ahn YM, Boshoff HI (2017) Elevation of fumarate levels compromise redox control and viability in Mycobacterium tuberculosis. Cell Chem Biol 24:243–245

Ahn S, Jung J, Jang IA, Madsen EL, Park W (2016) Role of glyoxylate shunt in oxidative stress response. J Biol Chem 291:11928–11938

Akhtar AA, Turner DPJ (2022) The role of bacterial ATP-binding cassette (ABC) transporters in pathogenesis and virulence: therapeutic and vaccine potential. Microb Pathog 171:105734

Akhtar S, Khan A, Sohaskey CD, Jagannath C, Sarkar D (2013) Nitrite reductase NirBD is induced and plays an important role during in vitro dormancy of mycobacterium tuberculosis. J Bacteriol 195(20):4592–4599

Albrecht AG, Netz DJA, Miethke M, Pierik AJ, Burghaus O, Peuckert F et al (2010) SufU is an essential iron-sulfur cluster scaffold protein in Bacillus subtilis. J Bacteriol 192:1643–1651

Almeida S, Sousa C, Abreu V, Diniz C, Dorneles EM, Lage AP, Barh D, Azevedo V (2017) Exploration of nitrate reductase metabolic pathway in Corynebacterium pseudotuberculosis. Int J Gen 2017(1):9481756

Amalia F, Syamsunarno MRA, Triatin RD, Fatimah SN, Chaidir L, Achmad TH (2022) The role of amino acids in tuberculosis infection: a literature review. Meta 12(10):933

Amon J, Bräu T, Grimrath A, Hänβler E, Hasselt K, Höller M, Jeβberger N, Ott L, Szököl J, Titgemeyer F (2008) Nitrogen control in mycobacterium smegmatis: nitrogen-dependent expression of ammonium transport and assimilation proteins depends on the OmpR-type regulator GlnR. J Bacteriol 190(21):7108–7116

Amon J, Titgemeyer F, Burkovski A (2009) A genomic view on nitrogen metabolism and nitrogen control in mycobacteria. J Mol Microbiol Biotechnol 17:20–29

Andersen ÅB, Andersen P, Ljungqvist L (1992) Structure and function of a 40,000-molecular-weight protein antigen of Mycobacterium tuberculosis. Infect Immun 60:2317–2323

Aravindhan V, Christy AJ, Roy S, Ajitkumar P, Narayanan PR, Narayanan S (2009) Mycobacterium tuberculosis groE promoter controls the expression of the bicistronic groESL1 operon and shows differential regulation under stress conditions. FEMS Microbiol Lett 292(1):42–49

Arcondéguy T, Jack R, Merrick M (2001) PII signal transduction proteins, pivotal players in microbial nitrogen control. Microbiol Mol Biol Rev 65(1):80–105

Arnold PK, Finley LWS (2023) Regulation and function of the mammalian tricarboxylic acid cycle. J Biol Chem 299:102838

Asthana P, Singh D, Pedersen JS, Hynönen MJ, Sulu R, Murthy AV, Laitaoja M, Jänis J, Riley LW, Venkatesan R (2021) Structural insights into the substrate-binding proteins Mce1A and Mce4A from Mycobacterium tuberculosis. IUCrJ 8:757–774

Atkinson MR, Ninfa AJ (1998) Role of the GlnK signal transduction protein in the regulation of nitrogen assimilation in Escherichia coli. Mol Microbiol 29:431–447

Balázsi G, van Oudenaarden A, Collins JJ (2011) Cellular decision making and biological noise: from microbes to mammals. Cell 144:910–925

Bandyopadhyay P, Pramanick I, Biswas R, Ps S, Sreedharan S, Singh S, Rajmani RS, Laxman S, Dutta S, Singh A (2022) S-Adenosylmethionine-responsive cystathionine β-synthase modulates sulfur metabolism and redox balance in Mycobacterium tuberculosis. Sci Adv 8:eabo0097

Banerjee S, Nandyala AK, Raviprasad P, Ahmed N, Hasnain SE (2007) Iron-dependent RNA-binding activity of mycobacterium tuberculosis aconitase. J Bacteriol 189(11):4046–4052

Barreto C, de Oliveira M, Balan A (2020) The ATP-binding cassette (ABC) transport systems in mycobacterium tuberculosis: structure, function, and possible targets for therapeutics. Biology 9(12):443

Bashiri G (2021) Lipid transport across the mycobacterial cell envelope. IUCr J 8(5):711–712

Basu P, Sandhu N, Bhatt A, Singh A, Balhana R, Gobe I, Crowhurst NA, Mendum TA, Gao L, Ward JL, Beale MH, McFadden J, Beste DJV (2018) The anaplerotic node is essential for the intracellular survival of Mycobacterium tuberculosis. J Biol Chem 293:5695–5704

Baughn AD, Rhee KY (2014) Metabolomics of central carbon metabolism in Mycobacterium tuberculosis. Microbiol Spectr 2(3):MGM2-0026-2013

Baughn AD, Garforth SJ, Vilchèze C, Jacobs WR Jr (2009) An anaerobic-type α-ketoglutarate ferredoxin oxidoreductase completes the oxidative tricarboxylic acid cycle of Mycobacterium tuberculosis. PLoS Pathog 5:e1000662

Becker SH. et al (2019) Regulation of protein quality control pathways by the mycobacterium tuberculosis proteasome. Doctoral dissertation, New York University

Beites T, Jansen RS, Wang R, Jinich A, Rhee KY, Schnappinger D, Ehrt S (2021) Multiple acyl-CoA dehydrogenase deficiency kills Mycobacterium tuber-

culosis in vitro and during infection. Nat Commun 12:6593
Beste DJ, McFadden J (2012) Metabolism of Mycobacterium tuberculosis. In: Systems biology of tuberculosis. Springer, pp 55–78
Beste DJ, Bonde B, Hawkins N, Ward JL, Beale MH, Noack S, Nöh K, Kruger NJ, Ratcliffe RG, McFadden J (2011) 13C metabolic flux analysis identifies an unusual route for pyruvate dissimilation in mycobacteria which requires isocitrate lyase and carbon dioxide fixation. PLoS Pathog 7:e1002091
Beste DJ, Nöh K, Niedenführ S, Mendum TA, Hawkins ND, Ward JL, Beale MH, Wiechert W, McFadden J (2013) 13C-flux spectral analysis of host-pathogen metabolism reveals a mixed diet for intracellular mycobacterium tuberculosis. Chem Biol 20(8):1012–1021
Bizior A, Williamson G, Harris T, Hoskisson PA, Javelle A (2023) Prokaryotic ammonium transporters: what has three decades of research revealed? Microbiology 169:001360
Blahut M, Sanchez E, Fisher CE, Outten FW (2020) Fe-S cluster biogenesis by the bacterial Suf pathway. Biochim Biophys Acta (BBA) Mol Cell Res 1867:118829
Bloch H, Segal W (1956) Biochemical differentiation of Mycobacterium tuberculosis grown in vivo and in vitro. J Bacteriol 72:132–141
Bode NJ, Darwin KH (2014) The pup-proteasome system of mycobacteria. In: Molecular Genetics of Mycobacteria, pp 667–680
Borah K, Beyß M, Theorell A, Wu H, Basu P, Mendum TA, Nöh K, Beste DJ, McFadden J (2019) Intracellular mycobacterium tuberculosis exploits multiple host nitrogen sources during growth in human macrophages. Cell Rep 29(11):3580–3591. e3584
Borah K, Mendum TA, Hawkins ND, Ward JL, Beale MH, Larrouy-Maumus G, Bhatt A, Moulin M, Haertlein M, Strohmeier G, Pichler H, Forsyth VT, Noack S, Goulding CW, McFadden J, Beste DJV (2021) Metabolic fluxes for nutritional flexibility of Mycobacterium tuberculosis. Mol Syst Biol 17:e10280
Braibant M, Gilot P, Content J (2000) The ATP binding cassette (ABC) transport systems of Mycobacterium tuberculosis. FEMS Microbiol Rev 24:449–467
Braymer JJ, Freibert SA, Rakwalska-Bange M, Lill R (2021) Mechanistic concepts of iron-sulfur protein biogenesis in biology. Biochim Biophys Acta (BBA) Mol Cell Res 1868:118863
Burns KE, Baumgart S, Dorrestein PC, Zhai H, McLafferty FW, Begley TP (2005) Reconstitution of a new cysteine biosynthetic pathway in Mycobacterium tuberculosis. J Am Chem Soc 127:11602–11603
Burns KE, Cerda-Maira FA, Wang T, Li H, Bishai WR, Darwin KH (2010) "Depupylation" of prokaryotic ubiquitin-like protein from mycobacterial proteasome substrates. Mol Cell 39(5):821–827
Burns-Huang K, Mundhra S (2019) Mycobacterium tuberculosis cysteine biosynthesis genes mec^{+}-cysO-cysM confer resistance to clofazimine. Tuberculosis 115:63–66
Bussi C, Gutierrez MG (2019) Mycobacterium tuberculosis infection of host cells in space and time. FEMS Microbiol Rev 43:341–361
Capodagli GC, Sedhom WG, Jackson M, Ahrendt KA, Pegan SD (2014) A noncompetitive inhibitor for Mycobacterium tuberculosis's class IIa fructose 1,6-bisphosphate aldolase. Biochemistry 53:202–213
Carroll KS, Gao H, Chen H, Stout CD, Leary JA, Bertozzi CR (2005) A conserved mechanism for sulfonucleotide reduction. PLoS Biol 3:e250
Casali N, Riley LW (2007) A phylogenomic analysis of the Actinomycetales mce operons. BMC Genomics 8:60
Castillo A, Taboada H, Mendoza A, Valderrama B, Encarnación S, Mora J (2000) Role of GOGAT in carbon and nitrogen partitioning in rhizobium etli. Microbiology 146(7):1627–1637
Cassio Barreto de Oliveira M, Balan A (2020) The ATP-binding cassette (ABC) transport Systems in Mycobacterium tuberculosis: structure, function, and possible targets for therapeutics. Biology 9:443
Chahal HK, Dai Y, Saini A, Ayala-Castro C, Outten FW (2009) The SufBCD Fe–S scaffold complex interacts with SufA for Fe–S cluster transfer. Biochemistry 48:10644–10653
Chalut C (2016) MmpL transporter-mediated export of cell-wall associated lipids and siderophores in mycobacteria. Tuberculosis 100:32–45
Chang DP, Guan XL (2021) Metabolic versatility of mycobacterium tuberculosis during infection and dormancy. Meta 11:88
Chang H-Y, Tsai W-C, Lee T-F, Sheng W-H (2021) Mycobacterium gordonae infection in immunocompromised and immunocompetent hosts: a series of seven cases and literature review. J Formos Med Assoc 120(1):524–532
Chartron J, Carroll KS, Shiau C, Gao H, Leary JA, Bertozzi CR, Stout CD (2006) Substrate recognition, protein dynamics, and iron-sulfur cluster in Pseudomonas aeruginosa adenosine 5′-phosphosulfate reductase. J Mol Biol 364:152–169
Chaturvedi S, Bhakuni V (2003) Unusual structural, functional, and stability properties of serine hydroxymethyltransferase from mycobacterium tuberculosis. J Biol Chem 278(42):40793–40805
Chaudhry R, Varacallo MA (2025) Biochemistry, Glycolysis. In: StatPearls. StatPearls Publishing Copyright © 2025, StatPearls Publishing LLC, Treasure Island (FL). https://www.ncbi.nlm.nih.gov/books/NBK482303/
Chen C, Yan Q, Tao M, Shi H, Han X, Jia L, Huang Y, Zhao L, Wang C, Ma X, Ma Y (2019) Characterization of serine acetyltransferase (CysE) from methicillin-resistant Staphylococcus aureus and inhibitory effect of two natural products on CysE. Microb Pathog 131:218–226
Chen J, Fruhauf A, Fan C, Ponce J, Ueberheide B, Bhabha G, Ekiert DC (2023a) Structure of an endogenous mycobacterial MCE lipid transporter. Nature 620:445–452

Chen Y, Wang Y, Chng SS (2023b) A conserved membrane protein negatively regulates Mce1 complexes in mycobacteria. Nat Commun 14:5897

Chen Z, Kong X, Ma Q, Chen J, Zeng Y, Liu H, Wang X, Lu S (2024) The impact of Mycobacterium tuberculosis on the macrophage cholesterol metabolism pathway. Front Immunol 15:1402024

Choi KH, Lai V, Foster CE, Morris AJ, Tolan DR, Allen KN (2006) New superfamily members identified for Schiff-base enzymes based on verification of catalytically essential residues. Biochemistry 45:8546–8555

Choudhary E, Sharma R, Pal P, Agarwal N (2022) Deciphering the proteomic landscape of Mycobacterium tuberculosis in response to acid and oxidative stresses. ACS Omega 7:26749–26766

Cole S, Brosch R, Parkhill J, Garnier T, Churcher C, Harris D, Gordon S, Eiglmeier K, Gas S, Barry C III (1998) Deciphering the biology of mycobacterium tuberculosis from the complete genome sequence. Nature 396(6707):190–190

Connor SE, Capodagli GC, Deaton MK, Pegan SD (2011) Structural and functional characterization of Mycobacterium tuberculosis triosephosphate isomerase. Acta Crystallogr D Biol Crystallogr 67:1017–1022

Conter C, Fruncillo S, Fernández-Rodríguez C, Martínez-Cruz LA, Dominici P, Astegno A (2020) Cystathionine β-synthase is involved in cysteine biosynthesis and H2S generation in toxoplasma gondii. Sci Rep 10(1):14657

Coutts G, Thomas G, Blakey D, Merrick M (2002) Membrane sequestration of the signal transduction protein GlnK by the ammonium transporter AmtB. EMBO J 21(4):536–545

Coux O, Tanaka K, Goldberg AL (1996) Structure and functions of the 20S and 26S proteasomes. Annu Rev Biochem 65(1):801–847

Cowley S, Ko M, Pick N, Chow R, Downing KJ, Gordhan BG, Betts JC, Mizrahi V, Smith DA, Stokes RW (2004) The mycobacterium tuberculosis protein serine/threonine kinase PknG is linked to cellular glutamate/glutamine levels and is important for growth in vivo. Mol Microbiol 52(6):1691–1702

Cumming BM, Steyn AJ (2015) Metabolic plasticity of central carbon metabolism protects mycobacteria. Proc Natl Acad Sci USA 112:13135–13136

D'Arcy P, Brnjic S, Olofsson MH, Fryknas M, Lindsten K, De Cesare M et al (2011) Inhibition of proteasome deubiquitinating activity as a new cancer therapy. Nat Med 17:1636–1640

Daniel J, Maamar H, Deb C, Sirakova TD, Kolattukudy PE (2011) Mycobacterium tuberculosis uses host triacylglycerol to accumulate lipid droplets and acquires a dormancy-like phenotype in lipid-loaded macrophages. PLoS Pathog 7:e1002093

Davidson AL, Dassa E, Orelle C, Chen J (2008) Structure, function, and evolution of bacterial ATP-binding cassette systems. Microbiol Mol Biol Rev 72:317–364

de Carvalho LP, Fischer SM, Marrero J, Nathan C, Ehrt S, Rhee KY (2010) Metabolomics of Mycobacterium tuberculosis reveals compartmentalized co-catabolism of carbon substrates. Chem Biol 17:1122–1131

de la Paz Santangelo M, Gest PM, Guerin ME, Coinçon M, Pham H, Ryan G, Puckett SE, Spencer JS, Gonzalez-Juarrero M, Daher R, Lenaerts AJ, Schnappinger D, Therisod M, Ehrt S, Sygusch J, Jackson M (2011) Glycolytic and non-glycolytic functions of mycobacterium tuberculosis fructose-1,6-bisphosphate aldolase, an essential enzyme produced by replicating and non-replicating bacilli. J Biol Chem 286:40219–40231

de la Torre LI, Vergara Meza JG, Cabarca S, Costa-Martins AG, Balan A (2021) Comparison of carbohydrate ABC importers from Mycobacterium tuberculosis. BMC Genomics 22:841

Devlin KL, Leach DT, Stratton KG, Lamichhane G, Lin VS, Beatty KE (2025) Proteomic characterization of Mycobacterium tuberculosis subjected to carbon starvation mSystems 10:e01530–e01524

Dolan SK, Welch M (2018) The glyoxylate shunt, 60 years on. Ann Rev Microbiol 72:309–330

Elchennawi I, Carpentier P, Caux C, Ponge M, Ollagnier de Choudens S (2023) Structural and biochemical characterization of mycobacterium tuberculosis zinc SufU-SufS complex. Biomolecules 13(5):732

Elharar Y, Roth Z, Hermelin I, Moon A, Peretz G, Shenkerman Y, Vishkautzan M, Khalaila I, Gur E (2014) Survival of mycobacteria depends on proteasome-mediated amino acid recycling under nutrient limitation. EMBO J 33(16):1802–1814

Ensinck D, Gerhardt EC, Rollan L, Huergo LF, Gramajo H, Diacovich L (2024) The PII protein interacts with the Amt ammonium transport and modulates nitrate/nitrite assimilation in mycobacteria. Front Microbiol 15:1366111

Eoh H, Rhee KY (2014) Methylcitrate cycle defines the bactericidal essentiality of isocitrate lyase for survival of Mycobacterium tuberculosis on fatty acids. Proc Natl Acad Sci USA 111:4976–4981

Eoh H, Liu R, Lim J, Lee JJ, Sell P (2022) Central carbon metabolism remodeling as a mechanism to develop drug tolerance and drug resistance in Mycobacterium tuberculosis. Front Cell Infect Microbiol 12:958240

Erickson AI, Sarsam RD, Fisher AJ (2015) Crystal structures of Mycobacterium tuberculosis CysQ, with substrate and products bound. Biochemistry 54:6830–6841

Esquilin-Lebron K, Dubrac S, Barras F, Boyd JM (2021) Bacterial approaches for assembling iron-sulfur proteins. MBio 12:e02425–e02421

Fan X, Zhao X, Wang R, Li M, Luan X, Wang R, Wan K, Liu H (2023) A novel multistage antigens ERA005f confer protection against mycobacterium tuberculosis by driving Th-1 and Th-17 type T cell immune responses. Front Immunol 14:1276887

Feng Z, Caceres NE, Sarath G, Barletta RG (2002) Mycobacterium smegmatis L-alanine dehydrogenase (Ald) is required for proficient utilization of alanine as a sole nitrogen source and sustained anaerobic growth. J Bacteriol 184(18):5001–5010

Fenn K, Wong CT, Darbari VC (2020) Mycobacterium tuberculosis uses Mce proteins to interfere with host cell signaling. Front Mol Biosci 6:149

Ferraris DM, Spallek R, Oehlmann W, Singh M, Rizzi M (2015) Structures of citrate synthase and malate dehydrogenase of Mycobacterium tuberculosis. Proteins 83:389–394

Fieweger RA, Wilburn KM, VanderVen BC (2019) Comparing the metabolic capabilities of bacteria in the Mycobacterium tuberculosis complex. Microorganisms 7:177

Furze CM, Delso I, Casal E, Guy CS, Seddon C, Brown CM, Parker HL, Radhakrishnan A, Pacheco-Gomez R, Stansfeld PJ (2021) Structural basis of trehalose recognition by the mycobacterial LpqY-SugABC transporter. J Biol Chem 296

Gago G, Diacovich L, Gramajo H (2018) Lipid metabolism and its implication in mycobacteria-host interaction. Curr Opin Microbiol 41:36–42

Gandotra S, Lebron MB, Ehrt S (2010) Mycobacterium tuberculosis proteasome active site threonine is essential for persistence yet dispensable for replication and resistance to nitric oxide. PLoS Pathog 6:e1001040

Gao B, Li G, Gu D, Wang J (2023) Research progress on GlnR-mediated regulation in Actinomycetes. Front Microbiol 14:1282523

Garcia PS, Gribaldo S, Py B, Barras F (2019) The SUF system: an ABC ATPase-dependent protein complex with a role in Fe–S cluster biogenesis. Res Microbiol 170(8):426–434

García-Gutiérrez Á, Cánovas FM, Ávila C (2018) Glutamate synthases from conifers: gene structure and phylogenetic studies. BMC Genomics 19(1):65

Gengenbacher M, Kaufmann SH (2012) Mycobacterium tuberculosis: success through dormancy. FEMS Microbiol Rev 36:514–532

Gerike U, Hough DW, Russell NJ, Dyall-Smith ML, Danson MJ (1998) Citrate synthase and 2-methylcitrate synthase: structural, functional and evolutionary relationships. Microbiology 144:929–935

Giffin MM, Modesti L, Raab RW, Wayne LG, Sohaskey CD (2012) Ald of mycobacterium tuberculosis encodes both the alanine dehydrogenase and the putative glycine dehydrogenase. J Bacteriol 194:1045–1054

Giffin MM, Shi L, Gennaro ML, Sohaskey CD (2016) Role of alanine dehydrogenase of mycobacterium tuberculosis during recovery from hypoxic nonreplicating persistence. PLoS One 11(5):e0155522

Goddard AD, Bali S, Mavridou DA, Luque-Almagro VM, Gates AJ, Dolores Roldán M, Newstead S, Richardson DJ, Ferguson SJ (2017) The P aracoccus denitrificans N ar K-like nitrate and nitrite transporters—probing nitrate uptake and nitrate/nitrite exchange mechanisms. Mol Microbiol 103(1):117–133

Goncharenko KV, Vit A, Blankenfeldt W, Seebeck FP (2015) Structure of the sulfoxide synthase EgtB from the ergothioneine biosynthetic pathway. Angew Chem Int Ed 54:2821–2824

Gouzy A, Poquet Y, Neyrolles O (2014) Nitrogen metabolism in Mycobacterium tuberculosis physiology and virulence. Nat Rev Microbiol 12:729–737

Gouzy A, Larrouy-Maumus G, Wu T-D, Peixoto A, Levillain F, Lugo-Villarino G, Guerquin-Kern J-L, De Carvalho LPS, Poquet Y, Neyrolles O (2013) Mycobacterium tuberculosis nitrogen assimilation and host colonization require aspartate. Nat Chem Biol 9(11):674–676

Griffin JE, Pandey AK, Gilmore SA, Mizrahi V, McKinney JD, Bertozzi CR, Sassetti CM (2012) Cholesterol catabolism by Mycobacterium tuberculosis requires transcriptional and metabolic adaptations. Chem Biol 19:218–227

Grosset J (2003) Mycobacterium tuberculosis in the extracellular compartment: an underestimated adversary. Antimicrob Agents Chemother 47:833–836

Guasch-Ferré M, Santos JL, Martínez-González MA, Clish CB, Razquin C, Wang D, Liang L, Li J, Dennis C, Corella D, Muñoz-Bravo C, Romaguera D, Estruch R, Santos-Lozano JM, Castañer O, Alonso-Gómez A, Serra-Majem L, Ros E, Canudas S, Asensio EM, Fitó M, Pierce K, Martínez JA, Salas-Salvadó J, Toledo E, Hu FB, Ruiz-Canela M (2020) Glycolysis/gluconeogenesis – and tricarboxylic acid cycle-related metabolites, Mediterranean diet, and type 2 diabetes. Am J Clin Nutr 111(4):835–844

Guida M (2024) Novel strategies against infectious diseases: leveraging small molecules and insights into pathogen biology

Gupta S, Kumar A, Singh K, Kumari R, Sharma A, Singh RK, Pandey SK, Anupurba S (2020) Rv1273c, an ABC transporter of mycobacterium tuberculosis promotes mycobacterial intracellular survival within macrophages via modulating the host cell immune response. Int J Biol Macromol 142:320–331

Hackett EE, Charles-Messance H, O'Leary SM, Gleeson LE, Muñoz-Wolf N, Case S, Wedderburn A, Johnston DGW, Williams MA, Smyth A, Ouimet M, Moore KJ, Lavelle EC, Corr SC, Gordon SV, Keane J, Sheedy FJ (2020) Mycobacterium tuberculosis limits host glycolysis and IL-1β by restriction of PFK-M via MicroRNA-21. Cell Rep 30:124–136.e4

Hakiem OR, Batra JK (2022) Role of HrcA in stress management in mycobacterium tuberculosis. J Appl Microbiol 132(4):3315–3326

Hards K, Adolph C, Harold LK, McNeil MB, Cheung CY, Jinich A, Rhee KY, Cook GM (2020) Two for the price of one: attacking the energetic-metabolic hub of mycobacteria to produce new chemotherapeutic agents. Prog Biophys Mol Biol 152:35–44

Harper C, Hayward D, Wiid I, van Helden P (2008) Regulation of nitrogen metabolism in mycobacterium tuberculosis: a comparison with mechanisms in Corynebacterium glutamicum and Streptomyces coelicolor. IUBMB Life 60(10):643–650

Harth G, Clemens DL, Horwitz MA (1994) Glutamine synthetase of Mycobacterium tuberculosis: extracellular release and characterization of its enzymatic activity. Proc Natl Acad Sci USA 91:9342–9346

Harth G, Masleša-Galić S, Tullius MV, Horwitz MA (2005) All four Mycobacterium tuberculosis glnA genes encode glutamine synthetase activities but only GlnA1 is abundantly expressed and essential for bacterial homeostasis. Mol Microbiol 58:1157–1172

Hartman T, Weinrick B, Vilchèze C, Berney M, Tufariello J, Cook GM, Jacobs WR Jr (2014) Succinate dehydrogenase is the regulator of respiration in Mycobacterium tuberculosis. PLoS Pathog 10:e1004510

Hatzios SK, Bertozzi CR (2011) The regulation of sulfur metabolism in mycobacterium tuberculosis. PLoS Pathog 7(7):e1002036

Hatzios SK, Schelle MW, Newton GL, Sogi KM, Holsclaw CM, Fahey RC, Bertozzi CR (2011) The mycobacterium tuberculosis CysQ phosphatase modulates the biosynthesis of sulfated glycolipids and bacterial growth. Bioorg Med Chem Lett 21(17):4956–4959

Hernández-Viveros D, Rechy-Iruretagoyena DA, Díaz-Molina R, Vique-Sánchez JL (2021) Triosephosphate isomerase from mMycobacterium tuberculosis as potential target to develop a new anti-TB drug. Biointerf Res Appl Chem 12:1–9

Hirschfield GR, McNeil M, Brennan PJ (1990) Peptidoglycan-associated polypeptides of Mycobacterium tuberculosis. J Bacteriol 172:1005–1013

Hirschler A, Carapito C, Maurer L, Zumsteg J, Villette C, Heintz D, Dahl C, Al-Nayal A, Sangal V, Mahmoud H (2021) Biodesulfurization induces reprogramming of sulfur metabolism in Rhodococcus qingshengii IGTS8: proteomics and untargeted metabolomics. Microbiol Spec 9(2):e00692–e00621

Howard NC, Khader SA (2020) Immunometabolism during Mycobacterium tuberculosis infection. Trends Microbiol 28:832–850

Huang L, Nazarova EV, Tan S, Liu Y, Russell DG (2018) Growth of Mycobacterium tuberculosis in vivo segregates with host macrophage metabolism and ontogeny. J Exp Med 215:1135–1152

Huang EY-W, Kwai BXC, Bhusal RP, Bashiri G, Leung IKH (2023) Mycobacterium tuberculosis Rv1916 is an acetyl-CoA-binding protein. Chembiochem 24:e202300162

Hudspeth JD, Boncella AE, Sabo ET, Andrews T, Boyd JM, Morrison CN (2022) Structural and biochemical characterization of Staphylococcus aureus cysteine desulfurase complex SufSU. ACS Omega 7:44124–44133

Huergo LF, Chandra G, Merrick M (2013) P(II) signal transduction proteins: nitrogen regulation and beyond. FEMS Microbiol Rev 37:251–283

Izumi A, Schnell R, Schneider G (2012) Crystal structure of NirD, the small subunit of the nitrite reductase NirbD from mycobacterium tuberculosis at 2.0 Å resolution. Proteins Structure Function Bioinfo 80(12):2799–2803

Jain S, Mandal RS, Anand S, Maiti S, Ramachandran S (2013) Probing the amino acids critical for protein oligomerisation and protein–nucleotide interaction in mycobacterium tuberculosis PII protein through integration of computational and experimental approaches. Biochimica et Biophysica Acta (BBA)-Proteins Proteomics 1834(12):2736–2749

Jeong J-A, Oh J-I (2019) Alanine dehydrogenases in mycobacteria. J Microbiol 57(2):81–92

Jeong J-A, Hyun J, Oh J-I (2015) Regulation mechanism of the ald gene encoding alanine dehydrogenase in mycobacterium smegmatis and mycobacterium tuberculosis by the Lrp/AsnC family regulator AldR. J Bacteriol 197(19):3142–3153

Jiang Q, Shi L (2021) Coordination of the uptake and metabolism of amino acids in mycobacterium tuberculosis-infected macrophages. Front Immunol 12:711462

Jiang Z, Wang M, Nicolas M, Ogé L, Pérez-Garcia MD, Crespel L, Li G, Ding Y, Le Gourrierec J, Grappin P, Sakr S (2022) Glucose-6-phosphate dehydrogenases: the hidden players of plant physiology. Int J Mol Sci 23:16128

Joshi P, Gupta A, Gupta V (2019) Insights into multifaceted activities of CysK for therapeutic interventions. 3 Biotech 9(2):44

Jurgenson CT, Burns KE, Begley TP, Ealick SE (2008) Crystal structure of a sulfur carrier protein complex found in the cysteine biosynthetic pathway of Mycobacterium tuberculosis. Biochemistry 47:10354–10364

Kalscheuer R, Weinrick B, Veeraraghavan U, Besra GS, Jacobs WR Jr (2010) Trehalose-recycling ABC transporter LpqY-SugA-SugB-SugC is essential for virulence of Mycobacterium tuberculosis. Proc Natl Acad Sci USA 107:21761–21766

Khan A, Sarkar D (2012) Nitrate reduction pathways in mycobacteria and their implications during latency. Microbiology 158(2):301–307

Khan TA, Mazhar H, Saleha S, Tipu HN, Muhammad N, Abbas MN (2016) Interferon-gamma improves macrophages function against M. tuberculosis in multidrug-resistant tuberculosis patients. Chem Res Pract 2016(1):7295390

Khan MZ, Bhaskar A, Upadhyay S, Kumari P, Rajmani RS, Jain P, Singh A, Kumar D, Bhavesh NS, Nandicoori VK (2017) Protein kinase G confers survival advantage to mycobacterium tuberculosis during latency-like conditions. J Biol Chem 292(39):16093–16108

Khan MZ, Singha B, Ali MF, Taunk K, Rapole S, Gourinath S, Nandicoori VK (2021) Redox homeostasis in mycobacterium tuberculosis is modulated by a novel actinomycete-specific transcription factor. EMBO J 40(14):e106111

Khan MF, Ali A, Khan SN, Obaid A, Awan FM, Khan AS, Jabbar A (2024) Investigating the anti-tubercular potential of novel non-antibiotic agents against drug-resistant mycobacterium tuberculosis. BioScientific Review 6(4):40–49

Kim JW, Lee JY, Oh M, Lee EW (2023) An integrated view of lipid metabolism in ferroptosis revisited via lipidomic analysis. Exp Mol Med 55:1620–1631

Klaassen CD, Boles JW (1997) Sulfation and sulfotransferases 5: the importance of 3′-phosphoadenosine 5′-phosphosulfate (PAPS) in the regulation of sulfation. FASEB J 11:404–418

Klepp LI, y Garcia JS (2022) Mycobacterial MCE proteins as transporters that control lipid homeostasis of the cell wall. Tuberculosis 132:102162

Krajewski WW, Jones TA, Mowbray SL (2005) Structure of mycobacterium tuberculosis glutamine synthetase in complex with a transition-state mimic provides functional insights. Proc Natl Acad Sci 102(30):10499–10504

Krishnan A, Burroughs AM, Iyer LM, Aravind L (2020) Comprehensive classification of ABC ATPases and their functional radiation in nucleoprotein dynamics and biological conflict systems. Nucleic Acids Res 48:10045–10075

Kumar A, Bose M, Brahmachari V (2003) Analysis of expression profile of mammalian cell entry (mce) operons of Mycobacterium tuberculosis. Infect Immun 71:6083–6087

Lee W, VanderVen BC, Fahey RJ, Russell DG (2013) Intracellular Mycobacterium tuberculosis exploits host-derived fatty acids to limit metabolic stress. J Biol Chem 288:6788–6800

Leigh J, Dodsworth J (2007) Nitrogen regulation in bacteria and archaea. Ann Rev Microbiol 61: 349–377

Li P, Gu Y, Li J, Xie L, Li X, Xie J (2017) Mycobacterium tuberculosis major facilitator superfamily transporters. J Membr Biol 250(6):573–585

Limenitakis J, Oppenheim RD, Creek DJ, Foth BJ, Barrett MP, Soldati-Favre D (2013) The 2-methylcitrate cycle is implicated in the detoxification of propionate in Toxoplasma gondii. Mol Microbiol 87:894–908

Lin G, Hu G, Tsu C, Kunes YZ, Li H, Dick L, Parsons T, Li P, Chen Z, Zwickl P, Weich N, Nathan C (2006) Mycobacterium tuberculosis prcBA genes encode a gated proteasome with broad oligopeptide specificity. Mol Microbiol 59:1405–1416

Lin H, Xing J, Wang H, Wang S, Fang R, Li X, Li Z, Song N (2024) Roles of lipolytic enzymes in Mycobacterium tuberculosis pathogenesis. Front Microbiol 15:1329715

Lisa M-N, Sogues A, Barilone N, Baumgart M, Gil M, Graña M, Durán R, Biondi RM, Bellinzoni M, Bott M (2021) A tetratricopeptide repeat scaffold couples signal detection to OdhI phosphorylation in metabolic control by the protein kinase PknG. MBio 12(5). https://doi.org/10.1128/mbio.01717-01721

Liu F, Ma N (2020) Multicriteria ABC inventory classification using the social choice theory. Sustainability (Switzerland) 12(1)

Lofthouse EK (2013) Systems-based investigation of species-specific metabolism in the Mycobacterium tuberculosis complex [PhD dissertation]. University of Surrey, Guildford

Long ER (1954) Chemistry and chemotherapy of tuberculosis, 3rd edn. Williams & Wilkins, Baltimore

Lovewell RR, Sassetti CM, VanderVen BC (2016) Chewing the fat: lipid metabolism and homeostasis during M. tuberculosis infection. Curr Opin Microbiol 29:30–36

Lyon RH, Rogers P, Hall WH, Lichstein HC (1967) Inducible glutamate transport in mycobacteria and its relation to glutamate oxidation. J Bacteriol 94(1):92–100

Mackenzie JS, Lamprecht DA, Asmal R, Adamson JH, Borah K, Beste DJ, Lee BS, Pethe K, Rousseau S, Krieger I, Sacchettini JC (2020) Bedaquiline reprograms central metabolism to reveal glycolytic vulnerability in Mycobacterium tuberculosis. Nat Commun 11:6092

Madej MG (2014) Function, structure, and evolution of the major facilitator superfamily: the LacY manifesto. Adv Biol 2014:523591

Madej MG, Kaback HR (2013) Evolutionary mix-and-match with MFS transporters II. Proc Natl Acad Sci 110(50):E4831–E4838

Madej MG, Dang S, Yan N, Kaback HR (2013) Evolutionary mix-and-match with MFS transporters. Proc Natl Acad Sci 110(15):5870–5874

Maharjan R, Zhang Z, Klenotic PA, Gregor WD, Tringides ML, Cui M, Purdy GE, Yu EW (2024) Structures of the mycobacterial MmpL4 and MmpL5 transporters provide insights into their role in siderophore export and iron acquisition. PLoS Biol 22(10):e3002874

Maksymiuk C, Balakrishnan A, Bryk R, Rhee KY, Nathan CF (2015) E1 of α-ketoglutarate dehydrogenase defends Mycobacterium tuberculosis against glutamate anaplerosis and nitroxidative stress. Proc Natl Acad Sci USA 112:E5834–E5843

Malm S, Tiffert Y, Micklinghoff J, Schultze S, Joost I, Weber I, Horst S, Ackermann B, Schmidt M, Wohlleben W, Ehlers S, Geffers R, Reuther J, Bange FC (2009) The roles of the nitrate reductase NarGHJI, the nitrite reductase NirBD and the response regulator GlnR in nitrate assimilation of Mycobacterium tuberculosis. Microbiology 155:1332–1339

Marrero J, Trujillo C, Rhee KY, Ehrt S (2013) Glucose phosphorylation is required for Mycobacterium tuberculosis persistence in mice. PLoS Pathog 9:e1003116

Marrero J, Rhee KY, Schnappinger D, Pethe K, Ehrt S (2010) Gluconeogenic carbon flow of tricarboxylic acid cycle intermediates is critical for mycobacterium tuberculosis to establish and maintain infection. Proc Natl Acad Sci 107(21):9819–9824

Martínez-Reyes I, Chandel NS (2020) Mitochondrial TCA cycle metabolites control physiology and disease. Nat Commun 11:102

Mathur D, Malik G, Garg LC (2006) Biochemical and functional characterization of triosephosphate isomerase from Mycobacterium tuberculosis H37Rv. FEMS Microbiol Lett 263:229–235

Maurya RK, Bharti S, Krishnan MY (2019) Triacylglycerols: fuelling the hibernating mycobacterium tuberculosis. Front Cell Infect Microbiol 8:450

McKinney JD, Höner Bentrup K, Muñoz-Elías EJ, Miczak A, Chen B, Chan WT, Swenson D, Sacchettini JC, Jacobs WR, Russell D (2000) Persistence of Mycobacterium tuberculosis in macrophages and mice requires the glyoxylate shunt enzyme isocitrate lyase. Nature 406:735–738

Mekonnen D, Derbie A, Mihret A, Yimer SA, Tønjum T, Gelaw B, Nibret E, Munshae A, Waddell SJ, Aseffa A (2021) Lipid droplets and the transcriptome of Mycobacterium tuberculosis from direct sputa: a literature review. Lipids Health Dis 20:129

Melly G, Purdy GE (2019) MmpL proteins in physiology and pathogenesis of Mycobacterium tuberculosis. Microorganisms 7:70

Miner M, Chang J, Pandey A, Sassetti C, Sherman D (2009) Role of cholesterol in Mycobacterium tuberculosis infection. Indian J Exp Biol 47:407–411

Mistry R, Byrne DP, Starns D, Barsukov IL, Yates EA, Fernig DG (2024) Polysaccharide sulfotransferases: the identification of putative sequences and respective functional characterisation. Essays Biochem 68(4):431–447

Mohn WW, van der Geize R, Stewart GR, Okamoto S, Liu J, Dijkhuizen L, Eltis LD (2008) The actinobacterial mce4 locus encodes a steroid transporter. J Biol Chem 283:35368–35374

Moreira-Teixeira L, Redford PS, Stavropoulos E, Ghilardi N, Maynard CL, Weaver CT, Freitas do Rosário AP, Wu X, Langhorne J, O'Garra A (2017) T cell–derived IL-10 impairs host resistance to mycobacterium tuberculosis infection. J Immunol 199(2):613–623

Mulholland CV, Wiggins TJ, Cui J, Vilchèze C, Rajagopalan S, Shultis MW, Reyes-Fernández EZ, Jacobs WR Jr, Berney M (2024) Propionate prevents loss of the PDIM virulence lipid in Mycobacterium tuberculosis. Nat Microbiol 9:1607–1618

Müller T, Walter B, Wirtz A, Burkovski A (2006) Ammonium toxicity in bacteria. Curr Microbiol 52:400–406

Muñoz-Elías EJ, McKinney JD (2005) Mycobacterium tuberculosis isocitrate lyases 1 and 2 are jointly required for in vivo growth and virulence. Nat Med 11(6):638–644

Muñoz-Elías EJ, Upton AM, Cherian J, McKinney JD (2006) Role of the methylcitrate cycle in Mycobacterium tuberculosis metabolism, intracellular growth, and virulence. Mol Microbiol 60:1109–1122

Nachin L, Loiseau L, Expert D, Barras F (2003) SufC: an unorthodox cytoplasmic ABC/ATPase required for [Fe-S] biogenesis under oxidative stress. EMBO J 22:427–437

Negi A, Sharma R (2024) The significance of persisters in tuberculosis drug discovery: exploring the potential of targeting the glyoxylate shunt pathway. Eur J Med Chem 265:116058

Negishi M, Pedersen LG, Petrotchenko E, Shevtsov S, Gorokhov A, Kakuta Y, Pedersen LC (2001) Structure and function of sulfotransferases. Arch Biochem Biophys 390:149–157

Noy T, Vergnolle O, Hartman TE, Rhee KY, Jacobs WR Jr, Berney M, Blanchard JS (2016) Central role of pyruvate kinase in carbon co-catabolism of Mycobacterium tuberculosis. J Biol Chem 291:7060–7069

O'Hare HM, Durán R, Cerveñansky C, Bellinzoni M, Wehenkel AM, Pritsch O, Obal G, Baumgartner J, Vialaret J, Johnsson K (2008) Regulation of glutamate metabolism by protein kinases in mycobacteria. Mol Microbiol 70(6):1408–1423

O'Leary SE, Jurgenson CT, Ealick SE, Begley TP (2008) O-phospho-L-serine and the thiocarboxylated sulfur carrier protein CysO-COSH are substrates for CysM, a cysteine synthase from mycobacterium tuberculosis. Biochemistry 47 (44):11606–11615

Olaniyan MF, Muhibi MA, Olaniyan TB (2023) Oxidative stress and inflammatory response interplay. J Prevent Diagnost Treat Strateg Med 2(2):94–100

Pandey AK, Sassetti CM (2008) Mycobacterial persistence requires the utilization of host cholesterol. Proc Natl Acad Sci USA 105:4376–4380

Pao SS, Paulsen IT, Saier MH Jr (1998) Major facilitator superfamily. Microbiol Mol Biol Rev 62(1):1–34

Paritala H, Carroll K (2013) New targets and inhibitors of mycobacterial sulfur metabolism. Infect Disord Drug Targets 13:85–115

Parkin LA, Becker KL, Maceren JP, Palande A, Previti ML, Seeliger JC. 2025. Metabolic tagging reveals surface-associated lipoproteins in mycobacteria. bioRxiv

Parveen S, Shen J, Lun S, Zhao L, Alt J, Koleske B, Leone RD, Rais R, Powell JD, Murphy JR (2023) Glutamine metabolism inhibition has dual immunomodulatory and antibacterial activities against mycobacterium tuberculosis. Nat Commun 14(1):7427

Patil V, Jain V (2021) Understanding metabolic remodeling in Mycobacterium smegmatis to overcome energy exigency and reductive stress under energy-compromised state. Front Microbiol 12:722229

Perkowski EF, Miller BK, McCann JR, Sullivan JT, Malik S, Allen IC, Godfrey V, Hayden JD, Braunstein M (2016) An orphaned Mce-associated membrane protein of Mycobacterium tuberculosis is a virulence factor that stabilizes Mce transporters. Mol Microbiol 100:90–107

Petrilli JD, Müller I, Araújo LE, Cardoso TM, Carvalho LP, Barros BC, Teixeira M, Arruda S, Riley LW, Queiroz A (2020) Differential host pro-inflammatory response to mycobacterial cell wall lipids regulated by the Mce1 operon. Front Immunol 11:1848

Pham TV, Murkin AS, Moynihan MM, Harris L, Tyler PC, Shetty N, Sacchettini JC, Huang HL, Meek TD (2017) Mechanism-based inactivator of isocitrate lyases 1 and 2 from Mycobacterium tuberculosis. Proc Natl Acad Sci USA 114:7617–7622

Phong WY, Lin W, Rao SPS, Dick T, Alonso S, Pethe K (2013) Characterization of phosphofructokinase activity in Mycobacterium tuberculosis reveals that a functional glycolytic carbon flow is necessary to limit the accumulation of toxic metabolic intermediates under hypoxia. PLoS One 8:e56037

Pimentel-Schmitt EF, Thomae AW, Amon J, Klieber MA, Roth HM, Muller YA, Jahreis K, Burkovski A, Titgemeyer F (2007) A glucose kinase from Mycobacterium smegmatis. J Mol Microbiol Biotechnol 12:75–81

Pinto R, Tang QX, Britton WJ, Leyh TS, Triccas JA (2004) The mycobacterium tuberculosis cysD and cysNC genes form a stress-induced operon that encodes a tri-functional sulfate-activating complex. Microbiology 150(6):1681–1686

Plaitakis A, Kalef-Ezra E, Kotzamani D, Zaganas I, Spanaki C (2017) The glutamate dehydrogenase pathway and its roles in cell and tissue biology in health and disease. Biology 6(1):11

Planck KA, Rhee K (2021) Metabolomics of mycobacterium tuberculosis. In: Mycobacteria Protocols. Springer, pp 579–593

Poyraz Ö, Brunner K, Lohkamp B, Axelsson H, Hammarström LG, Schnell R, Schneider G (2015) Crystal structures of the kinase domain of the sulfate-activating complex in mycobacterium tuberculosis. PLoS One 10(3):e0121494

Qiang L, Wang J, Zhang Y, Ge P, Chai Q, Li B, Shi Y, Zhang L, Gao GF, Liu CH (2019) Mycobacterium tuberculosis Mce2E suppresses the macrophage innate immune response and promotes epithelial cell proliferation. Cell Mol Immunol 16:380–391

Qi N, She G-L, Du W, Ye B-C (2021) Mycobacterium smegmatis GlnR regulates the Glyoxylate cycle and the Methylcitrate cycle on fatty acid metabolism by repressing icl transcription. Front Microbiol 12:603835

Rai R, Mathew BJ, Chourasia R, Singh AK, Chaurasiya SK (2025) Glutamate decarboxylase confers acid tolerance and enhances survival of mycobacteria within macrophages. J Biol Chem 301(4):108338

Rathor N, Chandolia A, Saini NK, Sinha R, Pathak R, Garima K et al (2013) An insight into the regulation of mce4 operon of Mycobacterium tuberculosis. Tuberculosis 93:389–397

Read R, Pashley CA, Smith D, Parish T (2007) The role of GlnD in ammonia assimilation in Mycobacterium tuberculosis. Tuberculosis 87:384–390

Reddy VS, Shlykov MA, Castillo R, Sun EI, Saier MH Jr (2012) The major facilitator superfamily (MFS) revisited. FEBS J 279(11):2022–2035

Rhee SG (2001) Regulation of phosphoinositide-specific phospholipase C. Annu Rev Biochem 70:281–312

Rhee KY, de Carvalho LP, Bryk R, Ehrt S, Marrero J, Park SW, Schnappinger D, Venugopal A, Nathan C (2011) Central carbon metabolism in Mycobacterium tuberculosis: an unexpected frontier. Trends Microbiol 19:307–314

Rieck B, Degiacomi G, Zimmermann M, Cascioferro A, Boldrin F, Lazar-Adler NR, Bottrill AR, le Chevalier F, Frigui W, Bellinzoni M, Lisa M-N, Alzari PM, Nguyen L et al (2014) PknG senses amino acid availability to control metabolism and virulence of Mycobacterium tuberculosis. PLoS Pathog 10:e1006399

Rodriguez GM, Sharma N, Biswas A, Sharma N (2022) The iron response of mycobacterium tuberculosis and its implications for tuberculosis pathogenesis and novel therapeutics. Front Cell Infect Microbiol 12:876667

Rojas-Pirela M, Andrade-Alviárez D, Rojas V, Kemmerling U, Cáceres AJ, Michels PA et al (2020) Phosphoglycerate kinase: structural aspects and functions, with special emphasis on the enzyme from Kinetoplastea. Open Biol 10:200302

Ruecker N, Jansen R, Trujillo C, Puckett S, Jayachandran P, Piroli GG, Frizzell N, Molina H, Rhee KY, Ehrt S (2017) Fumarase deficiency causes protein and metabolite succination and intoxicates Mycobacterium tuberculosis. Cell Chem Biol 24:306–315

Saini NK, Sharma M, Chandolia A, Pasricha R, Brahmachari V, Bose M (2008) Characterization of Mce4A protein of Mycobacterium tuberculosis: role in invasion and survival. BMC Microbiol 8:200

Samanovic MI, Li H, Darwin KH (2013) The pup-proteasome system of mycobacterium tuberculosis. In: Regulated proteolysis in microorganisms, pp 267–295

Sambou T, Dinadayala P, Stadthagen G, Barilone N, Bordat Y, Constant P, Levillain F, Neyrolles O, Gicquel B, Lemassu A, Daffé M (2008) Capsular glucan and intracellular glycogen of Mycobacterium tuberculosis: biosynthesis and impact on the persistence in mice. Mol Microbiol 70:762–774

Samuels AN, Wang ER, Harrison GA, Valenta JC, Stallings CL (2022) Understanding the contribution of metabolism to Mycobacterium tuberculosis drug tolerance. Front Cell Infect Microbiol 12:958555

Sarkar B, Singh J, Yadav M, Sharma P, Sharma RD, Singh S, Chandramouli A, Mehdiratta K, Kumar A, Kamat SS, Ghorpade DS, Mohanty D, Kumar D, Gokhale RS (2025) PPARγ mediated enhanced lipid biogenesis fuels Mycobacterium tuberculosis growth in a drug-tolerant hepatocyte environment. elife 14:RP103817

Sassetti CM, Rubin EJ (2003) Genetic requirements for mycobacterial survival during infection. Proc Natl Acad Sci USA 100:12989–12994

Savvi S, Warner DF, Kana BD, McKinney JD, Mizrahi V, Dawes SS (2008) Functional characterization of a vitamin B12-dependent methylmalonyl pathway in Mycobacterium tuberculosis: implications for propionate metabolism during growth on fatty acids. J Bacteriol 190:3886–3895

Saxena P, Asthana AK, Madan M (2014) Comparison of microscopy and PCR in detection of mycobacterium tuberculosis. J Microbiol Infectious Disease 4(04):141–144

Schelle MW, Bertozzi CR (2006) Sulfate metabolism in mycobacteria. Chembiochem 7(10):1516–1524

Schelle B, Karl N, Ludewig B, Siddell SG, Thiel V (2006) Nucleocapsid protein expression facilitates coronavirus replication. In: The Nidoviruses: toward control of SARS and other Nidovirus diseases. Springer, pp 43–48

Schnell R, Schneider G (2010) Structural enzymology of Sulphur metabolism in Mycobacterium tuberculosis. Biochem Biophys Res Commun 396:33–38

Schnell R, Sandalova T, Hellman U, Lindqvist Y, Schneider G (2005) Siroheme-and [Fe4-S4]-dependent NirA from mycobacterium tuberculosis is a sulfite reductase with a covalent Cys-Tyr bond in the active site. J Biol Chem 280(29):27319–27328

Selim KA, Zimmer E, Yehia H, Doello S (2021) Molecular and cellular mechanisms underlying the microbial survival strategies: insights into temperature and nitrogen adaptations. In: Climate change and the microbiome: sustenance of the ecosphere. Springer, pp 717–748

Serafini A, Tan L, Horswell S, Howell S, Greenwood DJ, Hunt DM, Phan MD, Schembri M, Monteleone M, Montague CR, Britton W (2019) Mycobacterium tuberculosis requires glyoxylate shunt and reverse methylcitrate cycle for lactate and pyruvate metabolism. Mol Microbiol 112:1284–1307

Shetty ND, Reddy MC, Palaninathan SK, Owen JL, Sacchettini JC (2010) Crystal structures of the apo and ATP bound mycobacterium tuberculosis nitrogen regulatory PII protein. Protein Sci 19(8):1513–1524

Sheridan KJ, Lechner BE, Keeffe GO, Keller MA, Werner ER, Lindner H, Jones GW, Haas H, Doyle S (2016) Ergothioneine biosynthesis and functionality in the opportunistic fungal pathogen. Aspergillus Fumigatus Sci Rep 6(1):35306

Shi L, Sohaskey CD, Pfeiffer C, Datta P, Parks M, McFadden J et al (2010) Carbon flux rerouting during Mycobacterium tuberculosis growth arrest. Mol Microbiol 78:1199–1215

Shi J, Feng Z, Song Q, Wen A, Liu T, Xu L, Ye Z, Xu S, Gao F, Xiao L, Zhu J, Das K, Zhao G, Li J, Feng Y, Lin W (2025) Structural insights into transcription regulation of the global OmpR/PhoB family regulator PhoP from Mycobacterium tuberculosis. Nat Commun 16:1573

Simcox BS, Rohde KH (2024) Orphan response regulator Nnar is critical for nitrate and nitrite assimilation in mycobacterium abscessus. Front Cell Infect Microbiol 14:1411333

Sirakova TD, Dubey VS, Deb C, Daniel J, Korotkova TA, Abomoelak B, Kolattukudy PE (2006) Identification of a diacylglycerol acyltransferase gene involved in accumulation of triacylglycerol in Mycobacterium tuberculosis under stress. Microbiology 152:2717–2725

Snášel J, Pichová I (2019) Allosteric regulation of pyruvate kinase from Mycobacterium tuberculosis by metabolites. Biochimica et Biophysica Acta (BBA) Prot Proteom 1867:125–139

Snášel J, Machová I, Šolínová V, Kašička V, Krečmerová M, Pichová I (2021) Phosphofructokinases A and B from Mycobacterium tuberculosis display different catalytic properties and allosteric regulation. Int J Mol Sci 22:1483

Sohaskey CD, Wayne LG (2003) Role of narK2X and narGHJI inHypoxic upregulation of nitrate reduction by Mycobacterium tuberculosis. J Bacteriol 185(24):7247–7256

Sohaskey CD, Modesti L (2009) Differences in nitrate reduction between mycobacterium tuberculosis and Mycobacterium bovis are due to differential expression of both narGHJI and narK2. FEMS Microbiol Lett 290(2):129–134

Soni DK, Dubey SK, Bhatnagar R (2020) ATP-binding cassette (ABC) import systems of Mycobacterium tuberculosis: target for drug and vaccine development. Emerging Microbes Infect 9:207–220

Sritharan M (2016) Iron homeostasis in mycobacterium tuberculosis: mechanistic insights into siderophore-mediated iron uptake. J Bacteriol 198(18):2399–2409

Steiner DJ, Furuya Y, Metzger DW (2014) Host-pathogen interactions and immune evasion strategies in Francisella tularensis pathogenicity. Infect Drug Resist 7:239–251

Steiner-Rebrova EM, Böth D, Lössl P, Vilaplana F, Schnell R, Schneider G (2014) CysK2 from Mycobacterium tuberculosis is an O-phospho-L-serine-dependent S-sulfocysteine synthase. J Bacteriol 196:2979–2986

Sturm A, Sun P, Avila-Pacheco J, Clatworthy AE, Bloom-Ackermann Z, Wuo MG, Gomez JE, Jin S, Clish CB, Kiessling LL, Hung DT (2024) Genetic factors affecting storage and utilization of lipids during dormancy in Mycobacterium tuberculosis. MBio 15:e03208–e03223

Sutter M, Damberger FF, Imkamp F, Allain FH, Weber-Ban E (2010) Prokaryotic ubiquitin-like protein (pup) is coupled to substrates via the side chain of its C-terminal glutamate. J Am Chem Soc 132:5610–5612

Takahashi-Iñiguez T, García-Hernández E, Arreguín-Espinosa R, Flores ME (2012) Role of vitamin B12 on methylmalonyl-CoA mutase activity. J Zhejiang Univ Sci B 13:423–437

Takayama K, Wang C, Besra GS (2005) Pathway to synthesis and processing of mycolic acids in Mycobacterium tuberculosis. Clin Microbiol Rev 18:81–101

Tang S, Hicks ND, Cheng Y-S, Silva A, Fortune SM, Sacchettini JC (2019) Structural and functional insight into the mycobacterium tuberculosis protein PrpR reveals a novel type of transcription factor. Nucleic Acids Res 47(18):9934–9949

Tian J, Bryk R, Itoh M, Suematsu M, Nathan C (2005) Variant tricarboxylic acid cycle in Mycobacterium tuberculosis: identification of alpha-ketoglutarate decarboxylase. Proc Natl Acad Sci USA 102:10670–10675

Tovilla-Coutiño DB, Momany C, Eiteman MA (2020) Engineered citrate synthase alters acetate accumulation in Escherichia coli. Metab Eng 61:171–180

Tripathi A, Anand K, Das M, O'Niel RA, P S S, Thakur C, R L RR, Rajmani RS, Chandra N, Laxman S, Singh A. (2022) Mycobacterium tuberculosis requires SufT for Fe-S cluster maturation, metabolism, and survival in vivo. PLoS Pathog 18:e1010475

Tripathi SM, Ramachandran R (2008) Overexpression, purification, crystallization and preliminary X-ray analysis of Rv2780 from mycobacterium tuberculosis H37Rv. Struct Biology Crystal Commun 64(5):367–370

Trujillo C, Blumenthal A, Marrero J, Rhee KY, Schnappinger D, Ehrt S (2014) Triosephosphate isomerase is dispensable in vitro yet essential for mycobacterium tuberculosis to establish infection. MBio 5:e00085–e00014

Tullius MV, Harth G, Horwitz MA (2003) Glutamine synthetase GlnA1 is essential for growth of Mycobacterium tuberculosis in human THP-1 macrophages and Guinea pigs. Infect Immun 71:3927–3936

Upton AM, McKinney JD (2007) Role of the methylcitrate cycle in propionate metabolism and detoxification in mycobacterium smegmatis. Microbiology 153:3973–3982

Velayati AA, Mitaria S, Farnia P, Farnia P, Ghanavi J, Maleknia M, Murase Y, Hoffner S (2024) Exploring the genetic landscape of *Mycobacterium tuberculosis*: unlocking the differences between latent and active tuberculosis. Int J Mycobacteriol 13:410–419

Von Rosen T, Keller LM, Weber-Ban E (2021) Survival in hostile conditions: pupylation and the proteasome in actinobacterial stress response pathways. Front Mol Biosci 8:685757

Voskuil MI, Visconti KC, Schoolnik GK (2004) Mycobacterium tuberculosis gene expression during adaptation to stationary phase and low-oxygen dormancy. Tuberculosis 84:218–227

Wang J, Li B-X, Ge P-P, Li J, Wang Q, Gao GF, Qiu X-B, Liu CH (2015) Mycobacterium tuberculosis suppresses innate immunity by coopting the host ubiquitin system. Nat Immunol 16(3):237–245

Warner DF (2014) Mycobacterium tuberculosis metabolism. Cold Spring Harb Perspect Med 5:a021121

Watanabe M, Chiba Y, Hirai MY (2021) Metabolism and regulatory functions of O-acetylserine, Sadenosylmethionine, homocysteine, and serine in plant development and environmental responses. Front Plant Sci 12:643403

Watanabe S, Kita A, Miki K (2005) Crystal structure of atypical cytoplasmic ABC-ATPase SufC from Thermus thermophilus HB8. J Mol Biol 353:1043–1054

Watzer B, Spät P, Neumann N, Koch M, Sobotka R, Macek B, Hennrich O, Forchhammer K (2019) The signal transduction protein PII controls ammonium, nitrate and urea uptake in cyanobacteria. Front Microbiol 10:1428

Weber I, Fritz C, Ruttkowski S, Kreft A, Bange FC (2000) Anaerobic nitrate reductase (narGHJI) activity of Mycobacterium bovis BCG in vitro and its contribution to virulence in immunodeficient mice. Mol Microbiol 35:1017–1025

Wei L, Liu L, Gong W (2024) Structure of mycobacterial ergothioneine-biosynthesis C-S lyase EgtE. J Biol Chem 300:105539

Wheeler PR, Coldham NG, Keating L, Gordon SV, Wooff EE, Parish T, Hewinson RG (2005) Functional demonstration of reverse transsulfuration in the Mycobacterium tuberculosis complex reveals that methionine is the preferred sulfur source for pathogenic mycobacteria. J Biol Chem 280:8069–8078

Wilburn KM, Fieweger RA, VanderVen BC (2018) Cholesterol and fatty acids grease the wheels of Mycobacterium tuberculosis pathogenesis. Pathogens and Disease 76:fty021

Williams M, Mizrahi V, Kana BD (2014) Molybdenum cofactor: a key component of Mycobacterium tuberculosis pathogenesis? Crit Rev Microbiol 40:18–29

Willson BJ, Dalzell L, Chapman LN, Thomas GH (2019) Enhanced functionalisation of major facilitator superfamily transporters via fusion of C-terminal protein domains is both extensive and varied in bacteria. Microbiology 165(4):419–424

Wipperman MF, Sampson NS, Thomas ST (2014) Pathogen roid rage: cholesterol utilization by Mycobacterium tuberculosis. Crit Rev Biochem Mol Biol 49:269–293

Wolber JM, Urbanek BL, Meints LM, Piligian BF, Lopez-Casillas IC, Zochowski KM et al (2017) The trehalose-specific transporter LpqY-SugABC is required for antimicrobial and anti-biofilm activity of trehalose analogues in Mycobacterium smegmatis. Carbohydr Res 450:60–66

Wolfson-Stofko B, Hadi T, Blanchard JS (2013) Kinetic and mechanistic characterization of the glyceraldehyde 3-phosphate dehydrogenase from Mycobacterium tuberculosis. Arch Biochem Biophys 540:53–61

Xia H, van Den Hof S, Cobelens F, Zhou Y, Zhao B, Wang S, Zhao Y (2020) Value of pyrazinamide for composition of new treatment regimens for multidrug-resistant mycobacterium tuberculosis in China. BMC Infect Dis 20(1):19

Xu Y, Liu Y, Chen S, Li X, Xu L, Qi Y, Jiang D, Jin S (2014) The B subfamily of plant ATP binding cassette transporters and their roles in auxin transport. Biol Plant 58(3):401–410

Xu Y, Ma S, Huang Z, Wang L, Raza SHA, Wang Z (2023) Nitrogen metabolism in mycobacteria: the key genes and targeted antimicrobials. Front Microbiol 14:1149041

Yan N (2013) Structural advances for the major facilitator superfamily (MFS) transporters. Trends Biochem Sci 38:151–159

Yelamanchi SD, Surolia A (2021) Targeting amino acid metabolism of mycobacterium tuberculosis for developing inhibitors to curtail its survival. IUBMB Life 73(4):643–658

Zaychikova MV, Danilenko VN (2020) The Actinobacterial mce operon: structure and functions. Biol Bull Rev 10:520–525

Zhang F, Xie JP (2011) Mammalian cell entry gene family of Mycobacterium tuberculosis. Mol Cell Biochem 352:1–10

Zhang Y, Li J, Li B, Wang J, Liu CH (2018) Mycobacterium tuberculosis Mce3C promotes mycobacteria entry into macrophages through activation of β2 integrin-mediated signalling pathway. Cell Microbiol 20:e12800

Zhang L, Hendrickson RC, Meikle V, Lefkowitz EJ, Ioerger TR, Niederweis M (2020) Comprehensive analysis of iron utilization by mycobacterium tuberculosis. PLoS Pathog 16(2):e1008337

Exploring the Protein Landscape of Dormant *Mycobacterium tuberculosis* Through In Vitro Functional Studies

7

Abstract

The investigation of functional proteins during dormancy in *Mycobacterium tuberculosis* (Mtb) through in vitro experiments provides crucial insights into the bacterium's survival mechanisms and pathogenicity. Dormancy constitutes a pivotal phase in the Mtb life cycle, enabling the pathogen to persist within hostile host environments, evade immune defenses, and establish latent tuberculosis infections. Although dormancy is often conceptually divided into early, mid, and late stages, these phases frequently overlap, representing a continuum of physiological and molecular adaptations rather than distinct, isolated steps. During early dormancy, which typically occurs within the initial days to weeks following exposure to stressors such as hypoxia and nutrient limitation, Mtb initiates rapid transcriptional and proteomic reprogramming. This reprogramming involves the upregulation of genes associated with stress response, anaerobic metabolism, and cell wall remodeling. These molecular changes drive initial morphological adaptations, including cell wall thickening and a marked reduction in replication rates, signaling the transition from active growth to a nonreplicating persistent state. Mid-dormancy, spanning several weeks to months, is characterized by further metabolic downregulation coupled with enhanced resistance to environmental stresses. Proteins involved in energy conservation, detoxification processes, and maintenance of cellular integrity become increasingly prominent during this phase. Morphologically, Mtb bacilli undergo size reduction, exhibit altered cell division patterns such as budding, and display structural transformations including the folding of rod-shaped cells into ovoid forms. These adaptations collectively support the bacterium's long-term survival under sustained hostile conditions. Late dormancy, which may extend over months or even years, is defined by the stabilization of a metabolically quiescent state accompanied by profound physiological modifications. Morphological hallmarks of this stage include the formation of specialized spore-like cells and filterable nonacid-fast forms. Despite these altered physical characteristics, the cells remain metabolically active, enabling Mtb to withstand prolonged immune pressure and antibiotic exposure. This stage underlies the bacterium's capacity for latent infection and potential reactivation. While the tripartite framework of early, mid, and late dormancy provides a useful conceptual model, the transitions between these stages are naturally gradual and overlapping. Overlapping genetic and phenotypic changes underscore the dynamic and continuous nature of dormancy in Mtb. A comprehensive understanding of the

P. Farnia et al., *Proteins in Mycobacterium Tuberculosis*, Advances in Experimental Medicine and Biology 1484, https://doi.org/10.1007/978-3-031-96883-9_7

functional proteins and regulatory networks operative throughout these stages is essential not only for elucidating the mechanisms that sustain Mtb during latency but also for identifying novel therapeutic targets. Targeting these pathways holds promise for preventing reactivation and improving the efficacy of tuberculosis treatment.

Keywords

Mycobacterium tuberculosis · Dormancy · Molecular chaperones · Dormancy survival regulator (DosR) · Heat shock protein X · Sigma factor · WhiB-like (Wbl) family

7.1 Introduction

Dormancy in *Mycobacterium tuberculosis* (Mtb) refers to a specialized physiological state in which the bacilli markedly reduce their metabolic activity and cease replication. This adaptation enables Mtb to survive under harsh conditions such as hypoxia (low oxygen), nutrient deprivation, and immune system pressure. Dormancy is critical for the bacterium's persistence within the host, often resulting in latent tuberculosis infection, where the bacteria remain viable but metabolically inactive for prolonged periods (Mehrotra et al. 2014; Ehrt et al. 2018). During this dormant phase, Mtb cells exhibit reduced susceptibility to antibiotics that target actively dividing bacteria, complicating treatment and often necessitating extended therapy durations. The transition into dormancy is tightly regulated by genetic mechanisms that enable the bacterium to adjust to environmental stresses by modulating gene expression related to energy conservation, stress response, and detoxification of harmful compounds. Importantly, the dormant state is reversible; when environmental conditions improve, Mtb can reactivate, resume growth, and potentially cause active disease. This ability to switch between active and dormant states underlies the persistent challenge of tuberculosis eradication and underscores the need for therapies targeting both bacterial populations (Naran et al. 2018). In vitro models have been developed to simulate dormancy and study Mtb's metabolic and physiological adaptations during latent infection. These models replicate key host-pathogen interactions and stress conditions, facilitating research into drug tolerance, resuscitation mechanisms, and novel therapeutic strategies. One of the earliest and most widely used models is the nutrient starvation model, first described by Loebel et al. (1933). This model mimics the nutrient-deprived environment within granulomas by transferring Mtb cultures into phosphate-buffered saline. Another prominent in vitro model is the hypoxic Wayne model, which simulates gradual oxygen depletion to replicate the hypoxic conditions of granulomas (Wayne 1976; Wayne and Sramek 1979, 1994; Wayne and Lin 1982; Wayne and Hayes 1996). Under this model, Mtb enters distinct nonreplicating persistence phases (NRP1 and NRP2), characterized by markedly reduced metabolic activity and replication rates. These phases closely mimic the physiological state of Mtb during latent infection, providing valuable insights into the mechanisms governing dormancy and reactivation. Several genes, including Rv0251c, Rv0841c, Rv1874, Rv2332, Rv2660c, and Rv3290c (*lat*), are highly induced during the nonreplicating persistence phases of Mtb, reflecting their roles in adaptation to hypoxia and survival during dormancy (Muttucumaru et al. 2004; Bacon et al. 2014; Joshi et al. 2021). The gene Rv0251c encodes an alpha-crystallin heat shock protein involved in stress protection and is part of the SigG regulon, which contributes to bacterial persistence under stress conditions (Lee et al. 2008; Keren et al. 2011). Rv3290c (*lat*) functions as an L-lysine-epsilon aminotransferase, playing a key role in amino acid metabolism during dormancy. Rv2332 encodes a malate oxidoreductase involved in intermediary metabolism, supporting energy adaptation under hypoxia and starvation. Although some genes, such as Rv1874 and Rv2660c, encode hypothetical proteins with uncharacterized functions, their consistent upregulation across persistence models suggests involvement in dormancy maintenance. Rv0841c is implicated in stress response, though its precise role remains unclear (Joshi et al. 2021).

The enduring hypoxic response (EHR) model, which extends hypoxic exposure to seven days, identified approximately 230 genes induced during prolonged oxygen deprivation. Comparative analyses revealed that 47 of these genes overlap with those induced in the nutrient starvation model, and five overlap with those in the Wayne hypoxic model. Genes such as Rv0251c, Rv1805c, Rv1152, Rv2517c, and Rv3290c are consistently upregulated across all three models, underscoring their likely significance in Mtb persistence under diverse stress conditions. These findings highlight the central role of these genes in enabling Mtb to survive and adapt during dormancy, making them potential targets for therapeutic intervention (Murphy and Brown 2007; Rustad et al. 2008, 2009; Joshi et al. 2021). Numerous experimental studies have significantly advanced our understanding of Mtb drug persisters, subpopulations of bacteria capable of surviving prolonged antibiotic exposure without acquiring genetic resistance. Early foundational in vivo work by McCune and Trompsett (1956) and McCune et al. (1966) established that a fraction of nonreplicating Mtb bacilli could persist in mouse models despite extended antibiotic treatment, introducing the concept of bacterial persistence. Building on these findings, Hu et al. (2000) and McKinney et al. (2000) demonstrated through in vitro and macrophage infection models that Mtb can enter a viable but nonreplicating persistent state under various stress conditions, including antibiotic pressure. These studies revealed that Mtb maintains viability despite growth arrest, underscoring its capacity to evade antibiotic killing via phenotypic tolerance. A pivotal in vitro drug persister model was developed by Keren et al. (2011), who enriched antibiotic-tolerant persister cells by exposing actively growing Mtb cultures to D-cycloserine for 14 days. Transcriptomic analyses of these persisters identified 282 genes with increased expression and 1408 genes with decreased expression, with many downregulated genes associated with growth and energy metabolism. Importantly, five genes, Rv0251c, Rv1805c, Rv1152, Rv2517c, and Rv3290c, were consistently upregulated across multiple dormancy and persistence models, indicating their central role in mediating antibiotic tolerance (Keren et al. 2011; Torrey et al. 2016; Joshi et al. 2021). Furthermore, Zhang et al. (2012) reviewed and experimentally characterized the molecular mechanisms underlying persister formation, emphasizing the critical roles of metabolic remodeling, activation of the stringent response, and toxin-antitoxin systems in promoting phenotypic antibiotic tolerance. Transcriptomic analyses of nutrient-starved Mtb reveal extensive gene expression changes, including upregulation of stress response genes and downregulation of genes involved in energy metabolism, lipid biosynthesis, and virulence factors. These adaptations enable Mtb to endure hostile conditions and contribute to its long-term persistence during latent infection (Betts et al. 2002, 2012; Gengenbacher and Kaufmann 2012; Jamet et al. 2015). Notably, the ATP-binding cassette (ABC) transporter gene cluster responsible for enzymes essential in mycolic acid biosynthesis is downregulated during nutrient deprivation, suggesting its role in Mtb's adaptation and survival under starvation (Jamet et al. 2015). The granuloma model serves as an advanced ex vivo system designed to replicate the complex in vivo environment encountered by Mtb during infection. This model typically involves infecting peripheral blood mononuclear cells (PBMCs) ex vivo to simulate the cellular diversity and microenvironmental complexity characteristic of granulomas (Puissegur et al. 2004; Parasa et al. 2013; Kapoor et al. 2013; Guirado et al. 2015; Marakalala et al. 2016; Nangpal et al. 2025). Granulomas are highly organized structures composed of infected macrophages and activated lymphocytes that create a specialized niche marked by hypoxia, oxidative stress, nutrient scarcity, acidic pH, and antibiotic pressure. These combined stressors critically influence Mtb's adaptive responses and capacity for persistence within the host (Russell et al. 2010; Guirado et al. 2015). Utilizing this granuloma model, researchers have investigated the multifactorial survival strategies employed by Mtb under complex stress conditions. Transcriptomic analyses performed in these systems have revealed that Mtb upregulates genes involved in key metabolic and regulatory pathways. For instance, genes related to carbon metabolism, including *pfkB* (phosphofructokinase B), *sugA* (sugar transporter), and *glpX* (fructose-1,6-bisphosphatase) are induced, reflecting metabolic reprogramming to optimize the utilization of limited carbon sources available within the granuloma microenvironment (Kapoor et al. 2013). Concurrently, genes involved in lipid biosynthesis, such as *fas* (fatty acid synthase) and *accD3* (acetyl-CoA carboxylase subunit D3), show increased expression, underscoring the critical role of lipid metabolism and cell wall remodeling in sustaining bacterial persistence (Kapoor et al. 2013). Toxin-antitoxin system genes, including *vapC* and *mazF*, are expressed at elevated levels within the granuloma environment, suggesting their involvement in regulating dormancy and enhancing stress tolerance (Kapoor et al. 2013). These findings indicate that *Mycobacterium tuberculosis* (Mtb) orchestrates a coordinated response that integrates metabolic shifts and regulatory mechanisms to survive the hostile conditions imposed by granulomas (Guirado et al. 2015). The granuloma model thus provides a valuable experimental platform for identifying genes essential for Mtb persistence and offers promising targets for the development of novel therapeutic interventions (Gengenbacher and Kaufmann 2012; Kapoor et al. 2013; Guirado et al. 2015; Torrey et al. 2016; Joshi et al. 2021). The lysosomal in vitro exposure (LivE) model exposes Mtb to lysosomal extracts derived from activated macrophages, simulating intracellular stress conditions such as nitric oxide exposure and iron limitation (Vandal et al. 2009). This model is designed to provide a controlled environment that closely mimics the intracellular stresses Mtb encounters to establish persistence. By exposing the bacteria to actual lysosomal contents, the LivE model offers a more physiologically relevant system compared to models that simulate individual stress factors alone (Rohde et al. 2007; Bhat et al. 2018). Transcriptomic analyses under these conditions reveal significant changes in gene expression, including upregulation of lipid transporter genes (mmpL4, mmpL11) and polyketide synthase (pks12), alongside downregulation of cell wall synthesis genes such as *fbpA* and *fbpB* (Wright et al. 2017). These expression patterns indicate substantial remodeling of the cell envelope, a critical adaptation enabling Mtb to withstand the harsh lysosomal environment and persist intracellularly (Vandal et al. 2009; Joshi et al. 2021). The lipid-rich dormancy model combines hypoxic conditions with the presence of cholesterol and fatty acids to closely replicate the lipid-rich environment Mtb encounters within granulomas during latent infection (Pandey and Sassetti 2008; Daniel et al. 2011). Transcriptomic analyses under these conditions reveal a distinct gene expression profile marked by pronounced induction of genes involved in cholesterol and fatty acid catabolism, including *fadA5*, *fadE23*, and *hsaD*. This reflects Mtb's metabolic adaptation to exploit host-derived lipids as primary carbon and energy sources when conventional nutrients are scarce (Rodríguez et al. 2014; Aguilar-Ayala et al. 2017). Concurrently, genes within the hypoxia-responsive DosR regulon, such as Rv3133c and Rv0081, are upregulated, demonstrating a tightly coordinated regulatory response that integrates oxygen limitation with metabolic stress (Park et al. 2003; Daniel et al. 2011). Additionally, transcriptional regulators like whiB3 play a crucial role in maintaining redox balance and modulating lipid metabolism during dormancy, further supporting Mtb's survival in this hostile environment (Pacl et al. 2018).

These metabolic and regulatory adaptations enable Mtb to persist in a dormant state by efficiently utilizing host lipids and managing oxidative stress, highlighting potential targets for therapeutic intervention aimed at eradicating latent tuberculosis infection. Recent studies have expanded on these findings: Sturm et al. (2024) identified the lipid transporter gene omamC as essential for Mtb survival during carbon starvation and dormancy, demonstrating that disruption of lipid metabolism pathways can effectively eradicate nonreplicating bacilli. Similarly, Kim and Shin (2023) showed that Mtb accumulates intrabacterial lipid inclusions and upregulates triacylglycerol synthesis, promoting a drug-tolerant, slow-growing phenotype. These adaptations are accompanied by induction of reductive stress response genes and specific tRNAs, suggesting novel mechanisms involved in dormancy and antibiotic tolerance (Singha et al. 2024). These studies reveal a complex network of metabolic and regulatory pathways enabling Mtb to persist under hostile, nutrient-limited, and hypoxic conditions. Key genes such as Rv0251c, Rv3290c (*lat*), and members of the DosR regulon are consistently identified across lipid-rich and other dormancy models, highlighting their potential as promising targets for novel therapeutics aimed at eradicating persistent tuberculosis infections (Chen et al. 2013). Among the various experimental approaches to studying Mtb dormancy, the Wayne hypoxia model (Wayne 1976; 1977) has gained significant attention due to its ability to closely replicate the gradual oxygen depletion that Mtb experiences within granulomas. In this model, Mtb cultures are sealed in tubes where oxygen is progressively consumed by bacterial respiration, inducing the bacteria to enter two distinct nonreplicating persistence (NRP) stages. The first stage, NRP-1, occurs approximately between days 7 and 10 and represents an early dormancy phase characterized by slowed bacterial growth while maintaining viability. The second stage, NRP-2, develops between days 14 and 20 and reflects a deeper dormant state marked by metabolic shutdown and increased tolerance to antibiotics (Wayne 1976, 1994; Wayne and Lin 1982; Wayne and Hayes 1996). Variations in this model have been instrumental in identifying Mtb genes that contribute to the establishment and maintenance of latency (Imboden and Schoolnik 1998; Yuan et al. 1998; Hu and Coates 1999). One of the key genes induced during hypoxia is *acr* (also known as hspX, Rv2031), which encodes the alpha-crystallin protein. This protein is a dominant antigen in vivo and is recognized by the majority of tuberculosis patient sera, underscoring its immunological significance (Lee et al. 1992; Verbon et al. 1992). Another critical hypoxia-induced gene is Rv3133c (DosR), a putative transcription factor that orchestrates the expression of a regulon essential for Mtb's adaptation to oxygen limitation (Park et al. 2003). Together, these discoveries have provided a mechanistic framework elucidating how Mtb persists within the hypoxic environment of granulomas. Building upon this foundational work, Velayati and Farnia employed atomic force microscopy to examine Mtb bacilli subjected to prolonged latency, lasting up to 48 months, using a modified Wayne model (Velayati et al. 2011, 2016; Velayati and Farnia 2016). These investigations revealed profound morphological changes in dormant Mtb, including cell wall thickening, reduction in cell size, and the emergence of spore-like and nonacid-fast forms (Velayati et al. 2011, 2016; Velayati and Farnia 2016). Importantly, these dormant bacilli retained viability and were capable of reactivation under favorable conditions, providing direct evidence of actual long-term persistence beyond the typical durations studied in vitro. More recently, Velayati et al. (2024) further advanced this understanding by exploring the genetic adaptations of latent Mtb bacilli subjected to combined stressors of oxygen, nitrogen, and nutrient starvation, shedding light on the complex molecular strategies employed during extended dormancy. Whole-genome sequencing of clinical isolates subjected to a double-stress dormancy model combining oxygen and nutrient deprivation over 12 months identified frameshift mutations in genes such as *fbiC*, *PPE35*, and *ald*, along with upstream regulatory mutations in *whiB6* (Velayati et al. 2024). These genetic changes affect pathways critical for redox balance, cell wall integrity, and stress response, providing a molecular basis for Mtb's remarkable ability to persist and adapt during latency. These studies collectively underscore the utility of the Wayne hypoxia model and its derivatives in unraveling the physiological and genetic basis of Mtb dormancy, offering critical insights into the pathogen's ability to survive hostile host environments and evade eradication. Proteins induced early in Mtb dormancy, such as DosR, HspX (Rv2031c), Rv2626c, and the acyl carrier protein Rv0100, continue to be expressed throughout late dormancy. These proteins play critical roles in maintaining stress resistance, conserving energy, and preserving cell envelope integrity during prolonged dormancy (Hudock et al. 2017; Kundu and Basu 2021; Wang et al. 2021; Gutka et al. 2024). In contrast, proteins like Rv1738, which are minimally expressed during early dormancy, become highly abundant in late dormancy. Rv1738 promotes ribosome hibernation and translational arrest, thereby intensifying the metabolic shutdown essential for long-term persistence (Bunker et al. 2015). Additionally, universal stress proteins and toxin-antitoxin systems show increased expression during late dormancy, reinforcing growth arrest and enhancing antibiotic tolerance (Peddireddy et al. 2017; Zhang et al. 2022). This chapter organizes the physiological, genomic, and proteomic changes associated with Mtb dormancy into three distinct stages: early, mid, and late dormancy, providing a clearer framework for understanding how tuberculosis bacilli adapt and persist under hostile conditions. This staged approach reflects the sequential metabolic slowdown, gene regulation, and protein expression patterns observed in both experimental models and clinical studies. It facilitates a comprehensive understanding of the mechanisms by which Mtb maintains long-term viability, develops antibiotic tolerance, and retains the potential for reactivation.

7.2 Complex Adaptations of *Mycobacterium tuberculosis* from Infection to Long-Term Persistence

Mycobacterium tuberculosis (Mtb) exhibits an extraordinary ability to persist within the human host for extended periods without causing active disease, a condition known as latent tuberculosis (LTB). This phenomenon was first recognized nearly a century ago (Opie and Aronson 1927; Feldman and Baggenstoss 1938) and has since been extensively studied to elucidate how Mtb evades immune clearance and antimicrobial therapy. The concept of "latency" in tuberculosis encompasses two interrelated but distinct phenomena, both involving Mtb bacilli entering a dormant state under different circumstances. The classical definition of latency refers to the state established shortly after initial infection. Following inhalation, Mtb bacilli reach the lungs and are phagocytosed by macrophages (Vynnycky and Fine 1997; Magombedze et al. 2013; Getahun et al. 2015; Boom et al. 2021; Yang et al. 2023; Behr et al. 2024). The host immune response leads to the formation of granulomas, organized aggregates of infected macrophages and activated lymphocytes that contain the bacilli and prevent their replication and dissemination. Within this microenvironment, Mtb enters a dormant state characterized by markedly reduced metabolic activity. Individuals with latent tuberculosis infection (LTBI) are asymptomatic and noncontagious, typically testing positive on immune sensitization assays such as the tuberculin skin test (TST) or interferon-gamma release assays (IGRA), while sputum smears and cultures remain negative (Flynn and Chan 2022; Metcalfe et al. 2011; Trajman et al. 2013; Wlodarczyk et al. 2014). Epidemiological data estimate that approximately one-quarter to one-third of the global population, around 2 to 2.5 billion people, harbor latent infection, representing a substantial reservoir for potential reactivation. Without intervention, 5–15% of these individuals will develop active tuberculosis during their lifetime, with risk markedly increased among immunocompromised populations or those with comorbidities such as HIV, diabetes, or malnutrition (WHO 2024). A second context for latency involves bacilli that enter a dormant or nonreplicating persistent state in response to antituberculosis chemotherapy. Since most anti-TB drugs preferentially target actively dividing bacteria, specific subpopulations of Mtb respond to drug pressure by slowing or halting replication and reducing metabolic activity, thereby adopting a "persister" phenotype (McDermott and Tompsett 1954; Mitchison 1954, 1985, 1998; Zhang et al. 2012). These dormant bacilli exhibit phenotypic drug tolerance—not due to genetic resistance but as a consequence of their low metabolic state, which diminishes antibiotic efficacy. This form of dormancy is observed in experimental models such as the Cornell model of latency and in clinical settings where bacilli survive despite prolonged chemotherapy, contributing to treatment failure or relapse (Andersson and Levin 1999; Cardona and Ruiz-Manzano 2004; Andersson and Hughes 2011; Gengenbacher and Kaufmann 2012). Notably, these bacilli are often undetectable by standard culture methods unless resuscitated, underscoring the necessity for extended TB treatment regimens to achieve complete eradication and prevent recurrence. Thus, the first form of latency arises naturally during initial infection when the immune system successfully contains the bacteria, resulting in a clinically silent state with dormant bacilli residing within granulomas. The second form develops as a survival strategy during or after drug treatment, wherein bacilli enter a nonreplicating, drug-tolerant state that complicates therapeutic efforts (Manabe and Bishai 2000; Magombedze et al. 2013; Grant and Hung 2013; Campaniço et al. 2020). Both forms are critical to understanding TB pathogenesis and control: the former constitutes the large global reservoir fueling new cases, while the latter explains persistent challenges related to treatment duration and relapse prevention. Recognizing these distinctions is essential for developing improved diagnostic tools, therapeutic strategies, and public health interventions aimed at ultimately eliminating tuberculosis. The notion of latent TB has evolved through foundational studies in pathology and bacteriology, progressively refining our understanding of Mtb's complex interactions with the host and its remarkable capacity for persistence. Initial observations revealed that Mtb can persist in a dormant state within host tissues, often remaining undetectable by conventional culture techniques but visible microscopically as acid-fast bacilli (AFB) (Beck and Yegian 1952; Medlar et al. 1952). These "viable but nonculturable" bacilli raised important questions regarding the true nature of bacterial viability during latency. The landmark Cornell model, developed in the 1950s by McDermott and colleagues, provided the first experimental evidence of latent bacilli in vivo (McCune and Tompsett 1956; McCune et al. 1966). In this model, mice infected intravenously with approximately 10^5 colony-forming units (CFU) of Mtb were treated with isoniazid and pyrazinamide for 12 weeks. Following cessation of drug therapy, no culturable bacilli were detected for 4–6 weeks, defining a "sterile" latent state. However, culturable bacilli reemerged in about 30% of mice after antibiotic withdrawal, demonstrating the presence of latent infection. Immunosuppression with corticosteroids increased relapse rates to as high as 83%, highlighting the critical role of host immunity in controlling latent infection. Clinical studies have also shown that rifampicin outperforms rifapentine in symptom relief and lesion absorption, although it is associated with higher toxicity (Waksman 1964; Mitchison 1979, 1985). Mitchison and Dickinson (1978) further advanced the field by proposing that tuberculosis lesions harbor heterogeneous bacillary populations with distinct metabolic activities and drug susceptibilities. These include (I) rapidly replicating bacilli, highly sensitive to most anti-TB drugs; (II) slow or intermittently growing bacilli, susceptible to pyrazinamide or rifampicin but resistant to isoniazid; and (III) latent bacilli, which are metabolically inactive and phenotypically tolerant to chemotherapy (Mitchison 1979; Grosset 1980; Fox et al. 1999; Cardona and Ruiz-Manzano 2004). This physiological and metabolic heterogeneity accounts for the difficulty in sterilizing tuberculosis infections and the prolonged duration of TB therapy. Pathological studies suggest that the microenvironments within granulomas, characterized by hypoxia, nutrient deprivation, and acidic pH, drive Mtb into nonreplicating persistence and confer drug tolerance (Wayne and Sohaskey 2001; Gomez and McKinney 2004; Barry et al. 2009). The Wayne hypoxia model was pivotal in elucidating the metabolic adaptations that occur during dormancy. In this model, gradual oxygen depletion in sealed cultures induces Mtb to shift its metabolism toward fermentative pathways, increasing the activity of enzymes such as isocitrate lyase and glycine dehydrogenase to maintain redox balance and energy production under low-oxygen conditions. Kaprelyants and Kell (1993, 2003) described three physiological states of Mtb in vitro: (I) viable, culturable bacilli capable of division; (II) dormant bacilli with low metabolic activity, unable to divide without resuscitation;

and (III) nonviable, nonculturable bacilli incapable of replication. These states likely mirror the dynamic spectrum of bacterial physiology in vivo, although direct extrapolation remains complex due to the intricacies of the host environment. Historical and contemporary studies have documented the presence of nonacid-fast, cell-wall-deficient variants of Mtb resembling L-forms described in other bacteria (Mattman 1970, 2000; Khomenko 1987; Chandrasekhar and Ratnam 1992; Smith 2003). These forms may arise under antibiotic pressure or other hostile host conditions, representing a survival strategy during dormancy. Drawing from current review literature, proteins associated with Mtb dormancy are classified into three distinct stages: early, middle, and late dormancy phases. This framework reflects the dynamic physiological and molecular adaptations the bacterium undergoes as it transitions from active replication to a long-term persistent state under hostile conditions such as hypoxia, nutrient limitation, and immune pressure. During the early dormancy stage, typically occurring within days to weeks following stress exposure, Mtb rapidly activates the dormancy survival regulator regulon DosR (DevR) (Voskuil et al. 2003; Magombedze et al. 2013). Key proteins induced during the early dormancy phase of Mtb include the transcription factor DosR and the sensor kinases DosS and DosT, which detect environmental signals. Highly upregulated effectors such as the molecular chaperone HspX (Acr, Rv2031c), the translation inhibitor DATIN (Rv0079), and latency-associated proteins Rv2623 and Rv2626c contribute to redox balance and immune modulation (Kumar et al. 2007; Devasundaram et al. 2016; Glass et al. 2017). These proteins facilitate metabolic downshift, energy conservation, and protection against oxidative and reactive nitrogen species (RNS) stress. Additionally, sigma factors such as *SigE* and *SigH* are upregulated early, coordinating the expression of stress response genes (Raman et al. 2001). Molecular chaperones DnaK and GroEL2 increase in abundance to maintain protein homeostasis during this phase (Rustad et al. 2008; Mogk et al. 1997; Xiao et al. 2024). The mid-dormancy stage corresponds to a deeper nonreplicating persistence phase, often modeled in vitro by prolonged hypoxia lasting weeks to months. Proteins involved in lipid metabolism and cell wall remodeling, such as Rv0100 (an acyl carrier protein), and enzymes supporting anaerobic respiration become more prominent (Hampshire et al. 2004). Regulatory proteins like the MprAB two-component system gain increased activity, integrating signals from the cell envelope and redox stress to sustain dormancy (He et al. 2006; Pang and Howard 2007). Members of the WhiB family of proteins—including WhiB3, WhiB5, and WhiB7—play critical roles in maintaining redox balance, antibiotic tolerance, and metabolic adaptation during this phase (Burian et al. 2012, 2013; Smith et al. 2012a; Lilic et al. 2021; Miotto et al. 2022; Guiza Beltran et al. 2024; Gutka et al. 2024). In the late dormancy stage, which may persist for months to years, Mtb maintains minimal basal metabolic activity to ensure long-term survival. Genes involved in basal metabolism, such as those encoding isocitrate lyase (*icl*), superoxide dismutase (*sodC*), and persistent sigma factors (*sigA*, *sigF*, *sigH*), continue to be expressed to preserve cellular integrity and readiness for resuscitation. Some latency-associated antigens remain expressed to modulate host immunity (McKinney et al. 2000; Sharma et al. 2000; Gengenbacher and Kaufmann 2012; Lee et al. 2015; Iona et al. 2016; Verma et al. 2022). Proteins such as HspX, DnaK, GroEL2, DosR, and sigma factors are expressed throughout all dormancy stages, reflecting their essential roles in survival and persistence (Joshi et al. 2021). The continued presence of these proteins enables Mtb to withstand prolonged stress and rapidly reactivate when environmental conditions become favorable (Nambi et al. 2015; Trutneva et al. 2020; Shleeva et al. 2020). Early dormancy is thus characterized by the rapid induction of stress response and metabolic downshift proteins; mid-dormancy involves sustained metabolic adaptation and redox regulation; and late dormancy maintains basal functions necessary for long-term survival. This three-stage model provides a comprehensive understanding of the temporal regulation and functional roles of various proteins that facilitate Mtb's transition from exponential growth to a persistent, nonreplicating state. Alongside these proteomic and genomic adaptations, morphological changes—including alterations in cell size, shape, cell wall composition, and acid-fastness—reflect the bacterium's dynamic physiological adjustments to hostile environments such as hypoxia, nutrient limitation, and immune pressure.

7.3 Environmental Sensing and Early Signaling in Dormancy Initiation of *Mycobacterium tuberculosis*

Mycobacterium tuberculosis (Mtb) activates several critical metabolic pathways during early dormancy to adapt effectively to hypoxic (low oxygen) and nutrient-limited conditions while conserving energy. A hallmark of this adaptation is the metabolic shift from aerobic respiration to a fermentative-like state, prominently involving the glyoxylate shunt and a reversed tricarboxylic acid (TCA) cycle. These pathways enable Mtb to bypass carbon loss and facilitate the accumulation and secretion of succinate, which plays a vital role in maintaining redox balance and sustaining the proton motive force under low-oxygen conditions (Watanabe et al. 2011; Eoh and Rhee 2014; Borah et al. 2021). Concurrently, fatty acid β-oxidation is upregulated, allowing the bacterium to degrade host-derived lipids and utilize them as essential carbon sources for survival during both infection and dormancy. Complementing this, the synthesis and accumulation of triacylglycerol serve as energy reserves and contribute to remodeling the bacterial cell envelope. These metabolic adaptations have been consistently demonstrated through transcriptomic, metabolomic, and lipidomic analyses, which reveal increased expression of genes and pathways involved in fatty acid degradation and storage under stress conditions such as hypoxia and nutrient limitation (Daniel et al. 2011; Rodríguez et al. 2014; Raghunandanan et al. 2019). Global systems biology studies underscore that β-oxidation is indispensable for Mtb's persistence within the host, while triacylglycerol accumulation supports survival during dormancy and environmental stress. In addition to these metabolic shifts, Mtb tightly regulates its intracellular redox state, notably modulating the ratio of reduced nicotinamide adenine dinucleotide (NADH) to oxidized nicotinamide adenine dinucleotide (NAD^+) to preserve cellular homeostasis. This metabolic reprogramming underpins a low-energy, nonreplicating state that enhances bacterial survival and persistence during dormancy. Investigations employing genetically encoded fluorescent biosensors have revealed that inhibition of the electron transport chain (ETC), disruption of membrane potential, and exposure to antimycobacterial drugs lead to the accumulation of NADH and elevated NADH/NAD^+ ratios within Mtb, particularly inside macrophages. These findings indicate metabolic stress and adaptive responses to hostile intracellular environments (Vilchèze et al. 2010; Hung et al. 2011; Zhao et al. 2011; Bilan et al. 2014; Boshoff et al. 2008; Bhat et al. 2016). Maintaining NAD(H) homeostasis is therefore critical for Mtb's survival, as perturbations in the NADH/NAD^+ balance are closely linked to the bacterium's capacity to enter and sustain dormancy (Watanabe et al. 2016). Wayne and Hayes (1996) developed an in vitro hypoxia model that closely mimics the oxygen-limited environment found within granulomas, demonstrating that Mtb enters a nonreplicating, drug-tolerant state under low oxygen tension. This dormant state is accompanied by the induction of specific dormancy-associated genes. These foundational studies laid the groundwork for understanding how Mtb persists within the host and evades immune clearance, highlighting the critical role of stress response mechanisms and specialized proteins during latency. Extensive investigations into the rapid production of key proteins during the early stages of Mtb dormancy have revealed that several proteins are specifically induced within days to weeks following exposure to dormancy-inducing stressors such as hypoxia and nitric oxide. As oxygen becomes limited, Mtb shifts from aerobic respiration to glycolytic and anaerobic metabolic pathways to generate energy under hypoxic conditions. This metabolic transition is tightly regulated by the dormancy survival regulator DosR (Sherman et al. 2001; Voskuil et al. 2003; Roberts et al. 2004; Rustad et al. 2008; Leistikow et al. 2010; Salina et al. 2022). Activation of DosR is mediated by two heme-containing sensor histidine kinases, DosS (Rv3132c) and DosT (Rv2027c), which detect environmental cues by binding gases such as oxygen, nitric oxide (NO), and carbon monoxide (CO) through their heme prosthetic groups (Cho et al. 2009). Both kinases share several structural domains: a histidine kinase domain and an ATP-binding domain located in the C-terminal region, and two tandem GAF (cyclic guanosine monophosphate [cGMP], adenylyl cyclase, *FhlA*) domains designated GAF-A and GAF-B in the N-terminal region. GAF domains are regulatory modules capable of binding small molecules and are broadly distributed among prokaryotes and eukaryotes (Martinez et al. 2002). The three-dimensional structures of the GAF-A and GAF-B domains have been elucidated, revealing that a b-type heme is embedded within the GAF-A domain, which consists of a five-strand antiparallel β-sheet and four α-helices. The heme is oriented nearly perpendicular to the β-sheet (Sardiwal et al. 2005; Ioanoviciu et al. 2007; Podust et al. 2008). Histidine residues H149 and H147 act as the proximal axial ligands for DosS and DosT, respectively (Sardiwal et al. 2005). The kinase activity of DosS and DosT is regulated by the binding state of the distal ligand and the redox state of the heme iron: when the heme iron is in the ferric (Fe^{3+}) form, the kinase remains inactive; reduction in the iron to the ferrous (Fe^{2+}) state or binding of NO or CO restores kinase activity. Flavin nucleotides have been reported to facilitate the reduction in DosS heme iron. The heme iron of DosT can resist oxidation from Fe^{2+} to Fe^{3+} in the presence of oxygen (Honaker et al. 2009), suggesting that DosT plays a more prominent role than DosS during the early phase of hypoxia, particularly during the transition from aerobic to hypoxic conditions. DosT functions primarily as an acute sensor that rapidly detects the onset of hypoxia as well as the presence of nitric oxide (NO) or carbon monoxide (CO). Under normal oxygen conditions, the heme group of DosT exists in the oxy form (Fe^{2+}-O_2), rendering it inactive. However, as oxygen tension decreases, the heme iron converts to the deoxy form (Fe^{2+}), enabling DosT to sense and respond promptly to hypoxic signals. DosT has an oxygen binding affinity (dissociation constant, Kd) of approximately 3.3 micromolar (μM), corresponding to oxygen levels at the onset of the nonreplicating persistence phase 1 (NRP-1), highlighting its critical role in early adaptation to hypoxic conditions (Sousa et al. 2007). This high sensitivity makes DosT more responsive to initial decreases in oxygen tension compared to DosS, allowing DosT to act first when Mtb transitions from aerobic to anaerobic conditions (Kim et al. 2010). In contrast, DosS functions as a chronic sensor that sustains dormancy signaling during prolonged hypoxia, thereby maintaining bacterial adaptation to long-term low-oxygen environments (Kumar et al. 2008). The heme group of DosS acts as both a redox and oxygen sensor; under hypoxic or reducing conditions, DosS transitions from the inactive ferric form (Fe^{3+}) to the active ferrous form (Fe^{2+}), initiating downstream signaling (Chao et al. 2015). DosS and DosT are components of a broader two-component signal transduction system (TCSS), a common bacterial mechanism for environmental adaptation. In this system, the sensor kinases DosS and DosT are membrane-associated, while the response regulator DosR (dormancy survival regulator, also known as DevR) is located in the cytoplasm (Rustad et al. 2008; Sivaramakrishnan 2013; Chauhan et al. 2011; Liu et al. 2024). Following autophosphorylation, DosS and DosT transfer the phosphate group to DosR. Phosphorylation induces a conformational change in DosR, enabling it to bind promoter regions upstream of DosR regulon genes. This binding initiates transcription of approximately 48 to 50 genes, many encoding proteins essential for metabolic adaptation, stress response, and persistence during dormancy or other hostile conditions (Wisedchaisri et al. 2008; Gautam et al. 2019; Sharma and Tyagi 2016; Apiche et al. 2024). The DosR regulon, also known as the dormancy survival regulator regulon, plays a central role in Mtb's adaptation to nonreplicating persistence (NRP), a physiological state widely believed to underlie latent tuberculosis infection in humans. Activation of the DosR regulon shifts Mtb's metabolism away from aerobic respiration, enabling the bacterium to survive under anaerobic or microaerophilic (low oxygen) conditions by maintaining essential energy levels and redox balance (Leistikow et al. 2010; Apiche et al. 2024). Functional studies have established that this regulon is critical for Mtb's survival during anaerobic dormancy and for rapid resumption of growth when favorable conditions return. Mutants deficient in DosR regulon induction fail to persist in hypoxic lesions within animal models such as guinea pigs and rabbits, which develop granulomas resembling those in human tuberculosis. In mouse infection models, where granulomas are typically non-hypoxic, DosR mutants do not exhibit attenuated virulence, highlighting the regulon's specific role in adaptation to hypoxic and anaerobic conditions observed in more physiologically relevant models such as guinea pigs and nonhuman primates (Aly et al. 2006; Via et al. 2008; Mehra et al. 2015). The genes within the DosR regulon encode a diverse set of proteins, including metabolic enzymes, stress response factors, and proteins involved in immune evasion and translation inhibition. Key members include Rv0079, which encodes the dormancy-associated translation inhibitor (DATIN) that suppresses protein synthesis; Rv2031c (also known as hspX), encoding the α-crystallin heat shock protein that stabilizes proteins under hypoxic stress; and Rv2623, an ATP-binding protein involved in regulating bacterial growth and persistence (Rustad et al. 2008; Leistikow et al. 2010; Kumar et al. 2012; Rodríguez et al. 2014; Peddireddy et al. 2017; Liu et al. 2024). Additional important genes include Rv1733c, encoding a hypoxia-induced protein involved in adaptation to low oxygen; Rv2029c, encoding 6-phosphofructokinase, a key enzyme for metabolic remodeling during hypoxia; Rv2626c, a DosR regulon protein contributing to dormancy and immune modulation; Rv1998c, a transcriptional regulator that aids in redox homeostasis under stress; Rv2032, encoding a citrate lyase subunit important for survival during nutrient deprivation; and Rv3134c, a stress response protein involved in persistence under hostile conditions (Bagchi et al. 2005; Muñoz-Elías et al. 2006; Kapoor et al. 2015).

The genes of the DosR regulon are organized into several genomic clusters within the Mtb genome, including Rv0079-Rv0081, Rv1733c-Rv1738, Rv2028c-Rv2032, Rv2623-Rv2631, and Rv3126c-Rv3134c. These clusters exhibit co-regulation and functional grouping, reflecting a coordinated bacterial response to environmental stimuli and underscoring their essential roles in survival during latent infection. The DosR regulon is specifically activated under hypoxic and other stress conditions to facilitate bacterial persistence and dormancy, ensuring Mtb's long-term survival within the host (Roupie et al. 2007; Gerasimova et al. 2011; Kumar et al. 2012; Voskuil and Schlesinger 2015). Clinically, the DosR regulon holds significant relevance in tuberculosis infection. Its antigens are specifically recognized by T cells in individuals with latent tuberculosis infection, indicating an active immune response against these proteins during latency. Moreover, expression of the DosR regulon has been detected in human sputum samples from infected patients and various animal models of tuberculosis, providing strong evidence for its involvement in bacterial latency and persistence within the host (Rustad et al. 2008; Honaker et al. 2009; Leistikow et al. 2010; Mehra et al. 2015). Due to their immunogenic properties during latent infection, proteins encoded by the DosR regulon are actively investigated as potential vaccine candidates targeting latent tuberculosis (Singh et al. 2014). Beyond promoting Mtb survival under hostile, hypoxic conditions, activation of the DosR regulon also modulates the host immune response, contributing to the pathogen's remarkable ability to persist for extended periods, often years, within the human host (Sherman et al. 2001; Voskuil et al. 2003; Zheng and Abramovitch 2020). Although some studies indicate that transcription of the DosR regulon gradually diminishes following the initial induction phase and is not sustained indefinitely during prolonged dormancy, the proteins produced early remain critical for survival throughout the mid and late stages of dormancy. This suggests that continuous transcriptional expression of the regulon is unnecessary once these proteins are synthesized and functional. Rustad et al. (2008, 2009) investigated the role of the DosR regulon using an Mtb mutant (DorKO) lacking the entire three-gene operon spanning Rv3134c-Rv3133c-dosR-Rv3132c-dosS. Under hypoxic conditions modeled by the Wayne hypoxia system and the defined hypoxia resazurin assay for dormancy (RAD) model, the DorKO mutant exhibited a marked survival defect compared to wild-type and complemented strains. While wild-type Mtb maintained viability for over 200 days under anaerobic dormancy, the DorKO mutant rapidly lost culturability on solid media within weeks, although some viability was initially retained in liquid culture. These results underscore the essential role of the DosR regulon in facilitating long-term survival and persistence of Mtb under oxygen-limited conditions that mimic the host environment during latent infection. Transcriptomic analyses reveal that the DosR regulon is strongly induced early during hypoxia but declines significantly during prolonged dormancy. This pattern indicates that sustained transcription is not required for long-term survival once essential proteins have been produced. Supporting this, measurements of oxygen consumption demonstrate that the DosR regulon modulates bacterial metabolism by slowing respiration, a critical adaptation facilitating Mtb's transition into and maintenance of a nonreplicating persistent (NRP) state. These findings establish the DosR regulon as a preparatory response, producing key proteins early to enable Mtb to endure hypoxic, nonrespiring states even after gene expression subsides (Rustad et al. 2008, 2009). This pre-adaptive strategy allows Mtb to synthesize DosR regulon proteins while energy availability is still sufficient, equipping the bacterium to survive extended periods of anaerobiosis or other stresses without requiring continuous regulon activation. The presence of these proteins during latent infection is corroborated by immune responses detected in latently infected individuals (Rustad et al. 2009). Consequently, only a subset of DosR regulon-encoded proteins may remain functionally active throughout dormancy, while the regulon's gene expression is transient and primarily important during early adaptation. Thus, the DosR regulon is transiently expressed upon entry into dormancy, with many proteins produced early persisting and functioning during prolonged stress, but continuous activation of all regulon genes is not maintained throughout the entire dormant period. Proteins encoded by the DosR regulon exhibit structural diversity, including small molecular chaperones, enzymes, and regulatory proteins. A key protein is Heat Shock Protein X (HspX), also known as alpha-crystallin and encoded by Rv2031c. HspX is a small heat shock protein chaperone that is upregulated over 1000-fold within 24–48 hours of hypoxic exposure. It functions to stabilize the proteome under hypoxic stress by preventing protein aggregation, thereby preserving protein function during dormancy (Yuan et al. 1996; Chang et al. 1996; Cunningham and Spreadbury 1998; Sherman et al. 2001; Kennaway et al. 2005; Mushtaq et al. 2015). Another important protein is Rv0079, known as Dormancy Associated Translation Inhibitor (DATIN). DATIN interacts with the 30S ribosomal subunit to suppress protein synthesis, conserving energy during dormancy by reducing metabolic demand (Kumar et al. 2012, 2013). Computational modeling and docking studies have confirmed DATIN's binding to ribosomal subunits, while in vitro translation assays demonstrated its inhibitory effect on protein synthesis. Beyond translation inhibition, DATIN stimulates immune responses by engaging Toll-like receptor 2 (TLR2), inducing proinflammatory cytokines such as tumor necrosis factor-alpha (TNF-α), interleukin-1 beta (IL-1β), interleukin-8 (IL-8), and interferon-gamma (IFN-γ) in macrophages and peripheral blood mononuclear cells (PBMCs). This immune activation may contribute to granuloma formation and maintenance during latent tuberculosis infection (Sharma and Bose 2001; Kumar et al. 2013). Other notable proteins encoded by the DosR regulon include Rv2623, Rv2626c, and Rv2004c, which are involved in redox balance, energy conservation, and adaptation to hypoxic conditions. For example, Rv2623 is a universal stress protein that binds ATP and regulates bacterial growth rate and persistence. It is highly induced under hypoxia and nitrosative stress and is required for establishing chronic infection in animal models. Loss of Rv2623 results in failure to establish latent infection and leads to hypervirulence, while overexpression attenuates growth, highlighting its role in energy regulation and persistence (Florczyk et al. 2001; O'Toole and Williams 2003; Drumm et al. 2009; Glass et al. 2017). The DosR regulon also encodes enzymes facilitating alternative metabolic pathways such as glycolysis and anaerobic respiration, supporting energy production in the absence of oxygen (Leistikow et al. 2010). Recent proteomic and transcriptomic analyses reveal that the early dormancy response involves at least four major regulatory systems beyond the DosR regulon. Sigma factor *SigE* is induced under envelope and oxidative stress, regulating genes involved in maintaining cell envelope integrity and stress adaptation (Manganelli et al. 2002). Similarly, *SigH* responds to oxidative and heat stress, controlling the expression of genes that protect against redox imbalance and protein damage (Sun et al. 2004). The WhiB-like (*Wbl*) family proteins participate in redox sensing and gene regulation and are activated in early dormancy, continuing into mid-dormancy, overlapping with DosR and sigma factor activities. For example, WhiB3 is induced by hypoxia and nitric oxide and plays a crucial role in maintaining redox homeostasis and lipid metabolism during stress (Singh et al. 2009). WhiB7 is upregulated in response to antibiotic stress and interacts with sigma factors to regulate genes involved in antibiotic resistance and stress adaptation (Burian et al. 2012). Toxin-antitoxin (TA) systems are induced from early to mid-dormancy, contributing to growth arrest and persistence by promoting reversible growth arrest under stress conditions. These systems are upregulated during hypoxia and nutrient starvation, functionally overlapping with DosR, Wbl proteins, and sigma factors (Ramage et al. 2009; Sala et al. 2014). Additionally, the MprAB two-component system, which modulates envelope stress adaptation and interacts with the DosR network, is primarily active from early to mid-dormancy, further overlapping with other regulatory systems. As dormancy progresses into the late phase, the expression of DosR-dependent genes is generally repressed, while other regulatory systems, such as toxin-antitoxin (TA) modules and WhiB-like (*Wbl*) proteins, continue to contribute to long-term persistence. Among these, the response regulator MtrA plays a critical role throughout dormancy and into resuscitation in Mtb. MtrA controls genes involved in DNA replication, cell division, and stress responses, thereby orchestrating key processes essential for bacterial survival and recovery. Li et al. (2010) and Gorla et al. (2018) demonstrated that MtrA directly regulates a broad set of genes, including dnaA, which is crucial for initiating DNA replication, and wag31, a gene involved in cell division. MtrA also modulates DNA repair pathways, ensuring genomic integrity under stress. Further studies by Sharma et al. (2015) and Chatterjee and Banta (2018) showed that MtrA regulates the expression of resuscitation-promoting factors (*rpfA*, *rpfB*, and *rpfC*), which are essential for transitioning Mtb from dormancy back to active growth. Additionally, MtrA controls stress response genes such as whiB4, which is involved in oxidative stress regulation, and fbpB (Ag85B), important for cell adhesion and cell wall remodeling (Chatterjee and Banta 2018; Rajagopalan et al. 2010). Functional studies by Fol et al. (2006) and Peterson et al. (2023) revealed that altering MtrA levels significantly affects bacterial survival, drug sensitivity, and the capacity to persist and resuscitate. These findings highlight MtrA's intersection with multiple dormancy pathways and its essential role in bacterial adaptation and persistence.

7.3.1 Heat Shock Protein X (HspX)

Heat Shock Protein X (HspX), also known as alpha-crystallin (*Acr*) and encoded by the Rv2031c gene, is a member of the small heat shock protein (sHsp) family, which is evolutionarily ancient and ubiquitous across all domains of life. These proteins are characterized by their relatively low molecular mass, typically ranging from 12 to 43 kilodaltons (kDa), and a distinctive tripartite domain structure (Verbon et al. 1992; Yuan et al. 1998; Cunningham and Spreadbury 1998; Stewart et al. 2005; Sun and MacRae 2005; Vaubourgeix et al. 2015). Central to this family is the conserved α-crystallin domain (ACD), approximately 90–100 amino acids in length, which forms the structural core essential for chaperone activity (Sun et al. 2005). This domain is flanked by two variable and flexible regions: an N-terminal segment and a short C-terminal extension, both of which contribute to the protein's ability to oligomerize and diversify its functional capabilities (Sun et al. 2005). Structurally, the ACD adopts a compact β-sandwich fold composed of two antiparallel β-sheets, a hallmark feature of sHsps (Sun et al. 2005). HspX assembles into oligomeric complexes of varying sizes, predominantly forming dodecameric oligomers arranged in a tetrahedral geometry. This quaternary structure is crucial for its chaperone function, providing a dynamic platform that interacts with a wide variety of substrate proteins. Unlike many chaperones that require adenosine triphosphate (ATP) to function, sHsps, including HspX, operate in an ATP-independent manner. They bind to nonnative or misfolded proteins, preventing their irreversible aggregation and maintaining them in a state competent for refolding or degradation by ATP-dependent chaperone systems (Stewart et al. 2005; Vaubourgeix et al. 2015; Jee et al. 2018). Functionally, HspX plays a pivotal role in the dormancy and latency of *M. tuberculosis* by stabilizing the proteome under stressful conditions and preventing protein damage, thereby aiding bacterial survival (Stewart et al. 2005; Hu et al. 2006; Vaubourgeix et al. 2015; Jee et al. 2018). Its expression is strongly upregulated early during the transition into dormancy, typically within the first 24 to 48 hours after exposure to hypoxia (low oxygen), nitric oxide, or other host-like stress signals. This rapid induction is tightly controlled by the dormancy survival regulator (DosR) regulon. Upon activation, DosR binds to promoter regions of dormancy-associated genes, including hspX, driving their transcription and ensuring a swift and robust response to environmental stress (Park et al. 2003; Hu et al. 2011; Mehra et al. 2015; Hutter et al. 2017). HspX's chaperone activity is essential for maintaining protein homeostasis during the early stages of dormancy, a period characterized by a rapid metabolic downshift and exposure to environmental challenges (Vaubourgeix et al. 2015; Jee et al. 2018). Beyond proteome stabilization, HspX also contributes to maintaining cell envelope integrity during dormancy and exhibits immunomodulatory effects by inhibiting the differentiation of monocytes into dendritic cells, thereby facilitating immune evasion (Stewart et al. 2005; Hu et al. 2006). Quantitatively, HspX expression increases dramatically by approximately 80-fold within the first 24–48 hours of exposure to hypoxia or other granuloma-like stressors (Voskuil et al. 2003; Rustad et al. 2008; Hu et al. 2011; Mehra et al. 2015; Hutter et al. 2017). As an ATP-independent molecular chaperone, HspX sequesters nonnative or misfolded proteins, preventing their irreversible aggregation under proteotoxic stress conditions typical of dormancy (Stewart et al. 2005). This activity serves as a first line of defense in protein quality control, complementing ATP-dependent chaperones such as Hsp70 and ClpB, which subsequently refold or degrade the sequestered substrates. Additionally, HspX facilitates the polar localization of protein aggregates within the bacterial cell, aiding in their management and clearance (Zhang et al. 2019). Experimental evidence confirms that HspX actively drives growth arrest and persistence, beyond serving as a mere marker of dormancy. Deletion of hspX increases replication rates in macrophage infection models and impairs the bacterium's ability to maintain long-term dormancy. Conversely, constitutive overexpression of HspX slows bacterial growth in vitro and enhances survival under oxygen-limited conditions, underscoring its critical role in adapting to and maintaining the dormant state (Stewart et al. 2005; Hu et al. 2006). Clinically, HspX protein and antibodies are detectable in patients with latent tuberculosis infection, confirming its immunodominance during latency.

During exponential growth under normoxic and nutrient-rich conditions, HspX protein is typically undetectable as the bacterium prioritizes rapid replication. However, low levels of hspX mRNA persist during this phase, with protein expression initiating only in the stationary phase or under oxygen limitation, indicating robust posttranscriptional regulation (Geluk et al. 2007; Taylor et al. 2012; Castro-Garza et al. 2017). Upon transition from the stationary to logarithmic growth phase, hspX transcription increases rapidly, but protein synthesis remains low, further supporting the existence of posttranscriptional control mechanisms (Yuan et al. 1999). The functional dichotomy between exponential and dormant phases is pronounced: during dormancy, HspX accumulates and assumes a central role in stress adaptation, protein quality control, and growth arrest, whereas in the exponential phase, its expression is repressed and its chaperone functions are largely unnecessary (Yuan et al. 1999; Hu et al. 2006). This tightly regulated expression pattern ensures that HspX is available precisely when needed to support Mtb survival under hostile conditions encountered during latent infection.

7.3.2 Deoxyribonucleic Acid-Binding Protein (KDnaK) and 60 Kilodalton Chaperonin 2 (GroEL2 is a Paralog of GroEL, a Molecular Chaperone Involved in Protein Folding GroEL2)

Deoxyribonucleic acid-binding protein K (DnaK) and the 60 kilodalton chaperonin 2 (GroEL2) are two essential molecular chaperones prominently upregulated during the early stages of dormancy in Mtb. Proteomic and transcriptomic analyses consistently demonstrate their increased expression under stress conditions such as hypoxia, underscoring their critical roles in protein folding and protection against stress-induced damage (Stewart et al. 2002; Ojha et al. 2005; Henderson et al. 2010; Sharma et al. 2016; Park et al. 2024). DnaK belongs to the highly conserved heat shock protein 70 (Hsp70) family of ATP-dependent molecular chaperones. Structurally, it comprises three main domains: an N-terminal ATPase domain that hydrolyzes ATP to regulate substrate binding and release; a substrate-binding domain that recognizes and binds exposed hydrophobic regions of nascent or stress-denatured proteins; and a C-terminal lid domain that assists in substrate encapsulation and release (Mayer et al. 2005; Hickey et al. 2009; Meimaridou et al. 2009; Clerico et al. 2015). Functionally, DnaK acts as a central hub within the bacterial cytosolic chaperone network, engaging numerous protein substrates during both co-translational and post-translational folding. Its primary functions include preventing protein aggregation, refolding misfolded proteins, and buffering the effects of mutations on protein structure and function, thereby maintaining proteome integrity under stress. DnaK operates cooperatively with its co-chaperones *DnaJ* (a Heat Shock Protein 40 homologue) and GrpE, forming the DnaK-DnaJ-GrpE (*KJE*) chaperone system, which coordinates efficient protein folding and quality control (Stewart et al. 2002; Mayer et al. 2005; Perales-Calvo et al. 2018). Structural studies have elucidated the asymmetric complex formed between DnaK and the GrpE dimer, revealing how GrpE allosterically regulates DnaK to simultaneously release ADP and substrate peptides, a mechanism essential for protein folding activity both in vitro and in vivo (Stewart et al. 2002). In Mtb, DnaK is indispensable for bacterial viability and growth, playing a unique role in the mycobacterial proteostasis network, especially under host-induced stress, where it cooperates with ClpB to dissolve and refold protein aggregates, contributing to bacterial recovery during stress (Rosenzweig et al. 2013). GroEL2, encoded by the *groEL2* gene, is a member of the Hsp60 family of chaperonins and a paralog of GroEL in Mtb. It assembles into a large double-ring oligomeric complex, typically a tetradecamer composed of two heptameric rings stacked back-to-back, forming a barrel-shaped nano-cage with a central cavity (Chilukoti et al. 2016). This complex functions in concert with the co-chaperonin GroES, encoded by *groES* (Rv3418c), which forms a lid over the central cavity. Unfolded or misfolded substrate proteins are encapsulated within this cavity, where ATP binding and hydrolysis drive conformational changes in GroEL2, providing an isolated environment for proper folding (Tang et al. 2007; Rao et al. 2009; Gardner et al. 2021). GroEL2 is the major chaperonin in Mtb and is particularly important under stress conditions that increase the risk of protein misfolding, such as oxidative and nitrosative stress encountered within macrophages (Qamra et al. 2004; Chilukoti et al. 2016). Recent studies have revealed that GroEL2 localizes to the bacterial surface, facilitating association with macrophages, and can be released into the host cell cytosol, where it interacts with host proteins such as mortalin to modulate apoptosis and immune responses (Georgieva et al. 2018; Vinod et al. 2021). The transcriptional regulation of these chaperones involves specific promoter elements.

For dnaK, two promoters have been identified: a housekeeping promoter (P1) that maintains basal expression and a stress-inducible promoter (P2) recognized by the alternative sigma factor SigH, which mediates expression during stress (Manganelli et al. 2002; Stewart et al. 2002; Sharp et al. 2016). The groEL2 gene promoter contains regulatory elements such as the Controlling Inverted Repeat of Chaperone Expression (CIRCE) and HspR-Associated Inverted Repeat (HAIR) boxes, which participate in heat shock regulation (Narberhaus 1999; Stewart et al. 2002; Jäger et al. 2004; Sato et al. 2007; Zhuo et al. 2018). The groEL2 promoter activity is strongly induced under stress conditions, including oxidative and nitrosative stress encountered within macrophages (Aravindhan et al. 2009; Joseph et al. 2012; Choudhary et al. 2014). Functional and genetic studies have demonstrated that disruption of these chaperone genes markedly reduces bacterial survival by impairing protein folding. Specifically, loss of dnaK increases vulnerability to environmental stresses and diminishes the bacterium's ability to persist under dormancy-like conditions (Genevaux et al. 2004; Warnecke 2012; Fourie and Wilson 2020). Similarly, mutations affecting groEL2 expression or processing alter Mtb's ability to modulate host immune responses and survive intracellularly (Rengarajan et al. 2008; Naffin-Olivos et al. 2014). GroEL2 is also subject to proteolytic cleavage by the Hip1 protease, generating a truncated monomeric form (GroEL2(cl)) that predominates during infection and dampens proinflammatory responses by dendritic cells and macrophages, thereby limiting cytokine production and antigen presentation (Naffin-Olivos et al. 2014). In contrast, the full-length GroEL2 protein is highly immunostimulatory, promoting strong T cell responses, but its continuous presence can be detrimental to bacterial growth, as observed in hip1 mutant strains. Both DnaK and GroEL2 are upregulated in response to dormancy-inducing stresses such as hypoxia, oxidative stress, and nutrient deprivation (Stewart et al. 2002; Chauhan et al. 2011; Rao et al. 2021; Liu et al. 2024). Their expression is controlled by heat shock and stress-responsive regulatory pathways, including alternative sigma factors and stress-sensing transcriptional regulators. Functionally, DnaK acts as an initial responder by binding exposed hydrophobic regions of unfolded proteins, either refolding them directly or transferring them to GroEL2 for encapsulated folding when necessary. GroEL2 provides a protected environment for proteins that cannot be properly refolded by DnaK alone, thereby maintaining protein homeostasis. This cooperation enables Mtb to survive the metabolic downshift and proteotoxic stress encountered during the transition to and maintenance of dormancy (Sharma et al. 2008; Kumar et al. 2015; Fourie and Wilson 2020). Additionally, GroEL2's surface localization facilitates bacterial association with macrophages, potentially influencing host-pathogen interactions.

7.3.3 Sigma Factor

In Mtb, sigma factors are specialized subunits of RNA polymerase that direct the enzyme to specific promoters, enabling precise regulation of gene expression in response to environmental stresses. Among these, the alternative sigma factor SigB, encoded by the sigB (Rv2710) gene, plays a critical role early in the onset of dormancy by orchestrating the bacterial stress response (Sachdeva et al. 2010; Liu et al. 2024). SigB belongs to the sigma-70 family of sigma factors, which are primarily responsible for promoter recognition and transcription initiation. Unlike the primary sigma factor SigA, SigB lacks the region 1-a domain that restricts promoter recognition, allowing SigB to target a distinct subset of stress-responsive genes. This structural difference enables SigB to regulate genes essential for adaptation and survival under stress conditions (Rodrigue et al. 2006; Ayala et al. 2020). Functionally, SigB acts as a master regulator during dormancy by directing RNA polymerase to transcribe a broad array of genes involved in stress adaptation. Transcriptomic studies by Betts et al. (2002) demonstrated that sigB expression is upregulated within 24 hours of nutrient starvation and hypoxia, coinciding with increased expression of genes encoding molecular chaperones such as GroEL2 and DnaK, enzymes involved in cell envelope remodeling, and ribosomal proteins. Voskuil et al. (2004) further confirmed that under hypoxic conditions, SigB-dependent transcription includes early secreted antigens such as ESAT-6-like proteins, members of the PE-PGRS family, and regulatory proteins including WhiB2 and IdeR. Complementary transcriptomic analyses by Salina et al. (2014, 2019a, b) and proteomic profiling by Gopinath et al. (2015) revealed elevated expression of stress-responsive proteins, including enzymes like ketoacyl synthase KasA, which SigB likely regulates. Collectively, these findings underscore the pivotal role of SigB in facilitating dormancy adaptation. Chromatin immunoprecipitation sequencing (ChIP-seq) has been effectively employed to identify direct SigB binding sites in mycobacteria, notably in *Mycobacterium smegmatis*, where SigB was shown to bind promoters of essential housekeeping and stress-responsive genes (Mundade et al. 2014; Hurst-Hess et al. 2019). This evidence highlights SigB's broad regulatory influence across diverse gene networks. Recently, Kundu and Basu (2021) emphasized the extensive regulatory network controlled by SigB, encompassing genes encoding stress response chaperones, enzymes involved in cell envelope remodeling, secreted antigens, ribosomal proteins, and various transcriptional regulators critical for survival during dormancy. SigB also exhibits autoregulatory behavior by enhancing its own expression under specific stress conditions, functioning as a terminal regulator within the sigma factor cascade (Rodrigue et al. 2006). During exponential growth, SigB is expressed at low basal levels and operates alongside the primary sigma factor SigA to maintain transcription of essential housekeeping genes, effectively serving as a backup when SigA activity is compromised (Voskuil et al. 2003). However, upon entry into dormancy or exposure to environmental stress, SigB assumes a dominant role in reshaping the transcriptional landscape, promoting gene expression programs that facilitate survival, adaptation, and growth arrest. Experimental studies have demonstrated that overexpression of SigB leads to reduced intracellular growth rates, highlighting its critical function in growth arrest and persistence (Manganelli et al. 2002; Stewart et al. 2002). In summary, SigB is a key alternative sigma factor in Mtb that integrates environmental signals to regulate a wide array of genes essential for the bacterial stress response, dormancy initiation, and long-term persistence within the host.

7.3.4 Sulfur Mobilization Operon

During the early phase of dormancy, Mtb undergoes substantial metabolic reprogramming, prominently characterized by the increased expression of iron-sulfur (Fe-S) cluster assembly proteins encoded by the suf operon (Rv1460-Rv1466) (Saini et al. 2012; Anand et al. 2021; Elchennawi et al. 2022). These proteins belong to the iron-sulfur cluster biosynthesis family and contain conserved cysteine-rich motifs that coordinate iron and sulfur atoms to form Fe-S clusters. Such clusters serve as essential cofactors for a diverse range of enzymes, especially those involved in anaerobic respiration and electron transport chains, which are critical for sustaining energy production under dormancy conditions (Willemse et al. 2018; Tripathi et al.

2022; Gorityala et al. 2024). The SUF (sulfur mobilization) system, encoded by the suf operon, represents the primary and exclusive machinery for Fe-S cluster assembly in Mtb and is indispensable for bacterial survival and persistence (Huet et al. 2005; Gorityala et al. 2024). This system comprises several key proteins with specialized functions: SufR (Rv1460), a zinc-finger transcriptional regulator, modulates operon expression in response to intracellular iron availability and oxidative stress; SufB (Rv1461), SufC (Rv1462), and SufD (Rv1463) assemble into a scaffold complex that facilitates Fe-S cluster formation and transfer; SufS (Rv1464), a cysteine desulfurase, mobilizes sulfur from L-cysteine; and SufU (Rv1465), an accessory sulfur transfer protein, with recent studies confirming a functional interaction between SufS and SufU (Elchennawi et al. 2023; Gorityala et al. 2024). SufT (Rv1466), containing a DUF59 domain, is critical for Fe-S cluster maturation and physically interacts with other SUF components and Fe-S-dependent enzymes such as aconitase and SufR (Tripathi et al. 2022). Disruption of any SUF component severely impairs Fe-S cluster biosynthesis, leading to metabolic dysfunction, increased susceptibility to oxidative and nitrosative stress, and diminished bacterial survival within macrophages and animal infection models, thereby underscoring the system's vital role in Mtb persistence (Huet et al. 2005; Gorityala et al. 2024). The early induction of Fe-S assembly proteins during dormancy supports key anaerobic respiration and electron transport processes that sustain energy production under hypoxic conditions (Saini et al. 2012; Cook et al. 2014; Birhanu et al. 2022; Elchennawi et al. 2022). Given that Fe-S clusters are highly sensitive to damage by reactive oxygen and nitrogen species, efficient assembly and maintenance of these clusters are critical for bacterial survival. Experimental knockdown of SufT leads to reduced activity of Fe-S-dependent enzymes, disruption of central metabolic pathways, impaired glycolytic and respiratory functions, and decreased survival under stress and within host cells (Tripathi et al. 2022). Strains deficient in SufT exhibit growth defects and altered levels of tricarboxylic acid (TCA) cycle intermediates, amino acids, and sulfur metabolites, highlighting the importance of Fe-S proteins in metabolic adaptation. This includes roles in the TCA cycle, amino acid and sulfur metabolism, and the methylcitrate cycle, reflecting Mtb's shift to alternative energy sources during dormancy (Tripathi et al. 2022). Fe-S proteins also play a crucial role in maintaining electron transport and energy metabolism by supporting respiratory enzymes and redox proteins that sustain ATP synthesis and redox balance, particularly under the low-oxygen conditions characteristic of dormancy (Huet et al. 2005; Gorityala et al. 2024). Proper Fe-S cluster assembly is essential for resistance to oxidative and nitrosative stress and for long-term survival within the host. Genetic disruptions of suf genes such as sufT or sufR result in impaired growth, altered metabolism, reduced intracellular survival, and compromised persistence in animal infection models, emphasizing the critical role of the SUF system in Mtb pathogenesis (Anand et al. 2021). Moreover, recent computational and experimental studies have identified SUF pathway proteins, including SufD, as promising drug targets. This highlights the pathway's essential functions in bacterial growth, stress resistance, and persistence, underscoring its potential for therapeutic intervention against tuberculosis (Gorityala et al. 2024).

7.3.5 Methylcitrate Synthase (*PrpC*) and Methylcitrate Dehydratase (*PrpD*)

Central to early adaptations in Mtb is the methylcitrate cycle, a specialized metabolic pathway essential for processing propionyl-coenzyme A (propionyl-CoA), a potentially toxic intermediate generated primarily through the catabolism of odd-chain fatty acids and cholesterol—key carbon sources during infection and dormancy (Upton et al. 2008; Griffin et al. 2012; Lee et al. 2013; Liu et al. 2018; Liu and Jiang 2021). The methylcitrate cycle detoxifies propionyl-CoA by converting it into metabolites that feed into central metabolic pathways such as the tricarboxylic acid (TCA) cycle, thereby maintaining cellular homeostasis and ensuring efficient energy production (Huang et al. 2023). Two enzymes, methylcitrate synthase and methylcitrate dehydratase, encoded by the genes *prpC* (Rv1131) and *prpD* (Rv1130), respectively, catalyze key steps in this cycle. Methylcitrate synthase initiates the cycle by catalyzing the condensation of propionyl-CoA with oxaloacetate, a four-carbon dicarboxylic acid, to produce 2-methylcitrate, a pivotal intermediate. This step is critical as it commits propionyl-CoA to further metabolism. Subsequently, methylcitrate dehydratase converts 2-methylcitrate into 2-methyl-cis-aconitate through a dehydration reaction, facilitating downstream enzymatic processes that ultimately yield pyruvate and succinate. These products integrate into central carbon metabolism, supporting both energy generation and biosynthetic functions (Muñoz-Elías et al. 2006; Masiewicz et al. 2012; Pei et al. 2021; Huang et al. 2023). The functionality of the methylcitrate cycle is indispensable for Mtb's survival and persistence, particularly during infection when the bacterium relies heavily on host-derived lipids and cholesterol as carbon sources (Muñoz-Elías and McKinney 2005; Griffin et al. 2012). Accumulation of propionyl-CoA, if not properly metabolized, can inhibit key metabolic enzymes and disrupt cellular processes, underscoring the necessity of this detoxification pathway (Marrero et al. 2010; Eoh and Rhee 2013). Thus, the methylcitrate cycle not only detoxifies propionyl-CoA but also enables Mtb to efficiently utilize lipid-derived substrates to sustain growth and maintenance under nutrient-limited and hypoxic conditions characteristic of dormancy (Muñoz-Elías and McKinney 2005; Griffin et al. 2012; Eoh and Rhee 2013). Genetic studies have demonstrated that disruption of the prpDC operon, encoding both methylcitrate dehydratase and methylcitrate synthase, severely impairs Mtb's ability to grow when propionate serves as the sole carbon source. Moreover, such disruption diminishes bacterial survival within macrophages, highlighting the operon's essential role in metabolic adaptation and pathogenesis (Savvi et al. 2008). Transcriptomic and proteomic analyses further reveal that expression of the prpDC operon is strongly induced early during the transition into dormancy. In vitro models simulating dormancy conditions, including hypoxia and nutrient limitation, show that within the first 48 hours of exposure, the prpDC operon is upregulated, supporting the metabolic shift that enables Mtb to exploit host-derived lipids and cholesterol as alternative carbon sources. This adaptation allows the bacterium to meet its energy demands and maintain metabolic flexibility under the nutrient-poor and low-oxygen conditions typical of dormancy (Muñoz-Elías and McKinney 2005). Proteomic profiling corroborates the increased abundance of methylcitrate cycle enzymes during early dormancy stages, reinforcing the pathway's significance in stress adaptation. Regulation of the prpDC operon is tightly controlled by the transcriptional regulator PrpR, encoded by Rv1129c. PrpR senses the intracellular accumulation of propionate or propionyl-CoA and directly activates transcription of the prpDC operon, ensuring timely expression of enzymes responsible for detoxifying propionyl-CoA during lipid metabolism. Functional studies of Mtb mutants lacking prpDC or prpR reveal that these strains fail to grow on propionate or cholesterol as sole carbon sources and exhibit significantly reduced survival within macrophages, underscoring the critical role of the methylcitrate cycle and its regulatory network in Mtb's persistence within the host (Tang et al. 2019).

Hence, the methylcitrate cycle, driven by the enzymatic activities of methylcitrate synthase (PrpC) and methylcitrate dehydratase (PrpD), is fundamental for detoxifying propionyl-CoA and facilitating metabolic adaptation during infection and dormancy in Mtb. The early and robust induction of the prpDC operon, regulated by the transcription factor PrpR, enables efficient metabolism of host-derived lipids, maintaining metabolic flexibility essential for the bacterium's survival, persistence, and pathogenicity under the nutrient-limited and hypoxic conditions encountered during latent tuberculosis infection.

7.4 Mid-Stage of Dormancy in *Mycobacterium tuberculosis*

Mid-dormancy in Mtb is characterized by a profound state of growth arrest, which is orchestrated through a sophisticated interplay of regulatory mechanisms, notably including toxin-antitoxin (TA) systems and the maintenance of redox homeostasis. During this phase, Mtb bacilli enter a nonreplicating persistent state, marked by a significant reduction or complete cessation of both cell division and metabolic activity. This physiological adaptation is crucial for the bacterium's ability to endure prolonged exposure to hostile environmental conditions, such as hypoxia (low oxygen levels) and nutrient limitation, which are commonly encountered within the granulomatous lesions of the host (Wayne and Sohaskey 2001; Sala et al. 2014) .Transcriptomic analyses have demonstrated a global downregulation of transcriptional activity during dormancy, consistent with the cessation of cellular growth. This transcriptional quiescence is followed by a rapid activation, or transcriptional burst, upon resuscitation from dormancy, enabling the bacterium to quickly resume metabolic activity and proliferation when favorable conditions return (Ignatov et al. 2015; Iona et al. 2016). These studies collectively underscore the pivotal roles that TA systems and redox balance play in maintaining the dormant state and facilitating the long-term persistence of Mtb within the host (Sala et al. 2014). A key molecular adaptation during mid-dormancy is the upregulation of Triacylglycerol Synthase-1 (*Tgs1*), encoded by the Rv3130c gene. Tgs1 is an acyltransferase enzyme that catalyzes the synthesis of triacylglycerol (TAG) by esterifying diacylglycerol with acyl-coenzyme A (acyl-CoA). As a member of the diacylglycerol acyltransferase (DGAT) family, Tgs1 facilitates the final and rate-limiting step in TAG biosynthesis, transferring acyl groups from acyl-CoA to diacylglycerol to form TAG (Sirakova et al. 2003; Daniel et al. 2004; Low et al. 2010; Chen et al. 2022). Structurally, Tgs1 contains conserved acyltransferase motifs that are characteristic of mycobacterial lipid metabolism enzymes. Experimental studies have shown that Tgs1 is primarily responsible for the accumulation of TAG in Mtb under stress conditions. Daniel et al. (2004) and Sirakova et al. (2006) demonstrated that when Tgs1 is expressed in *Escherichia coli*, it preferentially utilizes C26:0-CoA, a long-chain fatty acyl-CoA, for TAG synthesis. This substrate specificity is reflected in the predominance of C26 fatty acids in the TAG molecules produced by Tgs1, confirming its role in incorporating long-chain acyl-CoA substrates during TAG biosynthesis (Sirakova et al. 2006). Importantly, Tgs1 is a crucial component of the DosR (dormancy survival regulator) regulon, a regulatory network that is activated in response to hypoxia and nitric oxide (NO) and is essential for the adaptation of Mtb to dormancy (Voskuil et al. 2003). Expression of *tgs1* is strongly induced under various dormancy-inducing conditions, including hypoxia, nutrient starvation, acidic pH, and exposure to nitric oxide, as observed in both clinical isolates and in vitro dormancy models (Daniel et al. 2004; Chauhan and Tyagi 2009; Singh et al. 2011). This upregulation drives the accumulation of triacylglycerol (TAG) within intracellular lipid droplets, which serve as long-term energy reserves critical for sustaining bacterial viability during periods of metabolic inactivity (Sirakova et al. 2003; Mallick et al. 2021). Functional disruption of the *tgs1* gene nearly abolishes TAG synthesis under stress conditions, while genetic complementation restores this capability, firmly establishing Tgs1's central role in TAG biosynthesis (Sirakova et al. 2003; Deb et al. 2006). The storage of TAG mediated by Tgs1 supports metabolic reprogramming during mid-dormancy by diverting fatty acids away from membrane lipid synthesis toward storage. This shift contributes to reduced growth rates and enhances tolerance to antibiotics, a hallmark of persistent Mtb populations (Deb et al. 2009). Overexpression of *tgs1* in *Mycobacterium marinum*, a close relative of Mtb, increases lipid droplet formation and improves bacterial survival under nutrient starvation, underscoring the adaptive advantage conferred by this enzyme (Liu et al. 2020). Furthermore, TAG reserves accumulated during dormancy provide a rapid and accessible energy source that can be mobilized during reactivation, enabling Mtb to resume growth efficiently when environmental conditions become favorable (Low et al. 2009; Galagan et al. 2013; Chang and Guan 2021). Thus, Tgs1 orchestrates a critical metabolic shift toward lipid storage, enforces growth arrest, and primes the bacterium for long-term persistence and eventual reactivation. This makes Tgs1 a hallmark enzyme of mid-dormancy adaptation in Mtb (Daniel et al. 2004; Voskuil et al. 2003).

The WhiB family in Mtb comprises seven transcriptional regulators, each characterized by the presence of a [4Fe-4S] iron-sulfur cluster. These proteins play distinct yet sometimes overlapping roles in bacterial physiology, stress adaptation, and pathogenesis (Steyn et al. 2002; Larsson et al. 2012; Wu et al. 2017; Lilic et al. 2023). WhiB1 is essential for mycobacterial growth and exhibits high sensitivity to nitric oxide (NO). Exposure to NO modifies its iron-sulfur cluster, enhancing its DNA-binding ability. Upon sensing NO, WhiB1 represses genes such as those in the espA operon, which impacts the ESX-1 secretion system and virulence, and also participates in regulating cell division (Smith et al. 2012b). WhiB2 contributes to bacterial cell division by interacting with the cell division machinery and is required for septum formation. Additionally, it may play a role in drug resistance mechanisms (Konar et al. 2012). WhiB3 functions as a redox-sensitive transcriptional regulator that is particularly important during mid-dormancy and stress adaptation (Mehta and Singh 2019). It senses intracellular redox changes through its iron-sulfur cluster and modulates gene expression accordingly. Experimental evidence from Singh et al. (2007) and Steyn et al. (2002) demonstrated that WhiB3 acts as an intracellular redox sensor, maintaining redox homeostasis by regulating the production of virulence-associated lipids in response to environmental cues. It also interacts with the principal sigma factor RpoV (σA), underscoring its role in virulence regulation (Steyn et al. 2002; Singh et al. 2007). Further studies by Saini et al. (2012) confirmed that WhiB3 directly senses fluctuations in the intracellular redox environment during infection and dormancy, coordinating metabolic responses through its iron-sulfur cluster. WhiB4 serves as a redox sensor regulating genes linked to antioxidant defense, thereby helping Mtb survive oxidative stress imposed by the host. Studies by Chawla et al. (2012) showed that WhiB4 responds to oxygen (O_2) and nitric oxide (NO) through its [4Fe-4S] cluster and controls oxidative stress responses. Deletion of the whiB4 gene (MtbΔwhiB4) led to altered redox balance, reduced membrane potential, and overexpression of antioxidant systems such as alkyl hydroperoxidase (*ahpC-ahpD*) and rubredoxins (*rubA-rubB*). DNA binding assays revealed that oxidation of WhiB4's cysteine thiols induces disulfide-linked oligomerization, which activates DNA binding and transcriptional repression; this effect is reversed upon reduction. The MtbΔwhiB4 mutant displayed increased resistance to oxidative stress in vitro and enhanced survival inside macrophages. In animal models, this mutant showed hypervirulence in guinea pig lungs but impaired dissemination to the spleen (Chawla et al. 2012). These findings indicate that WhiB4 finely tunes the oxidative stress response to maintain redox balance and modulate virulence. Further mechanistic insights by Chawla et al. (2018) revealed that WhiB4 dynamically links genome condensation with oxidative stress response; under oxidative conditions, its oligomerization promotes nucleoid condensation and global changes in gene expression, coupling structural and transcriptional adaptations to oxidative stress. WhiB5 is associated with persistence and adaptation during nutrient starvation, contributing to long-term survival under stress and playing a vital role in reactivation and full virulence after chronic infection (Chawla et al. 2012).WhiB6 is a pivotal transcriptional regulator in Mtb that directs the expression of virulence-associated genes, including those integral to the ESX-1 secretion system, which is essential for bacterial adaptation to stress and survival within host cells (Solans et al. 2014; Chen et al. 2016). A recent investigation by Velayati et al. (2024) identified a consistent upstream mutation, designated whiB6:c.-75delG, in studied clinical Mtb isolates subjected to 12 months of in vitro dormancy, independent of their drug susceptibility profile. Remarkably, this mutation was also detected at a high frequency (95%) among drug-resistant tuberculosis isolates during the exponential growth phase, indicating a strong association with both dormancy and drug resistance. The whiB6:c.-75delG mutation is located in the promoter region of the gene and can disrupt normal gene expression control, potentially leading to altered levels of WhiB6 protein. This alteration may have downstream effects on cellular processes, including antibiotic susceptibility. Such regulatory changes could impair Mtb's ability to adapt to stress and maintain dormancy, which are critical factors in its persistence and pathogenicity. The presence of the same mutation in both long-term dormant cells (even after 24 months) and drug-resistant strains during the exponential phase of growth suggests a convergent adaptive mechanism. This implies that the mutation not only enhances survival during prolonged dormancy but may also confer a selective advantage under antibiotic pressure in actively replicating bacteria. Therefore, the whiB6:c.-75delG mutation likely plays a dual role in facilitating both dormancy persistence and the development of drug resistance, underscoring its significance in Mtb's long-term survival and posing challenges for clinical management. Mechanistically, WhiB6 functions as a redox-responsive transcription factor that regulates ESX-1 secretion and the DosR regulon, modulating aerobic and anaerobic metabolism, cell division, and virulence (Solans et al. 2014; Chen et al. 2016; Velayati et al. 2024). Its expression is tightly controlled by the virulence master regulator PhoP, which binds upstream promoter regions, integrating environmental signals such as nitric oxide and hypoxia to fine-tune WhiB6 levels and downstream gene expression (Solans et al. 2014; Chen et al. 2016; Bhattacharya et al. 2023). WhiB6's regulation of ESX-1 genes is critical for the secretion of ESAT-6, a major virulence factor required for phagosomal escape and host cell cytolysis. Mutations or polymorphisms in the whiB6 promoter region, such as those observed in laboratory strains like $H_{37}Rv$, can alter ESAT-6 expression and secretion, affecting virulence and immune evasion (Solans et al. 2014). Experimental models demonstrate that WhiB6 modulates mycobacterial infection, granuloma formation, and dissemination, highlighting its central role in disease progression (Solans et al. 2014; Chen et al. 2016). WhiB7 is a central transcriptional regulator that plays a critical role in intrinsic antibiotic resistance and redox homeostasis (Burian et al. 2012; Morris et al. 2005). Structurally, WhiB7 contains a unique C-terminal AT-hook DNA-binding motif, which enables it to bind AT-rich sequences located upstream of the −35-promoter region of its target genes (Lilic et al. 2021; Burian et al. 2012). This DNA-binding mechanism is reminiscent of eukaryotic chromosomal proteins and is rare among bacterial regulators, highlighting the distinctiveness of WhiB7 within bacterial transcription factors. Functionally, WhiB7 is activated in response to antibiotic exposure and various cellular stresses (Morris et al. 2005; Burian et al. 2012). Upon induction, WhiB7 interacts specifically with domain 4 of the primary sigma factor sigma A (σ^A4) within the RNA polymerase holoenzyme. This interaction facilitates the transcription of a suite of genes that confer resistance to multiple antibiotic classes (Wan et al. 2021). High-resolution structural studies, including cryo-electron microscopy and X-ray crystallography, have elucidated how WhiB7 simultaneously engages with σ^A4 and the AT-rich promoter DNA via its AT-hook, stabilizing the transcriptional complex and activating gene expression critical for antibiotic resistance (Lilic et al. 2021; Wan et al. 2021; Chen et al. 2016). WhiB7 belongs to the Actinobacteria-specific WhiB-like (Wbl) family of iron-sulfur ([4Fe-4S]) cluster-containing transcription factors, which are exclusive to this phylum and serve diverse regulatory functions (Steyn et al. 2002; Lilic et al. 2023). The redox-sensitive [4Fe-4S] cluster within WhiB7 enables it to respond dynamically to intracellular redox changes, linking environmental stress signals to transcriptional regulation. This redox sensing is crucial for modulating thiol redox balance shortly after antibiotic treatment, thereby contributing to the bacterium's adaptive responses (Burian et al. 2012; Morris et al. 2005). Overall, the WhiB family of [4Fe-4S] cluster-containing transcriptional regulators in Mtb organizes a wide range of physiological processes, including growth, cell division, redox sensing, stress adaptation, virulence, and antibiotic resistance. Each WhiB protein contributes uniquely to the bacterium's ability to survive and persist within the hostile host environment, making them critical factors in Mtb pathogenesis and dormancy.

During the mid-dormancy phase, the acyl carrier protein (ACP), encoded by the gene Rv0100, is significantly upregulated, reflecting its critical role in fatty acid biosynthesis and bacterial adaptation to dormancy (Gutka et al. 2024). ACPs are highly conserved proteins that function as central carriers of acyl groups during fatty acid synthesis. The Rv0100 protein consists of 78 amino acids and adopts a characteristic four-helix bundle structure. A pivotal feature of Rv0100 is the covalent attachment of a phosphopantetheine prosthetic group to a specific serine residue, which acts as a flexible arm that shuttles acyl intermediates between enzymatic complexes, facilitating the elongation and modification of fatty acid chains (Gutka et al. 2024).The marked upregulation of Rv0100 during dormancy has been demonstrated in nonreplicating persistence models such as the Wayne hypoxia model, underscoring its essential role in the biosynthesis of long-chain fatty acids and complex cell wall lipids, including phthiocerol dimycocerosates (PDIMs). PDIMs are specialized lipids that enhance Mtb's virulence and persistence by reinforcing the cell envelope and facilitating immune evasion (Gutka et al. 2024; Mulholland et al. 2024). Genomically, Rv0100 is part of an operon that includes genes such as Rv0101, encoding a peptide synthase, and Rv0099 (also known as FadD10), a fatty acid coenzyme A (CoA) ligase, illustrating its integration into broader lipid biosynthetic and degradation pathways vital for metabolic adaptation during dormancy (Sassetti and Rubin 2003; Gutka et al. 2024). Functionally, Rv0100 supports both energy storage and cell wall remodeling during mid-dormancy. By facilitating fatty acid biosynthesis, it enables the accumulation of triacylglycerol (TAG) within intracellular lipid droplets, which serve as essential energy reserves sustaining Mtb's long-term survival under nutrient-limited and hypoxic conditions typical of dormancy (Gutka et al. 2024). Rv0100 also provides critical precursors for PDIM synthesis, thereby strengthening the bacterial cell envelope and enhancing the pathogen's ability to evade host immune defenses (Hotter et al. 2005). Functional studies using knockout mutants have shown that loss of Rv0100 impairs bacterial growth, reduces lipid synthesis, and compromises survival both in vitro and in vivo, confirming its indispensable role in metabolic and structural adaptations necessary for dormancy and latent infection (Gutka et al. 2024). Thus, Rv0100 is a key factor in the metabolic reprogramming that underpins Mtb persistence within the host.

Resuscitation-promoting factors (Rpfs) constitute another critical group of proteins involved in Mtb's transition from dormancy to active growth. The Rpf family comprises five secreted proteins, among which RpfB and RpfE are the most extensively characterized. These proteins exhibit muralytic activity as peptidoglycan hydrolases, cleaving β-(1,4)-glycosidic bonds within the peptidoglycan layer of the bacterial cell wall (Kana et al. 2008; Ruggiero et al. 2009; Kana and Mizrahi 2010; Tanriver et al. 2025). This enzymatic action remodels the rigid peptidoglycan matrix, which is essential for maintaining cell wall integrity during dormancy and enabling the physical and biochemical changes required for resuscitation (Mukamolova et al. 2006). During dormancy, the Mtb cell wall undergoes structural modifications that contribute to growth arrest and increased resistance to environmental stresses (Bhat et al. 2022). Rpfs hydrolyze specific bonds within the peptidoglycan, alleviating the mechanical constraints imposed by the thickened and highly cross-linked cell wall characteristic of dormant cells (Mukamolova et al. 2006). This remodeling is critical for resuscitation, as it permits re-expansion of the cell wall and reactivation of the cell division machinery once favorable environmental conditions, such as increased oxygen and nutrient availability, return (Kana et al. 2008). Beyond their enzymatic role, Rpfs are implicated in signaling pathways that detect environmental cues and initiate molecular cascades leading to bacterial reactivation (Biketov et al. 2007; Kana and Mizrahi 2010). By modulating peptidoglycan structure, Rpfs generate specific muropeptide fragments that act as signaling molecules, activating transcriptional regulators and other factors involved in exiting dormancy (Mukamolova et al. 2006; Kana and Mizrahi 2010). This dual functionality positions Rpfs as pivotal mediators bridging the dormant and active states of Mtb (Kana et al. 2008). Thus, Rpfs such as RpfB and RpfE are secreted peptidoglycan hydrolases containing conserved lysozyme-like domains that maintain cell wall plasticity during dormancy and catalyze the remodeling necessary for resuscitation. Their enzymatic activity relieves the physical constraints imposed by dormancy and participates in signaling mechanisms that trigger the return to active growth, making them essential for the successful persistence and reactivation of Mtb (Kana and Mizrahi 2010; Mukamolova et al. 2006).

During the mid-dormancy phase, both type I and type II nicotinamide adenine dinucleotide (NADH) dehydrogenases perform essential yet distinct roles to sustain minimal energy metabolism and maintain redox balance under hypoxic and nutrient-limited conditions. Type I NADH dehydrogenase (NDH-1), encoded by the nuo operon (nuoA through nuoN, Rv3145-Rv3158), forms a large multisubunit membrane-bound complex homologous to mitochondrial complex I (Falk-Krzesinski et al. 1998; Rao et al. 2008; Hegde et al. 2012; Bhat et al. 2016; Płocińska et al. 2022). This complex, consisting of at least 14 subunits, functions as a proton-pumping NADH: menaquinone oxidoreductase, catalyzing the oxidation of NADH to NAD^+ while transferring electrons to menaquinone. Concurrently, it pumps protons across the membrane to generate a proton motive force necessary for ATP synthesis (Rao et al. 2008; Bhat et al. 2022). Although NDH-1 supports basal respiration and helps maintain the cellular $NADH/NAD^+$ ratio, its activity is diminished during dormancy compared to type II NADH dehydrogenases (NDH-2), especially under low oxygen conditions, as demonstrated in Wayne's hypoxic dormancy model and corroborated by gene expression and proteomic analyses (Shi et al. 2005; Rao et al. 2008; Berney and Cook 2010; Cumming et al. 2017; Płocińska et al. 2022; Bhat et al. 2022). In contrast, type II NADH dehydrogenases, collectively referred to as NDH-2, are encoded by the ndh (Rv1854c) and ndhA (Rv0392c) genes. These enzymes are single-subunit, membrane-associated flavoproteins that catalyze NADH oxidation and electron transfer to menaquinone without coupled proton translocation (Weinstein et al. 2005; Awasthy et al. 2014; Vilchèze et al. 2018). Both Ndh and NdhA share high amino acid similarity and play significant roles in maintaining redox homeostasis during dormancy, compensating for the reduced activity of NDH-1 under hypoxic conditions. Their activity ensures continued electron flow through the respiratory chain, enabling the bacterium to sustain essential metabolic functions despite limited oxygen availability. Together, the coordinated functions of resuscitation-promoting factors (Rpfs) and NADH dehydrogenases enable Mtb to maintain cell wall integrity, regulate energy metabolism, and balance redox states. These processes are critical for bacterial survival during dormancy and for successful reactivation when environmental conditions become favorable, highlighting the intricate metabolic adaptations that underpin Mtb's persistence and pathogenicity (Voskuil et al. 2011; Cook et al. 2017; Beites et al. 2019; Xu et al. 2023).

7.5 Late Stage of Dormancy in *Mycobacterium tuberculosis*

Late dormancy in Mtb, typically defined as the period extending beyond three months of non-replicative persistence, is characterized by a highly selective and adaptive proteomic profile that enables the bacterium to endure extreme and prolonged stress. During this phase, proteins that are abundantly expressed in the early and mid-stages of dormancy continue to play vital roles. Central among these is the dormancy survival regulator DosR (Rv3133c), which acts as the master regulator of the dormancy survival regulon. DosR is rapidly induced by hypoxia and other environmental stresses and sustains the expression of genes essential for stress resistance and energy conservation, thereby maintaining Mtb's metabolic adaptation to the hostile conditions of dormancy (Jiang et al. 2018; Leistikow et al. 2010; Nikitushkin et al. 2022; Sivaramakrishnan 2013; Gutka et al. 2024). Another prominent protein throughout all dormancy phases is HspX (Rv2031c), a molecular chaperone that stabilizes proteins and cellular structures against stress-induced damage. Its high-level expression persists well into late dormancy, underscoring its indispensable role in maintaining protein homeostasis during extended metabolic quiescence (Trutneva et al. 2020). Similarly, Rv2626c, a member of the DosR regulon, remains highly expressed under hypoxic conditions and is detectable in late dormancy, supporting bacterial adaptation and survival under sustained stress (Roupie et al. 2007). The acyl carrier protein Rv0100, pivotal for fatty acid biosynthesis and complex cell wall lipid production, also continues to be upregulated during dormancy, playing a critical role in maintaining cell envelope integrity and metabolic flexibility (Gutka et al. 2024). As dormancy progresses, specific proteins become increasingly expressed that are either absent or minimally present during earlier stages. Notably, Rv1738 is highly expressed in late dormancy. Structural studies reveal its similarity to bacterial ribosome hibernation-promoting factors (HPFs), suggesting a role in suppressing protein synthesis and facilitating the maintenance of non-replicative persistence (Deb et al. 2009; Shleeva et al. 2015; Bunker et al. 2015). Proteomic analyses indicate that while proteins involved in central metabolism, chaperone activity, and stress defense are retained, the overall diversity and abundance of the proteome decline significantly in late dormancy. Only a select group of highly stable and essential proteins, including Rv1738, remains to preserve cellular stability and viability under extended non-culturable conditions (Nikitushkin et al. 2022). A key finding by Bespyatykh et al. (2022) is the retention of proteins from the pH-dependent regulatory system PhoP even after five months in a dormant state. The transcriptional regulator PhoP (Rv0757) is significantly upregulated in dormant cells. The PhoP-PhoR two-component system enhances mycobacterial fitness under acidic conditions, which are critical during dormancy and persistence within host macrophages (Gonzalo-Asensio et al. 2008; Baker et al. 2019; Bespyatykh et al. 2022; Khan et al. 2024). Bespyatykh et al. identified the preservation of 1379 proteins in dormant cells after prolonged storage, with 468 proteins exhibiting statistically significant positive fold changes relative to actively growing cells, reflecting broad biochemical adaptation. The PhoP regulatory system persists throughout extended dormancy, regulating genes involved in stress response, adaptation, and virulence, thereby playing a vital role in enhancing mycobacterial survival under the acidic and hostile conditions typical of the granuloma environment. This contributes to the bacterium's ability to maintain prolonged non-replicative states (Gonzalo-Asensio et al. 2008; Baker et al. 2018; Bespyatykh et al. 2022). In addition, Bespyatykh et al. (2022) detected proteins involved in mycothiol and uridine monophosphate (UMP) biosynthesis salvage pathways, confirming their presence and functional relevance during late dormancy. This aligns with broader omics-based analyses that highlight the importance of these metabolic salvage pathways for long-term survival in dormant cells (Nikitushkin et al. 2022; Kumar et al. 2015; Chakraborty et al. 2021). Furthermore, specific DosR regulon proteins such as Rv2031c (α-crystallin/HspX) and Rv2626c persist at high levels in late dormancy, with α-crystallin/HspX accounting for up to 25% of the total protein content in dormant cells after extended storage. This observation is consistent with earlier reports identifying HspX as one of the most abundant proteins under anaerobic (dormant) conditions, alongside other stress response proteins (Bespyatykh et al. 2022). Chaperones and stress defense proteins, including DnaJ1, HtpG, GroEL2, DnaK, GroES, GroEL1, and HspX, form a "defending shell" that protects other proteins and maintains overall cell stability during prolonged dormancy (Bespyatykh et al. 2022). Nikitushkin et al. (2022) also reported the retention of chaperones and oxidative stress defense enzymes in dormant, non-culturable Mtb cells stored for four months, supporting sustained cell vitality under dormancy.Recent cryo-electron microscopy structural analyses and transcriptional investigations have further elucidated PhoP's role in regulating intracellular signaling molecules such as cyclic AMP (cAMP) and controlling gene expression programs essential for survival and virulence under stress conditions (Shi et al. 2025). These findings highlight genetic and proteomic adaptations as hallmarks of late-stage dormancy in Mtb, underpinning the bacterium's remarkable ability to persist in a metabolically quiescent yet viable state within the host for extended periods. Velayati et al. (2024) employed whole-genome sequencing combined with protein structure analysis to identify consistent frameshift mutations in the *fbiC*, *PPE35*, and *ald* genes across clinical TB isolates that were maintained under in vitro dormancy for 12 months. These mutations impact critical bacterial functions such as redox balance, cell wall integrity, and metabolic flexibility, thereby enhancing Mtb's survival under prolonged stress (Fig. 7.1).

The *fbiC* gene encodes a protein essential for the biosynthesis of coenzyme F420, a vital cofactor involved in redox reactions that detoxify harmful compounds within Mtb. Genomically, *fbiC* is located within a region flanked by genes such as *PE12*, *fadH*, and *TB8.4*, with a repeat region noted near *TB8.4*, highlighting the functional context of *fbiC* within the bacterial genome (Fig. 7.2). A frameshift mutation identified in the *fbiC* gene of dormant Mtb isolates (Fig. 7.3) alters the gene's reading frame, disrupting highly conserved protein domains as demonstrated by conservation scores and structural mapping. This frameshift likely impairs the synthesis or function of coenzyme F420, which may reduce the bacterium's metabolic activity and its capacity to manage oxidative stress. During dormancy, when Mtb encounters hostile conditions such as hypoxia and nutrient deprivation, this metabolic downregulation could confer an adaptive advantage by lowering susceptibility to oxidative damage and conserving energy, thereby enhancing persistence in a latent state. This adaptation enables the bacilli to evade immune responses and withstand conventional therapies until conditions favor reactivation. Thus, the *fbiC* frameshift mutation not only affects protein structure and function but also contributes to Mtb's long-term survival strategy via metabolic modulation and stress resistance. In the same regard, a duplication mutation (c.1798dupA) in the PPE35 gene causes a frameshift, resulting in a truncated, nonfunctional protein. PPE35 belongs to the PPE protein family, which plays a crucial role in maintaining bacterial cell wall integrity and modulating host immune responses, thereby facilitating immune evasion and persistence. The structural organization and key features of PPE35 are illustrated in Fig. 7.4, while Fig. 7.5 depicts the molecular impact of the c.1798dupA variant. This frameshift mutation disrupts the protein sequence starting at amino acid position 600, likely compromising PPE35's structural role in the cell wall. The weakened cell envelope may alter interactions with the host immune system, potentially enhancing Mtb's ability to survive in dormancy by evading immune detection and resisting hostile conditions. Hence, disruption of PPE35 through this frameshift mutation contributes to bacterial persistence during latency by affecting both cell wall integrity and immune modulation (Velayati et al. 2024). The *ald* gene mutation (c.266delA) leads to loss of function in L-alanine dehydrogenase, an enzyme pivotal for nitrogen metabolism (Figs. 7.6 and 7.7). This mutation restricts metabolic flexibility by limiting the bacterium's ability to utilize alternative nitrogen sources during nutrient scarcity typical of dormancy. Such metabolic constraints expose a critical vulnerability in dormant Mtb, suggesting that targeting nitrogen metabolism or amino acid assimilation pathways could effectively reduce bacterial persistence and the risk of reactivation. Collectively, these mutations reveal adaptive strategies by which Mtb balances sur-

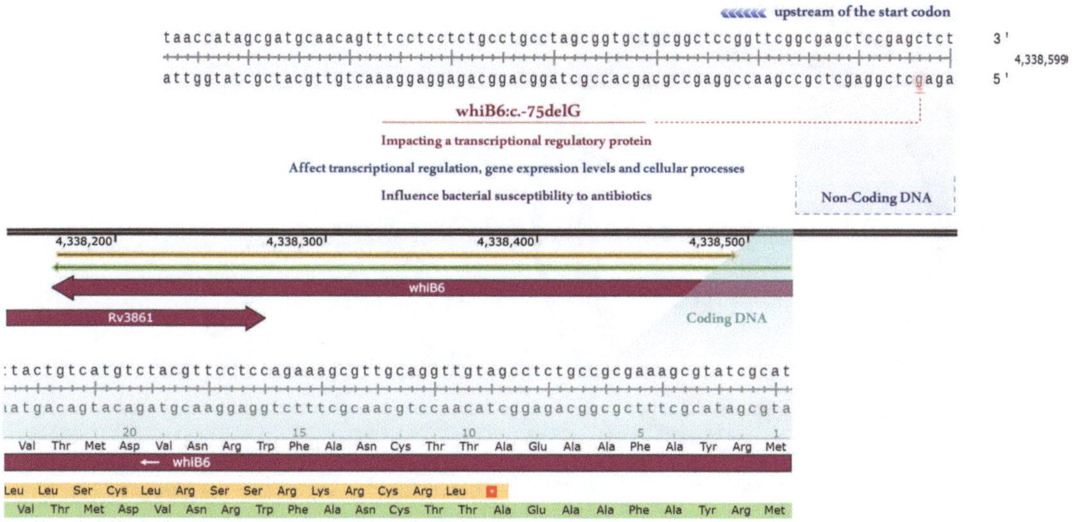

Fig. 7.1 The whiB6:c.-75delG mutation in dormant and drug-resistant *M.tuberculosis* isolate. The figure presents a comprehensive view of a genetic variant in the whiB6 gene region, highlighting both its genomic context and potential biological consequences. At the top, the DNA sequence is shown with a specific focus on a single-nucleotide deletion, labeled as whiB6:c.-75delG. This mutation occurs 75 bases upstream of the start codon, placing it within the noncoding, regulatory region of the gene. The upstream location is emphasized with an arrow and annotation, underlining its potential to impact gene regulation rather than directly altering the protein sequence. Beneath the sequence, the figure explains the functional significance of this variant. Since whiB6 encodes a transcriptional regulatory protein, the deletion could disrupt normal transcriptional regulation, gene expression levels, and various cellular processes. The annotation also notes that such regulatory changes may influence the bacterium's susceptibility to antibiotics, highlighting the clinical relevance of the mutation.The middle section of the figure provides a genomic map, displaying the positions of the whiB6 gene and its neighboring gene, Rv3861, along the chromosome. The coordinates along the top axis (in base pairs) help situate these genes within the broader genome. The arrows indicate the orientation and extent of each gene, with whiB6 and Rv3861 overlapping slightly, which may suggest potential regulatory interactions or shared promoter regions. The lower portion of the figure shifts focus to the coding region of whiB6. Here, both the DNA and corresponding amino acid sequences are displayed, with color-coded boxes representing the translation of nucleotide triplets into amino acids. The wild-type and mutated sequences are aligned for comparison, and the figure highlights how the deletion could affect the reading frame or protein structure if it were within the coding region. However, since the actual deletion is upstream, the primary impact is regulatory rather than structural

vival under prolonged stress with metabolic trade-offs, identifying potential targets for novel therapeutic interventions against latent and drug-resistant tuberculosis.

In conclusion, late dormancy in Mtb is characterized by the selective retention and upregulation of central metabolic enzymes, salvage pathway proteins, DosR regulon members, molecular chaperones, and specific genetic mutations. These features collectively enable the pathogen to withstand extended metabolic quiescence and resist diverse environmental stresses, underpinning its remarkable persistence within the host.

7.6 Morphological Changes and Their Correlation with Genomic Changes

Over the past several decades, extensive research using well-controlled laboratory models has elucidated the morphological adaptations of Mtb in response to dormancy-inducing stress (Wayne and Sramek 1964; Youmans 1979; Wayne and Lin 1982; Wayne and Hayes 1996; Betts et al. 2002; Kapoor et al. 2013). One of the most influential of these is Wayne's hypoxia model, which involves culturing Mtb in sealed, stationary liquid media where oxygen levels gradually

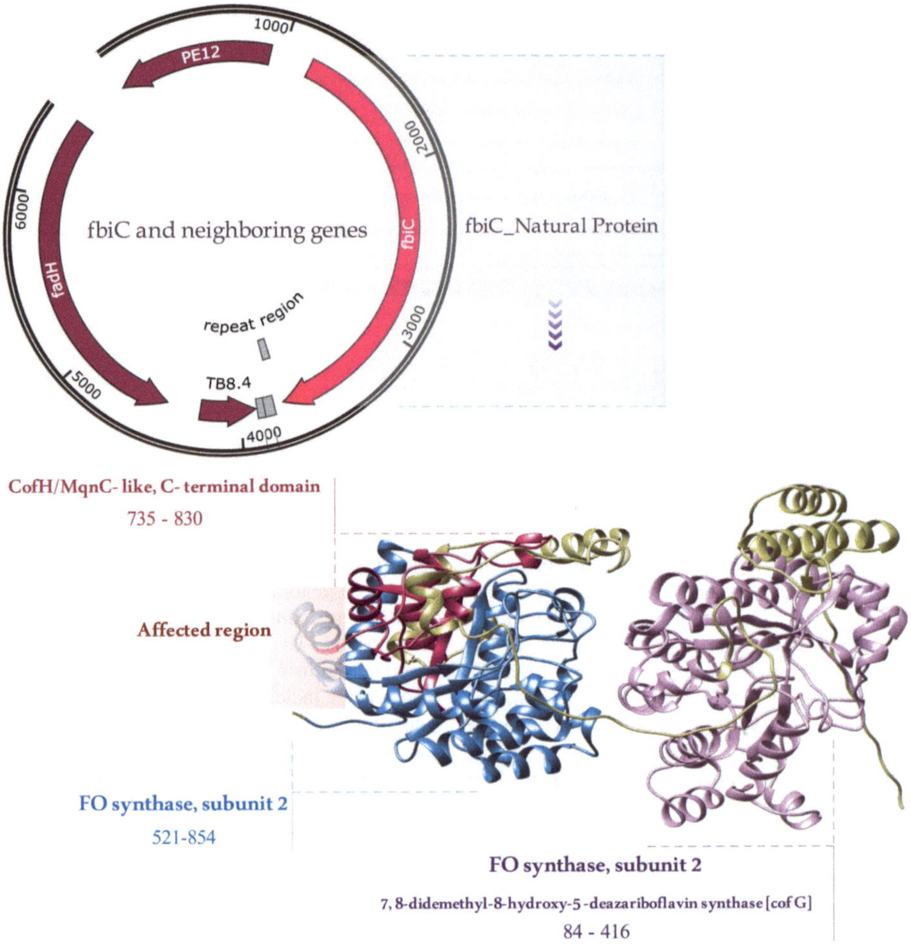

Fig. 7.2 Mapping fbiC domains in *Mycobacterium tuberculosis*. The figure provides an integrated overview of the fbiC gene and its neighboring genomic context, as well as the structural domains of the fbiC protein. In the upper left section, a circular map illustrates the location of the fbiC gene alongside adjacent genes such as PE12, fadH, and TB8.4, highlighting their arrangement and the presence of a repeat region near TB8.4. On the right side, the structural organization of the natural fbiC protein is depicted, with distinct domains highlighted in different colors. The C-terminal region of the protein contains a CofH/MqnC-like domain, spanning amino acids 735–830, which is marked in dark pink. Another prominent feature is the FO synthase, subunit two domain, which appears in two segments: one from amino acids 521 to 854 (shown in blue) and another from 84 to 416 (shown in purple). These domains are associated with the enzymatic activity involved in the biosynthesis of 7,8-didemethyl-8-hydroxy-5-deazariboflavin, an essential metabolic intermediate. A specific region within the protein structure is labeled as the "affected region," suggesting an area of functional or structural significance, possibly due to mutation or interaction. *Drawing by Mohsen Maleknia; Mycobacteriology Research Center, National Research Institute of Tuberculosis and Lung Disease, Shahid Beheshti University of Medical Sciences, Tehran, Iran

decline, thereby simulating the hypoxic environment encountered within the host (Wayne and Hayes 1996). Over incubation periods ranging from one to four weeks, light and electron microscopy revealed significant morphological transformations, including a reduction in cell size, a shift from the typical rod-shaped form to ovoid or coccoid shapes, and thickening of the cell wall (Wayne and Hayes 1996; Shleeva et al. 2011). Additionally, the bacteria accumulated intracellular lipid inclusions and exhibited a progressive loss of acid-fast staining, hallmarks of dormancy onset and progression (Shleeva et al. 2004; Wayne and Hayes 1996). Expanding

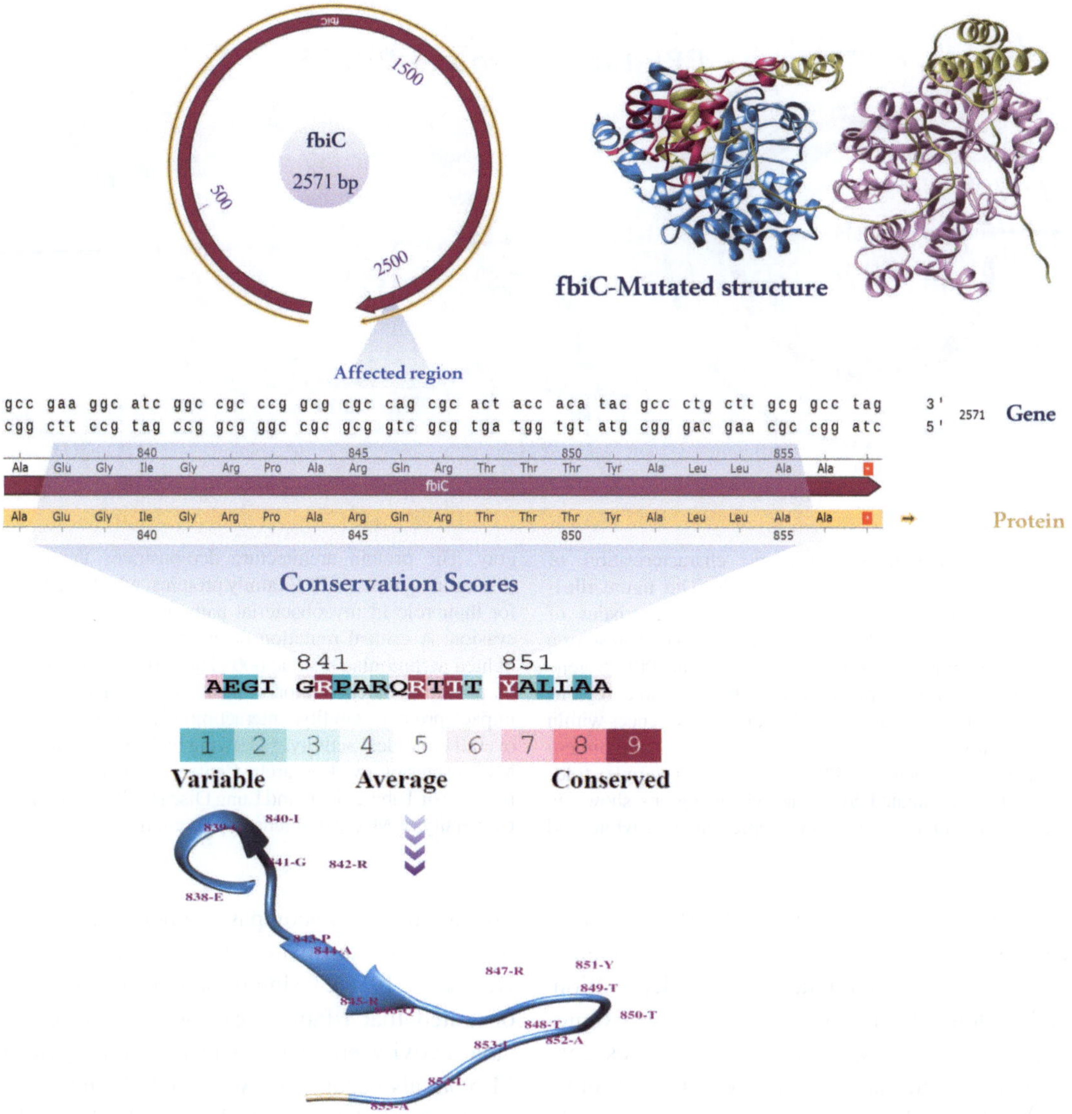

Fig. 7.3 Impact of a frameshift mutation in fbiC on conserved protein domains during dormancy in *Mycobacterium tuberculosis*. The figure provides a comprehensive visualization of the frameshift mutation in the fbiC gene and its consequences at the nucleotide, protein, and structural levels. At the top left, a circular map highlights the full length of the fbiC gene, which spans 2571 base pairs and pinpoints the specific region affected by the mutation. This affected region is then shown in greater detail, displaying both the nucleotide sequence and the corresponding amino acid translation. The frameshift mutation alters the reading frame, resulting in a different sequence of amino acids from position 840 onward. This change is evident in the comparison between the wild-type and mutated protein sequences, where a new set of amino acids replaces the normal sequence due to the frameshift. To the right, the three-dimensional structure of the mutated fbiC protein is shown, illustrating how the frameshift impacts the overall protein conformation. The affected region is mapped onto the protein structure, emphasizing the potential disruption to its standard folding and function. Below, conservation scores for the amino acids in the affected region are presented, with a color scale ranging from variable (blue) to highly conserved (pink). The altered amino acid sequence introduced by the frameshift includes several residues that are typically highly conserved, suggesting that the mutation could have serious functional consequences. Finally, the lower part of the figure provides a detailed view of the mutated region within the protein structure, labeling the specific amino acids that have been changed. This visualization underscores the importance of the frameshift mutation, as it not only alters the primary sequence but also affects conserved residues critical for protein function. * Drawing by Mohsen Maleknia; Mycobacteriology Research Center, National Research Institute of Tuberculosis and Lung Disease, Shahid Beheshti University of Medical Sciences, Tehran, Iran

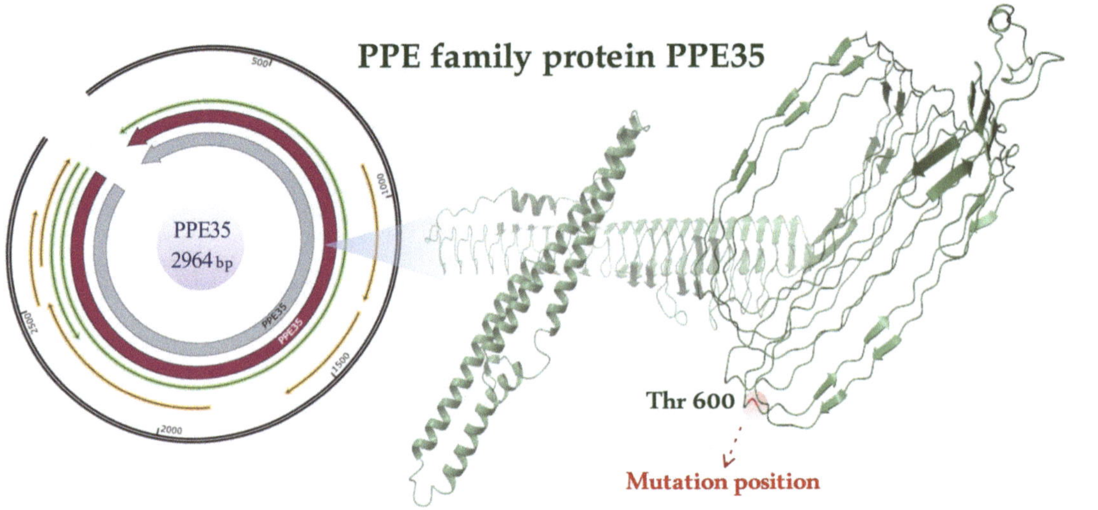

Fig. 7.4 Structural and genomic characterization of PPE35 in *Mycobacterium tuberculosis*. The figure illustrates the structural organization and characteristics of PPE family protein PPE35 in Mtb. The left panel presents a circular genomic representation showing the PPE35 gene spanning 2964 base pairs, with distinct colored regions indicating different functional domains or sequences within the gene structure. The right panel depicts the three-dimensional protein structure of PPE35, revealing a complex folding pattern dominated by alpha-helical regions shown in green and beta-sheet structures represented in white and gray. The protein architecture demonstrates the typical structural features of PPE family proteins, which are known for their role in mycobacterial pathogenesis and immune evasion. A critical mutation position is specifically highlighted at threonine residue 600 (Thr 600), indicating a site of potential functional or clinical significance that may impact protein stability, interaction with host factors, or overall biological activity. * Drawing by Mohsen Maleknia; Mycobacteriology Research Center, National Research Institute of Tuberculosis and Lung Disease, Shahid Beheshti University of Medical Sciences, Tehran, Iran

on these findings, Youmans (1979) systematically subjected Mtb to various environmental stresses such as hypoxia, nitrogen deprivation, and prolonged starvation. Over periods extending up to a month, microscopic analyses confirmed that the bacteria adopted smaller, more rounded morphologies accompanied by thickened cell walls and increased lipid inclusion formation. These observations reinforced the concept that dormancy in Mtb involves distinct morphological adaptations triggered by environmental stressors, including a loss of acid-fastness detectable by conventional staining methods.

Concurrently, pioneering studies by Dickinson and Mitchison in 1978 and 1979 significantly advanced understanding of tuberculosis chemotherapy, particularly regarding drug efficacy against dormant Mtb populations. Their 1978 work introduced in vitro models mimicking the heterogeneous bacterial populations present in infected tissues, encompassing both actively replicating and dormant bacilli (Mitchison and Dickinson 1978). Using these models, they demonstrated that rifampin exhibits potent bactericidal activity not only against actively growing Mtb but also against slowly metabolizing or dormant subpopulations, which typically show tolerance to drugs such as isoniazid. By inducing dormancy or slowed bacterial growth under controlled conditions, they evaluated rifampin's killing kinetics relative to other antitubercular agents, providing a mechanistic explanation for rifampin's critical role in shortening tuberculosis treatment by effectively targeting persistent bacterial cells. Following this, nutrient deprivation models were developed by cultivating Mtb in media lacking nitrogen and carbon sources (Betts et al. 2002; Deb et al. 2009). These models revealed similar morphological adaptations, including cell wall thickening, cell size reduction, and formation of intracellular lipid bodies, as

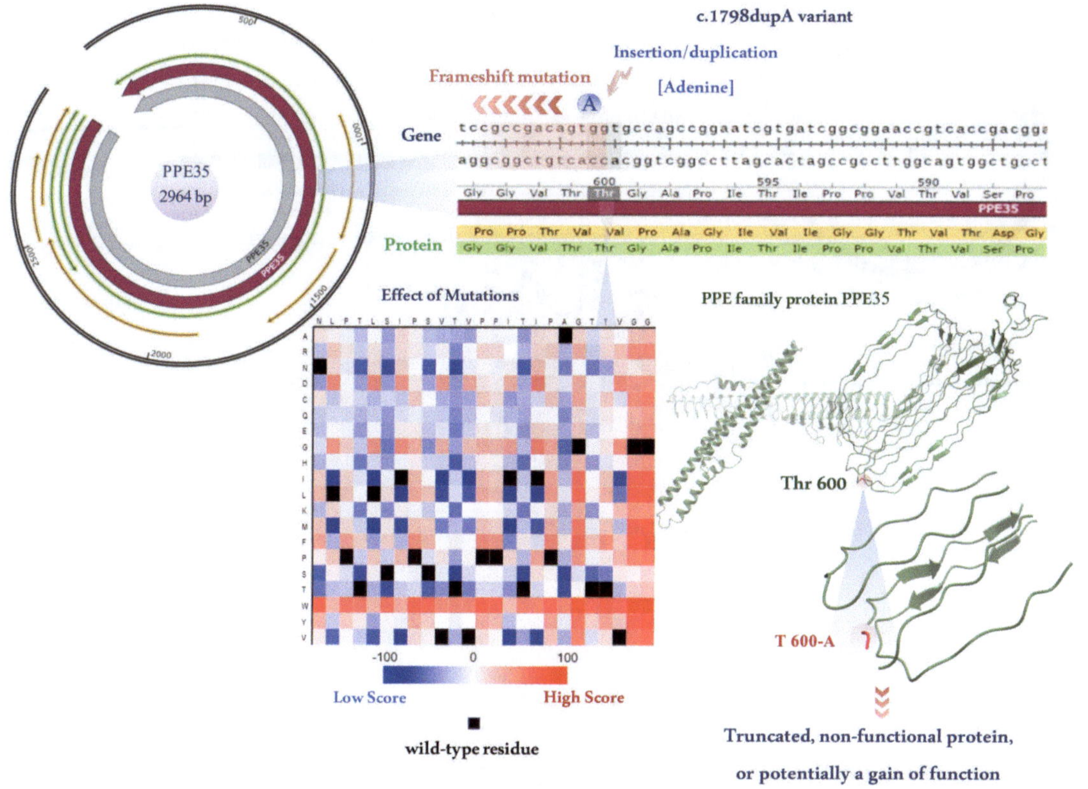

Fig. 7.5 Structural insights into PPE35 dysregulation by a c.1798dupA mutation in dormant. This figure demonstrates the molecular impact of the c.1798dupA variant in the PPE35 gene through multiple complementary visualizations. On the left circular diagram, the PPE35 gene is represented. This circular map visually locates the PPE35 gene within a larger genomic context, highlighting its position and orientation with colored arrows. Moving to the right, a magnified gene sequence is shown, pinpointing the c.1798dupA variant. Here, an adenine insertion is marked, demonstrating how this single-nucleotide duplication causes a frameshift mutation in the DNA sequence. Directly below the gene sequence, the corresponding protein sequence is displayed. This section shows how the frameshift alters the amino acid sequence starting from position 600, where the normal sequence is disrupted and replaced by a new, incorrect sequence due to the shifted reading frame. Beneath the protein sequence, a heatmap is presented, labeled "Effect of Mutations." This matrix uses a color scale from blue to red to indicate the predicted impact of different amino acid substitutions at each position, with blue representing low impact and red indicating high impact. Black squares denote the wild-type residues. To the right of the heatmap, a ribbon diagram illustrates the three-dimensional structure of the PPE35 protein. The position of threonine at residue 600 is highlighted, emphasizing where the frameshift mutation occurs. The structural model suggests that the mutation leads to a truncated protein, as shown by the abrupt end of the ribbon diagram after position 600. The annotation at the bottom of the figure explains that this truncation likely results in a nonfunctional protein or, less commonly, could produce a protein with a new, abnormal function. *Drawing by Mohsen Maleknia; Mycobacteriology Research Center, National Research Institute of Tuberculosis and Lung Disease, Shahid Beheshti University of Medical Sciences, Tehran, Iran

observed through light and electron microscopy (Deb et al. 2009; Shleeva et al. 2011). Conducted over several weeks, these experiments allowed detailed observation of the gradual transition of Mtb cells into dormant phenotypes in response to nutrient limitation (Betts et al. 2002; Deb et al. 2009). Paramasivan et al. (1998) further utilized Wayne's in vitro system to demonstrate that dormant Mtb populations reside in oxygen-limited regions of static cultures. They observed that these bacilli resist the bactericidal effects of anaerobic conditions and exhibit partial or complete tolerance to frontline drugs such as isoniazid and rifampicin. Using a murine tuberculosis

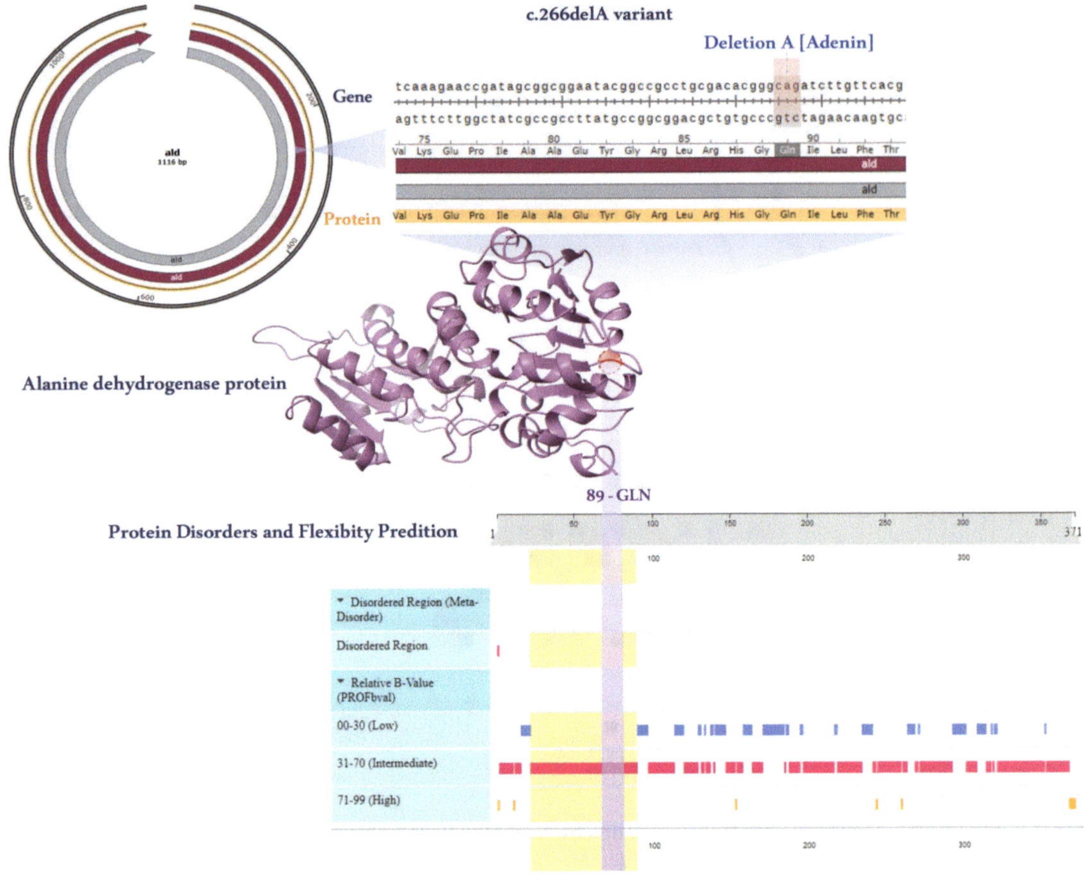

Fig. 7.6 The c.266delA frameshift mutation in the alanine dehydrogenase (ald) gene of dormant *Mycobacterium tuberculosis*. This figure outlines the molecular effects of the c.266delA variant in the ald gene, which encodes the alanine dehydrogenase protein. On the left, a circular genomic map shows the location and orientation of the ald gene within the Mtb genome. Moving to the right, a detailed view of the gene sequence pinpoints the exact site of the adenine deletion at nucleotide position 266. This deletion causes a frameshift in the coding sequence, as illustrated by the comparison of the wild-type and mutated protein sequences, where the normal amino acid sequence is disrupted starting at residue 89 (glutamine). At the central portion, the three-dimensional structure of the alanine dehydrogenase protein is shown, with the site of the frameshift mutation clearly marked. The bottom showed a graphical prediction of protein disorder and flexibility. This analysis maps regions of intrinsic disorder and flexibility across the protein's length, using color-coded bars to indicate areas with varying degrees of structural disorder and mobility. The frameshift site at residue 89 is highlighted, suggesting that the mutation may impact both the stability and dynamic properties of the protien. *Drawing by Mohsen Maleknia; Mycobacteriology Research Center, National Research Institute of Tuberculosis and Lung Disease, Shahid Beheshti University of Medical Sciences, Tehran, Iran

model, they showed that combining metronidazole with rifampicin was more effective at eradicating persistent Mtb bacilli than rifampicin alone, suggesting promising therapeutic strategies for targeting dormant bacterial populations (Paramasivan et al. 1998). Lim and Dick (2001) developed a plate-based dormancy culture system for *Mycobacterium smegmatis* using anoxic jars as an alternative to the liquid-based Wayne model. They demonstrated the system's utility for genetic studies of dormancy by successfully isolating metronidazole-resistant mutants. Building on Wayne and Hayes' observation that gradual oxygen depletion allows bacteria to adapt to anoxia, whereas rapid oxygen loss leads to cell death, Lim and Dick's work supported the con-

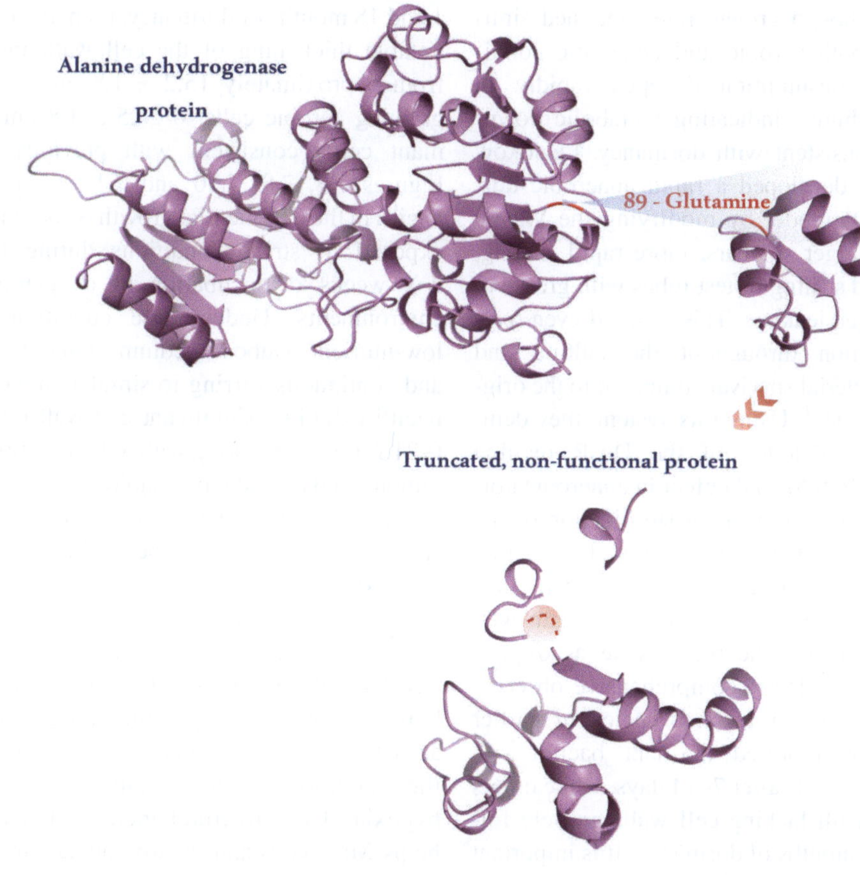

Fig. 7.7 Loss of function in alanine dehydrogenase caused by frameshift-induced truncation in dormant *Mycobacterium tuberculosis*. This figure demonstrates the structural consequences of a frameshift mutation in the alanine dehydrogenase protein caused by the c.266delA variant in dormant TB bacilli. The upper part of the image shows the complete, functional structure of the alanine dehydrogenase enzyme, with the location of glutamine at position 89 highlighted in red. This residue marks the site where the frameshift mutation occurs due to the deletion of an adenine nucleotide. The lower part of the figure depicts the resulting truncated protein, which is significantly shorter and lacks the complex folding seen in the full-length enzyme. This truncated form is labeled as nonfunctional, indicating that the frameshift mutation disrupts the typical structure and likely abolishes the enzyme's activity. The visual comparison between the intact and truncated proteins clearly illustrates how the mutation leads to a loss of function, which may have significant effects on the survival and metabolic processes of dormant Mtb

* Drawing by Mohsen Maleknia; Mycobacteriology Research Center, National Research Institute of Tuberculosis and Lung Disease, Shahid Beheshti University of Medical Sciences, Tehran, Iran

cept of a time-dependent "dormancy program" necessary for survival under low-oxygen conditions. Starck et al. (2004) adapted the Wayne dormancy model to investigate differences in the proteome of Mtb grown under aerobic versus anaerobic conditions. They cultured the Harlingen strain aerobically until it reached the transition phase between exponential growth and stationary phase (approximately 7 days), then transferred a portion of the cultures to anaerobic conditions for an additional 16 days. Using two-dimensional polyacrylamide gel electrophoresis (2D-PAGE), they identified 50 protein spots uniquely expressed or significantly more abundant under anaerobic conditions. Sixteen of these proteins were characterized by matrix-assisted laser desorption/ionization time-of-flight (MALDI-TOF) mass spectrometry, including the α-crystallin homologue HspX, elongation factor Tu (*Tuf*), GroEL2, and several metabolic enzymes. Although growth rates declined similarly under both aerobic and anaerobic conditions, ATP consumption dropped rapidly in anaerobic cultures, indicating metabolic downregulation consistent with dormancy. Leistikow et al. (2010) developed a rapid anaerobic dormancy (RAD) model by modifying the Wayne model with larger stir bars, more rapid stirring, and improved sealing of test tubes with grease to prevent oxygen leakage. This ensured even oxygen distribution throughout the culture and improved bacterial survival compared to the original Wayne model. Using this system, they demonstrated that deletion of the DosR regulon resulted in a 10,000-fold defect in anaerobic dormancy survival and eventual sterilization of the mutant culture. Their experiments showed that the DosR regulon is essential for Mtb to shift away from oxygen consumption, maintain ATP levels, and balance the redox state as oxygen becomes scarce. These comprehensive observations help reconcile inconsistencies in earlier studies: some reported dormant bacilli with thickened cell walls after 7–14 days, while others described bacilli lacking cell walls entirely following 4–13 months of dormancy. It is important to note that most previous research maintained Mtb under hypoxic conditions for durations ranging from a few days up to 12 months, reflecting the dynamic and heterogeneous nature of dormancy-associated morphological changes. Velayati et al. (2011, 2016), Velayati and Farnia (2016) pioneered a prolonged in vitro culture model exposing Mtb bacilli (standard laboratory strains, i.e., H_{37}RV and various clinical isolates) to combined oxygen limitation and nutrient scarcity, sustaining these conditions for up to 48 months. This extended model enabled a comprehensive analysis of the full spectrum of morphological changes associated with Mtb latency, with sequential observations ranging from as early as three days to as long as 48 months. Throughout this prolonged period, two distinct morphological stages of dormant Mtb bacilli were identified, providing valuable insights into the dynamic adaptations of the pathogen during latency (Velayati et al. 2011, 2016; Velayati and Farnia 2016). The first stage, occurring between 1 and 18 months of dormancy, is characterized by notable thickening of the cell wall, increasing from approximately 15.2 ± 1.8 nm in actively growing aerobic cells to 20.5 ± 1.8 nm in dormant cells, consistent with previous reports. Figures 7.8, 7.9, 7.10 and 7.11 compare Mtb bacilli in the exponential growth phase with those exposed to stress conditions during the first 2–4 weeks of incubation in oxygen-deprived environments. Under these conditions, using low-nutrient Dubos medium without glycerol and continuous stirring to simulate hypoxia, the bacilli exhibited significant cell wall thickening (~21.5 ± 1.8 nm) along with a denser, more voluminous, and irregular morphology. These changes reflect adaptive responses to environmental stressors and represent the earliest observable morphological alterations during dormancy. This cell wall thickening likely serves as a crucial adaptive mechanism, acting as a protective barrier that reduces permeability and restricts metabolic exchanges. By minimizing metabolic activity, the bacteria enhance survival under hostile conditions such as nutrient scarcity and hypoxia. This structural modification not only helps Mtb withstand environmental stresses but may also contribute to persistence within the host by increasing resistance to immune defenses and antibiotic treatment. Understanding these morphological changes provides critical insights into the mechanisms by which Mtb enters dormancy and maintains long-term survival, informing the development of more effective therapeutic strategies against tuberculosis. Continuing observations, illustrated in Figs. 7.12, 7.13, 7.14, 7.15 and 7.16, reveal that Mtb undergoes distinct morphological transformations during prolonged in vitro dormancy spanning 3–12 months. A striking feature is the tendency of bacilli to fold, often at their midpoints, although in some cases only the head or tail bends to wrap around the opposite end (Fig. 7.13). During the early dormancy phase (3–4 months), approximately 10–15% of the bacterial population exhibits this folding behavior. As dormancy extends to 9–12 months, the proportion of folded bacilli increases dramatically to 65–70%. This folding brings the two ends closer, reducing cell length to

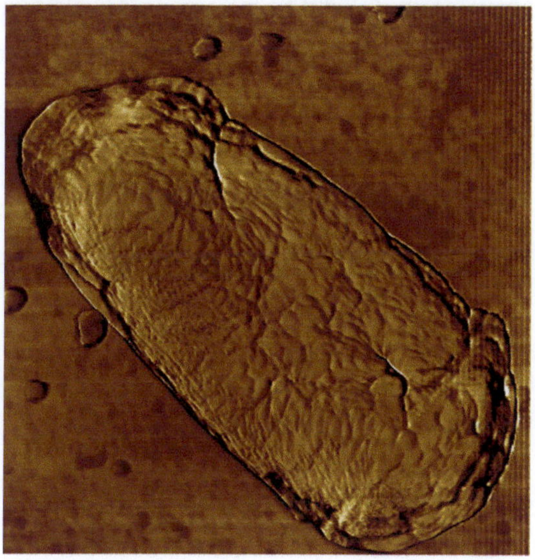

Fig. 7.8 Atomic force microscopy (AFM) friction mode image of *Mycobacterium tuberculosis* bacillus in exponential phase
The image shows an atomic force microscopy (AFM) friction mode micrograph of a *Mycobacterium tuberculosis* (Mtb) bacillus in its exponential growth phase. The rod-shaped bacterium exhibits a textured surface with characteristic wrinkles and ridges visible across its cell envelope. This high-resolution visualization reveals the ultrastructural details of the bacterial cell wall, which measures approximately 15.6 ± 1.6 nm in thickness. The bacillus displays the typical elongated morphology of Mtb, with a length ranging from 1 to 3 micrometers
*Experimental work by Prof. Parissa -Farnia and Prof. Gennady Konstantinovich Zhavnerko; Mycobacteriology Research Center, National Research Institute of Tuberculosis and Lung Disease, Shahid Beheshti University of Medical Sciences, Tehran, Iran & The Republican Research and Practical Centre for Epidemiology & and Microbiology, Filimonova 23, Minsk, Belarus

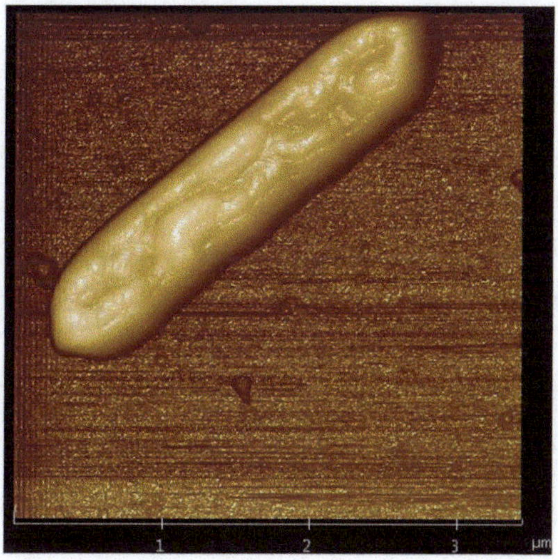

Fig. 7.9 Atomic force microscopy (AFM) height mode image of *Mycobacterium tuberculosis* bacillus in exponential phase
This atomic force microscopy (AFM) image, captured in height mode, depicts a single *Mycobacterium tuberculosis* (MTB) bacillus during its exponential growth phase, just as it is beginning to divide. The bacterium appears as an elongated rod, measuring between 1 and 3 micrometers in length. The surface of the cell displays distinct ridges and textures, which are characteristic features of Mtb during active replication. Height mode AFM provides a detailed three-dimensional view of the cell's surface topography, allowing for precise visualization of its structural features
*Experimental work by Prof. Parissa -Farnia and Prof. Gennady Konstantinovich Zhavnerko; Mycobacteriology Research Center, National Research Institute of Tuberculosis and Lung Disease, Shahid Beheshti University of Medical Sciences, Tehran, Iran & The Republican Research and Practical Centre for Epidemiology & and Microbiology, Filimonova 23, Minsk, Belarus

approximately 0.8 μm and causing a shape shift from rod-like to ovoid or spherical forms (Figs. 7.15 and 7.16). After a full year of dormancy, most bacilli appear nearly round or oval, often with thickened cell walls (Fig. 7.16). This transformation likely reflects a purposeful adaptation to conserve resources under nutrient limitation, minimizing surface area to reduce metabolic demands. This adaptation enhances the ability of dormant bacilli to withstand metabolic stress and environmental challenges by lowering metabolic requirements and increasing resistance to adverse conditions. Interestingly, within this first dormancy stage, a smaller subset (~20–25%) continues to divide irregularly over 3–18 months, often via asymmetrical splitting or budding. Despite this limited replication, the majority remain viable without division, highlighting a complex balance between dormancy maintenance and occasional replication during latent infection.

As dormancy progresses beyond 18 months and extends up to 48 months under complete oxygen and nutrient deprivation, Mtb bacilli begin to lose cell wall integrity. Specialized dormant forms emerge, notably spore-like cells mea-

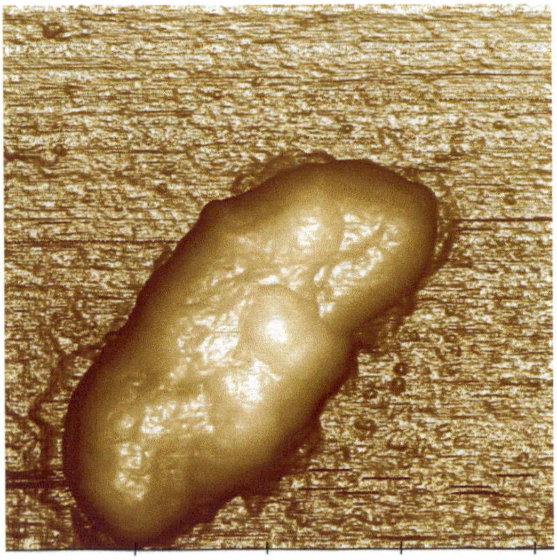

Fig. 7.10 Atomic force microscopy image of a *Mycobacterium tuberculosis* following two weeks of in vitro dormancy
The atomic force microscopy (AFM) image presented here, captured in height-friction mode, shows a single bacillus bacterium after two weeks of dormancy. The bacillus appears as an elongated, rod-shaped structure with a slightly thickened cell wall (19 ± 1.5 nm). This increased thickness is visible as a pronounced elevation in the AFM topography, indicating that the bacterium has reinforced its cell envelope, likely to enhance resistance to environmental stressors and preserve cellular integrity during inactivity. Surface details in the image reveal a relatively smooth but dense texture, with subtle undulations and bulges along the cell body. These features suggest internal reorganization and possibly the accumulation of storage lipids or protective macromolecules, which are commonly synthesized as bacteria prepare for or endure dormancy. The overall appearance of the bacillus is more robust compared to actively growing cells, supporting the idea that dormancy triggers significant structural changes. The use of height-friction mode in AFM not only provides topographical information but also highlights differences in material properties across the cell surface. This dual contrast allows for a more apparent distinction between the thickened cell wall and the surrounding substrate, as well as finer details of the cell's surface architecture
*Experimental work by Prof. Parissa -Farnia and Prof. Gennady Konstantinovich Zhavnerko; Mycobacteriology Research Center, National Research Institute of Tuberculosis and Lung Disease, Shahid Beheshti University of Medical Sciences, Tehran, Iran & The Republican Research and Practical Centre for Epidemiology & and Microbiology, Filimonova 23, Minsk, Belarus

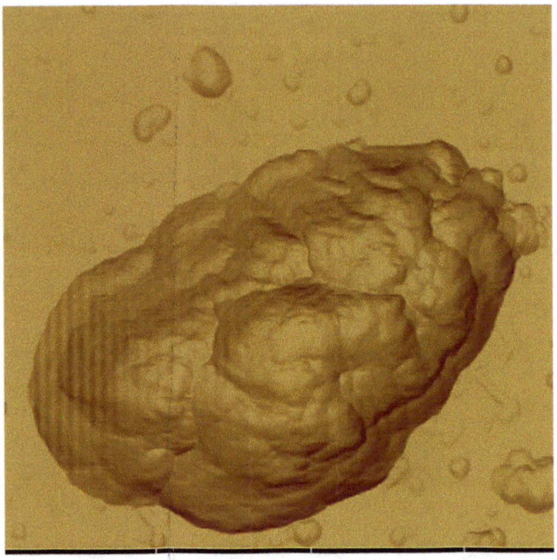

Fig. 7.11 Atomic force microscopy image of a *Mycobacterium tuberculosis* following one month of in vitro dormancy
The image shows a single bacillus after one month of in vitro dormancy. The TB bacilli appear as a dense, irregularly contoured structure, with a pronounced thickening of the cell wall, measured at approximately 20 ± 1 nm. This thickened cell wall is clearly visible in the AFM topography, where the outer envelope appears more robust and layered compared to actively growing cells. The thickening of the cell wall likely acts as a protective coat, providing enhanced resistance to hostile environmental conditions. By reinforcing the cell envelope, the bacillus can restrict the exchange of metabolites and limit metabolic activity, which is crucial for survival during prolonged periods of dormancy. This adaptation not only helps the cell withstand physical and chemical stressors but also preserves its internal environment until favorable growth conditions return
*Experimental work by Prof. Parissa -Farnia and Prof. Gennady Konstantinovich Zhavnerko; Mycobacteriology Research Center, National Research Institute of Tuberculosis and Lung Disease, Shahid Beheshti University of Medical Sciences, Tehran, Iran & The Republican Research and Practical Centre for Epidemiology & and Microbiology, Filimonova 23, Minsk, Belarus

suring approximately 0.1 to 0.5 μm in diameter (Figs. 7.17 and 7.18). Although these cells exhibit spore-like morphology and are predominantly nonacid-fast, they lack heat resistance and therefore are not true spores. Importantly, while these cell wall-deficient forms do not grow on standard culture media, they can regain acid-fastness and rod-shaped morphology when inoculated into the peritoneal cavity of guinea pigs, subsequently causing infection after prolonged incubation of 40–45 days. These findings indicate that dormant Mtb exists in two morphologi-

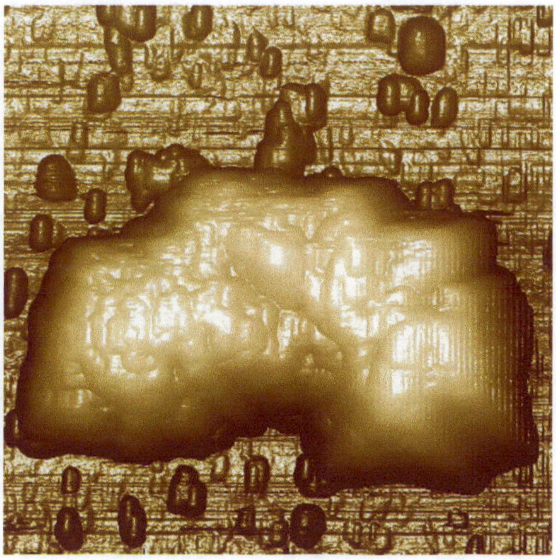

Fig. 7.12 Atomic force microscopy image of a *Mycobacterium tuberculosis* following 3-4 months of in vitro dormancy
The height-friction mode atomic force microscopy (AFM) image captures a Mtb bacilli after three months of in vitro dormancy. During this dormancy phase (3 – 4 months), approximately 10 – 1 5% of the bacilli population display a distinct folding phenomenon. Usually, this structural change is most prominent at the midpoint of the cell, where the bacillus bends or folds inward, departing from the characteristic straight, rod-shaped morphology seen in actively replicating TB cells
*Experimental work by Prof. Parissa -Farnia and Prof. Gennady Konstantinovich Zhavnerko; Mycobacteriology Research Center, National Research Institute of Tuberculosis and Lung Disease, Shahid Beheshti University of Medical Sciences, Tehran, Iran & The Republican Research and Practical Centre for Epidemiology & and Microbiology, Filimonova 23, Minsk, Belarus

Fig. 7.13 AFM contact-friction imaging reveals asymmetrical folding in dormant *Mycobacterium tuberculosis*
The AFM contact-friction image reveals that structural folding in dormant Mtb bacilli is not limited to the midpoint of the cell. In some cases, as illustrated in the image, one side of the bacilli can fold or wrap around the cell body, resulting in pronounced asymmetrical deformation. The observed folding and irregular surface features underscore the remarkable morphological plasticity of Mtb during extended dormancy
*Experimental work by Prof. Parissa -Farnia and Prof. Gennady Konstantinovich Zhavnerko; Mycobacteriology Research C enter, National Research Institute of Tuberculosis and Lung Disease, Shahid Beheshti University of Medical Sciences, Tehran, Iran & The Republican Research and Practical Centre for Epidemiology & and Microbiology, Filimonova 23, Minsk, Belarus

cally and physiologically distinct phases during latency, closely mirroring the pathogen's behavior within the host. A thorough understanding of these phases is crucial for advancing knowledge of Mtb persistence and guiding the development of targeted therapies aimed at eradicating latent tuberculosis infections.

Today, several studies have deepened our understanding of Mtb adaptation to stress and dormancy by integrating detailed morphological analyses with comprehensive transcriptomic and proteomic profiling (Raghunandanan et al. 2019; Kundu and Basu 2021). These investigations demonstrate that Mtb's response to hostile environments such as hypoxia and nutrient limitation is rapid and complex, involving significant molecular and structural changes within the first week of exposure. For example, transcriptomic analyses in macrophage infection models reveal major gene expression shifts as early as two days postinfection. Over 100 genes related to fatty acid metabolism, stress response proteins (including heat shock proteins and chaperones), anaerobic respiration, and metabolic adaptation are swiftly upregulated in response to environmental stressors. These early transcriptional changes correlate closely with morphological adaptations such as cell wall thickening, cell size reduction, and altered cell division patterns, all contributing to enhanced bacterial survival under adverse conditions. Proteomic analyses corroborate these findings by revealing increased syn-

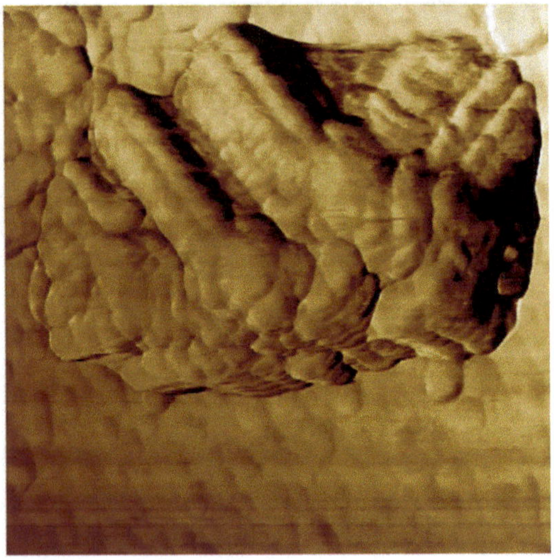

Fig. 7.14 AFM contact-friction image revealing folding in dormant *Mycobacterium tuberculosis*. The AFM contact-friction image reveals that structural "folding phenomena" exist in dormant Mtb bacilli. The imaging clearly shows multiple folded regions with varying depths and orientations, suggesting that the cell wall architecture of dormant TB exhibits distributed structural complexity across its entire surface. These folded patterns appear as overlapping, wave-like formations that create intricate surface topography, providing evidence for a more distributed pattern of structural organization in dormant Mtb bacilli than previously understood
*Experimental work by Prof. Parissa -Farnia and Prof. Gennady Konstantinovich Zhavnerko; Mycobacteriology Research Center, National Research Institute of Tuberculosis and Lung Disease, Shahid Beheshti University of Medical Sciences, Tehran, Iran & The Republican Research and Practical Centre for Epidemiology & and Microbiology, Filimonova 23, Minsk, Belarus

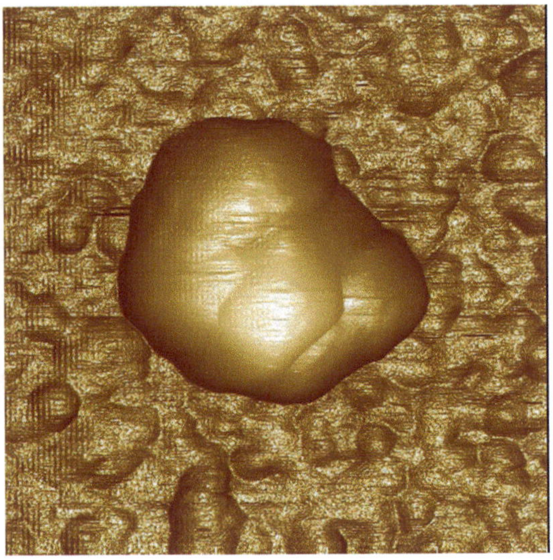

Fig. 7.15 Unique folding phenomenon in dormant *Mycobacterium tuberculosis* observed by atomic force microscopy
This image presents a high-resolution visualization of a dormant MTB bacilli captured after an extended period of in vitro dormancy. A single MTB cell exhibiting a distinctive folding phenomenon, where the head and tail regions of the bacillus curve inward and approach each other, resulting in a rounded or oval appearance rather than the typical rod shape seen in actively growing cells. This morphological adaptation is commonly observed after 6 to 7 months of dormancy, with more than 50% of the bacterial population displaying this compact form. The folding and compaction of the cell are thought to be survival strategies, reducing the cell's metabolic activity and surface area, thereby enhancing resistance to environmental stresses and nutrient deprivation
*Experimental work by Prof. Parissa -Farnia and Prof. Gennady Konstantinovich Zhavnerko; Mycobacteriology Research Center, National Research Institute of Tuberculosis and Lung Disease, Shahid Beheshti University of Medical Sciences, Tehran, Iran & The Republican Research and Practical Centre for Epidemiology & and Microbiology, Filimonova 23, Minsk, Belarus

thesis of proteins essential for preserving cell wall integrity, mitigating oxidative stress, and organizing metabolic downregulation during early dormancy. Key regulatory networks, including the DosR regulon and various transcription factors, coordinate these adaptive responses and facilitate Mtb's transition into dormancy. Recent high-resolution time-course studies have delineated these regulatory mechanisms across multiple distinct transcriptional phases, demonstrating that dormancy entry is a highly coordinated event involving modulation of over 60% of the Mtb genome. These findings emphasize that Mtb's transition into dormancy is a tightly regulated process initiated within days of exposure to environmental stressors and can be sustained as long as the stress persists. The integration of morphological, transcriptomic, and proteomic data has been instrumental in linking specific genetic programs to observable physical transformations, thereby providing a comprehensive understanding of how Mtb adapts, survives, and persists under conditions that closely resemble latent infection in the human host.

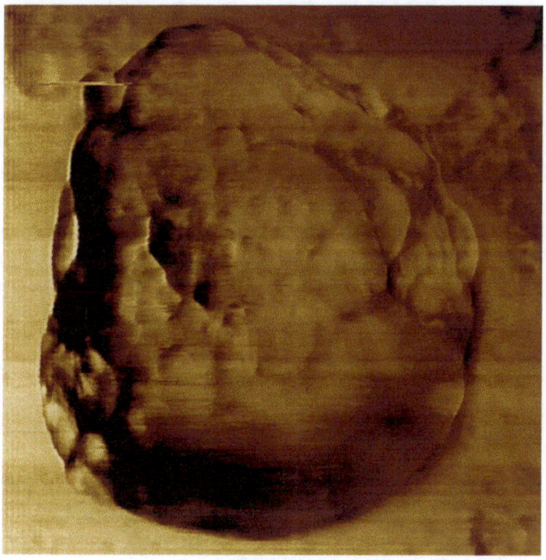

Fig. 7.16 Oval or round dormant *Mycobacterium tuberculosis*. Using atomic force microscopy (AFM), it was observed that 65–70% of Mtb bacilli develop a round or oval morphology as a result of a folding phenomenon following 9 to 12 months of in vitro dormancy. This adaptation likely enhances the survival of dormant TB bacilli by reducing metabolic demands and increasing resistance to environmental stressors
*Experimental work by Prof. Parissa -Farnia and Prof. Gennady Konstantinovich Zhavnerko; Mycobacteriology Research Center, National Research Institute of Tuberculosis and Lung Disease, Shahid Beheshti University of Medical Sciences, Tehran, Iran & The Republican Research and Practical Centre for Epidemiology & and Microbiology, Filimonova 23, Minsk, Belarus

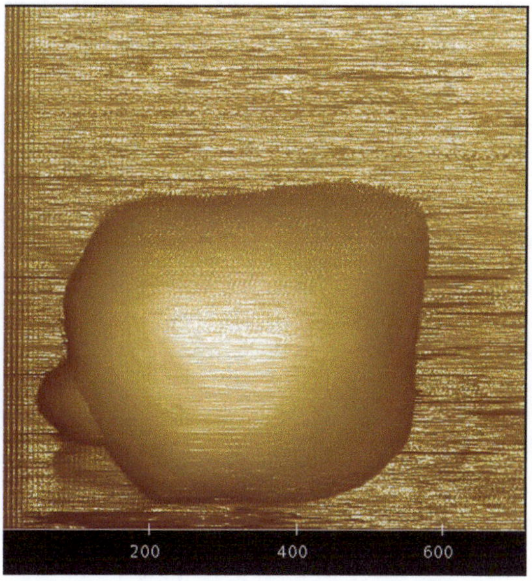

Fig. 7.17 Cell wall-deficient forms of dormant *Mycobacterium tuberculosis*. Dormant Mtb cells lose their cell wall and become smear-negative when stained with carbol fuchsin after remaining for 18 months under conditions of zero oxygen and nutrient deprivation. These adapted Mtb cells have an average diameter ranging from 0.1 to 0.5 µm. Although they do not grow on standard culture media, inoculation of these cell suspensions into the peritoneal cavity of guinea pigs results in infection after 40 to 45 days
*Experimental work by Prof. Parissa -Farnia and Prof. Gennady Konstantinovich Zhavnerko; Mycobacteriology Research Center, National Research Institute of Tuberculosis and Lung Disease, Shahid Beheshti University of Medical Sciences, Tehran, Iran & The Republican Research and Practical Centre for Epidemiology & and Microbiology, Filimonova 23, Minsk, Belarus

Conflict of Interest The authors declare that they have no conflict of interest related to the content of this chapter.

Ethical Consideration This chapter is based on conclusions drawn from the authors' research as well as data and information available in the literature, all of which have been appropriately cited. The figures and images presented originate from the authors' studies, which did not involve experiments on humans, animals, or cell cultures. All referenced experimental work was conducted using laboratory and wild-type strains of *Mycobacterium tuberculosis* (Mtb), following established standard protocols. The research project received approval from the Scientific Committee of the NRITLD, Tehran, Iran. The findings and interpretations presented are grounded in a rigorous analysis of existing research combined with the authors' experimental work.

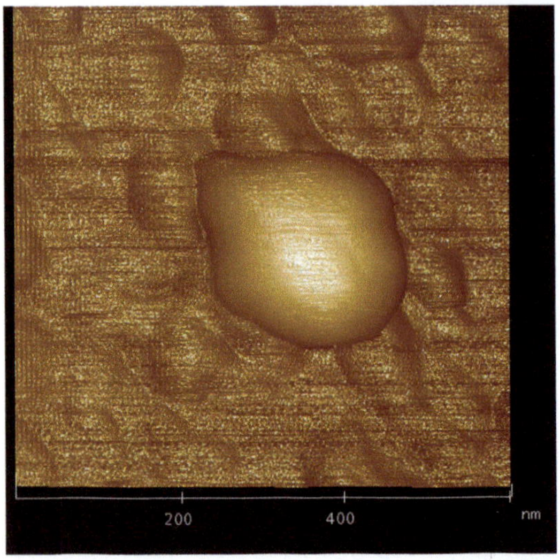

Fig. 7.18 Nonacid-fast persister forms of *Mycobacterium tuberculosis* after 26 months of dormancy

Nonacid-fast bacilli were observed using atomic force microscopy in height-friction mode. These cells are smear-negative when stained with carbol fuchsin and have an average diameter ranging from 0.1 to 0.5 μm. Inoculation of these cell suspensions into the peritoneal cavity of guinea pigs resulted in infection after 40 – 4 5 days.

*Experimental work by Prof. Parissa -Farnia and Prof. Gennady Konstantinovich Zhavnerko; Mycobacteriology Research Center, National Research Institute of Tuberculosis and Lung Disease, Shahid Beheshti University of Medical Sciences, Tehran, Iran & The Republican Research and Practical Centre for Epidemiology & and Microbiology, Filimonova 23, Minsk, Belarus

References

Aguilar-Ayala DA, Tilleman L, Van Nieuwerburgh F, Deforce D, Palomino JC, Vandamme P et al (2017) The transcriptome of Mycobacterium tuberculosis in a lipid-rich dormancy model through RNAseq analysis. Sci Rep 7:17665

Aly S, Wagner K, Keller C, Malm S, Malzan A, Brandau S, Bange FC, Ehlers S (2006) Oxygen status of lung granulomas in Mycobacterium tuberculosis-infected mice. J Pathol 210(3):298–305

Anand K, Tripathi A, Shukla K, Malhotra N, Jamithireddy AK, Jha RK, Chaudhury SN, Rajmani RS, Ramesh A, Nagaraja V, Gopal B, Nagaraju G, Seshayee ASN, Singh A (2021) Mycobacterium tuberculosis SufR responds to nitric oxide via its 4Fe–4S cluster and regulates Fe–S cluster biogenesis for persistence in mice. Redox Biol 46:102062

Andersson DI, Hughes D (2011) Persistence of antibiotic resistance in bacterial populations. FEMS Microbiol Rev 35(5):901–911

Andersson DI, Levin BR (1999) The biological cost of antibiotic resistance. Curr Opin Microbiol 2(5):489–493

Apiche EA, Yee E, Damodaran AR, Bhagi-Damodaran A (2024) Oxygen affinities of DosT and DosS sensor kinases with implications for hypoxia adaptation in Mycobacterium tuberculosis. bioRxiv [preprint]

Aravindhan V, Christy AJ, Roy S, Ajitkumar P, Narayanan PR, Narayanan S (2009) Mycobacterium tuberculosis groE promoter controls the expression of the bicistronic groESL1 operon and shows differential regulation under stress conditions. FEMS Microbiol Lett 292(1):42–49

Awasthy D, Ambady A, Narayana A, Morayya S, Sharma U (2014) Roles of the two type II NADH dehydrogenases in the survival of Mycobacterium tuberculosis in vitro. Gene 550(1):110–116

Ayala RF, Bartolini M, Grau R (2020) The stress-responsive alternative sigma factor SigB of Bacillus subtilis and its relatives: an old friend with new functions. Front Microbiol 11:1761

Bacon J, Alderwick LJ, Allnutt JA, Gabasova E, Watson R, Hatch KA et al (2014) Non-replicating Mycobacterium tuberculosis elicits a reduced infectivity profile with corresponding modifications to the cell wall and extracellular matrix. PLoS One 9(2):e87329

Bagchi G, Chauhan S, Sharma D, Tyagi JS (2005) Transcription and autoregulation of the *Rv3134c-devR-devS* operon of Mycobacterium tuberculosis. Microbiology 151(Pt 12):4057–4065

Baker JJ, Dechow SJ, Abramovitch RB (2019) Acid fasting: modulation of Mycobacterium tuberculosis metabolism at acidic pH. Trends Microbiol 27(11):942–953

Baker JJ, Abramovitch RB (2018) Genetic and metabolic regulation of mycobacterium tuberculosis acid growth arrest. Sci Rep 8(1):4168

Barry CE, Boshoff HI, Dartois V, Dick T, Ehrt S, Flynn J et al (2009) The spectrum of latent tuberculosis: rethinking the biology and intervention strategies. Nat Rev Microbiol 7(12):845–855

Beck F, Yegian D (1952) A study of the tubercle bacillus in resected pulmonary lesions. Am Rev Tuberc 66(1):44–51

Behr MA, Edelstein PH, Ramakrishnan L (2024) Rethinking the burden of latent tuberculosis to reprioritize research. Nat Microbiol 9(5):1157–1158

Beites T, O'Brien K, Tiwari D et al (2019) Plasticity of the Mycobacterium tuberculosis respiratory chain and its impact on tuberculosis drug development. Nat Commun 10:4970

Berney M, Cook GM (2010) Unique flexibility in energy metabolism allows mycobacteria to combat starvation and hypoxia. PLoS One 5(1):e8614

Bespyatykh JA, Basmanov DV (2022) Omics technologies in the diagnostics of mycobacterium tuberculosis. Med Extreme Situat 24(2):29–36. https://doi.org/10.47183/mes.2022.013

Betts JC, Lukey PT, Robb LC, McAdam RA, Duncan K (2002) Evaluation of a nutrient starvation model of mycobacterium tuberculosis persistence by gene and protein expression profiling. Mol Microbiol 43(3):717–731

Betts G, Poyntz H, Stylianou E, Reyes-Sandoval A, Cottingham M, Hill A, McShane H (2012) Optimising immunogenicity with viral vectors: mixing MVA and HAdV-5 expressing the mycobacterial antigen Ag85A in a single injection. PLoS One 7(12):e50447

Bhat SA, Iqbal IK, Kumar A (2016) Imaging the NADH:NAD+ homeostasis for understanding the metabolic response of Mycobacterium to physiologically relevant stresses. Front Cell Infect Microbiol 6:145

Bhat ZS, Rather MA, Maqbool M, Ahmad Z (2018) Drug targets exploited in mycobacterium tuberculosis: pitfalls and promises on the horizon. Biomed Pharmacother 103:1733–1747

Bhatt P, Sharma M, Sharma PP, Rathi B, Sharma S (2022) Mycobacterium tuberculosis dormancy regulon proteins Rv2627c and Rv2628 as Toll-like receptor agonists and potential adjuvants. Int Immunopharmacol 112:109238. https://doi.org/10.1016/j.intimp.2022.10923

Bhattacharya A, Hegde SR, Mande SC, Santoni D et al (2023) Mycobacterium tuberculosis PhoP integrates stress response to intracellular survival by regulating cAMP level. elife 12:RP92136

Biketov S, Potapov V, Ganina E, Downing K, Kana BD, Kaprelyants A (2007) The role of resuscitation promoting factors in pathogenesis and reactivation of Mycobacterium tuberculosis during intra-peritoneal infection in mice. BMC Infect Dis 7:146

Bilan DS, Matlashov ME, Gorokhovatsky AY, Schultz C, Enikolopov G, Belousov VV (2014) Genetically encoded fluorescent indicator for imaging NAD(+)/NADH ratio changes in different cellular compartments. Biochim Biophys Acta 1840(2):951–957

Birhanu AG, Gomez-Munoz M, Kalayou S, Riaz T, Lutter T, Yimer SA et al (2022) Proteome profiling of mycobacterium tuberculosis cells exposed to nitrosative stress. ACS Omega 7(4):3470–3482

Boom WH, Schaible UE, Achkar JM (2021) The knowns and unknowns of latent Mycobacterium tuberculosis infection. J Clin Invest 131(3):e136222

Borah K, Mendum TA, Hawkins ND, Ward JL, Beale MH, Larrouy-Maumus G, Bhatt A, Moulin M, Haertlein M, Strohmeier G, Pichler H (2021) Metabolic fluxes for nutritional flexibility of Mycobacterium tuberculosis. Mol Syst Biol 17(5):e10280

Boshoff HI, Xu X, Tahlan K, Dowd CS, Pethe K, Camacho LR et al (2008) Biosynthesis and recycling of nicotinamide cofactors in Mycobacterium tuberculosis. An essential role for NAD in nonreplicating bacilli. J Biol Chem 283(27):19329–19341

Bunker RD, Mandal K, Bashiri G, Chaston JJ, Pentelute BL, Lott JS, Kent SB, Baker EN (2015) A functional role of Rv1738 in Mycobacterium tuberculosis persistence suggested by racemic protein crystallography. Proc Natl Acad Sci USA 112(14):4310–4315

Burian J, Ramón-García S, Sweet G, Gómez-Velasco A, Av-Gay Y, Thompson CJ (2012) The mycobacterial transcriptional regulator whiB7 gene links redox homeostasis and intrinsic antibiotic resistance. J Biol Chem 287(1):299–310

Burian J, Yim G, Hsing M, Axerio-Cilies P, Cherkasov A, Spiegelman GB, Thompson CJ (2013) The mycobacterial antibiotic resistance determinant WhiB7 acts as a transcriptional activator by binding the primary sigma factor SigA (RpoV). Nucleic Acids Res 41(22):10062–10076

Campaniço A, Harjivan SG, Warner DF, Moreira R, Lopes F (2020) Addressing latent tuberculosis: new advances in mimicking the disease, discovering key targets, and designing hit compounds. Int J Mol Sci 21(22):8854

Cardona PJ, Ruiz-Manzano J (2004) On the nature of Mycobacterium tuberculosis-latent bacilli. Eur Respir J 24(6):1044–1051

Castro-Garza J, García-Jacobo P, Rivera-Morales LG, Quinn FD, Barber J, Karls R, Haas D, Helms S, Gupta T, Blumberg H, Tapia J, Luna-Cruz I, Rendon A, Vargas-Villarreal J, Vera-Cabrera L, Rodríguez-Padilla C (2017) Detection of anti-HspX antibodies and HspX protein in patient sera for the identification of recent latent infection by Mycobacterium tuberculosis. PLoS One 12(8):e0181714

Chakraborty D, Ghosh R, Halder A, Mondal D, Misra D (2021) Endoscopic ultrasonographic diagnosis of pancreatic tuberculosis in immunocompetent patients—a case series from eastern India. Indian J Gastroenterol 40(6):636–640

Chandrasekhar S, Ratnam S (1992) Studies on cell-wall deficient non-acid fast variants of Mycobacterium tuberculosis. Tuber Lung Dis 73(5):273–279

Chang DP, Guan XL (2021) Metabolic versatility of Mycobacterium tuberculosis during infection and dormancy. Meta 11(2):88

Chang Z, Primm TP, Jakana J, Lee IH, Serysheva I, Chiu W, Gilbert HF, Quiocho FA (1996) Mycobacterium tuberculosis 16-kDa antigen (Hsp16.3) functions as an oligomeric structure in vitro to suppress thermal aggregation. J Biol Chem 271(12):7218–7223

Chao W-C, Yen C-L, Wu Y-H, Chen S-Y, Hsieh C-Y, Chang T-C et al (2015) Increased resistin may suppress reactive oxygen species production and inflammasome activation in type 2 diabetic patients with pulmonary tuberculosis infection. Microbes Infect 17(3):195–204

Chatterjee S, Banta A (2018) The spectrum of tuberculosis of the spine in pediatric age group: a review. Childs Nerv Syst 34(10):1937–1945

Chauhan S, Tyagi JS (2009) Powerful induction of divergent tgs1-Rv3131 genes in mycobacterium tuberculosis is mediated by DevR interaction with a high-affinity site and an adjacent cryptic low-affinity site. J Bacteriol 191(19):6075–6081

Chauhan S, Sharma D, Singh A, Surolia A, Tyagi JS (2011) Comprehensive insights into Mycobacterium tuberculosis DevR (DosR) regulon activation switch. Nucleic Acids Res 39(17):7400–7414

Chawla M, Parikh P, Saxena A, Munshi M, Mehta M, Mai D et al (2012) Mycobacterium tuberculosis

WhiB4 regulates oxidative stress response to modulate survival and dissemination in vivo. Mol Microbiol 85(6):1148–1165

Chawla M, Mishra S, Anand K, Parikh P, Mehta M, Vij M, Verma T, Singh P, Jakkala K, Verma HN, AjitKumar P, Ganguli M, Seshasayee ASN, Singh A (2018) Redox-dependent condensation of the mycobacterial nucleoid by WhiB4. Redox Biol 19:116–133

Chen ZW (2013) Multifunctional immune responses of HMBPP-specific Vγ2Vδ2 T cells in M. tuberculosis and other infections. Cell Mol Immunol 10(1):58–64

Chen C-H, Chen Y-M, Lee C-W, Chang Y-J, Cheng C-Y, Hung J-K (2016) Early diagnosis of spinal tuberculosis. J Formos Med Assoc 115(10):825–836

Chen M, Yang C, Deng L, Wang F, Liu J (2022) Production of 1, 3-medium chain-2-long chain (MLM) triacylglycerols by metabolically engineered Escherichia coli. Biochem Eng J 184:108498

Chilukoti N, Kumar CM, Mande SC (2016) GroEL2 of mycobacterium tuberculosis reveals the importance of structural pliability in chaperonin function. J Bacteriol 198(3):486–497

Cho HY, Cho HJ, Kim YM, Oh JI, Kang BS (2009) Structural insight into the heme-based redox sensing by DosS from Mycobacterium tuberculosis. J Biol Chem 284(19):13057–13067

Choudhary E, Bishai W, Agarwal N (2014) Expression of a subset of heat stress induced genes of mycobacterium tuberculosis is regulated by 3′, 5′-cyclic AMP. PLoS One 9(2):e89759

Clerico EM, Tilitsky JM, Meng W, Gierasch LM (2015) How Hsp70 molecular machines interact with their substrates to mediate diverse physiological functions. J Mol Biol 427(7):1575–1588

Cook GM, Hards K, Vilchèze C, Hartman T, Berney M (2014) Energetics of respiration and oxidative phosphorylation in mycobacteria. In: Molecular Genetics of Mycobacteria, pp 389–409

Cook GM, Hards K, Dunn E, Heikal A, Nakatani Y, Greening C, Crick DC, Fontes FL, Pethe K, Hasenoehrl E, Berney M (2017) Oxidative phosphorylation as a target space for tuberculosis: success, caution, and future directions. In: Tuberculosis and the tubercle bacillus. ASM Press, pp 295–316

Cumming BM, Rahman MA, Lamprecht DA, Rohde KH, Saini V, Adamson JH, Russell DG, Steyn AJ (2017) Mycobacterium tuberculosis arrests host cycle at the G1/S transition to establish long term infection. PLoS Pathog 13(5):e1006389

Cunningham AF, Spreadbury CL (1998) Mycobacterial stationary phase induced by low oxygen tension: cell wall thickening and localization of the 16-kilodalton α-crystallin homolog. J Bacteriol 180(4):801–808

Daniel J, Deb C, Dubey VS, Sirakova TD, Abomoelak B, Morbidoni HR, Kolattukudy PE (2004) Induction of a novel class of diacylglycerol acyltransferases and triacylglycerol accumulation in Mycobacterium tuberculosis as it goes into a dormancy-like state in culture. J Bacteriol 186(15):5017–5030

Daniel J, Maamar H, Deb C, Sirakova TD, Kolattukudy PE (2011) Mycobacterium tuberculosis uses host triacylglycerol to accumulate lipid droplets and acquires a dormancy-like phenotype in lipid-loaded macrophages. PLoS Pathog 7(6):e1002093

Deb C, Daniel J, Sirakova TD, Abomoelak B, Dubey VS, Kolattukudy PE (2006) A novel lipase belonging to the hormone-sensitive lipase family induced under starvation to utilize stored triacylglycerol in mycobacterium tuberculosis. J Biol Chem 281(7):3866–3875

Deb C, Lee C-M, Dubey VS, Daniel J, Abomoelak B, Sirakova TD et al (2009) A novel in vitro multiplestress dormancy model for mycobacterium tuberculosis generates a lipid-loaded, drug-tolerant, dormant pathogen. PLoS One 4(6):e6077

Devasundaram S, Gopalan A, Das SD, Raja A (2016) Proteomics analysis of three different strains of Mycobacterium tuberculosis under in vitro hypoxia and evaluation of hypoxia associated antigen's specific memory T cells in healthy household contacts. Front Microbiol 7:1275

Drumm JE, Mi K, Bilder P, Sun M, Lim J, Bielefeldt-Ohmann H, Basaraba R, So M, Zhu G, Tufariello JM, Izzo AA, Orme IM, Almo SC, Leyh TS, Chan J (2009) Mycobacterium tuberculosis universal stress protein Rv2623 regulates bacillary growth by ATP-binding: requirement for establishing chronic persistent infection. PLoS Pathog 5(5):e1000460

Ehrt S, Schnappinger D, Rhee KY (2018) Metabolic principles of persistence and pathogenicity in Mycobacterium tuberculosis. Nat Rev Microbiol 16(8):496–507

Elchennawi I, Carpentier P, Caux C, Ponge M, Ollagnier de Choudens S (2023) Structural and biochemical characterization of Mycobacterium tuberculosis zinc SufU-SufS complex. Biomolecules 13(5):732

Elchennawi I, Ollagnier de Choudens S (2022) Iron–sulfur clusters toward stresses: implication for understanding and fighting tuberculosis. Inorganics 10(10):174

Eoh H, Rhee KY (2013) Multifunctional essentiality of succinate metabolism in adaptation to hypoxia in Mycobacterium tuberculosis. Proc Natl Acad Sci USA 110(16):6554–6559

Eoh H, Rhee KY (2014) Methylcitrate cycle defines the bactericidal essentiality of isocitrate lyase for survival of Mycobacterium tuberculosis on fatty acids. Proc Natl Acad Sci USA 111(13):4976–4981

Falk-Krzesinski HJ, Wolfe AJ (1998) Genetic analysis of the nuo locus, which encodes the protontranslocating NADH dehydrogenase in Escherichia coli. J Bacteriol 180(5):1174–1184

Feldman WH, Baggenstoss AH (1938) The residual infectivity of the primary complex of tuberculosis. Am J Pathol 14(4):473–490

Florczyk MA, McCue LA, Stack RF, Hauer CR, McDonough KA (2001) Identification and characterization of mycobacterial proteins differentially expressed under standing and shaking culture conditions, including Rv2623 from a novel class of putative ATP-binding proteins. Infect Immun 69(9):5777–5785

Flynn JL, Chan J (2001) Immunology of tuberculosis. Annu Rev Immunol 19(1):93–129

Flynn JL, Chan J (2022) Immune cell interactions in tuberculosis. Cell 185(25):4682–4702

Fol M, Chauhan A, Nair NK, Maloney E, Moomey M, Jagannath C, Madiraju MV, Rajagopalan M (2006) Modulation of Mycobacterium tuberculosis proliferation by MtrA, an essential two-component response regulator. Mol Microbiol 60(3):643–657

Fourie KR, Wilson HL (2020) Understanding GroEL and DnaK stress response proteins as antigens for bacterial diseases. Vaccines (Basel) 8(4):773

Fox W, Ellard GA, Mitchison DA (1999) Studies on the treatment of tuberculosis undertaken by the British Medical Research Council tuberculosis units, 1946–1986, with relevant subsequent publications. Int J Tuberc Lung Dis 3(10):S231–SS79

Galagan JE, Minch K, Peterson M, Lyubetskaya A, Azizi E, Sweet L et al (2013) The mycobacterium tuberculosis regulatory network and hypoxia. Nature 499(7457):178–183

Gautam US, Mehra S, Kumari P, Alvarez X, Niu T, Tyagi JS, Kaushal D (2019) Mycobacterium tuberculosis sensor kinase DosS modulates the autophagosome in a DosR-independent manner. Commun Biol 2:349

Geluk A, Lin MY, van Meijgaarden KE, Leyten EM, Franken KL, Ottenhoff TH, Klein MR (2007) T-cell recognition of the HspX protein of Mycobacterium tuberculosis correlates with latent M. Tuberculosis infection but not with M. bovis BCG vaccination. Infect Immun 75(6):2914–2921

Genevaux P, Keppel F, Schwager F, Langendijk-Genevaux PS, Hartl FU, Georgopoulos C (2004) In vivo analysis of the overlapping functions of DnaK and trigger factor. EMBO Rep 5(2):195–200

Gengenbacher M, Kaufmann SH (2012) Mycobacterium tuberculosis: success through dormancy. FEMS Microbiol Rev 36(3):514–532

Gerasimova A, Kazakov AE, Arkin AP, Dubchak I, Gelfand MS (2011) Comparative genomics of the dormancy regulons in mycobacteria. J Bacteriol 193(14):3446–3452

Georgieva M, Sia JK, Bizzell E, Madan-Lala R, Rengarajan J (2018) Mycobacterium tuberculosis GroEL2 modulates dendritic cell responses. Infect Immun 86(2). https://doi.org/10.1128/iai.00387-17

Getahun H, Matteelli A, Abubakar I, Aziz MA, Baddeley A, Barreira D et al (2015) Management of latent mycobacterium tuberculosis infection: WHO guidelines for low tuberculosis burden countries. Eur Respir J 46(6):1563–1576

Glass LN, Swapna G, Chavadi SS, Tufariello JM, Mi K, Drumm JE et al (2017) Mycobacterium tuberculosis universal stress protein Rv2623 interacts with the putative ATP binding cassette (ABC) transporter Rv1747 to regulate mycobacterial growth. PLoS Pathog 13(7):e1006515

Gomez JE, McKinney JD (2004) M. Tuberculosis persistence, latency, and drug tolerance. Tuberculosis (Edinb) 84(1–2):29–44

Gonzalo-Asensio J, Mostowy S, Harders-Westerveen J, Huygen K, Hernández-Pando R, Thole J, Behr M, Gicquel B, Martín C (2008) PhoP: a missing piece in the intricate puzzle of Mycobacterium tuberculosis virulence. PLoS One 3(10):e3496

Gopinath V, Raghunandanan S, Gomez RL, Jose L, Surendran A, Ramachandran R, Pushparajan AR, Mundayoor S, Jaleel A, Kumar RA (2015) Profiling the proteome of Mycobacterium tuberculosis during dormancy and reactivation. Mol Cell Proteomics 14(8):2160–2176

Gorityala N, Baidya AS, Sagurthi SR (2024) Genome mining of Mycobacterium tuberculosis: targeting SufD as a novel drug candidate through in silico characterization and inhibitor screening. Front Microbiol 15:1369645

Gorla P, Plocinska R, Sarva K, Satsangi AT, Pandeeti E, Donnelly R et al (2018) MtrA response regulator controls cell division and cell wall metabolism and affects susceptibility of mycobacteria to the first line antituberculosis drugs. Front Microbiol 9:2839

Grant SS, Hung DT (2013) Persistent bacterial infections, antibiotic tolerance, and the oxidative stress response. Virulence 4(4):273–283

Griffin JE, Pandey AK, Gilmore SA, Mizrahi V, McKinney JD, Bertozzi CR, Sassetti CM (2012) Cholesterol catabolism by Mycobacterium tuberculosis requires transcriptional and metabolic adaptations. Chem Biol 19(2):218–227

Grosset J (1980) Bacteriologic basis of short-course chemotherapy for tuberculosis. Clin Chest Med 1(2):231–241

Guirado E, Mbawuike U, Keiser TL, Arcos J, Azad AK, Wang SH et al (2015) Characterization of host and microbial determinants in individuals with latent tuberculosis infection using a human granuloma model. MBio 6(1):e02537–e02514

Gutka HJ, Bondoc JMG, Patwell R, Khan S, Grzelak EM, Goswami R, Voskuil MI, Movahedzadeh F (2024) Rv0100: an essential acyl carrier protein from M. Tuberculosis important in dormancy. PLoS One 19(6):e0304876

Hampshire T, Soneji S, Bacon J, James BW, Hinds J, Laing K, Stabler RA, Marsh PD, Butcher PD (2004) Stationary phase gene expression of Mycobacterium tuberculosis following a progressive nutrient depletion: A model for persistent organisms? Tuberculosis 84(3–4):228–238

He H, Hovey R, Kane J, Singh V, Zahrt TC (2006) MprAB is a stress-responsive two-component system that directly regulates expression of sigma factors SigB and SigE in Mycobacterium tuberculosis. J Bacteriol 188(6):2134–2143

Hegde SR, Rajasingh H, Das C, Mande SS, Mande SC (2012) Understanding communication signals during mycobacterial latency through predicted genome-wide protein interactions and boolean modeling. PLoS One 7(3):e33893

Henderson B, Lund PA, Coates ARM (2010) Multiple moonlighting functions of mycobacterial molecular chaperones. Tuberculosis 90(2):119–124

Hickey TB, Thorson LM, Speert DP, Daffé M, Stokes RW (2009) Mycobacterium tuberculosis Cpn60.2 and DnaK are located on the bacterial surface, where Cpn60.2 facilitates efficient bacterial association with macrophages. Infect Immun 77(8):3389–3401

Honaker RW, Leistikow RL, Bartek IL, Voskuil MI (2009) Unique roles of DosT and DosS in DosR regulon induction and mycobacterium tuberculosis dormancy. Infect Immun 77(8):3258–3263

Hotter GS, Wards BJ, Mouat P, Besra GS, Gomes J, Singh M et al (2005) Transposon mutagenesis of Mb0100 at the ppe1-nrp locus in Mycobacterium bovis disrupts phthiocerol dimycocerosate (PDIM) and glycosylphenol-PDIM biosynthesis, producing an avirulent strain with vaccine properties at least equal to those of M. bovis BCG. J Bacteriol 187(7):2267–2277

Hu Y, Coates AR (1999) Transcription of the stationary-phase-associated hspX gene of Mycobacterium tuberculosis is inversely related to synthesis of the 16-kilodalton protein. J Bacteriol 181(5):1380–1387

Hu Y, Mangan JA, Dhillon J, Sole KM, Mitchison DA, Butcher PD et al (2000) Detection of mRNA transcripts and active transcription in persistent Mycobacterium tuberculosis induced by exposure to rifampin or pyrazinamide. J Bacteriol 182(22):6358–6365

Hu Y, Movahedzadeh F, Stoker NG, Coates AR (2006) Deletion of the Mycobacterium tuberculosis alpha-crystallin-like hspX gene causes increased bacterial growth in vivo. Infect Immun 74(2):861–868

Hu Y, Mathema B, Jiang W, Kreiswirth B, Wang W, Xu B (2011) Transmission pattern of drug-resistant tuberculosis and its implication for tuberculosis control in eastern rural China. PLoS One 6(5):e19548

Huang Z, Zhang G, Lyon CJ, Hu TY, Lu S (2023) Outlook for CRISPR-based tuberculosis assays now in their infancy. Front Immunol 14:1172035

Hudock TA, Foreman TW, Bandyopadhyay N, Gautam US, Veatch AV, LoBato DN, Gentry KM, Golden NA, Cavigli A, Mueller M, Hwang SA, Hunter RL, Alvarez X, Lackner AA, Bader JS, Mehra S, Kaushal D (2017) Hypoxia sensing and persistence genes are expressed during the intragranulomatous survival of Mycobacterium tuberculosis. Am J Respir Cell Mol Biol 56(5):637–647

Huet G, Daffé M, Saves I (2005) Identification of the Mycobacterium tuberculosis SUF machinery as the exclusive mycobacterial system of [Fe-S] cluster assembly: evidence for its implication in the pathogen's survival. J Bacteriol 187(17):6137–6146

Hurst-Hess K, Biswas R, Yang Y, Rudra P, Lasek-Nesselquist E, Ghosh P (2019) Mycobacterial SigA and SigB cotranscribe essential housekeeping genes during exponential growth. MBio 10(3). https://doi.org/10.1128/mbio.00273-19

Hutter S, Villiger TK, Brühlmann D, Stettler M, Broly H, Soos M, Gunawan R (2017) Glycosylation flux analysis reveals dynamic changes of intracellular glycosylation flux distribution in Chinese hamster ovary fed-batch cultures. Metab Eng 43(Pt A):9–20

Ignatov DV, Salina EG, Fursov MV, Skvortsov TA, Azhikina TL, Kaprelyants AS (2015) Dormant non-culturable Mycobacterium tuberculosis retains stable low-abundant mRNA. BMC Genomics 16:954

Imboden P, Schoolnik GK (1998) Construction and characterization of a partial Mycobacterium tuberculosis cDNA library of genes expressed at reduced oxygen tension. Gene 213(1–2):107–117

Ioanoviciu A, Podust LM, Ortiz de Montellano PR (2007) The three-dimensional structures of the GAF-A and GAF-B domains of DevS/DosS, heme-based sensors from Mycobacterium tuberculosis. Biochemistry 46(7):1746–1755

Iona E, Pardini M, Mustazzolu A, Piccaro G, Nisini R, Fattorini L, Giannoni F (2016) Mycobacterium tuberculosis gene expression at different stages of hypoxia-induced dormancy and upon resuscitation. J Microbiol 54(8):565–572

Iza Beltran D, Wan T, Zhang L (2024) WhiB-like proteins: diversity of structure, function and mechanism. Biochim Biophys Acta Mol Cell Res 1871(7):119787. https://doi.org/10.1016/j.bbamcr.2024.119787. Epub 2024 Jun 13. PMID: 38879133; PMCID: PMC11365794

Jäger S, Jäger A, Klug G (2004) CIRCE is not involved in heat-dependent transcription of groESL but in stabilization of the mRNA 5′-end in Rhodobacter capsulatus. Nucleic Acids Res 32(1):386–396. https://doi.org/10.1093/nar/gkh174. PMID: 14729923; PMCID: PMC373284

Jamet S, Quentin Y, Coudray C, Texier P, Laval F, Daffé M et al (2015) Evolution of mycolic acid biosynthesis genes and their regulation during starvation in Mycobacterium tuberculosis. J Bacteriol 197(24):3797–3811

Jee B, Kumar S, Yadav R, Singh Y, Katoch V, Sharma N (2018) Small heat shock protein 16.3 of Mycobacterium tuberculosis: two decades of functional characterization. Cell Physiol Biochem 49(1):368–380

Jiang Q, Lu L, Wu J, Yang C, Prakash R, Zuo T et al (2018) Assessment of tuberculosis contact investigation in Shanghai, China: an 8-year cohort study. Tuberculosis 108:10–15

Joseph SV, Madhavilatha GK, Kumar RA, Mundayoor S (2012) Comparative analysis of mycobacterial truncated hemoglobin promoters and the groEL2 promoter in free-living and intracellular mycobacteria. Appl Environ Microbiol 78(18):6499–6506

Joshi H, Kandari D, Bhatnagar R (2021) Insights into the molecular determinants involved in Mycobacterium tuberculosis persistence and their therapeutic implications. Virulence 12(1):2721–2749

Kana BD, Mizrahi V (2010) Resuscitation-promoting factors as lytic enzymes for bacterial growth and signaling. FEMS Immunol Med Microbiol 58(1):39–50

Kana BD, Gordhan BG, Downing KJ, Sung N, Vostroktunova G, Machowski EE, Tsenova L, Young M, Kaprelyants A, Kaplan G, Mizrahi V (2008) The resuscitation-promoting factors of Mycobacterium tuberculosis are required for virulence and resuscitation from dormancy but are collectively dispensable for growth in vitro. Mol Microbiol 67(3):672–684

Kapoor N, Pawar S, Sirakova TD, Deb C, Warren WL, Kolattukudy PE (2013) Human granuloma in vitro model, for TB dormancy and resuscitation. PLoS One 8(1):e53657

Kapoor D, Sharma A, Singh K (2015) Redox homeostasis in plants under abiotic stress: role of reactive oxygen species and antioxidants. Front Environ Sci 3:13

Kaprelyants AS, Kell DB (1993) Dormancy in non-sporulating bacteria. FEMS Microbiol Rev 10(3–4):271–286

Kaprelyants AS, Shleeva MO, Mukamolova GV, Young DI, Kell DB, Young M (2003) Intercellular signaling and the multiplication of prokaryotes: bacterial cytokines. Trends Microbiol 11(9):398–403

Kennaway CK, Benesch JL, Gohlke U, Wang L, Robinson CV, Orlova EV et al (2005) Dodecameric structure of the small heat shock protein Acr1 from mycobacterium tuberculosis. J Biol Chem 280(39):33419–33425

Keren I, Minami S, Rubin E, Lewis K (2011) Characterization and transcriptome analysis of mycobacterium tuberculosis persisters. MBio 2(3). https://doi.org/10.1128/mbio.00100-11

Khan MT, Zaheer S, Amar W, Shafique K (2024) Effect of smoking cessation interventions on abstinence and tuberculosis treatment outcomes among newly diagnosed patients: a randomized controlled trial. Microbiol Spec 12(4):e03878–e03823

Khomenko A (1987) The variability of mycobacterium tuberculosis in patients with cavitary pulmonary tuberculosis in the course of chemotherapy. Tubercle 68(4):243–253

Kim H, Shin SJ (2023) Revolutionizing control strategies against Mycobacterium tuberculosis infection through selected targeting of lipid metabolism. Cell Mol Life Sci 80(10):291

Kim MJ, Park KJ, Ko IJ, Kim YM, Oh JI (2010) Different roles of DosS and DosT in the hypoxic adaptation of mycobacteria. J Bacteriol 192(19):4868–4875

Konar M, Alam MS, Arora C, Agrawal P (2012) WhiB2/Rv3260c, a cell division-associated protein of mycobacterium tuberculosis H37Rv, has properties of a chaperone. FEBS J 279(15):2781–2792

Kumar A, Toledo JC, Patel RP, Lancaster JR Jr, Steyn AJ (2007) Mycobacterium tuberculosis DosS is a redox sensor and DosT is a hypoxia sensor. Proc Natl Acad Sci USA 104(28):11568–11573

Kumar A, Deshane JS, Crossman DK, Bolisetty S, Yan BS, Kramnik I, Agarwal A, Steyn AJ (2008) Heme oxygenase-1-derived carbon monoxide induces the Mycobacterium tuberculosis dormancy regulon. J Biol Chem 283(26):18032–18039

Kumar A, Lewin A, Rani PS, Qureshi IA, Devi S, Majid M, Kamal E, Marek S, Hasnain SE, Ahmed N (2013) Dormancy associated translation inhibitor (DATIN/Rv0079) of Mycobacterium tuberculosis interacts with TLR2 and induces proinflammatory cytokine expression. Cytokine 64(1):258–264

Kumar A, Singh N, Yadav R, Kumar RP, Sharma S, Arora A, Singh T (2012) Crystal structure of peptidyl-tRNA hydrolase from mycobacterium smegmatis reveals novel features related to enzyme dynamics. Int J Biochem Mol Biol 3(1):58

Kumar P, Singh A, Sharma U, Singh D, Dobhal M, Singh S (2013) Anti-mycobacterial activity of plumericin and isoplumericin against MDR mycobacterium tuberculosis. Pulm Pharmacol Ther 26(3):332–335

Kumar MK, Kumar P, Singh A (2015) Recent advances in the diagnosis and treatment of childhood tuberculosis. J Nat Sci Biol Med 6(2):314

Kundu M, Basu J (2021) Applications of transcriptomics and proteomics for understanding dormancy and resuscitation in mycobacterium tuberculosis. Front Microbiol 12:642487

Larsson C, Luna B, Ammerman NC, Maiga M, Agarwal N, Bishai WR (2012) Gene expression of mycobacterium tuberculosis putative transcription factors whiB1-7 in redox environments. PLoS One 7(7):e37516

Lee BY, Hefta SA, Brennan PJ (1992) Characterization of the major membrane protein of virulent Mycobacterium tuberculosis. Infect Immun 60(5):2066–2074

Lee JH, Karakousis PC, Bishai WR (2008) Role of stress response sigma factor SigG in Mycobacterium tuberculosis. J Bacteriol 190(11):3973–3980

Lee W, VanderVen BC, Fahey RJ, Russell DG (2013) Intracellular Mycobacterium tuberculosis exploits host-derived fatty acids to limit metabolic stress. J Biol Chem 288(10):6788–6800

Lee JY, Jung YW, Jeong I, Joh JS, Sim SY, Choi B, Jee HG, Lim DG (2015) Immune parameters differentiating active from latent tuberculosis infection in humans. Tuberculosis 95(6):758–763

Leistikow RL, Morton RA, Bartek IL, Frimpong I, Wagner K, Voskuil MI (2010) The Mycobacterium tuberculosis DosR regulon assists in metabolic homeostasis and enables rapid recovery from nonrespiring dormancy. J Bacteriol 192(6):1662–1670. https://doi.org/10.1128/JB.00926-09. Epub 2009 Dec 18. PMID: 20023019; PMCID: PMC2832541

Lilic M, Darst SA, Campbell EA (2021) Structural basis of transcriptional activation by the mycobacterium tuberculosis intrinsic antibiotic-resistance transcription factor WhiB7. Mol Cell 81(14):2875–2886. e5

Lilic M, Holmes NA, Bush MJ, Marti AK, Widdick DA, Findlay KC et al (2023) Structural basis of dual activation of cell division by the actinobacterial transcription factors WhiA and WhiB. Proc Natl Acad Sci 120(11):e2220785120

Liu Q, Ma A, Wei L, Pang Y, Wu B, Luo T et al (2018) China's tuberculosis epidemic stems from historical expansion of four strains of mycobacterium tuberculosis. Nat Ecol Evol 2(12):1982–1992

Lim A, Dick T (2001) Plate-based dormancy culture system for mycobacterium smegmatis and isolation of metronidazole-resistant mutants. FEMS Microbiol Lett 200(2):215–219

Li Y, Zeng J, He ZG (2010) Characterization of a functional C-terminus of the Mycobacterium tuberculosis MtrA responsible for both DNA binding and interac-

tion with its two-component partner protein. MtrB J Biochem 148(5):549–556
Liu Y, Jiang H (2021) Directed evolution of Propionyl-CoA carboxylase for succinate biosynthesis. Trends Biotechnol 39(4):330–331
Liu DQ, Zhang JL, Pan ZF, Mai JT, Mei HJ, Dai Y, Zhang L, Wang QZ (2020) Over-expression of Tgs1 in Mycobacterium marinum enhances virulence in adult zebrafish. Int J Med Microbiol 310(1):151378
Liu Y, Li H, Dai D, He J, Liang Z (2024) Gene regulatory mechanism of Mycobacterium tuberculosis during dormancy. Curr Issues Mol Biol 46(6):5825–5844
Loebel RO, Shorr E, Richardson HB (1933) The influence of adverse conditions upon the respiratory metabolism and growth of human tubercle bacilli. J Bacteriol 26:167–200
Low S, Ang L, Cutter J, James L, Chee C, Wang Y, Chew S (2009) Mortality among tuberculosis patients on treatment in Singapore. Int J Tuberc Lung Dis 13(3):328–334
Low KL, Shui G, Natter K, Yeo WK, Kohlwein SD, Dick T, Rao SP, Wenk MR (2010) Lipid droplet-associated proteins are involved in the biosynthesis and hydrolysis of triacylglycerol in Mycobacterium bovis bacillus Calmette-Guerin. J Biol Chem 285(28):21662–21670
Magombedze G, Dowdy D, Mulder N (2013) Latent tuberculosis: models, computational efforts and the pathogen's regulatory mechanisms during dormancy. Front Bioeng Biotechnol 1:4
Mallick I, Santucci P, Poncin I, Point V, Kremer L, Cavalier JF, Canaan S (2021) Intrabacterial lipid inclusions in mycobacteria: unexpected key players in survival and pathogenesis? FEMS Microbiol Rev. 45(6):fuab029 45
Manabe YC, Bishai WR (2000) Latent Mycobacterium tuberculosis—persistence, patience, and winning by waiting. Nat Med 6(12):1327–1329
Manganelli R, Voskuil MI, Schoolnik GK, Dubnau E, Gomez M, Smith I (2002) Role of the extracytoplasmic-function σ factor σH in mycobacterium tuberculosis global gene expression. Mol Microbiol 45(2):365–374
Marakalala MJ, Raju RM, Sharma K, Zhang YJ, Eugenin EA, Prideaux B et al (2016) Inflammatory signaling in human tuberculosis granulomas is spatially organized. Nat Med 22(5):531–538
Marrero J, Rhee KY, Schnappinger D, Pethe K, Ehrt S (2010) Gluconeogenic carbon flow of tricarboxylic acid cycle intermediates is critical for Mycobacterium tuberculosis to establish and maintain infection. Proc Natl Acad Sci USA 107(21):9819–9824
Martinez SE, Beavo JA, Hol WG (2002) GAF domains: two-billion-year-old molecular switches that bind cyclic nucleotides. Mol Interv 2(5):317–323
Masiewicz P, Brzostek A, Wolański M, Dziadek J, Zakrzewska-Czerwińska J (2012) A novel role of the PrpR as a transcription factor involved in the regulation of methylcitrate pathway in mycobacterium tuberculosis. PLoS ONE 7
Mattman LH (1970) Cell wall-deficient forms of mycobacteria. Ann N Y Acad Sci 174:852–859
Mattman LH (2000) Cell Wall deficient forms: stealth pathogens, 3rd edn. CRC Press
Mayer MP, Bukau B (2005) Hsp70 chaperones: cellular functions and molecular mechanism. Cell Mol Life Sci 62(6):670–684
McCune RM Jr, Tompsett R (1956) Fate of mycobacterium tuberculosis in mouse tissues as determined by the microbial enumeration technique. I. The persistence of drug-susceptible tubercle bacilli in the tissues despite prolonged antimicrobial therapy. J Exp Med 104(5):737–762
McCune RM, Feldmann FM, Lambert HP, McDermott W (1966) Microbial persistence: I. The capacity of tubercle bacilli to survive sterilization in mouse tissues. J Exp Med 123(3):445–468
McDermott W, Tompsett R (1954) Microbial persistence. Yale J Biol Med 26(3):221–245
McKinney JD, Höner zu Bentrup K, Muñoz-Elías EJ, Miczak A, Chen B, Chan WT, Swenson D, Sacchettini JC, Jacobs WR, Russell DG (2000) Persistence of Mycobacterium tuberculosis in macrophages and mice requires the glyoxylate shunt enzyme isocitrate lyase. Nature 406(6797):735–738
Medlar EM, McDermott W, McAdams AJ (1952) The persistence of tubercle bacilli in human tissues. Am Rev Tuberc 66(1):1–16
Mehra S, Foreman TW, Didier PJ, Ahsan MH, Hudock TA, Kissee R et al (2015) The DosR regulon modulates adaptive immunity and is essential for mycobacterium tuberculosis persistence. Am J Respir Crit Care Med 191(10):1185–1196
Mehrotra P, Jamwal SV, Saquib N, Sinha N, Siddiqui Z, Manivel V, Chatterjee S, Rao KVS (2014) Pathogenicity of Mycobacterium tuberculosis is expressed by regulating metabolic thresholds of the host macrophage. PLoS Pathog 10(7):e1004265
Mehta M, Singh A (2019) Mycobacterium tuberculosis WhiB3 maintains redox homeostasis and survival in response to reactive oxygen and nitrogen species. Free Radic Biol Med 135:50–58
Meimaridou E, Gooljar SB, Chapple JP (2009) From hatching to dispatching: the multiple cellular roles of the Hsp70 molecular chaperone machinery. J Mol Endocrinol 42(1):1–9
Metcalfe JZ, Everett CK, Steingart KR, Cattamanchi A, Huang L, Hopewell PC, Pai M (2011) Interferon-γ release assays for active pulmonary tuberculosis diagnosis in adults in low-and middle-income countries: systematic review and meta-analysis. J Infect Dis 204(suppl_4):S1120–S11S9
Mitchison D (1954) Tubercle bacilli resistant to isoniazid. Br Med J 1(4854):128
Mitchison DA (1979) Basic mechanisms of chemotherapy. Chest 76(6 Suppl):771–781
Mitchison DA (1985) The action of antituberculosis drugs in short-course chemotherapy. Tubercle 66(3):219–225

Mitchison D (1998) Basic concepts in the chemotherapy of tuberculosis. In: Mycobacteria: II Chemotherapy. Springer, pp 15–50

Mitchison D, Dickinson J (1978) Short term chemotherapy of tuberculosis. Bactericidal mechanisms in short term chemotherapy. Bulletin of the international union against Tuberculosis 53(4):270–275

Miotto P, Sorrentino R, De Giorgi S, Provvedi R, Cirillo DM, Manganelli R (2022) Transcriptional regulation and drug resistance in Mycobacterium tuberculosis. Front Cell Infect Microbiol 12:990312

Mogk A, Homuth G, Scholz C, Kim L, Schmid FX, Schumann W (1997) The GroE chaperonin machine is a major modulator of the CIRCE heat shock regulon of Bacillus subtilis. EMBO J 16(15):4579–4590

Morris RP, Nguyen L, Gatfield J, Visconti K, Nguyen K, Schnappinger D, Ehrt S, Liu Y, Heifets L, Pieters J, Schoolnik G, Thompson CJ (2005) Ancestral antibiotic resistance in Mycobacterium tuberculosis. Proc Natl Acad Sci USA 102(34):12200–12205

Murphy DJ, Brown JR (2007) Identification of gene targets against dormant phase Mycobacterium tuberculosis infections. BMC Infect Dis 7:84

Mushtaq K, Sheikh JA, Amir M, Khan N, Singh B, Agrewala JN (2015) Rv2031c of Mycobacterium tuberculosis: a master regulator of Rv2028-Rv2031 (HspX) operon. Front Microbiol 6:351

Muttucumaru DG, Roberts G, Hinds J, Stabler RA, Parish T (2004) Gene expression profile of Mycobacterium tuberculosis in a non-replicating state. Tuberculosis 84(3–4):239–246

Mulholland CV, Wiggins TJ, Cui J, Vilchèze C, Rajagopalan S, Shultis MW et al (2024) Propionate prevents loss of the PDIM virulence lipid in mycobacterium tuberculosis. Nat Microbiol 9(6):1607–1618

Mundade R, Ozer HG, Wei H, Prabhu L, Lu T (2014) Role of ChIP-seq in the discovery of transcription factor binding sites, differential gene regulation mechanism, epigenetic marks and beyond. Cell Cycle 13(18):2847–2852

Muñoz-Elías EJ, Upton AM, Cherian J, McKinney JD (2006) Role of the methylcitrate cycle in mycobacterium tuberculosis metabolism, intracellular growth, and virulence. Mol Microbiol 60(5):1109–1122

Muñoz-Elías EJ, JD MK (2005) Mycobacterium tuberculosis isocitrate lyases 1 and 2 are jointly required for in vivo growth and virulence. Nat Med 11(6):638–644

Mukamolova GV, Murzin AG, Salina EG, Demina GR, Kell DB, Kaprelyants AS, Young M (2006) Muralytic activity of Micrococcus luteus Rpf and its relationship to physiological activity in promoting bacterial growth and resuscitation. Mol Microbiol 59(1):84–98

Naffin-Olivos JL, Georgieva M, Goldfarb N, Madan-Lala R, Dong L, Bizzell E et al (2014) Mycobacterium tuberculosis Hip1 modulates macrophage responses through proteolysis of GroEL2. PLoS Pathog 10(5):e1004132

Nambi S, Long JE, Mishra BB, Baker R, Murphy KC, Olive AJ et al (2015) The oxidative stress network of Mycobacterium tuberculosis reveals coordination between radical detoxification systems. Cell Host Microbe 17(6):829–837

Nangpal P, Nagpal NL, Angrish N, Khare G (2025) Model systems to study Mycobacterium tuberculosis infections: an overview of scientific potential and impediments. Front Cell Infect Microbiol 15:1572547

Naran K, Moosa A, Barry CE 3rd, Mizrahi V, Warner DF (2018) Toward eradication of Mycobacterium tuberculosis: the way forward. Microbiol Spectr 6(4). https://doi.org/10.1128/microbiolspec.TBTB2-0007-2016

Narberhaus F (1999) Negative regulation of bacterial heat shock genes. Mol Microbiol 31(1):1–8

Nikitushkin V, Shleeva M, Loginov D, Dyčka FF, Sterba J, Kaprelyants A (2022) Shotgun proteomic profiling of dormant,'non-culturable'Mycobacterium tuberculosis. PLoS One 17(8):e0269847

Ojha A, Anand M, Bhatt A, Kremer L, Jacobs WR, Hatfull GF (2005) GroEL1: a dedicated chaperone involved in mycolic acid biosynthesis during biofilm formation in mycobacteria. Cell 123(5):861–873

Opie EL, Aronson JD (1927) Tubercle bacilli in latent tuberculosis lesions and in lung tissue without tuberculosis lesions. Arch Pathol 4(1):1–21

O'Toole R, Williams HD (2003) Universal stress proteins and Mycobacterium tuberculosis. Res Microbiol 154(6):387–392

Pacl HT, Reddy VP, Saini V, Chinta KC, Steyn AJ (2018) Host-pathogen redox dynamics modulate mycobacterium tuberculosis pathogenesis. Pathog Dis 76(5):fty036

Pandey AK, Sassetti CM (2008) Mycobacterial persistence requires the utilization of host cholesterol. Proc Natl Acad Sci 105(11):4376–4380

Pang X, Howard ST (2007) Regulation of the alpha-crystallin gene acr2 by the MprAB two-component system of Mycobacterium tuberculosis. J Bacteriol 189(17):6213–6221

Park HD, Guinn KM, Harrell MI, Liao R, Voskuil MI, Tompa M et al (2003) Rv3133c/dosR is a transcription factor that mediates the hypoxic response of mycobacterium tuberculosis. Mol Microbiol 48(3):833–843

Park E-J, Kim I, Jo E-K (2024) Host immune pathways to mycobacterium tuberculosis infection. J Bacteriol Virol 54(3):167–190

Paramasivan C (1998) An overview on drug resistant tuberculosis in India. Indian J Tuberculo 45(2):73–81

Peddireddy V, Doddam SN, Ahmed N (2017) Mycobacterial dormancy systems and host responses in tuberculosis. Front Immunol 8:84

Pei J-F, Qi N, Li Y-X, Wo J, Ye B-C (2021) RegX3-mediated regulation of methylcitrate cycle in mycobacterium smegmatis. Front Microbiol 12:619387

Perales-Calvo J, Giganti D, Stirnemann G, Garcia-Manyes S (2018) The force-dependent mechanism of DnaK-mediated mechanical folding. Sci Adv. 4(2):eaaq0243 4:eaaq0243

Peterson EJR, Brooks AN, Reiss DJ, Kaur A, Do J, Pan M, Wu WJ, Morrison R, Srinivas V, Carter W, Arrieta-Ortiz ML, Ruiz RA, Bhatt A, Baliga NS (2023) MtrA modulates Mycobacterium tuberculosis cell division in host microenvironments to mediate intrinsic resistance and drug tolerance. Cell Rep 42(8):112875

Płocińska R, Wasik K, Płociński P, Lechowicz E, Antczak M, Błaszczyk E et al (2022) The orphan response regulator Rv3143 modulates the activity of the NADH dehydrogenase complex (Nuo) in mycobacterium tuberculosis via protein–protein interactions. Front Cell Infect Microbiol 12:909507

Podust LM, Ioanoviciu A, Ortiz de Montellano PR (2008) 2.3 Å X-ray structure of the heme-bound GAF domain of sensory histidine kinase DosT of Mycobacterium tuberculosis. Biochemistry 47(47):12523–12531

Puissegur MP, Botanch C, Duteyrat JL, Delsol G, Caratero C, Altare F (2004) An in vitro dual model of mycobacterial granulomas to investigate the molecular interactions between mycobacteria and human host cells. Cell Microbiol 6(5):423–433

Qamra R, Srinivas V, Mande SC (2004) Mycobacterium tuberculosis GroEL homologues unusually exist as lower oligomers and retain the ability to suppress aggregation of substrate proteins. J Mol Biol 342(2):605–617

Raghunandanan S, Jose L, Gopinath V, Kumar RA (2019) Comparative label-free lipidomic analysis of Mycobacterium tuberculosis during dormancy and reactivation. Sci Rep 9(1):3660

Rajagopalan M, Dziedzic R, Al Zayer M, Stankowska D, Ouimet M-C, Bastedo DP et al (2010) Mycobacterium tuberculosis origin of replication and the promoter for immunodominant secreted antigen 85B are the targets of MtrA, the essential response regulator. J Biol Chem 285(21):15816–15827

Ramage HR, Connolly LE, Cox JS (2009) Comprehensive functional analysis of mycobacterium tuberculosis toxinantitoxin systems: implications for pathogenesis, stress responses, and evolution. PLoS Genet 5(12):e1000767

Raman S, Song T, Puyang X, Bardarov S, Jacobs WR Jr, Husson RN (2001) The alternative sigma factor SigH regulates major components of oxidative and heat stress responses in Mycobacterium tuberculosis. J Bacteriol 183(20):6119–6125

Rao S (2009) Tuberculosis and patient gender: an analysis and its implications in tuberculosis control. Lung India 26(2):46–47

Rao SP, Alonso S, Rand L, Dick T, Pethe K (2008) The protonmotive force is required for maintaining ATP homeostasis and viability of hypoxic, nonreplicating Mycobacterium tuberculosis. Proc Natl Acad Sci USA 105(33):11945–11950

Rao SD, Datta P, Gennaro ML, Igoshin OA (2021) Chaperone-mediated stress sensing in Mycobacterium tuberculosis enables fast activation and sustained response. mSystems 6(1):e00979–e00920

Rengarajan J, Murphy E, Park A, Krone CL, Hett EC, Bloom BR, Glimcher LH, Rubin EJ (2008) Mycobacterium tuberculosis Rv2224c modulates innate immune responses. Proc Natl Acad Sci USA 105(1):264–269

Roberts DM, Liao RP, Wisedchaisri G, Hol WG, Sherman DR (2004) Two sensor kinases contribute to the hypoxic response of Mycobacterium tuberculosis. J Biol Chem 279(22):23082–23087

Rodrigue S, Provvedi R, Jacques PÉ, Gaudreau L, Manganelli R (2006) The σ factors of Mycobacterium tuberculosis. FEMS Microbiol Rev 30(6):926–941

Rodríguez JG, Hernández AC, Helguera-Repetto C, Aguilar Ayala D, Guadarrama-Medina R, Anzóla JM, Bustos JR, Zambrano MM, González-Y-Merchand J, García MJ, Del Portillo P (2014) Global adaptation to a lipid environment triggers the dormancy-related phenotype of Mycobacterium tuberculosis. MBio 5(3):e01125–e01114

Rohde K, Yates RM, Purdy GE, Russell DG (2007) Mycobacterium tuberculosis and the environment within the phagosome. Immunol Rev 219(1):37–54

Rosenzweig R, Moradi S, Zarrine-Afsar A, Glover JR, Kay LE (2013) Unraveling the mechanism of protein disaggregation through a ClpB-DnaK interaction. Science 339(6123):1080–1083

Roupie V, Romano M, Zhang L, Korf H, Lin MY, Franken KL, Ottenhoff TH, Klein MR, Huygen K (2007) Immunogenicity of eight dormancy regulon-encoded proteins of Mycobacterium tuberculosis in DNA-vaccinated and tuberculosis-infected mice. Infect Immun 75(2):941–949

Ruggiero A, Tizzano B, Pedone E, Pedone C, Wilmanns M, Berisio R (2009) Crystal structure of the resuscitation-promoting factor (ΔDUF)RpfB from M. tuberculosis. J Mol Biol 385(1):153–162

Russell DG, VanderVen BC, Lee W, Abramovitch RB, Kim MJ, Homolka S et al (2010) Mycobacterium tuberculosis wears what it eats. Cell Host Microbe 8(1):68–76

Rustad TR, Harrell MI, Liao R, Sherman DR (2008) The enduring hypoxic response of Mycobacterium tuberculosis. PLoS One 3(1):e1502

Rustad TR, Sherrid AM, Minch KJ, Sherman DR (2009) Hypoxia: a window into Mycobacterium tuberculosis latency. Cell Microbiol 11(8):1151–1159

Sachdeva P, Misra R, Tyagi AK, Singh Y (2010) The sigma factors of Mycobacterium tuberculosis: regulation of the regulators. FEBS J 277(3):605–626

Saini V, Farhana A, Steyn AJ (2012) Mycobacterium tuberculosis WhiB3: a novel iron-sulfur cluster protein that regulates redox homeostasis and virulence. Antioxid Redox Signal 16(7):687–697

Sala A, Bordes P, Genevaux P (2014) Multiple toxin-antitoxin systems in Mycobacterium tuberculosis. Toxins (Basel) 6(3):1002–1020

Salina EG, Makarov V (2022) Mycobacterium tuberculosis dormancy: how to fight a hidden danger. Microorganisms 10(12):2334

Salina EG, Bychenko OS, Mamedov IZ, Azhikina TL, Kaprelyants AS (2019a) Resuscitation of dormant "non-culturable" Mycobacterium tuberculosis is characterized by immediate transcriptional burst. Front Cell Infect Microbiol 9:272

Salina EG, Grigorov AS, Skvortsova YV, Majorov KB, Bychenko OS, Ostrik AA, Logunova NN, Ignatov DV, Kaprelyants AS, Apt AS, Azhikina TL (2019b) MTS1338, a small Mycobacterium tuberculosis RNA, regulates transcriptional shifts consistent with bacterial adaptation for entering into dormancy and survival within host macrophages. Front Cell Infect Microbiol 9:405

Salina EG, Waddell SJ, Hoffmann N, Rosenkrands I, Butcher PD, Kaprelyants AS (2014) Potassium availability triggers mycobacterium tuberculosis transition to, and resuscitation from, non-culturable (dormant) states. Open Biol 4(10):140106

Sardiwal S, Kendall SL, Movahedzadeh F, Rison SC, Stoker NG, Djordjevic S (2005) A GAF domain in the hypoxia/NO-inducible mycobacterium tuberculosis DosS protein binds haem. J Mol Biol 353(5):929–936

Sassetti CM, Rubin EJ (2003) Genetic requirements for mycobacterial survival during infection. Proc Natl Acad Sci USA 100(22):12989–12994

Sato N, Fujimura T, Masuzawa M, Yogi Y, Matsuoka M, Kanoh M et al (2007) Recombinant mycobacterium leprae protein associated with entry into mammalian cells of respiratory and skin components. J Dermatol Sci 46(2):101–110

Savvi S, Warner DF, Kana BD, McKinney JD, Mizrahi V, Dawes SS (2008) Functional characterization of a vitamin B12-dependent methylmalonyl pathway in Mycobacterium tuberculosis: implications for propionate metabolism during growth on fatty acids. J Bacteriol 190(11):3886–3895

Sharma S, Bose M (2001) Role of cytokines in immune response to pulmonary tuberculosis. Asian Pac J Allergy Immunol 19(3):213–219

Sharma S, Tyagi JS (2016) Mycobacterium tuberculosis DevR/DosR dormancy regulator activation mechanism: dispensability of phosphorylation, cooperativity and essentiality of α10 helix. PLoS One 11(8):e0160723

Sharma V, Sharma S, zu Bentrup KH, McKinney JD, Russell DG, Jacobs WR, Sacchettini JC (2000) Structure of isocitrate lyase, a persistence factor of mycobacterium tuberculosis. Nat Struct Biol 7(8):663–668

Sharma JB, Roy KK, Pushparaj M, Gupta N, Jain SK, Malhotra N, Mittal S (2008) Genital tuberculosis: an important cause of Asherman's syndrome in India. Arch Gynecol Obstet 277(1):37–41

Sharma AK, Chatterjee A, Gupta S, Banerjee R, Mandal S, Mukhopadhyay J, Basu J, Kundu M (2015) MtrA, an essential response regulator of the MtrAB two-component system, regulates the transcription of resuscitation-promoting factor B of Mycobacterium tuberculosis. Microbiology 161(6):1271–1281

Sharma A, Bloss E, Heilig CM, Click ES (2016) Tuberculosis caused by Mycobacterium africanum, United States, 2004–2013. Emerg Infect Dis 22(3):396

Sharp JD, Singh AK, Park ST, Lyubetskaya A, Peterson MW, Gomes AL, Potluri LP, Raman S, Galagan JE, Husson RN (2016) Comprehensive definition of the SigH regulon of Mycobacterium tuberculosis reveals transcriptional control of diverse stress responses. PLoS One 11(3):e0152145

Sherman DR, Voskuil M, Schnappinger D, Liao R, Harrell MI, Schoolnik GK (2001) Regulation of the mycobacterium tuberculosis hypoxic response gene encoding α-crystallin. Proc Natl Acad Sci 98(13):7534–7539

Shi L, Sohaskey CD, Kana BD, Dawes S, North RJ, Mizrahi V et al (2005) Changes in energy metabolism of Mycobacterium tuberculosis in mouse lung and under in vitro conditions affecting aerobic respiration. Proc Natl Acad Sci USA 102(43):15629–15634

Shi X-J, Shi K-X, Han F, Wang L, Cai X, Zhao G-P, et al. Genetic determinants of mycobacterium tuberculosis adaptation and drug efficacy during stationary phase growth. Microbiol Spectr 2025;13(9):e01096–25

Shleeva M, Mukamolova GV, Young M, Williams HD, Kaprelyants AS (2004) Formation of 'non culturable'cells of mycobacterium smegmatis in stationary phase in response to growth under suboptimal conditions and their Rpf-mediated resuscitation. Microbiology 150(6):1687–1697

Shleeva MO, Kudykina YK, Vostroknutova GN, Suzina NE, Mulyukin AL, Kaprelyants AS (2011) Dormant ovoid cells of mycobacterium tuberculosis are formed in response to gradual external acidification. Tuberculosis 91(2):146–154

Shleeva M, Kondratieva T, Rubakova E, Vostroknutova G, Kaprelyants A, Apt A (2015) Reactivation of dormant "non-culturable" Mycobacterium tuberculosis developed in vitro after injection in mice: both the dormancy depth and host genetics influence the outcome. Microb Pathog 78:63–66

Shleeva M, Savitsky A, Nikitushkin V, Soloviev I, Trutneva K, Keruchenko YS, Kaprelyants A (2020) Effect of photodynamic inactivation against dormant forms and active growing cells of mycobacterium smegmatis. Appl Biochem Microbiol 56(3):285–291

Singh A, Guidry L, Narasimhulu KV, Mai D, Trombley J, Redding KE, Giles GI, Lancaster JR Jr, Steyn AJ (2007) Mycobacterium tuberculosis WhiB3 responds to O2 and nitric oxide via its [4Fe-4S] cluster and is essential for nutrient starvation survival. Proc Natl Acad Sci USA 104(28):11562–11567

Singh A, Crossman DK, Mai D, Guidry L, Voskuil MI, Renfrow MB, Steyn AJ (2009) Mycobacterium tuberculosis WhiB3 maintains redox homeostasis by regulating virulence lipid anabolism to modulate macrophage response. PLoS Pathog 5(8):e1000545

Singh S, Saraav I, Sharma S (2014) Immunogenic potential of latency associated antigens against Mycobacterium tuberculosis. Vaccine 32(6):712–716

Singha B, Murmu S, Nair T, Rawat RS, Sharma AK, Soni V (2024) Metabolic rewiring of Mycobacterium tuberculosis upon drug treatment and antibiotics resistance. Meta 14(1):63

Singh P, Cole ST (2011) Mycobacterium leprae: genes, pseudogenes and genetic diversity. Future Microbiol 6(1):57–71

Sirakova TD, Dubey VS, Cynamon MH, Kolattukudy PE (2003) Attenuation of mycobacterium tuberculosis by disruption of a mas-like gene or a chalcone synthase-like gene, which causes deficiency in dimycocerosyl phthiocerol synthesis. J Bacteriol 185(10):2999–3008

Sirakova TD, Dubey VS, Deb C, Daniel J, Korotkova TA, Abomoelak B, Kolattukudy PE (2006) Identification of a diacylglycerol acyltransferase gene involved in accumulation of triacylglycerol in Mycobacterium tuberculosis under stress. Microbiology 152(9):2717–2725

Sivaramakrishnan S (2013) Ortiz de Montellano PR. The dosS-dosT/dosR mycobacterial sensor system. Biosensors 3(3):259–282

Smith I (2003) Mycobacterium tuberculosis pathogenesis and molecular determinants of virulence. Clin Microbiol Rev 16(3):463–496

Smith LJ, Stapleton MR, Buxton RS, Green J (2012a) Structure-function relationships of the Mycobacterium tuberculosis transcription factor WhiB1. PLoS One 7(7):e40407

Smith T, Wolff KA, Nguyen L (2012b) Molecular biology of drug resistance in mycobacterium tuberculosis. In: Pathogenesis of Mycobacterium tuberculosis and its Interaction with the Host Organism, pp 53–80

Solans L, Aguiló N, Samper S, Pawlik A, Frigui W, Martín C et al (2014) A specific polymorphism in mycobacterium tuberculosis H37Rv causes differential ESAT-6 expression and identifies WhiB6 as a novel ESX-1 component. Infect Immun 82(8):3446–3456

Sousa EH, Tuckerman JR, Gonzalez G, Gilles-Gonzalez MA (2007) DosT and DevS are oxygen-switched kinases in Mycobacterium tuberculosis. Protein Sci 16(8):1708–1719

Steyn AJ, Collins DM, Hondalus MK, Jacobs WR Jr, Kawakami RP, Bloom BR (2002) Mycobacterium tuberculosis WhiB3 interacts with RpoV to affect host survival but is dispensable for in vivo growth. Proc Natl Acad Sci 99(5):3147–3152

Starck J, Kallenius G, Marklund B-I, Andersson DI, Akerlund T (2004) Comparative proteome analysis of mycobacterium tuberculosis grown under aerobic and anaerobic conditions. Microbiology 150(11):3821–3829

Sturm A, Sun P, Avila-Pacheco J, Clatworthy AE, Bloom-Ackermann Z, Wuo MG et al (2024) Genetic factors affecting storage and utilization of lipids during dormancy in Mycobacterium tuberculosis. MBio 15(2):e0320823

Stewart GR, Wernisch L, Stabler R, Mangan JA, Hinds J, Laing KG, Young DB, Butcher PD (2002) Dissection of the heat-shock response in Mycobacterium tuberculosis using mutants and microarrays. Microbiology 148(Pt 10):3129–3138

Stewart GR, Newton SM, Wilkinson KA, Humphreys IR, Murphy HN, Robertson BD, Wilkinson RJ, Young DB (2005) The stress-responsive chaperone alpha-crystallin 2 is required for pathogenesis of Mycobacterium tuberculosis. Mol Microbiol 55(4):1127–1137

Sun Y, MacRae TH (2005) Small heat shock proteins: molecular structure and chaperone function. Cell Mol Life Sci 62(21):2460–2476

Sun R, Converse PJ, Ko C, Tyagi S, Morrison NE, Bishai WR (2004) Mycobacterium tuberculosis ECF sigma factor sigC is required for lethality in mice and for the conditional expression of a defined gene set. Mol Microbiol 52(1):25–38

Sun M, Andreassi JL, Liu S, Pinto R, Triccas JA, Leyh TS (2005) The trifunctional sulfate-activating complex (SAC) of mycobacterium tuberculosis. J Biol Chem 280(9):7861–7866

Tang Y, Wu W-j, Yang S, Wang D-G, Zhang Q, Liu X et al (2019) Surgical treatment of thoracolumbar spinal tuberculosis—a multicentre, retrospective, case-control study. J Orthop Surg Res 14(1):233

Tanriver G, Khan SA, Góra A, Chegou NN, Mahmoudi S (2025) Exploring the multifaceted roles of resuscitation-promoting factors in tuberculosis: implications for diagnosis, vaccine development, and drug targeting. Biotechnol Rep (Amst) 46:e00886

Taylor JL, Wieczorek A, Keyser AR, Grover A, Flinkstrom R, Karls RK, Bielefeldt-Ohmann H, Dobos KM, Izzo AA (2012) HspX-mediated protection against tuberculosis depends on its chaperoning of a mycobacterial molecule. Immunol Cell Biol 90(10):945–954

Torrey HL, Keren I, Via LE, Lee JS, Lewis K (2016) High persister mutants in mycobacterium tuberculosis. PLoS One 11(5):e0155127

Trajman A, Steffen R, Menzies D (2013) Interferon-gamma release assays versus tuberculin skin testing for the diagnosis of latent tuberculosis infection: an overview of the evidence. Pulmonary medicine 2013(1):601737

Tripathi A, Anand K, Das M, O'Niel RA, Sabarinath PS, Thakur C et al (2022) Mycobacterium tuberculosis requires SufT for Fe-S cluster maturation, metabolism, and survival in vivo. PLoS Pathog 18(4):e1010475

Trutneva KA, Shleeva MO, Demina GR, Vostroknutova GN, Kaprelyans AS (2020) One-year old dormant, "non-culturable" Mycobacterium tuberculosis preserves significantly diverse protein profile. Front Cell Infect Microbiol 10:26

Vandal OH, Roberts JA, Odaira T, Schnappinger D, Nathan CF, Ehrt S (2009) Acid-susceptible mutants of Mycobacterium tuberculosis share hypersusceptibility to cell wall and oxidative stress and to the host environment. J Bacteriol 191(2):625–631

Vaubourgeix J, Lin G, Dhar N, Chenouard N, Jiang X, Botella H, Lupoli T, Mariani O, Yang G, Ouerfelli O, Unser M, Schnappinger D, McKinney J, Nathan C (2015) Stressed mycobacteria use the chaperone ClpB to sequester irreversibly oxidized proteins asymmetrically within and between cells. Cell Host Microbe 17(2):178–190

Velayati AA, Farnia P (2016) Atlas of Mycobacterium tuberculosis, 1st edn. Academic, London. Hardback ISBN: 9780128038086. eBook ISBN: 978012803880

Velayati AA, Farnia P, Masjedi MR, Zhavnerko GK, Merza MA, Ghanavi J, Tabarsi P, Farnia P, Poleschuyk NN, Ignatyev G (2011) Sequential adaptation in latent tuberculosis bacilli: observation by atomic force microscopy (AFM). Int J Clin Exp Med 4(3):193–199

Velayati AA, Abeel T, Shea T, Konstantinovich Zhavnerko G, Birren B, Cassell GH, Earl AM, Hoffner S, Farnia P (2016) Populations of latent Mycobacterium tuberculosis lack a cell wall: isolation, visualization, and whole-genome characterization. Int J Mycobacteriol 5(1):66–73

Velayati AA, Mitaria S, Farnia P, Farnia P, Ghanavi J, Maleknia M, Murase Y, Hoffner S (2024) Exploring the genetic landscape of Mycobacterium tuberculosis: unlocking the differences in between latent and active tuberculosis. Int J Mycobacteriol 13(4):410–419

Verbon A, Kuijper S, Jansen H, Speelman P, Kolk A (1992) Antibodies against secreted and non-secreted antigens in mice after infection with live mycobacterium tuberculosis. Scand J Immunol 36(3):371–384

Verma A, Ghoshal A, Dwivedi VP, Bhaskar A (2022) Tuberculosis: the success tale of less explored dormant Mycobacterium tuberculosis. Front Cell Infect Microbiol 12:1079569

Via LE, Lin PL, Ray SM, Carrillo J, Allen SS, Eum SY et al (2008) Tuberculous granulomas are hypoxic in Guinea pigs, rabbits, and nonhuman primates. Infect Immun 76(6):2333–2340

Vilchèze C, Weinrick B, Wong KW, Chen B, Jacobs J, William R (2010) NAD+ auxotrophy is bacteriocidal for the tubercle bacilli. Mol Microbiol 76(2):365–377

Vilchèze C, Weinrick B, Leung LW, Jacobs WR Jr (2018) Plasticity of Mycobacterium tuberculosis NADH dehydrogenases and their role in virulence. Proc Natl Acad Sci USA 115(7):1599–1604

Vinod V, Pushkaran AC, Kumar A, Mohan CG, Biswas R (2021) Interaction mechanism of mycobacterium tuberculosis GroEL2 protein with macrophage lectin-like, oxidized low-density lipoprotein receptor-1: an integrated computational and experimental study. Biochimica et Biophysica Acta (BBA) Gen Sub 1865(1):129758

Voskuil MI, Schlesinger LS (2015) Toward resolving the paradox of the critical role of the DosR regulon in mycobacterium tuberculosis persistence and active disease. American Thoracic Society; p. 1103–1105, Am J Respir Crit Care Med, 191

Voskuil MI, Schnappinger D, Visconti KC, Harrell MI, Dolganov GM, Sherman DR, Schoolnik GK (2003) Inhibition of respiration by nitric oxide induces a Mycobacterium tuberculosis dormancy program. J Exp Med 198(5):705–713

Voskuil M, Schnappinger D, Rutherford R, Liu Y, Schoolnik G (2004) Regulation of the mycobacterium tuberculosis PE/PPE genes. Tuberculosis 84(3–4):256–262

Voskuil MI, Bartek IL, Visconti K, Schoolnik GK (2011) The response of Mycobacterium tuberculosis to reactive oxygen and nitrogen species. Front Microbiol 2:105

Vynnycky E, Fine PEM (1997) The natural history of tuberculosis: the implications of age-dependent risks of disease and the role of reinfection. Epidemiol Infect 119(2):183–201

Waksman SA (1964) The role of chemotherapy in the treatment of tuberculosis. Am Rev Respir Dis 89:47–62

Wan T, Horová M, Beltran DG, Li S, Wong H-X, Zhang L-M (2021) Structural insights into the functional divergence of WhiB-like proteins in mycobacterium tuberculosis. Mol Cell 81(14):2887–2900. e5

Wang H, Wan L, Shi J, Zhang T et al (2021) Quantitative proteomics reveals that dormancy-related proteins mediate the attenuation in mycobacterium strains. Virulence 12(1):2228–2246. https://doi.org/10.1080/21505594.2021.1965703

Warnecke T (2012) Loss of the DnaK-DnaJ-GrpE chaperone system among the Aquificales. Mol Biol Evol 29(11):3485–3495

Watanabe S, Zimmermann M, Goodwin MB, Sauer U, Barry CE 3rd, Boshoff HI (2011) Fumarate reductase activity maintains an energized membrane in anaerobic mycobacterium tuberculosis. PLoS Pathog 7(10):e1002287

Watanabe A, Matsumoto T, Igari H, Sawa J, Yamaguchi Y, Sakatani M (2016) Risk of developing active tuberculosis in rheumatoid arthritis patients on adalimumab in Japan. Int J Tuberc Lung Dis 20(1):101–108

Wright CC, Hsu FF, Arnett E, Dunaj JL, Davidson PM, Pacheco SA et al (2017) The mycobacterium tuberculosis MmpL11 cell wall lipid transporter is important for biofilm formation, intracellular growth, and nonreplicating persistence. Infect Immun 85(8). https://doi.org/10.1128/iai.00131-17

Wayne LG (1976) Dynamics of submerged growth of Mycobacterium tuberculosis under aerobic and microaerophilic conditions. Am Rev Respir Dis 114(4):807–811

Wayne LG (1994) Dormancy of Mycobacterium tuberculosis and latency of disease. Eur J Clin Microbiol Infect Dis 13(11):908–914

Wayne LG, Sramek HA (1979) Antigenic differences between extracts of actively replicating and synchronized resting cells of Mycobacterium tuberculosis. Infect Immun 24(2):363–370

Wayne LG, Sramek HA (1994) Metronidazole is bactericidal to dormant cells of Mycobacterium tuberculosis. Antimicrob Agents Chemother 38(9):2054–2058

Wayne LG, Hayes LG (1996) An in vitro model for sequential study of shiftdown of Mycobacterium tuberculosis through two stages of nonreplicating persistence. Infect Immun 64(6):2062–2069

Wayne LG, Sohaskey CD (2001) Nonreplicating persistence of mycobacterium tuberculosis. Ann Rev Microbiol 55:139–163

Weinstein EA, Yano T, Li LS, Avarbock D, Avarbock A, Helm D et al (2005) Inhibitors of type II NADH:menaquinone oxidoreductase represent a class of Antitubercular drugs. Proc Natl Acad Sci USA 102(12):4548–4553

Willemse D, Weber B, Masino L, Warren RM, Adinolfi S, Pastore A, Williams MJ (2018) Rv1460, a SufR homologue, is a repressor of the suf operon in mycobacterium tuberculosis. PLoS One 13(7):e0200145

Wisedchaisri G, Wu M, Sherman DR, Hol WG (2008) Crystal structures of the response regulator DosR from Mycobacterium tuberculosis suggest a helix rearrangement mechanism for phosphorylation activation. J Mol Biol 378(1):227–242

Wlodarczyk M, Rudnicka W, Janiszewska-Drobinska B, Kielnierowski G, Kowalewicz-Kulbat M, Fol M, Druszczynska M (2014) Interferon-gamma assay in combination with tuberculin skin test are insufficient for the diagnosis of culture-negative pulmonary tuberculosis. PLoS One 9(9):e107208

World Health Organization (2024) Global Tuberculosis Report 2024. World Health Organization, Geneva

Wu J, Wang S, Lu C, Shao L, Gao Y, Zhou Z et al (2017) Multiple cytokine responses in discriminating between active tuberculosis and latent tuberculosis infection. Tuberculosis 102:68–75

Xiao X, Fay A, Santos Molina P, Kovach A, Glickman MS, Li H (2024) Structure of the Mycobacterium tuberculosis DnaK–GrpE complex reveals how key DnaK roles are controlled. Nat Commun 15:660

Xu Y, Ehrt S, Schnappinger D, Beites T (2023) Synthetic lethality of Mycobacterium tuberculosis NADH dehydrogenases is due to impaired NADH oxidation. MBio 14(3):e01045–e01023

Yang J, Zhang L, Qiao W, Luo Y (2023) Mycobacterium tuberculosis: pathogenesis and therapeutic targets. MedComm 4(5):e353

Yuan Y, Crane DD, Barry CE 3rd. (1996) Stationary phase-associated protein expression in Mycobacterium tuberculosis: function of the mycobacterial alpha-crystallin homolog. J Bacteriol 178(15): 4484–4492

Yuan Y, Crane DD, Simpson RM, Zhu Y, Hickey MJ, Sherman DR et al (1998) The 16-kDa α-crystallin (Acr) protein of Mycobacterium tuberculosis is required for growth in macrophages. Proc Natl Acad Sci USA 95(16):9578–9583

Zhang Y, Yew WW, Barer MR (2012) Targeting persisters for tuberculosis control. Antimicrob Agents Chemother 56(5):2223–2230

Zhang K, Wang X, Tu J, Rong H, Werz O, Chen X (2019) The interplay between depression and tuberculosis. J Leukoc Biol 106(3):749–757

Zhang LY, Wang CL, Yan MY, Geng YM, Yin H, Jia HY, Zhu CZ, Li ZH, Ren GX, Pan LP, Sun YC, Zhang ZD (2022) Toxin-antitoxin systems Alter adaptation of Mycobacterium smegmatis to environmental stress. Microbiol Spectr 10(6):e0281522

Zheng H, Abramovitch RB (2020) Inhibiting DosRST as a new approach to tuberculosis therapy. Future Med Chem 12(5):457–467

Zhao Y, Jin J, Hu Q, Zhou HM, Yi J, Yu Z et al (2011) Genetically encoded fluorescent sensors for intracellular NADH detection. Cell Metab 14(5):555–566

Zhuo L, Zhang Z, Pan Z, Sheng D-h, Hu W, Li Y-z (2018) CIRCE element evolved for the coordinated transcriptional regulation of bacterial duplicate groELs. Biochimica et Biophysica Acta (BBA)-Gen Regul Mech 1861(10):928–937

Structural and Functional Characterization of Protein Adhesins and Pili in *Mycobacterium tuberculosis*

8

Abstract

Mycobacterium tuberculosis (Mtb) is a highly adaptable pathogen whose ability to establish infection and persist within the host is primarily driven by a diverse array of adhesins, surface proteins, and structures that mediate attachment to host cells and tissues. These adhesins play critical roles in bacterial colonization, invasion, immune evasion, and long-term persistence, making them central to Mtb pathogenesis. Mtb adhesins can be broadly categorized based on their structural features, functional roles, and binding specificities. Protein adhesins such as LpqH (the 19-kDa lipoprotein), PstS-1 (phosphate-binding protein), Cpn60.2 (chaperonin GroEL2), Apa (alanine- and proline-rich antigen), members of the PE-PGRS protein family, the antigen 85 complex (Ag85A, Ag85B, and Ag85C), glutamine synthetase A1, malate synthase, N-acetylmuramoyl-L-alanine amidase, the histone-like HU protein, protein kinase D, ESAT-6, Mce1, and HBHA (heparin-binding hemagglutinin adhesin) exhibit specific affinities for host cell receptors and extracellular matrix components such as fibronectin, laminin, and epithelial surfaces. Many of these proteins are multifunctional "moonlighting" enzymes, performing both metabolic and adhesive roles, thereby enhancing bacterial survival and virulence. In addition to protein adhesins, Mtb expresses a variety of pili and fimbrial structures that further augment its capacity to interact with the host. Recent atomic force microscopy studies have identified seven distinct pili types on the mycobacterial surface: Type III (secretion needle pili), Type IV secretion pili, Type IV-like pili, curli-like pili (MTP), and the newly described Types V (relief funnel pili), VI (adhesion tapering pili), and VII (adhesion flap pili). These pili exhibit remarkable structural diversity, with Types VI and VII found exclusively in extensively drug-resistant (XDR) and totally drug-resistant (TDR) Mtb strains, suggesting a role in adaptation to antibiotic pressure and enhanced survival. Functionally, curli-like and Type IV pili contribute to biofilm formation, bacterial aggregation, and robust adhesion to macrophages and epithelial cells, while the novel pili types (V, VI, and VII) may mediate specialized adhesion and intercellular interactions, particularly in drug-resistant isolates. The adhesins and pili target diverse host molecules: some facilitate macrophage binding (e.g., LpqH, PstS-1, Cpn60.2, Apa), others bind fibronectin and laminin (e.g., PE-PGRS proteins, antigen 85 complex, malate synthase, curli-like pili), and yet others adhere to epithelial cells (e.g., Mce1, HBHA, Type IV pili). This multifaceted adhesion strategy enables Mtb to colonize a wide range of tissue environments, evade immune

P. Farnia et al., *Proteins in Mycobacterium Tuberculosis*, Advances in Experimental Medicine and Biology 1484, https://doi.org/10.1007/978-3-031-96883-9_8

defenses, and establish latent or persistent infections. The redundancy and multifunctionality of these adhesins and pili underscore their evolutionary significance in Mtb's lifecycle. Collectively, the surface adhesins and pili constitute a complex and versatile toolkit that underpins Mtb pathogenicity. The discovery of multiple distinct pili types, especially those associated with drug resistance, opens new avenues for research into bacterial adaptation, persistence, and immune evasion mechanisms. A comprehensive understanding of these structures and their interactions with host tissues holds promise for the development of innovative diagnostics, vaccines, and antiadhesin therapies aimed at combating both drug-sensitive and drug-resistant tuberculosis.

Keywords

Mycobacterium tuberculosis · Protein adhesins · Secretion needle pili · Curli-like pili (MTP) · Types V (relief funnel pili) · Types VI (adhesion tapering pili) · Types VII (adhesion flap pili)

8.1 Introduction

When an infected individual releases *Mycobacterium tuberculosis* (Mtb) into the air, others inhale it as tiny airborne droplets. These infectious droplets vary in size, typically ranging from about 0.65 micrometers for the smallest particles to more than 7 micrometers for medium and larger droplets (Fennelly and Jones-López 2015). Smaller aerosol particles are expected to bypass the nasopharyngeal and tracheobronchial regions, depositing in the distal airways, whereas larger particles tend to be trapped in the upper airway or oropharynx. This entrapment can potentially lead to tuberculosis (TB) manifestations in the oropharynx or cervical lymph nodes (Laal 2012; Fennelly and Jones-López 2015). Upon reaching the lower respiratory tract, Mtb is primarily phagocytosed by macrophages and dendritic cells. However, analyses of sputum samples from TB patients have identified neutrophils as the predominant phagocytic cells infected with Mtb (Mehta et al. 2006; Bermudez and Goodman 1996; Eum et al. 2010; Cohen et al. 2018). Furthermore, evidence indicates that Mtb can infect non-myelocytic cells within TB patients (Lerner et al. 2015; Randall et al. 2015; Ganbat et al. 2016). Bacilli that evade clearance mechanisms within the upper respiratory tract are transported to the alveolar spaces, which are composed of type I and type II alveolar epithelial cells (AECs) in addition to a diverse population of immune cells, including alveolar macrophages, dendritic cells, and neutrophils. These cellular constituents collectively constitute the initial microenvironment encountered by Mtb during pulmonary infection and play critical roles in host-pathogen interactions and subsequent immune responses (Lerner et al. 2015). Notably, Mtb possesses the ability to invade alveolar epithelial cells (AECs), which are increasingly recognized as pivotal contributors to the initial establishment of infection and subsequent disease progression (Mehta et al. 2006; Hernández-Pando et al. 1996; Smith 2003; Ashiru et al. 2012). The alveolar epithelial layer predominantly consists of type I and type II AECs. Type I AECs are thin, squamous cells covering the majority of the alveolar surface and are essential for gas exchange. In contrast, type II AECs are cuboidal cells responsible for surfactant production and serve as progenitors for type I cells, thereby contributing to epithelial repair and homeostasis (Saini et al. 2015; Hadifar et al. 2019). Alveolar epithelial cells (AECs) outnumber alveolar macrophages by several hundredfold within the alveolus, with estimates indicating approximately 30,000 AECs compared to only about 50 macrophages per alveolus. This substantial numerical disparity suggests that AECs are likely the initial host cell population encountered by Mtb upon reaching the pulmonary alveolar space (Crandall and Matthay 2001,; Crystal et al. 2008; Schneeberger and Lynch 1992; Mehta et al. 2006; Saini et al. 2015; Thacker et al. 2020; Pisu et al. 2020). Initial evidence for Mtb infection of alveolar epithelial cells (AECs) in humans was demonstrated by detecting Mtb DNA within type II AECs obtained from autopsied lung tissue of individuals who were latently infected with Mtb, but had died from causes unrelated to tuberculosis (Hernández-Pando et al. 1996; Ryndak et al. 2015). More recent studies have identified both Mtb DNA and viable bacteria in non-macrophage cells not only in the lungs but also in extra-pulmonary sites such as the kidney and spleen of latently infected individuals from TB-endemic regions (Barrios-Payán et al. 2012). Experimental infection models utilizing the A549 human type II AEC carcinoma cell line and primary bronchial epithelial cells have demonstrated that Mtb can adhere to and invade these epithelial cells, albeit generally at lower frequencies than macrophages (Bermudez and Goodman 1996; Birkness et al. 1999; García-Pérez et al. 2003; Ryndak et al. 2015; Reuschl et al. 2017; Matsuyama et al. 2018; Thacker et al. 2020; de Waal et al. 2022). In vitro analyses demonstrate that laboratory strains, i.e., notably HRv, effectively invade and proliferate within type II alveolar epithelial cells (AECs). This process relies on both microtubule- and microfilament-dependent pathways for cellular entry (Mehta et al. 2006; Bermudez and Goodman 1996). The intracellular replication capacity within AECs suggests these cells provide a protected niche conducive to bacterial survival and initial infection establishment (Agarwal et al. 2018). Advanced imaging techniques, including lung-on-chip technology, have shown that pulmonary surfactant regulates Mtb growth in both AECs and macrophages; however, AECs are more permissive to Mtb replication. When surfactant production is impaired, Mtb growth becomes uncontrolled in both cell types, highlighting the protective role of an intact alveolar barrier (Thacker et al. 2020). Transcriptional profiling of Mtb replicating within type II AECs reveals upregulation of genes associated with active metabolism and virulence, further indicating that AECs could provide a supportive environment for bacterial proliferation prior to the full activation of adaptive immune responses (Saini et al. 2015; Pisu et al. 2020). Recent investigations employing air-liquid interface (ALI) models of human primary bronchial epithelial cells have demonstrated that airway epithelial cells mount a rapid and robust immune response upon Mtb infection. This response is characterized by the upregulation of antimicrobial peptides and pro-inflammatory cytokines, which serve to recruit neutrophils and modulate the local immune environment (Nouailles et al. 2014; Leiva-Juárez et al. 2018; Johnston et al. 2021; de Waal et al. 2022). These epithelial responses are initiated not only by direct bacterial infection but are also potentiated by paracrine signaling molecules secreted by infected macrophages, including interleukin-1 beta (IL-1β) and type I interferons, which amplify the inflammatory milieu and coordinate innate immune defenses (Pisu et al. 2020; Saini et al. 2015).

A key feature of Mtb pathogenesis is the disruption of the epithelial barrier, primarily driven by virulence factors such as the early secreted antigenic target 6 kDa protein (ESAT-6). ESAT-6 has been demonstrated to impair the integrity of tight junctions between epithelial cells, facilitating bacterial passage across the epithelial layer and enabling dissemination beyond the initial infection site (Kinhikar et al. 2010). Furthermore, Mtb's ability to adhere to and invade alveolar epithelial cells (AECs) is primarily mediated by bacterial surface adhesins that specifically recognize and bind to receptors on the host epithelial surface (Bermudez and Goodman 1996; Ashiru et al. 2010). Adhesins are generally surface-exposed molecules that facilitate cell-to-cell or cell-to-extracellular matrix (ECM) interactions. These include intercellular adhesion molecules (ICAMs) and surface adhesion molecules (SAMs), which mediate bacterial attachment to host cells and ECM components such as fibronectin and laminin (da Silva Neto et al. 2009; Hallstrom and McCormick 2015; Viljoen et al. 2022). Both bacterial factors and host determinants intricately contribute to these interactions, with adhesins playing a pivotal role in orchestrating the complex process of host cell invasion (Klemm and Schembri 2000; Pizarro-Cerdá and Cossart 2006). Among the key adhesins identified, the 19 kDa lipoprotein antigen is notable for its ability to bind mannose receptors on macrophages, thereby facilitating phagocytosis (Diaz-Silvestre et al. 2005). LpqH, a glycosylated lipoprotein characterized by a β-sandwich fold, acts as a potent agonist of Toll-like receptor 2 (TLR2), initiating downstream immune activation pathways. Structural and functional studies have demonstrated that this 19 kDa lipoprotein antigen engages TLR2-dependent signaling and activates the NLRP3 inflammasome, leading to cytokine secretion and increased bacterial internalization (Brightbill et al. 1999; Liu et al. 2021). Beyond these adhesins, Mtb also utilizes metabolic enzymes such as malate synthase (MS), a key component of the glyoxylate shunt pathway, which is essential for bacterial persistence and intracellular survival in vivo (McKinney et al. 2000). Mtb malate synthase (MS) has been shown to function as an adhesin by binding extracellular matrix components such as laminin and fibronectin. When heterologously expressed in *Mycobacterium smegmatis*, MS localizes to the bacterial cell wall and significantly increases adherence to A549 human alveolar epithelial cells. Moreover, antibodies directed against the C-terminal domain of MS markedly inhibit this binding, highlighting its critical role in mediating host-cell interactions (Kinhikar et al. 2006). Pilus structures also contribute substantially to Mtb adhesion. The mycobacterial pilus (MTP), encoded by the mtp gene (Rv3312A), facilitates bacterial binding to laminin, macrophages, and epithelial cells. Functional studies employing pili-deficient mutants reveal a marked decrease in adhesion and invasion of both macrophages and epithelial cells, whereas overexpression of MTP enhances these capacities. Similarly, Type IV pili (encoded by Rv3654c-Rv3660c) are implicated in promoting initial attachment and colonization of host tissues (Ramsugit and Pillay 2014; Ramsugit et al. 2016). Following attachment, Mtb invades alveolar epithelial cells through specific interactions between bacterial adhesins, such as heparin-binding hemagglutinin adhesin (HBHA), and host cell surface receptors, including heparan sulfate proteoglycans like syndecan-4, as well as ECM components such as fibronectin and laminin (Pethe et al. 2001; Kinhikar et al. 2006; Zimmermann et al. 2016). Syndecan-4 (Sdc4) has been identified as a critical receptor mediating Mtb attachment and internalization in AECs; knockdown experiments targeting Sdc4 result in significantly diminished bacterial uptake, highlighting its essential role in the invasion process (Zimmermann et al. 2016). HBHA specifically binds to the heparan sulfate chains present on syndecan-4, facilitating bacterial adherence and subsequent internalization (Pethe et al. 2001; Zimmermann et al. 2016). In vitro infection models commonly utilize the A549 human alveolar epithelial cell line, which is incubated with either wild-type or mutant Mtb strains. Following incubation, non-adherent bacteria are removed by washing, and the number of cell-associated bacteria is quantified either by plating for colony-forming units (CFU) or by immunofluorescence microscopy (Glickman and Jacobs 2001; Smith 2003; Zumla et al. 2013). Recent methodological advancements include the application of fluorescent probes such as DLF-1, which enable rapid and quantitative detection of intracellular Mtb, thereby enhancing the sensitivity and throughput of infection assays (Zhang et al. 2016). Functional studies involving knockout or knockdown of host receptors, particularly syndecan-4, have confirmed the specificity of Mtb interactions with alveolar epithelial cells. Reduced expression or functional blockade of Sdc4 significantly decreases mycobacterial attachment and internalization, further underscoring the receptor's importance in mediating bacterial invasion (Zimmermann et al. 2016). Importantly, once internalized, Mtb replicates more efficiently within epithelial cells than within macrophages. Mehta et al. (2006) demonstrated that although Mtb invades macrophages more readily, its intracellular replication rate is substantially higher in epithelial cells. Moreover, bacteria that have undergone replication within alveolar epithelial cells exhibit markedly increased invasiveness toward other non-phagocytic cells (Bermudez et al. 2002). These findings suggest that epithelial cells not only serve as an early intracellular niche for Mtb replication but may also facilitate subsequent bacterial dissemination across the alveolar barrier and into the systemic circulation, thereby contributing to the establishment and progression of tuberculosis infection (Zimmermann et al. 2016;

Bermudez et al. 2002). In summary, the initial attachment and invasion of Mtb are orchestrated by a diverse repertoire of bacterial adhesins, including heparin-binding hemagglutinin adhesin (HBHA), LpqH, malate synthase, mycobacterial pili (MTP), and Type IV pili, which mediate precise interactions with host cell surface receptors and extracellular matrix components (Pethe et al. 2001; Kinhikar et al. 2006; Alteri et al. 2007; Zimmermann et al. 2016). These molecular interactions are critical for successful colonization, intracellular survival, and dissemination of Mtb, as demonstrated by numerous in vitro and in vivo experimental studies (Bermudez et al. 2002; Zimmermann et al. 2016). A comprehensive understanding of these early host-pathogen interactions provides valuable insights into the pathogenesis of tuberculosis and identifies potential molecular targets for therapeutic intervention.

8.2 Adhesin Proteins in *Mycobacterium tuberculosis*

The cell envelope of Mtb exhibits a uniquely intricate architecture, characterized by a thick, lipid-rich outer layer comprising mycolic acids, glycolipids, lipoproteins, and polysaccharides. This complex structural organization facilitates the presentation of a diverse repertoire of surface adhesins, encompassing proteins, lipoproteins, glycoproteins, glycolipids, and glycopolymers, each contributing distinct chemical properties essential for host-pathogen interactions. (Squeglia et al. 2018; Singh et al. 2022; Daffé and Marrakchi 2019; Jacobo-Delgado et al. 2023). Unlike many bacterial pathogens that rely primarily on classical surface proteins for adhesion, Mtb utilizes a multifaceted adhesin repertoire that extends beyond proteins to include lipid and carbohydrate molecules, reflecting its evolutionary adaptation to the intracellular lifestyle and complex host environment. Therefore, adhesins in Mtb are specialized surface molecules that enable the bacterium to attach to host cells and extracellular matrix (ECM) components, representing a critical initial step for colonization, immune evasion, and pathogenesis. They interact with a broad spectrum of host receptors, including Toll-like receptors, C-type lectin receptors, scavenger receptors, as well as various ECM proteins such as fibronectin, collagen, elastin, laminin, fibrillin, and vitronectin (Singh et al. 2022). The multifaceted adhesin repertoire facilitates the bacterium's ability to bind diverse host cell types and ECM components, promoting successful colonization, intracellular survival, and dissemination during infection (Singh et al. 2022). Understanding these interactions at the molecular level is essential for developing novel therapeutic strategies aimed at blocking bacterial adhesion and preventing tuberculosis pathogenesis. They also include enzymes and chaperones that have evolved secondary adhesive functions, expanding the repertoire of molecules involved in host interaction. These adhesins may be anchored in the cell envelope with or without signal peptides and can exist as either monomeric or multimeric complexes. The interaction between mycobacterial adhesins and host receptors not only mediates bacterial attachment but also triggers immune responses and significantly influences disease progression. A deeper understanding of these interactions is essential for the development of new diagnostics, vaccines, and antiadhesion therapies targeting tuberculosis (Menozzi et al. 1996; Pethe et al. 2001; Ramsugit and Pillay 2014; Ramsugit et al. 2016; Zhao et al. 2020). Based on their structure and mode of presentation, Mtb adhesins can be broadly classified into two main categories: fimbrial (pilus-associated) and non-fimbrial adhesins (Squeglia et al. 2018). Fimbrial adhesins, or pili, are short, hair-like appendages on the bacterial cell surface that mediate attachment to host cells and extracellular matrix components, facilitating infection and representing key virulence factors. These pilus structures promote bacterial adherence by physically bridging the pathogen to host receptors, thereby enhancing colonization and invasion. Non-fimbrial adhesins, in contrast, are typically surface-exposed proteins or protein complexes that mediate adhesion without forming filamentous structures. These include a diverse set of molecules, such as lipoproteins, enzymes, chaperones, and glycoproteins, that have evolved secondary adhesive functions in addition to their primary cellular role.

8.2.1 Fimbrial Adhesins or Pili in *Mycobacterium tuberculosis*

Prior to discussing the pili of Mtb, it is essential first to provide an overview of fimbrial adhesins, commonly referred to as pili. These filamentous appendages may be distributed uniformly across the bacterial surface or localized to one or both cellular poles. While pili are traditionally associated with Gram-negative bacteria, they are also present in certain Gram-positive bacteria and archaea. Functionally, these structures are integral to bacterial adhesion, facilitating colonization and contributing significantly to pathogenicity (Jefferson 2004; Scott and Zähner 2006). Pili structures enhance bacterial interaction with their surrounding environment by maximizing surface contact. These appendages are antigenic, with the genes encoding pili typically residing either on the bacterial chromosome or on plasmids. Various types of pili have been characterized across both Gram-positive and Gram-negative bacteria, reflecting their widespread functional significance in microbial physiology and pathogenesis. Pili were initially identified in Gram-negative bacteria as receptors for bacteriophages, highlighting their role in bacterial surface interactions (Costerton et al. 1999; Telford et al. 2006). Gram-negative bacteria are distinguished by a dual membrane system consisting of an inner cytoplasmic membrane and an outer membrane, separated by a thin peptidoglycan layer that maintains cell wall integrity (Telford et al. 2006; Kline et al. 2009). Several classes of pili have been characterized in these organisms, including Type I pili, Type III secretion system needles, Type IV pili, Type IV secretion system pili, curli pili, and chaperone-usher (CU) pili, each contributing uniquely to bacterial adhesion, motility, and pathogenicity (Fernandez and Berenguer 2000; Van Gerven et al. 2015; Ramirez et al. 2020; Dodson et al. 1993; Proft and Baker 2009; Kline et al. 2009; Costa et al. 2015; Hospenthal et al. 2017). In 1965, Brinton employed electron microscopy, crystallography, and X-ray diffraction techniques to elucidate the structure of the Type I pilus (Brinton Jr. 1965). These pili are widely distributed among Enterobacteriaceae species, including the majority of *Escherichia coli* strains (Proft and Baker 2009; Schmidt et al. 2022). Under electron microscopy, Type I pili appear as rigid, rod-shaped filaments approximately 1–2 μm in length, exhibiting a peritrichous distribution. Notably, they possess a flexible distal tip that mediates specific interactions with host cell surface receptors (Hahn et al. 2002). Type I pili represent the predominant adhesins in uropathogenic *Escherichia coli* (UPEC) strains, playing a critical role in the pathogenesis of bladder infections (cystitis) by facilitating bacterial adhesion, colonization, and evasion of urinary clearance mechanisms (Connell et al. 1996; Nykvist and Pettersson 2003; Wright et al. 2007; Schwartz et al. 2013). Structurally, Type I pili are composed of over 1000 pilin subunits arranged in a helical fashion. Atomic force microscopy studies have demonstrated that these pili can withstand mechanical forces up to 60 piconewtons (pN) through a mechanism of helical unwinding, thereby maintaining adhesion under shear stress (Miller et al. 2006). The Type I pilus has a short fibrillar tip, 3 nm wide, made up of three subunits: FimH, FimG, and FimF (Jones et al. 1995; Nishiyama et al. 2005; Miller et al. 2006). FimH, located at the distal tip, is responsible for attaching the pilus to mannosylated receptors found on various cell types and is a significant factor in cystitis development (Jones et al. 1995; Schmidt et al. 2022). FimH has two β-sandwich-shaped domains: a distal N-terminal lectin domain containing the mannose-binding pocket, and a C-terminal pilin domain that anchors the adhesin to the fimbrial tip (Jones et al. 1995; Schilling et al. 2001). The rigid pilus rod is made up of repeating FimA subunits forming a helical quaternary structure, 6–7 nm thick, with a helical cavity 20 Å wide and 3.1 subunits per turn (Hultgren et al. 1986; Miller et al. 2006; Spaulding et al. 2017). P pili are critical virulence factors in uropathogenic *Escherichia coli* (UPEC), facilitating the establishment of pyelonephritis through specific recognition and adhesion to kidney epithelial tissues (Bullitt and Makowski 1995; Mulvey et al. 2001). The P pilus is a complex organelle encoded by an 11-gene cluster known as the pyelonephritis-associated pili (pap) operon (Lillington et al. 2014). Structurally, the pilus extends several micrometers in length and measures approximately 6.8 nm in diameter, forming a right-handed helical cylinder composed predominantly of repeating PapA subunits arranged with 3.28 subunits per helical turn (Soto and Hultgren 1999; Lillington et al. 2014). The distal tip fibrillum, thinner at 2–3 nm, contains the adhesin PapG along with minor pilus proteins PapE, PapF, and PapK, which collectively mediate host receptor binding (Lee et al. 2004). Three allelic variants of PapG (PapG I, II, and III) exhibit distinct receptor specificities, all recognizing a conserved Gal(α1–4) Gal moiety linked to ceramide via a β-glucose residue, but differing in additional sugar modifications such as N-acetylgalactosamine or sialic acid residues. PapG binds to Gal(α1–4) Gal-containing glycolipids of the globoseries (GbO4) present on erythrocytes and renal epithelial cells, underscoring the pivotal role of P pili in UPEC-mediated pyelonephritis (Lane and Mobley 2007). The tip fibrillum's architecture is stabilized by PapF, which links PapG to a polymer primarily composed of PapE subunits, with PapK serving as the connector between the tip fibrillum and the pilus rod. The pilus rod itself consists of thousands of PapA subunits arranged in a right-handed helix approximately 6.8 nm thick, featuring a helical pitch of 24.9 Å and 3.28 subunits per turn. Anchoring the pilus to the bacterial outer membrane is the minor subunit PapH (Fällman et al. 2005). Both P and Type I pili share a head-to-tail arrangement of their major pilin subunits; however, mechanical stress or exposure to 50% glycerol can disrupt these interactions, causing the pilus rod to unwind into a linear fiber resembling the tip fibrillum (Miller et al. 2006). This unwinding mechanism enables the pili to absorb tensile forces over a broader range of lengths, thereby enhancing their resilience against shear stress in the urinary tract (Lillington et al. 2014; Busch and Waksman 2012). Genetically, the pap operon encodes P pili, whereas the fim operon encodes Type I pili, which, despite architectural similarities, tend to be more rigid and susceptible to mechanical breakage (Lillington et al. 2014; Busch and Waksman 2012). Curli are a unique class of adhesive extracellular amyloid fibers produced by Gram-negative bacteria, first discovered in 1989 in fibronectin-binding *Escherichia coli* isolates from bovine fecal samples (Olsén et al. 1989). They consist of repeating subunits of the major pilin CsgA (or curlin), which lacks structural homology to other pilins. Curli play crucial roles in bacterial interactions, biofilm formation, and inflammation (Olsén et al. 1989; Epstein and Chapman 2008; Perov et al. 2019; Yan et al. 2020). Curli fibers are extracellular amyloid structures that play a significant role in bacterial adhesion by binding to host extracellular matrix components such as fibronectin, thereby facilitating colonization (Epstein and Chapman 2008). Beyond adhesion, curli fibers contribute to pathogenesis by exacerbating amyloidosis, inducing proinflammatory responses, and aggravating hypotension and bleeding disorders during sepsis. Structurally, amyloid fibers possess a core of β-strands, conferring remarkable resistance to chemical and thermal denaturation as well as proteolytic degradation (Epstein and Chapman 2008).

Curli fibers protect bacteria against environmental stresses, including desiccation and antimicrobial agents. In *Escherichia coli*, curli are primarily composed of two proteins, CsgA and CsgB, each approximately 100 amino acids in length (Shu et al. 2012). The mature CsgA protein contains an N-terminal domain responsible for secretion and an amyloidogenic core domain that is resistant to protease digestion. Both CsgA and CsgB feature five repeating units characterized by the consensus sequence Ser-X_5-Gln-X_4-Asn-X_5-Gln, a motif typical of amyloidogenic peptides. Assembly of curli fibers involves the head-to-tail stacking of curlin subunits into an extended β-helical structure (Sleutel et al. 2023). Curli fibers share biophysical properties with amyloid aggregates implicated in neurodegenerative diseases, including their non-branching morphology, β-sheet-rich composition, and resistance to protease digestion and sodium dodecyl sulfate (SDS) treatment (Sleutel et al. 2023). Notably, curli expression has been detected at physiological temperature (37 °C), and these fibers function as pathogen-associated molecular patterns (PAMPs) recognized by Toll-like receptor 2 (TLR2), thereby activating innate immune responses (Kikuchi et al. 2005). Functionally, curli can bind plasminogen and tissue-type plasminogen activator (t-PA), facilitating the conversion of plasminogen to plasmin. This enzymatic activity promotes fibrin degradation, enhancing bacterial dissemination during infection (Sjöbring et al. 1994). Curli biogenesis proceeds via an extracellular nucleation-precipitation pathway. Soluble CsgA and CsgB are secreted through the outer membrane lipoprotein CsgG, with the assistance of accessory proteins CsgE and CsgF. The minor subunit CsgB nucleates the polymerization of CsgA into amyloid fibers, although CsgB itself does not polymerize (Bhoite et al. 2019; Yan et al. 2020). Genetically, curli production is governed by two divergently transcribed operons: csgBAC and csgDEFG in *E. coli*, and agfBAC and agfDEFG in Salmonella species (Bordeau and Felden 2014). The csgBAC operon encodes the major (CsgA) and minor (CsgB) curli subunits, while the csgDEFG operon encodes the transcriptional regulator CsgD and the assembly factors CsgE, CsgF, and CsgG. The curli assembly system spatially segregates nucleation and elongation functions between CsgB and CsgA, respectively, tightly regulating amyloid fiber formation (Bhoite et al. 2019; Yan et al. 2020).

Type III secretion-related organelles are surface appendages found in Gram-negative bacteria (Roine et al. 1997; Kubori et al. 1998; Blocker et al. 2003). First observed in Salmonella typhimurium as needle-like structures and in Pseudomonas syringae as pilus-like structures, these organelles are also known as "injectisomes" due to their ability to translocate proteins into host cells in a contact-dependent manner (Cornelis 2006; Proft and Baker 2009). T3SSs provide Gram-negative pathogens with a unique virulence mechanism, injecting bacterial effector proteins directly into the host cell cytoplasm and bypassing the extracellular space. Although they have no adhesive role, T3SSs are genetically, structurally, and functionally related to the bacterial flagellum assembly system (Blocker et al. 2003). T3SS filaments have been identified in a range of animal and plant pathogens, allowing bacteria to translocate substrates directly into the host cytoplasm to exert a broad range of virulence functions (Cornelis 2006; Proft and Baker 2009).

Unlike Gram-negative bacteria, Gram-positive bacteria lack an outer membrane and are characterized by a thick peptidoglycan cell wall. Despite these structural differences, many Gram-positive species also produce pili (also called fimbriae), which play crucial roles in adhesion, colonization, and pathogenesis (Proft and Baker 2009; Telford et al. 2006). The discovery of pili in Gram-positive bacteria was relatively recent compared to Gram-negative bacteria. Early electron microscopy studies identified pilus-like structures in *Corynebacterium diphtheriae*, *Streptococcus pyogenes*, *Streptococcus agalactiae*, and *Enterococcus faecalis*. The chaperone-usher or type IV assembly systems in assembled through a sortase-dependent mechanism. Sortase enzymes catalyze covalent cross-linking of pilin subunits to one another as well as to the peptidoglycan layer of the bacterial cell wall, thereby anchoring the pilus structure (Ton-That and Schneewind 2004; Telford et al. 2006). A typical Gram-positive pilus comprises a shaft formed by polymerization of major pilin subunits, complemented by one or more minor pilins localized at the tip or base. The tip pilin frequently functions as the adhesin, mediating specific interactions with host tissues.

For instance, the SpaA-type pilus of *Corynebacterium diphtheriae* consists of the major pilin SpaA, the tip adhesin SpaC, and the base pilin SpaB (Ton-That et al. 2004; Telford et al. 2006). In *Streptococcus pyogenes*, pilus operons located within the fibronectin-collagen-T antigen (FCT) region encode structural subunits such as FctA, FctB, and Cpa, with Cpa functioning as the tip adhesin mediating host cell attachment (Nakata et al. 2020). Similarly, *Streptococcus agalactiae* expresses pilus subunits PilA, PilB, and PilC, where PilC serves as the adhesin responsible for binding to host surfaces (Lauer et al. 2005). These pili facilitate adhesion to a broad spectrum of host molecules, including extracellular matrix (ECM) proteins such as fibronectin, collagen, and laminin, as well as to epithelial cells, thereby promoting colonization and biofilm formation (Telford et al. 2006; Mandlik et al. 2008;Proft and Baker 2009). For example, the SpaCBA pilus of *Corynebacterium diphtheriae* binds epithelial cells and contributes to colonization of the respiratory tract (Ton-That et al. 2004). In *Streptococcus pneumoniae*, the pilus tip adhesin RrgA interacts with the polymeric immunoglobulin receptor and ECM components, facilitating nasopharyngeal colonization and enhancing virulence (Barocchi et al. 2006). Beyond adhesion, pili in Gram-positive bacteria play critical roles in immune evasion and persistence. Their covalent attachment to the peptidoglycan cell wall and robust structural integrity confer resistance to mechanical shear forces and enzymatic degradation (Telford et al. 2006; Proft and Baker 2009). Furthermore, pili are often immunodominant antigens and are under investigation as promising vaccine candidates against streptococcal and enterococcal infections (Barocchi et al. 2006; Proft and Baker 2009).

Alteri et al. (2005, 2007) employed transmission electron microscopy (TEM) to identify thin, hair-like fibers approximately 2–3 nanometers in diameter on the surfaces of *Mycobacterium tuberculosis* strains H37Rv, H37Ra, and CDC1551. These structures, termed *Mycobacterium tuberculosis* pili (MTP), were observed to mediate bacterial aggregation. To elucidate the genetic basis of MTP, Alteri et al. (2007) investigated the mtp gene (Rv3312A), which encodes a low molecular weight pilin subunit. Through the generation of isogenic mtp deletion mutants, they demonstrated the essential role of this gene in pilus formation, as mutants lacking mtp failed to produce pili. Crucially, these mutants exhibited a marked decrease in adherence to laminin, a significant component of the host extracellular matrix, implicating MTP in host cell adhesion. Furthermore, the study revealed that MTP facilitates bacterial aggregation and biofilm development, processes integral to Mtb persistence and pathogenesis. The clinical significance of MTP was further supported by the detection of anti-MTP antibodies in the sera of tuberculosis patients, indicating in vivo expression of these pili during natural human infection. Chemically, the MTP pilin is a low molecular weight protein, with its mature form comprising approximately 103 amino acids and a molecular mass of approximately 5 kDa. When expressed recombinantly in *Escherichia coli*, the protein exhibits an apparent molecular weight of around 14.5 kDa, attributable to the absence of native posttranslational cleavage in the heterologous host. The amino acid sequence of MTP pilin is characterized by a high degree of hydrophobicity, constituting roughly 60% of its residues. It contains two cysteine residues located at the C-terminus, which are presumed to facilitate disulfide bond formation, thereby contributing to structural stability. Structurally, MTP pilin subunits polymerize into thin, flexible, coiled-coil fibers measuring 2–3 nm in diameter that can extend several microns from the bacterial surface. These fibers exhibit strong aggregative properties, forming dense meshworks that promote bacterial clumping and biofilm formation. Although MTP lacks significant sequence homology to other characterized pilins, it shares key biochemical and structural features with the curli family of bacterial amyloids, such as those produced by *Escherichia coli* (Alteri et al. 2007; Ramsugit and Pillay 2014). Like curli, MTP fibers are aggregative, resistant to protease degradation, and involved in adhesion to both abiotic surfaces and host extracellular matrix proteins, particularly laminin. Functionally, the MTP pilin is essential for several virulence-associated processes in Mtb. It mediates adhesion to host tissues, promotes bacterial aggregation and biofilm development, and enhances the bacterium's ability to adhere to and invade epithelial cells. The presence of anti-MTP antibodies in the sera of tuberculosis patients further confirms that these pili are expressed during human infection and are immunologically relevant. Thereby, the MTP pilin subunit, which is a curli-like, amyloidogenic protein that forms the core of the MTP fibers, can play a pivotal role in the pathogenesis and persistence of Mtb within the host. Between 2000 and 2024, several studies by Velayati et al. (2009, 2012, 2016, 2024), Velayati and Farnia (2016) and Farnia et al. (2023a), Farnia and Velayati (2023) utilized advanced imaging techniques, including atomic force microscopy (AFM), to examine pili-like structures on the surfaces of various clinical isolates of Mtb. By applying AFM with a range of positively and negatively charged substrates (Zhavnerko and Poleschuyk 2012; Zhavnerko et al. 2017), these studies revealed significant variability in pili morphology that correlated with the drug susceptibility profiles of the isolates. Based on these findings, Farnia et al. (2023a), Farnia and Velayati (2023) classified Mtb pili into seven distinct types. Four of these, Type III pili, Type IV secretion pili, Type IV-like pili, and curli-like pili (MTP), share structural similarities with pili previously characterized in Gram-negative and Gram-positive bacteria. The other three types, designated Type V, Type VI, and Type VII, represent novel pilus forms not reported before. Earlier, Type IV and curli-like pili (MTP) had been described and characterized by Alteri et al. (2005, 2007) using negative staining and transmission electron microscopy (TEM). Velayati and Farnia (2024) further emphasized that pili expression varies markedly under stress and drug-resistant conditions compared to drug-susceptible isolates, suggesting that pili may play crucial roles in bacterial adhesion, invasion, and survival within the host. Certain pili types, such as Type III secretion needles, Type IV secretion pili, Type IV-like pili, curli-like pili (MTP), and Type V pili (also known as funnel relief pili), are consistently observed across Mtb populations regardless of drug susceptibility. In contrast, Type VI and VII pili are rare and primarily found in some drug-resistant isolates. This diversity in pili morphology and distribution likely reflects adaptive mechanisms linked to drug resistance and pathogenicity in Mtb (Farnia et al. 2023a; Farnia and Velayati 2023). Other research analyses have demonstrated that curli-like pili, also known as MTPs, facilitate biofilm formation by promoting cell-to-cell adhesion (Ramsugit et al. 2013). Later on, it was shown that these pili play a pivotal role in host infection through specific binding to laminin, a key component of the extracellular matrix (ECM), thereby mediating bacterial attachment to host tissues (Ramsugit and Pillay 2014; Vinod et al. 2020). MTPs exhibit biochemical properties characteristic of amyloid fibers, which have been implicated in various human diseases (Alteri et al. 2007). These pili specifically bind to laminin, a key extracellular matrix (ECM) protein, facilitating bacterial adhesion to host tissues. The detection of IgG antibodies reactive to MTP in the sera of patients with active tuberculosis provides strong evidence that these pili are expressed during natural infection (Alteri et al. 2007). Immunological studies have also localized MTP within bacterial aggregates, indicating their role in promoting interbacterial interactions and biofilm formation, which are critical for persistence and pathogenesis (Ramsugit et al. 2013; Ramsugit and Pillay 2014). In vitro assays further confirm the high affinity of isolated MTP for laminin, supporting the hypothesis that MTP-mediated adhesion is essential for Mtb colonization within the host (Alteri et al. 2007; Ramsugit et al. 2013). The Rv3312A gene, commonly referred to as mtp, encodes the pilin subunits that constitute the MTP pili, which are critical for bacterial adhesion and pathogenesis (Ramsugit et al. 2013; Nyawo 2016). At the genomic level, mtp is located within a chromosomal region flanked upstream by Rv3312c, a gene of currently unknown function, and downstream by genes such as add, deoA, and cdd, which are implicated in intermediary metabolism pathways. This genomic context suggests that mtp resides within a locus potentially linked to both structural and metabolic functions, although its precise regulatory interactions remain to be elucidated. In a comprehensive molecular epidemiological study, Naidoo et al. (2014) employed amplicon sequencing to analyze the presence and conservation of the *mtp* gene across a diverse collection of clinical Mtb complex isolates. They found that all 91 isolates harbored the mtp gene, and 89 of them exhibited 100% sequence conservation, with only two isolates showing point mutations, underscoring the gene's ubiquity and high conservation in Mtb populations. Complementing these observations, Hosseini et al. (2014) reported complete conservation of both the mtp and flp genes in a cohort of 36 clinical isolates, further supporting the notion that these pilin-encoding genes are stable and conserved features within the Mtb genome (Hosseini et al. 2014). Together, these studies underscore the evolutionary conservation of the mtp gene, highlighting its fundamental role in the biology and pathogenicity of Mtb. The high degree of conservation observed across geographically and genetically diverse clinical isolates suggests that MTP-mediated adhesion mechanisms are critical for successful host colonization and infection.

Type IV pili, which have been documented in Mtb, are similar to other proteinaceous appendages commonly found on the surfaces of numerous bacterial pathogens infecting humans, animals, and plants (Craig et al. 2004; Govender et al. 2014; Berry and Pelicic 2015; Kuhn 2019). Type IV pili are multifunctional structures implicated in adherence to host cells, biofilm formation, and motility, such as the gliding movement observed in *Clostridia* species (Rakotoarivonina et al. 2002, 2005; Varga et al. 2006; Rodgers et al. 2011). In Mtb, Type IV pili correspond to the Flp (fimbrial low molecular weight protein) pilus system, encoded by a seven-gene operon comprising tadZ, tadA, tadB, tadC, flp, tadE, and tadF (Govender et al. 2014; Alteri et al. 2022). Each gene product is essential for pilus assembly and function. The flp gene (Rv3656c) encodes the major pilin subunit, a low molecular weight, highly hydrophobic protein that polymerizes into thin, flexible fibers characteristic of Type IV pili. Structurally, this pilin subunit, like other Type IV pilins, contains a conserved N-terminal α-helix that forms the core of the pilus filament. At the same time, its C-terminal region is more variable and surface-exposed, potentially mediating specific interactions with host cells (Berry and Pelicic 2015). Biochemically, Type IV pilins share conserved features critical for their maturation and assembly, including an invariant glutamate residue at the fifth position of the mature amino terminus and a glycine residue immediately preceding the cleavage site recognized by prepilin peptidase (Craig et al. 2004). These conserved residues are essential for the proper processing and polymerization of pilin subunits into functional pili. The specific roles of proteins encoded by the Mtb Tad (tight adherence) pilus operon have been characterized as follows: TadZ functions in directing the localization of the pilus assembly machinery within the bacterial cell; TadA is an ATPase that provides the energy necessary for pilus assembly and extrusion; TadB and TadC are integral membrane proteins forming the core assembly platform that anchors the pilus in the membrane and facilitates export of pilin subunits; TadE and TadF are minor pilins incorporated into the pilus structure, influencing pilus stability, function, and potentially host interaction specificity. Additionally, the pilD gene, located downstream of the operon, encodes a pre-pilin peptidase responsible for processing leader peptides from major and minor pilin subunits, a critical step for assembly of mature, functional pili (Alteri et al. 2022). This operon organization and functional assignment are consistent with the canonical Tad system described in other bacteria, where these components collectively mediate pilus biogenesis, surface localization, and host cell adherence. In Mtb, expression of the tad/flp operon genes (tadZ, tadA, tadB, tadC, flp, tadE, and tadF) is induced upon interaction with macrophages and alveolar epithelial cells, highlighting their role in pathogenesis and host colonization. Although Mtb harbors a reduced set of Tad genes compared to typical Tad loci in other species, the conserved functions of these proteins in pilus assembly and function remain critical for bacterial adherence and virulence (Mann et al. 2016; Alteri et al. 2022; Tekedar et al. 2024). The expression of the Mtb Flp Type IV pilus operon is tightly regulated and is significantly upregulated during the stationary phase of bacterial growth as well as upon contact with host cells, including macrophages and alveolar epithelial cells. In contrast, its expression remains low during exponential growth and dormancy phases (Viljoen et al. 2022). Functional studies have demonstrated that Flp pili are critical for adhesion to and invasion of A549 alveolar epithelial cells but do not significantly contribute to interactions with macrophages, indicating a degree of host-cell specificity (Viljoen et al. 2022). Mutants deficient in flp or other genes within the operon exhibit reduced biofilm formation and impaired adhesion to epithelial cells, underscoring the essential role of this pilus system in colonization, persistence, and the overall pathogenesis of Mtb (Alteri et al. 2022). Expression analyses reveal that the tad/flp pilus genes (tadZ, tadA, tadB, tadC, flp, tadE, and tadF) are significantly more expressed in stationary phase cultures compared to exponential phase or nonreplicating persistence stages (NRP1 and NRP2) induced by hypoxia. This suggests that Flp pili are dispensable during dormancy but play a crucial role during active growth and host interaction (Viljoen et al. 2022). Upon 4 h of contact with macrophages or alveolar epithelial cells, Flp pilus gene expression is markedly elevated relative to bacteria grown in culture alone. Immunodetection further confirmed the presence of Flp pili on both intracellular and extracellular bacteria during infection. Collectively, these findings highlight the role of Flp pili in mediating bacterial adherence and invasion of epithelial cells, facilitating biofilm formation, and promoting bacterial survival during infection. The host-cell specificity of Flp pili suggests their involvement in the initial colonization of the alveolar epithelium, a critical step in tuberculosis pathogenesis. Given their importance, Flp pili represent promising targets for the development of novel antiadhesion therapies and vaccines against tuberculosis (Alteri et al. 2022).

As illustrated in Fig. 8.1, atomic force microscopy (AFM) has revealed the presence of Type III secretion needle-like pili in both drug-susceptible and drug-resistant strains of Mtb (Velayati and Farnia 2016, 2024; Farnia et al. 2023a; Farnia and Velayati 2023). These pili are characterized as short or long, rigid, needle-shaped structures that protrude either singly or from different sides of the bacterial cell (Fig. 8.2). They measure approximately 50 μm ± 10 nm in length, with some exhibiting a tapered tip, particularly in drug-resistant isolates. The needle-like pili were observed protruding from the bacterial surface and, in certain instances, appeared to bridge adjacent bacilli, suggesting a potential role in intercellular interaction or secretion (Fig. 8.3). Structurally, the Type III pili identified in Mtb closely resemble the Type III secretion system (T3SS) needles found in Gram-negative pathogens such as *Salmonella* and *Shigella* (Notti and Stebbins 2016). In these organisms, the T3SS

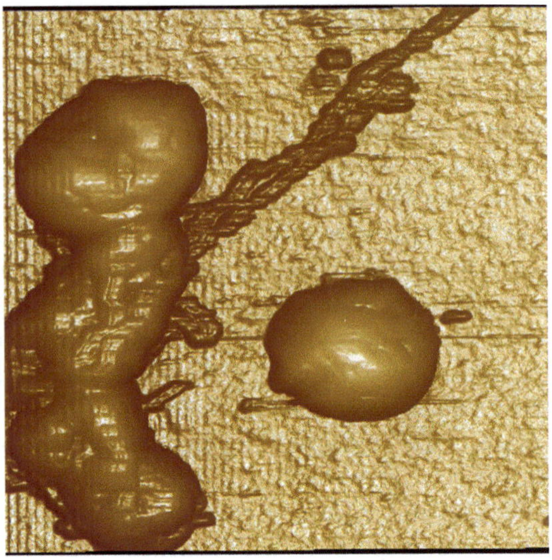

Fig. 8.1 Visualization of Type III pili on the surface of *Mycobacterium tuberculosis* by atomic force microscopy This atomic force microscopy (AFM) image shows a long, filamentous structure extending from the surface of Mycobacterium tuberculosis (Mtb), identified as a Type III pilus. Type III pili are present in both drug-susceptible and drug-resistant Mtb strains. However, in drug-resistant strains, these pili can reach lengths of up to 50 μm ± 10 nm and may occasionally exhibit tapering at the distal end. Such structural adaptations may be linked to the bacteria's response to drug resistance and could play a role in pathogenicity. The high-resolution AFM technique used here enables detailed visualization of these surface appendages, highlighting their potential significance in bacterial adhesion and survival mechanisms

*Experimental work by Prof. Parissa -Farnia and Prof. Gennady Konstantinovich Zhavnerko; Mycobacteriology Research Center, National Research Institute of Tuberculosis and Lung Disease, Shahid Beheshti University of Medical Sciences, Tehran, Iran & The Republican Research and Practical Centre for Epidemiology & and Microbiology, Filimonova 23, Minsk, Belarus

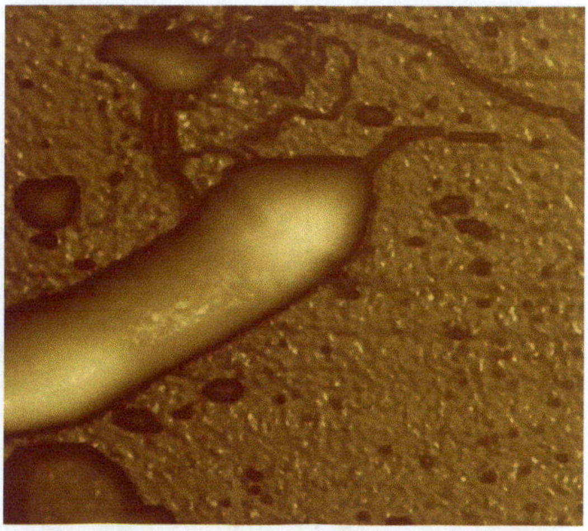

Fig. 8.2 High-resolution AFM image of Type III pilus extending from *Mycobacterium tuberculosis*
Atomic force microscopy (AFM) image showing a Mtb cell with a Type III pilus. Type III pili are visible as filamentous structures that can protrude from the head, tail, or sides of the bacterial cell. Their length is variable, as illustrated in the image
*Experimental work by Prof. Parissa -Farnia and Prof. Gennady Konstantinovich Zhavnerko; Mycobacteriology Research Center, National Research Institute of Tuberculosis and Lung Disease, Shahid Beheshti University of Medical Sciences, Tehran, Iran & The Republican Research and Practical Centre for Epidemiology & and Microbiology, Filimonova 23, Minsk, Belarus

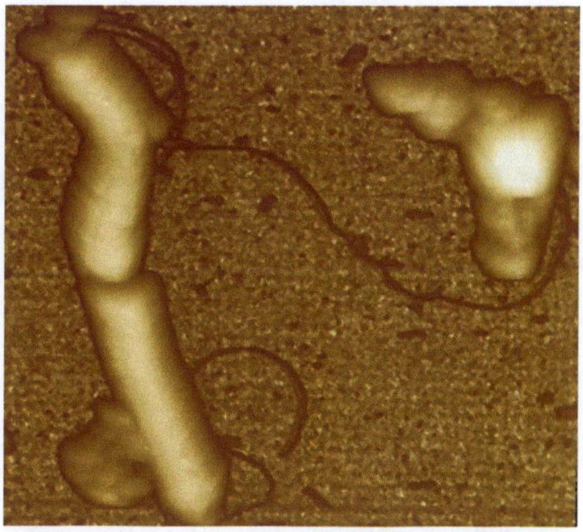

Fig. 8.3 Atomic force microscopy reveals Type III pili-mediated bridging between *Mycobacterium tuberculosis* Bacilli
This atomic force microscopy image shows two Mtb bacilli, with visible filamentous structures extending between them. These structures are likely Type III pili, which form bridges connecting the bacterial cells. Traditionally, pili are known for mediating attachment to host cells, but this image highlights their additional role in facilitating direct cell-to-cell contact between Mtb bacilli. Such connections may promote bacterial aggregation, biofilm formation, and potentially enable intercellular communication or exchange of molecules. The exact nature of what is exchanged, whether it involves proteins, signaling molecules, or even genetic material, remains to be fully understood and warrants further investigation
*Experimental work by Prof. Parissa -Farnia and Prof. Gennady Konstantinovich Zhavnerko; Mycobacteriology Research Center, National Research Institute of Tuberculosis and Lung Disease, Shahid Beheshti University of Medical Sciences, Tehran, Iran & The Republican Research and Practical Centre for Epidemiology & and Microbiology, Filimonova 23, Minsk, Belarus

needle is a helical polymer composed of small pilin subunits that assemble into a narrow, hollow conduit facilitating the translocation of effector proteins into host cells. Although the precise protein components of the Type III pili in Mtb remain unidentified, their overall architecture, a straight, rigid filament extending from the bacterial surface, is highly reminiscent of classical T3SS needles. This structural homology suggests that Mtb Type III pili may serve analogous functions in secretion and host interaction, despite the distinctive complexity of the mycobacterial cell envelope. Using atomic force microscopy (AFM), we identified two types of Type IV pili in Mtb: Type IV secretion pili and Type IV-like pili, the latter characterized by a tapering end. The main difference between these two pili types is their site of protrusion: Type IV secretion pili predominantly emerge from the head (pole) regions of the bacteria (96% ± 2%), whereas Type IV-like pili mostly protrude from the mid-side of the bacterial cell (6% ± 89%) (Fig. 8.4). Type IV secretion pili typically measure 20–70 nm in length and 7–15 nm in width, exhibiting a low surface roughness of 1.8 ± 1.9 nm. These pili are found in only 2–10% of drug-susceptible clinical isolates. In contrast, the Type IV-like pili present a short, cone-shaped tapering protrusion. As shown in Fig. 8.5, Type IV-like pili can form physical bridges between two Mtb bacilli, suggesting a role in bacterial aggregation or communication. Type IV pili, in general, are proteinaceous sur-

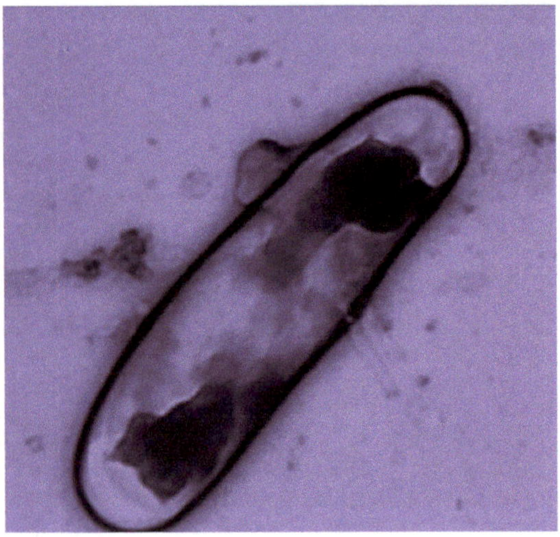

Fig. 8.4 Type IV-like pili in *Mycobacterium tuberculosis* Bacilli.
Type IV-like pili. Transmission electron microscopy shows a short tapering protrusion, typically seen on the left side of Mycobacterium tuberculosis. Sometimes, these pili protrude from the head (pole) region of the bacterium

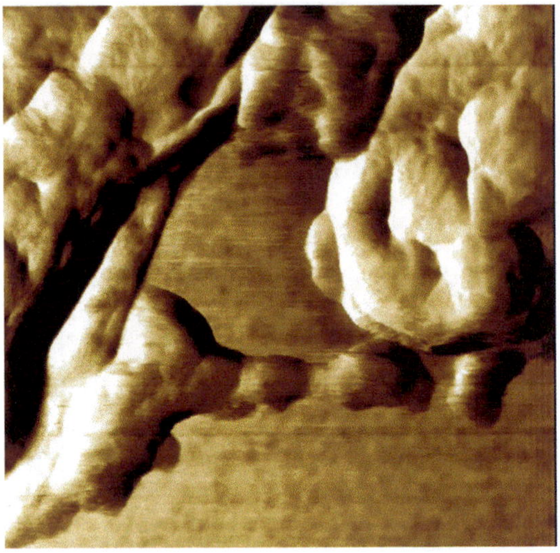

Fig. 8.5 Atomic force microscopy reveals Type IV-like pili bridging between *Mycobacterium tuberculosis*
This atomic force microscopy image presents a side view of two Mtb bacilli. A prominent filamentous structure, resembling a Type IV-like pilus, is seen extending from one bacterium and reaching the other, thereby forming a physical bridge between the two cells. This observation suggests that Type IV-like pili in Mtb are capable of mediating direct cell-to-cell contact, extending their known function beyond mere attachment to surfaces or host cells. The formation of such pili bridges may play a critical role in bacterial aggregation, biofilm development, and potentially intercellular communication. These connections could facilitate the exchange of signaling molecules or other factors between bacilli, which may influence their collective behavior, survival, and adaptation within host environments
*Experimental work by Prof. Parissa -Farnia and Prof. Gennady Konstantinovich Zhavnerko; Mycobacteriology Research Center, National Research Institute of Tuberculosis and Lung Disease, Shahid Beheshti University of Medical Sciences, Tehran, Iran & The Republican Research and Practical Centre for Epidemiology & and Microbiology, Filimonova 23, Minsk, Belarus

face structures involved in various functions such as adherence to host cells and biofilm formation. They are widely distributed among bacterial pathogens infecting humans, animals, and plants. In Mtb, a major proposed function of Type IV pili is the initiation of biofilm formation, which is important for bacterial survival and pathogenicity. The appearance of curli-like pili (MTP) under atomic force microscopy (Fig. 8.6) is consistent with previous reports (Alteri 2005, 2007).

Although three novel pilus types, Type V (relief funnel pili), Type VI (adhesion tapering pili), and Type VII (adhesion flap pili), each exhibiting distinct morphologies and distribution patterns, particularly in drug-resistant *Mycobacterium tuberculosis* strains, have been newly identified, their functional characterization and structural details remain to be resolved (Farnia et al. 2023a; Farnia and Velayati 2023). Type V pili, known as relief funnel pili, are considerably longer, measuring between 2 and 3 micrometers, and feature a characteristic funnel-shaped or tapered terminus. In drug-susceptible strains, these pili typically lie flat alongside the bacterial cell wall in a resting state. However, in drug-resistant strains, they detach from the cell surface and prominently display their funnel-shaped morphology. This unique structural adaptation may enhance the bacterium's ability to anchor to surfaces or host tissues and promote biofilm formation. Given the thickened cell wall characteristic of drug-resistant strains, these elongated pili could facilitate overcoming physical barriers to attachment and colonization, thereby providing a survival advantage during

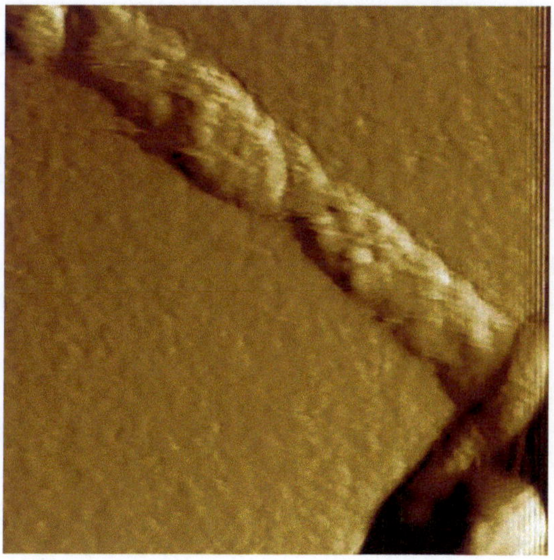

Fig. 8.6 Visualization of curli pili on the surface of totally drug-resistant *Mycobacterium tuberculosis* by atomic force microscopy
Atomic Force Microscopy (AFM) image illustrating the ultrastructural details of Curli pili protruding from the side of a Totally Drug-Resistant Tuberculosis strain. The Curli pili are visualized as prominent, fibrillar, tube-like structures extending outward from the bacterial cell surface. The high-resolution AFM imaging provides clear evidence of the physical architecture and spatial arrangement of Curli pili, highlighting their significance in the biology of drug-resistant *Mycobacterium tuberculosis* strains
*Experimental work by Prof. Parissa -Farnia and Prof. Gennady Konstantinovich Zhavnerko; Mycobacteriology Research Center, National Research Institute of Tuberculosis and Lung Disease, Shahid Beheshti University of Medical Sciences, Tehran, Iran & The Republican Research and Practical Centre for Epidemiology & and Microbiology, Filimonova 23, Minsk, Belarus

antibiotic treatment or immune responses (Fig. 8.7a, b). Type VI pili, termed adhesion tapering pili, are approximately 25 nanometers in length, whereas Type VII pili, known as adhesion flap pili, have diameters ranging from 1.4 to 3 nanometers (Fig. 8.8). Both types have been detected exclusively in extensively drug-resistant (XDR) and totally drug-resistant (TDR) Mtb strains. Type VI pili are distinguished by a pronounced tapering end, while Type VII pili exhibit a distinctive flap-like structure(Fig. 8.9). These adhesion pili likely play critical roles in mediating tight binding to host cells or between bacterial cells, processes essential for robust biofilm formation and persistence. To facilitate the study of these pili, Farnia and Velayati (2023) developed a novel light microscopy-based method for detecting and visualizing pili on clinical Mtb isolates. This approach involves culturing diverse isolates and applying specialized staining techniques that enhance the visibility of pili structures under light microscopy. Unlike electron microscopy, this method allows for direct, accessible observation of pili on the bacterial surface, enabling efficient analysis of pilus morphology and distribution across both drug-susceptible and drug-resistant strains. The study confirmed the presence of pili in XDR and TDR isolates, underscoring their potential involvement in bacterial adhesion, biofilm development, and pathogenesis (Farnia and Velayati 2023). This light microscopy technique thus represents a valuable tool for advancing research into the structural diversity and functional significance of pili in tuberculosis infection and drug resistance. Additionally, it is hypothesized that in the latent or resistant state, which has a thick cell wall with less permeability, these pili may serve as molecular bridges, enabling the bacilli to maintain essential interactions with their environment, evade host defenses, and resist antibiotic penetration. As it was already demonstrated that the thickness of the cell wall is significantly different in various TB isolates, susceptible, MDR, XDR, and TDR-TB or untreatable tuberculosis isolates (Velayati et al. 2009). The average thickness of the Mtb cell wall varies significantly between drug-susceptible and drug-resistant strains. In drug-susceptible isolates, the cell wall thickness averages approximately 15.6 ± 1.3 nm, whereas in multidrug-resistant (MDR) strains, it increases to 17.1 ± 1.03 nm. In extensively drug-resistant (XDR) and totally drug-resistant (TDR) strains, the thickness can reach up to 20.2 ± 1.5 nm, and in some cases, even as high as 26 ± 1.0 nm (Velayati et al. 2009). This thickening primarily occurs in the intermediate and outer electron-transparent layers of the cell envelope. Notably, studies have shown that in XDR and TDR strains, the basal peptidoglycan (PG) layer becomes denser and extends into the intermediate layer, resulting in a thickened, dense

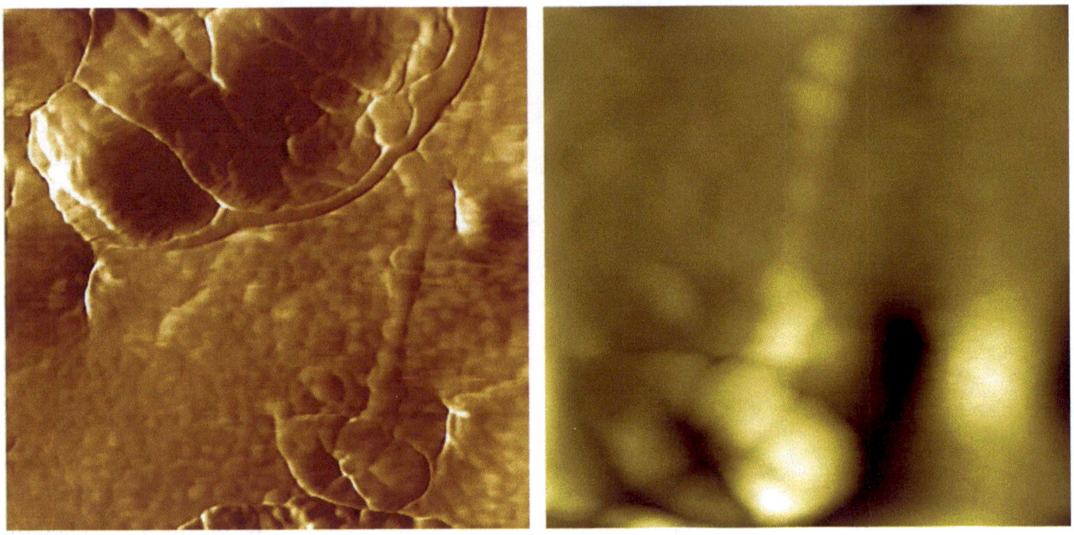

Fig. 8.7 (**a**) Visualization of Type V on the surface of drug-resistant *Mycobacterium tuberculosis* by atomic force microscopy. Atomic force microscopy (AFM) reveals the presence of Type V (relief funnel) pili on the mycobacterial cell surface as distinct, rod-shaped structures that terminate in tapering or funnel-shaped tips. They can be seen protruding from the cell surface, and their origin from the cytoplasmic membrane is supported by their low surface membrane thickness, measured at approximately 3.5 ± 1.2 nm. The length of these pili ranges from 8.9 nm to 1.5 µm while the diameter of their funnel-shaped tips is about 700 to 890 nm. Type V pili are particularly notable in drug-resistant tuberculosis (TB). In drug-susceptible bacilli, these pili are typically in a resting or nonfunctional state, lying alongside the cell wall. In contrast, in extensively drug-resistant (XDR) and totally drug-resistant (TDR) TB strains, the pili are more prominent and may be actively involved in cellular adaptation. It is possible that these bacilli utilize Type V pili to enhance their survival under harsh conditions or to facilitate attachment to other bacteria or host cells, representing a significant adaptation mechanism in drug-resistant TB. * Experimental work by Prof. Parissa -Farnia and Prof. Gennady Konstantinovich Zhavnerko; Mycobacteriology Research Center, National Research Institute of Tuberculosis and Lung Disease, Shahid Beheshti University of Medical Sciences, Tehran, Iran & The Republican Research and Practical Centre for Epidemiology & and Microbiology, Filimonova 23, Minsk, Belarus. (**b**) Visualization of Type V pili by Atomic Force Microscopy. This image presents a height-mode atomic force microscopy (AFM) scan highlighting the Type V (relief funnel) pili. The pili appear as rod-shaped structures with a distinct tapering or funnel-shaped tip. Notably, there is a central hole visible in the funnel, suggesting a possible channel or functional opening within the pilus. The observed pili exhibit low surface roughness (3.5 ± 1.2 nm), supporting the hypothesis that they originate from the cytoplasmic membrane. These structural features are characteristic of Type V pili and may be linked to their unique roles in mycobacterial adaptation and drug resistance. * Experimental work by Prof. Parissa -Farnia and Prof. Gennady Konstantinovich Zhavnerko; Mycobacteriology Research Center, National Research Institute of Tuberculosis and Lung Disease, Shahid Beheshti University of Medical Sciences, Tehran, Iran & The Republican Research and Practical Centre for Epidemiology & and Microbiology, Filimonova 23, Minsk, Belarus

layer surrounding the plasma membrane (Velayati et al. 2009). The peptidoglycan layer is a critical structural component of the bacterial cell envelope, providing both shape and mechanical integrity to the cell wall. The importance of mycolic acids for maintaining the structural and functional integrity of the mycobacterial cell wall has been well documented. Wang et al. (2000) demonstrated that mutant strains of *M. smegmatis* (155NS1), which are unable to synthesize full-length mycolic acids, exhibit abnormal cell wall ultrastructure and are temperature-sensitive, failing to grow at 37 °C (Liu and Nikaido 1999; Wang et al. 2000). Beyond determining cell shape and acting as a permeability barrier, the mycobacterial cell wall plays a vital role in regulating pathogen-host interactions (Watanabe et al. 2001). Alsteens et al. (2008) reported that

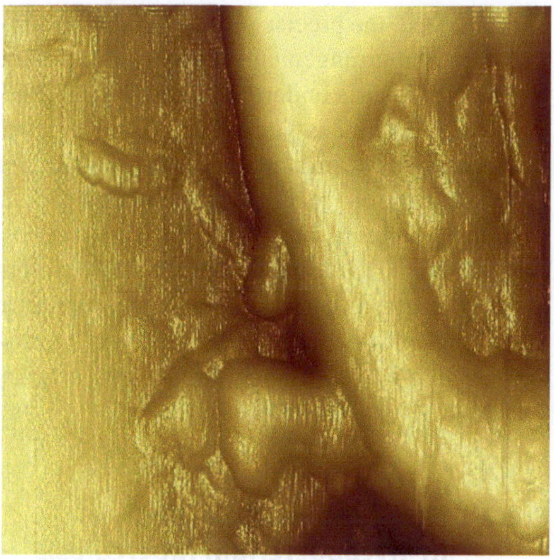

Fig. 8.8 Visualization of Type VI pili by atomic force microscopy
This atomic force microscopy (AFM) image displays a cross-sectional view of extensively drug-resistant tuberculosis (XDR-TB) cells. The image highlights the exceptionally thick cell wall characteristic of XDR and TDR-TB bacilli, measuring approximately 25 ± 2.5 nm. This increased cell wall thickness is a notable adaptation associated with drug resistance, potentially contributing to the bacterium's ability to withstand harsh environmental conditions and antibiotic pressure. In addition to the prominent cell wall, the image also reveals the presence of Type VI pili, which are specialized adhesion structures. These pili are observed as short, tapering projections extending from the cell surface, with lengths ranging from 15 to 25 nm and ending in a characteristic taper. Although only a small percentage of cells display these Type VI pili, their presence suggests a role in cellular adhesion, possibly facilitating interactions between bacteria or attachment to host tissues. The combination of extreme cell wall thickness and the emergence of Type VI adhesion pili underscores the remarkable structural adaptations of XDR and TDR-TB bacilli, which may enhance their survival, persistence, and pathogenicity in the face of multidrug therapy
*Experimental work by Prof. Parissa -Farnia and Prof. Gennady Konstantinovich Zhavnerko; Mycobacteriology Research Center, National Research Institute of Tuberculosis and Lung Disease, Shahid Beheshti University of Medical Sciences, Tehran, Iran & The Republican Research and Practical Centre for Epidemiology & and Microbiology, Filimonova 23, Minsk, Belarus

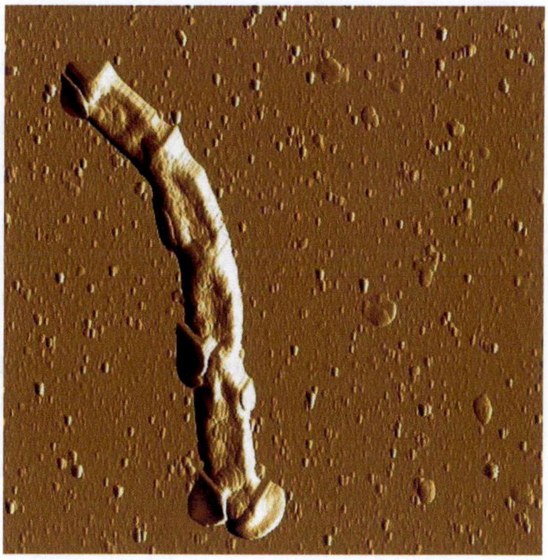

Fig. 8.9 Visualization of Type VII pili by atomic force microscopy
This atomic force microscopy (AFM) image captures the surface architecture of a totally drug-resistant (TDR) Mycobacterium tuberculosis, highlighting the presence of the newly identified Type VII adhesion flap pili. These pili are observed as short, flap-like projections extending from the bacterial surface. Type VII pili are notably minor, with a length of only 2 – 5 nm and a tip diameter ranging from 1.4 to 3 nm. Their unique morphology distinguishes them from other known pili types. Importantly, Type VII pili were detected exclusively in TDR-TB bacilli and were absent in drug-susceptible or less-resistant strains, suggesting a possible association with extreme drug resistance. The precise function of these pili remains under investigation; however, their structure and strategic positioning on the cell surface suggest a potential role in mediating cell-to-cell or cell-to-host attachment, which may contribute to the persistence and virulence of TDR-TB. The discovery of Type VII pili adds a new dimension to our understanding of mycobacterial adaptation mechanisms under intense antibiotic pressure
*Experimental work by Prof. Parissa -Farnia and Prof. Gennady Konstantinovich Zhavnerko; Mycobacteriology Research Center, National Research Institute of Tuberculosis and Lung Disease, Shahid Beheshti University of Medical Sciences, Tehran, Iran & The Republican Research and Practical Centre for Epidemiology & and Microbiology, Filimonova 23, Minsk, Belarus

exposure to various antimycobacterial drugs induces a marked increase in surface roughness in *M. bovis BCG strains* (1173P2, WHO, Stockholm, Sweden). They further suggested that the progressive erosion of the mycobacterial cell wall during therapy may alter treatment efficacy, with potentially adverse effects. Investigations by Velayati et al. (2009) into the cell wall thickening and associated biological adaptations in XDR and TDR isolates revealed noteworthy changes in pili expression. Their studies documented that 2–4% of MDR-TB bacilli simultaneously express both Type V and Type VI pili, while the majority (6–7%) express either Type V or Type VI pili alone (Velayati and Farnia 2024; Farnia et al. 2023a; Farnia and Velayati 2023). Additionally, approximately 2–3% of XDR and TDR bacilli express three distinct pilus types concurrently. It is hypothesized that latent or highly drug-resistant Mtb bacilli compensate for the reduced permeability and increased rigidity of their thickened cell walls by deploying diverse pili structures that extend beyond the cell envelope. These pili facilitate critical processes such as adhesion, biofilm formation, and molecular exchange, which are essential for bacterial survival under hostile conditions. This adaptive strategy is supported by recent findings demonstrating variability in pilus morphology and abundance between drug-susceptible and drug-resistant strains. Drug-resistant bacilli exhibit unique pilus structures that likely enhance persistence and pathogenicity (Velayati et al. 2009, 2012, 2016; Alteri et al. 2007; Singh et al. 2017a, b). The distinctive morphologies and increased abundance of pili in drug-resistant strains likely represent evolutionary adaptations that enable the bacilli to persist, communicate, and thrive despite antibiotic pressure and host immune defenses. Therefore, it is reasonable to propose that during resistant or latent states, the thickened and less permeable cell wall of Mtb is functionally compensated by the expression of diverse pili. These pili extend beyond the cell envelope, facilitating critical processes such as adhesion, biofilm formation, and molecular exchange, mechanisms essential for bacterial survival under stressful conditions. The structural diversity and functional specialization of Mtb pili underscore their pivotal role in enabling the bacterium to adapt, persist, and maintain infection, particularly during drug resistance and latency. Although these newly identified pili in drug-resistant Mtb strains have been observed using various microscopic techniques, including light microscopy, electron microscopy, and atomic force microscopy, their detailed protein structures remain unresolved. While their morphology and spatial distribution are well characterized, the precise molecular architecture, including the arrangement of pilin subunits and the structural basis underlying their unique functional properties, has yet to be fully elucidated. Deciphering these protein structures is critical for understanding the mechanisms governing pilus assembly, stability, and interaction with host cells. Such insights could inform the development of targeted therapeutic strategies aimed at disrupting pili-mediated adhesion and persistence, thereby enhancing treatment efficacy against drug-resistant tuberculosis.

8.2.2 Non-fimbrial Adhesins in *Mycobacterium tuberculosis*

Mtb expresses a diverse repertoire of adhesins, including proteins, lipoproteins, glycoproteins, and pili, that are essential for mediating attachment to host cells, interacting with extracellular matrix components, and facilitating the establishment of infection.

8.2.2.1 Alanine-Proline-Rich Antigen (Apa); Rv1860

Alanine-proline-rich antigenic glycoprotein (Apa; Rv1860), also known as the 45–47 kDa antigenic complex, is a heavily mannosylated protein of *Mycobacterium tuberculosis* composed of 325 amino acids with a characteristic abundance of alanine and proline residues (Dobos et al. 1996). Structurally, Apa belongs to a family of glycoproteins distinguished by extensive O-mannosylation at specific threonine residues, Thr49 and Thr57 (bearing α-1,2-mannobiose), Thr76 (α-mannose), and Thr316 (mannose, α-1,2-mannobiose, or triose),which critically define its recognition by host cell receptors (Ragas et al. 2007). The protein folds into a large seven-stranded antiparallel β-sheet flanked by three α-helices, forming a stable scaffold that supports its multifunctional roles (Zhang et al. 2020). Functionally, Apa acts as an important adhesin and an immune-dominant antigen. Its extensive mannosylation enables specific binding to pulmonary surfactant protein A (SP-A)—a C-type lectin of the innate immune system, and other C-type lectin receptors such as DC-SIGN and the macrophage mannose receptor, facilitating early pathogen recognition and host-pathogen interactions (Ragas et al. 2007). Additionally, Apa binds fibronectin, a key extracellular matrix protein, which promotes bacterial adhesion to host tissues (Nandakumar et al. 2013). This fibronectin-binding capacity is mediated primarily through interactions with the FnIII-4 and FnIII-5 modules of fibronectin's cell-binding domain, although the previously proposed Fn-binding motif (residues 258–261) appears structurally buried and inaccessible, suggesting alternative binding interfaces (Zhang et al. 2020). Immunologically, Apa elicits a robust Th1-type T cell response, inducing interferon-gamma production from both CD4+ and CD8+ T cells, highlighting its role in host defense (UniProt P9WIR7; Dobos et al. 1996). Its antigenicity and surface exposure make Apa a promising vaccine candidate. Indeed, Nandakumar et al. (2013) demonstrated that Apa significantly enhances protection against virulent *M. tuberculosis* in mice when used as a booster following Bacillus Calmette-Guérin (BCG) vaccination. Therefore, Apa is a structurally distinct, heavily glycosylated adhesin that mediates critical interactions with host immune lectins and extracellular matrix components, contributing to *M. tuberculosis* pathogenesis and immune modulation. Its detailed structural characterization and multifunctional binding properties provide valuable insights for the development of novel antituberculosis vaccines and therapeutics.

8.2.2.2 Fibronectin (FN)-Binding Protein (FnBP)

Fibronectin-binding proteins (FnBPs), initially identified as the 30-kDa or α-antigen (Abou-Zeid et al. 1988), are enzymes classified as mycolyltransferases (Belisle et al. 1997) that possess the capacity to bind fibronectin, a key extracellular matrix (ECM) glycoprotein (Ratliff et al. 1988). This fibronectin-binding property is believed to facilitate tissue colonization by Mtb (Kuo et al. 2012, 2019). The FnBP complex comprises antigen-85A, antigen-85B, antigen-85C (Ag85A, Ag85B, Ag85C), as well as FnBP A, B, and C, all of which play pivotal roles in the pathogenesis of Mtb (Abou-Zeid et al. 1988; Kuo et al. 2012, 2019). Their ability to bind fibronectin, a multifunctional glycoprotein abundantly present on host cell surfaces, has been well documented (Cole et al. 1998; Henderson et al. 2011; Meena et al. 2016). Fibronectin functions as a ligand for various integrin receptors, suggesting that the interaction between FnBPs and host integrins is mediated through fibronectin itself. In the context of *Staphylococcus aureus*, fibronectin facilitates bacterial adhesion to endothelial cells, which subsequently triggers phagocytosis and cytokine activation, thereby eliciting an appropriate immune response (Meena et al. 2016). Similarly, FnBPs in Mtb recognize and bind adhesive matrix molecules on microbial surfaces, contributing to infection (Henderson et al. 2011). These proteins typically contain LPxTG motifs in their C-terminal regions, which enable covalent attachment to the ECM (Naito et al. 2000; Massey et al. 2001). Upon binding fibronectin, FnBPs mediate interactions between bacterial virulence factors and host cell integrin receptors, initiating intracellular signaling cascades. This includes reorganization of the actin cytoskeleton and activation of tyrosine kinases such as focal adhesion kinase (FAK) and Src family kinases—non-receptor protein tyrosine kinases. The activation of these kinases stimulates phospholipase activity, which promotes recruitment of the tryptophan aspartate-containing coat protein (TACO, also known as CORO1A or coronin-1) to the phagosomal membrane. The presence of TACO inhibits phagosomal maturation, thereby preventing bacterial degradation (Meena et al. 2016). This mechanism illustrates how fibronectin, acting as an adhesion molecule, modulates host cell signaling pathways to facilitate the prolonged intracellular survival of Mtb within granulomas (Bisht and Meena 2019).

8.2.2.3 19-kilodalton Lipoprotein

The 19 kDa protein, also known as LpqH and encoded by the lpqH gene (Rv3763), is a well-characterized glycolipoprotein adhesin prominently expressed on the surface of Mtb. Structurally, LpqH belongs to the β-fold protein family and exists as a monomer composed of 119 amino acids, with a theoretical molecular weight of approximately 12 kDa. Its crystal structure, resolved at 2.85 Å resolution, reveals a compact fold dominated by β-sheets, which confers stability and facilitates diverse molecular interactions. LpqH is both anchored to the mycobacterial cell envelope and secreted extracellularly, positioning it strategically to mediate host-pathogen interactions (Chatterjee and Khoo 1998; Chatterjee et al. 2022). Functionally, the 19 kDa protein acts as a pathogen-associated molecular pattern (PAMP) recognized by host pattern recognition receptors (PRRs), including Toll-like receptors (TLRs) such as TLR2, as well as the macrophage mannose receptor (López et al. 2003; Mogensen 2009; Naser et al. 2013; Bisht and Meena 2019). Engagement of these receptors by LpqH initiates signaling cascades that lead to the production of pro-inflammatory cytokines, including interleukins and tumor necrosis factor-alpha (TNF-α), thereby eliciting a robust immune response (López et al. 2003; Bisht and Meena 2019). Experimental evidence demonstrates that peptides derived from LpqH specifically bind to monocyte-derived macrophages and effectively inhibit mycobacterial entry, underscoring its critical role as an adhesin. Moreover, LpqH induces apoptosis in differentiated THP-1 cells and monocyte-derived macrophages in a dose- and time-dependent manner. The polypeptide component of LpqH mediates this apoptotic effect through TLR2 signaling pathways, independent of its lipid modification (Diaz-Silvestre et al. 2005; Ocampo et al. 2014). Consistent with this, other studies have shown the involvement of LpqH in host cell signaling via TLR2 (Basu et al. 2012). Additionally, Neufert et al. (2001) reported that the 19 kDa protein serves as a potent activator of neutrophils. Beyond its role in immune activation, LpqH and related lipoproteins stimulate both humoral and cellular immune responses against Mtb, highlighting their importance in disease control. However, these lipoproteins also modulate host immune functions by inhibiting the expression of major histocompatibility complex class II (MHC-II) molecules and impairing antigen processing in a TLR2-dependent manner (Noss et al. 2001). Gehring et al. (2003) further demonstrated that the 19 kDa protein interferes with interferon-gamma (IFN-γ) signaling by downregulating HLA-DR expression, a key MHC-II cell surface receptor regulated by IFN-γ. This downregulation consequently reduces the expression of Fc gamma receptor I (Fcγ-R1), which belongs to the immunoglobulin superfamily, and diminishes the processing and presentation of bacterial antigens to CD4+ T cells (Gehring et al. 2003). Such immune evasion strategies facilitate the prolonged survival and persistence of Mtb within host cells (Neufert et al. 2001). Notably, the inhibitory effects of the 19 kDa protein are specific to MHC-II expression, without affecting MHC class I molecules. In addition to its immunomodulatory functions, the 19 kDa protein contributes to nutrient transport across the mycobacterial cell wall, supporting bacterial survival (Noss et al. 2001; Bisht and Meena 2019). Ciaramella et al. (2000) reported that LpqH primarily induces apoptosis in monocytes and macrophages by activating TLR2, which leads to the upregulation of death receptors and associated signaling molecules, thus initiating a death receptor-mediated apoptotic cascade. These findings collectively emphasize the multifaceted role of the 19 kDa protein in mediating Mtb invasion, immune modulation, and intracellular persistence, warranting further investigation into its potential as a therapeutic target.

8.2.2.4 Lipoarabinomannan (LAM)

Mannose-capped lipoarabinomannan (Man-LAM) is abundantly expressed in pathogenic mycobacteria such as *M. tuberculosis* and *M. ulcerans*. In contrast, arabinosylated lipoarabinomannan (Ara-LAM) predominates in non-pathogenic species like *M. fortuitum* and *M. smegmatis* (Torrelles and Chatterjee 2023). The phosphatidylinositol anchor of LAM facilitates attachment of the pathogen to the host cell membrane, playing a critical role in host-pathogen interactions (Brennan and Nikaido 1995). It has been proposed that LAM, together with lipomannans, forms a complex glycoconjugate with immunomodulatory functions, comprising diverse chemical structures that contribute to the pathogenesis of Mtb infection (Koul et al. 2004; Meena and Rajni 2010). These molecules engage specific receptors on host cell surfaces to modulate immune responses. For instance, LAM disrupts intracellular calcium homeostasis by chelating calcium ions or inhibiting Ca2+/Calmodulin-dependent protein kinase II (CaMKII), thereby impairing phagosome maturation. Man-LAM further promotes intracellular survival of mycobacteria by inhibiting apoptosis; it achieves this through phosphorylation of the pro-apoptotic protein BAD, which activates protein kinase B (PKB/Akt), ultimately blocking apoptotic signaling pathways (Malik et al. 2001). These mechanisms underscore the role of Man-LAM in facilitating mycobacterial invasion and persistence within host cells, thereby advancing disease progression. Beyond macrophages, receptors such as Toll-like receptors (TLRs) and C-type lectins are also expressed on dendritic cells (DCs) and other immune cells (Basu 2004; Wojtas et al. 2011; Zhou et al. 2019). TLRs are evolutionarily conserved pattern recognition receptors that detect pathogen-associated molecular patterns (PAMPs), initiating both innate and adaptive immune responses. Man-LAM's interaction with C-type lectins enables recognition of diverse pathogens bearing distinct carbohydrate motifs, thereby modulating host-pathogen dynamics. Moreover, LAM's interaction with C-type lectin receptors on dendritic cells modulates the cross-talk between TLRs and other signaling pathways, further influencing cytokine production. This results in the downregulation of IL-12 synthesis, which is crucial for effective antigen presentation and activation of cell-mediated immunity (Tailleux et al. 2003; Geijtenbeek et al. 2009). By suppressing IL-12, LAM effectively attenuates the host's immune response, facilitating immune evasion by Mtb. Collectively, these mechanisms illustrate how LAM subverts LPS-induced TLR signaling pathways, leading to impaired IL-12 production and a weakened immune response. This immunomodulation contributes to the persistence of Mtb within host cells and highlights the sophisticated strategies employed by the pathogen to evade immune clearance.

8.2.2.5 Heparin-Binding Hemagglutinin (HBHA)

Heparin-binding hemagglutinin adhesin (HBHA) is a key adhesin of Mtb to epithelial cells and is also implicated in the pathogen's dissemination beyond the lungs (Pethe et al. 2001). The adherence of mycobacteria to epithelial cells is inhibited explicitly by substances like heparin and sulfated carbohydrates, suggesting that HBHA binds to sulfated receptors on the host cell surface (Menozzi et al. 2006; Pethe et al. 2001; Sohn et al. 2011; Lomino et al. 2011). Furthermore, HBHA functions as a methylated antigen that elicits a specific T cell response (Temmerman et al. 2004). Pethe et al. (2001) demonstrated that a *M. bovis BCG* mutant strain with a disrupted HBHA gene showed no significant difference in phagocytic cell invasion compared to the wild-type, but exhibited a 60% reduction in adherence to pneumocytes. This observation suggests that HBHA-mediated adherence is specific to non-phagocytic cells and is crucial for mycobacterial dissemination, which is essential for the pathogen's ability to spread infection within the host. HBHA functions as a multifunctional adhesion molecule, promoting both bacterial aggregation and interaction with eukaryotic cells. The 28-kDa HBHA is the most extensively studied major adhesin in Mtb, playing a vital role in the pathogen's capacity to disseminate from the primary infection site through its interaction with host epithelial cells. HBHA facilitates two critical steps in tuberculosis pathogenesis: bacterial aggregation and cell adhesion (Esposito et al. 2011). The aggregation mediated by HBHA is pivotal for the formation of bacterial clumps, which enhances the ability of Mtb to adhere to and more effectively invade host cells (Esposito et al. 2011; Pethe et al. 2001). Moreover, the potential of HBHA in detecting latent TB infection has been investigated (Hougardy et al. 2007). The humoral immune response triggered by HBHA, along with other Mtb adhesins, holds promise for protective immunity by preventing the extrapulmonary spread of the pathogen from the lungs (Locht and Hougardy 2006; Dutt and Choreño-Parra 2025).

8.2.2.6 Glyceraldehyde-3-Phosphate Dehydrogenase (GAPDH)

Glyceraldehyde-3-phosphate dehydrogenase (GAPDH) in Mtb is a multifunctional enzyme belonging to the glyceraldehyde-3-phosphate dehydrogenase family (Malhotra et al. 2017). It is characterized by a conserved NAD^+-binding domain and a catalytic site responsible for its canonical glycolytic function—the oxidative phosphorylation of glyceraldehyde 3-phosphate to 1,3-bisphosphoglycerate (Sirover 2014; Gani et al. 2021). Beyond this metabolic role, Mtb GAPDH acts as a moonlighting protein, performing diverse functions essential for bacterial virulence and survival. Structurally, the recently resolved X-ray crystal structure of Mtb GAPDH reveals a tetrameric assembly, with each monomer adopting a Rossmann fold typical of dehydrogenases, which facilitates NAD^+ binding and enzymatic activity (Kumar et al. 2024). The protein surface exposes evolutionarily conserved residues and flexible loop regions that mediate interactions with host molecules, enabling its multifunctionality. Functionally, Mtb GAPDH localizes to the bacterial surface and secretome, where it serves as a receptor for key host proteins such as plasminogen, transferrin, and lactoferrin (Gani et al. 2021; Kumar et al. 2024). Binding to plasminogen promotes its conversion to plasmin, facilitating tissue invasion and bacterial dissemination. Importantly, GAPDH's interaction with transferrin and lactoferrin, major iron-binding proteins of the host, enables Mtb to acquire iron, an essential nutrient often limited during infection (Monfeli and Beeson 2007; Malhotra et al. 2017). Notably, GAPDH exhibits a higher affinity for lactoferrin (Kd ~31.7 nM) compared to transferrin (Kd ~160 nM), reflecting its adaptation to lung environments where lactoferrin is abundant (Malhotra et al. 2017). GAPDH also interacts with various extracellular matrix proteins, including fibronectin, fibrinogen, laminin, and collagen, as well as with human epithelial and endothelial cells, thereby facilitating bacterial adhesion and colonization (Dumke et al. 2011; Wang et al. 2021). GAPDH also binds cytoskeletal proteins such as actin and myosin, suggesting a role in modulating host cell structure and signaling pathways to promote bacterial invasion. Beyond its adhesive functions, GAPDH participates in cellular signaling networks. For instance, it is involved in the phosphorylation of the E3 ubiquitin ligase Siah-1 by apoptosis signal-regulating kinase 1 (ASK1), integrating stress responses that may influence bacterial survival under host-imposed stresses (Seidler 2013; Tristan et al. 2011). Furthermore, GAPDH has been implicated in apoptosis induction and transcriptional regulation, underscoring its broad impact on host-pathogen interactions (Tajima et al. 1999; Malhotra et al. 2017). Experimental mutagenesis studies have demonstrated that the enzymatic activity and ligand-binding functions of GAPDH are mediated by distinct residues, indicating functional independence between its metabolic and moonlighting roles (Malhotra et al. 2017). For example, C-terminal lysine residues modulate both enzymatic activity and binding to host ligands such as plasminogen and transferrin (Kumar et al. 2024). Given its central role in adhesion, iron acquisition, immune modulation, and stress response, GAPDH represents a promising target for novel antituberculosis therapies. Inhibiting GAPDH functions or disrupting its interactions with host proteins could impair Mtb's ability to colonize, acquire essential nutrients, and evade immune defenses, offering new avenues for drug development (Malhotra et al. 2017).

8.2.2.7 Antigen-85 (Ag85) Complex

The Antigen-85 (Ag85) complex of Mtb is a vital protein family consisting of three major secreted components: Ag85A (*FbpA*; Rv3804c, 338 amino acids), Ag85B (*FbpB*; Rv1886c, 325 amino acids), and Ag85C (*FbpC*; Rv0129c, 340 amino acids). These proteins share significant sequence homology and belong to the α/β-hydrolase fold superfamily, characterized by a conserved catalytic triad, typically serine, glutamate, and histidine residues (Ser124, Glu228, His260 in Ag85C), which underpins their enzymatic activity (Ronning et al. 2000; Kremer et al. 2002). Structurally, X-ray crystallographic analyses of Ag85C reveal a compact α/β-hydrolase fold featuring a hydrophobic pocket and an extended tunnel approximately 21 Å in length, which likely accommodates the substrate trehalose monomycolate (TMM). The active site is stabilized by an oxyanion hole formed by residues such as Leu40 and Met125, facilitating catalysis (Ronning et al. 2000). Conserved surface residues across Ag85A, B, and C form putative binding sites for host extracellular matrix proteins, including fibronectin, supporting their role in host-pathogen interactions (Ronning et al. 2000; Kremer et al. 2002). Functionally, the Ag85 complex serves dual roles in Mtb physiology and pathogenesis. Primarily, these proteins exhibit mycolyltransferase activity, catalyzing the transfer of mycolic acids from TMM to the cell wall arabinogalactan and to other TMM molecules, thereby synthesizing trehalose dimycolate (TDM or cord factor), a key glycolipid essential for cell wall integrity and virulence (Goins et al. 2018). This enzymatic activity is critical for maintaining the impermeability and robustness of the mycobacterial outer membrane, contributing to antibiotic resistance and immune evasion. In addition to their enzymatic function, Ag85 proteins mediate adhesion to host tissues by binding fibronectin on host cell surfaces, facilitating bacterial uptake and colonization (Kuo et al. 2012; Yang et al. 2025). This fibronectin-binding property implicates the Ag85 complex in the early stages of infection, promoting internalization by macrophages and epithelial cells. Genetic disruption of the fbp genes encoding Ag85 proteins impairs cell wall biosynthesis and attenuates virulence, underscoring their essential role in Mtb survival and pathogenicity (Armitige et al. 2000; Kremer et al. 2002). Immunologically, the Ag85 complex is highly antigenic, eliciting robust cellular and humoral immune responses. The 32 kDa Ag85 protein isolated from *M. bovis* stimulates significant proliferation and interferon-gamma (IFN-γ) secretion in peripheral blood leukocytes from tuberculosis patients, highlighting its potential as a diagnostic marker and vaccine candidate (Launois et al. 1994; Sai Priya et al. 2010; Huygen 2014). Elevated levels of circulating Ag85 proteins complexed with fibronectin or immunoglobulin G have been detected in the serum of patients with active tuberculosis, supporting their utility as biomarkers for disease diagnosis and monitoring (Druszczynska et al. 2017). Given their central role in cell wall biosynthesis, host adhesion, and immune modulation, Ag85 proteins represent attractive targets for antitubercular drug development. Recent studies have identified inhibitors such as ebselen, which covalently modify critical catalytic residues (e.g., Cys209 in Ag85C), disrupting the active site and enzymatic function, thereby impairing mycolyltransferase activity and bacterial viability (Favrot et al. 2013;Goins et al. 2017). The high conservation and essentiality of the Ag85 complex reduce the likelihood of resistance development, making these proteins promising candidates for novel therapeutic interventions (Iskandar et al. 2025). In summary, the Ag85 complex proteins of Mtb are structurally conserved mycolyltransferases with crucial roles in maintaining cell wall integrity, mediating host tissue adhesion, and modulating immune responses. Their multifunctional nature and surface accessibility underscore their significance in tuberculosis pathogenesis and their potential as diagnostic markers, vaccine antigens, and drug targets.

8.2.2.8 Malate Synthase (*glcB*; Rv1837c)

Malate synthase, encoded by the *glcB* gene (Rv1837c), plays a pivotal role in the survival and pathogenicity of Mtb. A distinctive feature of this enzyme is its unique C-terminal region, which mediates binding to host extracellular matrix proteins such as laminin and fibronectin, as well as to A549 lung epithelial cells, facilitating bacterial adherence and colonization (Kinhikar et al. 2006; Singh et al. 2017a, b; Ryndak and Laal 2019). This adhesive property is crucial for establishing infection, especially in the lung environment. Within the host, Mtb often encounters hypoxic or low-oxygen conditions. Under such stress, malate synthase undergoes adaptive changes that enhance bacterial survival. It functions as a key enzyme in the glyoxylate pathway—a metabolic bypass of the tricarboxylic acid (TCA) cycle—which allows the bacterium to assimilate carbon compounds derived from fatty acids and convert them into intermediates that feed into the TCA cycle, thereby supporting replication within macrophages (Ahn et al. 2016; Ehrt et al. 2018). This pathway is essential during persistent infection when glucose availability is limited, and fatty acids become the primary carbon source. Malate synthase also contributes to maintaining redox balance by regenerating oxidizing cofactors and preserving carbon skeletons during hypoxic stress, which is critical for bacterial growth and survival in the hostile intracellular environment. Its immunogenic nature is evidenced by the detection of specific serum antibodies in tuberculosis patients, highlighting its potential utility in serodiagnostic assays (Achkar et al. 2006; Mayer-Barber and Barber 2015; Singh et al. 2017a, b). Thus, malate synthase serves a dual function as both an adhesin and a stress-resistance factor. The specificity of malate synthase for laminin binding is a key virulence trait, with antibodies from tuberculosis patient sera recognizing this interaction (Samanich et al. 2001; Smith et al. 2003; Squeglia et al. 2018). The glyoxylate shunt, in which malate synthase participates, is indispensable for Mtb's fatty acid metabolism during persistence and has been identified as a promising target for antitubercular drug development (Puckett et al. 2017). The glcB-encoded malate synthase is essential for pathogenicity, with in vivo studies confirming its expression during active infection and cytoplasmic localization (Kinhikar et al. 2006). Interestingly, malate synthase is secreted despite lacking conventional secretion signals or cell wall anchoring motifs, leading to its classification as an "anchorless adhesin." The secreted enzyme enhances bacterial adherence to lung epithelial cells by binding to laminin and fibronectin, thereby promoting host infection and dissemination (Kinhikar et al. 2006). This unconventional secretion and adhesion mechanism underscores malate synthase's multifaceted role in Mtb pathogenesis.

8.2.2.9 Chaperonins as Adhesins (*dnaK* and *groEL2*; Rv0350 and Rv0440)

Among the various classes of heat shock proteins, GroESL and DnaK represent the primary bacterial stress-response systems essential for maintaining protein homeostasis under adverse conditions (Fourie and Wilson 2020). These highly conserved molecular chaperones are encoded by the essential groEL and groES genes and play critical roles in protein folding, assembly, and protection against stress-induced damage. In Mtb, two paralogs of the cpn60 gene family exist: cpn60.1, which is organized in an operon with cpn10, and cpn60.2, transcribed independently from the myc28 locus (Qamra et al. 2004; Shahar et al. 2011). Both Cpn60 proteins have been detected on the bacterial surface, specifically localized within the mycobacterial capsule, a protective layer surrounding the cell envelope (Kong et al. 1993; Qamra et al. 2004). Notably, Cpn60.2 and DnaK have been identified as integral components of this capsule, implicating them in host-pathogen interactions beyond their canonical intracellular chaperone functions. Cpn60.2 plays a pivotal role in mediating bacterial adherence to host macrophages by binding to specific macrophage surface receptors. This interaction is critical for efficient bacterial association and internalization. Experimental evidence shows that blocking surface-exposed Cpn60.2 with specific antibodies significantly reduces Mtb adhesion to macrophages (Hickey et al. 2009; Sharma et al. 2012). Mechanistically, Cpn60.2 stabilizes the interaction between Mtb and alveolar macrophages through binding to the host receptor CD43, functioning as a major adhesin facilitating bacterial colonization (Hickey et al. 2009). This extracellular adhesive function complements the intracellular role of Cpn60.2 in managing cellular stress, suggesting a coordinated network that supports bacterial survival and persistence within the host (Henderson 2010). Despite being expressed independently from the classical GroEL/ES operon, Cpn60.2 is indispensable for Mtb viability. Genetic deletion of the cpn60.2 gene results in complete growth inhibition, and attempts to generate knockout mutants via recombineering have been unsuccessful, underscoring its essentiality (Shahar et al. 2011; Hu et al. 2008). Although both Cpn60.2 and DnaK proteins are present on the bacterial surface, only Cpn60.2 is required for effective macrophage adherence, highlighting its unique role in host-pathogen interactions (Joseph et al. 2017). In summary, GroESL and DnaK chaperones in Mtb not only perform fundamental intracellular stress-response functions but also serve as critical adhesins on the bacterial surface. The dual localization and multifunctionality of these proteins underscore their importance in bacterial pathogenesis and represent promising targets for therapeutic intervention.

8.2.2.10 Mycobacterium Cell Entry-1 Protein (*Mce1*, Rv0169)

The Mycobacterium cell entry-1 protein (Mce1), encoded by the Rv0169 gene, is a critical component of the Mce1 complex—a multi-protein assembly encoded by the mce1 operon in Mtb (Ahmad et al. 1999; Chitale et al. 2001; Casali et al. 2006;Yan et al. 2019; Chen et al. 2023). This complex plays a multifaceted role in bacterial physiology and pathogenesis, integrating functions related to lipid transport, host cell adhesion, and immune modulation. Structurally, the Mce1 complex comprises six substrate-binding proteins (Mce1A through Mce1F, with Rv0169 specifically encoding Mce1A), two permease subunits (YrbE1A and YrbE1B), and additional associated membrane proteins. Together, these components form an ATP-binding cassette (ABC) transporter system that spans the mycobacterial cell envelope (Asthana et al. 2021). Recent advances in cryo-electron microscopy have elucidated the architecture of this complex, revealing an elongated, curved, needle-like structure predominantly composed of helical domains from the Mce1A-F subunits. This unique conformation creates a hydrophobic channel traversing the cell envelope, which facilitates the import of essential lipid molecules such as fatty acids and mycolic acids. These lipids are vital for maintaining cell envelope integrity and supporting energy metabolism, especially under the nutrient-limited conditions encountered during infection (Chen et al. 2023). Functionally, Mce1 serves a dual purpose: it acts as a lipid transporter and as an adhesin mediating bacterial entry into mammalian epithelial cells. This adhesive property was first demonstrated by heterologous expression of mce1A in nonpathogenic *Escherichia coli*, which conferred the ability to invade HeLa epithelial cells, a capability absent in wild-type *Escherichia coli*. Complementary experiments using latex beads coated with purified Mce1A protein further confirmed its adhesin role, showing significantly enhanced binding and uptake by epithelial cells compared to control beads. These findings underscore Mce1A's direct involvement in host cell adhesion and invasion (Casali et al. 2002, 2006; Kohwiwattanagun et al. 2007; Vinod et al. 2020). Genetic disruption of the mce1 operon results in diminished bacterial entry into host cells and attenuated virulence in animal models, highlighting the operon's critical contribution to Mtb pathogenicity (Shimono et al. 2003). The mce1 operon is tightly regulated by the transcriptional repressor Mce1R, which features a dynamic N-terminal DNA-binding domain and a stable C-terminal domain containing a conserved ligand-binding cavity (Casali et al. 2006). This structural configuration suggests that Mce1R may be amenable to pharmacological targeting, offering a potential strategy to interfere with lipid uptake and virulence. Beyond its role in lipid transport and adhesion, the Mce1 complex influences host immune responses and facilitates bacterial persistence within the host. By modulating lipid composition and cell wall homeostasis, Mce1 contributes to the bacterium's ability to evade immune detection and establish chronic infection. This multifaceted functionality makes the Mce1 complex an attractive target for novel therapeutic interventions aimed at disrupting Mtb survival and dissemination (Maity et al. 2023).

8.2.2.11 PE-PGRS Adhesins

The PE-PGRS family represents a unique group of glycine-rich, surface-exposed proteins in Mtb, characterized by a conserved N-terminal PE domain and a polymorphic C-terminal PGRS domain containing multiple GGA-GGX repeats. These repeats assemble into fibrillar structures that protrude from the bacterial surface, facilitating critical interactions with host tissues and the extracellular matrix (ECM) (Brennan and Delogu

2002; Sampson 2011; De Maio et al. 2020; D'Souza et al. 2023; Theorupun et al. 2025). Functionally, PE-PGRS proteins act as adhesins, mediating attachment to ECM components such as fibronectin, thereby promoting bacterial adhesion, colonization, and persistence within the host (Espitia et al. 1999; Brennan et al. 2001; Akhter et al. 2012). For instance, PE_PGRS17 and Wag22 (Rv1759c) have been demonstrated to bind fibronectin, thereby enhancing Mtb adherence and colonization of host tissues. Biochemical assays and immunofluorescence studies confirm their localization on the mycobacterial surface, underscoring their direct involvement in host-pathogen interactions (Qian et al. 2020). The glycine-rich PGRS domains form flexible, amyloid-like fibrils that facilitate stable attachment to ECM components, promoting bacterial aggregation and biofilm formation, both critical for Mtb persistence. By acting as molecular bridges to host integrins via fibronectin binding, PE-PGRS proteins also enhance phagocytic uptake and intracellular survival, supporting infection establishment. Among these, Wag22 (Rv1759c) is a prominent PE-PGRS protein distinguished by its large size (914 amino acids) and glycine-rich composition. Structurally, Wag22 comprises an N-terminal PE domain responsible for cell wall localization and a C-terminal PGRS domain that confers protein stability and protection from proteolysis (Fig. 8.10). Functionally, Wag22's fibronectin-binding capacity suggests a pivotal role in mediating host ECM interactions during infection. It is actively expressed during infection, localizes on the mycobacterial surface, influences cell surface properties, and modulates host immune responses. Studies in Mycobacterium marinum, a close relative of Mtb, reveal that inactivation of wag22 homologues reduces bacterial persistence within granulomas due to impaired macrophage replication, highlighting Wag22's essential role in bacterial survival within host tissues (Solomon et al. 2003). Moreover, Wag22 elicits immune responses that may contribute to sustaining latent tuberculosis infection, indicating its involvement in chronic infection stages. Recent whole genome sequencing (WGS) analyses reveal no sequence alterations in Wag22 among drug-susceptible, mono-drug resistant, or MDR-TB isolates compared to the reference strain H37Rv (Farnia et al. 2023b). However, significant deletions have been documented in XDR and TDR-TB clinical isolates. These deletions span amino acids 699 (tyrosine) to 710 (asparagine) and 725 (proline) to 914 (threonine), reducing the protein length from 914 to 714 amino acids (Fig. 8.11). Other TDR-TB isolates exhibit deletions affecting amino acids 1 (methionine) to 83 (glycine) and 725 to 914, resulting in a markedly truncated Wag22 protein (Fig. 8.12). Such extensive deletions may induce frameshift mutations, potentially leading to loss or abnormal function. These findings warrant further investigation to determine whether these alterations contribute to immune evasion and promote persistence in highly resistant Mtb strains (Farnia et al. 2023b). Similarly, PE_PGRS31, encoded by Rv1768, plays a crucial role in enhancing bacterial survival within host macrophages by modulating immune signaling pathways. Its PE domain directly interacts with the host protein S100A9, preventing S100A9 from binding to Toll-like receptor 4 (TLR4). This inhibition suppresses the downstream TLR4-MyD88-NF-κB-TNF-α signaling cascade, reducing pro-inflammatory responses that typically aid bacterial clearance (Liu et al. 2020). Additionally, PE_PGRS31 binding to S100A9 disrupts arachidonic acid metabolism by activating 5-lipoxygenase, increasing lipoxin A4 levels while downregulating cyclooxygenase-2 and prostaglandin E2 expression. These biochemical changes create an anti-inflammatory environment conducive to bacterial persistence within macrophages, a hallmark of dormancy and latent infection (Yuan et al. 2019). Structurally, PE_PGRS31 consists of 618 amino acids with a molecular mass of approximately 51,962.8 Da (Fig. 8.13). Whole genome sequencing (WGS) analyses show that the domain organization and structural integrity of PE_PGRS31 are conserved in all studied clinical isolates and laboratory strains (H37Rv) during both exponential and latent phases. However, in totally drug-resistant (TDR) isolates, a significant deletion reduces the protein length from 618

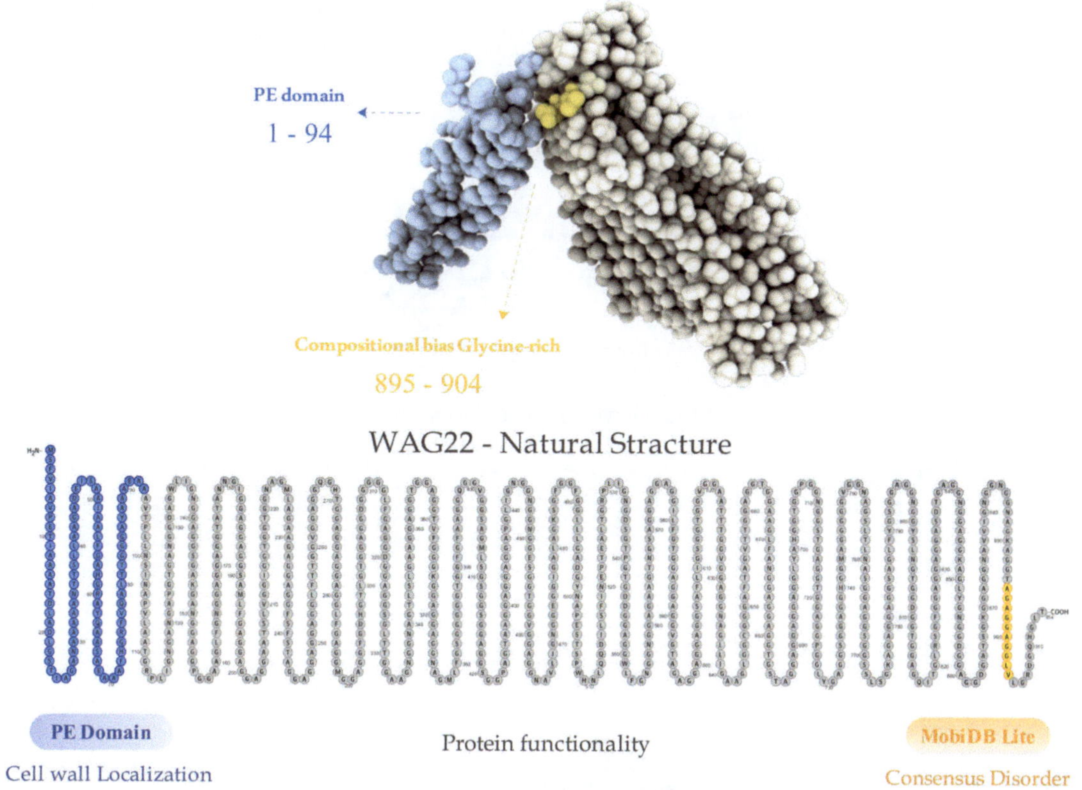

Fig. 8.10 Structural features and domain organization of the PE-PGRS family protein Wag22 (Rv1759c) in H37RV The figure shows the natural structure of the PE-PGRS family protein Wag22 (Rv1759c), which is 914 amino acids in length with a molecular mass of 74,353.8 daltons (Da). The N-terminal PE domain (residues 1 – 9 4, shown in blue) is responsible for cell wall localization. The central region is associated with protein functionality, while the C-terminal region (residues 895 – 9 04, highlighted in yellow) displays a glycine-rich compositional bias and is annotated as a consensus disorder region by MobiDB Lite. Structural representations include a space-filling model and a linear schematic, emphasizing the spatial and functional domains. Whole genome sequencing (WGs) documented no deletions, insertions, or substitutions in the amino acid sequence (n = 914) in susceptible, mono-drug resistant, or even MDR-TB isolates during the exponential growth state compared to the reference strain $H_{37}Rv$"
*Drawing by Mohsen Maleknia; Mycobacteriology Research Center, National Research Institute of Tuberculosis and Lung Disease, Shahid Beheshti University of Medical Sciences, Tehran, Iran

to 194 amino acids, corresponding to the loss of residues 195 to 618 (Fig. 8.14).Moreover, all extensively drug-resistant isolates examined exhibit a common deletion between amino acids 404 and 618 (Farnia et al. 2023b; Velayati and Farnia 2024). These deletions result in truncated proteins that may lose or alter function, potentially representing an adaptive mechanism to evade host immunity while preserving essential bacterial survival domains. The truncation of the C-terminal region, important for stability and host interaction, may reduce immune recognition and contribute to dormancy and latency in resistant bacilli. Further research is required to elucidate the precise impact of these deletions on PE_PGRS31 function and drug-resistant tuberculosis pathogenesis. Among the most thoroughly studied PE_PGRS proteins is PE_PGRS33 (Rv1818c), which localizes on the Mtb surface and mediates adhesion to macrophages and epithelial cells primarily through fibronectin binding and Toll-like receptor 2 (TLR2) interaction. Engagement of TLR2 by PE_PGRS33 activates pro-inflammatory cytokine responses, modulat-

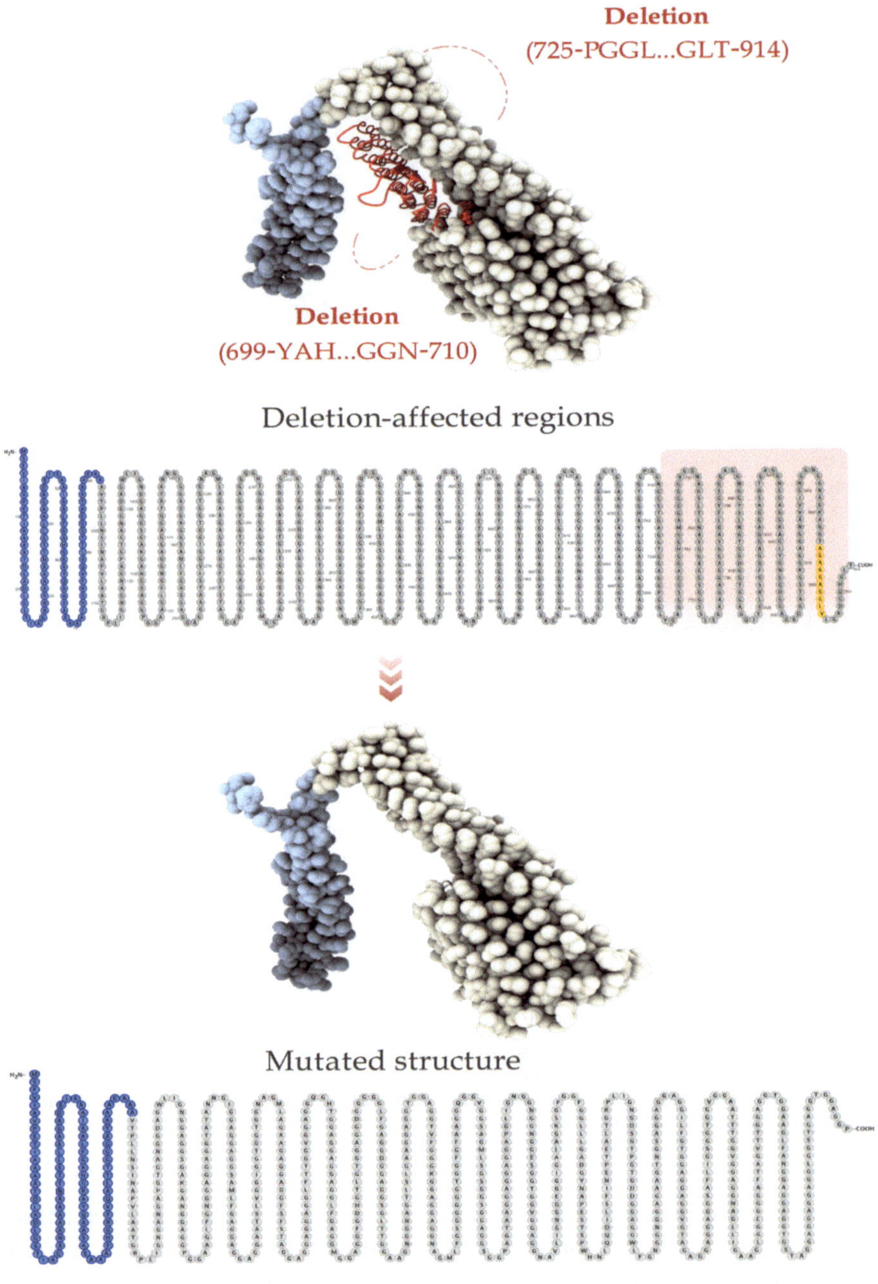

Fig. 8.11 Structural features and domain organization of Wag22 (Rv1759c) in extensively drug-resistant tuberculosis clinical isolates after prolonged treatment

The figure illustrates the structural consequences of large deletions in the proline-glutamate-proline-glutamate-polymorphic GC-rich sequences (PE-PGRS) family protein Wag22 (Rv1759c) in Mycobacterium tuberculosis. At the top, the native three-dimensional structure and corresponding linear schematic of Wag22 are presented, with the N-terminal PE domain highlighted in blue and the glycine-rich C-terminal region in yellow. Regions affected by deletions are marked in red, specifically spanning amino acids 699 – 7 10 (YAH...GGN) and 725 – 9 14 (PGGL...GLT). These extensive deletions are frequently observed in persister cells of extensively drug-resistant (XDR) tuberculosis, particularly in clinical isolates subjected to prolonged treatment exceeding 18 months. Such large deletions can cause a frameshift, potentially resulting in the absence or abnormal function of Wag22, a phenomenon that warrants further investigation to determine whether this alteration facilitates immune evasion by the bacilli. The middle panel highlights these deletion-affected regions within the linear sequence. The lower part of the figure depicts the mutated structure and its

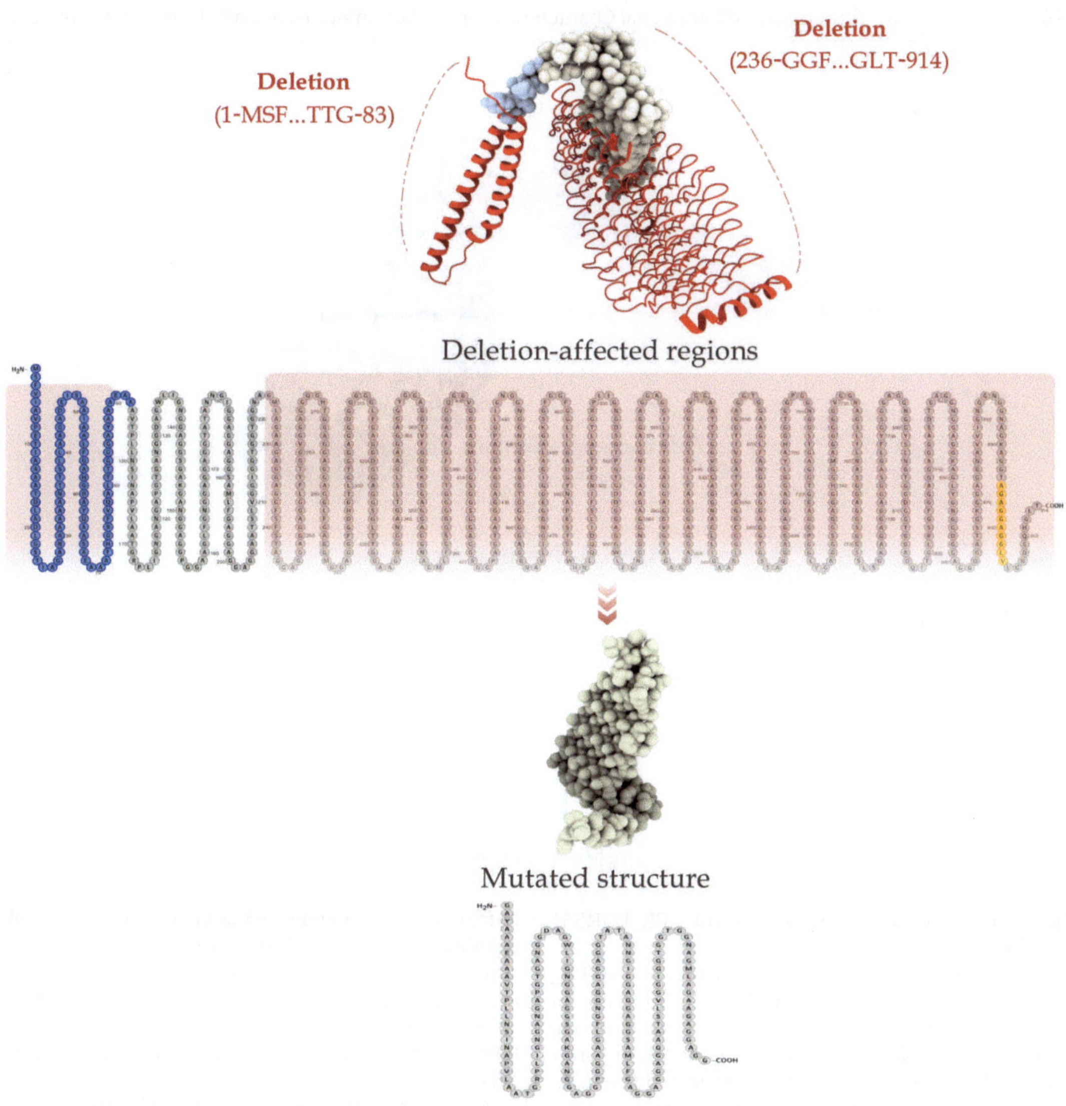

Fig. 8.12 Structural features and domain organization of Wag22 (Rv1759c) in totally drug-resistant tuberculosis clinical isolates after prolonged treatment
This figure illustrates the structural consequences of large deletions in the PE-PGRS family protein Wag22 (Rv1759c. At the top, the protein structure is shown, highlighting two major deletion regions: one at the N-terminus (amino acids 1 – 8 3, labeled as 1-MSF...TTG-83) and another spanning amino acids 236 – 9 14 (236-GGF... GLT-914). The central schematic maps these deletions onto the protein's linear topology, with the deletion-affected regions shaded in pink. The blue and orange segments represent unaffected regions at the N- and C-termini, respectively. The lower part of the figure shows the mutated structure resulting from these deletions, which is drastically truncated compared to the native protein. This visualization emphasizes how such extensive deletions, observed in totally drug-resistant (TDR) isolates, can disrupt the typical structure and likely function of Wag22, which warrants further investigation of this protein
*Drawing by Mohsen Maleknia; Mycobacteriology Research Center, National Research Institute of Tuberculosis and Lung Disease, Shahid Beheshti University of Medical Sciences, Tehran, Iran

schematic representation, demonstrating how these deletions result in a truncated and structurally altered protein with loss of the C-terminal domains. This visualization underscores that while Wag22 remains stable and conserved in drug-susceptible and MDR strains, significant structural changes can occur in XDR clinical isolates following extended treatment, which may impact the protein's function, immunogenicity, and detectability in clinical samples
*Drawing by Mohsen Maleknia; Mycobacteriology Research Center, National Research Institute of Tuberculosis and Lung Disease, Shahid Beheshti University of Medical Sciences, Tehran, Iran

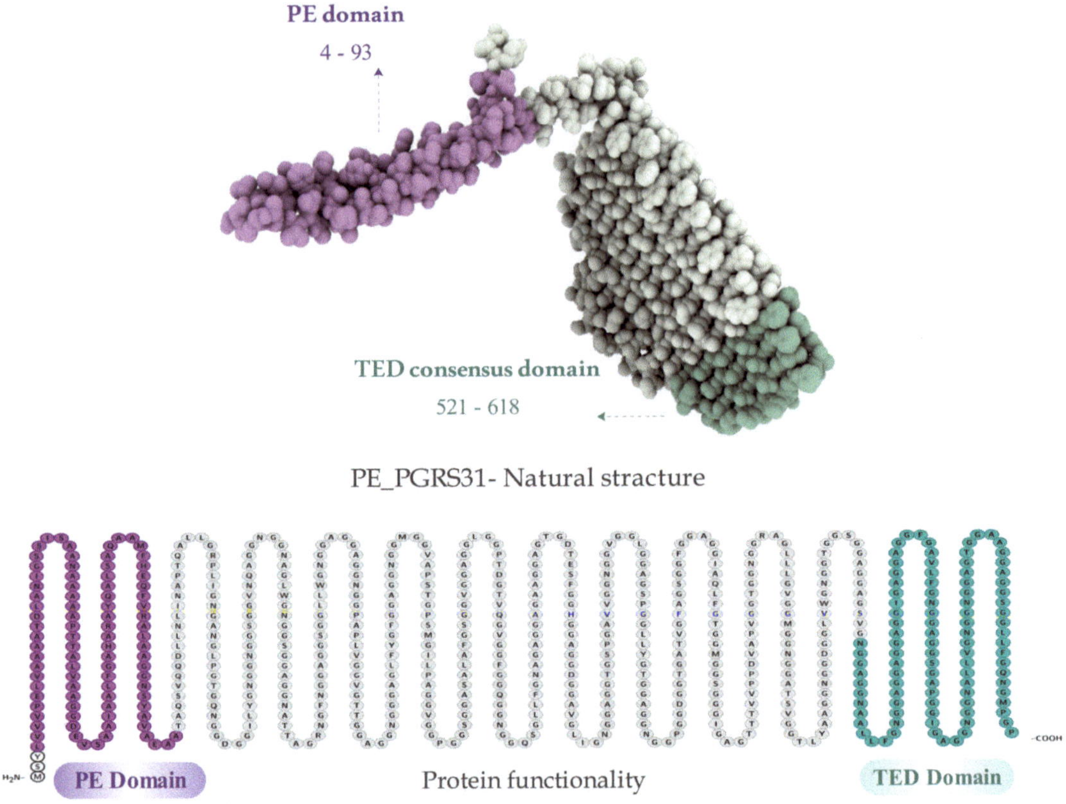

Fig. 8.13 Structural organization of PE_PGRS31 (Rv1768)
The figure displays the natural structure of the PE_PGRS31 protein (Rv1768) from Mtb, which consists of 618 amino acids. The three-dimensional model at the top highlights the PE domain (amino acids 4 to 93) in purple and the TED consensus domain (amino acids 521 to 618) in teal, with the central region in gray representing the variable PGRS domain. Below, a schematic linear representation maps these domains, with the PE domain at the N-terminus, the TED domain at the C-terminus, and the repetitive glycine-rich PGRS region in between. The structural integrity of PE_PGRS31 remained unchanged in susceptible, mono-drug resistant, and MDR-TB isolates compared to the $H_{37}Rv$ reference strain, "no deletions, insertions, or substitutions were detected using WGS"
*Drawing by Mohsen Maleknia; Mycobacteriology Research Center, National Research Institute of Tuberculosis and Lung Disease, Shahid Beheshti University of Medical Sciences, Tehran, Iran

ing host immunity and significantly contributing to tuberculosis pathogenesis (Delogu et al. 2004; Palucci et al. 2016). Functional studies demonstrate that Mtb mutants lacking PE_PGRS33 exhibit reduced macrophage entry and attenuated virulence, underscoring its essential role as an adhesin (Brodin et al. 2010; Palucci et al. 2016). The gene encoding PE_PGRS33 spans 1497 base pairs, located at 2,061,178 bp on the Mtb genome, and encodes a 498-amino-acid protein with a molecular mass of approximately 40,755.1 Da and an isoelectric point of 3.94 (Fig. 8.15).

Farnia et al. (2023b) reported that 66.6% of the clinical *Mycobacterium tuberculosis* isolates analyzed exhibited genetic variability in the PE_PGRS33 gene, consistent with previous

tion represents a significant loss of protein function, as the deleted region likely contains important functional domains essential for the protein's biological activity in the mycobacterial cell wall

*Drawing by Mohsen Maleknia; Mycobacteriology Research Center, National Research Institute of Tuberculosis and Lung Disease, Shahid Beheshti University of Medical Sciences, Tehran, Iran

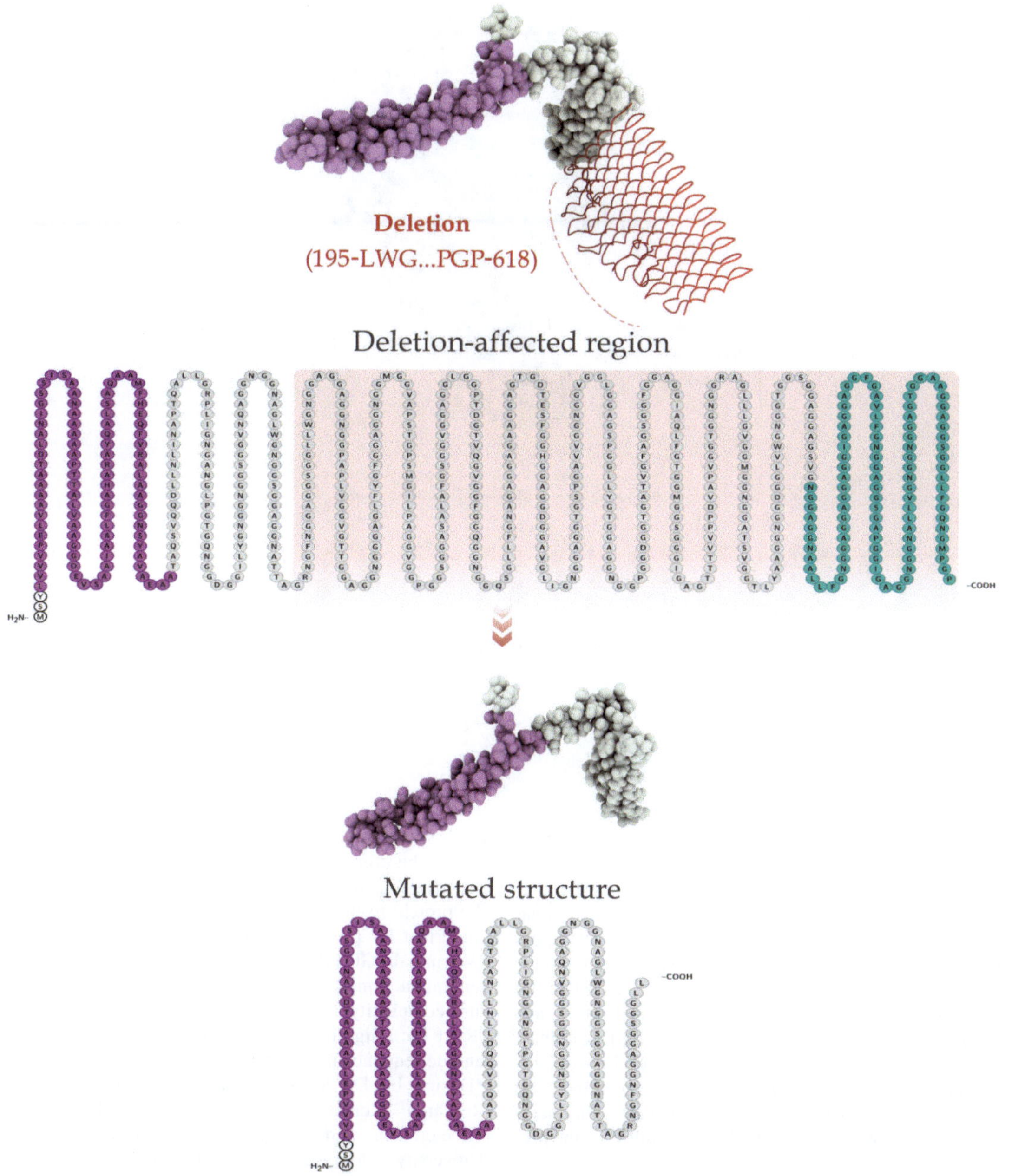

Fig. 8.14 Structural consequences of a large internal deletion in the PE_PGRS33 (Rv1768) protein of TDR-TB isolate

The figure demonstrates the structural impact of a large deletion mutation in the PE_PGRS33 protein (Rv1768) found in one studied TDR-TB isolates. The wild-type protein structure shown in the upper panel reveals the full-length protein consisting of 618 amino acids, with multiple transmembrane domains represented by the purple, gray, and teal colored segments in the schematic diagram. The critical deletion spans from amino acid position 195 (leucine) to position 618 (proline), resulting in the removal of 424 amino acids and reducing the protein length to only 194 amino acids. The middle schematic illustrates the deletion-affected region highlighted in pink, which encompasses the majority of the protein's central and C-terminal domains. This extensive deletion fundamentally alters the protein architecture, as demonstrated in the lower panel showing the mutated structure. The resulting truncated protein retains only the N-terminal transmembrane domains (shown in purple) and a small portion of the adjacent region, while completely losing the large C-terminal portion that would normally extend beyond the membrane. This dramatic structural modifica-

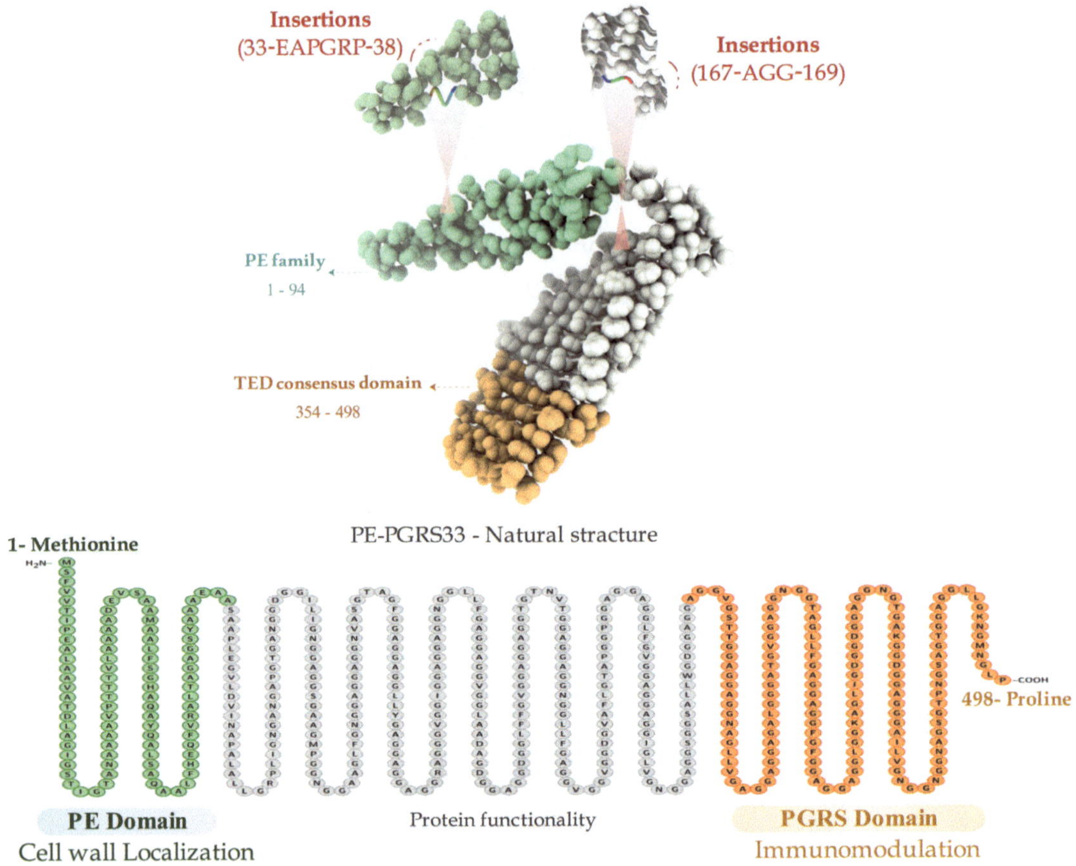

Fig. 8.15 Structural organization and key sequence insertions in the PE_PGRS33 (Rv1818c) protein
The figure illustrates the natural structure of the PE_PGRS33 (Rv1818c) protein, emphasizing its modular organization and important sequence features. The upper part of the figure displays a three-dimensional model, highlighting the PE family domain (amino acids 1 to 94) in green, which is involved in cell wall localization, and the PGRS domain (amino acids 354 to 498) in orange, known for its role in immunomodulation. Two distinct sequence insertions are marked: one between amino acids 33 and 38 (EAPGRP) and another between amino acids 167 and 169 (AGG), both of which may influence the protein's structure and function. The schematic below further details the domain organization, showing the N-terminal methionine and the PE domain responsible for anchoring the protein to the cell wall, followed by a central region important for protein functionality, and ending with the C-terminal PGRS domain that contributes to immune system interactions. Overall, the figure underscores the structural complexity and functional significance of PE_PGRS33 (Rv1818c), highlighting both its conserved domains and unique sequence insertion
*Design by Dr. Mohsen Maleknia; Mycobacteriology Research Center, National Research Institute of Tuberculosis and Lung Disease, Shahid Beheshti University of Medical Sciences, Tehran, Iran

smaller deletion from position 281 (glycine) to 283 (leucine) in the middle region. The "Deletion-affected regions" section of the figure shows the linear representation of the original protein with distinct functional domains colored in green, gray, and orange. These deletions fundamentally alter the protein architecture, as demonstrated in the "nonfunctional structure" at the bottom, where the resulting truncated protein retains only the gray and orange domains while completely losing the green N-terminal region. Beyond the major deletions, the isolate also underwent insertion events at two specific locations: amino acids 33 – 38 (EAPGRP) and positions 167 – 169 (AGG). Additionally, 17 amino acid substitutions occurred throughout the protein at various positions, including 1, 28, 32, 41, 44, 47, 49, 50, and positions 51 – 59
*Drawing by Mohsen Maleknia; Mycobacteriology Research Center, National Research Institute of Tuberculosis and Lung Disease, Shahid Beheshti University of Medical Sciences, Tehran, Iran

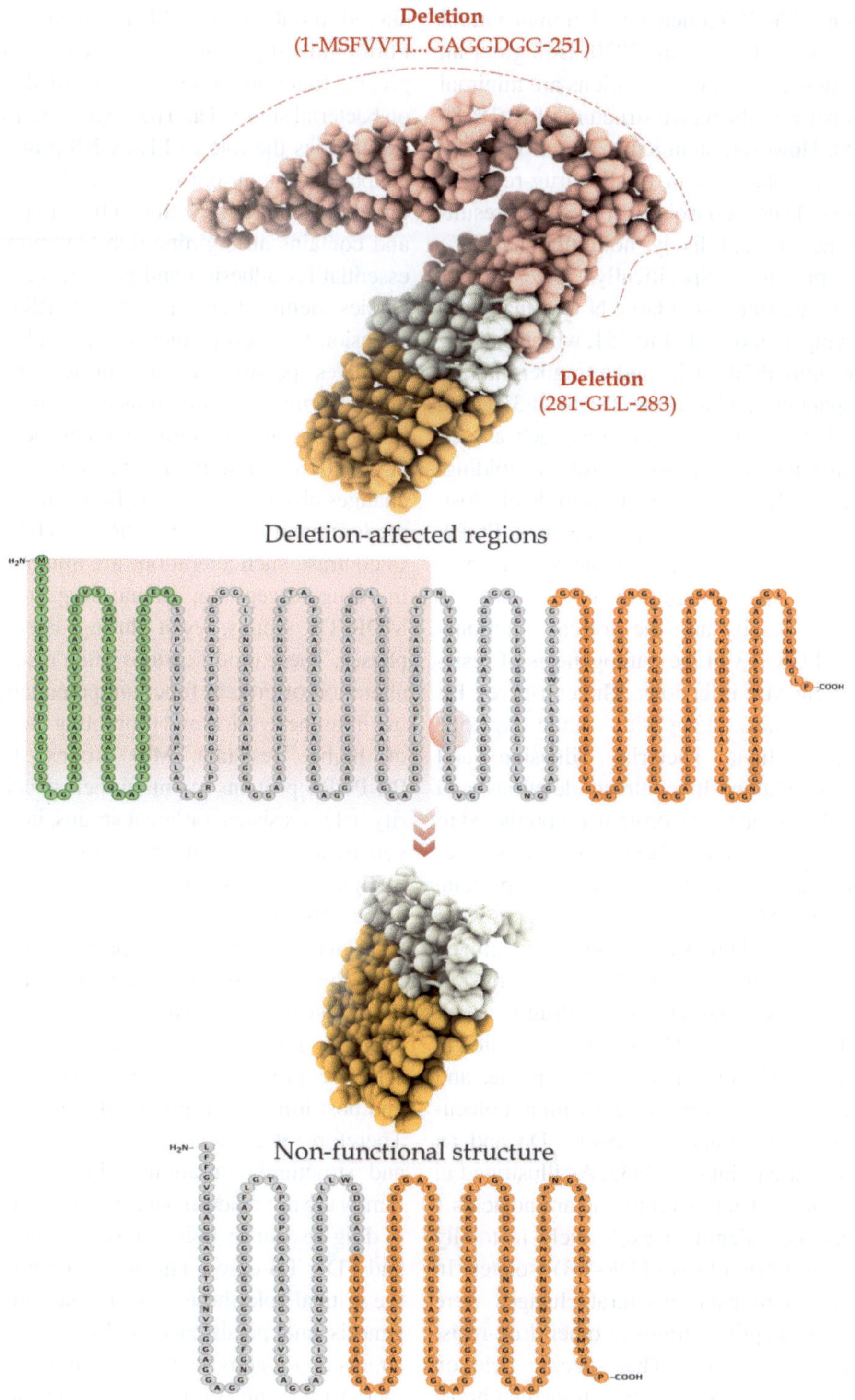

Fig. 8.16 Structural consequences of major deletions in PE_PGRS33 (Rv1818c) in a dormant totally drug-resistant tuberculosis isolate.

The most striking changes involve major deletions that dramatically reduce the Rv1818c from its original 498 amino acids to just 253 amino acids. The figure illustrates two primary deletion events: a large N-terminal deletion spanning from the first amino acid (methionine) to position 251 (glycine), effectively removing the entire pink-colored domain visible in the upper 3D structure, and a

observations of high-frequency variation in natural Mtb populations (Qian et al. 2020; Delogu et al. 2004). In most cases, these variations are minimal when compared to the native structure of Rv1818c (Fig. 8.15). However, significant alterations were predominantly observed in totally drug-resistant tuberculosis (TDR-TB) isolates (Fig. 8.16), resulting in truncated and likely nonfunctional PE_PGRS33 proteins. Specifically, two major deletions were identified: a large N-terminal deletion removing amino acids 1 to 251, which encompasses the entire PE domain, and a smaller internal deletion spanning amino acids 281 to 283. These deletions disrupt essential functions such as cell wall anchoring and proper protein folding. Consequently, the PE domain is completely lost, leaving only the central and C-terminal PGRS domains intact, which likely abolishes the protein's function. These findings underscore both the structural complexity and the critical functional role of PE_PGRS33 in the pathogenesis of resistant and latent Mtb infections. The conserved PE and PGRS domains, along with unique sequence insertions, facilitate bacterial adhesion and immune modulation. In contrast, deletions and mutations that truncate the protein are predicted to reduce virulence and alter host-pathogen interactions. Another important PE_PGRS protein, encoded by Rv1803c (PE_PGRS32), is less extensively characterized but is increasingly recognized for its role in *Mycobacterium tuberculosis* physiology and virulence, particularly in drug-resistant strains. The gene spans 1920 base pairs, located at position 2,044,923 bp on the Mtb genome, and encodes a 639-amino acid protein with a molecular mass of approximately 54,849.5 Da and an acidic isoelectric point of 3.7853. As illustrated in Fig. 8.17, a large deletion removing amino acids 1 to 212 has been identified exclusively in totally drug-resistant tuberculosis (TDR-TB) isolates. In contrast, no significant structural changes were observed in susceptible strains or other drug-resistant isolates (Fig. 8.18). This specific deletion encompasses key N-terminal transmembrane domains, resulting in a truncated protein with altered topology and likely impaired function. Further investigation is warranted to elucidate the precise functional consequences of this mutation on bacterial survival and pathogenesis. PE_PGRS3 exemplifies the role of PE-PGRS proteins in Mtb adaptation to nutrient limitation. It is upregulated under phosphate starvation, a host-imposed stress, and contains an arginine-rich C-terminal domain essential for adhesion and persistence. Functional studies demonstrate that PE_PGRS3 enhances adhesion to macrophages and epithelial cells and improves persistence in murine spleen tissue, underscoring its importance in host-pathogen interactions and infection maintenance (De Maio et al. 2018). In summary, the primary structural changes observed in the PE-PGRS family proteins predominantly occur in XDR and TDR-TB cases. In contrast, such alterations are minimal or absent in drug-susceptible, mono-drug resistant, or MDR-TB strains, even during latent infection phases. These modifications often result in loss or alteration of protein function, potentially facilitating immune evasion and promoting the persistence of highly resistant Mtb strains. Conversely, PE-PGRS proteins maintain their structural integrity in less resistant or latent strains, indicating that genetic and structural variability is closely linked to high levels of drug resistance rather than latency alone. This pattern strongly suggests that structural changes in PE-PGRS proteins are associated with the development and maintenance of high drug resistance in Mtb, likely representing an adaptive mechanism to survive under intense selective pressure from antimicrobial treatment and host immune responses (Farnia et al. 2023b). Therefore, it is clear that the genetic variability and structural alterations within the PE-PGRS family are not random but correlate with the degree of drug resistance, being most pronounced in XDR and TDR TB cases. This association underscores the critical role these proteins play in the pathogenesis and persistence of highly resistant Mtb strains and highlights their potential as targets for therapeutic intervention and diagnostic markers for resistance profiling.

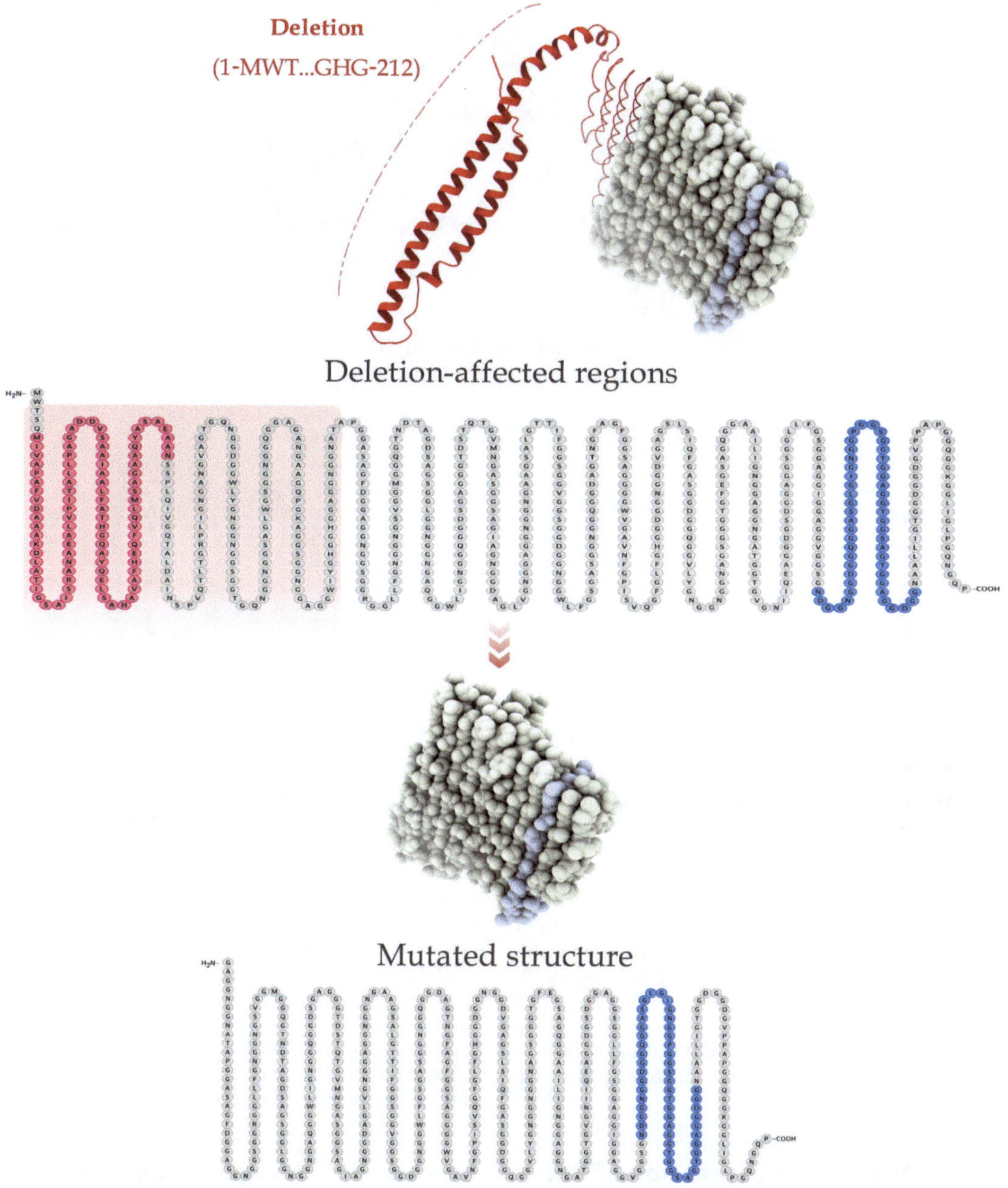

Fig. 8.17 Structural and topological consequences of a large N-terminal deletion in the PE_PGRS32 (Rv1803) protein
The figure demonstrates the structural and topological impact of a large N-terminal deletion in the PE_PGRS32 (Rv1803) protein. At the top, the deletion is specified as the removal of amino acids 1 to 212, labeled as (1-MWT... GHG-212). It is highlighted in red on the protein structure, corresponding to the N-terminal region. The middle section presents a schematic of the protein's transmembrane topology, where the deleted N-terminal transmembrane domains are shaded in pink, indicating the regions lost due to the deletion. A distinct segment, marked in blue, represents a transmembrane domain that remains intact. At the bottom, the mutated protein structure is shown, illustrating the absence of the N-terminal domains and the preservation of the blue-marked region, resulting in a truncated protein topology
*Drawing by Mohsen Maleknia; Mycobacteriology Research Center, National Research Institute of Tuberculosis and Lung Disease, Shahid Beheshti University of Medical Sciences, Tehran, Iran

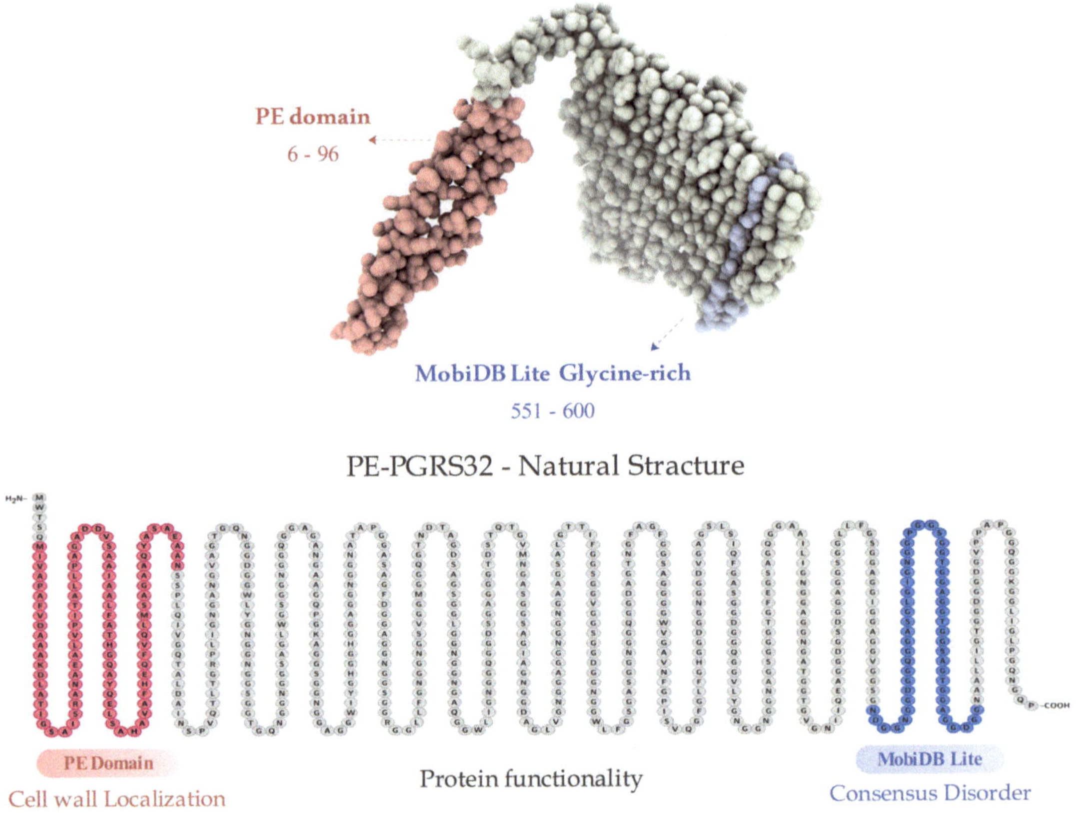

Fig. 8.18 Structural features and functional domains of the PE_PGRS32 (Rv1803) protein in *Mycobacterium tuberculosis*
The image illustrates the natural structure and topology of the PE_PGRS32 family protein (Rv1803), which is composed of 639 amino acids and has a molecular mass of 54,849.5 Da. The top part of the image shows the three-dimensional structure of the protein, highlighting the PE domain (amino acids 6 – 9 6) in pink, which is associated with cell wall localization, and a glycine-rich region (amino acids 551 – 6 00) in blue, identified as a consensus disorder region by MobiDB Lite. The lower part of the image presents a schematic diagram of the protein's amino acid sequence, again marking the PE domain and the glycine-rich region, emphasizing their roles in protein functionality and structural disorder, respectively. Notably, the analysis revealed that no changes in amino acid sequences, such as deletions, insertions, or substitutions, were observed in susceptible, mono-drug resistant, MDR-TB, or XDR-TB isolates, both in the exponential growth phase and under an in vitro model of latency
*Design by Dr. Mohsen Maleknia; Mycobacteriology Research Center, National Research Institute of Tuberculosis and Lung Disease, Shahid Beheshti University of Medical Sciences, Tehran, Iran

8.3 Differential Expression of Pili and Adhesins During the Exponential and Latent Phases of *Mycobacterium tuberculosis*

During the exponential or active growth phase, Mtb employs a diverse array of adhesins to colonize and invade host tissues effectively (Yang et al. 2025. Among these, the heparin-binding hemagglutinin adhesin (HBHA) is prominently expressed and plays a pivotal role in mediating attachment to epithelial cells, thereby facilitating extrapulmonary dissemination of the pathogen (Raze et al. 2018). Another critical group of adhesins is the antigen 85 (Ag85) complex, composed of Ag85A, Ag85B, and Ag85C. This complex not only binds fibronectin within the extracellular matrix but also contributes to cell wall biosynthesis, thereby supporting both the structural integrity of the bacterium and its interactions with the host during active infection

(Świerzko et al. 2016). Members of the PE-PGRS protein family are also abundantly expressed during this phase, mediating attachment to fibronectin and enhancing tissue colonization (Zhang et al. 2025). LpqH, commonly referred to as the 19-kDa antigen, is a glycosylated lipoprotein that interacts with monocytes and macrophages via mannose receptors and Toll-like receptors, promoting bacterial uptake and modulating host immune responses (Ramon-Luing et al. 2023). Similarly, PstS-1, a component of the phosphate transport system, functions as an adhesin by interacting with macrophages and is essential for nutrient acquisition during rapid bacterial proliferation (Oliveira and Balan 2020). The chaperonin Cpn60.2 (GroEL2) exhibits a dual role: it facilitates protein folding intracellularly. It acts as an adhesin on the bacterial surface by binding to macrophage CD43, thereby contributing to host-pathogen interactions (Piplani et al. 2023). Additionally, the Mycobacterium cell entry-1 protein (Mce1) is integral to the early stages of infection, facilitating bacterial entry into epithelial cells (Fenn et al. 2020). Regarding pili, the active phase is characterized by the expression of curli-like pili, also known as mycobacterial pili (MTP), encoded by Rv3312A. These pili mediate binding to laminin, macrophages, and epithelial cells, promoting bacterial aggregation and biofilm formation—processes essential for colonization and establishment of infection (Alteri et al. 2007; Ramsugit and Pillay 2016). Furthermore, Type IV pili, including Flp pili, are expressed during this phase and play crucial roles in adhesion, colonization, and invasion, particularly upon contact with host cells. As Mtb transitions into the persistence or latent phase, a state characterized by reduced metabolic activity and adaptation to the hostile microenvironment within granulomas, the expression profile of adhesins and pili undergoes significant modulation. Many adhesins and pili that are abundantly expressed during active growth are downregulated during latency, reflecting the bacterium's strategic need to evade host immune detection and conserve energy resources (Gengenbacher and Kaufmann 2012; Magombedze et al. 2013; Abbasnia et al. 2024).

Despite this overall downregulation, specific adhesins remain expressed or are selectively upregulated to support bacterial survival under these stress conditions. The antigen 85 (Ag85) complex, for instance, continues to be expressed at lower levels, maintaining essential interactions with host cells and contributing to cell wall integrity and maintenance during persistence (Abbasnia et al. 2024). Similarly, specific members of the PE-PGRS protein family are expressed during latency, likely facilitating the formation of microcolonies and biofilms that protect host immune responses and antibiotic treatment. Malate synthase, a metabolic enzyme with moonlighting adhesive functions, is notably upregulated during the latent phase. Its increased expression supports metabolic adaptation to nutrient-limited conditions and sustains reduced but critical adhesion to host tissues, thereby aiding bacterial persistence. Additionally, the chaperonin Cpn60.2 (GroEL2) remains functionally important during latency by inhibiting apoptosis in infected macrophages, which promotes intracellular survival and persistence of Mtb within host cells. Recent studies (Velayati and Farnia 2024) have reported the discovery of novel pili types in Mtb, designated as Types V, VI, and VII. These pili exhibit distinctive morphologies, including relief funnel, adhesion tapering, and adhesion flap structures, respectively. Notably, these newly identified pili have been observed predominantly in persistent or latent bacterial populations. These pili likely play crucial roles in facilitating nutrient uptake from the extracellular environment into the bacterial cytoplasm and in mediating the adhesion of persister cells to epithelial surfaces within the host. Such functions are vital for bacterial survival and long-term persistence under the hostile conditions characteristic of latency, including nutrient scarcity and immune system pressure. Overall, pili and adhesins are differentially regulated between the active and latent phases of *Mycobacterium tuberculosis*, reflecting the bacterium's shift from aggressive host colonization and dissemination during active infection to stealthy persistence and survival during latency. Given these potential roles, further research is essential to clarify the presence, regulatory mechanisms, and functional significance of these newly identified pili types in latent Mtb. A deeper understanding of their contribution to pathogenesis and persistence may uncover novel targets for therapeutic interventions aimed at eradicating latent tuberculosis infections.

Conflict of Interest The authors declare that they have no conflict of interest related to the content of this chapter.

Ethical Consideration This chapter is based on conclusions drawn from the authors' research as well as data and information available in the literature, all of which have been appropriately cited. The figures and images presented originate from the authors' studies, which did not involve experiments on humans, animals, or cell cultures. All referenced experimental work was conducted using laboratory and wild-type strains of *Mycobacterium tuberculosis* (Mtb), following established standard protocols. The research project received approval from the Scientific Committee of the NRITLD, Tehran, Iran. The findings and interpretations presented are grounded in a rigorous analysis of existing research combined with the authors' experimental work.

References

Abbasnia S, Hashem Asnaashari AM, Sharebiani H, Soleimanpour S, Mosavat A, Rezaee SA (2024) Mycobacterium tuberculosis and host interactions in the manifestation of tuberculosis. J Clin Tuberc Other Mycobact Dis 36:100458. https://doi.org/10.1016/j.jctube.2024.100458. PMID: 38983441; PMCID: PMC11231606

Abou-Zeid C, Ratliff TL, Wiker HG, Harboe M, Bennedsen J, Rook GA (1988) Characterization of fibronectin-binding antigens released by Mycobacterium tuberculosis and Mycobacterium bovis BCG. Infect Immun 56(12):3046–3051. https://doi.org/10.1128/iai.56.12.3046-3051.1988. PMID: 3141278; PMCID: PMC259698

Achkar JM, Dong Y, Holzman RS, Belisle J, Kourbeti IS, Sherpa T, Condos R, Rom WN, Laal S (2006) Mycobacterium tuberculosis malate synthase- and MPT51-based serodiagnostic assay as an adjunct to rapid identification of pulmonary tuberculosis. Clin Vaccine Immunol 13(11):1291–1293

Agarwal S, Ghosh S, Sharma S, Kaur K, Verma I (2018) Mycobacterium tuberculosis H37Rv expresses differential proteome during intracellular survival within alveolar epithelial cells compared with macrophages. Pathogens Dis 76(6):fty058

Ahmad S, Akbar PK, Wiker HG, Harboe M, Mustafa AS (1999) Cloning, expression and immunological reactivity of two mammalian cell entry proteins encoded by the mce1 operon of Mycobacterium tuberculosis. Scand J Immunol 50(5):510–518

Ahn S, Jung J, Jang IA, Madsen EL, Park W (2016) Role of glyoxylate shunt in oxidative stress response. J Biol Chem 291(22):11928–11938

Akhter Y, Ehebauer MT, Mukhopadhyay S, Hasnain SE (2012) The PE/PPE multigene family codes for virulence factors and is a possible source of mycobacterial antigenic variation: perhaps more? Biochimie 94(1):110–116

Alsteens D, Verbelen C, Dague E, Raze D, Baulard AR, Dufrêne YF (2008) Organization of the mycobacterial cell wall: a nanoscale view. Pflugers Arch 456(1): 117–125. https://doi.org/10.1007/s00424-007-0386-0. Epub 2007 Nov 28. PMID: 18043940

Alteri CJ (2005) Novel pili of mycobacterium tuberculosis [dissertation]. The University of Arizona, Tucson

Alteri CJ, Xicohténcatl-Cortes J, Hess S, Caballero-Olín G, Girón JA, Friedman RL (2007) Mycobacterium tuberculosis produces pili during human infection. Proc Natl Acad Sci USA 104(12):5145–5150. https://doi.org/10.1073/pnas.0602304104 PMID: 17360408; PMCID: PMC1817835

Alteri CJ, Rios-Sarabia N, De la Cruz MA, González-y-Merchand JA, Soria-Bustos J, Maldonado-Bernal C, Cedillo ML, Yáñez-Santos JA, Martínez-Laguna Y, Torres J, Friedman RL, Girón JA, Ares MA (2022) The Flp type IV pilus operon of *Mycobacterium tuberculosis* is expressed upon interaction with macrophages and alveolar epithelial cells. Front Cell Infect Microbiol 12:916247

Armitige LY, Jagannath C, Wanger AR, Norris SJ (2000) Disruption of the genes encoding antigen 85A and antigen 85B of Mycobacterium tuberculosis H37Rv: effect on growth in culture and in macrophages. Infect Immun 68(2):767–778

Ashiru OT, Pillay M, Sturm AW (2010) Adhesion to and invasion of pulmonary epithelial cells by the F15/LAM4/KZN and Beijing strains of Mycobacterium tuberculosis. J Med Microbiol 59(Pt 5):528–533

Ashiru OT, Pillay M, Sturm AW (2012) *Mycobacterium tuberculosis* isolates grown under oxygen deprivation invade pulmonary epithelial cells. Anaerobe 18(4):471–474

Asthana P, Singh D, Pedersen JS, Hynönen MJ, Sulu R, Murthy AV, Laitaoja M, Jänis J, Riley LW, Venkatesan R (2021) Structural insights into the substrate-binding proteins Mce1A and Mce4A from *Mycobacterium tuberculosis*. IUCrJ 8(Pt 5):757–774

Barocchi MA, Ries J, Zogaj X, Hemsley C, Albiger B, Kanth A, Dahlberg S, Fernebro J, Moschioni M, Masignani V, Hultenby K, Taddei AR, Beiter K, Wartha F, von Euler A, Covacci A, Holden DW, Normark S, Rappuoli R, Henriques-Normark B (2006) A pneumococcal pilus influences virulence and host inflammatory responses. Proc Natl Acad Sci USA 103(8):2857–2862

Barrios-Payán J, Saqui-Salces M, Jeyanathan M, Alcántara-Vazquez A, Castañon-Arreola M, Rook G, Hernandez-Pando R (2012) Extrapulmonary locations of mycobacterium tuberculosis DNA during latent infection. J Infect Dis 206(8):1194–1205

Basu J (2004) Mycobacteria within its intracellular niche: survival of the pathogen or its host? Curr Sci 86(1):103–110

Basu J, Shin DM, Jo EK (2012) Mycobacterial signaling through toll-like receptors. Front Cell Infect Microbiol 2:145

Belisle JT, Vissa VD, Sievert T, Takayama K, Brennan PJ, Besra GS (1997) Role of the major antigen of Mycobacterium tuberculosis in cell wall biogenesis. Science 276(5317):1420–1422

Bermudez LE, Goodman J (1996) Mycobacterium tuberculosis invades and replicates within type II alveolar cells. Infect Immun 64(4):1400–1406

Bermudez LE, Sangari FJ, Kolonoski P, Petrofsky M, Goodman J (2002) The efficiency of the translocation of Mycobacterium tuberculosis across a bilayer of epithelial and endothelial cells as a model of the alveolar wall is a consequence of transport within mononuclear phagocytes and invasion of alveolar epithelial cells. Infect Immun 70(1):140–146

Berry JL, Pelicic V (2015) Exceptionally widespread nanomachines composed of type IV pilins: the prokaryotic Swiss Army knives. FEMS Microbiol Rev 39(1):134–154

Bhoite S, van Gerven N, Chapman MR, Remaut H (2019) Curli biogenesis: bacterial amyloid assembly by the type VIII secretion pathway. EcoSal Plus 8(2). https://doi.org/10.1128/ecosalplus.ESP-0037-2018. PMID: 30892177; PMCID: PMC6428212

Birkness KA, Deslauriers M, Bartlett JH, White EH, King CH, Quinn FD (1999) An in vitro tissue culture bilayer model to examine early events in Mycobacterium tuberculosis infection. Infect Immun 67(2):653–658

Bisht D, Meena LS (2019) Adhesion molecules facilitate host-pathogen interaction & mediate *Mycobacterium tuberculosis* pathogenesis. Indian J Med Res 150(1):23–32

Blocker A, Komoriya K, Aizawa S (2003) Type III secretion systems and bacterial flagella: insights into their function from structural similarities. Proc Natl Acad Sci USA 100(6):3027–3030

Bordeau V, Felden B (2014) Curli synthesis and biofilm formation in enteric bacteria are controlled by a dynamic small RNA module made up of a pseudoknot assisted by an RNA chaperone. Nucleic Acids Res 42(7):4682–4696

Brennan PJ, Nikaido H (1995) The envelope of mycobacteria. Annu Rev Biochem 64:29–63

Brennan MJ, Delogu G (2002) The PE multigene family: a 'molecular mantra' for mycobacteria. Trends Microbiol 10(5):246–249

Brennan MJ, Delogu G, Chen Y, Bardarov S, Kriakov J, Alavi M, Jacobs WR Jr (2001) Evidence that mycobacterial PE_PGRS proteins are cell surface constituents that influence interactions with other cells. Infect Immun 69(12):7326–7333

Brightbill HD, Libraty DH, Krutzik SR, Yang RB, Belisle JT, Bleharski JR, Maitland M, Norgard MV, Plevy SE, Smale ST, Brennan PJ, Bloom BR, Godowski PJ, Modlin RL (1999) Host defense mechanisms triggered by microbial lipoproteins through toll-like receptors. Science 285(5428):732–736

Brinton CC Jr (1965) The structure, function, synthesis and genetic control of bacterial pili and a molecular model for DNA and RNA transport in gram negative bacteria. Trans N Y Acad Sci 27(8):1003–1054

Brodin P, Poquet Y, Levillain F, Peguillet I, Larrouy-Maumus G, Gilleron M, Ewann F, Christophe T, Fenistein D, Jang J, Jang MS, Park SJ, Rauzier J, Carralot JP, Shrimpton R, Genovesio A, Gonzalo-Asensio JA, Puzo G, Martin C, Brosch R et al (2010) High content phenotypic cell-based visual screen identifies Mycobacterium tuberculosis acyltrehalose-containing glycolipids involved in phagosome remodeling. PLoS Pathog 6(9):e1001100

Bullitt E, Makowski L (1995) Structural polymorphism of bacterial adhesion pili. Nature 373(6510):164–167

Busch A, Waksman G (2012) Chaperone-usher pathways: diversity and pilus assembly mechanism. Philos Trans R Soc Lond Ser B Biol Sci 367(1592):1112–1122. https://doi.org/10.1098/rstb.2011.0206. PMID: 22411982; PMCID: PMC3297437

Casali N, Konieczny M, Schmidt MA, Riley LW (2002) Invasion activity of a mycobacterium tuberculosis peptide presented by the Escherichia coli AIDA autotransporter. Infect Immun 70(12):6846–6852. https://doi.org/10.1128/IAI.70.12.6846-6852.2002. PMID: 12438361; PMCID: PMC133103

Casali N, White AM, Riley LW (2006) Regulation of the Mycobacterium tuberculosis mce1 operon. J Bacteriol 188(2):441–449. https://doi.org/10.1128/JB.188.2.441-449.2006. PMID: 16385033; PMCID: PMC1347267

Chatterjee D, Khoo KH (1998) Mycobacterial lipoarabinomannan: an extraordinary lipoheteroglycan with profound physiological effects. Glycobiology 8(2):113–120

Chatterjee S, Kundapura SV, Basak AJ, Mukherjee D, Dash S, Ganguli N, Das AK, Mukherjee G, Samanta D, Ramagopal UA (2022) High-resolution crystal structure of LpqH, an immunomodulatory surface lipoprotein of Mycobacterium tuberculosis reveals a distinct fold and a conserved cleft on its surface. Int J Biol Macromol 210:494–503

Chen J, Fruhauf A, Fan C, Ponce J, Ueberheide B, Bhabha G, Ekiert DC (2023) Structure of an endogenous mycobacterial MCE lipid transporter. Nature 620(7973):445–452

Chitale S, Ehrt S, Kawamura I, Fujimura T, Shimono N, Anand N, Lu S, Cohen-Gould L, Riley LW (2001) Recombinant *Mycobacterium tuberculosis* protein associated with mammalian cell entry. Cell Microbiol 3(4):247–254

Ciaramella A, Martino A, Cicconi R, Colizzi V, Fraziano M (2000) Mycobacterial 19-kDa lipoprotein mediates Mycobacterium tuberculosis-induced apoptosis in monocytes/macrophages at early stages of infection. Cell Death Different 7(12):1270–1272

Cohen SB, Gern BH, Delahaye JL, Adams KN, Plumlee CR, Winkler JK, Sherman DR, Gerner MY, Urdahl KB (2018) Alveolar macrophages provide an early Mycobacterium tuberculosis niche and initiate dissemination. Cell Host Microbe 24(3):439–446.e4

Cole ST, Brosch R, Parkhill J, Garnier T, Churcher C, Harris D, Gordon SV, Eiglmeier K, Gas S, Barry CE 3rd, Tekaia F, Badcock K, Basham D, Brown D, Chillingworth T, Connor R, Davies R, Devlin K, Feltwell T, Gentles S et al (1998) Deciphering the biology of Mycobacterium tuberculosis from the complete genome sequence. Nature 393(6685):537–544

Connell I, Agace W, Klemm P, Schembri M, Mărild S, Svanborg C (1996) Type 1 fimbrial expression enhances Escherichia coli virulence for the urinary tract. Proc Natl Acad Sci USA 93(18):9827–9832

Cornelis GR (2006) The type III secretion injectisome. Nat Rev Microbiol 4(11):811–825

Costa TR, Felisberto-Rodrigues C, Meir A, Prevost MS, Redzej A, Trokter M, Waksman G (2015) Secretion systems in gram-negative bacteria: structural and mechanistic insights. Nat Rev Microbiol 13(6):343–359

Costerton JW, Stewart PS, Greenberg EP (1999) Bacterial biofilms: a common cause of persistent infections. Science 284(5418):1318–1322

Craig L, Pique ME, Tainer JA (2004) Type IV pilus structure and bacterial pathogenicity. Nat Rev Microbiol 2(5):363–378

Crandall ED, Matthay MA (2001) Alveolar epithelial transport. Basic science to clinical medicine. Am J Respir Crit Care Med 163(4):1021–1029

Crystal RG, Randell SH, Engelhardt JF, Voynow J, Sunday ME (2008) Airway epithelial cells: current concepts and challenges. Proc Am Thorac Soc 5(7):772–777

da Silva Neto BR, Fátima da Silva J, Mendes-Giannini MJ, Lenzi HL, de Almeida Soares CM, Pereira M (2009) The malate synthase of Paracoccidioides brasiliensis is a linked surface protein that behaves as an anchorless adhesin. BMC Microbiol 9:1–2

Daffé M, Marrakchi H (2019) Unraveling the structure of the mycobacterial envelope. Microbiology Spectrum 7(4)

Delogu G, Pusceddu C, Bua A, Fadda G, Brennan M, Zanetti S (2004) Rv1818c-encoded PE_PGRS protein of mycobacterium tuberculosis is surface exposed and influences bacterial cell structure. Mol Microbiol 52(3):725–733. https://doi.org/10.1111/j.1365-2958.2004.04007.x

D'Souza C, Kishore U, Tsolaki AG (2023) The PE-PPE family of *Mycobacterium tuberculosis*: proteins in disguise. Immunobiology 228(2):152321

De Maio F, Battah B, Palmieri V, Petrone L, Corrente F, Salustri A, Palucci I, Bellesi S, Papi M, Rubino S, Sali M, Goletti D, Sanguinetti M, Manganelli R, De Spirito M, Delogu G (2018) PE_PGRS3 of Mycobacterium tuberculosis is specifically expressed at low phosphate concentration, and its arginine-rich C-terminal domain mediates adhesion and persistence in host tissues when expressed in Mycobacterium smegmatis. Cell Microbiol 20(12):e12952

De Maio F, Berisio R, Manganelli R, Delogu G (2020) PE_PGRS proteins of *Mycobacterium tuberculosis*: A specialized molecular task force at the forefront of host-pathogen interaction. Virulence 11(1):898–915

de Waal AM, Hiemstra PS, Ottenhoff TH, Joosten SA, van der Does AM (2022) Lung epithelial cells interact with immune cells and bacteria to shape the microenvironment in tuberculosis. Thorax 77(4):408–416

Diaz-Silvestre H, Espinosa-Cueto P, Sanchez-Gonzalez A, Esparza-Ceron MA, Pereira-Suarez AL, Bernal-Fernandez G, Espitia C, Mancilla R (2005) The 19-kDa antigen of Mycobacterium tuberculosis is a major adhesin that binds the mannose receptor of THP-1 monocytic cells and promotes phagocytosis of mycobacteria. Microb Pathog 39(3):97–107

Dobos KM, Khoo KH, Swiderek KM, Brennan PJ, Belisle JT (1996) Definition of the full extent of glycosylation of the 45-kilodalton glycoprotein of Mycobacterium tuberculosis. J Bacteriol 178(9):2498–2506

Dodson KW, Jacob-Dubuisson F, Striker RT, Hultgren SJ (1993) Outer-membrane PapC molecular usher discriminately recognizes periplasmic chaperone-pilus subunit complexes. Proc Natl Acad Sci USA 90(8):3670–3674

Druszczynska M, Wawrocki S, Szewczyk R, Rudnicka W (2017) Mycobacteria-derived biomarkers for tuberculosis diagnosis. Indian J Med Res 146(6):700–707

Dumke R, Hausner M, Jacobs E (2011) Role of Mycoplasma pneumoniae glyceraldehyde-3-phosphate dehydrogenase (GAPDH) in mediating interactions with the human extracellular matrix. Microbiology 157(Pt 8):2328–2338

Dutt TS, Choreño-Parra JA (2025) Editorial: tuberculosis and humoral immunity. Front Immunol 16:1562567

Ehrt S, Schnappinger D, Rhee KY (2018) Metabolic principles of persistence and pathogenicity in *Mycobacterium tuberculosis*. Nat Rev Microbiol 16(7):496–507

Epstein EA, Chapman MR (2008) Polymerizing the fibre between bacteria and host cells: the biogenesis of functional amyloid fibres. Cell Microbiol 10(7):1413–1420

Espitia C, Laclette JP, Mondragón-Palomino M, Amador A, Campuzano J, Martens A, Singh M, Cicero R, Zhang Y, Moreno C (1999) The PE-PGRS glycine-rich proteins of Mycobacterium tuberculosis: a new family of fibronectin-binding proteins? Microbiology 145(Pt 12):3487–3495

Esposito C, Marasco D, Delogu G, Pedone E, Berisio R (2011) Heparin-binding hemagglutinin HBHA from Mycobacterium tuberculosis affects actin polymerisation. Biochem Biophys Res Commun 410(2):339–344

Eum SY, Kong JH, Hong MS, Lee YJ, Kim JH, Hwang SH, Cho SN, Via LE, Barry CE 3rd. (2010) Neutrophils are the predominant infected phagocytic cells in the airways of patients with active pulmonary TB. Chest 137(1):122–128

Fällman E, Schedin S, Jass J, Uhlin BE, Axner O (2005) The unfolding of the P pili quaternary structure by stretching is reversible, not plastic. EMBO Rep 6(1):52–56. https://doi.org/10.1038/sj.embor.7400310. PMID: 15592451; PMCID:PMC1299220

Farnia P, Velayati AA (2023) Visualization of pili in extensively and totally drug-resistant *Mycobacterium tuberculosis* using light microscopy. Intl J Mycobacteriol 12(4):369–371

Farnia P, Zhavnerko GK, Farnia P, Poleschuyk NN, Ghanavi J, Velayati AA (2023a) Identification of seven types of pili in *Mycobacterium tuberculosis* using atomic force microscopy. Intl J Mycobacteriol 12(4):478–485

Farnia P, Farnia P, Ghanavi J, Besharati S, Velayati AA (2023b) Comparison of Proline-glutamate-proline-glutamate-polymorphic GC-rich sequences family protein Wag22 (Rv1759c), PE_PGRS31 (Rv1768), PE_PGRS32 (Rv1803), and PE_PGRS33 gene (Rv1818c) in exponential state and under in vitro model of latency in same clinical isolates of Mycobacterium tuberculosis: frameshift mutation in extensively drug-resistant and totally drug-resistant tuberculosis bacilli. Biomed Biotechnol Res J (BBRJ) 7(4):621–632

Favrot L, Grzegorzewicz AE, Lajiness DH, Marvin RK, Boucau J, Isailovic D, Jackson M, Ronning DR (2013) Mechanism of inhibition of Mycobacterium tuberculosis antigen 85 by ebselen. Nat Commun 4:2748

Fenn K, Wong CT, Darbari VC (2020) Mycobacterium tuberculosis uses Mce proteins to interfere with host cell signaling. Front Mol Biosci 8(6):149. https://doi.org/10.3389/fmolb.2019.00149. PMID: 31998747; PMCID: PMC6961568

Fennelly KP, Jones-López EC (2015) Quantity and quality of inhaled dose predicts immunopathology in tuberculosis. Front Immunol 6:313

Fernandez LA, Berenguer J (2000) Secretion and assembly of regular surface structures in gram-negative bacteria. FEMS Microbiol Rev 24:21–44

Fourie KR, Wilson HL (2020) Understanding GroEL and DnaK stress response proteins as antigens for bacterial diseases. Vaccine 8(4):773

Ganbat D, Seehase S, Richter E, Vollmer E, Reiling N, Fellenberg K, Gaede KI, Kugler C, Goldmann T (2016) Mycobacteria infect different cell types in the human lung and cause species dependent cellular changes in infected cells. BMC Pulm Med 16:19

Gani Z, Boradia VM, Kumar A, Patidar A, Talukdar S, Choudhary E, Singh R, Agarwal N, Raje M, Iyengar Raje C (2021) Mycobacterium tuberculosis glyceraldehyde-3-phosphate dehydrogenase plays a dual role-As an adhesin and as a receptor for plasmin(ogen). Cell Microbiol 23(5):e13311

García-Pérez BE, Mondragón-Flores R, Luna-Herrera J (2003) Internalization of Mycobacterium tuberculosis by macropinocytosis in non-phagocytic cells. Microb Pathog 35(2):49–55

Gehring AJ, Rojas RE, Canaday DH, Lakey DL, Harding CV, Boom WH (2003) The Mycobacterium tuberculosis 19-kilodalton lipoprotein inhibits gamma interferon-regulated HLA-DR and fc gamma R1 on human macrophages through toll-like receptor 2. Infect Immun 71(8):4487–4497

Geijtenbeek TB, den Dunnen J, Gringhuis SI (2009) Pathogen recognition by DC-SIGN shapes adaptive immunity. Future Microbiol 4(7):879–890

Gengenbacher M, Kaufmann SH (2012) Mycobacterium tuberculosis: success through dormancy. FEMS Microbiol Rev 36(3):514–532

Glickman MS, Jacobs WR (2001) Microbial pathogenesis of Mycobacterium tuberculosis: dawn of a discipline. Cell 104(4):477–485

Goins CM, Dajnowicz S, Thanna S, Sucheck SJ, Parks JM, Ronning DR (2017) Exploring covalent allosteric inhibition of antigen 85C from Mycobacterium tuberculosis by Ebselen derivatives. ACS Infect Dis 3(5):378–387

Goins CM, Dajnowicz S, Smith MD, Parks JM, Ronning DR (2018) Mycolyltransferase from *Mycobacterium tuberculosis* in covalent complex with tetrahydrolipstatin provides insights into antigen 85 catalysis. J Biol Chem 293(10):3651–3662

Govender VS, Ramsugit S, Pillay M (2014) Mycobacterium tuberculosis adhesins: potential biomarkers as anti-tuberculosis therapeutic and diagnostic targets. Microbiology 160(Pt 9):1821–1831

Hadifar S, Behrouzi A, Fateh A, Khatami S, Rahimi Jamnani F, Siadat SD, Vaziri F (2019) Comparative study of interruption of signaling pathways in lung epithelial cell by two different Mycobacterium tuberculosis lineages. J Cell Physiol 234(4):4739–4753

Hahn E, Wild P, Hermanns U, Sebbel P, Glockshuber R, Häner M, Taschner N, Burkhard P, Aebi U, Müller SA (2002) Exploring the 3D molecular architecture of Escherichia coli type 1 pili. J Mol Biol 323(5):845–857

Hallstrom KN, McCormick BA (2015) Pathogenicity islands: origins, structure, and roles in bacterial pathogenesis. In: Tang Y-W, Sussman M, Liu D, Poxton I, Schwartzman J (eds) Molecular medical microbiology, 2nd edn. Academic Press, pp 303–314

Henderson B (2010) Integrating the cell stress response: a new view of molecular chaperones as immunological and physiological homeostatic regulators. Cell Biochem Funct 28(1):1–14

Henderson B, Nair S, Pallas J, Williams MA (2011) Fibronectin: a multidomain host adhesin targeted by bacterial fibronectin-binding proteins. FEMS Microbiol Rev 35(1):147–200

Hernández-Pando R, Orozcoe H, Sampieri A, Pavón L, Velasquillo C, Larriva-Sahd J, Alcocer JM, Madrid MV (1996) Correlation between the kinetics of Th1, Th2 cells and pathology in a murine model of experimental pulmonary tuberculosis. Immunology 89(1):26–33

Hickey TB, Thorson LM, Speert DP, Daffé M, Stokes RW (2009) Mycobacterium tuberculosis Cpn60.2 and DnaK are located on the bacterial surface, where Cpn60.2 facilitates efficient bacterial association with macrophages. Infect Immun 77(8):3389–3401

Hospenthal MK, Costa TRD, Waksman G (2017) A comprehensive guide to pilus biogenesis in gram-negative bacteria. Nat Rev Microbiol 15(6):365–379. https://doi.org/10.1038/nrmicro.2017.40. PMID: 28496159

Hosseini H, Fooladi AA, Arjomandzadegan M, Emami N, Bornasi H, Fateh A (2014) Genetics study and transmission electron microscopy of pili in susceptible and resistant clinical isolates of Mycobacterium tuberculosis. Asian Pac J Trop Med 7S1:S199–S203

Hougardy JM, Schepers K, Place S, Drowart A, Lechevin V, Verscheure V, Debrie AS, Doherty TM, Van Vooren JP, Locht C, Mascart F (2007) Heparin-binding-hemagglutinin-induced IFN-gamma release as a diagnostic tool for latent tuberculosis. PLoS One 2(10):e926

Hu Y, Henderson B, Lund PA, Tormay P, Ahmed MT, Gurcha SS, Besra GS, Coates AR (2008) A mycobacterium tuberculosis mutant lacking the groEL homologue cpn60. 1 is viable but fails to induce an inflammatory response in animal models of infection. Infect Immun 76(4):1535–1546

Hultgren SJ, Schwan WR, Schaeffer AJ, Duncan JL (1986) Regulation of production of type 1 pili among urinary tract isolates of Escherichia coli. Infect Immun 54(3):613–620

Huygen K (2014) The Immunodominant T-cell epitopes of the Mycolyl-transferases of the antigen 85 complex of M. Tuberculosis. Front Immunol 5:321

Iskandar IW, Nurhasanah A, Hatta M, Hamid F, Handayani I, Chaera U, Yusriyyah AA, Jamaluddin BD, Zaenab S, Hidayah N, Karimah N, Permana AD, Massi MN (2025) Computational drug repurposing for tuberculosis by inhibiting Ag85 complex proteins. Narra J 5(1):e1130. https://doi.org/10.52225/narra.v5i1.1130. PMID: 40352212; PMCID:PMC12059857

Jacobo-Delgado YM, Rodríguez-Carlos A, Serrano CJ, Rivas-Santiago B (2023) *Mycobacterium tuberculosis* cell-wall and antimicrobial peptides: a mission impossible? Front Immunol 14:1194923

Jefferson KK (2004) What drives bacteria to produce a biofilm? FEMS Microbiol Lett 236(2):163–173

Johnston SL, Goldblatt DL, Evans SE, Tuvim MJ, Dickey BF (2021) Airway epithelial innate immunity. Front Physiol 12:749077

Jones CH, Pinkner JS, Roth R, Heuser J, Nicholes AV, Abraham SN, Hultgren SJ (1995) FimH adhesin of type 1 pili is assembled into a fibrillar tip structure in the Enterobacteriaceae. Proc Natl Acad Sci USA 92(6):2081–2085

Joseph S, Yuen A, Singh V, Hmama Z (2017) *Mycobacterium tuberculosis* Cpn60.2 (GroEL2) blocks macrophage apoptosis via interaction with mitochondrial mortalin. Biology Open 6(4):481–488

Kikuchi T, Mizunoe Y, Takade A, Naito S, Yoshida S (2005) Curli fibers are required for development of biofilm architecture in Escherichia coli K-12 and enhance bacterial adherence to human uroepithelial cells. Microbiol Immunol 49(9):875–884

Kinhikar AG, Vargas D, Li H, Mahaffey SB, Hinds L, Belisle JT, Laal S (2006) Mycobacterium tuberculosis malate synthase is a laminin-binding adhesin. Mol Microbiol 60(4):999–1013

Kinhikar AG, Verma I, Chandra D, Singh KK, Weldingh K, Andersen P, Hsu T, Jacobs WR Jr, Laal S (2010) Potential role for ESAT6 in dissemination of mycobacterium tuberculosis via human lung epithelial cells. Mol Microbiol 75(1):92–106. https://doi.org/10.1111/j.1365-2958.2009.06959.x. Epub 2009 Nov 10. PMID: 19906174; PMCID: PMC2846543

Klemm P, Schembri MA (2000) Bacterial adhesins: function and structure. Int J Med Microbiol 290(1):27–35

Kline KA, Fälker S, Dahlberg S, Normark S, Henriques-Normark B (2009) Bacterial adhesins in host-microbe interactions. Cell Host Microbe 5(6):580–592

Kohwiwattanagun J, Kawamura I, Fujimura T, Mitsuyama M (2007) Mycobacterial mammalian cell entry protein 1A (Mce1A)-mediated adherence enhances the chemokine production by A549 alveolar epithelial cells. Microbiol Immunol 51(2):253–261

Kong TH, Coates AR, Butcher PD, Hickman CJ, Shinnick TM (1993) Mycobacterium tuberculosis expresses two chaperonin-60 homologs. Proc Nat Acad Sci 90(7):2608–2612

Koul A, Herget T, Klebl B, Ullrich A (2004) Interplay between mycobacteria and host signalling pathways. Nat Rev Microbiol 2(3):189–202

Kremer L, Maughan WN, Wilson RA, Dover LG, Besra GS (2002) The M. Tuberculosis antigen 85 complex and mycolyltransferase activity. Lett Appl Microbiol 34(4):233–237

Kubori T, Matsushima Y, Nakamura D, Uralil J, Lara-Tejero M, Sukhan A, Galán JE, Aizawa SI (1998) Supramolecular structure of the salmonella typhimurium type III protein secretion system. Science 280(5363):602–605

Kuhn S (2019) The role of type IV pili in Mycobacterium tuberculosis pathogenesis. Doctoral dissertation, University of Cape Town

Kumar A, Kumar R, Boradia VM, Malhotra H, Kumar A, Seth S, Garg P, Karthikeyan S, Raje M, Iyengar Raje C (2024) Stoichiometry of ligand binding and role of C-terminal lysines in Mycobacterium tuberculosis and human GAPDH multifunctionality. FEBS J 291(23):5236–5255

Kuo CJ, Bell H, Hsieh CL, Ptak CP, Chang YF (2012) Novel mycobacteria antigen 85 complex binding motif on fibronectin. J Biol Chem 287(3):1892–1902

Kuo CJ, Gao J, Huang JW, Ko TP, Zhai C, Ma L, Liu W, Dai L, Chang YF, Chen TH, Hu Y, Yu X, Guo RT, Chen CC (2019) Functional and structural investigations of fibronectin-binding protein Apa from *Mycobacterium tuberculosis*. Biochim Biophys Acta Gen Subj 1863(9):1351–1359

Laal S (2012) How does *Mycobacterium tuberculosis* establish infection? J Infect Dis 206(8):1157–1159

Lane MC, Mobley HL (2007) Role of P-fimbrial-mediated adherence in pyelonephritis and persistence of uropathogenic *Escherichia coli* (UPEC) in the mammalian kidney. Kidney Int 72(1):19–25

Lauer P, Rinaudo CD, Soriani M, Margarit I, Maione D, Rosini R, Taddei AR, Mora M, Rappuoli R, Grandi G, Telford JL (2005) Genome analysis reveals pili in group B streptococcus. Science 309(5731):105

Launois P, Niang N'Diaye M, Sarthou JL, Drowart A, Van Vooren JP, Cartel JL, Huygen K (1994) T cell reactivity against antigen 85 but not against the 18- and 65-kD heat shock proteins in the early stages of acquired immunity against Mycobacterium leprae. Clin Exp Immunol 96(1):86–90

Lee YM, DiGiuseppe PA, Silhavy TJ, Hultgren SJ (2004) P pilus assembly motif necessary for activation of the CpxRA pathway by PapE in *Escherichia coli*. J Bacteriol 186(13):4326–4337

Leiva-Juárez MM, Kolls JK, Evans SE (2018) Lung epithelial cells: therapeutically inducible effectors of antimicrobial defense. Mucosal Immunol 11(1):21–34

Lerner TR, Borel S, Gutierrez MG (2015) The innate immune response in human tuberculosis. Cell Microbiol 17(9):1277–1285

Lillington J, Geibel S, Waksman G (2014) Biogenesis and adhesion of type 1 and P pili. Biochim Biophys Acta Gen Subj 1840(9):2783–2793

Liu J, Nikaido H (1999) A mutant of Mycobacterium smegmatis defective in the biosynthesis of mycolic acids accumulates meromycolates. Proc Natl Acad Sci USA 96(7):4011–4016

Liu S, Xie Y, Luo W, Dou Y, Xiong H, Xiao Z, Zhang XL (2020) PE_PGRS31-S100A9 interaction promotes mycobacterial survival in macrophages through the regulation of NF-κB-TNF-α signaling and arachidonic acid metabolism. Front Microbiol 8(11):845

Liu L, Zhai K, Chen Y, Chen X, Wang G, Wu L (2021) Effect and mechanism of *Mycobacterium tuberculosis* lipoprotein LpqH in NLRP3 Inflammasome activation in mouse Ana-1 macrophage. Biomed Res Int 2021:8239135

Locht C, Hougardy JM (2006) Heparin-binding hemagglutinin, from an extrapulmonary dissemination factor to a powerful diagnostic and protective antigen against tuberculosis. Tuberculosis (Edinb) 86(3–4):303–309

Lomino JV, Tripathy A, Redinbo MR (2011) Triggered Mycobacterium tuberculosis heparin-binding hemagglutinin adhesin folding and dimerization. J Bacteriol 193(9):2089–2096

López M, Sly LM, Luu Y, Young D, Cooper H, Reiner NE (2003) The 19-kDa Mycobacterium tuberculosis protein induces macrophage apoptosis through toll-like receptor-2. J Immunol 170(5):2409–2416

Magombedze G, Dowdy D, Mulder N (2013) Latent tuberculosis: models, computational efforts and the pathogen's regulatory mechanisms during dormancy. Front Bioeng Biotechnol 1:4

Maity D, Singh D, Bandhu A (2023) Mce1R of Mycobacterium tuberculosis prefers long-chain fatty acids as specific ligands: a computational study. Mol Divers 27(6):2523–2543

Malik Z, Iyer S, Kusner D (2001) *Mycobacterium tuberculosis* phagosomes exhibit altered calmodulin-dependent signal transduction: contribution to inhibition of phagosome-lysosome fusion and intracellular survival in human macrophages. J Immunol 166(5):3392–3401

Malhotra H, Patidar A, Boradia VM, Kumar R, Nimbalkar RD, Kumar A, Gani Z, Kaur R, Garg P, Raje M, Raje CI (2017) Mycobacterium tuberculosis glyceraldehyde-3-phosphate dehydrogenase functions as a receptor for human lactoferrin. Front Cell Infect Microbiol 7:245. https://doi.org/10.3389/fcimb.2017.00245. PMID: 28642848; PMCID:PMC5462994

Mandlik A, Swierczynski A, Das A, Ton-That H (2008) Pili in gram-positive bacteria: assembly, involvement in colonization and biofilm development. Trends Microbiol 16(1):33–40

Mann KM, Pride AC, Flentie K, Kimmey JM, Weiss LA, Stallings CL (2016) Analysis of the contribution of MTP and the predicted Flp pilus genes to Mycobacterium tuberculosis pathogenesis. Microbiology 162(10):1784–1796

Massey RC, Kantzanou MN, Fowler T, Day NP, Schofield K, Wann ER, Berendt AR, Höök M, Peacock SJ (2001) Fibronectin-binding protein A of Staphylococcus aureus has multiple, substituting, binding regions that mediate adherence to fibronectin and invasion of endothelial cells. Cell Microbiol 3(12):839–851

Matsuyama M, Martins AJ, Shallom S, Kamenyeva O, Kashyap A, Sampaio EP, Kabat J, Olivier KN, Zelazny AM, Tsang JS, Holland SM (2018) Transcriptional response of respiratory epithelium to nontuberculous mycobacteria. Am J Respir Cell Mol Biol 58(2):241–252

Mayer-Barber KD, Barber DL (2015) Innate and adaptive cellular immune responses to mycobacterium tuberculosis infection. Cold Spring Harb Perspect Med 5(12):a018424

McKinney JD, Hönerzu Bentrup K, Muñoz-Elías EJ, Miczak A, Chen B, Chan WT, Swenson D, Sacchettini JC, Jacobs WR, Russell DG (2000) Persistence of Mycobacterium tuberculosis in macrophages and mice requires the glyoxylate shunt enzyme isocitrate lyase. Nature 406(6797):735–738

Meena LS, Rajni. (2010) Survival mechanisms of pathogenic mycobacterium tuberculosis H37Rv. FEBS J 277(11):2416–2427. https://doi.org/10.1111/j.1742-4658.2010.07666.x

Meena PR, Monu, Meena LS (2016) Fibronectin binding protein and Ca^{2+} play an access key role to mediate pathogenesis in Mycobacterium tuberculosis: an overview. Biotechnol Appl Biochem 63(6):820–826

Mehta PK, Karls RK, White EH, Ades EW, Quinn FD (2006) Entry and intracellular replication of Mycobacterium tuberculosis in cultured human microvascular endothelial cells. Microb Pathog 41(2–3):119–124

Menozzi FD, Rouse JH, Alavi M, Laude-Sharp M, Muller J, Bischoff R, Brennan MJ, Locht C (1996)

Identification of a heparin-binding hemagglutinin present in mycobacteria. J Exp Med 184(3):993–1001
Menozzi FD, Reddy VM, Cayet D, Raze D, Debrie AS, Dehouck MP, Cecchelli R, Locht C (2006) *Mycobacterium tuberculosis* heparin-binding haemagglutinin adhesin (HBHA) triggers receptor-mediated transcytosis without altering the integrity of tight junctions. Microbes Infect 8(1):1–9
Miller E, Garcia T, Hultgren S, Oberhauser AF (2006) The mechanical properties of E. Coli type 1 pili measured by atomic force microscopy techniques. Biophys J 91(10):3848–3856
Mogensen TH (2009) Pathogen recognition and inflammatory signaling in innate immune defenses. Clin Microbiol Rev 22(2):240–273
Monfeli RR, Beeson C (2007) Targeting iron acquisition by Mycobacterium tuberculosis. Infect Disord Drug Targets 7(3):213–220
Mulvey MA, Schilling JD, Hultgren SJ (2001) Establishment of a persistent Escherichia coli reservoir during the acute phase of a bladder infection. Infect Immun 69(7):4572–4579
Naidoo N, Ramsugit S, Pillay M (2014) Mycobacterium tuberculosis pili (MTP), a putative biomarker for a tuberculosis diagnostic test. Tuberculosis (Edinb) 94(3):338–345
Naito M, Fukuda T, Sekiguchi K, Yamada T (2000) The domains of human fibronectin mediating the binding of alpha antigen, the most immunopotent antigen of mycobacteria that induces protective immunity against mycobacterial infection. Biochem J 347(Pt 3):725–731
Nakata M, Sumitomo T, Patenge N, Kreikemeyer B, Kawabata S (2020) Thermosensitive pilus production by FCT type 3 streptococcus pyogenes controlled by Nra regulator translational efficiency. Mol Microbiol 113(1):173–189
Nandakumar S, Kannanganat S, Dobos KM, Lucas M, Spencer JS, Fang S, McDonald MA, Pohl J, Birkness K, Chamcha V, Ramirez MV, Plikaytis BB, Posey JE, Amara RR, Sable SB (2013) O-mannosylation of the Mycobacterium tuberculosis adhesin Apa is crucial for T cell antigenicity during infection but is expendable for protection. PLoS Pathog 9(10):e1003705
Naser SA, Thanigachalam S, Spinelli N, Safavi MM, Naser N, Khan O (2013) The 19 kDa protein from Mycobacterium avium subspecies paratuberculosis is a glycolipoprotein. Adv Microbiol 3(7):520–528
Neufert C, Pai RK, Noss EH, Berger M, Boom WH, Harding CV (2001) Mycobacterium tuberculosis 19-kDa lipoprotein promotes neutrophil activation. J Immunol 167(3):1542–1549
Nishiyama M, Horst R, Eidam O, Herrmann T, Ignatov O, Vetsch M, Bettendorff P, Jelesarov I, Grütter MG, Wüthrich K, Glockshuber R (2005) Structural basis of chaperone--subunit complex recognition by the type 1 pilus assembly platform FimD. EMBO J 24(12):2075–2086
Noss EH, Pai RK, Sellati TJ, Radolf JD, Belisle J, Golenbock DT, Boom WH, Harding CV (2001) Toll-like receptor 2-dependent inhibition of macrophage class II MHC expression and antigen processing by 19-kDa lipoprotein of Mycobacterium tuberculosis. J Immunol 167(2):910–918
Notti RQ, Stebbins CE (2016) The structure and function of type III secretion systems. Microbiol Spectrum 4(1)
Nouailles G, Dorhoi A, Koch M, Zerrahn J, Weiner J 3rd, Faé KC, Arrey F, Kuhlmann S, Bandermann S, Loewe D, Mollenkopf HJ, Vogelzang A, Meyer-Schwesinger C, Mittrücker HW, McEwen G, Kaufmann SH (2014) CXCL5-secreting pulmonary epithelial cells drive destructive neutrophilic inflammation in tuberculosis. J Clin Invest 124(3):1268–1282
Nyawo GR (2016) The role of Mycobacterium tuberculosis pili (MTP) in the pathogenesis of tuberculosis. Doctoral dissertation, University of KwaZulu-Natal
Nykvist P, Pettersson RF (2003) Uropathogenic Escherichia coli express type 1 fimbriae only in surface adherent populations under physiological growth conditions. J Infect Dis 187(1):32–41
Ocampo M, Curtidor H, Vanegas M, Patarroyo MA, Patarroyo ME (2014) Specific interaction between Mycobacterium tuberculosis lipoprotein-derived peptides and target cells inhibits mycobacterial entry in vitro. Chem Biol Drug Des 84(6):626–641
Oliveira CB, Balan A (2020) The ATP-binding cassette (ABC) transport systems in mycobacterium tuberculosis: structure, function, and possible targets for therapeutics. Biology (Basel) 9(12):443. https://doi.org/10.3390/biology9120443. PMID: 33322523; PMCID: PMC7765882
Olsén A, Jonsson A, Normark S (1989) Fibronectin binding mediated by a novel class of surface organelles on Escherichia coli. Nature 338(6217):652–655
Palucci I, Camassa S, Cascioferro A, Sali M, Anoosheh S, Zumbo A, Minerva M, Iantomasi R, De Maio F, Di Sante G, Ria F, Sanguinetti M, Palù G, Brennan MJ, Manganelli R, Delogu G (2016) PE_PGRS33 contributes to Mycobacterium tuberculosis entry in macrophages through interaction with TLR2. PLoS One 11(3):e0150800
Pethe K, Alonso S, Biet F, Delogu G, Brennan MJ, Locht C, Menozzi FD (2001) The heparin-binding haemagglutinin of M. Tuberculosis is required for extrapulmonary dissemination. Nature 412(6843):190–194. https://doi.org/10.1038/35084083. PMID: 11449276
Piplani B, Kumar CMS, Lund PA, Chaudhuri TK (2023) Mycobacterial chaperonins in cellular proteostasis: evidence for chaperone function of Cpn60.1 and Cpn60.2-mediated protein folding. Mol Microbiol
Pisu D, Huang L, Grenier JK, Russell DG (2020) Dual RNA-Seq of Mtb-infected macrophages in vivo reveals ontologically distinct host-pathogen interactions. Cell Rep 30(2):335–350.e4 https://doi.org/10.1016/j.celrep.2019.12.033. PMID: 31940480; PMCID: PMC7032562
Pizarro-Cerdá J, Cossart P (2006) Bacterial adhesion and entry into host cells. Cell 124(4):715–727

Proft T, Baker EN (2009) Pili in gram-negative and gram-positive bacteria – structure, assembly and their role in disease. Cell Mol Life Sci 66(4):613–635

Puckett S, Trujillo C, Wang Z, Eoh H, Ioerger TR, Krieger I, Sacchettini J, Schnappinger D, Rhee KY, Ehrt S (2017) Glyoxylate detoxification is an essential function of malate synthase required for carbon assimilation in *Mycobacterium tuberculosis*. Proc Natl Acad Sci USA 114(11):E2225–E2232

Qamra R, Srinivas V, Mande SC (2004) Mycobacterium tuberculosis GroEL homologues unusually exist as lower oligomers and retain the ability to suppress aggregation of substrate proteins. J Mol Biol 342(2):605–617. https://doi.org/10.1016/j.jmb.2004.07.066

Qian J, Chen R, Wang H, Zhang X (2020) Role of the PE/PPE family in host–pathogen interactions and prospects for anti-tuberculosis vaccine and diagnostic tool design. Front Cell Infect Microbiol 10:594288

Ragas A, Roussel L, Puzo G, Rivière M (2007) The Mycobacterium tuberculosis cell-surface glycoprotein apa as a potential adhesin to colonize target cells via the innate immune system pulmonary C-type lectin surfactant protein A. J Biol Chem 282(8):5133–5142

Rakotoarivonina H, Jubelin G, Hebraud M, Gaillard-Martinie B, Forano E, Mosoni P (2002) Adhesion to cellulose of the gram-positive bacterium Ruminococcus albus involves type IV pili. Microbiology 148(Pt 6):1871–1880

Rakotoarivonina H, Larson MA, Morrison M, Girardeau JP, Gaillard-Martinie B, Forano E, Mosoni P (2005) The Ruminococcus albus pilA1-pilA2 locus: expression and putative role of two adjacent *pil* genes in pilus formation and bacterial adhesion to cellulose. Microbiology 151(Pt 4):1291–1299

Ramirez MS, Bonomo RA, Tolmasky ME (2020) Carbapenemases: transforming Acinetobacter baumannii into a yet more dangerous menace. Biomolecules 10(5):720. https://doi.org/10.3390/biom10050720. PMID: 32380691; PMCID:PMC7291396

Ramon-Luing LA, Palacios Y, Ruiz A, Téllez-Navarrete NA, Chavez-Galan L (2023) Virulence factors of mycobacterium tuberculosis as modulators of cell death mechanisms. Pathogens 12(6):839. https://doi.org/10.3390/pathogens12060839 PMID: 37375529; PMCID: PMC10304248

Ramsugit S, Pillay M (2014) Mycobacterium tuberculosis pili promote adhesion to and invasion of THP-1 macrophages. Jpn J Infect Dis 67(6):476–478. https://doi.org/10.7883/yoken.67.476

Ramsugit S, Guma S, Pillay B, Jain P, Larsen MH, Danaviah S, Pillay M (2013) Pili contribute to biofilm formation in vitro in Mycobacterium tuberculosis. Antonie Van Leeuwenhoek 104(5):725–735

Ramsugit S, Pillay B, Pillay M (2016) Evaluation of the role of mycobacterium tuberculosis pili (MTP) as an adhesin, invasin, and cytokine inducer of epithelial cells. Braz J Infect Dis 20(2):160–165. https://doi.org/10.1016/j.bjid.2015.11.002. Epub 2015 Dec 31. PMID: 26748229; PMCID: PMC9427661

Randall PJ, Hsu NJ, Quesniaux V, Ryffel B, Jacobs M (2015) *Mycobacterium tuberculosis* infection of the 'non-classical immune cell'. Immunol Cell Biol 93(9):789–795

Ratliff TL, McGARR JA, Abou-zeid C, Rook GA, Stanford JL, Aslanzadeh J, Brown EJ (1988) Attachment of mycobacteria to fibronectin-coated surfaces. Microbiology 134(5):1307–1313

Raze D, Verwaerde C, Deloison G, Werkmeister E, Coupin B, Loyens M, Brodin P, Rouanet C, Locht C (2018) Heparin-binding hemagglutinin Adhesin (HBHA) is involved in Intracytosolic lipid inclusions formation in mycobacteria. Front Microbiol 9:2258

Reuschl AK, Edwards MR, Parker R, Connell DW, Hoang L, Halliday A, Jarvis H, Siddiqui N, Wright C, Bremang S, Newton SM, Beverley P, Shattock RJ, Kon OM, Lalvani A (2017) Innate activation of human primary epithelial cells broadens the host response to Mycobacterium tuberculosis in the airways. PLoS Pathog 13(9):e1006577

Rodgers K, Arvidson CG, Melville S (2011) Expression of a Clostridium perfringens type IV pilin by Neisseria gonorrhoeae mediates adherence to muscle cells. Infect Immun 79(8):3096–3105

Roine E, Wei W, Yuan J, Nurmiaho-Lassila EL, Kalkkinen N, Romantschuk M, He SY (1997) Hrp pilus: an hrp-dependent bacterial surface appendage produced by pseudomonas syringae pv. Tomato DC3000. Proc Natl Acad Sci USA 94(7):3459–3464

Ronning DR, Vissa VD, Besra GS, Belisle JT, Sacchettini JC (2000) Structure of the cell wall mycolyl transferase from *Mycobacterium tuberculosis*, Ag85C, and its complex with trehalose 6-sulfate, a product analog. Nat Struct Biol 7(2):141–146

Ryndak MB, Laal S (2019) *Mycobacterium tuberculosis* primary infection and dissemination: a critical role for alveolar epithelial cells. Front Cell Infect Microbiol 9:299

Ryndak MB, Singh KK, Peng Z, Laal S (2015) Transcriptional profile of *Mycobacterium tuberculosis* replicating in type II alveolar epithelial cells. PLoS One 10(4):e0123745

Sai Priya VH, Latha GS, Hasnain SE, Murthy KJ, Valluri VL (2010) Enhanced T cell responsiveness to Mycobacterium bovis BCG r32-kDa ag correlates with successful anti-tuberculosis treatment in humans. Cytokine 52(3):190–193

Saini G, Porte J, Weinreb PH, Violette SM, Wallace WA, McKeever TM, Jenkins G (2015) αvβ6 integrin may be a potential prognostic biomarker in interstitial lung disease. Eur Respir J 46(2):486–494

Samanich K, Belisle JT, Laal S (2001) Homogeneity of antibody responses in tuberculosis patients. Infect Immun 69(7):4600–4609

Sampson SL (2011) Mycobacterial PE/PPE proteins at the host-pathogen interface. Clin Dev Immunol 2011:497203

Schilling JD, Mulvey MA, Hultgren SJ (2001) Structure and function of Escherichia coli type 1 pili: new insight into the pathogenesis of urinary tract infections. J Infect Dis 183(Suppl 1):S36–S40. https://doi.org/10.1086/318855. PMID:11171012

Schmidt AK, Fitzpatrick AD, Schwartzkopf CM, Faith DR, Jennings LK, Coluccio A, Hunt DJ, Michaels LA, Hargil A, Chen Q, Bollyky PL (2022) A filamentous bacteriophage protein inhibits type IV pili to prevent superinfection of Pseudomonas aeruginosa. MBio 13(1):e02441–e02421

Schneeberger EE, Lynch RD (1992) Structure, function, and regulation of cellular tight junctions. Am J Physiol Lung Cell Mol Physiol 262(6):L647–L661

Schwartz DJ, Kalas V, Pinkner JS, Chen SL, Spaulding CN, Dodson KW, Hultgren SJ (2013) Positively selected FimH residues enhance virulence during urinary tract infection by altering FimH conformation. Proc Natl Acad Sci USA 110(39):15530–15537

Scott JR, Zähner D (2006) Pili with strong attachments: gram-positive bacteria do it differently. Mol Microbiol 62(2):320–330

Seidler NW (2013) Basic biology of GAPDH. In: Seidler NW (ed) GAPDH: biological properties and diversity. Springer, pp 1–36

Shahar A, Melamed-Frank M, Kashi Y, Shimon L, Adir N (2011) The dimeric structure of the Cpn60.2 chaperonin of *Mycobacterium tuberculosis* at 2.8 Å reveals possible modes of function. J Mol Biol 412(2):192–203

Sharma M, Bose M, Abhimanyu S, L., Diwakar, A., Kumar, S., Gaur, S. N., & Banavalikar, J. N. (2012) Intracellular survival of Mycobacterium tuberculosis in macrophages is modulated by phenotype of the pathogen and immune status of the host. Intl J Mycobacteriol 1(2):65–74

Shimono N, Morici L, Casali N, Cantrell S, Sidders B, Ehrt S, Riley LW (2003) Hypervirulent mutant of Mycobacterium tuberculosis resulting from disruption of the mce1 operon. Proc Natl Acad Sci USA 100(26):15918–15923

Shu Q, Crick SL, Pinkner JS, Ford B, Hultgren SJ, Frieden C (2012) The E. Coli CsgB nucleator of curli assembles to β-sheet oligomers that alter the CsgA fibrillization mechanism. Proc Natl Acad Sci USA 109(17):6502–6507

Singh G, Kumar A, Maan P, Kaur J (2017a) Cell wall associated factors of Mycobacterium tuberculosis as major virulence determinants: current perspectives in drugs discovery and design. Curr Drug Targets 18(16):1904–1918

Singh KS, Sharma R, Keshari D, Singh N, Singh SK (2017b) Down-regulation of malate synthase in Mycobacterium tuberculosis H37Ra leads to reduced stress tolerance, persistence and survival in macrophages. Tuberculosis (Edinb) 106:73–81

Singh R, Sharma D, Sharma D, Gautam S, Gupta MK, Bisht D (2022) Unraveling major proteins of Mycobacterium tuberculosis envelope. Current Proteomics 19(5):372–379

Sirover MA (2014) Structural analysis of glyceraldehyde-3-phosphate dehydrogenase functional diversity. Int J Biochem Cell Biol 57:20–26

Sjöbring U, Pohl G, Olsén A (1994) Plasminogen, absorbed by Escherichia coli expressing curli or by salmonella enteritidis expressing thin aggregative fimbriae, can be activated by simultaneously captured tissue-type plasminogen activator (t-PA). Mol Microbiol 14(3):443–452

Sleutel M, Pradhan B, Volkov AN, Remaut H (2023) Structural analysis and architectural principles of the bacterial amyloid curli. Nat Commun 14(1):2822

Smith I (2003) Mycobacterium tuberculosis pathogenesis and molecular determinants of virulence. Clin Microbiol Rev 16(3):463–496

Smith CV, Huang CC, Miczak A, Russell DG, Sacchettini JC, Höner zu Bentrup, K. (2003) Biochemical and structural studies of malate synthase from Mycobacterium tuberculosis. J Biol Chem 278(3):1735–1743

Sohn H, Kim JS, Shin SJ, Kim K, Won CJ, Kim WS, Min KN, Choi HG, Lee JC, Park JK, Kim HJ (2011) Targeting of Mycobacterium tuberculosis heparin-binding hemagglutinin to mitochondria in macrophages. PLoS Pathog 7(12):e1002435

Solomon JM, Leung GS, Isberg RR (2003) Intracellular replication of Mycobacterium marinum within Dictyostelium discoideum: efficient replication in the absence of host coronin. Infect Immun 71(6):3578–3586. https://doi.org/10.1128/IAI.71.6.3578-3586.2003. PMID: 12761143; PMCID: PMC155778

Soto GE, Hultgren SJ (1999) Bacterial adhesins: common themes and variations in architecture and assembly. J Bacteriol 181(4):1059–1071

Spaulding CN, Klein RD, Ruer S, Kau AL, Schreiber HL, Cusumano ZT, Dodson KW, Pinkner JS, Fremont DH, Janetka JW, Remaut H, Gordon JI, Hultgren SJ (2017) Selective depletion of uropathogenic E. coli from the gut by a FimH antagonist. Nature 546(7659):528–532

Squeglia F, Ruggiero A, De Simone A, Berisio R (2018) A structural overview of mycobacterial adhesins: key biomarkers for diagnostics and therapeutics. Protein Sci 27(2):369–380. https://doi.org/10.1002/pro.3346. Epub 2017 Dec 27. PMID: 29139177; PMCID: PMC5775178

Świerzko AS, Bartłomiejczyk MA, Brzostek A, Łukasiewicz J, Michalski M, Dziadek J, Cedzyński M (2016) Mycobacterial antigen 85 complex (Ag85) as a target for ficolins and mannose-binding lectin. Int J Med Microbiol 306(4):212–221. https://doi.org/10.1016/j.ijmm.2016.04.004

Tailleux L, Schwartz O, Herrmann JL, Pivert E, Jackson M, Amara A, Legres L, Dreher D, Nicod LP, Gluckman JC, Lagrange PH, Gicquel B, Neyrolles O (2003) DC-SIGN is the major Mycobacterium tuberculosis receptor on human dendritic cells. J Exp Med 197(1):121–127

Tajima H, Tsuchiya K, Yamada M, Kondo K, Katsube N, Ishitani R (1999) Over-expression of GAPDH induces apoptosis in COS-7 cells transfected with cloned GAPDH cDNAs. Neuroreport 10(10):2029–2033

Tekedar HC, Patel F, Blom J, Griffin MJ, Waldbieser GC, Kumru S, Abdelhamed H, Dharan V, Hanson LA, Lawrence ML (2024) Tad pili contribute to the virulence and biofilm formation of virulent *Aeromonas hydrophila*. Front Cell Infect Microbiol 14:1425624

Telford JL, Barocchi MA, Margarit I, Rappuoli R, Grandi G (2006) Pili in gram-positive pathogens. Nat Rev Microbiol 4(7):509–519

Temmerman S, Pethe K, Parra M, Alonso S, Rouanet C, Pickett T, Drowart A, Debrie AS, Delogu G, Menozzi FD, Sergheraert C, Brennan MJ, Mascart F, Locht C (2004) Methylation-dependent T cell immunity to Mycobacterium tuberculosis heparin-binding hemagglutinin. Nat Med 10(9):935–941

Thacker VV, Dhar N, Sharma K, Barrile R, Karalis K, McKinney JD (2020) A lung-on-chip model of early *Mycobacterium tuberculosis* infection reveals an essential role for alveolar epithelial cells in controlling bacterial growth. elife 9:e59961

Theorupun SC, Massi MN, Nurhasanah A, Syamsuri F, Hatta M, Rasita YD et al (2025) PPE18 and PepA variations in Mycobacterium tuberculosis clinical isolates from Makassar, Indonesia: challenges for immune recognition and vaccine development. Intl J Mycobacteriol 14(2):191–200

Ton-That H, Schneewind O (2004) Assembly of pili in gram-positive bacteria. Trends Microbiol 12(5):228–234

Ton-That H, Marraffini LA, Schneewind O (2004) Pilus assembly in corynebacteria: sortases and pilin elements involved in pilus assembly of *Corynebacterium diphtheriae*. Mol Microbiol 52(3):251–261

Torrelles JB, Chatterjee D (2023) Collected thoughts on mycobacterial Lipoarabinomannan, a cell envelope Lipoglycan. Pathogens 12(11):1281

Tristan C, Shahani N, Sedlak TW, Sawa A (2011) The diverse functions of GAPDH: views from different subcellular compartments. Cell Signal 23(2):317–323

Van Gerven N, Klein RD, Hultgren SJ, Remaut H (2015) Bacterial amyloid formation: structural insights into curli biogenesis. Trends Microbiol 23(11):693–706

Varga JJ, Nguyen V, O'Brien DK, Rodgers K, Walker RA, Melville SB (2006) Type IV pili-dependent gliding motility in the gram-positive pathogen Clostridium perfringens and other clostridia. Mol Microbiol 62(3):680–694

Velayati AA, Farnia P (2016) Atlas of mycobacterium tuberculosis. Academic Press (Elsevier), London. ISBN: 9780128038086

Velayati AA, Farnia P (2024) Pili in mycobacterium tuberculosis: structure, function, and therapeutic advances, 1st ed. Academic. Press (Elsevier), London. ISBN: 9780443217128 (print); 9780443217135 (eBook)

Velayati AA, Abeel T, Shea T, Zhavnerko GK, Birren B, Cassell GH, Earl AM, Hoffner S, Farnia P (2016) Populations of latent mycobacterium tuberculosis lack a cell wall: isolation, visualization, and whole-genome characterization. Int J Mycobacteriol 5(1):66–73. https://doi.org/10.1016/j.ijmyco.2015.12.001

Velayati AA, Farnia P, Masjedi MR, Ibrahim T, Tabarsi P, Haroun R, Kuan H, Ghanavi J, Varahram M (2009) Totally drug-resistant tuberculosis strains: evidence of adaptation at the cellular level. Eur Respir J 34(5):1202–1203

Velayati AA, Farnia P, Masjedi MR (2012) Pili in totally drug resistant *Mycobacterium tuberculosis* (TDR-TB). Intl J Mycobacter 1(2):57–58

Viljoen A, Dufrêne YF, Nigou J (2022) Mycobacterial adhesion: from hydrophobic to receptor-ligand interactions. Microorganisms 10(2):454

Vinod V, Vijayrajratnam S, Vasudevan AK, Biswas R (2020) The cell surface adhesins of Mycobacterium tuberculosis. Microbiol Res 232:126392

Wang L, Slayden RA, Barry CE 3rd, Liu J (2000) Cell wall structure of a mutant of Mycobacterium smegmatis defective in the biosynthesis of mycolic acids. J Biol Chem 275(10):7224–7229

Wang J, Li Y, Pan L, Li J, Yu Y, Liu B, Zubair M, Wei Y, Pillay B, Olaniran AO, Chiliza TE, Shao G, Feng Z, Xiong Q (2021) Glyceraldehyde-3-phosphate dehydrogenase (GAPDH) moonlights as an adhesin in Mycoplasma hyorhinis adhesion to epithelial cells as well as a plasminogen receptor mediating extracellular matrix degradation. Vet Res 52(1):80

Watanabe M, Aoyagi Y, Ridell M, Minnikin DE (2001) Separation and characterization of individual mycolic acids in representative mycobacteria. Microbiology 147(Pt 7):1825–1837

Wojtas B, Fijalkowska B, Wlodarczyk A, Schollenberger A, Niemialtowski M, Hamasur B, Pawlowski A, Krzyzowska M (2011) Mannosylated lipoarabinomannan balances apoptosis and inflammatory state in mycobacteria-infected and uninfected bystander macrophages. Microb Pathog 51(1–2):9–21

Wright KJ, Seed PC, Hultgren SJ (2007) Development of intracellular bacterial communities of uropathogenic *Escherichia coli* depends on type 1 pili. Cell Microbiol 9(9):2230–2241

Yan S, Zhen J, Li Y, Zhang C, Stojkoska A, Lambert N, Li Q, Li P, Xie J (2019) Mce-associated protein Rv0177 alters the cell wall structure of *Mycobacterium smegmatis* and promotes macrophage apoptosis via regulating the cytokines. Int Immunopharmacol 66:205–214

Yan Z, Yin M, Chen J, Li X (2020) Assembly and substrate recognition of curli biogenesis system. Nat Commun 11:241

Yang H, Ma Y, Lei X, Chai S, Zhang S, Su G, Li S, Du L (2025) Stopping tuberculosis at the gate: the role of *Mycobacterium tuberculosis* adhesins in infection and intervention. Vaccines (Basel) 13(7):676

Yuan CH, Zhang S, Xiang F, Gong H, Wang Q, Chen Y, Luo W (2019) Secreted Rv1768 from RD14 of *Mycobacterium tuberculosis* activates macrophages and induces a strong IFN-γ-releasing of CD4+ T cells. Front Cell Infect Microbiol 9:341

Zhang R, Zhao J, Han G, Liu Z, Liu C, Zhang C, Liu B, Jiang C, Liu R, Zhao T, Han MY (2016) Real-time discrimination and versatile profiling of spontaneous reactive oxygen species in living organisms with a single fluorescent probe. J Am Chem Soc 138(11):3769–3778

Zhang W, Mo S, Liu M, Liu L, Yu L, Wang C (2020) Rationally designed protein building blocks for programmable hierarchical architectures. Front Chem 8:587975. https://doi.org/10.3389/fchem.2020.587975. PMID: 33195088;PMCID: PMC7658299

Zhang Z, Dong L, Li X, Deng T, Wang Q (2025) The PE/PPE family proteins of *Mycobacterium tuberculosis*: evolution, function, and prospects for tuberculosis control. Front Immunol 16:1606311

Zhao D, Lin D, Xu C (2020) A protein fragment of Rv3194c located on mycobacterial cell surface efficiently prevents adhesion of recombinant Mycobacterium smegmatis, and promises a new anti-adhesive drug. Microb Pathog 149:104498

Zhavnerko G, Poleschuyk NN (2012) Mycobacterium under AFM tip: Advantages of polyelectrolyte modified substrate. Int J Mycobacteriol 1(1):53-56. https://doi.org/10.1016/j.ijmyco.2012.01.009. Epub 2012 Feb 2. PMID: 26786951

Zhavnerko G, Farnia P, Poleschuk N (2017) Brief reports on the use of atomic force microscopy in visualization of mycobacterium tuberculosis. Biomed Biotechnol Res J 1:9. https://doi.org/10.4103/bbrj.bbrj_21_17

Zhou KL, Li X, Zhang XL, Pan Q (2019) Mycobacterial mannose-capped lipoarabinomannan: a modulator bridging innate and adaptive immunity. Emerg Microbes Infect 8(1):1168–1177

Zimmermann N, Saiga H, Houthuys E, Moura-Alves P, Koehler A, Bandermann S, Dorhoi A, Kaufmann SH (2016) Syndecans promote mycobacterial internalization by lung epithelial cells. Cell Microbiol 18(12):1846–1856

Zumla A, Nahid P, Cole ST (2013) Advances in the development of new tuberculosis drugs and treatment regimens. Nat Rev Drug Discov 12(5):388–404

Key Proteins and Enzymatic Mechanisms Underlying Drug Resistance in *Mycobacterium tuberculosis*

9

Abstract

Drug resistance in *Mycobacterium tuberculosis* (Mtb) is a complex phenomenon resulting from multiple coordinated molecular and genetic mechanisms that allow the bacterium to evade antimicrobial agents. Intrinsic resistance primarily stems from the unique architecture of the mycobacterial cell envelope, characterized by a dense, lipid-rich barrier that substantially limits antibiotic penetration and reduces intracellular drug concentrations. Additionally, enzymatic degradation mechanisms further diminish drug efficacy by inactivating antibiotics extracellularly. Acquired resistance is predominantly driven by spontaneous chromosomal mutations that modify drug targets or disrupt prodrug activation pathways, thereby reducing antibiotic binding affinity and therapeutic activity while preserving essential bacterial functions. Upregulation of drug targets and drug-modifying enzymes through genetic alterations leads to elevated thresholds for effective inhibition. Active efflux systems also play a critical role by transporting antibiotics out of the cell, lowering intracellular drug levels below therapeutic thresholds, and contributing to multidrug resistance. Collectively, these interrelated mechanisms establish a formidable defense network that complicates tuberculosis treatment and underscores the pressing need for novel therapeutic approaches to circumvent resistance and improve clinical outcomes.

Keywords

Mycobacterium tuberculosis · Drug-resistant TB · Acquired resistance · Prodrug activation · Drug-inactivating enzymes · Efflux pump

9.1 Introduction

Mycobacterium tuberculosis (Mtb), the causative agent of tuberculosis (TB), exhibits intrinsic resistance to many commonly used antibiotics and chemotherapeutic agents, primarily due to its distinctive cell wall architecture and metabolic adaptations. The treatment of TB generally necessitates a combination of three to four drugs, as only a limited number of antimycobacterial agents are efficacious against these pathogens. Despite these multidrug regimens, TB continues to impose substantial global health burdens marked by high infection rates, morbidity, and mortality. This problem is further compounded by the rising prevalence of drug-resistant strains, including monodrug-resistant TB, multidrug-resistant TB (MDR-TB), pre-extensively drug-resistant TB (pre-XDR), extensively drug-resistant TB (XDR-TB), and totally drug-resistant TB (TDR-TB), which present increasing therapeutic challenges and treatment failures

P. Farnia et al., *Proteins in Mycobacterium Tuberculosis*, Advances in Experimental Medicine and Biology 1484, https://doi.org/10.1007/978-3-031-96883-9_9

(Velayati et al. 2013a, b; Velayati and Farnia 2016; Seung et al. 2015; Joob and Wiwanitkit 2019; Conradie et al. 2020; Cobongela and Mgwenya 2024; Johnson et al. 2024; Matteelli et al. 2025). Drug resistance in Mtb is mediated through a variety of interrelated molecular mechanisms. These include genetic mutations that alter antibiotic targets or enzymes required for prodrug activation, enzymatic inactivation of drugs, active efflux of antimicrobials, and reduced cell wall permeability (Cocozaki et al. 2016; Ramasubban et al. 2024; Brammachary et al. 2025). For instance, mutations in the *rpoB* gene encoding the RNA polymerase β subunit reduce rifampicin binding and confer high-level resistance (Telenti et al. 1993; Campbell et al. 2001; Xu et al. 2022; Alifano et al. 2015; Yadav 2024). Similarly, mutations in *katG* prevent activation of the prodrug isoniazid, while alterations in inhA, ahpC, and fabG1 modify enoyl-ACP reductase and related enzymes, resulting in resistance to isoniazid and ethionamide (Vilchèze and Jacobs 2014; Vilcheze et al. 2018). Resistance to ethambutol arises from mutations in *embB* and *embC*, which affect arabinosyl transferases essential for cell wall biosynthesis (Goude et al. 2009). Mutations in *gyrA* and *gyrB* disrupt DNA gyrase activity, conferring fluoroquinolone resistance (Kanematsu et al. 1998; Lee et al. 2005). Aminoglycoside and capreomycin resistance have been linked to mutations in *rrs* (16S rRNA) and *rpsL* (ribosomal protein S12) genes (Alangaden et al. 1998; Springer et al. 2001). The *pncA* gene encoding pyrazinamidase is vital for pyrazinamide activation, and its mutation leads to drug resistance (Sreevatsan et al. 1997). Similarly, changes in *ethA* impair ethionamide activation, while mutations in *folC* and *thyA* confer resistance to p-aminosalicylic acid (Cuevas-Córdoba et al. 2013; Aono et al. 2014; Dookie et al. 2018). Resistance to newer anti-TB agents such as delamanid and pretomanid is associated with mutations in the *ddn* gene, and mutations in *tlyA* confer capreomycin resistance (Maus et al. 2005). Further resistance to cycloserine involves mutations in *alr*, and alterations in rplC and *rrl* contribute to linezolid resistance. Enzymatic drug inactivation also significantly contributes to resistance. β-lactamase enzymes hydrolyze β-lactam antibiotics, rendering them inactive, whereas the Eis acetyltransferase modifies aminoglycosides (e.g., kanamycin), reducing their antimicrobial efficacy (Labby et al. 2013; Labby and Garneau-Tsodikova 2013; Willby et al. 2016; Sanz-García et al. 2019). Loss-of-function mutations in *pncA* hinder pyrazinamide activation, further complicating treatment. Efflux pumps actively expel a broad range of antibiotics from the bacterial cell, lowering intracellular drug concentrations and contributing to multidrug resistance. Notably, members of the MmpL family, including MmpL5, mediate efflux of drugs such as bedaquiline and clofazimine, and their overexpression, along with regulatory proteins like Eis, amplifies resistance phenotypes (Van Bambeke et al. 2000; Martin et al. 2020; Remm et al. 2022). The mycobacterial cell wall, enriched with mycolic acids and complex lipids, acts as an effective physical barrier to many drugs. The antigen 85 complex (Ag85A, Ag85B, Ag85C), responsible for transferring mycolic acids, plays a crucial role in maintaining cell wall integrity, thereby contributing to intrinsic resistance (Favrot and Ronning 2012, Favrot et al. 2014). Other enzymes involved in cell wall biosynthesis, such as *KasA*, *AhpC*, *FabG1*, *EmbB*, and *EmbC*, influence susceptibility, while porin proteins like MspA and channel proteins such as CpnT regulate permeability and drug uptake (Slayden and Barry 3rd 2002; Brammachary et al. 2025; Gessese et al. 2025). Recent studies have identified additional proteins implicated in drug resistance and bacterial persistence. Rv0148, a putative short-chain dehydrogenase/reductase, is overexpressed in aminoglycoside-resistant strains and may contribute to antibiotic neutralization (Sharma et al. 2015, 2018). Upregulation of ABC transporter proteins, including Rv0933, Rv1273c, and Rv1819c, in resistant strains suggests roles in drug efflux (Målen et al. 2011). Membrane-associated proteins such as MmpL4, Rv1269c, Rv3137, and SseA demonstrate altered expression in virulent Beijing lineage strains, potentially elevating mutation rates and accelerating resistance development (de Keijzer et al. 2017). Additionally, components of the respiratory chain, including the cytochrome bc_1:aa_3 complex and F_1F_0 ATP synthase, are essential for bacterial energetics and survival. These complexes are targeted by novel drugs like Q203 and bedaquiline, which inhibit ATP synthesis and exhibit potent activity against drug-resistant strains (Ofori-Anyinam et al. 2020; Verma et al. 2025). Currently, over 20 promising antimycobacterial agents, both new chemical entities and derivatives targeting novel pathways, are in clinical development or have recently received approval. Notable examples include bedaquiline, which inhibits ATP synthase, and oxazolidinones such as sutezolid and delpazolid, which disrupt protein synthesis. Although these developments are encouraging, the emergence of resistance-associated variants targeting new drugs jeopardizes their long-term effectiveness. Therefore, combination therapies incorporating higher doses of rifamycins and fluoroquinolones (e.g., moxifloxacin) are under investigation to optimize treatment outcomes. Equally critical is the innovation of drug delivery systems capable of enhancing drug stability, targeted intracellular uptake, controlled release, biocompatibility, and affordability, particularly in resource-poor settings (Hoagland et al. 2016; Angula et al. 2021; Dhameliya et al. 2022; Gairola et al. 2022; Nair et al. 2023; Kufa et al. 2025; Omoteso et al. 2025). In summary, the drug resistance in Mtb resulted from a complex, interconnected network involving classical genetic mutations, enzymatic drug modification, efflux mechanisms, and the formidable barrier presented by the mycobacterial cell wall. Addressing this global challenge necessitates continued in-depth research into these molecular mechanisms, coupled with the development of innovative therapeutic agents and drug delivery platforms to improve clinical management and control of drug-resistant tuberculosis.

9.2 Molecular Mechanisms and Protein Classes in *Mycobacterium tuberculosis*

Drug resistance in Mtb arises through diverse molecular mechanisms, each involving distinct protein classes that enable the bacterium to evade the effects of antimicrobial agents. These mechanisms are broadly categorized into six groups, reflecting differences in their biochemical pathways and genetic underpinnings. Understanding these pathways is critical for addressing the global challenge of drug-resistant tuberculosis. One primary mechanism involves target alteration, where mutations in genes encoding drug targets prevent antibiotic binding. For example, mutations in the *rpoB* gene, which encodes the β-subunit of RNA polymerase, disrupt rifampicin binding, leading to resistance (Donnabella et al. 1994). Similarly, mutations in *inhA* (enoyl-ACP reductase) and *katG* (catalase-peroxidase) reduce isoniazid efficacy, while alterations in *gyrA* and *gyrB* (DNA gyrase subunits) confer fluoroquinolone resistance (Wade and Zhang 2004). Ribosomal targets are also affected: mutations in 16S rRNA (*rrs*) or ribosomal protein S12 (*rpsL*) cause resistance to aminoglycosides, and tlyA mutations impair capreomycin binding by altering rRNA methylation. These mutations often cluster in specific regions, such as the rifampicin resistance-determining region (RRDR) of rpoB, enabling targeted molecular diagnostics (Springer et al. 2001). A second mechanism, abolition of prodrug activation, involves mutations in enzymes required to convert inactive prodrugs into their active forms. For instance, *katG* mutations prevent the activation of isoniazid, while *pncA* mutations disrupt pyrazinamide activation by pyrazinamidase. Similarly, *ethA* mutations hinder ethionamide activation, and mutations in *ddn* (nitroreductase) or F420 biosynthesis genes (fgd1, fbiA/B/C) block the activation of nitroimidazoles like delamanid. These loss-of-function mutations highlight the reliance of certain drugs on bacterial metabolism for efficacy (Metcalfe et al. 2008). Enzymatic drug inactivation, while less prevalent in Mtb compared to other pathogens, involves the chemical modification of antibiotics that reduces their efficacy. For example, overexpression of the *eis* gene, which encodes an aminoglycoside acetyltransferase, leads to the acetylation and subsequent inactivation of kanamycin. Similarly, although β-lactamases are relatively uncommon in Mtb due to its intrinsic resistance mechanisms, they are capable of hydrolyzing β-lactam antibiotics.

These enzymes exemplify how bacterial proteins can directly neutralize antimicrobial agents, although enzymatic drug inactivation generally plays a secondary role in the development of tuberculosis drug resistance (Chen et al. 2012; Luthra et al. 2018). Efflux pumps constitute a crucial mechanism of drug resistance by actively transporting antibiotics out of the bacterial cell, thereby reducing intracellular drug concentrations. The major families of efflux pumps in Mtb include the Major Facilitator Superfamily (MFS), exemplified by LfrA and P55 pumps that expel fluoroquinolones and tetracyclines; the Small Multidrug Resistance (SMR) family, such as the Mmr pump that exports acriflavine; and the Resistance-Nodulation-Division (RND) family, including MmpL5, which is associated with clofazimine resistance. Additionally, ATP-Binding Cassette (ABC) transporters, like PstB, also contribute to the efflux of fluoroquinolones. The expression of these pumps is frequently upregulated in response to antibiotic exposure, facilitating the development of multidrug-resistant phenotypes (Szumowski et al. 2013). Cell wall impermeability acts as a physical barrier, limiting drug penetration. The mycobacterial cell wall, rich in mycolic acids, arabinogalactan, and trehalose dimycolate, is synthesized by enzymes encoded by the ag85 complex. Mutations in these pathways or reduced porin expression further restrict drug uptake, contributing to intrinsic resistance against many antibiotics, including rifampicin. This impermeability underscores the unique structural defenses of Mtb (Favrot and Ronning 2012; Sachan et al. 2023). Recent research has uncovered novel proteins and pathways that extend beyond the classical mechanisms of drug resistance in Mtb. One such protein, Rv0148, which is overexpressed in aminoglycoside-resistant strains, contains a short-chain dehydrogenase/reductase domain and is thought to bind or chemically modify aminoglycosides, thereby reducing their efficacy (Sharma et al. 2015, 2018; Sharma and Bisht 2017). Additionally, stress response proteins involved in maintaining redox balance, such as AhpC, and those responsible for DNA repair, including RecA, enhance bacterial survival under drug-induced stress. Regulatory modifications also influence the expression of efflux pumps, further contributing to resistance. Although less well understood, epigenetic adaptations may also play a role in modulating the expression of resistance-associated genes. These resistance strategies differ fundamentally in their mechanisms of action. Target alteration and prodrug inactivation directly interfere with the interaction between drugs and their targets, whereas efflux pumps and cell wall impermeability reduce intracellular drug concentrations by limiting access. Enzymatic drug inactivation chemically neutralizes antibiotics before they exert their effects, while emerging mechanisms, such as those involving Rv0148, suggest novel modes of antibiotic sequestration or modification. (Lata et al. 2015; Catalão et al. 2019). The proteins implicated in drug resistance encompass a wide range of functional classes, enzymes, transporters, structural components, and regulatory factors, highlighting the remarkable evolutionary adaptability and resourcefulness of Mtb.

9.3 Intrinsic Resistance via Cell Wall Impermeability

The intrinsic drug resistance of Mtb is primarily due to its uniquely complex and highly impermeable cell envelope, which acts as an effective barrier against many antibiotics. This cell envelope consists of multiple layers, including a plasma membrane, a thick cell wall core, and an outer lipid-rich layer, each contributing distinctive biochemical and structural characteristics that together confer robust resistance (Brennan 2003; Jankute et al. 2015; Poulton and Rock 2022; Jacobo-Delgado et al. 2023; Schami 2024). At the core of the Mtb cell envelope lies the peptidoglycan-arabinogalactan-mycolic acid complex (mAGP), a large macromolecular assembly essential for bacterial viability and the impermeability of the cell wall. The peptidoglycan (PG) layer is composed of linear glycan strands made up of alternating $\beta(1 \rightarrow 4)$-linked N-acetylglucosamine and N-acetylmuramic acid residues, cross-linked by peptide chains that provide mechanical strength and maintain cell shape (Brennan and Nikaido 1995; Alderwick et al. 2015). This PG is covalently attached through a linker unit (L-Rha-D-GlcNAc-P) to arabinogalactan (AG), a linear galactofuran polymer further branched with arabinofuranose residues (McNeil 1999; Daffe et al. 1990; Besra et al. 1995; Mishra et al. 2011; Jankute et al. 2012). AG serves as the scaffold for mycolic acids (MAs), which are very long-chain α-alkyl, β-hydroxy fatty acids oriented perpendicular to the membrane plane (Alderwick et al. 2015). This mAGP complex creates a dense, highly cross-linked matrix that acts as a physical barrier against hydrophilic antibiotics, substantially limiting their penetration (Minnikin et al. 1984). The mycolic acids form a thick, waxy, and hydrophobic outer membrane, known as the mycomembrane, which is unique to mycobacteria (Jarlier and Nikaido 1994). This lipid-rich layer is interspersed with free lipids and complex glycolipids including trehalose dimycolate (TDM, or cord factor), trehalose monomycolate, phosphatidylinositol mannosides (PIMs), lipomannan (LM), lipoarabinomannan (LAM), sulfolipids, and phthiocerol dimycocerosates (PDIM) (Chiaradia et al. 2017; Daffé and Marrakchi 2019; Daffé et al. 2019). These lipids significantly contribute to the hydrophobicity of the cell wall, diminishing permeability to numerous antibiotics. Furthermore, they play pivotal roles in immune evasion and virulence by modulating host immune responses (Vincent and Nyongesa 2018). The biosynthesis and maintenance of this complex cell wall depend on a multitude of specialized enzymes and structural proteins. Foremost among these is the antigen 85 complex (Ag85A, Ag85B, Ag85C), encoded by the *ag85* gene family. This complex comprises mycolyltransferases that catalyze the transfer of mycolic acids onto arabinogalactan and trehalose, synthesizing TDM and other mycolylated glycolipids essential for preserving the integrity and impermeability of the mycomembrane (Belisle et al. 1997; Ronning et al. 2004; Elamin et al. 2009; Goins et al. 2018). Disruption of *ag85* genes sensitizes Mtb to frontline drugs, highlighting their crucial role in intrinsic resistance.

β-Lactamases such as *BlaC* are extracellular enzymes capable of hydrolyzing the β-lactam ring of β-lactam antibiotics, rendering these drugs inactive prior to reaching their targets (Chambers et al. 1995; Wang et al. 2006; Hugonnet et al. 2009; Wivagg et al. 2014; Lehmann et al. 2018; Warrier et al. 2016; Mishra et al. 2017). BlaC is a broad-spectrum class A β-lactamase with a serine active site that efficiently degrades penicillins and cephalosporins, thereby contributing significantly to intrinsic resistance against these antibiotics (Hugonnet et al. 2009; Wivagg et al. 2014; Tooke et al. 2019; Jacobs et al. 2024).

CpnT, a channel protein containing a necrosis-inducing toxin domain, forms pores in the outer membrane that facilitate nutrient uptake and potentially influence drug susceptibility by modulating membrane permeability (Danilchanka et al. 2014). Structurally, CpnT integrates a pore-forming domain with a toxin domain, thereby linking nutrient transport and virulence. Lipoproteins such as LppX and LprG (Rv1411c) mediate the transport of complex lipids like PDIM and triacylglycerols across the cell envelope, contributing to lipid homeostasis and cell wall integrity (Sulzenbacher et al. 2006; Drage et al. 2010). These lipoproteins facilitate dynamic remodeling of the cell wall in response to environmental stresses and antibiotic exposure. Additionally, peptidoglycan-modifying enzymes, including transglycosylases and transpeptidases, preserve the structural scaffold of the cell wall; modifications in PG cross-linking can alter cell wall permeability and affect drug resistance (Kieser et al. 2015; Chiang and West 2024). Overall, the architecture of the Mtb cell wall is characterized by a multilayered, highly cross-linked matrix possessing both hydrophobic and hydrophilic domains. The covalent linkage of peptidoglycan to arabinogalactan and subsequently to mycolic acids produces a rigid and impermeable barrier. The outer lipid layer, enriched with mycolic acids and glycolipids, imparts extreme hydrophobicity, markedly reducing the penetration of hydrophilic antibiotics. Embedded proteins not only synthesize and transport these complex lipids but also regulate permeability and interact with host immune factors. Together, these structural components and associated proteins form a dynamic and resilient barrier that limits antibiotic access, enzymatically degrades or modifies drugs, and underlies the intrinsic drug resistance phenotype of Mtb (Nasirzadeh et al. 2022; Jacobo-Delgado et al. 2023; Liu et al. 2025). Understanding the detailed structure-function relationships of these proteins and their roles in cell wall assembly and maintenance is critical for the design of novel therapeutic strategies targeting the mycobacterial cell envelope.

9.4 Acquired Resistance Through Drug Target Alteration

Extensive experimental and clinical evidence demonstrates that resistance primarily arises from mutations in specific genes encoding drug targets or enzymes responsible for prodrug activation. These genetic alterations reduce drug binding or efficacy, thereby enabling the bacteria to survive antibiotic treatment. Studies have identified approximately 25 key genes responsible for resistance to specific antibiotics, with mutations in these genes accounting for 36–95% of observed resistance cases, depending on the drug. However, a significant proportion, ranging from 5% to 64%, of resistant isolates lacks classical mutations, indicating the presence of alternative or additional mechanisms yet to be fully elucidated (Zhang and Yew 2015; Sharma and Bisht 2017).

9.4.1 RNA Polymerase β-Subunit (RpoB) and RNA Polymerase Binding Protein A (RbpA)

RNA polymerase β-subunit (RpoB) and RNA polymerase binding protein A (RbpA) are critical contributors to Mtb resistance against rifampicin, a cornerstone antituberculosis drug (Heep et al. 2001; Borukhov and Nudler 2003; Herrera et al. 2003; Campbell et al. 2001; Floss and Yu 2005; Sidiq et al. 2019). A comprehensive understanding of their structural features, functional roles, and resistance-associated mutations is essential to effectively tackle rifampicin-resistant tuberculosis. RpoB encodes the β-subunit of the DNA-dependent RNA polymerase (RNAP), a multisubunit enzyme complex vital for bacterial transcription. The RNAP holoenzyme consists of five subunits (α_2ββ′ω), with RpoB (β subunit) and RpoC (β′ subunit) forming the catalytic core responsible for nucleotide polymerization, a process coordinated by Mg^{2+} ions (Miller et al. 1994; Sosunov et al. 2005; Stephanie et al. 2022). The *rpoB* gene spans approximately 3534 base pairs, encoding a protein of 1178 amino acids (Miller et al. 1994). During transcription initiation, the holoenzyme binds to promoter DNA outside the enzyme's active site channel, where the σ subunit mediates sequence-specific contacts and directs the template strand into the catalytic cleft (Murakami et al. 2002). The rifampicin binding pocket within RpoB is formed by conserved amino acid residues that stabilize the drug through hydrogen bonds, hydrophobic interactions, and van der Waals forces, thereby effectively inhibiting transcription elongation (Mustaev et al. 1994, 1997; Zhang et al. 2012; Campbell et al. 2001; Zenkin and Yuzenkova 2015; Molodtsov et al. 2017; Amusengeri et al. 2022; Khan et al. 2025). Rifampicin resistance predominantly arises from mutations clustered within an 81-base pair rifampicin resistance-determining region (RRDR), spanning codons 426–452 (Mtb numbering). These mutations alter residues that either directly contact rifampicin or maintain the structural integrity of its binding pocket, thus reducing drug affinity and therapeutic efficacy. The most prevalent mutation, Serine to Leucine at codon 531 (S531L), accounts for approximately 69% of rifampicin-resistant clinical isolates. This substitution introduces a bulky leucine side chain that causes steric hindrance within the binding pocket (Campbell et al. 2011; Tan et al. 2017; Jamieson et al. 2014; Jagielski et al. 2018; Li et al. 2022; Brandis and Hughes 2013; Molodtsov et al. 2017; Su et al. 2022). While S531L does not significantly alter RNAP conformation in the drug's absence, rifampicin binding is disrupted, resulting in disorder at the rifampicin interface and a reduction of approximately 40% in contact surface area (Lahiri et al. 2016; Molodtsov et al. 2017). This disruption diminishes van der Waals interactions, decreases binding free energy, and substantially lowers rifampicin affinity, thereby increasing the minimum inhibitory concentration (MIC) needed to inhibit bacterial growth (Muthaiah et al. 2017). The mutation further displaces structural elements such as fork loop 2, undermining drug binding and allowing transcription to proceed despite rifampicin's presence. Another common mutation, histidine to tyrosine at codon 526 (H526Y), induces more pronounced conformational changes within the rifampicin binding pocket. This substitution creates steric clashes that prevent rifampicin binding by disrupting critical hydrogen bonds and altering the electrostatic environment (Telenti et al. 1993; Campbell et al. 2001; Kumar and Jena 2014). Both S531L and H526Y mutations impart high-level rifampicin resistance, often correlated with MIC values exceeding 32 μg/mL. Other mutations, such as Aspartate to Valine at codon 516 (D516V), do not substantially change the overall structure but alter the electrostatic surface of the binding pocket, thereby reducing rifampicin affinity through modified charge interactions (Jamieson et al. 2014). Additional mutation-prone codons within the RRDR, such as 441, 451, and 456, critically influence rifampicin binding by disrupting electrostatic interactions and communication between key protein regions (Singh et al. 2017; Amusengeri et al. 2022). Molecular dynamics simulations and binding free energy calculations reveal that mutations like Asp441Val, Ser456Gln, and His451Pro markedly alter rifampicin binding dynamics (Amusengeri et al. 2022). These mechanistic insights are invaluable for the rational design of novel therapeutics and combination regimens aimed at overcoming rifampicin resistance (Vummidi 2023). Beyond antibiotic resistance alone, rpoB mutations may exert pleiotropic effects on Mtb physiology and pathogenesis (Koch et al. 2014). Complementing RpoB's role, RNA polymerase binding protein A (RbpA) is an essential transcriptional activator unique to actinobacteria, including Mtb (Bortoluzzi et al. 2013; Tabib-Salazar et al. 2013). RbpA directly associates with the RNAP core enzyme, predominantly interacting with the β subunit (Hu et al. 2012; Hubin et al. 2015), and selectively binds principal sigma factors σ^A and σ^B, excluding alternative sigma factors (Bortoluzzi et al. 2013; Tabib-Salazar et al. 2013). Structurally, RbpA consists of a well-defined central core flanked by flexible N- and C-terminal regions, with the C-terminal domain containing a sigma-interacting region critical for sigma factor engagement (Tabib-Salazar et al. 2013; Hubin et al. 2015). Moreover, RbpA may interact with promoter DNA during transcription initiation, stabilizing the transcription complex (Hubin et al. 2015). Functionally, RbpA stabilizes the σ^A-containing RNAP holoenzyme, promoting formation of the open promoter complex and driving transcription of housekeeping genes (Hu et al. 2012; Bortoluzzi et al. 2013). The basic linker and sigma-interacting domains suffice for transcriptional activation, while other regions refine regulatory specificity (Hubin et al. 2015). Structural studies reveal that RbpA engages both conserved and unique regions of σ^A, explaining its selective affinity for housekeeping sigma factors (Hubin et al. 2015). Although RbpA does not bind rifampicin directly, it indirectly influences rifampicin susceptibility by stabilizing the transcription complex and modulating RNAP conformation (Hu et al. 2012; Bortoluzzi et al. 2013; Stephanie et al. 2022). Moreover, RbpA can mitigate destabilizing effects caused by rpoB mutations, partially restoring transcriptional activity in rifampicin-resistant strains, thus impacting bacterial fitness and persistence during therapy (Newell et al. 2006; Zhu et al. 2018; Jensen et al. 2019).

In summary, mutations within the RRDR of RpoB alter the rifampicin binding pocket, reducing drug affinity and conferring resistance, while RbpA modulates RNAP stability and transcriptional activity under these conditions.

9.4.2 Enoyl-ACP Reductase (InhA)

Enoyl-acyl carrier protein (ACP) reductase (InhA) is a pivotal enzyme in Mtb, catalyzing a crucial step in the biosynthesis of mycolic acids, essential components of the mycobacterial cell wall. As a member of the short-chain dehydrogenase/reductase (SDR) family, InhA is specifically characterized as a tyrosine-dependent NADH-dependent enoyl-ACP reductase (Zhang et al. 1992; Heym et al. 1993; Rozwarski et al. 1998; Chollet et al. 2015). The *inhA* gene encodes a homotetrameric enzyme, with each subunit exhibiting a Rossmann fold domain that is responsible for NADH binding and catalytic function. Structurally, InhA adopts an α/β fold, featuring a central β-sheet flanked by α-helices (Rozwarski et al. 1998). Its active site encompasses a hydrophobic pocket designed to accommodate fatty acyl substrates in a U-shaped conformation adjacent to the nicotinamide ring of NADH (Rozwarski et al. 1998). Integral residues lining this pocket include Tyr158, which stabilizes the substrate through hydrogen bonding, alongside other hydrophobic amino acids (Rozwarski et al. 1998; Parikh et al. 1999). The enzyme's catalytic mechanism involves covalent adduct formation between NADH and the substrate (Vögeli et al. 2018), with Tyr158 playing a critical role in protonation stereospecificity and Thr196 participating in both hydride transfer and protonation events. Additionally, Lys165 is central to NADH cofactor binding (Parikh et al. 1999). Inhibitors targeting InhA commonly form ternary complexes with the enzyme and NAD cofactor, exhibiting slow-onset inhibition associated with ordering of the substrate-binding loop (Pan and Tonge 2012). The catalytic process culminates in hydride transfer from NADH to the substrate, followed by protonation, ultimately producing long-chain meromycolic acids indispensable for maintaining cell wall integrity.

InhA constitutes the principal target of the first-line antituberculosis drug isoniazid (INH) and the second-line agent ethionamide (ETH). Both INH and ETH function as prodrugs requiring enzymatic activation: INH by KatG catalase-peroxidase, and ETH by EthA monooxygenase. KatG is a heme-containing bifunctional enzyme with catalase and peroxidase activities (Banerjee et al. 1994; Morlock et al. 2003; Vilchèze and Jacobs 2014; Chollet et al. 2015; Sikandar and Xing 2025). Although KatG shares structural similarities primarily with peroxidases rather than catalases, it exhibits robust catalase activity due to the presence of a unique Met-Tyr-Trp (MYW) covalent adduct, which is essential for its catalase function (Njuma et al. 2014; Aziz and Goodwin 2021). This MYW cofactor forms autocatalytically through a compound I-mediated mechanism (Ghiladi et al. 2005). It undergoes redox cycling between radical and fully covalent states, enabling KatG to harness heme intermediates for catalytic oxygen production (Njuma et al. 2014). The catalytic reaction mechanism involves deprotonation of the active-site tryptophan residue, which functionally substitutes for the distal histidine found in monofunctional catalases (Loewen et al. 2014). A conformationally dynamic arginine residue, termed the "arginine switch" (R418), located approximately 20 Å from the heme center, modulates the delicate balance between catalase and peroxidase activities in a pH-dependent manner (Loewen et al. 2014). Collectively, these distinctive structural features enable KatG to perform catalase activity rivaling canonical monofunctional catalases (Liu 2024). Despite extensive study, the precise mechanistic roles of these structural elements in KatG's bifunctional enzymatic activity remain incompletely resolved (Hegde et al. 2021; Xu et al. 2024). KatG's bifunctional activity not only confers protection against oxidative stress but also enzymatically activates INH, converting it into reactive species such as the isonicotinoyl radical. This radical subsequently forms an INH-NAD adduct that potently inhibits InhA, disrupting mycolic acid synthesis and compromising cell wall integrity (Bertrand et al. 2004; Cade et al. 2010). Resistance to INH frequently arises through mutations in katG that diminish catalase-peroxidase activity, thereby impairing prodrug activation and enabling bacterial survival under antibiotic pressure. The most prevalent of these mutations is the substitution of serine to threonine at codon 315 (S315T), identified in approximately 61.9–87% of INH-resistant clinical isolates worldwide (Pym et al. 2002; Unissa et al. 2015; Rahim et al. 2012). This mutation introduces a bulkier threonine side chain proximal to the heme-containing active site, sterically hindering INH access and reducing the generation of reactive species crucial for antimicrobial activity. Importantly, the S315T mutation selectively attenuates catalase-peroxidase activity while preserving sufficient catalase function to mitigate oxidative stress, thus maintaining bacterial fitness despite conferring moderate- to high-level INH resistance (Wei et al. 2003; Mo et al. 2004; Marney et al. 2018; Rossini and Dias 2023). Other katG mutations, such as W341R and L398P, destabilize the tertiary structure, disrupt the heme pocket, or interfere with MYW cofactor formation, further impairing enzymatic function (Ghiladi et al. 2005). Conversely, mutations like R146P have minimal impact on drug susceptibility, underscoring the heterogeneity in functional consequences among katG variants (Hsu et al. 2020). Deletions or frameshifts causing partial or complete loss of katG expression abolish catalase-peroxidase activity, resulting in high-level INH resistance but often imposing a fitness cost due to compromised oxidative defense. Such mutations trigger conformational alterations, e.g., increased flexibility near the active site and disruption of key covalent bonds such as between heme and His270, that destabilize electron transfer, impair catalysis, and prevent INH activation (Pym et al. 2002; Mo et al. 2004; Ando et al. 2014; Hsu et al. 2020; Shukla et al. 2024).

Mutations in the *inhA* gene and its promoter region account for approximately 10–22% of INH resistance cases worldwide (Nono et al. 2024). *inhA* encodes the enoyl-ACP reductase enzyme, a key member of the fatty acid synthase II (FAS-II) system integral to mycolic acid biosynthesis (Sacco et al. 2007; Hopf et al. 2022). Structurally, InhA exists as a homotetramer with each monomer presenting a Rossmann fold facilitating NADH and enoyl substrate binding. The INH-NAD adduct formed upon KatG activation binds tightly within InhA's active site, inhibiting reductase activity and thereby blocking mycolic acid synthesis (Banerjee et al. 1998; Rawat et al. 2003; Galina et al. 2022). Resistance frequently arises from mutations in the *inhA* promoter, notably a cytosine-to-thymine substitution at position −15 (−15 C > T), which induces overexpression of *InhA*. Present in about 16.9–22% of INH-resistant isolates, this mutation results in elevated enzyme levels that sequester the inhibitory INH-NAD adduct, conferring low-level resistance (Vilchèze and Jacobs 2014; Valafar. 2021; Dokrungkoon et al. 2023; Nono et al. 2024). Though less common, coding region mutations in inhA alter amino acid residues within the active or substrate-binding site, attenuating drug binding affinity and allowing fatty acid synthesis to proceed despite drug presence (Vilchèze and Jacobs 2014, 2019).

Mutations in *ahpC* and its regulatory regions have also been implicated in INH resistance, primarily as compensatory mechanisms. *ahpC* encodes alkyl hydroperoxide reductase subunit C, a peroxiredoxin that assembles into a dodecameric ring stabilized by an extended N-terminal region. Conserved catalytic cysteine residues facilitate the reduction in peroxides, protecting Mtb from oxidative damage (Sherman et al. 1996; Rinder et al. 1998; Guimarães et al. 2005; Oh and Jeon 2014; Nieto et al. 2016; Hsu et al. 2020). Upregulation or mutations of *ahpC* can compensate for diminished KatG catalase activity, enhancing bacterial survival despite impaired INH activation, though *ahpC* mutations alone do not confer resistance (Rinder et al. 1998; Bakhtiyariniya et al. 2022). Another gene associated with INH resistance is *ndh*, encoding NADH dehydrogenase II, a flavoprotein involved in the respiratory electron transport chain that catalyzes NADH oxidation. Mutations in *ndh* increase the intracellular NADH/NAD^+ ratio, thus raising NADH levels that competitively inhibit INH-NAD adduct formation by competing for binding to *InhA*. Though less frequently reported, *ndh* mutations contribute indirectly to INH resistance by modulating cellular redox homeostasis and reducing drug efficacy (Miesel et al. 1998; Lee et al. 2001; Cardoso et al. 2007; Dalla Costa et al. 2009; Goossens et al. 2020). Collectively, mutations in *katG*, *inhA* (both promoter and coding regions), *ahpC*, and *ndh* constitute a complex and multifactorial resistance network. While *katG* mutations mainly impair prodrug activation by structurally destabilizing the catalase-peroxidase enzyme, *inhA* mutations reduce drug binding or increase target abundance. *ahpC* mutations compensate for oxidative stress defense loss, and *ndh* mutations alter redox balance, often coexisting and potentially acting synergistically to complicate therapeutic outcomes (Zhang and Yew 2015; Khan et al. 2023).

9.4.3 Arabinosyltransferase (EmbB, EmbC)

Arabinosyltransferases EmbB and EmbC are integral membrane glycosyltransferases essential for Mtb cell wall biosynthesis, specifically catalyzing the construction of the arabinan domains of arabinogalactan (AG) and lipoarabinomannan (LAM), respectively (Goude et al. 2009; Alderwick et al. 2018; Liu et al. 2025). Both enzymes belong to the GT-C family of glycosyltransferases, distinguished by multiple transmembrane helices and conserved catalytic motifs, sharing approximately 42% amino acid sequence identity that reflects their evolutionary and functional relatedness. EmbB, a critical arabinosyltransferase in mycobacterial cell wall assembly, has been deeply characterized through cryo-electron microscopy (cryo-EM) studies of its *Mycobacterium smegmatis* homologue. The protein contains 15 transmembrane helices arranged in a GT-C fold, accompanied by two periplasmic carbohydrate-binding modules (CBMs) that form a β-sheet structure (Tan et al. 2020; Zhang et al. 2020). The active site comprises a large cavity located at the membrane-periplasm interface, lined by conserved charged residues such as Asp285, Asp286, Arg389, and Glu391, which are critical for catalytic function (Zhang et al. 2020). This cavity accommodates both the donor substrate, decaprenyl-phosphate-arabinose (DPA), and the acceptor arabinan chain, enabling the transfer of arabinose residues necessary for arabinogalactan polymerization (Das et al. 2020; Zhang et al. 2020; Tan et al. 2020). Adjacent to the active site, a loop between transmembrane helices 13 and 14 regulates access to the acceptor substrate; mutations in this region—most notably at Met306—are frequently implicated in ethambutol resistance. Such mutations alter substrate specificity or drug binding while preserving catalytic activity, thereby reducing the drug's inhibitory effect. Structural analyses also reveal tightly bound phosphatidylglycerol lipids and calcium ions that contribute to EmbB's stabilization within the membrane environment. EmbC primarily catalyzes the polymerization of arabinose during lipoarabinomannan biosynthesis (Korkegian et al. 2014). This enzyme comprises a transmembrane N-terminal domain and a hydrophilic C-terminal domain vital for catalytic activity (Shi et al. 2006). The C-terminal domain contains two subdomains, with subdomain II functioning as a carbohydrate-binding module (CBM) (Alderwick et al. 2011). EmbC recognizes and binds arabinofuranoside acceptors through conserved tryptophan residues present within these CBMs, which are essential for substrate binding (Alderwick et al. 2011). The protein harbors two functional motifs: a GT-C motif containing critical aspartate residues and a proline-rich region necessary for arabinan synthesis (Berg et al. 2005; Korkegian et al. 2014). EmbC's activity extends arabinan chains in LAM, particularly adding linear termini to the branched Ara-(18–22) core structure (Shi et al. 2006). Mutations in embC can produce truncated forms of LAM lacking arabinose branching, emphasizing its essential role in mycobacterial cell wall construction (Korkegian et al. 2014; Amin et al. 2008; Goude et al. 2009; Das et al. 2020). Both EmbB and EmbC serve as targets of ethambutol, a first-line antituberculosis agent that inhibits arabinan biosynthesis by competitively binding near their active sites. This binding obstructs arabinose transfer, compromising cell wall formation (Amin et al. 2008; Goude et al. 2009; Das et al. 2020). Mutations in *embB* codon 306 (e.g., Met306Ile/Val/Leu), as well as in codons 406 and 497 and *embC*, alter arabinosyltransferase function, resulting in ethambutol resistance (Belanger et al. 1996; Radmacher et al. 2005; Safi et al. 2008; Sun et al. 2018). Ethambutol's antibacterial effect stems primarily from disrupting arabinosyltransferase-mediated cell wall assembly, competitively inhibiting substrate binding and thereby preventing arabinose incorporation into the cell wall, weakening its structural integrity. Mutations in *embB* are the predominant cause of ethambutol resistance, identified in approximately 85% of resistant clinical isolates, while *embC* mutations are less common, occurring in about 2.8% of cases (Goude et al. 2009). The most frequently observed *embB* mutation occurs at codon 306, with substitutions such as Met306Ile or Met306Val reported in 27–87% of ethambutol-resistant isolates, with many studies citing frequencies between 55% and 69% (Bakuła et al. 2013; Bwalya et al. 2022). Additional mutation hotspots include codons 354, 406, and 497 (Belanger et al. 1996; Alcaide et al. 1997; Radmacher et al. 2005; Brossier et al. 2006; Safi et al. 2008; Xu et al. 2015; Sun et al. 2018). These mutations induce conformational changes in or near the ethambutol-binding pocket, diminishing drug affinity. For example, alterations at *embB* codon 306 reshape the substrate-binding pocket's geometry and electrostatic profile, reducing ethambutol binding and allowing arabinan synthesis to continue despite drug presence. Although less frequent, *embC* mutations proximal to the drug-binding site may contribute to resistance, particularly when co-occurring with *embB* mutations (Mohammadi et al. 2019). Collectively, these alterations impair arabinosyltransferase enzymatic function, decreasing ethambutol's efficacy and underpinning clinical resistance (Belanger et al. 1996; Radmacher et al. 2005; Safi et al. 2008; Xu et al. 2015; Sun et al. 2018).

9.4.4 Pyrazinamidase (PncA)

Pyrazinamidase/nicotinamidase (PncA), the enzyme responsible for activating pyrazinamide (PZA) in Mtb, belongs to the amidohydrolase superfamily, a diverse group of enzymes catalyzing the hydrolytic cleavage of amide or ester bonds at carbon and phosphorus centers. Members of this superfamily share a characteristic (β/α)8-barrel, or TIM barrel, structural fold, enabling catalytic hydrolysis across a wide range of substrates (Seibert and Raushel 2005). Embedded within this fold is a mononuclear or binuclear metal center typically coordinated by conserved amino acid residues, essential for enzymatic activity. This metal center plays a dual catalytic role by activating the substrate's scissile bond and facilitating deprotonation of a bound water molecule, generating a nucleophilic hydroxide ion which attacks the amide bond (Scorpio et al. 1997; Todd et al. 2001; Seibert and Raushel 2005; Zhang et al. 2008; Petrella et al. 2011; Zapata-Pérez et al. 2016; Tunstall et al. 2021). Structurally, PncA is a monomeric protein containing Mn^{2+} and Fe^{2+} ions, with amino acid residues essential for catalysis (Asp8, Lys96, Cys138) and metal coordination (Asp49, His51, His57, His71) (Zhang et al. 2008). Enzymes within this superfamily often exhibit promiscuous activities, indicative of their evolutionary and mechanistic relationships (Roodveldt and Tawfik 2005). The crystal structure of PncA, resolved at 2.2 Å resolution (PDB ID: 3PL1), reveals a single α/β domain characterized by a six-stranded parallel β-sheet flanked by α-helices (Petrella et al. 2011). The active site contains a catalytic triad formed by Cys138, Asp8, and Lys96, consistent with a cysteine-based amidohydrolase mechanism. A Fe^{2+} ion is coordinated within the active site by conserved residues Asp49, His51, His57, and His71, playing a pivotal role in catalysis (Smith et al. 2012). The substrate-binding cavity is relatively small, approximately 10 Å deep and 7 Å wide, and lined by conserved residues such as Phe13 and Trp68, which are critical for substrate recognition and binding. Functionally, PncA catalyzes the conversion of the prodrug pyrazinamide into its active form, pyrazinoic acid (POA), which accumulates within the bacterial cytoplasm, especially under acidic conditions encountered in macrophage phagosomes (Petrella et al. 2011; Supo-Escalante et al. 2020; Tunstall et al. 2021; Carter et al. 2024). POA disrupts membrane energetics, inhibits fatty acid synthase I (FAS I), interferes with coenzyme A biosynthesis, and sensitizes Mtb to oxidative stress, collectively resulting in bacterial cell death. Consequently, PncA's enzymatic activity is indispensable for PZA's bactericidal effect (Ostrer et al. 2025). Resistance to pyrazinamide primarily arises from mutations in the *pncA* gene, which are highly heterogeneous and dispersed throughout the coding region. These mutations include missense substitutions, nonsense mutations, insertions, deletions, and frameshifts that impair or abolish *PncA* activity by disrupting the protein's three-dimensional structure, altering active site geometry, destabilizing metal coordination, or compromising protein folding and stability (Sreevatsan et al. 1997; Cheng et al. 2000; Modlin et al. 2021; Kim et al. 2023, 2024; Li et al. 2024). Such alterations reduce substrate binding or cause complete loss of catalytic function, preventing conversion of PZA into its active form and thereby rendering the drug ineffective. Given that *PncA* is nonessential under normal bacterial growth conditions, these mutations confer pyrazinamide resistance without significantly impairing bacterial fitness.

To date, many distinct *pncA* mutations have been identified in clinical isolates, including prevalent missense mutations at codons 12, 51, 57, 71, 85, 94, 132, and 142. Notably, substitutions such as His57Asp, Asp49Ala, and Gly97Asp, located near the active site or metal-binding residues, severely diminish enzymatic function (Zhang et al. 2003a, b, 2014; Ramirez-Busby and Valafar 2015; Rajendran and Palaniyandi 2022). Insertions or deletions leading to frameshifts and premature stop codons also frequently result in loss of function. Furthermore, mutations in the *pncA* promoter region have been reported to decrease gene expression, contributing to resistance. The diverse and scattered nature of these mutations reflects *PncA*'s structural flexibility and indicates multiple potential routes through which its function can be impaired (Ramirez-Busby and Valafar 2015; Shi et al. 2022).

9.4.5 DNA Gyrase Subunits (GyrA, GyrB)

DNA gyrase in Mtb is a type II topoisomerase enzyme that plays an indispensable role in maintaining DNA topology during key cellular processes such as replication, transcription, and recombination. By introducing negative supercoils into closed circular double-stranded DNA, DNA gyrase facilitates essential DNA transactions critical for bacterial survival. The enzyme functions as a tetrameric complex composed of two GyrA and two GyrB subunits, encoded by the *gyrA* and *gyrB* genes, respectively (Madhusudan et al. 1994; Drlica and Zhao 1997; Champoux 2001; Collin et al. 2011). This assembly is conventionally designated as $GyrA_2GyrB_2$.

The GyrA subunit (~90 kDa) contains the breakage-reunion active site responsible for catalyzing DNA strand cleavage and religation. Structurally, it comprises two distinct domains: an N-terminal domain that binds DNA and mediates strand cleavage and religation and a C-terminal domain (CTD) that promotes DNA wrapping around the enzyme (Manjunatha et al. 2001). Uniquely, the CTD of Mtb GyrA harbors two functional GyrA-box motifs, which confer a hybrid functionality that differentiates it from gyrases in other bacterial species (Takiff et al. 1994; Hooper 1999; Collin et al. 2011; Bush et al. 2015). The GyrB subunit (~80 kDa) harbors ATPase activity that provides the energy required for DNA strand passage and supercoiling through hydrolysis of ATP. It contains several functional domains, including the GHKL ATPase domain (named after gyrase, Hsp90, histidine kinase, and MutL proteins), a transducer domain, and a TOPRIM domain. The TOPRIM domain coordinates essential metal ions necessary for catalysis and mediates interactions with GyrA and DNA (Kamsri et al. 2024). Two structural elements within the ATPase domain, the ATP-lid loop (residues 104–125) and a Corynebacteriales-specific flexible C-loop (residues 214–245), stabilize the enzyme's open conformation and regulate its ATPase activity. The transducer domain transmits conformational changes induced by ATP hydrolysis to the DNA cleavage core, thereby facilitating strand passage (Piton et al. 2010; Collin et al. 2011; Bush et al. 2015). The catalytic cycle of DNA gyrase initiates with ATP binding and hydrolysis by the GyrB subunit, triggering conformational changes that enable DNA cleavage mediated by GyrA (Agrawal et al. 2013; Tretter et al. 2010). Crystal structures of the GyrB ATPase domain reveal a 12° domain rearrangement upon ATP hydrolysis, augmenting the distance between transducer domains by approximately 7 Å (Stanger et al. 2014). This conformational shift is indispensable for coupling ATP hydrolysis with DNA binding and strand passage. Moreover, dynamical-nonequilibrium molecular dynamics simulations of Mtb GyrB have identified two communication pathways linking the ATPase active site to the C-loop and the transducer domain's C-terminal region, which may underlie the unique functional properties of Mtb DNA gyrase (Kamsri et al. 2024). Upon ATP binding, the ATPase domains of GyrB dimerize, capturing the transported DNA segment (T-segment). Subsequent ATP hydrolysis and phosphate release drive conformational cycling essential for enzymatic function (Corbett et al. 2004). DNA gyrase is the principal target of fluoroquinolone antibiotics such as moxifloxacin and levofloxacin. These drugs exert bactericidal activity by stabilizing the DNA cleavage complex and inhibiting ligation of DNA strands, thereby inducing lethal double-strand breaks. Resistance to fluoroquinolones in Mtb primarily arises from mutations within the quinolone resistance-determining regions (QRDRs) of the *gyrA* (codons 74–113) and *gyrB* (codons 461–499) genes. These mutations alter amino acid residues critical for drug binding while generally preserving the catalytic activity of DNA gyrase (Willmott and Maxwell 1993; Maxwell 1997; Laponogov et al. 2009; Hooper and Jacoby 2016; Collins and Osheroff 2024). Within gyrA, resistance-associated mutations frequently cluster at codons corresponding to Ser83 and Asp87 (using *Escherichia coli* numbering), which directly interact with fluoroquinolone molecules. Substitutions such as Ser83Leu, Asp87Gly, and Asp87Asn induce conformational alterations in the DNA cleavage site that reduce drug binding affinity and confer phenotypic resistance. In Mtb, fluoroquinolone resistance is most commonly linked to mutations at codons 90, 91, and 94 within the gyrA QRDR. However, emerging evidence highlights additional mutations both within and outside this canonical region that may contribute to resistance. Positions such as Tyr50, Ala74, Glu21, Ser95, and Gly668 have been reported in clinical isolates, although their exact functional impact remains less well characterized and often requires further verification (Maruri et al. 2012; Malik et al. 2012; Kabir et al. 2020). For instance, the p.Ser95Thr substitution is frequently observed as a phylogenetic marker and appears to have a limited direct role in resistance, whereas mutations like p.Glu21Gln and p. Gly668Asp, found outside the classical QRDR, are hypothesized to affect fluoroquinolone binding or DNA supercoiling through allosteric or structural perturbations.

Recently, using whole-genome sequencing, we identified a consistent mutation pattern in the gyrA gene at three loci, p.Glu21Gln, p.Ser95Thr, and p.Gly668Asp, in XDR and TDR-TB clinical isolates. While the p.Ser95Thr mutation is primarily considered a lineage-specific marker with limited direct contribution to resistance, the p.Glu21Gln substitution is predicted to disrupt local charge distribution within DNA gyrase, thereby impairing fluoroquinolone binding. The p.Gly668Asp mutation, located in the C-terminal domain, may affect DNA supercoiling activity. Collectively, these mutations are likely contributors to the fluoroquinolone-resistant phenotype (Fig. 9.1). This constellation of mutations highlights the multifaceted mechanisms by which drug-resistant TB strains evade fluoroquinolone-mediated inhibition, involving both canonical QRDR mutations and less typical substitutions that modulate enzyme structure and function. A comprehensive understanding of these diverse mutations is essential to improve molecular diagnostic accuracy and inform effective therapeutic decision-making.

Mutations in *gyrB*, although less frequent, are significant and often involve residues within or

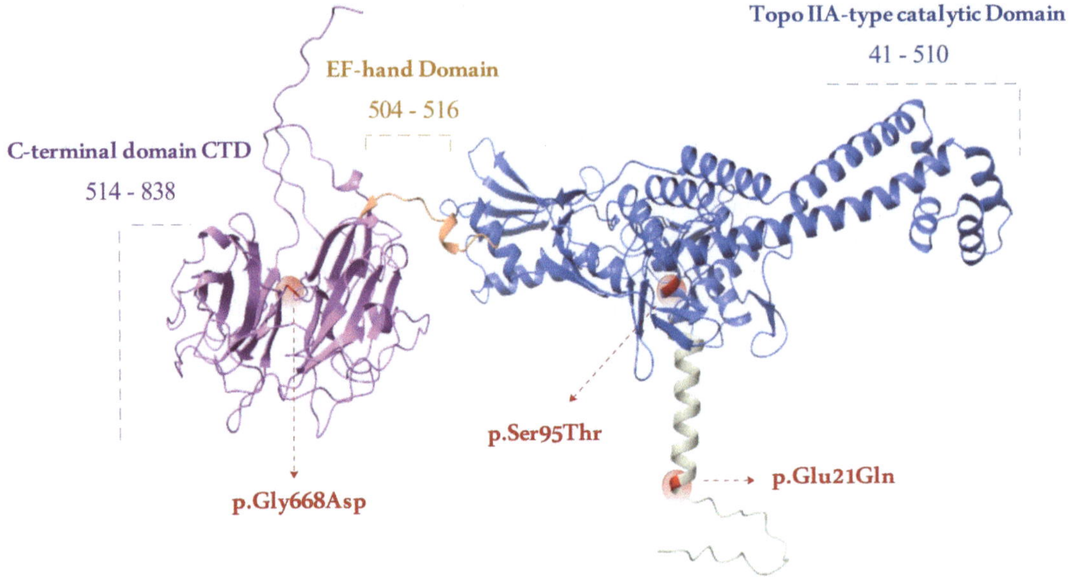

Fig. 9.1 Structural localization of *gyrA* missense mutations associated with fluoroquinolone resistance in XDR- and TDR-TB isolates
This figure presents the three-dimensional structure of the GyrA subunit of *Mycobacterium tuberculosis* DNA gyrase (comprising 838 amino acids), highlighting the missense mutations identified with 100% frequency among XDR and TDR-TB strains isolated from clinical isolates. The primary domains of the protein are annotated: the Topoisomerase IIA-type catalytic domain (residues 41 – 5 10, blue), the EF-hand domain (residues 504 – 5 16, orange), and the C-terminal domain (residues 514 – 8 38, purple). The p.Glu21Gln mutation, indicated by a red sphere near the N-terminus, may alter local charge distribution and influence fluoroquinolone binding. The p.Ser95Thr mutation, marked within the catalytic core, is associated with decreased enzymatic sensitivity to fluoroquinolones. The p.Gly668Asp mutation, also depicted as a red sphere, resides in the C-terminal domain and is linked to reduced DNA supercoiling activity. Collectively, these amino acid substitutions disrupt key functional regions of GyrA, thus contributing to high-level fluoroquinolone resistance by altering DNA-drug interaction dynamics and reducing enzymatic efficiency
*Drawing by Mohsen Maleknia; Mycobacteriology Research Center, National Research Institute of Tuberculosis and Lung Disease, Shahid Beheshti University of Medical Sciences, Tehran, Iran

proximal to the ATPase and TOPRIM domains, including Asp472 and Asn510. Substitutions such as Asp472His and Asn510Asp confer resistance by altering enzyme conformational dynamics or drug-binding pockets, resulting in diminished fluoroquinolone efficacy (Shafiq et al. 2024). These mutations may also influence ATP hydrolysis rates or DNA binding, subtly modulating enzyme activity to maintain bacterial viability while evading drug inhibition. Mechanistically, amino acid substitutions in DNA gyrase induce conformational alterations that reduce fluoroquinolone binding affinity to the DNA-enzyme-drug ternary complex. This prevents stabilization of the cleavage complex necessary for fluoroquinolone bactericidal action while preserving the essential catalytic functions of DNA gyrase. Consequently, Mtb can maintain effective DNA supercoiling even in the presence of fluoroquinolone therapy, underpinning clinical drug resistance (Collins et al. 2024).

9.4.6 Ribosomal Protein S12 (RpsL)

Ribosomal protein S12 (RpsL), encoded by the *rpsL* gene (Rv0682) in Mtb, is an essential and highly conserved component of the 30S ribosomal subunit that plays a crucial role in maintaining translational fidelity. This protein belongs to the universally conserved uS12 family and interacts closely with the 16S ribosomal RNA (rRNA) at the decoding center, particularly stabilizing helix 44 (Zhang et al. 2003a, b; Sharma et al. 2007; Sun et al. 2010; Datta et al. 2021). This interaction is vital for accurate codon-anticodon pairing during protein synthesis, ensuring the proper selection and accommodation of aminoacyl-tRNAs at the ribosomal A site. Over the years, extensive structural and functional studies have highlighted RpsL's role in modulating the GTPase activity of elongation factors and controlling the accuracy of decoding (Schuwirth et al. 2005; Demirci et al. 2013; Shrestha et al. 2020). Resistance to streptomycin, a key aminoglycoside antibiotic used in tuberculosis treatment, is mainly mediated by mutations in the *rpsL* gene. Early research by Honoré and Cole (1994) identified novel mutations in *rpsL* that confer high-level streptomycin resistance without significantly impairing ribosomal function. Among these, mutations at codons 43 and 88, both encoding lysine residues, are the most commonly observed and are critical for maintaining the ribosome's structural integrity and its interaction with the antibiotic. The K43R substitution is the most prevalent globally and strongly linked to high-level resistance, while the K88R mutation also contributes to resistance but is less frequent (Toivonen et al. 1999; Spies et al. 2011). Molecular analyses of clinical isolates have confirmed that mutations at codon 43 account for the majority of streptomycin-resistant cases, with the K43R substitution reducing streptomycin binding affinity by altering the ribosomal decoding site's conformation (Mieskes et al. 2000). Significantly, these mutations do not substantially affect bacterial fitness, which explains their persistence in clinical populations (Spies et al. 2011). Further investigations demonstrated that the K42R mutation in ribosomal protein S12 of *Mycobacterium smegmatis* does not influence susceptibility to other aminoglycosides such as kanamycin or amikacin, nor does it affect resistance patterns associated with 16S rRNA mutations (Kalapala et al. 2010). This finding underscores the specificity of *rpsL* mutations for streptomycin resistance. Moreover, resistance to other aminoglycosides like kanamycin and amikacin is primarily linked to mutations in the 16S rRNA gene (*rrs*), with the A1401G substitution being the principal cause of high-level resistance (Shrestha et al. 2020). Overexpression of wild-type RpsL does not alter susceptibility to these drugs, further emphasizing the distinct roles of *rpsL* and *rrs* mutations in aminoglycoside resistance (Kalapala et al. 2010). Collectively, mutations at codons 43 and 88 of *rpsL* are key determinants of streptomycin resistance in Mtb. These mutations modify the ribosomal protein structure in a way that diminishes drug binding while preserving translational fidelity, allowing resistant strains to survive and proliferate clinically (Villellas et al. 2013; Arjomandzadegan and Gravand 2015). Understanding these molecular mechanisms is crucial for developing rapid diagnostic assays and guiding effective treatment strategies to combat drug-resistant tuberculosis (Mao et al. 2025).

9.4.7 16S rRNA (Rrs)

The 16S ribosomal RNA (16S rRNA), encoded by the *rrs* gene in Mtb, is a fundamental structural and functional component of the 30S ribosomal subunit (Zhao et al. 2014). It forms the decoding center where messenger RNA (mRNA) codons are precisely read and matched with aminoacyl-tRNAs, thereby ensuring translational fidelity (Verma et al. 1994). The 16S rRNA adopts intricate secondary and tertiary structures, including helix 44, which harbors dynamic nucleotides A1492 and A1493 that play a critical role in monitoring codon-anticodon pairing during protein synthesis. Aminoglycoside antibiotics such as streptomycin, kanamycin, and amikacin target the aminoacyl (A) site within this decoding region, disrupting translation and inhibiting bacterial protein synthesis. Resistance to these drugs in Mtb is predominantly driven by mutations in the *rrs* gene that alter the ribosomal conformation, thereby reducing antibiotic binding affinity (Meier et al. 1994; Garneau-Tsodikova and Labby 2016; Krause et al. 2016). Early molecular analyses by Prammananan et al. (1999) identified the A1401G mutation (corresponding to position 1408 in Mtb numbering) as the principal genetic change responsible for high-level kanamycin resistance in clinical isolates. This single-nucleotide substitution induces conformational changes in the ribosomal A site, preventing stable drug binding and allowing protein synthesis to continue despite antibiotic pressure. Preceding this, Suzuki et al. (1998) confirmed the widespread occurrence of the A1401G mutation across diverse Mtb strains. Subsequent studies expanded the spectrum of *rrs* mutations linked to aminoglycoside resistance. Pfister et al. (2003) demonstrated that mutations within the bacterial A-site of 16S rRNA, particularly at position 1408 and the U1406-U1495 base pair, disrupt aminoglycoside binding. The A1408G mutation confers high-level resistance to 6'-NH3+ aminoglycosides by preventing the insertion of ring I into the A-site helix. Mutations in the U1406-U1495 pair generally confer low-to-moderate resistance, with the notable exception of the U1406C/U1495A double mutation. Additionally, mutations such as U1406C, C1496U, and U1498C in helix 44 confer resistance to hygromycin B in *Mycobacterium smegmatis* (Pfister et al. 2003). Nessar et al. (2011) identified several mutations in the *rrs* gene associated with kanamycin resistance, including A1408G, U1406A, C1409U, and G1491U. The A1401G mutation also confers high-level resistance to amikacin and reduced susceptibility to capreomycin (Sirgel et al. 2012). Capreomycin and viomycin bind across the ribosomal interface involving 16S rRNA helix 44 and 23S rRNA helix 69, with residues in helix 44 being key determinants of drug selectivity (Akbergenov et al. 2011). The U1406A mutation, while conferring kanamycin resistance, significantly attenuates Mtb virulence by impairing ribosome maturation and translation initiation. This mutation reduces the expression of numerous proteins, including key virulence factors and metabolic enzymes (Watanabe et al. 2016). It is important to highlight that these mutations vary in their dominance patterns; U1406A and A1408G mutations act dominantly to confer resistance, whereas C1409U and G1491U mutations exhibit recessive behavior (Nessar et al. 2011). Mutations in helix 44, especially A1401G and neighboring nucleotides, disrupt aminoglycoside and capreomycin binding pockets, thereby conferring resistance (Huang et al. 2015). The A1401G mutation is prevalent among Mtb strains resistant to amikacin, kanamycin, and capreomycin, though it is also occasionally found in susceptible strains (Georghiou et al. 2012). This mutation, along with others in the *rrs* gene, *eis* promoter, *tlyA*, and *gidB* genes, is associated with resistance to second-line injectable drugs (Khanna et al. 2010). Epidemiological studies reveal that Beijing lineage strains of Mtb exhibit a significantly higher prevalence of mutations in *rpoB* and *katG* genes compared to non-Beijing strains, potentially contributing to increased rifampicin and isoniazid resistance (Guo et al. 2022). While no general association has been found between streptomycin resistance and the Beijing genotype, these strains are significantly correlated with high-level streptomycin resistance and the *rpsL* K43R mutation (Sun et al. 2012, 2016). Systematic reviews of allelic exchange experiments have validated the causal relationship between many of these mutations and drug resistance, emphasizing the role of genetic background in modulating resistance phenotypes (Nebenzahl-Guimaraes et al. 2014; Watanabe et al. 2016). Structural studies further revealed that the *rrs* gene exists as part of a single rRNA operon in Mtb, limiting compensatory mutation options and amplifying the functional consequences of deleterious mutations on ribosomal activity and antibiotic susceptibility (Watanabe et al. 2016). Park et al. (2020) demonstrated that mutations in both 16S and 23S rRNA genes independently contribute to multidrug resistance, highlighting the complexity of rRNA-mediated resistance mechanisms. Recent advances in molecular diagnostics have leveraged these insights (Vemula and Gupta 2022). Overall, mutations in the *rrs* gene alter the ribosomal decoding site, reducing drug binding and conferring clinically relevant resistance.

9.4.8 23S rRNA (rrl Gene)

The 23S ribosomal RNA (23S rRNA), encoded by the rrl gene in Mtb, constitutes a vital structural and functional RNA component of the 50S large ribosomal subunit. Comprising approximately 2900 nucleotides, the 23S rRNA folds into six well-defined secondary structure domains (I–VI) that adopt intricate tertiary conformations essential for ribosomal function. Among these, key functional regions include the peptidyl transferase center (PTC) and the sarcin-ricin loop (SRL), located within domain VI (helix 95) (Noller and Woese 1981; Gutell et al. 1993; Ji et al. 1994; Kurabachew et al. 2003; Laughlin et al. 2022). The sarcin-ricin loop is a highly conserved RNA element critical for protein synthesis. Initially, it was considered essential for GTP hydrolysis catalyzed by elongation factors EF-Tu and EF-G; however, more recent evidence demonstrates its primary role in anchoring these factors to the ribosome (García-Ortega et al. 2010; Shi et al. 2012). Advances in cryo-electron microscopy (cryo-EM) have elucidated near-atomic structures of the Mtb 50S ribosomal subunit and the intact 70S ribosome, unveiling the elaborate architecture of 23S rRNA and its intra-ribosomal interactions. For instance, a 3 Å resolution structure of the *Mycobacterium smegmatis* 50S subunit revealed a unique conformational switch in 23S rRNA helices H67-H71, which may regulate 70S ribosome availability (Baid and Sengupta 2023). Additionally, the structure of the Mtb 50S subunit bound to the dual-specificity methyltransferase TlyA clarified the mechanism of rRNA recognition and modification via base flipping of the target nucleotide (Laughlin et al. 2022). Furthermore, a 3.3 Å resolution structure of the Mtb 50S subunit complexed with clarithromycin highlighted dynamic interactions within the nascent polypeptide exit tunnel, shedding light on the antibiotic's specificity and efficacy in tuberculosis treatment (Zhang et al. 2021). Resistance mechanisms involving 23S rRNA in Mtb primarily arise from mutations and post-translational modifications that alter antibiotic binding. Methylation of rRNA nucleotides represents a prominent defense, particularly against macrolides and related antibiotics (Madsen et al. 2005). In Mtb, the Erm (37) methyltransferase encoded by *erm* (37) targets nucleotide A2058 of the 23S rRNA and subsequently methylates adjacent nucleotides A2057 and A2059. This unique methylation pattern confers resistance to ketolides such as telithromycin and is attributed to Erm (37)'s truncated C-terminal domain, affecting its rRNA interaction dynamics (Madsen et al. 2005). Recent structural analyses have uncovered that dimethylation of A2058 imparts macrolide resistance through a previously unrecognized mechanism, underscoring the critical role of the desosamine sugar moiety in macrolide binding (Svetlov et al. 2021). Linezolid, a member of the oxazolidinone class of antibiotics, inhibits bacterial protein synthesis by binding to the 23S rRNA within the peptidyl transferase center of the 50S subunit (Kloss et al. 1999; Leach et al. 2011). It specifically blocks formation of the 70S initiation complex by disrupting the N-formylmethionyl-tRNA ribosome, mRNA ternary complex (Swaney et al. 1998). Resistance to linezolid in Mtb and other bacteria predominantly occurs via mutations in the *rrl* gene encoding 23S rRNA and the *rplC* gene encoding ribosomal protein L3 (Long and Vester 2008; Pi et al. 2019). The *rplC* T460C mutation, resulting in a Cys154Arg substitution, is prevalent among linezolid-resistant Mtb strains (Beckert et al. 2012; Pi et al. 2019). Mutations in *rrl*, notably G2814T and G2270T, also contribute to resistance by modifying the linezolid binding site in the peptidyl transferase center and reducing drug affinity (Pi et al. 2019; Long and Vester 2008). Remarkably, mutations distant from the binding site may confer resistance via allosteric structural rearrangements of rRNA (Belousoff et al. 2017). Understanding these mechanisms is imperative for developing novel oxazolidinones capable of circumventing existing resistance pathways (Long and Vester 2008; Belousoff et al. 2017). Clinical observations underscore the rapid emergence of linezolid resistance during prolonged therapy, often associated with suboptimal dosing; genomic studies confirm *rrl* and *rplC* mutations as primary determinants (Wasserman et al. 2019). Functional characterizations reveal that these mutations impair linezolid binding without significantly affecting ribosomal function, enabling resistant bacteria to persist despite drug exposure (Hu et al. 2020). Moreover, strong correlations between phenotypic resistance and mutations in *rrl* and *rplC* validate molecular diagnostics targeting these loci (Gan et al. 2023).

In addition to genetic mutations, toxin-antitoxin (TA) systems contribute significantly to the regulation of 23S rRNA integrity and overall ribosomal function in Mtb. A prominent example is the VapBC family of TA systems, which includes VapC toxins, PIN domain endoribonucleases that inhibit bacterial translation by cleaving specific RNA molecules. For instance, VapC20 specifically targets the sarcin-ricin loop of 23S rRNA, a critical region for protein synthesis, thereby disrupting ribosomal function (Winther et al. 2013). Other VapC toxins selectively cleave essential transfer RNAs (tRNAs); notably, VapC-mt4 cleaves three distinct tRNAs at their anticodon loops, impeding translation (Cruz et al. 2015). A broad study characterizing 12 distinct VapC toxins from Mtb demonstrated that 11 preferentially cleave different tRNAs, while one specifically targets the sarcin-ricin loop, reflecting functional diversity within this toxin family (Winther et al. 2013). Structural analysis of the VapBC26 complex revealed that the ribonuclease activity of VapC26 is tightly regulated and inhibited by the C-terminal region of its cognate VapB26 antitoxin, illustrating how TA systems control toxin activity (Kang et al. 2017). Together, these findings highlight the multifaceted strategies employed by VapC toxins to modulate RNA stability, thereby influencing Mtb's growth dynamics and persistence through targeted RNA cleavage. Complementing the VapBC family, MazF toxins represent a distinct class of sequence-specific endoribonucleases within toxin-antitoxin systems. Structural analyses of MazF toxins provide further insights into how Mtb modulates ribosomal function. Specifically, MazF-mt6 from Mtb features a PemK-like fold characterized by a shortened β1-β2 linker. This structural arrangement forms a deep, solvent-accessible active site capable of recognizing and binding structured RNA substrates, including helix 70 of the 23S rRNA (Hoffer et al. 2017). MazF-mt6 exhibits potent and selective endoribonuclease activity, cleaving mRNA specifically at 5′-UU↓CCU-3′ motifs. Crucially, it also uniquely targets a single UUCCU site located within the ribosomal A site of 23S rRNA. This precise and dual targeting effectively disrupts translation and destabilizes the association of ribosomal subunits, thereby exerting a profound inhibitory effect on overall protein synthesis (Schifano et al. 2013). Site-directed mutagenesis studies have pinpointed Asp10, Arg13, and Thr36 as essential residues for MazF-mt6's ribonuclease activity, suggesting a proton-relay mechanism in its catalysis (Hoffer et al. 2017; Dobbins et al. 2019).

Collectively, the *rrl* gene encodes 23S rRNA, a structurally and functionally complex RNA fundamental to ribosomal catalytic activity and antibiotic binding in Mtb. Mutations and methylation modifications, especially within domain V and the sarcin-ricin loop, confer resistance to macrolides and oxazolidinones by diminishing drug affinity. Concurrently, mutations in ribosomal proteins L3 and L4 further contribute to linezolid resistance. Additionally, toxin-mediated cleavage of 23S rRNA modulates translational efficiency and promotes bacterial persistence, thereby contributing to drug tolerance.

9.4.9 Ribosomal Proteins L3 and L4 (RplC, RplD)

Ribosomal proteins L3 (RplC) and L4 (RplD) are essential components of the 50S large ribosomal subunit in Mtb. The Mtb ribosome possesses species-specific features, including unique ribosomal proteins and rRNA expansion segments (Yang et al. 2017). These distinctive characteristics make the Mtb protein synthesis machinery a potential target for therapeutic interventions (Kumar et al. 2021). Notably, the *rplC* gene, encoding the L3 protein, has been identified as a novel target in linezolid-resistant Mtb strains, with the *rplC* T460C mutation found in multiple resistant isolates (Beckert et al. 2012). Additionally, specific Mtb ribosomal proteins, including S3 and L16, exhibit idiosyncratic behavior under various conditions, suggesting potential extraribosomal functions (Fan et al. 2014). Structurally, both proteins possess conserved domains that facilitate direct interaction with 23S rRNA and contribute to the formation of the peptidyl transferase center

(PTC), the catalytic core of the ribosome. L3 contains a globular domain with extensions that reach into the PTC region, directly influencing antibiotic binding sites. Similarly, L4 has an extended loop that interacts with the nascent peptide exit tunnel, impacting ribosome function and antibiotic susceptibility (Schuwirth et al. 2005; Wilson 2014; Klitgaard et al. 2015; Conkle-Gutierrez et al. 2024). Functionally, ribosomal proteins L3 and L4 play crucial roles in ribosome structure and function, particularly in the peptidyl transferase center (PTC) and peptide exit tunnel. L3 mutations can affect tRNA binding affinities, peptidyltransferase activity, and ribosome structure, influencing antibiotic susceptibility and cell growth (Petrov et al. 2004). Similarly, L4 mutations in the extended loop region can impact poly(Phe) synthesis, tRNA binding, and erythromycin sensitivity (Tsagkalia et al. 2005). Both proteins contribute to antibiotic resistance, with L3 mutations associated with reduced susceptibility to PTC-targeting antibiotics like linezolid and tiamulin (Klitgaard et al. 2015). However, the extended loops of L4 and L22 are not essential for ribosome assembly or L4-mediated autogenous control (Zengel et al. 2003). These findings highlight the complex interplay between ribosomal proteins, rRNA, and antibiotics, emphasizing the importance of L3 and L4 in ribosome function and antibiotic resistance mechanisms. Beckert et al. (2012) reported that the T460C mutation in the *rplC* gene, resulting in a cysteine-to-arginine substitution at position 154 (Cys154Arg) in ribosomal protein L3, is a dominant mutation strongly associated with linezolid resistance in Mtb. This mutation was consistently observed in both clinical isolates and in vitro-selected mutants, but was absent in susceptible strains (Beckert et al. 2012; Pi et al. 2019). The Cys154Arg substitution directly affects the drug-binding site near the peptidyl transferase center, conferring resistance to linezolid and sutezolid (Makafe et al. 2016). Whole-genome sequencing of linezolid-resistant isolates revealed that 88.3% harbored the *rplC* T460C mutation, while the remaining isolates had mutations in the rrl gene (Pi et al. 2019). Clinical studies in Moscow found that 13 out of 16 linezolid-resistant isolates bore the RplC C154R substitution, with these strains exhibiting higher linezolid MICs compared to isolates with *rrl* mutations (Ushtanit et al. 2021). Hence, mutations in *rplC*, particularly the T460C substitution leading to the Cys154Arg amino acid change, are strongly associated with linezolid resistance in Mtb, primarily by altering ribosomal structure and reducing drug binding affinity (Kumar et al. 2024). While mutations in *rplD* occur less frequently and their contribution to resistance is less clearly defined, combined mutations in ribosomal proteins and the 23S rRNA gene collectively contribute to varying levels of linezolid resistance. These insights are critical for improving molecular diagnostics and guiding the development of effective therapies against drug-resistant tuberculosis.

9.4.10 Ribosomal RNA Methyltransferase (TlyA)

Ribosomal RNA methyltransferase TlyA, encoded by the tlyA gene (Rv1694) in Mtb, is a SAM-dependent 2′-O-methyltransferase that modifies specific nucleotides in both 16S and 23S rRNAs. Structurally, TlyA contains conserved methyltransferase domains responsible for binding the methyl donor S-adenosylmethionine (SAM) and the rRNA substrate, catalyzing methylation at cytidine residues C1409 in 23S rRNA and C1402 in 16S rRNA (Parish and Stoke 2000; Maus et al. 2005; Johansen et al. 2006; Rahman et al. 2010; Witek et al. 2017; Laughlin et al. 2022). These methylations are critical for maintaining ribosomal function and antibiotic susceptibility, particularly influencing the binding of tuberactinomycin antibiotics such as capreomycin and viomycin, which target the ribosomal interface between the 30S and 50S subunits. Functionally, TlyA-mediated methylation enhances the affinity of capreomycin and related antibiotics to the ribosome, thereby sensitizing Mtb to these drugs (Maus et al. 2005; Johansen et al. 2006). Loss of TlyA activity, often due to mutations or deletions in the *tlyA* gene, results in the absence of these methylations, leading to reduced drug binding and consequent resistance. Maus et al. (2005) first demonstrated that *tlyA* mutations confer capreomycin resistance by abolishing 2′-O-methylation of rRNA. Subsequent studies confirmed that *tlyA* mutants lack methylation at C1409 and C1402, correlating with high-level capreomycin resistance. Recent structural and biochemical studies have elucidated how Mtb TlyA specifically recognizes and modifies the 50S ribosomal subunit. Using single-particle cryo-electron microscopy, the TlyA-50S complex in a postcatalytic state was captured, revealing that TlyA binds near the peptidyl transferase center on helix 69 of the 23S rRNA, employing a base-flipping mechanism to methylate the 2'-O position of the target nucleotide (Laughlin et al. 2022). This methylation induces conformational changes that propagate to the H69 tip, creating a more favorable binding pocket for capreomycin (Nandi et al. 2024). TlyA's unique dual substrate specificity is facilitated by a novel RXWV motif in its interdomain linker, which is crucial for S-adenosyl-L-methionine binding and may contribute to domain interactions and substrate recognition (Witek et al. 2017; Laughlin et al. 2022).

Rahman et al. (2015) revealed the multifunctional nature of TlyA, demonstrating that besides its established role as a 2′-O-rRNA methyltransferase, TlyA also exhibits hemolytic activity and localizes to the bacterial surface.

TlyA surface localization was confirmed by confocal microscopy and immunofluorescence in *Mycobacterium marinum*, *Mycobacterium smegmatis*, and *Escherichia coli* expressing TlyA, despite lacking classical secretion signals such as those from Sec or Tat pathways. Proteinase K accessibility assays demonstrated TlyA's surface exposure, correlating with its hemolytic function. Furthermore, TlyA was found in membrane vesicles secreted by these bacteria, indicating a non-classical secretion mechanism that facilitates its extracellular presence and contributes to virulence and host-pathogen interactions. Zhao et al. (2019) characterized methylation patterns on 16S and 23S rRNAs, emphasizing that methylation by enzymes like TlyA critically modifies ribosomal RNA, thereby affecting antibiotic binding and resistance mechanisms. These modifications influence the efficacy of antibiotics targeting the ribosome. Laughlin et al. (2022) assessed the impact of TlyA deficiency on capreomycin resistance in Mtb strains, finding that loss of TlyA function leads to significant resistance to capreomycin and viomycin. Phenotypic assays supported previous genetic and biochemical data, confirming that tlyA mutants exhibit reduced drug susceptibility, underscoring TlyA's essential role in mediating antibiotic action and resistance. Collectively, *tlyA* encodes a dual-specificity rRNA methyltransferase that methylates key cytidines in both 16S and 23S rRNAs, modifications essential for the effective binding of capreomycin and related antibiotics. Mutations or deletions in *tlyA* abolish these methylations, inducing conformational changes in the ribosome that reduce antibiotic binding and confer resistance. Additionally, TlyA functions as a surface-exposed hemolysin, underscoring its significant role in Mtb biology by contributing not only to antibiotic susceptibility but also to virulence and host-pathogen interactions.

9.4.11 ATP Synthase Subunit C (AtpE)

ATP synthase subunit c (AtpE) in Mtb is an integral membrane protein encoded by the atpE gene. It belongs to the F-type ATPase family, a complex enzyme comprising two main sections: the membrane-embedded FO domain and the cytoplasmic F1 domain (Lu et al. 2014; Kamariah et al. 2019; Sarathy et al. 2019). The FO domain includes subunits a, b, and the c-ring, which is formed by nine identical c subunits (Preiss et al. 2015). The F1 domain contains subunits γ, δ, ε, and the α3:β3 hexameric headpiece. The α3:β3 headpiece is connected to the c-ring by a central stalk, composed of ε and γ subunits, and a peripheral stalk formed by b and δ subunits (Kamariah et al. 2019). Proton translocation from the periplasmic space to the cytoplasm drives rotation of the c-ring. This rotational energy is transmitted via the ε and γ subunits to the catalytic α3:β3 headpiece, inducing conformational changes that power ATP synthesis (Nakamoto et al. 1999). Functionally, AtpE is essential for energy production in Mtb, supporting aerobic growth and survival under hypoxic or dormant conditions by maintaining cellular ATP levels (Preiss et al. 2015). Genetic knockdown studies have demonstrated that depletion of *atpE* is bactericidal, which underscores its critical role in Mtb viability (Harikishore and Grüber 2024).

Bedaquiline (BDQ), a diarylquinoline antibiotic, specifically targets AtpE by binding to the c subunit and inhibiting proton translocation. This action collapses the proton motive force and halts ATP synthesis, demonstrating high efficacy against both replicating and nonreplicating Mtb (Andries et al. 2005; Koul et al. 2014). However, resistance to BDQ has emerged, primarily due to mutations in the atpE gene that alter the drug-binding site, thereby reducing BDQ affinity (Petrella et al. 2006; Huitric et al. 2010; Biukovic et al. 2013; Nehru et al. 2024). Key resistance-conferring mutations in *atpE* have been identified at amino acid positions such as Ala63, Glu61, and Ile66, all located within the transmembrane helices forming the BDQ-binding pocket (Huitric et al. 2010; Snobre et al. 2023). For example, the atpE Ile66Val mutation has been experimentally shown to minimally disrupt BDQ-ATP synthase interactions without significantly elevating the minimum inhibitory concentration (MIC), whereas other mutations more profoundly reduce drug susceptibility (Otum et al. 2023). Site-directed mutagenesis and in silico analyses confirmed that the Ile66Val substitution does not increase MIC beyond the critical concentration (0.25–0.5 μg/mL), whereas mutations in regulatory genes such as Rv0678 cause higher levels of resistance (Otum et al. 2023). Clinical and laboratory isolates frequently reveal *atpE* mutations, including I66M and A63P, as determinants of BDQ resistance, often accompanied by mutations in Rv0678, which regulates efflux pump expression (Otum et al. 2023; Ismail et al. 2022, 2023). These mutations predominantly localize to the BDQ-binding site of the ATP synthase c subunit, disrupting critical drug interactions and diminishing binding affinity (Ismail et al. 2019). Recent surveillance studies in high-burden countries such as Russia report a high prevalence of *atpE* mutations among BDQ-resistant Mtb isolates, confirming the clinical relevance of these variants (Zimenkov et al. 2025). Additionally, mutations in the *atpB* gene, encoding the ATP synthase a subunit, have been identified; these may influence BDQ susceptibility by altering the subunit's structure near the drug-binding interface (Zimenkov et al. 2025). Functional investigations employing CRISPR interference have validated *atpE* as essential for bacterial survival, with knockdown inducing bactericidal effects at nanomolar BDQ concentrations (Harikishore and Grüber 2024). Structural studies utilizing cryo-electron microscopy have elucidated atomic-level interactions between BDQ and ATP synthase, demonstrating how specific mutations in AtpE impair drug binding without abolishing enzyme activity. These insights provide a foundation for structure-based drug design aimed at overcoming resistance (Laughlin et al. 2022). Intriguingly, some clinical isolates from regions such as East Java, Indonesia, exhibit BDQ resistance without detectable *atpE* mutations, implicating alternative mechanisms such as efflux pump regulation driven by Rv0678 mutations (Lestari et al. 2023). This underscores the complexity of BDQ resistance and the necessity of comprehensive molecular surveillance. Hence, the ATP synthase c subunit (AtpE), encoded by *atpE*, forms the proton-translocating c-ring essential for ATP synthesis in Mtb. Bedaquiline targets AtpE by inhibiting proton flow, but resistance mainly arises from mutations altering the drug-binding site, thereby reducing BDQ efficacy. Experimental, structural, and clinical data collectively emphasize the pivotal role of *atpE* mutations in BDQ resistance and highlight the urgent need for continued surveillance and development of novel inhibitors.

9.4.12 Alanine Racemase (Alr)

Alanine racemase (Alr) is a pyridoxal 5′-phosphate (PLP)-dependent enzyme encoded by the alr gene in Mtb. This gene is part of a complex heptacistronic operon containing multiple regulatory elements, indicating its involvement in stress adaptation (Bhat et al. 2017). Alr catalyzes the reversible conversion of L-alanine to D-alanine, a critical precursor for peptidoglycan cross-linking in the bacterial cell wall, making it essential for cell wall biosynthesis and bacterial viability (LeMagueres et al. 2005; Prosser and de Carvalho 2013; Asojo et al. 2014). Functional studies have confirmed that inactivation of Alr severely impairs Mtb growth, with mutants requiring external D-alanine supplementation and exhibiting reduced survival within macrophages and mouse models (Awasthy et al. 2012). Furthermore, Alr is the primary target of the anti-tuberculosis drug D-cycloserine (Awasthy et al. 2012).). Structurally, Alr is classified within the fold-type III PLP-dependent enzyme family and functions as a homodimer, with each monomer consisting of 384 amino acids. Each monomer includes an N-terminal α/β-barrel domain and a C-terminal domain mainly composed of β-strands. The PLP cofactor is covalently attached through an internal aldimine bond to a conserved lysine residue (Lys42 in Mtb Alr), forming the active site where substrate binding and catalysis take place (LeMagueres et al. 2005; Anthony et al. 2011). The catalytic mechanism proceeds via a Schiff base intermediate formed between PLP and the substrate, which facilitates proton abstraction and the racemization process. The geometry of the active site is highly conserved among bacterial alanine racemases, with critical residues lining the substrate-binding pocket and entrance channel that regulate substrate access and enzymatic activity (LeMagueres et al. 2005).

D-cycloserine (DCS), a structural analogue of D-alanine, acts as an irreversible inhibitor of Alr by creating a stable adduct with the PLP cofactor, thereby blocking the enzyme's function and disrupting peptidoglycan synthesis (Anthony et al. 2011). Resistance to cycloserine in Mtb is primarily associated with mutations in the *alr* gene that alter the enzyme's structure, reducing its affinity for the drug. Whole-genome sequencing of cycloserine-resistant clinical isolates has uncovered several nonsynonymous mutations in *alr*, which tend to cluster near the active site and substrate-binding regions. These mutations correlate with increased minimum inhibitory concentrations (MICs) for cycloserine. Notably, substitutions such as D322N, M319T, and Y364D affect residues essential for PLP binding or substrate access, thereby weakening cycloserine's inhibitory effect (Chen et al. 2017; Nakatani et al. 2017). Detailed structural and kinetic analyses of the D322N mutation demonstrate a roughly 240-fold decrease in cycloserine binding affinity. Crystallographic studies reveal that replacing aspartate with asparagine at position 322 disrupts hydrogen bonding networks and alters the conformation of the PLP-binding pocket, which impairs the formation of the inhibitory PLP-cycloserine adduct. This mutation allows the enzyme to continue producing D-alanine despite the presence of the drug, conferring resistance. Other mutations near the active site similarly reduce susceptibility by modifying substrate access or cofactor orientation (Nakatani et al. 2017; Evangelopoulos et al. 2019; de Chiara et al. 2023). Besides *alr*, mutations in other genes involved in D-alanine metabolism, such as *ddl* (D-alanine:D-alanine ligase) and *ald* (alanine dehydrogenase), as well as alterations affecting drug uptake and efflux, contribute to cycloserine resistance. Nevertheless, *alr* remains the principal genetic determinant of resistance in Mtb (Chen et al. 2017; Evangelopoulos et al. 2019).

9.4.13 D-Alanine Ligase (Ddl)

D-Alanine:D-alanine ligase (Ddl), encoded by the ddl gene in Mtb, is an essential ATP-dependent enzyme that catalyzes the ligation of two D-alanine molecules to form the D-alanyl-D-alanine dipeptide, a crucial precursor in peptidoglycan biosynthesis and bacterial cell wall formation (Galperin and Koonin 1997; Bruning et al. 2011, 2023; Prosser and de Carvalho 2013; Yang et al. 2018; Qin et al. 2021). Belonging to the ATP-grasp superfamily, Ddl exhibits a characteristic three-domain architecture that creates a ligand-binding cavity at the interface of these domains. The enzyme functions as a homodimer, with each monomer containing discrete domains responsible for ATP and D-alanine substrate binding (Bruning et al. 2011; Pederick et al. 2020). Notably, Mtb Ddl displays a unique structural conformation distinct from Ddl enzymes of other bacterial species. This conformation is characterized by highly flexible nucleotide and D-alanine binding pockets, which undergo substantial rearrangements to facilitate substrate entry and catalysis (Maitra et al. 2019). The active site encompasses two distinct D-alanine binding sites with differing affinities: a high-affinity site (Kd ~95 nM) and a low-affinity site (Kd ~1.2 mM), in addition to an ATP binding site. The catalytic mechanism initiates with D-alanine binding to the high-affinity site, followed by phosphorylation of its carboxyl group by the γ-phosphate of ATP. Subsequently, a second D-alanine molecule binds to the low-affinity site, and ligation proceeds through nucleophilic attack by the amine group of this second molecule on the phosphorylated intermediate, yielding the D-alanyl-D-alanine dipeptide (Huynh et al. 2015).

D-cycloserine (DCS), a structural analogue of D-alanine, acts as a second-line antituberculosis drug by competitively inhibiting Ddl. It binds to both D-alanine sites, obstructing dipeptide formation and thereby disrupting peptidoglycan synthesis (Evangelopoulos et al. 2019; Pederick et al. 2020; de Chiara and Prosser 2023). Inhibition studies indicate that DCS binding affinity increases with higher ATP concentrations, exhibiting an IC50 of approximately 0.37 mM under standard assay conditions. Significantly, ATP binding precedes the binding of either D-alanine or DCS, with DCS competitively inhibiting substrate binding at both D-alanine sites. The pH dependence of inhibition further suggests that DCS binds optimally in its zwitterionic form (Prosser and de Carvalho 2013). Resistance to cycloserine in Mtb is frequently associated with mutations in the *ddl* gene that reduce drug binding affinity or alter enzyme activity, though such mutations are observed less often than mutations in *alr* (Chen et al. 2017; Evangelopoulos et al. 2019; Rahman et al. 2020; de Chiara and Prosser 2023). These mutations may impact substrate affinity or enzyme kinetics, potentially diminishing DCS efficacy. Additionally, *ddl* overexpression or the presence of compensatory mutations can enhance enzymatic capacity to produce D-alanyl-D-alanine despite the presence of the drug (Desjardins et al. 2016).

Beyond direct genetic alterations, the intricate peptidoglycan biosynthesis pathway and the mycobacteria-specific lipid-rich cell wall contribute to intrinsic resistance by limiting drug penetration and modifying metabolic fluxes (Sachan et al. 2023; Chiang and West 2024). Given its indispensable role in cell wall integrity and bacterial viability, Ddl represents a promising target for novel antibiotic development. Structural studies have identified druggable pockets within the enzyme, offering avenues for rational design of more effective inhibitors (Maitra et al. 2019; Chauhan et al. 2024).

Overall, D-alanine:D-alanine ligase in Mtb is an ATP-dependent homodimeric enzyme with a three-domain structure that catalyzes the formation of the D-alanyl-D-alanine dipeptide, a key precursor in peptidoglycan biosynthesis (Pederick et al. 2020). Its adaptive ligand-binding sites accommodate ATP and D-alanine substrates, facilitating efficient catalysis. Cycloserine inhibits Ddl by mimicking D-alanine and blocking dipeptide synthesis; however, resistance arises through mutations that affect substrate or inhibitor binding as well as increased enzyme expression (Prosser and de Carvalho 2013). Detailed structural and kinetic elucidations of Ddl provide a foundation for the design and development of improved therapeutics against drug-resistant tuberculosis.

9.4.14 Erythromycin Resistance Methylase (Erm37)

Erythromycin resistance methyltransferase Erm (37) in Mtb is an S-adenosylmethionine (SAM)-dependent 23S rRNA methyltransferase responsible for intrinsic resistance to macrolide antibiotics. It achieves this by methylating specific adenine residues within the ribosomal RNA, thereby preventing antibiotic binding (Buriánková et al. 2004; Maravić 2004). The *erm* (37) gene encodes a protein of approximately 179 amino acids, characterized by conserved methyltransferase motifs that facilitate SAM binding and catalyze the transfer of methyl groups to the 23S rRNA, particularly at nucleotide A2058 (Escherichia coli numbering), located in the peptidyl transferase center (Buriánková et al. 2004; Madsen et al. 2005; Broderick et al. 2014). Madsen et al. (2005) resolved the crystal structure of Erm (37), unveiling a unique "slippage" mechanism whereby the enzyme translocates along the rRNA substrate, enabling methylation of multiple adjacent adenines (A2057, A2058, and A2059). This produces an atypical methylation pattern that distinguishes Erm (37) from other members of the Erm family. This broader methylation confers resistance not only to macrolides but also to lincosamides and streptogramin B (MLS) antibiotics. Functional assays identified essential arginine residues (R119, R140, R141, R142) critical for RNA binding; mutations at these positions markedly impair methyltransferase activity, underscoring their role in substrate recognition (Madsen et al. 2005). Extending these findings, Goh et al. (2022) employed biochemical and genetic approaches to demonstrate that Erm (37) exhibits relatively relaxed substrate specificity compared to other Erm enzymes. Their work confirmed that Erm (37)-mediated methylation significantly reduces macrolide binding affinity, thereby conferring intrinsic resistance in Mtb. Significantly, Erm (37) methylation impacts clinically relevant macrolides such as clarithromycin (Goh et al. 2022). Clinically, while most erythromycin resistance in Mtb arises from mutations in the 23S rRNA gene, a subset of resistant strains harbors erm (37), which serves as an intrinsic resistance determinant limiting macrolide efficacy (Andini and Nash 2006; Fyfe et al. 2016). Comparative structural studies of ErmE, a related 23S rRNA methyltransferase from Streptomyces, have provided valuable insights into conserved catalytic mechanisms and RNA-binding interactions across the Erm family. These investigations elucidated the SAM binding site, substrate recognition features, and conformational dynamics critical for methylation activity (Stsiapanava and Selmer 2019). Such structural knowledge has advanced our understanding of how Erm methyltransferases achieve target specificity and mediate antibiotic resistance. Ongoing research continues to refine the functional and regulatory landscape of Erm enzymes (Alexander et al. 2025).

In addition to conferring antibiotic resistance, Mtb influences host-pathogen interactions via secreted methyltransferases like Rv1988 (Yaseen et al. 2015). This enzyme specifically methylates histone H3 at arginine 42 (H3R42), a rare modification within the globular domain rather than the typical histone tail. This epigenetic alteration alters nucleosome structure and suppresses immune response gene expression, enabling bacterial evasion of host defenses and enhancing survival within macrophages. Experimental deletion of Rv1988 reduces Mtb's infectivity, while ectopic expression in nonpathogenic mycobacteria enhances survival, highlighting the enzyme's crucial role in manipulating host epigenetics to promote pathogenesis (Yaseen et al. 2015).

Overall, unlike other Erm methyltransferases that target only A2058 in 23S rRNA, Erm (37) employs a distinctive mechanism: it initially methylates A2058 and then proceeds to methylate adjacent nucleotides A2057 and A2059 through a slippage mechanism (Madsen et al. 2005). This atypical methylation pattern underlies resistance to a broader spectrum of antibiotics, including ketolides such as telithromycin (Madsen et al. 2005). The unique, truncated C-terminal domain of Erm (37) likely contributes to its distinctive interaction with rRNA (Madsen et al. 2005; Stsiapanava and Selmer 2019). Generally, Erm methyltransferases confer resistance by modifying the ribosomal binding site for macrolide antibiotics, thereby disrupting their interaction with rRNA (Maravić 2004). Understanding these mechanisms is critical for the development of new antimicrobial agents and potential Erm inhibitors aimed at overcoming macrolide resistance.

9.4.15 Pentapeptide Repeat Protein MfpA (Rv3361c)

Pentapeptide repeat proteins (PRPs) are characterized by repeating units of five amino acids that fold into a distinctive right-handed quadrilateral β-helix structure. This unique conformation enables them to mimic DNA or other macromolecules, facilitating interactions with diverse molecular targets. In Mtb, the PRP MfpA (Rv3361c) plays a pivotal role in conferring resistance to fluoroquinolone antibiotics. MfpA structurally resembles B-form DNA, allowing it to bind DNA gyrase, an essential bacterial enzyme targeted by fluoroquinolones, and competitively inhibit DNA binding to protect the enzyme from drug-induced damage, thus enhancing bacterial survival (Zhang et al. 2021; Saran and Saha 2022). Structural analyses have demonstrated that the stable β-helix fold formed by the pentapeptide repeats is integral to MfpA's DNA mimicry and interaction with gyrase (Vetting et al. 2006). Furthermore, MfpA exhibits remarkable stability of its secondary and tertiary structures even under denaturing conditions. This resilience is reinforced by MfpA's propensity to form dimers or oligomers, features likely influencing its regulatory functions (Khrapunov et al. 2008). Recent crystallographic studies of the *Mycobacterium smegmatis* homologue revealed that MfpA functions as a T-segment DNA mimic that stimulates gyrase ATPase activity and diminishes fluoroquinolone-induced DNA cleavage. This activity is explicitly mediated through the ATPase domain of the gyrase B subunit, as confirmed by mutational analyses (Feng et al. 2021). Genetically, mfpA resides within an operon comprising *mfpB*, *mfpC*, *mfpD*, and *mfpE*. Within this operon, *MfpB* operates as a small GTPase, *MfpC* functions as its guanine nucleotide exchange factor (GEF), and *MfpD* acts as a GTPase-activating protein (GAP) that downregulates *MfpB* activity. This intricate regulatory network finely tunes *MfpA*-mediated fluoroquinolone resistance, illustrating a sophisticated mechanism by which Mtb modulates its antibiotic defense (Huang et al. 2022). Functional studies conducted in *Mycobacterium smegmatis* demonstrate that alterations in *mfpD* expression significantly affect fluoroquinolone sensitivity, further underscoring the complexity of regulatory controls over *MfpA*-mediated resistance (Mayer and Takiff 2014; Huang et al. 2022, 2024; Zhang et al. 2025).

9.4.16 16S rRNA Methyltransferase (GidB)

The *gidB* gene in Mtb encodes a highly conserved S-adenosylmethionine (SAM)-dependent 16S rRNA methyltransferase responsible for methylating the N7 position of guanine residue G518 (corresponding to G527 in *Escherichia coli*) within the 530 loops of the 16S rRNA (Okamoto et al. 2007; Wong et al. 2013; Rodríguez-García et al. 2021; Rodríguez-Castillo et al. 2015, 2021). This methylation generates 7-methylguanosine (m^7G), a modification critical for maintaining ribosomal decoding fidelity during translation and ensuring accurate protein synthesis (Bi et al. 2024). Structurally, GidB belongs to the Rossmann fold methyltransferase family, characterized by a SAM-binding domain that facilitates the transfer of a methyl group to the rRNA substrate (Gana et al. 2013). This modification fine-tunes the ribosome's functional architecture, which is essential for normal ribosomal activity and susceptibility to antibiotics (Wong et al. 2013). Loss or disruption of this conserved m^7G methylation, frequently caused by mutations in *gidB*, results in low-level streptomycin resistance in Mtb and other bacteria (Okamoto et al. 2007; Wong et al. 2011, 2013; Verma et al. 2014). The absence of methylation induces structural alterations in the ribosome that reduce streptomycin binding affinity, without significantly compromising bacterial fitness, as demonstrated in isogenic gidB deletion mutants (Okamoto et al. 2007; Wong et al. 2011). Clinically, *gidB* mutations have been detected in a substantial proportion of streptomycin-resistant Mtb isolates, particularly in strains lacking mutations in primary resistance genes such as *rpsL* and *rrs* (Villellas et al. 2013).

Resistance conferred by *gidB* mutations typically manifests as a modest increase in the minimum inhibitory concentration (MIC) for streptomycin, producing a heterogeneous spectrum of phenotypes ranging from susceptibility to resistance. This phenotypic variability means that some *gidB* mutants may be classified as sensitive during routine clinical testing, complicating treatment decisions and affecting patient outcomes (Wong et al. 2011). The reported observations underscore the pivotal role of *gidB* in ribosome biology and antibiotic susceptibility, establishing it as a critical genetic marker for low-level streptomycin resistance with important implications for tuberculosis therapy.

9.4.17 Proteins in Ethionamide Resistance

Resistance to ethionamide (ETH) in Mtb primarily arises from mutations in several key proteins belonging to distinct families, each characterized by unique structures, functions, and regulatory mechanisms. A comprehensive understanding of these proteins and the mutations conferring resistance is critical for addressing ETH resistance effectively (Brossier et al. 2011; Vilchèze and Jacobs 2014; Zhang et al. 2017; da Silva et al. 2018; Anand et al. 2022; Cao et al. 2023). The principal enzyme responsible for ETH activation is EthA, encoded by the *ethA* gene (Rv3854c). EthA is a flavin-containing monooxygenase of the Baeyer-Villiger monooxygenase family. Structurally, EthA possesses a flavin adenine dinucleotide (FAD) cofactor at its active site, which facilitates the oxidation of ethionamide, thereby converting the prodrug into its active form (Vannelli et al. 2002). Functionally, EthA catalyzes the S-oxidation of ETH, generating reactive intermediates that form an ETH-NAD adduct. This adduct inhibits the enoyl-ACP reductase InhA, effectively blocking mycolic acid synthesis, crucial for bacterial cell wall integrity (Morì et al. 2015). Mutations in *ethA*, including missense, nonsense, frameshift mutations, and deletions, impair or abolish EthA enzymatic activity, preventing ETH activation and causing resistance (de Souza et al. 2020). These mutations represent the predominant cause of ETH resistance in clinical Mtb isolates, exhibiting considerable diversity among strains (Morlock et al. 2003).

Expression of *ethA* is tightly regulated by EthR, a TetR-family transcriptional repressor encoded by *ethR* (Rv3855). EthR binds to the *ethA* promoter region, repressing its transcription. Structurally, EthR forms a homodimer containing a helix-turn-helix DNA-binding domain typical of TetR repressors (Carette et al. 2012). Mutations or overexpression of EthR enhance repression of *eth*A, reducing EthA protein levels, which in turn decreases ETH activation and contributes to resistance. Additionally, small molecules such as cyclic di-GMP can bind EthR, increasing its DNA-binding affinity and repression activity, further modulating resistance (Zhang et al. 2017).

Another critical target affected by ETH is InhA, encoded by the *inhA* gene. InhA belongs to the short-chain dehydrogenase/reductase (SDR) family and functions as a homotetrameric NADH-dependent enoyl-ACP reductase essential for mycolic acid biosynthesis. It contains a Rossmann fold for NADH binding and an active site accommodating enoyl substrates. The activated ETH-NAD adduct binds tightly to InhA, inhibiting its reductase activity (Banerjee et al. 1994). Mutations in *inhA*, particularly in the promoter region (e.g., the −15 C > T substitution) or within the coding sequence, can lead to overexpression or amino acid changes that reduce binding affinity for the ETH-NAD adduct, resulting in low-to-moderate-level ETH resistance (Vilchèze and Jacobs 2014). Additional Baeyer-Villiger monooxygenases, such as MymA (Rv3083) and Rv0565c, have also been implicated in ETH activation, although their clinical significance remains less well defined (Grant et al. 2016). MymA shares structural features with EthA, including the FAD cofactor and conserved catalytic residues essential for enzymatic activity. Functionally, MymA can oxidize ETH, contributing to prodrug activation. Mutations in *mymA* that disrupt FAD binding or substrate interaction reduce enzymatic activity, diminishing ETH activation and contributing to resistance, albeit usually to a lesser extent than ethA mutations (Tomas et al. 2022).

Mutations in *ethA* and *mymA* primarily arise under selective pressure from ETH treatment. These often target amino acids critical for maintaining active-site conformation or cofactor binding, resulting in loss of enzyme function. Furthermore, regulatory mutations altering the expression levels of these enzymes influence ETH susceptibility (Grant et al. 2016). Recent multi-omic analyses have identified additional gene clusters, such as Rv3093c-Rv3095, encoding a putative monooxygenase/repressor pair analogous to the *ethA*/*ethR* system, which may also participate in ETH metabolism and resistance, expanding the current understanding of ETH resistance mechanisms (Wan et al. 2023). Moreover, intrinsic resistance of Mtb is modulated by its complex, lipid-rich cell wall, which limits drug permeability and reduces susceptibility to hydrophilic compounds like ETH. Mutations in genes involved in cell wall biosynthesis and regulation may indirectly affect ETH resistance by altering drug access (Goossens et al. 2020). Overall, Ethionamide resistance in Mtb primarily stems from *ethA* mutations that disrupt prodrug activation, with ethR mutations further reducing EthA expression. Mutations in *inhA* decrease drug binding to the target enzyme, while other monooxygenases like MymA contribute minimally. Understanding these mechanisms and cell wall permeability is crucial for improving diagnostics and treatments for ETH-resistant tuberculosis.

9.4.18 Ddn (Deazaflavin-Dependent Nitroreductase; Rv3547), Fgd1 (F420-Dependent Glucose-6-Phosphate Dehydrogenase; Rv0407), FbiA, FbiB, FbiC (F420 Biosynthesis Proteins), Rv0132c (FbiD Homologue))

Delamanid and pretomanid represent a novel class of nitroimidazole antimycobacterial agents developed to combat MDR and XDR tuberculosis. Delamanid received European Medicines Agency (EMA) approval in 2014, followed by the US Food and Drug Administration's (FDA) authorization of pretomanid in 2019. Pretomanid has been incorporated into combination treatment regimens, notably the bedaquiline-pretomanid-linezolid (BPaL) protocol, which is specifically designed to combat drug-resistant tuberculosis (TB) (Ignatius and Dooley 2019; Mudde et al. 2022; Occhineri et al. 2022; Nguyen et al. 2023; Zhao et al. 2024). While resistance to pretomanid remains relatively infrequent, its prevalence among newly diagnosed TB cases ranges from 2% to 5%, increasing to 10–15% in patients experiencing treatment failure or relapse. These statistics highlight the critical need for continual molecular surveillance and in-depth mechanistic studies to monitor and understand resistance patterns effectively (Bloemberg et al. 2015; Hoffmann et al. 2016; Liu et al. 2022; Timm et al. 2023; Labuda et al. 2024). Resistance to delamanid and pretomanid primarily arises from mutations that disrupt the F420-dependent prodrug bioactivation pathway. Key proteins involved include the nitroreductase Ddn, glucose-6-phosphate dehydrogenase Fgd1, and the F420 biosynthetic enzymes FbiA, FbiB, and FbiC (Fujiwara et al. 2018; Liu et al. 2022). Importantly, the spontaneous mutation frequency conferring resistance to delamanid is comparable to that observed for isoniazid and PA-824, highlighting the importance of ongoing surveillance (Fujiwara et al. 2018). Mutations in *ddn*, *fgd1*, *fbiA*, fbiB, and *fbiC* correspond with distinct F420 chromatographic profiles, indicative of disrupted cofactor biosynthesis or function (Haver et al. 2015; Fujiwara et al. 2018). Certain mutations selectively abolish pretomanid activation while preserving basal Ddn enzymatic activity essential for survival (Lee et al. 2020). Notably, delamanid and pretomanid exhibit differential binding to Ddn, which may underpin variations in clinical efficacy across resistant strains (Lee et al. 2020). Understanding these resistance mechanisms is critical for developing rapid diagnostics and optimizing therapeutic interventions (Haver et al. 2015; Gómez-González et al. 2021; Mudde et al. 2022).

Ddn, a deazaflavin-dependent nitroreductase, mediates the activation of bicyclic nitroimidazole prodrugs, including PA-824 (pretomanid) and OPC-67683 (delamanid), through the utilization of the F420 coenzyme. This enzyme catalyzes a hydride transfer reaction that leads to the generation of reactive nitrogen species, essential for the drugs' antimycobacterial activity (Cellitti et al. 2012; Mohamed et al. 2016). The flexible N-terminal region of Ddn plays a key role in substrate binding, although complete structural elucidation of the protein remains incomplete (Cellitti et al. 2012). Ddn exhibits stereoselectivity, preferentially reducing the (S)-isomer of PA-824 that carries bulky substituents (Gurumurthy et al. 2013). The catalytic mechanism proceeds via hydride transfer from reduced F420 ($F420H_2$), followed by protonation and spontaneous intramolecular rearrangements. This process is facilitated by a triad of tyrosine residues that form a hydrophobic barrier around the active site, transiently allowing water molecules to access the site for proton transfer (Mohamed et al. 2016). Notably, the ddn Trp20Stop mutation, characterized by Mansjö et al. (2022), introduces a premature stop codon, resulting in a truncated, nonfunctional enzyme. This loss-of-function mutation is strongly associated with high-level resistance to both delamanid and pretomanid and has been detected exclusively in resistant isolates from the Swedish strain collection.

Fgd1, encoded by the *fgd1* gene, is a member of the glucose-methanol-choline (GMC) oxidoreductase family and contains an FAD cofactor. It catalyzes the oxidation of glucose-6-phosphate to 6-phosphogluconolactone, simultaneously reducing coenzyme F420 to its active form, $F420H_2$. This reduced coenzyme serves as the essential electron donor for Ddn-mediated prodrug activation (Purwantini et al. 2016). Mutations in *fgd1* diminish the availability of $F420H_2$, thereby impairing Ddn's function and contributing to drug resistance (Cellitti et al. 2012). The biosynthesis of F420 in Mtb is orchestrated by several enzymes, including FbiA, FbiB, FbiC, and the recently identified FbiD homologue Rv0132c. FbiC, a radical S-adenosylmethionine (SAM) enzyme, catalyzes the formation of the F420 precursor Fo from 5-amino-6-ribitylamino-2,4(1H,3H)-pyrimidinedione and tyrosine (Decamps et al. 2012; Mehta et al. 2015). FbiD functions as a guanylyltransferase that synthesizes dehydro-F420-0 using phosphoenolpyruvate as a substrate, revising earlier understanding that implicated 2-phospho-L-lactate (Bashiri et al. 2019). FbiA and FbiB catalyze subsequent glutamylation steps that convert Fo into mature coenzyme F420. Specifically, FbiB's C-terminal domain reduces dehydro-F420-0 using $FMNH_2$, while its N-terminal domain acts as a γ-glutamyl ligase (Bashiri et al. 2019). This elucidated biosynthetic pathway has facilitated heterologous expression and production of F420 in *Escherichia coli*, advancing mechanistic understanding of the complex enzymatic transformations involved in F420 biosynthesis (Bashiri et al. 2019; Mehta et al. 2015). The comprehensive pathway highlights the integral role of Fgd1 and the Fbi enzyme family in maintaining coenzyme F420 levels, which are essential for the activation of key antitubercular prodrugs in Mtb.

Mutations in the F420 biosynthetic genes (*fbiA*, *fbiB*, *fbiC*, *fbiD*), alongside *ddn* and *fgd1*, represent primary genetic determinants of pretomanid and delamanid resistance (Rifat et al. 2020; Haver et al. 2015). A genome-wide association study encompassing over 6000 clinical isolates identified novel resistance-associated mutations spanning diverse antituberculosis agents, including cycloserine, ethionamide, and para-aminosalicylic acid (Coll et al. 2018). Intriguingly, natural polymorphisms such as the *ddn* Tyr29 deletion have been reported in delamanid-naïve patients, correlating with elevated delamanid MICs (Reichmuth et al. 2020). Recent studies have highlighted the role of *fbiC* mutations in bacterial dormancy and persistence, implicating them in treatment failure and pathogen survival (Farnia et al. 2024a; Velayati et al. 2024).

9.5 Abolition of Prodrug Activation

Prodrug activation constitutes a fundamental mechanism underlying the effective treatment of tuberculosis (TB), as several frontline and second-line anti-TB drugs are administered as inactive compounds requiring enzymatic conversion within Mtb to exert their bactericidal effects. Notable examples include isoniazid (INH) and ethionamide (ETH), both of which rely on specific bacterial enzymes for activation: KatG, a catalase-peroxidase, activates INH, whereas EthA, a Baeyer-Villiger monooxygenase, mediates ETH activation (Gygli et al. 2017; Sachan et al. 2023). Mutations that impair or abolish the function of these activating enzymes prevent the conversion of prodrugs into their active metabolites, thereby resulting in resistance via the mechanism known as abolition of prodrug activation (Mori et al. 2015, 2017). This resistance mechanism is a major contributor to treatment failure and the emergence of drug-resistant TB strains. Resistance to INH and ETH arises through multiple molecular pathways. Mutations in the *katG* gene reduce or eliminate the enzyme's catalytic activity, frequently leading to high-level resistance due to complete loss of catalase-peroxidase function. Alternatively, mutations in the *inhA* gene or its promoter region alter the drug target, decreasing the binding affinity of the activated drug and typically conferring low- to-moderate-level resistance (Vilchèze and Jacobs 2014; Purkan et al. 2016). For ETH, resistance similarly involves mutations in *ethA* that abrogate prodrug activation, as well as mutations in *inhA* that affect the drug target, reducing susceptibility (Zhang et al. 1992; Morlock et al. 2003; Sachan et al. 2023). Molecular diagnostic assays that simultaneously target the *ethA* and *inhA* genes enable accurate discrimination between resistance mechanisms arising from impaired prodrug activation and those due to target site modifications. Complementary functional assays evaluating enzymatic activity and phenotypic drug susceptibility further elucidate the molecular basis of resistance (Rodriguez et al. 2015; Machado et al. 2016). Specifically, strains harboring *ethA* mutations exhibit diminished activation of ethionamide (ETH), whereas those with *inhA* mutations maintain prodrug activation but display reduced drug susceptibility as a consequence of altered target binding affinity. A comparable pattern is observed in isoniazid (INH) resistance, where *katG* mutations abolish prodrug activation, while *inhA* mutations alter the drug target. Additionally, overexpression of EthR, a TetR-family transcriptional repressor of *ethA*, can suppress EthA expression, phenotypically mimicking loss of activation and thus complicating resistance profiles (Dookie et al. 2018). Resistance to pyrazinamide (PZA), a key first-line anti-TB prodrug, primarily arises from mutations in the *pncA* gene, which encodes the enzyme pyrazinamidase responsible for converting PZA into its active form, pyrazinoic acid. Loss-of-function mutations in *pncA* disrupt this critical enzymatic activation, resulting in resistance via abolition of prodrug activation. In contrast, target-based resistance to PZA is relatively rare, likely due to the elusive and poorly defined nature of its molecular target within Mtb (Sachan et al. 2023). Recent advances have elucidated key resistance mechanisms to nitroimidazole prodrugs such as pretomanid and delamanid. Activation of these agents depends on the nitroreductase Ddn and the coenzyme F420 biosynthesis pathway, involving genes such as *fbiA*, *fbiB*, *fbiC*, *fbiD*, and *fgd1*. Mutations in these genes disrupt prodrug activation, culminating in resistance (Haver et al. 2015; Rifat et al. 2020; Fujiwara et al. 2018). Notably, mutations in Rv2983 (*fbiD*) have recently been identified as novel contributors to nitroimidazole resistance (Rifat et al. 2020). While most mutations confer high-level resistance to both delamanid and pretomanid, certain mutations in *fbiB* and Rv2983 differentially influence the degree of resistance. Importantly, some of these resistance-associated mutations preserve the native enzymatic activity of Ddn, which is essential for bacterial survival. This allows resistant *Mycobacterium tuberculosis* strains to emerge and persist (Lee et al. 2020). Therefore, careful molecular monitoring of these mutations is critical to ensure the effective clinical use of nitroimidazole drugs and to manage the development of resistance. In summary, resistance arises either from mutations that disrupt prodrug activation, leading to high-level resistance, or from target-site modifications causing variable resistance levels. Distinguishing these mechanisms is essential for accurate diagnostics and effective treatment. A multifaceted approach, including DNA sequencing, enzymatic assays, phenotypic testing, and computational modeling, provides a comprehensive framework to identify mutations and predict their impact (Singh et al. 2020).

9.6 Drug Inactivation by Modifying Enzyme

Drug-inactivating enzymes in Mtb play a pivotal role in antibiotic resistance by chemically modifying antimicrobial agents, thereby neutralizing their therapeutic efficacy. Among these, the enhanced intracellular survival (*Eis*) acetyltransferase is one of the most extensively characterized. Eis is an intracellular protein belonging to the GCN5-related N-acetyltransferase (GNAT) superfamily, featuring a unique tripartite fold formed by the fusion of two GNAT domains and a C-terminal region, resulting in a distinctive heart-shaped structure (Chen et al. 2011; Favrot et al. 2016; Baumgartner et al. 2021). This structural complexity creates a considerable, bifurcated, and highly negatively charged substrate-binding pocket capable of accommodating aminoglycoside antibiotics such as kanamycin A and capreomycin. Eis enzymatically multiacetylates several amino groups on these drugs, effectively inactivating them (Fong and Berghuis 2002; Smith et al. 2017). The acetylation mechanism employed by Eis follows a random-sequential bisubstrate kinetic model characteristic of aminoglycoside acetyltransferases (Tsodikov et al. 2014). Contrastingly, other Mtb acetyltransferases such as homoserine transacetylase (HTA) operate via a ping-pong kinetic mechanism, regulated by substrate binding and feedback inhibition (Sharma et al. 2023). Additionally, enzymes like MbtA, which participate in siderophore biosynthesis, undergo reversible acetylation and deacetylation mediated by the acetyltransferase Pat (Rv0998) and the deacetylase DAc (Rv1151c), modulating function under iron-limiting conditions (Vergnolle et al. 2016). The bifunctional enzyme GlmU, essential for mycobacterial cell wall synthesis, exhibits a distinctive acetyl-CoA binding conformation and is regulated by phosphorylation via PknB. Key catalytic residues, His-374 and Asn-397, as well as substrate-binding residues Ser-416 and Trp-460, govern its acetyltransferase activity (Jagtap et al. 2012). Beyond direct drug inactivation, Eis significantly contributes to Mtb immune evasion. Eis acetylates DUSP16/MKP-7, a JNK-specific phosphatase, thereby inhibiting JNK-dependent autophagy, phagosome maturation, and reactive oxygen species (ROS) generation (Kim et al. 2012). This suppression of autophagic and inflammatory responses occurs through a redox-dependent pathway involving modulation of ROS production and proinflammatory cytokine release (Shin et al. 2010). Molecular docking studies highlight Eis's binding to DUSP16/MKP-7 via charge complementarity and favorable geometric interactions (Yoon et al. 2013). Eis further enhances interleukin-10 (IL-10) expression by promoting histone H3 acetylation at the IL-10 promoter, activating the Akt/mTOR/p70S6K pathway and thereby inhibiting autophagy, cumulatively fostering intracellular bacterial survival (Duan et al. 2016). Structure-guided drug design, leveraging crystal structures of Eis (e.g., PDB IDs 6X10, 6X6Y), has led to the development of potent thieno[2,3-d]pyrimidine derivatives that competitively inhibit Eis by targeting its substrate-binding pocket. Punetha et al. (2020) synthesized a series of Eis inhibitors that effectively block Eis-mediated acetylation of kanamycin, restoring drug susceptibility in resistant Mtb strains with Eis overexpression. These findings underscore Eis inhibitors' therapeutic potential as adjuvants against aminoglycoside resistance.

The other critical drug-inactivating enzyme is the β-lactamase BlaC, an Ambler class A serine β-lactamase that hydrolyzes β-lactam antibiotics, including penicillins, cephalosporins, and carbapenems, which inhibit bacterial cell wall synthesis (Luthra et al. 2018). BlaC's active site contains a conserved serine residue that mediates nucleophilic attack on the β-lactam ring, forming a covalent acyl-enzyme intermediate that is subsequently hydrolyzed to release an inactive antibiotic (Mehta et al. 2015, 2021). With broad substrate specificity and high catalytic efficiency, BlaC confers intrinsic β-lactam resistance in Mtb. However, β-lactamase inhibitors such as clavulanate effectively inhibit BlaC activity, restoring β-lactam efficacy and providing key strategies to combat resistance (Luthra et al. 2018). The l,d-transpeptidases LdtMt1 and LdtMt2 catalyze 3 → 3 cross-linking in Mtb peptidoglycan, a predominant linkage type in this pathogen's cell wall (Cordillot et al. 2013). Carbapenems, such as meropenem and imipenem, efficiently inactivate these enzymes by binding and acylating their catalytic cysteine residues. This inactivation proceeds through rapid formation of a non-covalent enzyme-antibiotic complex, followed by slower acylation that yields a highly stable acyl-enzyme intermediate (Dubée et al. 2012). Side chain variations among carbapenems influence binding affinity and acylation kinetics, with carbapenems outperforming cephalosporins in l,d-transpeptidase inhibition (Dubée et al. 2012; Sütterlin et al. 2018). Crystal structures of LdtMt2 complexed with carbapenems confirm covalent attachment to catalytic Cys-354, elucidating the molecular basis of inhibition (Gokulan et al. 2018). This unique carbapenem targeting of l,d-transpeptidases offers promising avenues for addressing drug-resistant TB (Cordillot et al. 2013). Nevertheless, Mtb can inactivate carbapenems indirectly via BlaC-mediated hydrolysis or slow hydrolysis of the acyl-enzyme complex, limiting their clinical efficacy (Luthra et al. 2018). Recently, the iron- and α-ketoglutarate-dependent sulfate ester dioxygenase Rv3406 was identified as capable of inactivating the experimental antitubercular agent RCB18350, expanding the repertoire of drug-inactivating enzymes (Recchia et al. 2025). In addition, cytochrome bd oxidase has emerged as a promising target for novel anti-TB drugs, with structural and mechanistic insights accelerating inhibitor development (Harikishore et al. 2023). Enzymes involved in fatty acid biosynthesis, cellular metabolism, and metal uptake are also under investigation as potential drug targets (Yan et al. 2022). Strategies to combat multidrug-resistant TB increasingly focus on designing drugs with novel mechanisms, phenotypic screening of small molecule libraries, and drug repositioning approaches (Singh and Chibale 2021). An emerging resistance mechanism involves the methyltransferase encoded by Rv0560c, which can N-methylate pyrido-benzimidazole compounds, such as the experimental drug "compound 14," preventing inhibition of the essential enzyme DprE1, a critical factor in arabinogalactan synthesis for the mycobacterial cell wall (Warrier et al. 2016). This methylation-mediated drug inactivation represents a novel resistance pathway, although its clinical relevance requires further validation. Thus, Mtb employs a range of drug-inactivating enzymes, including acetyltransferases, β-lactamases, dioxygenases, and methyltransferases, to chemically modify or degrade antibiotics, rendering them ineffective and facilitating resistance. The structural diversity of these enzymes underpins their ability to catalyze various resistance-conferring reactions. Recent advances in structural biology have enabled the development of targeted inhibitors, such as clavulanate and novel Eis inhibitors, which hold promise for overcoming enzymatic resistance and improving treatment outcomes for drug-resistant TB.

9.7 Efflux Pump-Mediated Drug Extrusion

Efflux pumps are critical contributors to drug resistance in Mtb, playing a significant role in both intrinsic and acquired resistance mechanisms by actively exporting a wide range of antimicrobial agents out of bacterial cells, thereby reducing intracellular drug concentrations and nullifying their efficacy (da Silva et al. 2011). These transmembrane proteins transport substrates, including first- and second-line anti-TB drugs, from the cytoplasm across the complex mycobacterial cell envelope to the extracellular milieu (Pal et al. 2014). Notably, Mtb possesses an extensive array of putative efflux pumps relative to the size of its genome. These efflux systems are classified into five major families based on their structural organization and energy transduction mechanisms: the Major Facilitator Superfamily (MFS), ATP-Binding Cassette (ABC) transporters, Resistance-Nodulation-Division (RND) superfamily, Small Multidrug Resistance (SMR) family, and the Multidrug and Toxic Compound Extrusion (MATE) family (Kanji et al. 2019; Chetri 2023; Saier and Paulsen 2001; Varela et al. 2023). Mechanistically, ABC transporters utilize the energy derived from ATP hydrolysis to facilitate the translocation of substrates across cellular membranes. In contrast, efflux pumps belonging to the MFS, RND, SMR, and MATE families operate primarily via electrochemical gradients of protons (H^+) or sodium ions (Na^+), employing drug:H^+ or drug:Na^+ antiport mechanisms (Saier and Paulsen 2001). This mechanistic diversity affords these transporters the capacity to recognize and extrude a broad spectrum of structurally diverse antibiotics, thereby contributing to resistance against multiple drug classes (Pal et al. 2014). The MFS family is predominantly implicated in antibiotic resistance within Gram-positive bacteria, including Mtb, whereas the RND family plays a more prominent role in resistance among Gram-negative organisms (Durães et al. 2018). Current investigative efforts focus on the identification and refinement of both natural and synthetic efflux pump inhibitors, aiming to counteract antimicrobial resistance by impairing efflux functionality (Durães et al. 2018). EfpA, the first multidrug efflux pump characterized in Mtb, belongs to the MFS and is essential for bacterial viability and drug resistance, particularly against isoniazid (INH) (Roberts et al. 2024; Wang et al. 2024). EfpA plays a crucial role in Mtb's recovery following INH exposure and may be involved in the development of resistance during prolonged therapy (Roberts et al. 2024). Structural investigations employing cryo-electron microscopy have revealed that EfpA functions as a lipid flippase, containing bound endogenous lipids spanning the membrane (Li et al. 2024; Wang et al. 2024). Inhibitors such as BRD-8000.3 function via competitive binding to a tunnel-shaped lateral fenestration within the EfpA transporter, displacing a phospholipid fatty acid chain and thereby obstructing access to the natural substrate binding site (Khandelwal et al. 2024; Wang et al. 2024). This mode of inhibition parallels that of darobactin, an antibiotic which targets the BamA insertase by occupying its lateral gate and substituting a bound lipid molecule (Kaur et al. 2021). These structural insights have provided a molecular basis for the observed synergistic effects of these inhibitors and underscore their potential in combination therapeutic strategies to potentiate antitubercular efficacy (Khandelwal et al. 2024; Chim et al. 2012). Additionally, BRD-9327 binds within the outer vestibule region of EfpA, likely interfering with its "alternate access" transport mechanism (Khandelwal et al. 2024). Complementary mutagenesis and transport studies have further substantiated EfpA's role as a multidrug efflux pump with broad substrate specificity (Rai and Mehra 2021). Other MFS transporters exhibiting diverse physiological functions include Rv1877 and Rv0191. Rv1877 acts as a multidrug efflux pump that is essential for the detoxification of spermidine, a polyamine involved in cellular growth and stress adaptation processes. In contrast, Rv0191 appears to facilitate the transport of low molecular weight thiols, implicating its role in defense against oxidative stress (Sao Emani and Reiling 2024). These functions correspond closely with the well-established protective role of mycothiol, a predominant low molecular weight thiol unique to mycobacteria, which safeguards the cell against reactive oxygen species, alkylating agents, and various antibiotics (Newton et al. 1999; Rawat et al. 2002). Significantly, mycothiol-deficient strains demonstrate heightened susceptibility to oxidative insults and multiple antimicrobials, yet paradoxically exhibit increased resistance to isoniazid (INH), suggesting the presence of intricate regulatory networks governing redox homeostasis within Mtb (Rawat et al. 2002). Collectively, these observations highlight the multifaceted roles of MFS transporters beyond classical drug efflux, potentially encompassing the regulation of membrane potential, intracellular pH balance, and modulation of stress response pathways (Dos Santos et al. 2014).

The Tap efflux pump, encoded by Rv1258c, represents another critical MFS transporter implicated in multidrug resistance through the active extrusion of a broad spectrum of antibiotics, including rifampicin, tetracycline, and aminoglycosides (Aínsa et al. 1998; Pal et al. 2014). Structurally, Tap functions as a proton-dependent antiporter comprising twelve transmembrane helices, underscoring its importance in mediating antibiotic tolerance (Aínsa et al. 1998). The discovery of a novel G442A mutation in Rv1258c within rifampicin-resistant clinical isolates implicates this pump in clinically relevant resistance phenotypes (Farzaneh et al. 2024). Functional ablation of Tap in *Mycobacterium bovis BCG* elicited widespread modulation of gene expression pertaining to cell wall biosynthesis and stress response pathways, thereby revealing its broader physiological contributions beyond drug transport (Ramón-García et al. 2012). Moreover, both overexpression and disruption of Tap significantly altered antibiotic susceptibility profiles, correlating with its confirmed role in active efflux as demonstrated by intracellular drug accumulation assays (Ramón-García et al. 2012). ABC transporters, defined by their two transmembrane and two nucleotide-binding domains, harness the energy from ATP hydrolysis to transport substrates across cell membranes. In Mtb, ABC transporters facilitate the uptake of nutrients and export of toxic compounds, influencing pathogenesis, virulence, and drug resistance (Ford and Beis 2019; Cassio Barreto de Oliveira and Balan 2020). The Mtb genome encodes at least 26 complete and 11 incomplete ABC transporters, equally divided between importers and exporters (Braibant et al. 2000). Unlike exporters, importers possess periplasmic substrate-binding domains critical for high-affinity nutrient acquisition (Oliveira and Balan 2020). Recent advances in structural biology have provided molecular frameworks elucidating ABC transport cycle mechanisms (Rees et al. 2009). Among importers and exporters, Rv0194 and Rv1273c have been identified as integral multidrug efflux pumps, capable of exporting β-lactams, streptomycin, and chloramphenicol, and mediating biofilm formation, thereby contributing to antibiotic tolerance (Danilchanka et al. 2008; Adhikary et al. 2023). Knockout studies of efflux pump genes Rv1218c, Rv3065, Rv0849, and Rv1258c further corroborate the substantial role of ABC and SMR transporters in mediating drug extrusion (Balganesh et al. 2012).

The RND family of efflux pumps, widely studied in Gram-negative bacteria for their multidrug resistance capacity, is also present in mycobacteria and contributes to intrinsic resistance and virulence. These transporters typically consist of proton motive force-driven tripartite complexes expelling diverse harmful compounds (Colclough et al. 2020; Kavanaugh et al. 2024; Zwama and Nishino 2021). Structural and evolutionary analyses have highlighted amino acid substitutions enhancing RND pump efficiency, though therapeutic targeting via efflux pump inhibitors has been limited due to issues with toxicity, formulation, and pharmacokinetics (Duffey et al. 2024). Computational approaches, including molecular docking, offer renewed potential for the design of specific inhibitors against RND pumps (Duffey et al. 2024). Understanding the regulation and physiological functions of these pumps remains critical in overcoming MDR challenges (Colclough et al. 2020; Georgieva et al. 2022; Kavanaugh et al. 2024). Within Mtb, the RND-type mycobacterial membrane protein large (MmpL) family represents a set of large multidomain transporters involved in both lipid transport and drug efflux, central to bacterial physiology and drug resistance (Viljoen et al. 2017). The MmpS5-MmpL5 efflux system, regulated negatively by the transcriptional repressor Rv0678, is clinically significant for mediating resistance to bedaquiline and clofazimine. Farnia et al. (2024b) demonstrated that mutations in the Rv0678 gene lead to derepression of the MmpS5-MmpL5 efflux pump, resulting in increased efflux activity. Whole-genome sequencing of multidrug-resistant Mtb clinical isolates revealed a high prevalence (92.3%) of unique insertion-deletion polymorphisms and amino acid substitutions clustered

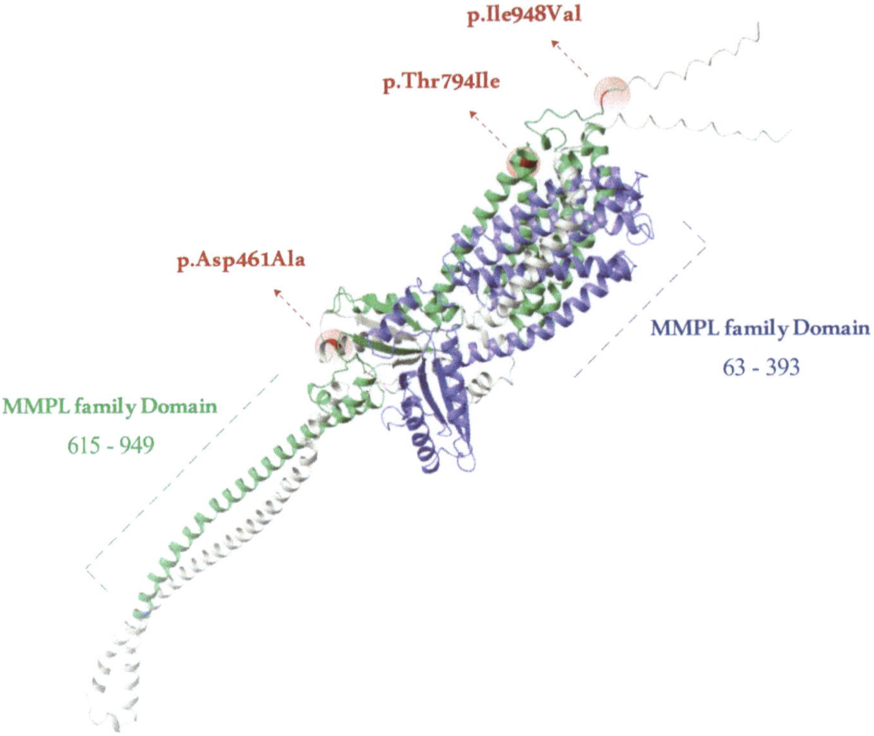

Fig. 9.2 Structural localization of high-frequency missense mutations in the MmpL5 efflux transporter of XDR- and TDR-TB isolates. This figure illustrates the modeled three-dimensional structure of the MmpL5 transmembrane efflux transporter (964 amino acids) from *Mycobacterium tuberculosis*. The protein comprises two MMPL family domains: residues 63–393 (blue) and 615–949 (green). High-frequency missense mutations observed in XDR- and TDR-TB isolates (≥96%), p.Asp461Ala, p.Thr794Ile, and p.Ile948Val are highlighted in red. The p.Asp461Ala mutation is located within the transmembrane region and may interfere with critical electrostatic interactions required for substrate transport. The p.Thr794Ile and p.Ile948Val mutations are situated in the C-terminal MMPL domain and are predicted to cause conformational changes that alter protein structure and modulate efflux pump activity, leading to reduced intracellular drug accumulation. These structural alterations are associated with enhanced efflux function. They would increase resistance to bedaquiline and other lipophilic antimicrobials, underscoring the role of MmpL5 as a key efflux pump contributing to multidrug resistance in these clinical isolates

*Drawing by Mohsen Maleknia; Mycobacteriology Research Center, National Research Institute of Tuberculosis and Lung Disease, Shahid Beheshti University of Medical Sciences, Tehran, Iran

within the MmpL5 (Rv0676) gene, reflecting in vivo adaptive evolution driven by subtherapeutic drug pressure. Notably, no significant mutations were detected in MmpS5 (Rv0677) or the Rv0678 regulatory gene itself, suggesting that structural variations in MmpL5 play a critical role in modulating efflux pump function and contributing to decreased drug susceptibility. Building on these findings, we identified three concurrent mutations, p.Ile948Val, p.Thr794Ile, and p. Asp461Ala, in XDR and TDR-TB isolates. The p.Ile948Val missense mutation is implicated in altering drug transport mechanisms, thereby promoting multidrug resistance, while the p. Thr794Ile substitution induces conformational changes affecting drug permeability across the bacterial cell envelope. The p.Asp461Ala mutation modulates transport efficiency, further enhancing resistance. Together, these mutations underscore the multifaceted ways in which alterations in bacterial transport and permeability pathways contribute to drug resistance in Mtb (Fig. 9.2).

The other efflux families, SMR and MATE, energized by proton or sodium gradients, also participate in intrinsic resistance mechanisms. The

MATE transporter encoded by the mmp gene in *Mycobacterium smegmatis* contributes to resistance against various antibiotics and is part of a constitutively expressed operon important for basal drug tolerance (Mishra and Daniels 2013). SMR transporters, among the smallest multidrug efflux proteins, utilize proton-drug antiport strategies and are ubiquitous among pathogenic bacteria including Mtb, contributing to resistance by significantly reducing intracellular drug accumulation (Ninio et al. 2001). Thus, efflux pumps are central to multidrug resistance in Mtb, as they effectively diminish intracellular antibiotic concentrations and promote the formation of biofilms, structured microbial communities that bolster bacterial persistence by increasing tolerance to antimicrobial agents and evading host immune responses. These pumps facilitate the export of signaling molecules and toxic metabolites essential for biofilm development and maintenance (Askoura et al. 2011; Zhang et al. 2023, 2024; Ren et al. 2024). Today, targeting efflux systems with inhibitors has emerged as a promising strategy to both restore antibiotic susceptibility and disrupt biofilm integrity, thereby addressing chronic, drug-resistant infections (Nishino et al. 2021). A thorough understanding of efflux pump regulation, substrate specificity, and inhibition is crucial for the rational development of novel efflux pump inhibitors (EPIs) aimed at overcoming drug resistance and improving tuberculosis treatment outcomes (Askoura et al. 2011; Zhang et al. 2023; Ren et al. 2024; Nishino et al. 2021).

In conclusion, drug resistance in Mtb arises from multiple, interrelated mechanisms that collectively enhance bacterial survival under antibiotic pressure. These mechanisms encompass genetic mutations that modify drug targets, thereby reducing antibiotic binding affinity and therapeutic efficacy. Furthermore, overexpression of target proteins can overwhelm drug action, diminishing its inhibitory potential. Enzymatic modification or inactivation of antimicrobial agents further attenuates their effectiveness. Crucially, active efflux of drugs via membrane transporters lowers intracellular drug concentrations, facilitating multidrug resistance. Alterations in the bacterial cell envelope also contribute by decreasing membrane permeability and limiting drug access to intracellular targets. These resistance mechanisms may operate independently or synergistically, posing significant challenges to effective treatment.

Conflict of Interest The authors declare that they have no conflict of interest related to the content of this chapter.

Ethical Consideration This chapter is based on conclusions drawn from the authors' research as well as data and information available in the literature, all of which have been appropriately cited. The figures and images presented originate from the authors' studies, which did not involve experiments on humans, animals, or cell cultures. All referenced experimental work was conducted using laboratory and wild-type strains of *Mycobacterium tuberculosis* (Mtb), following established standard protocols. The research project received approval from the Scientific Committee of the NRITLD, Tehran, Iran. The findings and interpretations presented are grounded in a rigorous analysis of existing research combined with the authors' experimental work.

References

Adhikary A, Chatterjee D, Ghosh AS (2023) ABC superfamily transporter Rv1273c of Mycobacterium tuberculosis acts as a multidrug efflux pump. FEMS Microbiol Lett 370:fnad114. https://doi.org/10.1093/femsle/fnad114. PMID: 37881010

Aínsa JA, Blokpoel MC, Otal I, Young DB, De Smet KA, Martín C (1998) Molecular cloning and characterization of tap, a putative multidrug efflux pump present in mycobacterium fortuitum and mycobacterium tuberculosis. J Bacteriol 180(22):5836–5843. https://doi.org/10.1128/JB.180.22.5836-5843.1998. PMID: 9811639; PMCID: PMC107655

Agrawal A, Roué M, Spitzfaden C, Petrella S, Aubry A, Hann M, Bax B, Mayer C (2013) Mycobacterium tuberculosis DNA gyrase ATPase domain structures suggest a dissociative mechanism that explains how

ATP hydrolysis is coupled to domain motion. Biochem J 456(2):263–273

Akbergenov R, Shcherbakov D, Matt T, Duscha S, Meyer M, Wilson DN, Böttger EC (2011) Molecular basis for the selectivity of antituberculosis compounds capreomycin and viomycin. Antimicrob Agents Chemother 55(10):4712–4717

Alangaden GJ, Kreiswirth BN, Aouad A, Khetarpal M, Igno FR, Moghazeh SL, Manavathu EK, Lerner SA (1998) Mechanism of resistance to amikacin and kanamycin in Mycobacterium tuberculosis. Antimicrob Agents Chemother 42(5):1295–1297

Alcaide F, Pfyffer GE, Telenti A (1997) Role of embB in natural and acquired resistance to ethambutol in mycobacteria. Antimicrobl Agents Chemother 41(10):2270-2273

Alderwick LJ, Lloyd GS, Ghadbane H, May JW, Bhatt A, Eggeling L, Fütterer K, Besra GS (2011) The C-terminal domain of the Arabinosyltransferase Mycobacterium tuberculosis EmbC is a lectin-like carbohydrate binding module. PLoS Pathog 7(2):e1001299

Alderwick LJ, Harrison J, Lloyd GS, Birch HL (2015) The mycobacterial cell wall--peptidoglycan and arabinogalactan. Cold Spring Harb Perspect Med 5(8):a021113

Alderwick LJ, Birch HL, Krumbach K, Bott M, Eggeling L, Besra GS (2018) AftD functions as an α1→ 5 arabinofuranosyltransferase involved in the biosynthesis of the mycobacterial cell wall core. Cell Surf 1:2-14

Alexander DC, Farquhar T, Adams JME, Suchan DM, Workman SD, Wallace RJ Jr, Brown-Elliott BA, El-Halfawy OM, Cameron ADS (2025) Macrolide resistance due to erm(55). Microbiol Spectr 13(3):e0239724

Alifano P, Palumbo C, Pasanisi D, Talà A (2015) Rifampicin-resistance, rpoB polymorphism and RNA polymerase genetic engineering. J Biotechnol 202:60–77

Amin AG, Goude R, Shi L, Zhang J, Chatterjee D, Parish T (2008) EmbA is an essential arabinosyltransferase in Mycobacterium tuberculosis. Microbiology 154(1):240–248

Amusengeri A, Khan A, Tastan BÖ (2022) The structural basis of Mycobacterium tuberculosis RpoB drug-resistant clinical mutations on rifampicin drug binding. Molecules 27(3):885

Anand PK, Kumar A, Saini A, Kaur J (2022) Mutation in EthA protein of Mycobacterium tuberculosis conferred drug tolerance against ethionamide in Mycobacterium smegmatis mc2155. Comput Biol Chem 98:107677

Andini N, Nash KA (2006) Intrinsic macrolide resistance of the Mycobacterium tuberculosis complex is inducible. Antimicrob Agents Chemother 50(7):2560–2562

Ando H, Miyoshi-Akiyama T, Watanabe S, Kirikae T (2014) A silent mutation in mabA confers isoniazid resistance on Mycobacterium tuberculosis. Mol Microbiol 91(3):538–547

Andries K, Verhasselt P, Guillemont J, Göhlmann HW, Neefs JM, Winkler H, Van Gestel J, Timmerman P, Zhu M, Lee E, Williams P, de Chaffoy D, Huitric E, Hoffner S, Cambau E, Truffot-Pernot C, Lounis N, Jarlier V (2005) A diarylquinoline drug active on the ATP synthase of Mycobacterium tuberculosis. Science 307(5707):223–227

Angula KT, Legoabe LJ, Beteck RM (2021) Chemical classes presenting novel antituberculosis agents currently in different phases of drug development: A 2010–2020 review. Pharmaceuticals 14(5):461

Anthony KG, Strych U, Yeung KR, Shoen CS, Perez O, Krause KL, Cynamon MH, Aristoff PA, Koski RA (2011) New classes of alanine racemase inhibitors identified by high-throughput screening show antimicrobial activity against Mycobacterium tuberculosis. PLoS One 6(5):e20374

Aono A, Chikamatsu K, Yamada H, Kato T, Mitarai S (2014) Association between pncA gene mutations, pyrazinamidase activity, and pyrazinamide susceptibility testing in Mycobacterium tuberculosis. Antimicrob Agents Chemother 58(8):4928–4930

Arjomandzadegan M, Gravand S (2015) Analysis of rpsL and rrs genes mutations related to streptomycin resistance in Mdr and Xdr clinical isolates of Mycobacterium tuberculosis. Tuberk Toraks 63(4):235–242

Askoura M, Mottawea W, Abujamel T, Taher I (2011) Efflux pump inhibitors (EPIs) as new antimicrobial agents against Pseudomonas aeruginosa. Libyan J Med 13:6

Asojo OA, Nelson SK, Mootien S, Lee Y, Rezende WC, Hyman DA, Matsumoto MM, Reiling S, Kelleher A, Ledizet M, Koski RA, Anthony KG (2014) Structural and biochemical analyses of alanine racemase from the multidrug-resistant Clostridium difficile strain 630. Acta Crystallogr D Biol Crystallogr 70(Pt 7):1922–1933

Awasthy D, Bharath S, Subbulakshmi V, Sharma U (2012) Alanine racemase mutants of Mycobacterium tuberculosis require D-alanine for growth and are defective for survival in macrophages and mice. Microbiology (Reading). 158(Pt 2):319–327

Aziz T, Goodwin D (2021) New insight into the protein-based cofactor of Mycobacterium tuberculosis KatG: toward better understanding an unusual catalase mechanism. FASEB J 35(S1):05396

Baid P, Sengupta J (2023) Cryo-EM captures a unique conformational rearrangement in 23S rRNA helices of the Mycobacterium 50S subunit. Int J Biol Macromol 253(Pt 3):126876

Bakhtiyariniya P, Khosravi AD, Hashemzadeh M, Savari M (2022) Detection and characterization of mutations in genes related to isoniazid resistance in Mycobacterium tuberculosis clinical isolates from Iran. Mol Biol Rep 49:6135–6143

Bakuła Z, Napiórkowska A, Bielecki J, Augustynowicz-Kopeć E, Zwolska Z, Jagielski T (2013) Mutations in the embB gene and their association with ethambutol resistance in multidrug-resistant Mycobacterium tuberculosis clinical isolates from Poland. Biomed Res Int 2013:167954

Balganesh M, Dinesh N, Sharma S, Kuruppath S, Nair AV, Sharma U (2012) Efflux pumps of Mycobacterium tuberculosis play a significant role in antituberculosis activity of potential drug candidates. Antimicrob Agents Chemother 56(5):2643–2651

Banerjee A, Dubnau E, Quemard A, Balasubramanian V, Um KS, Wilson T, Collins D, de Lisle G, Jacobs WR Jr (1994) inhA, a gene encoding a target for isoniazid and ethionamide in Mycobacterium tuberculosis. Science 263(5144):227–230

Banerjee A, Sugantino M, Sacchettini JC, Jacobs WR Jr (1998) The mabA gene from the inhA operon of Mycobacterium tuberculosis encodes a 3lketoacyl reductase that fails to confer isoniazid resistance. Microbiology 144(10):2697–2704

Bashiri G, Antoney J, Jirgis ENM, Shah MV, Ney B, Copp J, Stuteley SM, Sreebhavan S, Palmer B, Middleditch M, Tokuriki N, Greening C, Scott C, Baker EN, Jackson CJ (2019) A revised biosynthetic pathway for the cofactor F420 in prokaryotes. Nat Commun 10(1):1558

Baumgartner JT, Habeeb Mohammad TS, Czub MP, Majorek KA, Arolli X, Variot C, Anonick M, Minor W, Ballicora MA, Becker DP, Kuhn ML (2021) Gcn5-related N-acetyltransferases (GNATs) with a catalytic serine residue can play ping-pong too. Front Mol Biosci 8:646046

Beckert P, Hillemann D, Kohl TA, Kalinowski J, Richter E, Niemann S, Feuerriegel S (2012) rplC T460C identified as a dominant mutation in linezolid-resistant Mycobacterium tuberculosis strains. Antimicrob Agents Chemother 56(5):2743–2745

Belanger AE, Besra GS, Ford ME, Mikusová K, Belisle JT, Brennan PJ, Inamine JM (1996) The embAB genes of Mycobacterium avium encode an arabinosyl transferase involved in cell wall arabinan biosynthesis that is the target for the antimycobacterial drug ethambutol. Proc Natl Acad Sci USA 93(21):11919–11924

Belisle JT, Vissa VD, Sievert T, Takayama K, Brennan PJ, Besra GS (1997) Role of the major antigen of Mycobacterium tuberculosis in cell wall biogenesis. Science 276(5317):1420–1422

Belousoff MJ, Eyal Z, Radjainia M, Ahmed T, Bamert RS, Matzov D, Bashan A, Zimmerman E, Mishra S, Cameron D, Elmlund H, Peleg AY, Bhushan S, Lithgow T, Yonath A (2017) Structural basis for linezolid binding site rearrangement in the Staphylococcus aureus ribosome. MBio 8(3):e00395–e00317

Berg S, Starbuck J, Torrelles JB, Vissa VD, Crick DC, Chatterjee D, Brennan PJ (2005) Roles of conserved proline and glycosyltransferase motifs of EmbC in biosynthesis of lipoarabinomannan. J Biol Chem 280(7):5651–5663

Bertrand T, Eady NA, Jones JN, Nagy JM, Jamart-Gregoire B, Raven EL, Brown KA (2004) Crystal structure of Mycobacterium tuberculosis catalase-peroxidase. J Biol Chem 279(37):38991–38999

Besra GS, Khoo KH, McNeil MR, Dell A, Morris HR, Brennan PJ (1995) A new interpretation of the structure of the mycolyl-arabinogalactan complex of Mycobacterium tuberculosis as revealed through characterization of oligoglycosylalditol fragments by fast-atom bombardment mass spectrometry and 1H nuclear magnetic resonance spectroscopy. Biochemistry 34(13):4257–4266

Bhat AH, Pathak D, Rao A (2017) The alr-groEL1 operon in Mycobacterium tuberculosis: an interplay of multiple regulatory elements. Sci Rep 7:43772

Bi Z, Chen YX, Young ID, Su HW, Chen Y, Hong JY, Fraser JS, Javid B (2024) Ribosomal RNA methylation by GidB modulates discrimination of mischarged tRNA. elife 13:RP102752

Biukovic G, Basak S, Manimekalai MS, Rishikesan S, Roessle M, Dick T, Rao SP, Hunke C, Grüber G (2013) Variations of subunit {varepsilon} of the mycobacterium tuberculosis F1Fo ATP synthase and a novel model for mechanism of action of the tuberculosis drug TMC207. Antimicrob Agents Chemother 57(1):168-176. https://doi.org/10.1128/AAC.01039-12. Epub 2012 Oct 22. PMID: 23089752; PMCID: PMC3535943

Bloemberg GV, Keller PM, Stucki D, Trauner A, Borrell S, Latshang T, Coscolla M, Rothe T, Hömke R, Ritter C, Feldmann J, Schulthess B, Gagneux S, Böttger EC (2015) Acquired resistance to Bedaquiline and Delamanid in therapy for tuberculosis. N Engl J Med 373(20):1986–1988

Bortoluzzi A, Muskett FW, Waters LC, Addis PW, Rieck B, Munder T, Schleier S, Forti F, Ghisotti D, Carr MD, O'Hare HM (2013) Mycobacterium tuberculosis RNA polymerase-binding protein A (RbpA) and its interactions with sigma factors. J Biol Chem 288(20):14438–14450

Borukhov S, Nudler E (2003) RNA polymerase holoenzyme: structure, function and biological implications. Curr Opin Microbiol 6(2):93–100

Braibant M, Gilot P, Content J (2000) The ATP binding cassette (ABC) transport systems of Mycobacterium tuberculosis. FEMS Microbiol Rev 24(4):449–467

Brammachary U, Ramachandra V, Palavesam S, Raj Cuppusamy Kapalamurthy V, Muralidhar A, Muthaiah M (2025) Mechanisms and action of drug resistance on Mycobacterium tuberculosis [Internet]. In: New Insights in AntibioticResistance and Pneumococcal Vaccines. IntechOpen. Available from: https://doi.org/10.5772/intechopen.108943

Brandis G, Hughes D (2013) Genetic characterization of compensatory evolution in strains carrying rpoB Ser531Leu, the rifampicin resistance mutation most frequently found in clinical isolates. J Antimicrob Chemother 68(11):2493–2497

Brennan PJ (2003) Structure, function, and biogenesis of the cell wall of Mycobacterium tuberculosis. Tuberculosis (Edinb) 83(1–3):91–97

Brennan PJ, Nikaido H (1995) The envelope of mycobacteria. Annu Rev Biochem 64:29–63

Broderick JB, Duffus BR, Duschene KS, Shepard EM (2014) Radical S-adenosylmethionine enzymes. Chem Rev 114(8):4229–4317

Brossier F, Veziris N, Truffot-Pernot C, Jarlier V, Sougakoff W (2006) Performance of the genotype MTBDR line probe assay for detection of resistance to rifampin and isoniazid in strains of Mycobacterium tuberculosis with low- and high-level resistance. J Clin Microbiol 44(10):3659–3664

Brossier F, Veziris N, Truffot-Pernot C, Jarlier V, Sougakoff W (2011) Molecular investigation of resistance to the antituberculous drug ethionamide in multidrug-resistant clinical isolates of Mycobacterium tuberculosis. Antimicrob Agents Chemother 55(1):355–360

Bruning JB, Murillo AC, Chacon O, Barletta RG, Sacchettini JC (2011) Structure of the mycobacterium tuberculosis Dalanine: D-alanine ligase, a target of the antituberculosis drug D-cycloserine. Antimicrob Agents Chemother 55(1):291–301. https://doi.org/10.1128/AAC.00558-10. PMID: 21098251; PMCID: PMC3019691

Buriánková K, Doucet-Populaire F, Dorson O, Gondran A, Ghnassia JC, Weiser J, Pernodet JL (2004) Molecular basis of intrinsic macrolide resistance in the Mycobacterium tuberculosis complex. Antimicrob Agents Chemother 48(1):143–150

Bush NG, Evans-Roberts K, Maxwell A (2015) DNA Topoisomerases. EcoSal Plus 6:2. https://doi.org/10.1128/ecosalplus.ESP-0010-2014

Bwalya P, Solo E, Chizimu J, Shrestha D, Mbulo G, Thapa J, Nakajima C, Suzuki Y (2022) Characterization of embB mutations involved in ethambutol resistance in multi-drug resistant Mycobacterium tuberculosis isolates in Zambia. Tuberculosis 133:102184

Cade CE, Dlouhy AC, Medzihradszky KF, Salas-Castillo SP, Ghiladi RA (2010) Isoniazid-resistance conferring mutations in Mycobacterium tuberculosis KatG: catalase, peroxidase, and INH-NADH adduct formation activities. Protein Sci 19(3):458–474

Campbell EA, Korzheva N, Mustaev A, Murakami K, Nair S, Goldfarb A, Darst SA (2001) Structural mechanism for rifampicin inhibition of bacterial rna polymerase. Cell 104(6):901–912

Campbell PJ, Morlock GP, Sikes RD, Dalton TL, Metchock B, Starks AM, Hooks DP, Cowan LS, Plikaytis BB, Posey JE (2011) Molecular detection of mutations associated with first- and second-line drug resistance compared with conventional drug susceptibility testing of Mycobacterium tuberculosis. Antimicrob Agents Chemother 55(5):2032–2041

Cao B, Mijiti X, Deng LL, Wang Q, Yu JJ, Anwaierjiang A, Qian C, Li M, Fang DA, Jiang Y, Zhao LL, Zhao X, Wan K, Liu H, Li G, Yuan X (2023) Genetic characterization conferred co-resistance to isoniazid and Ethionamide in Mycobacterium tuberculosis isolates from southern Xinjiang, China. Infect Drug Resist. 16:3117–3135

Cardoso RF, Cardoso MA, Leite CQ, Sato DN, Mamizuka EM, Hirata RD et al (2007) Characterization of ndh gene of isoniazid resistant and susceptible mycobacterium tuberculosis isolates from Brazil. Mem Inst Oswaldo Cruz 102(1):59–61. https://doi.org/10.1590/S0074-02762007000100009

Carette X, Blondiaux N, Willery E, Hoos S, Lecat-Guillet N, Lens Z et al (2012) Structural activation of the transcriptional repressor EthR from Mycobacterium tuberculosis by single amino acid change mimicking natural and synthetic ligands. Nucleic Acids Res 40(7):3018–3030

Carter JJ, Walker TM, Walker AS, Whitfield MG, Morlock GP, Lynch CI, Adlard D, Peto TEA, Posey JE, Crook DW, Fowler PW (2024) Prediction of pyrazinamide resistance in Mycobacterium tuberculosis using structure-based machine-learning approaches. JAC Antimicrob Resist 6(2):dlae037

Cassio Barreto de Oliveira M, Balan A (2020) The ATP-binding cassette (ABC) transport Systems in Mycobacterium tuberculosis: structure, function, and possible targets for therapeutics. Biology (Basel) 9(12):443

Catalão MJ, Filipe SR, Pimentel M (2019) Revisiting anti-tuberculosis therapeutic strategies that target the peptidoglycan structure and synthesis. Front Microbiol 10:190

Cellitti SE, Shaffer J, Jones DH, Mukherjee T, Gurumurthy M, Bursulaya B, Boshoff HI, Choi I, Nayyar A, Lee YS, Cherian J, Niyomrattanakit P, Dick T, Manjunatha UH, Barry CE 3rd, Spraggon G, Geierstanger BH (2012) Structure of Ddn, the deazaflavin-dependent nitroreductase from Mycobacterium tuberculosis involved in bioreductive activation of PA-824. Structure 20(1):101–112

Chambers HF, Moreau D, Yajko D, Miick C, Wagner C, Hackbarth C, Kocagöz S, Rosenberg E, Hadley WK, Nikaido H (1995) Can penicillins and other beta-lactam antibiotics be used to treat tuberculosis?. Antimicrob Agents Chemother 39(12):2620–2624

Champoux JJ (2001) DNA topoisomerases: structure, function, and mechanism. Annu Rev Biochem 70:369–413

Chauhan M, Barot R, Yadav R, Joshi K, Mirza S, Chikhale R, Srivastava VK, Yadav MR, Murumkar PR (2024) The Mycobacterium tuberculosis cell wall: an alluring drug target for developing newer anti-TB drugs—a perspective. Chem Biol Drug Des 104(3):e14612

Chen W, Biswas T, Porter VR, Tsodikov OV, Garneau-Tsodikova S (2011) Unusual regioversatility of acetyltransferase Eis, a cause of drug resistance in XDR-TB. Proc Natl Acad Sci 108(24):9804–9808

Chen W, Green KD, Garneau-Tsodikova S (2012) Cosubstrate tolerance of the aminoglycoside resistance enzyme Eis from Mycobacterium tuberculosis. Antimicrob Agents Chemother 56(11):5831–5838

Chen J, Zhang S, Cui P, Shi W, Zhang W, Zhang Y (2017) Identification of novel mutations associated with cycloserine resistance in Mycobacterium tuberculosis. J Antimicrob Chemother 72(12):3272–3276

Cheng SJ, Thibert L, Sanchez T, Heifets L, Zhang Y (2000) pncA mutations as a major mechanism of pyrazinamide resistance in Mycobacterium tuberculosis: spread of a monoresistant strain in Quebec, Canada. Antimicr Agents Chemother 44(3):528–532

Chetri S (2023) The culmination of multidrug-resistant efflux pumps vs. meager antibiotic arsenal era: urgent need for an improved new generation of EPIs. Front

Microbiol 14:1149418. https://doi.org/10.3389/fmicb.2023.1149418. PMID: 37138605; PMCID: PMC10149990

Chiang CY, West NP (2024) The fall of the mycobacterial cell wall: interrogating peptidoglycan synthesis for novel anti-TB agents. PeerJ 12:e18404

Chiaradia L, Lefebvre C, Parra J, Marcoux J, Burlet-Schiltz O, Etienne G, Tropis M, Daffé M (2017) Dissecting the mycobacterial cell envelope and defining the composition of the native mycomembrane. Sci Rep 7:12807

Chim N, Owens CP, Contreras H, Goulding CW (2012) Withdrawn. Infect disord drug targets. Epub ahead of print. PMID: 23167715; PMCID: PMC3695056

Chollet A, Mourey L, Lherbet C, Delbot A, Julien S, Baltas M, Bernadou J, Pratviel G, Maveyraud L, Bernardes-Génisson V (2015) Crystal structure of the enoyl-ACP reductase of Mycobacterium tuberculosis (InhA) in the apo-form and in complex with the active metabolite of isoniazid pre-formed by a biomimetic approach. J Struct Biol 190(3):328–337

Cobongela S, Mgwenya T (2024) Treatment strategies for Mycobacterium tuberculosis infection. In: IntechOpen

Cocozaki AI, Altman RB, Huang J, Buurman ET, Kazmirski SL, Doig P, Prince DB, Blanchard SC, Cate JH, Ferguson AD (2016) Resistance mutations generate divergent antibiotic susceptibility profiles against translation inhibitors. Proc Natl Acad Sci USA 113(29):8188–8193

Colclough AL, Alav I, Whittle EE, Pugh HL, Darby EM, Legood SW, McNeil HE, Blair JM (2020) RND efflux pumps in gram-negative bacteria; regulation, structure and role in antibiotic resistance. Future Microbiol 15:143–157

Coll F, Phelan J, Hill-Cawthorne GA, Nair MB, Mallard K, Ali S, Abdallah AM, Alghamdi S, Alsomali M, Ahmed AO, Portelli S, Oppong Y, Alves A, Bessa TB, Campino S, Caws M, Chatterjee A, Crampin AC, Dheda K, Furnham N, Glynn JR, Grandjean L, Minh Ha D, Hasan R, Hasan Z, Hibberd ML, Joloba M, Jones-López EC, Matsumoto T, Miranda A, Moore DJ, Mocillo N, Panaiotov S, Parkhill J, Penha C, Perdigão J, Portugal I, Rchiad Z, Robledo J, Sheen P, Shesha NT, Sirgel FA, Sola C, Oliveira Sousa E, Streicher EM, Helden PV, Viveiros M, Warren RM, McNerney R, Pain A, Clark TG (2018) Genome-wide analysis of multi- and extensively drug-resistant Mycobacterium tuberculosis. Nat Genet 50(2):307–316

Collin F, Karkare S, Maxwell A (2011) Exploiting bacterial DNA gyrase as a drug target: current state and perspectives. Appl Microbiol Biotechnol 92(3):479–497

Collins JA, Osheroff N (2024) Gyrase and topoisomerase IV: recycling old targets for new antibacterials to combat fluoroquinolone resistance. ACS Infect Dis 10(4):1097–1115. https://doi.org/10.1021/acsinfecdis.4c00128

Conradie F, Diacon AH, Ngubane N, Howell P, Everitt D, Crook AM, Mendel CM, Egizi E, Moreira J, Timm J, McHugh TD, Wills GH, Bateson A, Hunt R, Van Niekerk C, Li M, Olugbosi M, Spigelman M, Nix-TB Trial Team (2020) Treatment of highly drug-resistant pulmonary tuberculosis. N Engl J Med 382(10):893–902

Corbett KD, Shultzaberger RK, Berger JM (2004) The C-terminal domain of DNA gyrase A adopts a DNA-bending β-pinwheel fold. Proc Natl Acad Sci USA 101(19):7293–7298

Cordillot M, Dubée V, Triboulet S, Dubost L, Marie A, Hugonnet JE, Arthur M, Mainardi JL (2013) In vitro cross-linking of Mycobacterium tuberculosis peptidoglycan by L,D-transpeptidases and inactivation of these enzymes by carbapenems. Antimicrob Agents Chemother 57(12):5940–5945

Cruz JW, Sharp JD, Hoffer ED, Maehigashi T, Vvedenskaya IO, Konkimalla A, Husson RN, Nickels BE, Dunham CM, Woychik NA (2015) Growth-regulating Mycobacterium tuberculosis VapC-mt4 toxin is an isoacceptor-specific tRNase. Nat Commun 6:7480

Cuevas-Córdoba B, Xochihua-González SO, Cuellar A, Fuentes-Domínguez J, Zenteno-Cuevas R (2013) Characterization of pncA gene mutations in pyrazinamide-resistant Mycobacterium tuberculosis isolates from Mexico. Infect Genet Evol 19:330–334

Daffé M, Marrakchi H (2019) Unraveling the structure of the mycobacterial envelope. Microbiol Spectrum 7(4):10–128

Daffe M, Brennan PJ, McNeil M (1990) Predominant structural features of the cell wall arabinogalactan of Mycobacterium tuberculosis as revealed through characterization of oligoglycosyl alditol fragments by gas chromatography/mass spectrometry and by 1H and 13C NMR analyses. J Biol Chem 265(12):6734–6743

Daffé M, Quémard A, Marrakchi H (2019) Mycolic acids: from chemistry to biology. In: Biogenesis of fatty acids, lipids and membranes. Springer, Cham, pp 181–216

Dalla Costa ER, Ribeiro MO, Silva MS, Arnold LS, Rostirolla DC, Cafrune PI et al (2009) Correlations of mutations in katG, oxyR-ahpC and inhA genes and in vitro susceptibility in Mycobacterium tuberculosis clinical strains segregated by spoligotype families from tuberculosis prevalent countries in South America. BMC Microbiol 9:39

Danilchanka O, Mailaender C, Niederweis M (2008) Identification of a novel multidrug efflux pump of Mycobacterium tuberculosis. Antimicrob Agents Chemother 52(7):2503–2511

Danilchanka O, Sun J, Pavlenok M, Maueröder C, Speer A, Siroy A, Marrero J, Trujillo C, Mayhew DL, Doornbos KS, Muñoz LE, Herrmann M, Ehrt S, Berens C, Niederweis M (2014) An outer membrane channel protein of Mycobacterium tuberculosis with exotoxin activity. Proc Natl Acad Sci USA 111(18):6750–6755

Das N, Jena PK, Pradhan SK (2020) Arabinosyltransferase C enzyme of Mycobacterium tuberculosis, a potential drug target: An insight from molecular docking study. Heliyon 6(2):e02693

Datta M, Pillai M, Modak MJ, Liiv A, Khaja FT, Hussain T, Remme J, Varshney U (2021) A mutation in the ribosomal protein uS12 reveals novel functions of its universally conserved PNSA loop. Mol Microbiol 115(6):1292–1308

Decamps L, Philmus B, Benjdia A, White R, Begley TP, Berteau O (2012) Biosynthesis of F0, precursor of the F420 cofactor, requires a unique two radical-SAM domain enzyme and tyrosine as substrate. J Am Chem Soc 134(44):18173–18176

da Silva PE, Von Groll A, Martin A, Palomino JC (2011) Efflux as a mechanism for drug resistance in mycobacterium tuberculosis. FEMS Immunol Med Microbiol 63(1):1–9. https://doi.org/10.1111/j.1574-695X.2011.00831.x

da Silva DA, Ferreira NV, Rego AM, Barbosa PCP, Machado RF, Pimentel A et al (2018) Integrated analysis of ethionamide resistance loci in mycobacterium tuberculosis clinical isolates. Tuberculosis (Edinb) 113:163–174. https://doi.org/10.1016/j.tube.2018.08.010

de Chiara C, Prosser GA, Ogrodowicz R, de Carvalho LPS (2023) Structure of the D-cycloserine-resistant variant D322N of alanine racemase from mycobacterium tuberculosis. ACS Bio Med Chem Au 3(3):233–239. https://doi.org/10.1021/acsbiomedchemau.2c00074. PMID: 37363078; PMCID: PMC10288493

Demirci H, Wang L, Murphy FV 4th, Murphy EL, Carr JF, Blanchard SC, Jogl G, Dahlberg AE, Gregory ST (2013) The central role of protein S12 in organizing the structure of the decoding site of the ribosome. RNA 19(12):1791–1801

de Keijzer J, Mulder A, de Ru AH, van Soolingen D, van Veelen PA (2017) Parallel reaction monitoring of clinical mycobacterium tuberculosis lineages reveals pre-existent markers of rifampicin tolerance in the emerging Beijing lineage. J Proteome 150:9–17. https://doi.org/10.1016/j.jprot.2016.08.022

Desjardins CA, Cohen KA, Munsamy V, Abeel T, Maharaj K, Walker BJ, Shea TP, Almeida DV, Manson AL, Salazar A, Padayatchi N, O'Donnell MR, Mlisana KP, Wortman J, Birren BW, Grosset J, Earl AM, Pym AS (2016) Genomic and functional analyses of Mycobacterium tuberculosis strains implicate ald in D-cycloserine resistance. Nat Genet 48(5):544–551

de Souza VC, Antunes D, LHS S, PVZC G, Caffarena ER, ACR G et al (2020) Insights into the mechanism of ethionamide resistance in mycobacterium tuberculosis through an in silico structural evaluation of EthA and mutants identified in clinical isolates. Catalysts 10(5):543. https://doi.org/10.3390/catal10050543

Dhameliya TM, Devani AA, Patel KA, Shah KC (2022) Comprehensive coverage on anti-mycobacterial Endeavour reported in 2021. ChemistrySelect 7(19):e202200921

Dobbins A, Guzmán K, Herm C, Phillips E, Russell K (2019) A physical model of Mycobacterium tuberculosis MazF-mt6 illustrates the catalytic residues needed for 23S rRNA cleavage via a proton-relay mechanism. FASEB J 33(1 Suppl):493.3

Dokrungkoon T, Tulyaprawat O, Suwannakarn K, Ngamskulrungroj P (2023) In vitro modeling of isoniazid resistance mechanisms in mycobacterium tuberculosis H37Rv. Front Microbiol 10(14):1171861. https://doi.org/10.3389/fmicb.2023.1171861. PMID: 37581473; PMCID: PMC10417620

Donnabella V et al (1994) Isolation of the gene for the beta subunit of RNA polymerase from rifampin-resistant Mycobacterium tuberculosis and identification of new mutations. Am J Respir Cell Mol Biol 11:639–643

Dookie N, Rambaran S, Padayatchi N, Mahomed S, Naidoo K (2018) Evolution of drug resistance in Mycobacterium tuberculosis: a review on the molecular determinants of resistance and implications for personalized care. J Antimicrob Chemother 73(5):1138–1151

Dos Santos SC, Teixeira MC, Dias PJ, Sá-Correia I (2014) MFS transporters required for multidrug/multixenobiotic (MD/MX) resistance in the model yeast: understanding their physiological function through post-genomic approaches. Front Physiol 5:180

Drage MG, Tsai HC, Pecora ND, Cheng TY, Arida AR, Shukla S, Rojas RE, Seshadri C, Moody DB, Boom WH, Sacchettini JC, Harding CV (2010) Mycobacterium tuberculosis lipoprotein LprG (Rv1411c) binds triacylated glycolipid agonists of toll-like receptor 2. Nat Struct Mol Biol 17(9):1088–1095

Drlica K, Zhao X (1997) DNA gyrase, topoisomerase IV, and the 4-quinolones. Microbiol Mol Biol Rev 61(3):377–392. https://doi.org/10.1128/mmbr.61.3.377-392.1997. PMID: 9293187; PMCID: PMC232616

Duan L, Yi M, Chen J, Li S, Chen W (2016) Mycobacterium tuberculosis EIS gene inhibits macrophage autophagy through up-regulation of IL-10 by increasing the acetylation of histone H3. Biochem Biophys Res Commun 473(4):1229–1234

Dubée V, Triboulet S, Mainardi JL, Ethève-Quelquejeu M, Gutmann L, Marie A, Dubost L, Hugonnet JE, Arthur M (2012) Inactivation of Mycobacterium tuberculosis l,d-transpeptidase $LdtMt_1$ by carbapenems and cephalosporins. Antimicrob Agents Chemother 56(8):4189–4195

Duffey M, Jumde RP, da Costa RMA, Ropponen HK, Blasco B, Piddock LJV (2024) Extending the potency and lifespan of antibiotics: inhibitors of gram-negative bacterial efflux pumps. ACS Infect Dis. 10(5):1458–1482

Durães F, Pinto M, Sousa E (2018) Medicinal chemistry updates on bacterial efflux pump modulators. Curr Med Chem 25(42):6030–6069. https://doi.org/10.2174/0929867325666180209142612

Elamin AA, Stehr M, Oehlmann W, Singh M (2009) The mycolyltransferase 85A, a putative drug target of Mycobacterium tuberculosis: development of a novel assay and quantification of glycolipid-status of the mycobacterial cell wall. J Microbiol Methods 79(3):358–363

Evangelopoulos D, Prosser GA, Rodgers A, Dagg BM, Khatri B, Ho MM, Gutierrez MG, Cortes T, de Carvalho LP (2019) Comparative fitness analysis of

D-cycloserine resistant mutants reveals both fitness-neutral and high-fitness cost genotypes. Nat Commun 10(1):4177

Fan X, Tang X, Yan J, Xie J (2014) Identification of idiosyncratic Mycobacterium tuberculosis ribosomal protein subunits with implications in extraribosomal function, persistence, and drug resistance based on transcriptome data. J Biomol Struct Dyn 32(10):1546–1551

Farnia P, Maleknia M, Farnia P, Ghanavi J (2024a) Adaptive mechanisms of Mycobacterium tuberculosis: role of fbiC mutations in dormancy and survival. Int J Mycobacteriol 13(4):355–361

Farnia P, Besharati S, Farina P, Ayoubi S, Marjani M, Ghanavi J, Tabarsi P, Velayati AA (2024b) The role of efflux pumps transporter in multi-drug resistant tuberculosis: mycobacterial memberane protein (MmpL5). Intl J Mycobacteriol 13(1):7–14

Farzaneh S, Norouzi F, Fazeli H, Salehi M, Safari M, Nasr EB (2024) Novel mutation in efflux pump Rv1258c (tap) gene in drug resistant clinical isolates of Mycobacterium tuberculosis in Iran. J Infect Dev Ctries 18(2):243–250

Favrot L, Ronning DR (2012) Targeting the mycobacterial envelope for tuberculosis drug development. Expert Rev AntiInfect Ther 10(9):1023–1036. https://doi.org/10.1586/eri.12.91. PMID: 23106277; PMCID: PMC3571691

Favrot L, Lajiness DH, Ronning DR (2014) Inactivation of the Mycobacterium tuberculosis antigen 85 complex by covalent, allosteric inhibitors. J Biol Chem 289(36):25031–25040

Favrot L, Blanchard JS, Vergnolle O (2016) Bacterial GCN5-related N-acetyltransferases: from resistance to regulation. Biochemistry 55(7):989–1002

Feng L, Mundy JEA, Stevenson CEM, Mitchenall LA, Lawson DM, Mi K, Maxwell A (2021) The pentapeptide-repeat protein, MfpA, interacts with mycobacterial DNA gyrase as a DNA T-segment mimic. Proc Natl Acad Sci USA 118(11):e2016705118

Floss HG, Yu TW (2005) Rifamycin-mode of action, resistance, and biosynthesis. Chem Rev 105(2):621–632

Fong DH, Berghuis AM (2002) Substrate promiscuity of an aminoglycoside antibiotic resistance enzyme via target mimicry. EMBO J 21(10):2323–2331. https://doi.org/10.1093/emboj/21.10.2323. PMID: 12006485; PMCID: PMC126009

Ford RC, Beis K (2019) Learning the ABCs one at a time: structure and mechanism of ABC transporters. Biochem Soc Trans 47(1):23–36

Fujiwara M, Kawasaki M, Hariguchi N, Liu Y, Matsumoto M (2018) Mechanisms of resistance to delamanid, a drug for Mycobacterium tuberculosis. Tuberculosis (Edinb) 108:186–194

Fyfe C, Grossman TH, Kerstein K, Sutcliffe J (2016) Resistance to macrolide antibiotics in public health pathogens. Cold Spring Harb Perspect Med 6(10):a025395

Gairola A, Benjamin A, Weatherston JD, Cirillo JD, Wu HJ (2022) Recent developments in drug delivery for treatment of tuberculosis by targeting macrophages. Adv Therapeut 5(6):2100193

Galina L, Hopf FS, Abbadi BL, Sperotto ND, Czeczot AM, Duque-Villegas MA, Perello MA, Matter LB, de Souza EV, Parish T, Machado P (2022) Evaluation of 3-Deoxy-D-Arabino-Heptulosonate 7-phosphate synthase (DAHPS) as a vulnerable target in Mycobacterium tuberculosis. Microbiol Spectrum 10(4):e00728–e00722

Galperin MY, Koonin EV (1997) A diverse superfamily of enzymes with ATP-dependent carboxylate-amine/thiol ligase activity. Protein Sci 6(12):2639–2643. https://doi.org/10.1002/pro.5560061218. PMID: 9416615; PMCID: PMC2143612

Gan WC, Ng HF, Ngeow YF (2023) Mechanisms of linezolid resistance in mycobacteria. Pharmaceuticals (Basel) 16(6):784

Gana R, Rao S, Huang H, Wu C, Vasudevan S (2013) Structural and functional studies of S-adenosyl-L-methionine binding proteins: a ligand-centric approach. BMC Struct Biol 13:6

García-Ortega L, Álvarez-García E, Gavilanes JG, Martínez-del-Pozo Á, Joseph S (2010) Cleavage of the sarcin–ricin loop of 23S rRNA differentially affects EF-G and EF-Tu binding. Nucleic Acids Res 38(12):4108–4119

Garneau-Tsodikova S, Labby KJ (2016) Mechanisms of resistance to aminoglycoside antibiotics: overview and perspectives. Medchemcomm 7(1):11–27. https://doi.org/10.1039/C5MD00344J. PMID: 26877861; PMCID: PMC4752126

Georghiou SB, Magana M, Garfein RS, Catanzaro DG, Catanzaro A, Rodwell TC (2012) Evaluation of genetic mutations associated with Mycobacterium tuberculosis resistance to amikacin, kanamycin and capreomycin: a systematic review. PLoS One 7(3):e33275

Georgieva ER, Ahmad AB, Adetuyi O, Majeed S (2022) Production of recombinant Mtb membrane efflux pump for structural and functional studies to reveal mechanisms of drug resistance. FASEB J 36(S1):R2532

Gessese AT, Kinde MZ, Eshetu T, Kerisew B (2025) Whole-genome sequencing analysis to identify antimicrobial resistance regions and virulence factors in Mycobacterium tuberculosis isolates from the Amhara region. Ethiopia Scientif Reports 15(1):1–5

Ghiladi RA, Knudsen GM, Medzihradszky KF, Ortiz de Montellano PR (2005) The met-Tyr-Trp cross-link in Mycobacterium tuberculosis catalase-peroxidase (KatG): autocatalytic formation and effect on enzyme catalysis and spectroscopic properties. J Biol Chem 280(24):22651–22663

Goh BC, Xiang X, Lescar J, Dedon PC (2022) Crystal structure and functional analysis of mycobacterial erythromycin resistance methyltransferase Erm38 reveals its RNA-binding site. J Biol Chem 298(2):101571

Goins CM, Dajnowicz S, Smith MD, Parks JM, Ronning DR (2018) Mycolyltransferase from Mycobacterium tuberculosis in covalent complex with tetrahydrolipstatin provides insights into antigen 85 catalysis. J Biol Chem 293(10):3651–3662

Gokulan K, Khare S, Cerniglia CE, Foley SL, Varughese KI (2018) Structure and inhibitor specificity of L, D-transpeptidase (LdtMt2) from mycobacterium tuberculosis and antibiotic resistance: calcium binding promotes dimer formation. AAPS J 20(2):44

Goossens SN, Sampson SL, Van Rie A (2020) Mechanisms of drug-induced tolerance in Mycobacterium tuberculosis. Clin Microbiol Rev 34(1):e00141–e00120

Gómez-González PJ, Perdigão J, Gomes P, Puyen ZM, Santos-Lázaro D, Napier G et al (2021) Genetic diversity of candidate loci linked to mycobacterium tuberculosis resistance to bedaquiline, delamanid and pretomanid. Sci Rep 11:19431. https://doi.org/10.1038/s41598-021-98862-4. PMID: 34561159; PMCID: PMC8460363

Goude R, Amin AG, Chatterjee D, Parish T (2009) The arabinosyltransferase EmbC is inhibited by ethambutol in Mycobacterium tuberculosis. Antimicrob Agents Chemother 53(10):4138–4146

Grant SS, Wellington S, Kawate T, Desjardins CA, Silvis MR, Wivagg C, Thompson M, Gordon K, Kazyanskaya E, Nietupski R, Haseley N, Iwase N, Earl AM, Fitzgerald M, Hung DT (2016) Baeyer-Villiger monooxygenases EthA and MymA are required for activation of replicating and non-replicating Mycobacterium tuberculosis inhibitors. Cell Chem Biol 23(6):666–677

Guimarães BG, Souchon H, Honoré N, Saint-Joanis B, Brosch R, Shepard W, Cole ST, Alzari PM (2005) Structure and mechanism of the alkyl Hydroperoxidase AhpC, a key Elementof the Mycobacterium tuberculosis defense system against OxidativeStress. J Biol Chem 280(27):25735–25742

Guo S, Chongsuvivatwong V, Lei S (2022) Comparison on major gene mutations related to rifampicin and isoniazid resistance between Beijing and nonBeijing strains of Mycobacterium tuberculosis: a systematic review and Bayesian metaanalysis. Genes (Basel) 13(10):1849

Gurumurthy M, Rao M, Mukherjee T, Rao SP, Boshoff HI, Dick T, Barry CE 3rd, Manjunatha UH (2013) A novel F(420) -dependent anti-oxidant mechanism protects mycobacterium tuberculosis against oxidative stress and bactericidal agents. Mol Microbiol 87(4):744-755. https://doi.org/10.1111/mmi.12127. Epub 2012 Dec 28. PMID: 23240649; PMCID: PMC356724

Gutell RR, Gray MW, Schnare MN (1993) A compilation of large subunit (23S and 23S-like) ribosomal RNA structures: 1993. Nucleic Acids Res 21(13):3055–3074

Gygli SM, Borrell S, Trauner A, Gagneux S (2017) Antimicrobial resistance in Mycobacterium tuberculosis: mechanistic and evolutionary perspectives. FEMS Microbiol Rev 41(3):354–373

Harikishore A, Grüber G (2024) Mycobacterium tuberculosis F-ATP synthase inhibitors and targets. Antibiotics (Basel) 13(12):1169

Harikishore A, Mathiyazakan V, Pethe K, Grüber G (2023) Novel targets and inhibitors of the Mycobacterium tuberculosis cytochrome bd oxidase to foster anti-tuberculosis drug discovery. Expert Opin Drug Discov 18(8):917–927

Haver HL, Chua A, Ghode P, Lakshminarayana SB, Singhal A, Mathema B, Wintjens R, Bifani P (2015) Mutations in genes for the F420 biosynthetic pathway and a nitroreductase enzyme are the primary resistance determinants in spontaneous in vitro-selected PA-824-resistant mutants of Mycobacterium tuberculosis. Antimicrob Agents Chemother 59(9): 5316–5323

Heep M, Brandstätter B, Rieger U, Lehn N, Richter E, Rüsch-Gerdes S, Niemann S (2001) Frequency of rpoB mutations inside and outside the cluster I region in rifampin-resistant clinical mycobacterium tuberculosis isolates. J Clin Microbiol 39(1):107–110. https://doi.org/10.1128/JCM.39.1.107-110.2001. PMID: 11136755; PMCID: PMC87688

Hegde P, Boshoff H, Rusman Y, Aragaw W, Salomon C, Dick T, Aldrich C (2021) Reinvestigation of the structure-activity relationships of isoniazid. Tuberculosis (Edinb) 129:102100

Herrera L, Jiménez S, Valverde A (2003) García-Aranda MA, Sáez-Nieto JA. Molecular analysis of rifampicin-resistant Mycobacterium tuberculosis isolated in Spain (1996–2001). Description of new mutations in the rpoB gene and review of the literature. Int J Antimicrob Agents 21(5):403–408

Heym B, Zhang Y, Poulet S, Young D, Cole ST (1993) Characterization of the katG gene encoding a catalase-peroxidase required for the isoniazid susceptibility of mycobacterium tuberculosis. J Bacteriol 175(13):4255-4259

Hoagland DT, Liu J, Lee RB, Lee RE (2016) New agents for the treatment of drug-resistant Mycobacterium tuberculosis. Adv Drug Deliv Rev 102:55–72

Hoffer ED, Miles SJ, Dunham CM (2017) The structure and function of Mycobacterium tuberculosis MazF-mt6 toxin provide insights into conserved features of MazF endonucleases. J Biol Chem 292(19):7718–7726

Hoffmann H, Kohl TA, Hofmann-Thiel S, Merker M, Beckert P, Jaton K, Nedialkova L, Sahalchyk E, Rothe T, Keller PM, Niemann S (2016) Delamanid and Bedaquiline resistance in Mycobacterium tuberculosis ancestral Beijing genotype causing extensively drug-resistant tuberculosis in a Tibetan refugee. Am J Respir Crit Care Med 193(3):337–340

Honoré N, Cole ST (1994) Streptomycin resistance in mycobacteria. Antimicrob Agents Chemother 38(2):238–242. https://doi.org/10.1128/AAC.38.2.238. PMID: 8192450; PMCID: PMC284433

Hooper DC (1999) Mechanisms of fluoroquinolone resistance. Drug Resist Updat 2(1):38–55. https://doi.org/10.1054/drup.1998.0068

Hooper DC, Jacoby GA (2016) Topoisomerase inhibitors: fluoroquinolone mechanisms of action and resistance. Cold Spring Harb Perspect Med 6(9):a025320. https://doi.org/10.1101/cshperspect.a025320. PMID: 27581430; PMCID: PMC5287101

Hopf FS, Roth CD, de Souza EV, Galina L, Czeczot AM, Machado P, Basso LA, Bizarro CV (2022) Bacterial enoyl-reductases: the ever-growing list of fabs, their mechanisms and inhibition. Front Microbiol 13:891610

Hsu LY, Lai LY, Hsieh PF, Lin TL, Lin WH, Tasi HY, Lee WT, Jou R, Wang JT (2020) Two novel katG mutations conferring isoniazid resistance in Mycobacterium tuberculosis. Front Microbiol 11:1644

Huang Y, Yan S, Li Y, Ai X, Yu X, Ge Y, Lv X, Fan L, Xie J (2022) Mycobacterium fluoroquinolone resistance protein D (MfpD), a GTPase-activating protein of GTPase MfpB, is involved in fluoroquinolones potency. Microbiol Spectr. 10(6):e0209822

Huang Y, Shen Q, Xu H, Huang L, Xiang S, Li P, Fan L, Xie J (2024) Mycobacterium smegmatis MfpC is a GEF that regulates mfpA translationally to alter the fluoroquinolone efficacy. Commun Biol. 7(1):1035

Hubin EA, Tabib-Salazar A, Humphrey LJ, Flack JE, Olinares PD, Darst SA, Campbell EA, Paget MS (2015) Structural, functional, and genetic analyses of the actinobacterial transcription factor RbpA. Proc Natl Acad Sci USA 112(23):7171–7176

Hugonnet JE, Tremblay LW, Boshoff HI, Barry CE 3rd, Blanchard JS (2009) Meropenem-clavulanate is effective against extensively drug-resistant mycobacterium tuberculosis. Science 323(5918):1215–1218. https://doi.org/10.1126/science.1167498. PMID: 19251630; PMCID: PMC2679150

Hu Y, Morichaud Z, Chen S, Leonetti JP, Brodolin K (2012) Mycobacterium tuberculosis RbpA protein is a new type of transcriptional activator that stabilizes the σ A-containing RNA polymerase holoenzyme. Nucleic Acids Res 40(14):6547–6557

Hu M, Fu L, Wang B, Xu J, Guo S, Zhao J, Li Y, Chen X, Lu Y (2020) Genetic and virulence characteristics of linezolid and Pretomanid dual drug-resistant strains induced from Mycobacterium tuberculosis in vitro. Infect Drug Resist. 13:1751–1761

Huitric E, Verhasselt P, Koul A, Andries K, Hoffner S, Andersson DI (2010) Rates and mechanisms of resistance development in Mycobacterium tuberculosis to a novel diarylquinoline ATP synthase inhibitor. Antimicrob Agents Chemother 54(3):1022–1028

Huynh KH, Hong MK, Lee C, Tran HT, Lee SH, Ahn YJ, Cha SS, Kang LW (2015) The crystal structure of the D-alanine-D-alanine ligase from Acinetobacter baumannii suggests a flexible conformational change in the central domain before nucleotide binding. J Microbiol 53(11):776–782

Ignatius EH, Dooley KE (2019) New drugs for the treatment of tuberculosis. Clin Chest Med 40(4):811–828. https://doi.org/10.1016/j.ccm.2019.08.010

Ismail NA, Omar SV, Moultrie H, Bhyat Z, Conradie F, Enwerem M et al (2022) Assessment of epidemiological and genetic characteristics and clinical outcomes of resistance to bedaquiline in patients treated for rifampicin-resistant tuberculosis: a cross-sectional and longitudinal study. Lancet Infect Dis 22(4):496–506. https://doi.org/10.1016/S1473-3099(21)00470-9

Ismail N, Dippenaar A, Warren RM, Peters RP, Omar SV (2023) Emergence of canonical and noncanonical genomic variants following in vitro exposure of clinical mycobacterium tuberculosis strains to bedaquiline or clofazimine. Antimicrob Agents Chemother 67(4):e01368-22. https://doi.org/10.1128/AAC.01368-22. PMID: 37071516; PMCID: PMC10102159

Jacobo-Delgado YM, Rodríguez-Carlos A, Serrano CJ, Rivas-Santiago B (2023) Mycobacterium tuberculosis cell-wall and antimicrobial peptides: a mission impossible? Front Immunol 14:1194923

Jacobs LMC, Consol P, Chen Y (2024) Drug discovery in the field of β-lactams: an academic perspective. Antibiotics (Basel). 13(1):59

Jagielski T, Bakuła Z, Brzostek A, Minias A, Stachowiak R, Kalita J, Napiórkowska A, Augustynowicz-Kopeć E, Żaczek A, Vasiliauskiene E, Bielecki J (2018) Characterization of mutations conferring resistance to rifampin in Mycobacterium tuberculosis clinical strains. Antimicrob Agents Chemother 62(10):10–128

Jagtap PK, Soni V, Vithani N, Jhingan GD, Bais VS, Nandicoori VK, Prakash B (2012) Substrate-bound crystal structures reveal features unique to Mycobacterium tuberculosis N-acetyl-glucosamine 1-phosphate uridyltransferase and a catalytic mechanism for acetyl transfer. J Biol Chem 287(47):39524–39537

Jamieson FB, Guthrie JL, Neemuchwala A, Lastovetska O, Melano RG, Mehaffy C (2014) Profiling of rpoB mutations and MICs for rifampin and rifabutin in Mycobacterium tuberculosis. J Clin Microbiol 52(6):2157–2162

Jankute M, Grover S, Rana AK, Besra GS (2012) Arabinogalactan and lipoarabinomannan biosynthesis: structure, biogenesis and their potential as drug targets. Future Microbiol 7(1):129–147

Jankute M, Cox JA, Harrison J, Besra GS (2015) Assembly of the mycobacterial cell wall. Ann Rev Microbiol 69:405–423

Jarlier V, Nikaido H (1994) Mycobacterial cell wall: structure and role in natural resistance to antibiotics. FEMS Microbiol Lett 123(1–2):11–18

Jensen MR, Houldsworth M, Brown K, Murakami KS, Rutherford ST, Hochschild A et al (2019) RbpA modulates transcription and rifampicin susceptibility in mycobacterium tuberculosis. J Bacteriol 201(5):e00661-18. https://doi.org/10.1128/JB.00661-18. PMID: 30523160; PMCID: PMC6377349

Ji YE, Kempsell KE, Colston MJ, Cox RA (1994) Nucleotide sequences of the spacer-1, spacer-2 and trailer regions of the rrn operons and secondary structures of precursor 23S rRNAs and precursor 5S rRNAs of slow-growing mycobacteria. Microbiology (Reading) 140(Pt 7):1763–1773. https://doi.org/10.1099/13500872-140-7-1763

Johansen SK, Maus CE, Plikaytis BB, Douthwaite S (2006) Capreomycin binds across the ribosomal subunit interface using tlyA-encoded 2'-O-methylations in 16S and 23S rRNAs. Mol Cell 23(2):173–182

Johnson TM, Rivera CG, Lee G, Zeuli JD (2024) Pharmacology of emerging drugs for the treatment of multi-drug resistant tuberculosis. J Clin Tuberc Other Mycobact Dis. 37:100470

Joob B, Wiwanitkit V (2019) Tuberculosis case detection rate: An observation on the change in 12-year period (2005–2016). Biomed Biotechnol Res J (BBRJ). 3(3):192–195

Kabir S, Tahir Z, Mukhtar N, Sohail M, Saqalein M, Rehman A (2020) Fluoroquinolone resistance and mutational profile of gyrA in pulmonary MDR tuberculosis patients. BMC Pulm Med 20(1):138

Kalapala SK, Hobbie SN, Böttger EC, Shcherbakov D (2010) Mutation K42R in ribosomal protein S12 does not affect susceptibility of Mycobacterium smegmatis 16S rRNA A-site mutants to 2-deoxystreptamines. PLoS One 5(8):e11960

Kampranis SC, Bates AD, Maxwell A (1999) A model for the mechanism of strand passage by DNA gyrase. Proc Natl Acad Sci USA 96(15):8414–8419

Kamsri B, Kamsri P, Punkvang A, Chimprasit A, Saparpakorn P, Hannongbua S et al (2024) Signal propagation in the ATPase domain of Mycobacterium tuberculosis DNA gyrase from dynamical-nonequilibrium molecular dynamics simulations. Biochemistry 63(11):1493–1504

Kanematsu E, Deguchi T, Yasuda M, Kawamura T, Nishino Y, Kawada Y (1998) Alterations in the GyrA subunit of DNA gyrase and the ParC subunit of DNA topoisomerase IV associated with quinolone resistance in enterococcus faecalis. Antimicrob Agents Chemother 42(2):433–435

Kang SM, Kim DH, Lee KY, Park SJ, Yoon HJ, Lee SJ, Im H, Lee BJ (2017) Functional details of the Mycobacterium tuberculosis VapBC26 toxin-antitoxin system based on a structural study: insights into unique binding and antibiotic peptides. Nucleic Acids Res 45(14):8564–8580

Kanji A, Hasan R, Hasan Z (2019) Efflux pump as alternate mechanism for drug resistance in Mycobacterium tuberculosis. Indian J Tuberc 66(1):20–25

Kaur H, Jakob RP, Marzinek JK, Green R, Imai Y, Bolla JR, Agustoni E, Robinson CV, Bond PJ, Lewis K, Maier T, Hiller S (2021) The antibiotic darobactin mimics a β-strand to inhibit outer membrane insertase. Nature 593(7857):125–129

Kavanaugh LG, Dey D, Shafer WM, Conn GL (2024) Structural and functional diversity of resistance-nodulation-division (RND) efflux pump transporters with implications for antimicrobial resistance. Microbiol Mol Biol Rev 88(3):e0008923

Khan AS, Phelan JE, Khan MT, Ali S, Qasim M, Mohammad N, Napier G, Ahmad S, Alam J, Khattak B, Campino S, Clark TG, Khan TA (2023) Genetic mutations underlying isoniazid-resistant Mycobacterium tuberculosis in Khyber Pakhtunkhwa, Pakistan. Tuberculosis (Edinb) 138:102286

Khan MT, Dumont E, Zhu J, Sahibzada KI, Lodhi MS, Qadir M, Wei D (2025) Insight into rifampicin resistance behind discordant mutations in RpoB of mycobacterium tuberculosis: a molecular dynamics simulations study. Results Chem 16:102380. https://doi.org/10.1016/j.rechem.2025.102380

Khandelwal NK, Gupta M, Gomez JE, Barkho S, Guan Z, Eng AY, Kawate T, Balasubramani SG, Sali A, Hung DT, Stroud RM (2024) Structure and inhibition mechanisms of Mycobacterium tuberculosis essential transporter efflux protein A. bioRxiv [Preprint] 5:2024.09.04.611325

Khanna A, Raj VS, Tarai B, Sood R, Pareek PK, Upadhyay DJ, Sharma P, Rattan A, Saini KS, Singh H (2010) Emergence and molecular characterization of extensively drug-resistant Mycobacterium tuberculosis clinical isolates from the Delhi region in India. Antimicrob Agents Chemother 54(11):4789–4793

Khrapunov S, Cheng H, Hegde S, Blanchard J, Brenowitz M (2008) Solution structure and refolding of the mycobacterium tuberculosis pentapeptide repeat protein MfpA. J Biol Chem 283(52):36290–36299. https://doi.org/10.1074/jbc.M804702200. PMID: 18945684; PMCID: PMC2605971

Kieser KJ, Baranowski C, Chao MC, Long JE, Sassetti CM, Waldor MK, Sacchettini JC, Ioerger TR, Rubin EJ (2015) Peptidoglycan synthesis in Mycobacterium tuberculosis is organized into networks with varying drug susceptibility. Proc Natl Acad Sci USA 112(42):13087–13092

Kim KH, An DR, Song J, Yoon JY, Kim HS, Yoon HJ, Im HN, Kim J, Kim DJ, Lee SJ, Kim KH, Lee HM, Kim HJ, Jo EK, Lee JY, Suh SW (2012) Mycobacterium tuberculosis Eis protein initiates suppression of host immune responses by acetylation of DUSP16/MKP-7. Proc Natl Acad Sci USA 109(20):7729–7734

Kim NY, Kim DY, Chu J, Jung SH (2023) pncA large deletion is the characteristic of pyrazinamide-resistant Mycobacterium tuberculosis belonging to the east Asian lineage. Infect Chemother 55(2):247–256

Kim CG, Jose J, Hay MP, Choi PJ (2024) Novel prodrug strategies for the treatment of tuberculosis. Chem Asian J 19(23):e202400944

Klitgaard RN, Ntokou E, Nørgaard K, Biltoft D, Hansen LH, Trædholm NM, Kongsted J, Vester B (2015) Mutations in the bacterial ribosomal protein l3 and their association with antibiotic resistance. Antimicrob Agents Chemother 59(6):3518–3528

Kloss P, Xiong L, Shinabarger DL, Mankin AS (1999) Resistance mutations in 23 S rRNA identify the site of action of the protein synthesis inhibitor linezolid in the ribosomal peptidyl transferase center. J Mol Biol 294(1):93–101

Ko EM, Min J, Kim H, Jeong JA, Lee S, Kim S (2024) Molecular characteristics of drug-susceptible Mycobacterium tuberculosis clinical isolates based on treatment duration. Osong Public Health Res Perspect 15(5):385

Koch A, Mizrahi V, Warner DF (2014) The impact of drug resistance on Mycobacterium tuberculosis physiology: what can we learn from rifampicin? Emerg Microbes Infect 3(3):e17

Kong TH, Coates AR, Butcher PD, Hickman CJ, Shinnick TM (1993) Mycobacterium tuberculosis expresses two chaperonin-60 homologs. Proc Nat Acad Sci 90(7):2608-2612

Korkegian A, Roberts DM, Blair R, Parish T (2014) Mutations in the essential arabinosyltransferase EmbC lead to alterations in Mycobacterium tuberculosis lipoarabinomannan. J Biol Chem 289(51):35172–35181

Koul A, Vranckx L, Dhar N, Göhlmann HW, Özdemir E, Neefs JM et al (2014) Delayed bactericidal response of mycobacterium tuberculosis to bedaquiline involves remodelling of bacterial metabolism. Nat Commun 5:3369. https://doi.org/10.1038/ncomms4369. PMID: 24569628; PMCID: PMC3948051

Krause KM, Serio AW, Kane TR, Connolly LE (2016) Aminoglycosides: An overview. Cold Spring Harb Perspect Med 6(6):a027029

Kufa M, Finger V, Kovar O, Soukup O, Torruellas C, Roh J, Korabecny J (2025) Revolutionizing tuberculosis treatment: breakthroughs, challenges, and hope on the horizon. Acta Pharm Sin B 15(3):1311–1332

Kumar S, Jena L (2014) Understanding rifampicin resistance in tuberculosis through a computational approach. Genomics Inform 12(4):276–282

Kumar N, Sharma S, Kaushal PS (2021) Protein synthesis in Mycobacterium tuberculosis as a potential target for therapeutic interventions. Mol Asp Med 81:101002

Kumar V, Jain P, Ratnam R, Singh U, Jain A, Pandey BK et al (2024) Agreement between phenotypically detected linezolid resistance and mutations in rrl and rplC genes of Mycobacterium tuberculosis isolates using nanopore sequencing. Int J Mycobacteriol. 13(3):331–336

Kurabachew M, Sandaa RA, Enger Ø, Bjorvatn B (2003) Sequence analysis in the 23S rDNA region of Mycobacterium tuberculosis and related species. J Microbiol Methods 54(3):373–380

Labby KJ, Garneau-Tsodikova S (2013) Strategies to overcome the action of aminoglycoside-modifying enzymes for treating resistant bacterial infections. Future Med Chem 5(11):1285–1309. https://doi.org/10.4155/fmc.13.80. PMID:23859208; PMCID: PMC3819198

Labuda SM, Seaworth B, Dasgupta S, Goswami ND (2024) BPaL accelerated monitoring project team. Bedaquiline, pretomanid, and linezolid with or without moxifloxacin for tuberculosis. Lancet. Respir Med 12(2):e5–e6

Lahiri N, Shah RR, Layre E, Young D, Ford C, Murray MB, Fortune SM, Moody DB (2016) Rifampin resistance mutations are associated with broad chemical remodeling of Mycobacterium tuberculosis. J Biol Chem 291(27):14248–14256

Laponogov I, Sohi MK, Veselkov DA, Pan XS, Sawhney R, Thompson AW, McAuley KE, Fisher LM, Sanderson MR (2009) Structural insight into the quinolone-DNA cleavage complex of type IIA topoisomerases. Nat Struct Mol Biol 16(6):667–669

Lata M, Sharma D, Deo N, Tiwari PK, Bisht D, Venkatesan K (2015) Proteomic analysis of ofloxacin-mono resistant Mycobacterium tuberculosis isolates. J Proteome 127:114–121

Laughlin ZT, Nandi S, Dey D, Zelinskaya N, Witek MA, Srinivas P et al (2022) 50S subunit recognition and modification by the mycobacterium tuberculosis ribosomal RNA methyltransferase TlyA. Proc Natl Acad Sci U S A 119(14):e2120352119. https://doi.org/10.1073/pnas.2120352119. PMID: 35357969; PMCID: PMC9168844

Leach KL, Brickner SJ, Noe MC, Miller PF (2011) Linezolid, the first oxazolidinone antibacterial agent. Ann N Y Acad Sci 1222:49–54

Lee AS, Teo AS, Wong SY (2001) Novel mutations in ndh in isoniazid-resistant Mycobacterium tuberculosis isolates. Antimicrob Agents Chemother 45(7):2157–2159

Lee JK, Lee YS, Park YK, Kim BS (2005) Alterations in the GyrA and GyrB subunits of topoisomerase II and the ParC and ParE subunits of topoisomerase IV in ciprofloxacin-resistant clinical isolates of Pseudomonas aeruginosa. Int J Antimicrob Agents 25(4):290–295

Lee BM, Harold LK, Almeida DV, Afriat-Jurnou L, Aung HL, Forde BM et al (2020) Predicting nitroimidazole antibiotic resistance mutations in mycobacterium tuberculosis with protein engineering. PLoS Pathog 16(2):e1008287. https://doi.org/10.1371/journal.ppat.1008287. PMID: 32032366; PMCID: PMC7032734

Lehmann J, Cheng TY, Aggarwal A, Park AS, Zeiler E, Raju RM, Akopian T, Kandror O, Sacchettini JC, Moody DB, Rubin EJ, Sieber SA (2018) An antibacterial β-lactone kills Mycobacterium tuberculosis by disrupting mycolic acid biosynthesis. Angew Chem Int Ed Engl 57(1):348–353

LeMagueres P, Im H, Ebalunode J, Strych U, Benedik MJ, Briggs JM, Kohn H, Krause KL (2005) The 1.9 A crystal structure of alanine racemase from Mycobacterium tuberculosis contains a conserved entryway into the active site. Biochemistry 44(5):1471–1481

Lestari BW, Nijman G, Larasmanah A, Soeroto AY, Santoso P, Alisjahbana B, Chaidir L, Andriyoko B, van Crevel R, Hill PC (2023) Management of drug-resistant tuberculosis in Indonesia: a four-year cascade of care analysis. Lancet Reg Health Southeast Asia. 22:100294. https://doi.org/10.1016/j.lansea.2023.100294. PMID: 38482149; PMCID: PMC10934335

Li MC, Wang XY, Xiao TY, Lin SQ, Liu HC, Qian C, Xu D, Li GL, Zhao XQ, Liu ZG, Zhao LL (2022) rpoB mutations are associated with variable levels of rifampin and rifabutin resistance in mycobacterium tuberculosis. Infect Drug Resist 15:6853–6861. https://doi.org/10.2147/IDR.S382580. PMID: 36381836; PMCID: PMC9660061

Li D, Zhang X, Yao Y, Sun X, Sun J, Ma X, Yuan K, Bai G, Pang X, Hua R, Guo T, Mi Y, Wu L, Zhang J, Wu Y, Liu Y, Wang P, Wong CCL, Chen X, Gao F (2024) Structure and function of Mycobacterium tuberculosis EfpA as a lipid transporter and its inhibition by BRD-8000.3. Proc Natl Acad Sci USA 121(44):e2412653121

Liu A (2024) Catalase-peroxidase (KatG): a potential frontier in tuberculosis drug development. Crit Rev Biochem Mol Biol 59(6):434–446

Liu Y, Shi J, Li L, Wu T, Chu P, Pang Y, Guo Y, Gao M, Lu J (2022) Spontaneous mutational patterns and novel mutations for Delamanid resistance in Mycobacterium tuberculosis. Antimicrob Agents Chemother 66(12):e0053122

Liu B, Su P, Hu P, Yan M, Li W, Yi S, Chen Z, Zhang X, Guo J, Wan X, Wang J, Gong D, Bai H, Wan K, Liu H, Li G, Tan Y (2024) Prevalence, transmission and genetic diversity of pyrazinamide resistance among multidrug-resistant Mycobacterium tuberculosis isolates in Hunan. China Infect Drug Resist 17:403–416

Liu Y, Brown CM, Erramilli S, Su Y-C, Guu S-Y, Tseng P-S et al (2025) Structural insights into terminal arabinosylation of mycobacterial cell wall arabinan. Nat Commun 16:3973

Loewen PC, Carpena X, Vidossich P, Fita I, Rovira C (2014) An ionizable active-site tryptophan imparts catalase activity to a peroxidase core. J Am Chem Soc 136(20):7249–7252

Long KS, Vester B (2008) Antibiotic resistance mechanisms, with an emphasis on those related to the ribosome. EcoSal Plus 3(1)

Long KS, Vester B (2012) Resistance to linezolid caused by modifications at its binding site on the ribosome. Antimicrob Agents Chemother 56(2):603–612

Lu P, Lill H, Bald D (2014) ATP synthase in mycobacteria: special features and implications for a function as drug target. Biochim Biophys Acta Bioenerg 1837(7):1208–1218

Luthra S, Rominski A, Sander P (2018) The role of antibiotic-target-modifying and antibiotic-modifying enzymes in Mycobacterium abscessus drug resistance. Front Microbiol 9:2179

Machado D, Pires D, Perdigao J, Couto I, Portugal I, Martins M, Amaral L, Anes E, Viveiros M (2016_ Ion channel blockers as antimicrobial agents, efflux inhibitors, and enhancers of macrophage killing activity against drug resistant mycobacterium tuberculosis. PloS One 11(2):e0149326

Madhusudan K, Ramesh V, Nagaraja V (1994) Cloning and sequence analysis of DNA gyrase genes from Mycobacterium tuberculosis. Curr Sci 66(9):664–667

Madsen CT, Jakobsen L, Buriánková K, Doucet-Populaire F, Pernodet JL, Douthwaite S (2005) Methyltransferase Erm(37) slips on rRNA to confer atypical resistance in Mycobacterium tuberculosis. J Biol Chem 280(47):38942–38947

Maitra A, Munshi T, Healy J, Martin LT, Vollmer W, Keep NH, Bhakta S (2019) Cell wall peptidoglycan in Mycobacterium tuberculosis: An Achilles' heel for the TB-causing pathogen. FEMS Microbiol Rev 43(5):548–575

Makafe GG, Cao Y, Tan Y, Julius M, Liu Z, Wang C, Njire MM, Cai X, Liu T, Wang B, Pang W, Tan S, Zhang B, Yew WW, Lamichhane G, Guo J, Zhang T (2016) Role of the Cys154Arg substitution in ribosomal protein L3 in Oxazolidinone resistance in Mycobacterium tuberculosis. Antimicrob Agents Chemother 60(5):3202–3206

Målen H, De Souza GA, Pathak S, Søfteland T, Wiker HG (2011) Comparison of membrane proteins of mycobacterium tuberculosis H37Rv and H37Ra strains. BMC Microbiol 11:18. https://doi.org/10.1186/1471-2180-11-18

Malik S, Willby M, Sikes D, Tsodikov OV, Posey JE (2012) New insights into fluoroquinolone resistance in Mycobacterium tuberculosis: functional genetic analysis of gyrA and gyrB mutations. PLoS One 7(6):e39754

Manjunatha UH, Mahadevan S, Visweswariah SS, Nagaraja V (2001) Monoclonal antibodies to mycobacterial DNA gyrase A inhibit DNA supercoiling activity. Eur J Biochem 268(7):2038–2046

Mansjö M, Karlsson Lindsjö O, Grönfors Seeth C, Groenheit R, Werngren J (2022) The ddn Trp20Stop mutation and its association with lineage 4.5 and resistance to Delamanid and Pretomanid in Mycobacterium tuberculosis. Antimicrob Agents Chemother 66(12):e0102622

Mao X, Wang J, Xu J, Xu P, Hu H, Li L, Zhang Z, Song Y (2025) Current diagnosing strategies for Mycobacterium tuberculosis and its drug resistance: a review. J Appl Microbiol 136(5):lxaf100

Maravić G (2004) Macrolide resistance based on the Erm-mediated rRNA methylation. Curr Drug Targets Infect Disord 4(3):193–202. https://doi.org/10.2174/1568005043340777

Marney MW, Metzger RP, Hecht D, Valafar F (2018) Modeling the structural origins of drug resistance to isoniazid via key mutations in Mycobacterium tuberculosis catalase-peroxidase. KatG Tuberculosis 108:155–162

Martin A, Bouyakoub Y, Soumillion K, Mantu EN, Colmant A, Rodriguez-Villalobos H (2020) Targeting bedaquiline mycobacterial efflux pump to potentially enhance therapy in Mycobacterium abscessus. Intl J Mycobacteriol 9(1):71–75

Maruri F, Sterling TR, Kaiga AW, Blackman A, van der Heijden YF, Mayer C, Cambau E, Aubry A (2012) A systematic review of gyrase mutations associated with fluoroquinolone-resistant Mycobacterium tuberculosis and a proposed gyrase numbering system. J Antimicrob Chemother 67(4):819–831

Matteelli A, Lovatti S, Rossi B, Rossi L (2025) Update on multidrug-resistant tuberculosis preventive therapy toward the global tuberculosis elimination. Int J Infect Dis 1:155

Maus CE, Plikaytis BB, Shinnick TM (2005) Mutation of tlyA confers capreomycin resistance in

Mycobacterium tuberculosis. Antimicrob Agents Chemother 49(2):571–577
Maxwell A (1997) DNA gyrase as a drug target. Trends Microbiol 5(3):102–109. https://doi.org/10.1016/S0966-842X(96)10085-8
Mayer C, Takiff H (2014) The molecular genetics of fluoroquinolone resistance in Mycobacterium tuberculosis. Microbiol Spectr.:2
McNeil M (1999) Arabinogalactan in mycobacteria: structure, biosynthesis, and genetics. In: Genetics of bacterial polysaccharides. CRC, Washington, DC, pp 207–223
Mehta AP, Abdelwahed SH, Mahanta N, Fedoseyenko D, Philmus B, Cooper LE et al (2015) Radical Sadenosylmethionine (SAM) enzymes in cofactor biosynthesis: a treasure trove of complex organic radical rearrangement reactions. J Biol Chem 290(7):3980–3996. https://doi.org/10.1074/jbc.R114.6237938. PMID: 25477515; PMCID: PMC432680
Mehta SC, Furey IM, Pemberton OA, Boragine DM, Chen Y, Palzkill T (2021) KPC-2 β-lactamase enables carbapenem antibiotic resistance through fast deacylation of the covalent intermediate. J Biol Chem 296. 100155
Meier A, Kirschner P, Bange FC, Vogel U, Böttger EC (1994) Genetic alterations in streptomycin-resistant Mycobacterium tuberculosis: mapping of mutations conferring resistance. Antimicrob Agents Chemother 38(2):228–233
Metcalfe C, Macdonald IK, Murphy EJ, Brown KA, Raven EL, Moody PC (2008) The tuberculosis prodrug isoniazid bound to activating peroxidases. J Biol Chem 283(10):6193–6200
Miesel L, Weisbrod TR, Marcinkeviciene JA, Bittman R, Jacobs WR Jr (1998) NADH dehydrogenase defects confer isoniazid resistance and conditional lethality in Mycobacterium smegmatis. J Bacteriol 180(9):2459–2467
Mieskes KT, Rüsch-Gerdes S, Truffot-Pernot C, Feldmann K, Tortoli E, Casal M, Löscher T, Rinder H (2000) Rapid, simple, and culture-independent detection of rpsL codon 43 mutations that are highly predictive of streptomycin resistance in Mycobacterium tuberculosis. Am J Trop Med Hyg 63(1–2):56–60
Miller LP, Crawford JT, Shinnick TM (1994) The rpoB gene of Mycobacterium tuberculosis. Antimicrob Agents Chemother 38(4):805–811
Minnikin DE, Minnikin SM, Parlett JH, Goodfellow M, Magnusson M (1984) Mycolic acid patterns of some species of mycobacterium. Arch Microbiol 139:225–231. https://doi.org/10.1007/BF00402005
Mishra MN, Daniels L (2013) Characterization of the MSMEG_2631 gene (mmp) encoding a multidrug and toxic compound extrusion (MATE) family protein in Mycobacterium smegmatis and exploration of its polyspecific nature using biolog phenotype microarray. J Bacteriol 195(7):1610–1621
Mishra AK, Driessen NN, Appelmelk BJ, Besra GS (2011) Lipoarabinomannan and related glycoconjugates: structure, biogenesis and role in mycobacterium tuberculosis physiology and host–pathogen interaction. FEMS Microbiol Rev 35(6):1126–1157. https://doi.org/10.1111/j.1574-6976.2011.00276.x
Mo L, Zhang W, Wang J, Weng XH, Chen S, Shao LY, Pang MY, Chen ZW (2004) Three-dimensional model and molecular mechanism of Mycobacterium tuberculosis catalase-peroxidase (KatG) and isoniazid-resistant KatG mutants. Microb Drug Resist 10(4):269–279
Modlin SJ, Marbach T, Werngren J, Mansjö M, Hoffner SE, Valafar F (2021) Atypical genetic basis of pyrazinamide resistance in Monoresistant Mycobacterium tuberculosis. Antimicrob Agents Chemother 65(6):e01916–e01920
Mohamed AE, Condic-Jurkic K, Ahmed FH, Yuan P, O'Mara ML, Jackson CJ, Coote ML (2016) Hydrophobic shielding drives catalysis of hydride transfer in a family of F420H2-dependent enzymes. Biochemistry 55(49):6908–6918
Mohammadi B, Ramazanzadeh R, Rouhi S, Mohajeri P, Nouri B (2019) Point-mutations in embB306 gene and their association with resistance to ethambutol in Mycobacterium tuberculosis in clinical isolates. J Adv Med Biomed Res 27(125):43–48
Molodtsov V, Scharf NT, Stefan MA, Garcia GA, Murakami KS (2017 Mar) Structural basis for rifamycin resistance of bacterial RNA polymerase by the three most clinically important RpoB mutations found in Mycobacterium tuberculosis. Mol Microbiol 103(6):1034–1045
Mori G, Chiarelli LR, Esposito M, Makarov V, Bellinzoni M, Hartkoorn RC et al (2015) Thiophenecarboxamide derivatives activated by EthA kill Mycobacterium tuberculosis by inhibiting the CTP synthetase PyrG. Chem Biol 22(7):917–927
Mori G, Chiarelli LR, Riccardi G, Pasca MR (2017) New prodrugs against tuberculosis. Drug Discov Today 22(3):519–525
Morlock GP, Metchock B, Sikes D, Crawford JT, Cooksey RC (2003) ethA, inhA, and katG loci of ethionamide-resistant clinical Mycobacterium tuberculosis isolates. Antimicrob Agents Chemother 47(12):3799–3805
Mudde SE, Upton AM, Lenaerts A, Bax HI, De Steenwinkel JEM (2022) Delamanid or pretomanid? A Solomonic judgement! J Antimicrob Chemother 77(4):880–902
Murakami KS, Masuda S, Campbell EA, Muzzin O, Darst SA (2002) Structural basis of transcription initiation: an RNA polymerase holoenzyme-DNA complex. Science 296(5571):1285–1290
Mustaev A, Zaychikov E, Severinov K, Kashlev M, Polyakov A, Nikiforov V, Goldfarb A (1994) Topology of the RNA polymerase active center probed by chimeric rifampicin-nucleotide compounds. Proc Natl Acad Sci USA 91(23):12036–12040
Mustaev A, Kozlov M, Markovtsov V, Zaychikov E, Denissova L, Goldfarb A (1997) Modular organization of the catalytic center of RNA polymerase. Proc Natl Acad Sci USA 94(12):6641–6645
Muthaiah M, Shivekar SS, Cuppusamy Kapalamurthy VR, Alagappan C, Sakkaravarthy A, Brammachary

U (2017) Prevalence of mutations in genes associated with rifampicin and isoniazid resistance in Mycobacterium tuberculosis clinical isolates. J Clin Tuberc Other Mycobact Dis 8:19–25

Nair A, Greeny A, Nandan A, Sah RK, Jose A, Dyawanapelly S, Junnuthula V, Sadanandan P (2023) Advanced drug delivery and therapeutic strategies for tuberculosis treatment. J Nanobiotechnol 21(1):414

Nakamoto RK, Ketchum CJ, al-Shawi MK (1999) Rotational coupling in the F0F1 ATP synthase. Annu Rev Biophys Biomol Struct 28:205–234

Nakatani Y, Opel-Reading HK, Merker M, Machado D, Andres S, Kumar SS, Moradigaravand D, Coll F, Perdigão J, Portugal I, Schön T, Nair D, Devi KRU, Kohl TA, Beckert P, Clark TG, Maphalala G, Khumalo D, Diel R, Klaos K, Aung HL, Cook GM, Parkhill J, Peacock SJ, Swaminathan S, Viveiros M, Niemann S, Krause KL, Köser CU (2017) Role of alanine racemase mutations in Mycobacterium tuberculosis d-Cycloserine resistance. Antimicrob Agents Chemother 61(12):e01575–e01517

Nandi S, Dey D, Srinivas P, Dunham CM, Conn GL (2024) Distant ribose 2′-o-methylation of 23s rrna helix 69 pre-orders the capreomycin drug binding pocket at the ribosome subunit interface. bioRxiv [Preprint] 5:2024.11.05.619916

Nasirzadeh Z, Farnia P, Nowroozi J, Farnia P, Velayati AA (2022) Comparing pili producing gene (mtp-flp) in susceptible and resistant dormant Mycobacterium tuberculosis strains with active clinical isolates. Biomed Biotechnol Res J (BBRJ). 6(2):224–229

Nebenzahl-Guimaraes H, Jacobson KR, Farhat MR, Murray MB (2014) Systematic review of allelic exchange experiments aimed at identifying mutations that confer drug resistance in Mycobacterium tuberculosis. J Antimicrob Chemother 69(2):331–342

Nehru VJ, Brammacharry U, Ramkumar SS, Khusro A, Vadakunnel MJ, Gunasekaran S, David E, Priya VV, Aljowaie RM, Almutairi SM (2024) Molecular analysis of bedaquiline resistance and genetic variations in clinical isolates of multidrug and fluoroquinolones-resistant Mycobacterium tuberculosis. J King Saud Univ Sci 36(6):103226

Nessar R, Reyrat JM, Murray A, Gicquel B (2011) Genetic analysis of new 16S rRNA mutations conferring aminoglycoside resistance in Mycobacterium abscessus. J Antimicrob Chemother 66(8):1719–1724

Newell KV, Thomas DP, Collins D, Romero R, Zhang G, Darst SA et al (2006) Overexpression of RbpA partially restores transcriptional activity in rifampicin-resistant Mycobacterium tuberculosis. Antimicrob Agents Chemother 50(12):4107–4113

Newton GL, Unson MD, Anderberg SJ, Aguilera JA, Oh NN, delCardayre SB, Av-Gay Y, Fahey RC (1999) Characterization of Mycobacterium smegmatis mutants defective in 1-d-myo-inosityl-2-amino-2-deoxy-alpha-d-glucopyranoside and mycothiol biosynthesis. Biochem Biophys Res Commun 255(2):239–244

Nguyen TVA, Nguyen QH, Nguyen TNT, Anthony RM, Vu DH, Alffenaar JC (2023) Pretomanid resistance: An update on emergence, mechanisms and relevance for clinical practice. Int J Antimicrob Agents 62(4):106953

Nieto RLM, Mehaffy C, Creissen E, Troudt J, Troy A, Bielefeldt-Ohmann H, Burgos M, Izzo A, Dobos KM (2016) Virulence of Mycobacterium tuberculosis after acquisition of isoniazid resistance: individual nature of katG mutants and the possible role of AhpC. PLoS One 11(11):e0166807

Ninio S, Rotem D, Schuldiner S (2001) Functional analysis of novel multidrug transporters from human pathogens. J Biol Chem 276(51):48250–48256

Nishino K, Yamasaki S, Nakashima R, Zwama M, Hayashi-Nishino M (2021) Function and inhibitory mechanisms of multidrug efflux pumps. Front Microbiol 12:737288

Njuma OJ, Ndontsa EN, Goodwin DC (2014) Catalase in peroxidase clothing: interdependent cooperation of two cofactors in the catalytic versatility of KatG. Arch Biochem Biophys 544:27–39

Noller HF, Woese CR (1981) Secondary structure of 16S ribosomal RNA. Science 212(4493):403–411. https://doi.org/10.1126/science.6163215

Nono VN, Nantia EA, Mutshembele AM, Teagho SN, Simo YW, Takong BS et al (2024) Prevalence of katG and inhA mutations in genes associated with isoniazid resistance in Mycobacterium tuberculosis clinical isolates in Cameroon. medRxiv [Preprint]. https://doi.org/10.1101/2024.05.20.24308014

Occhineri S, Matucci T, Rindi L, Tiseo G, Falcone M, Riccardi N, Besozzi G (2022) Pretomanid for tuberculosis treatment: an update for clinical purposes. Curr Res Pharmacol Drug Discov 3:100128

Ofori-Anyinam B, Riley AJ, Jobarteh T, Gitteh E, Sarr B, Faal-Jawara TI et al (2020) Comparative genomics shows differences in the electron transport and carbon metabolic pathways of Mycobacterium africanum relative to Mycobacterium tuberculosis and suggests an adaptation to low oxygen tension. Tuberculosis (Edinb) 120:101899

Oh E, Jeon B (2014) Role of alkyl hydroperoxide reductase (AhpC) in the biofilm formation of campylobacter jejuni. PLoS One 9(1):e87312. https://doi.org/10.1371/journal.pone.0087312. PMID: 24489935; PMCID: PMC3904908

Okamoto S, Tamaru A, Nakajima C, Nishimura K, Tanaka Y, Tokuyama S, Suzuki Y, Ochi K (2007) Loss of a conserved 7-methylguanosine modification in 16S rRNA confers low-level streptomycin resistance in bacteria. Mol Microbiol 63(4):1096–1106

Omoteso OA, Fadaka AO, Walker RB, Khamanga SM (2025) Innovative strategies for combating multidrug-resistant tuberculosis: advances in drug delivery systems and treatment. Microorganisms 13(4):722

Ostrer L, Crooks TA, Howe MD, Vo S, Jia Z, Hegde P, Schacht N, Aldrich CC, Baughn AD (2025) Mechanism of the dual action self-potentiating antitubercular drug morphazinamide. bioRxiv [Preprint] 31:2024.10.08.617272

Otum C, Rivière E, Barnard M, Loubser J, Williams M, Streicher E, Rie A, Warren R, Klopper M (2023) Site-directed mutagenesis of Mycobacterium tuberculosis and functional validation to investigate potential bedaquiline resistance-causing mutations. Sci Rep 13:35563

Pal R, Fatima Z, Hameed S (2014) Efflux pumps in drug resistance of mycobacterium tuberculosis: a panoramic view. Int J Curr Microbiol Appl Sci 3(8):528–546

Pan P, Tonge PJ (2012) Targeting InhA, the FASII enoyl-ACP reductase: SAR studies on novel inhibitor scaffolds. Curr Top Med Chem 12(7):672–693

Parikh S, Moynihan DP, Xiao G, Tonge PJ (1999) Roles of tyrosine 158 and lysine 165 in the catalytic mechanism of InhA, the enoyl-ACP reductase from Mycobacterium tuberculosis. Biochemistry 38(41):13623–13634

Park HE, Kim S, Shim S, Park HT, Park WB, Im YB, Yoo HS (2020) 16S and 23S rRNA gene mutation independent multidrug resistance of non-tuberculous mycobacteria isolated from south Korean soil. Microorganisms. 8(8):1114

Parish T, Stoker NG (2000) Use of a flexible cassette method to generate a double unmarked mycobacterium tuberculosis tlyA plcABC mutant by gene replacement. Microbiology (Reading). 146(Pt 8):1969–1975. https://doi.org/10.1099/00221287-146-8-1969

Pederick JL, Thompson AP, Bell SG, Bruning JB (2020) D-alanine–d-alanine ligase as a model for the activation of ATP-grasp enzymes by monovalent cations. J Biol Chem 295(23):7894–7904

Petrella S et al (2006) Genetic basis for natural and acquired resistance to the diarylquinoline R207910 in mycobacteria. Antimicrob Agents Chemother 50:2853–2856

Petrella S, Gelus-Ziental N, Maudry A, Laurans C, Boudjelloul R, Sougakoff W (2011) Crystal structure of the pyrazinamidase of Mycobacterium tuberculosis: insights into natural and acquired resistance to pyrazinamide. PLoS One 6(1):e15785

Petrov A, Meskauskas A, Dinman JD (2004) Ribosomal protein L3: influence on ribosome structure and function. RNA Biol 1(1):59–65

Pfister P, Hobbie S, Vicens Q, Böttger EC, Westhof E (2003) The molecular basis for A-site mutations conferring aminoglycoside resistance: relationship between ribosomal susceptibility and X-ray crystal structures. Chembiochem 4(10):1078–1088

Pi R, Liu Q, Jiang Q, Gao Q (2019) Characterization of linezolid-resistance-associated mutations in Mycobacterium tuberculosis through WGS. J Antimicrob Chemother 74(7):1795–1798

Piton J, Petrella S, Delarue M, André-Leroux G, Jarlier V, Aubry A, Mayer C (2010) Structural insights into the quinolone resistance mechanism of Mycobacterium tuberculosis DNA gyrase. PLoS One 5(8):e12245

Poulton NC, Rock JM (2022) Unraveling the mechanisms of intrinsic drug resistance in mycobacterium tuberculosis. Front Cell Infect Microbiol 12:997283. https://doi.org/10.3389/fcimb.2022.997283. PMID: 36325467; PMCID: PMC9618640

Prammananan T, Sander P, Springer B, Böttger EC (1999) RecA-mediated gene conversion and aminoglycoside resistance in strains heterozygous for rRNA. Antimicrob Agents Chemother 43:447–453

Preiss L, Langer JD, Yildiz Ö, Eckhardt-Strelau L, Guillemont JE, Koul A et al (2015) Structure of the mycobacterial ATP synthase Fo rotor ring in complex with the anti-TB drug bedaquiline. Sci Adv 1(4):e1500106. https://doi.org/10.1126/sciadv.1500106. PMID: 26601184; PMCID: PMC4640650

Prosser GA, de Carvalho LP (2013) Metabolomics reveal D-alanine:D-alanine ligase as the target of D-cycloserine in mycobacterium tuberculosis. ACS Med Chem Lett 4(12):1233–1237. https://doi.org/10.1021/ml400349n. PMID: 24478820; PMCID: PMC3903091

Punetha A, Ngo H, Holbrook S, Green K, Willby M, Bonnett S et al (2020) Structure-guided optimization of inhibitors of acetyltransferase Eis from Mycobacterium tuberculosis. ACS Chem Biol 15(6):1434–1443

Purkan I, Natalia D, Syah YM, Retnoningrum DS, Kusuma HS (2016) Mutation of katG in a clinical isolate of Mycobacterium tuberculosis: effects on catalase-peroxidase for isoniazid activation. Ukr Biochem J 88(5):71–81

Purwantini E, Daniels L, Mukhopadhyay B (2016) F420H2 is required for phthiocerol dimycocerosate synthesis in mycobacteria. J Bacteriol 198(15):2020–2028. https://doi.org/10.1128/JB.01035-15. PMID: 27185825; PMCID: PMC4944228

Pym AS, Saint-Joanis B, Cole ST (2002) Effect of katG mutations on the virulence of Mycobacterium tuberculosis and the implication for transmission in humans. Infect Immun 70(9):4955–4960

Qin Y, Xu L, Teng Y, Wang Y, Ma P (2021) Discovery of novel antibacterial agents: recent developments in D-alanyl-D-alanine ligase inhibitors. Chem Biol Drug Des 98(3):305–322

Radmacher E, Stansen KC, Besra GS, Alderwick LJ, Maughan WN, Hollweg G, Sahm H, Wendisch VF, Eggeling L (2005) Ethambutol, a cell wall inhibitor of Mycobacterium tuberculosis, elicits L-glutamate efflux of Corynebacterium glutamicum. Microbiology 151(5):1359–1368

Rahim Z, Nakajima C, Raqib R, Zaman K, Endtz H, van der Zanden A, Suzuki Y (2012) Molecular mechanism of rifampicin and isoniazid resistance in Mycobacterium tuberculosis from Bangladesh. Tuberculosis (Edinb) 92(6):529–534

Rahman A, Srivastava SS, Sneh A, Ahmed N, Krishnasastry MV (2010) Molecular characterization of tlyA gene product, Rv1694 of Mycobacterium tuberculosis: a non-conventional hemolysin and a ribosomal RNA methyl transferase. BMC Biochem 11:35

Rahman MA, Sobia P, Dwivedi VP, Bhawsar A, Singh DK, Sharma P, Moodley P, Van Kaer L, Bishai WR, Das G (2015) Mycobacterium tuberculosis TlyA protein negatively regulates T Helper (Th) 1 and Th17 differentiation and promotes tuberculosis pathogenesis. J Biol Chem 290(23):14407-14417. https://doi.org/10.1074/jbc.M115.653600. Epub 2015 Apr 6. PMID: 25847237; PMCID: PMC4505508

Rahman MA, Kaiser F, Jamshidi S, Freitas Monteiro M, Rahman KM, Mullany P, Roberts AP (2020) Integron gene cassettes harboring novel variants of d-alanine-d-alanine ligase confer high-level resistance to d-cycloserine. Sci Rep 10(1):20709

Rai D, Mehra S (2021) The mycobacterial efflux pump EfpA can inducez high drug tolerance to many Antituberculosis drugs, including Moxifloxacin, in Mycobacterium smegmatis. Antimicrob Agents Chemother 65(11):e0026221

Rajendran A, Palaniyandi K (2022) Mutations associated with pyrazinamide resistance in Mycobacterium tuberculosis: A review and update. Curr Microbiol 79(11):348

Ramasubban G, Michael JS, Gupta R, Venkatesan M, Hoffner S, Asalapuram P (2024) Rapid detection of M. Tuberculosis and its resistance to rifampicin and isoniazid with the mfloDx™ MDR-TB test. Intl J Mycobacteriol 13(1):91–95

Ramirez-Busby SM, Valafar F (2015) Systematic review of mutations in pyrazinamidase associated with pyrazinamide resistance in Mycobacterium tuberculosis clinical isolates. Antimicrob Agents Chemother 59(9):5267–5277

Ramón-García S, Mick V, Dainese E, Martín C, Thompson CJ, De Rossi E, Manganelli R, Aínsa JA (2012) Functional and genetic characterization of the tap efflux pump in Mycobacterium bovis BCG. Antimicrob Agents Chemother 56(4):2074–2083

Rawat M, Newton GL, Ko M, Martinez GJ, Fahey RC, Av-Gay Y (2002) Mycothiol-deficient Mycobacterium smegmatis mutants are hypersensitive to alkylating agents, free radicals, and antibiotics. Antimicrob Agents Chemother 46(11):3348–3355

Rawat R, Whitty A, Tonge PJ (2003) The isoniazid-NAD adduct is a slow, tight-binding inhibitor of InhA, the Mycobacterium tuberculosis enoyl reductase: adduct affinity and drug resistance. Proc Natl Acad Sci 100(24):13881–13886

Recchia D, Stelitano G, Egorova A, Batisti Biffignandi G, Savková K, Kafková R, Huszár S, Marino Cerrato A, Slayden RA, Cummings JE, Whittel N, Bauman AA, Robertson GT, Rank L, Urbina F, Lane TR, Ekins S, Riabova O, Kazakova E, Mikušová K, Sassera D, Degiacomi G, Chiarelli LR, Makarov V, Pasca MR (2025) Mycobacterium tuberculosis sulfate Ester dioxygenase Rv3406 is able to inactivate the RCB18350 compound. ACS Infect Dis. 11(4):986–997

Reece RJ, Maxwell A (1991) DNA gyrase: structure and function. Crit Rev Biochem Mol Biol 26(3–4): 335–375

Rees DC, Johnson E, Lewinson O (2009) ABC transporters: the power to change. Nat Rev Mol Cell Biol 10(3):218–227

Reichmuth ML, Hömke R, Zürcher K, Sander P, Avihingsanon A, Collantes J, Loiseau C, Borrell S, Reinhard M, Wilkinson RJ, Yotebieng M, Fenner L, Böttger EC, Gagneux S, Egger M, Keller PM (2020) Natural polymorphisms in Mycobacterium tuberculosis conferring resistance to Delamanid in drug-naive patients. Antimicrob Agents Chemother 64(11):e00513-20

Remm S, Earp JC, Dick T, Dartois V, Seeger MA (2022) Critical discussion on drug efflux in mycobacterium tuberculosis. FEMS Microbiol Rev 46(1):fuab050. https://doi.org/10.1093/femsre/fuab050

Ren J, Wang M, Zhou W, Liu Z (2024) Efflux pumps as potential targets for biofilm inhibition. Front Microbiol 15:1315238. https://doi.org/10.3389/fmicb.2024.1315238. PMID: 38596384; PMCID: PMC11002903

Rifat D, Li SY, Ioerger T, Shah K, Lanoix JP, Lee J, Bashiri G, Sacchettini J, Nuermberger E (2020) Mutations in fbiD (Rv2983) as a novel determinant of resistance to Pretomanid and Delamanid in Mycobacterium tuberculosis. Antimicrob Agents Chemother 65(1):e01948–e01920

Rinder H, Thomschke A, Rüsch-Gerdes S, Bretzel G, Feldmann K, Rifai M, Löscher T (1998) Significance of ahpC promoter mutations for the prediction of isoniazid resistance in Mycobacterium tuberculosis. Eur J Clin Microbiol Infect Dis 17:508–511

Roberts AH, Moon CW, Faulkner V, Kendall SL, Waddell SJ, Bacon J (2024) EfpA is required for regrowth of mycobacterium tuberculosis following isoniazid exposure. Antimicrob Agents Chemother 7:68

Rodríguez-Castillo JA, Arce-Mendoza AY, Quintanilla-Siller A, Rendon A, Salinas-Carmona MC, Rosas-Taraco AG (2015) Possible association of rare polymorphism in the ABCB1 gene with rifampin and ethambutol drug-resistant tuberculosis. Tuberculosis (Edinb) 95(5):532–537

Rodríguez-García Á, Mares-Alejandre RE, Muñoz-Muñoz PLA, Ruvalcaba-Ruiz S, González-Sánchez RA, Bernáldez-Sarabia J et al (2021) Molecular analysis of streptomycin resistance genes in clinical strains of Mycobacterium tuberculosis and biocomputational analysis of the MtGidB L101F variant. Antibiotics 10(7):807

Rodríguez JG, Pino C, Tauch A, Murcia MI (2015) Complete genome sequence of the clinical Beijing-like strain mycobacterium tuberculosis 323 using the PacBio real-time sequencing platform. Genome Announc 3(2):e00371-15. https://doi.org/10.1128/genomeA.00371-15. PMID: 25931600; PMCID: PMC4417696

Ronning DR, Vissa V, Besra GS, Belisle JT, Sacchettini JC (2004) Mycobacterium tuberculosis antigen 85A and 85C structures confirm binding orientation and conserved substrate specificity. J Biol Chem 279(35):36771–36777

Roodveldt C, Tawfik DS (2005) Shared promiscuous activities and evolutionary features in various members of the amidohydrolase superfamily. Biochemistry 44(38):12728–12736

Rossini ND, Dias MV (2023) Mutations and insights into the molecular mechanisms of resistance of mycobacterium tuberculosis to first-line. Genet Mol Biol 23(46):e20220261. https://doi.org/10.1590/1678-4685-GMB-2022-0261

Rozwarski DA, Grant GA, Barton DH, Jacobs WR Jr, Sacchettini JC (1998) Modification of the NADH of the isoniazid target (InhA) from mycobacterium tuberculosis. Science 279(5347):98–102. https://doi.org/10.1126/science.279.5347.98

Sacco E, Covarrubias AS, O'Hare HM, Carroll P, Eynard N, Jones TA, Parish T, Daffé M, Bäckbro K, Quémard A (2007) The missing piece of the type II fatty acid synthase system from Mycobacterium tuberculosis. Proc Natl Acad Sci 104(37):14628–14633

Sachan RS, Mistry V, Dholaria M, Rana A, Devgon I, Ali I et al (2023) Overcoming mycobacterium tuberculosis drug resistance: novel medications and repositioning strategies. ACS Omega 8(36):32244–32257. https://doi.org/10.1021/acsomega.3c05336. PMID: 37773624; PMCID: PMC10537019

Safi H, Sayers B, Hazbón MH, Alland D (2008) Transfer of embB codon 306 mutations into clinical mycobacterium tuberculosis strains alters susceptibility to ethambutol, isoniazid, and rifampin. Antimicrob Agents Chemother 52(6):2027–2034. https://doi.org/10.1128/AAC.01486-07. PMID: 18378710; PMCID: PMC2415778

Saier MH, Paulsen IT (2001) Phylogeny of multidrug transporters. Semin Cell Dev Biol 12(3):205–213

Sanz-García F, Anoz-Carbonell E, Pérez-Herrán E, Martín C, Lucía A, Rodrigues L, Aínsa JA (2019) Mycobacterial aminoglycoside acetyltransferases: a little of drug resistance, and a lot of other roles. Front Microbiol 10:46

Sao Emani C, Reiling N (2024) The efflux pumps Rv1877 and Rv0191 play differential roles in the protection of Mycobacterium tuberculosis against chemical stress. Front Microbiol 15:1359188

Saran N, Saha S (2022) Survey of mycobacterial fluoroquinolone resistance protein conservon (mfp conservon) in Mycobacteriaceae and identification of its promoter activity. Gene Rep 29:101684

Sarathy JP, Ragunathan P, Shin J, Cooper CB, Upton AM, Grüber G, Dick T (2019) TBAJ-876 retains Bedaquiline's activity against subunits c and ε of Mycobacterium tuberculosis F-ATP synthase. Antimicrob Agents Chemother 63(10):e01191–e01119

Schami AM (2024) Variation of phthiocerol dimycocerosate and phosphatidyl-myo-inositol mannoside levels on the cell envelope of mycobacterium tuberculosis influences infection outcomes in macrophages. Doctoral dissertation, The University of Texas Health Science Center at San Antonio

Schifano JM, Edifor R, Sharp JD, Ouyang M, Konkimalla A, Husson RN, Woychik NA (2013) Mycobacterial toxin MazF-mt6 inhibits translation through cleavage of 23S rRNA at the ribosomal A site. Proc Natl Acad Sci USA 110(21):8501–8506

Schuwirth BS, Borovinskaya MA, Hau CW, Zhang W, Vila-Sanjurjo A, Holton JM, Cate JH (2005) Structures of the bacterial ribosome at 3.5 A resolution. Science 310(5749):827–834

Scorpio A, Lindholm-Levy P, Heifets L, Gilman R, Siddiqi S, Cynamon M, Zhang Y (1997) Characterization of pncA mutations in pyrazinamide-resistant Mycobacterium tuberculosis. Antimicrob Agents Chemother 41(3):540–543

Seibert CM, Raushel FM (2005) Structural and catalytic diversity within the amidohydrolase superfamily. Biochemistry 44(17):6383–6391

Seung KJ, Keshavjee S, Rich ML (2015) Multidrug-resistant tuberculosis and extensively drug-resistant tuberculosis. Cold Spring Harb Perspect Med 5(9):a017863

Shafiq A, Khan H, Shahid M, Fatima N, Khan M (2024) Mutational insights into gyrA and gyrB genes in Mycobacterium tuberculosis: a genetic basis for fluoroquinolone resistance in multidrug-resistant tuberculosis. J Pure Appl Microbiol 18(4)

Sharma D, Bisht D (2017) Mycobacterium tuberculosis hypothetical proteins and proteins of unknown function: hope for exploring novel resistance mechanisms as well as future targets of drug resistance. Front Microbiol 8:465. https://doi.org/10.3389/fmicb.2017.00465

Sharma D, Cukras AR, Rogers EJ, Southworth DR, Green R (2007) Mutational analysis of S12 protein and implications for the accuracy of decoding by the ribosome. J Mol Biol 374(4):1065–1076

Sharma D, Kumar B, Lata M, Joshi B, Venkatesan K, Shukla S, Bisht D (2015) Comparative proteomic analysis of aminoglycosides resistant and susceptible mycobacterium tuberculosis clinical isolates for exploring potential drugtargets. PLoS One 10(10):e0139414. https://doi.org/10.1371/journal.pone.0139414

Sharma D, Singh R, Deo N, Bisht D (2018) Interactome analysis of Rv0148 to predict potential targets and their pathways linked to aminoglycosides drug resistance: an in silico approach. Microb Pathog 121:179–183. https://doi.org/10.1016/j.micpath.2018.05.034

Sharma S, Jayasinghe YP, Mishra NK, Orimoloye MO, Wong TY, Dalluge JJ, Ronning DR, Aldrich CC (2023) Structural and functional characterization of Mycobacterium tuberculosis Homoserine Transacetylase. ACS Infect Dis. 9(3):540–553

Sherman DR, Mdluli K, Hickey MJ, Arain TM, Morris SL, Barry CE III, Stover CK (1996) Compensatory ahp C gene expression in isoniazid-resistant Mycobacterium tuberculosis. Science 272(5268):1641–1643

Shi L, Berg S, Lee A, Spencer JS, Zhang J, Vissa V, McNeil MR, Khoo KH, Chatterjee D (2006) The carboxy terminus of EmbC from Mycobacterium smegmatis mediates chain length extension of the arabinan in lipoarabinomannan. J Biol Chem 281(28):19512–19526

Shi X, Khade PK, Sanbonmatsu KY, Joseph S (2012) Functional role of the sarcin-ricin loop of the 23S rRNA in the elongation cycle of protein synthesis. J Mol Biol 419(3–4):125–138

Shi D, Zhou Q, Xu S, Zhu Y, Li H, Xu Y (2022) Pyrazinamide resistance and pncA mutation profiles in multidrug resistant Mycobacterium tuberculosis. Infect Drug Resist 15:4985–4994

Shin DM, Jeon BY, Lee HM, Jin HS, Yuk JM, Song CH, Lee SH, Lee ZW, Cho SN, Kim JM, Friedman RL, Jo EK (2010) Mycobacterium tuberculosis eis regulates autophagy, inflammation, and cell death through redox-dependent signaling. PLoS Pathog 6(12):e1001230

Shrestha D, Maharjan B, Oo NA, Isoda N, Nakajima C, Suzuki Y (2020) Molecular analysis of streptomycin-resistance associating genes in Mycobacterium tuberculosis isolates from Nepal. Tuberculosis 125:101985

Shukla S, Bhardwaj N, Singh A (2024) Drug resistance in mycobacterium tuberculosis: an evolutionary perspective and its adaptation to the lung microenvironment. The Microbe 10:100189

Sidiq Z, Hanif M, Chopra KK, Khanna A, Jadhav I, Dwivedi KK (2019) Second-line drug susceptibilities of multidrug- and rifampicin-resistant Mycobacterium tuberculosis isolates in Delhi. Biomed Biotechnol Res J (BBRJ) 3(2):87–91

Sikandar M, Xing L (2025) Advances in the mechanisms of drug resistance of Mycobacterium tuberculosis. Microb Pathog 206:107861

Simanshu DK, Yamaguchi Y, Park JH, Inouye M, Patel DJ (2013) Structural basis of mRNA recognition and cleavage by toxin MazF and its regulation by antitoxin MazE in Bacillus subtilis. Mol Cell 52(3):447–458

Singh V, Chibale K (2021) Strategies to combat multi-drug resistance in tuberculosis. Acc Chem Res 54(10):2361–2376

Singh A, Grover S, Sinha S, Das M, Somvanshi P, Grover A (2017) Mechanistic principles behind molecular mechanism of rifampicin resistance in mutant RNA polymerase Beta subunit of Mycobacterium tuberculosis. J Cell Biochem 118(12):4594–4606

Singh R, Dwivedi SP, Gaharwar US, Meena R, Rajamani P, Prasad T (2020) Recent updates on drug resistance in Mycobacterium tuberculosis. J Appl Microbiol 128(6):1547–1567

Sirgel FA, Tait M, Warren RM, Streicher EM, Böttger EC, van Helden PD, Gey van Pittius NC, Coetzee G, Hoosain EY, Chabula-Nxiweni M, Hayes C, Victor TC, Trollip A (2012) Mutations in the rrs A1401G gene and phenotypic resistance to amikacin and capreomycin in Mycobacterium tuberculosis. Microb Drug Resist 18(2):193–197

Slayden RA, Barry CE 3rd (2002) The role of KasA and KasB in the biosynthesis of meromycolic acids and isoniazid resistance in mycobacterium tuberculosis. Tuberculosis 82(4–5):149–160

Slayden RA, Barry CE 3rd (2002) The role of KasA and KasB in the biosynthesis of meromycolic acids and isoniazid resistance in Mycobacterium tuberculosis. Tuberculosis 82(4–5):149–160

Smith BC, Anderson MA, Hoadley KA, Keck JL, Cleland WW, Denu JM (2012) Structural and kinetic isotope effect studies of nicotinamidase (Pnc1) from Saccharomyces cerevisiae. Biochemistry 51(1):243–256

Smith CA, Bhattacharya M, Toth M, Stewart NK, Vakulenko SB (2017) Aminoglycoside resistance profile and structural architecture of the aminoglycoside acetyltransferase AAC(6′)-Im. Microb Cell 4(12):402–410

Snobre J, Villellas MC, Coeck N, Mulders W, Tzfadia O, de Jong BC, Andries K, Rigouts L (2023) Bedaquiline- and clofazimine-selected Mycobacterium tuberculosis mutants: further insights on resistance driven largely by Rv0678. Sci Rep 13:10444

Sosunov V, Zorov S, Sosunova E, Nikolaev A, Zakeyeva I, Bass I, Goldfarb A, Nikiforov V, Severinov K, Mustaev A (2005) The involvement of the aspartate triad of the active center in all catalytic activities of multisubunit RNA polymerase. Nucleic Acids Res 33(13):4202–4211

Spies FS, Ribeiro AW, Ramos DF, Ribeiro MO, Martin A, Palomino JC, Rossetti ML, da Silva PE, Zaha A (2011) Streptomycin resistance and lineage-specific polymorphisms in Mycobacterium tuberculosis gidB gene. J Clin Microbiol 49(7):2625–2630

Springer B, Kidan YG, Prammananan T, Ellrott K, Böttger EC, Sander P (2001) Mechanisms of streptomycin resistance: selection of mutations in the 16S rRNA gene conferring resistance. Antimicrob Agents Chemother 45(10):2877–2884

Sreevatsan S, Pan X, Zhang Y, Kreiswirth BN, Musser JM (1997) Mutations associated with pyrazinamide resistance in pncA of Mycobacterium tuberculosis complex organisms. Antimicrob Agents Chemother 41(3):636–640

Stanger FV, Dehio C, Schirmer T (2014) Structure of the N-terminal gyrase B fragment in complex with ADP·Pi reveals rigid-body motion induced by ATP hydrolysis. PLoS One 9(9):e107289

Stephanie F, Tambunan USF, Siahaan TJ (2022) Mycobacterium tuberculosis transcription machinery: a review on the mycobacterial RNA polymerase and drug discovery efforts. Life (Basel) 12(11):1774. https://doi.org/10.3390/life12111774. PMID: 36362929; PMCID: PMC9695777

Stsiapanava A, Selmer M (2019) Crystal structure of ErmE 23S rRNA methyltransferase in macrolide resistance. Sci Rep 9(1):14607. https://doi.org/10.1038/s41598-019-51174-0. PMID: 31601908; PMCID: PMC6787224

Sütterlin L, Edoo Z, Hugonnet JE, Mainardi JL, Arthur M (2018) Peptidoglycan cross-linking activity of L,Dtranspeptidases from Clostridium difficile and inactivation of these enzymes by β-lactams. Antimicrob Agents Chemother 62(1):e01607–17. https://doi.org/10.1128/AAC.01607-17. PMID: 29061738; PMCID: PMC5740366

Su F, Cao L, Ren X, Hu J, Wang X, Fu Y, Zhou Y, Wu H, Wen Y, Jiang M, Tavengana G (2022) RpoB gene mutation characteristics of rifampicin-resistant tuberculosis in Anqing, China. Jundishapur J Microbiol 15(9):e127306

Sulzenbacher G, Canaan S, Bordat Y, Neyrolles O, Stadthagen G, Roig-Zamboni V, Rauzier J, Maurin D, Laval F, Daffé M, Cambillau C, Gicquel B, Bourne Y, Jackson M (2006) LppX is a lipoprotein required for the translocation of phthiocerol dimycocerosates to the surface of Mycobacterium tuberculosis. EMBO J 25(7):1436–1444

Sun YJ, Luo JT, Wong SY, Lee AS (2010) Analysis of rpsL and rrs mutations in Beijing and non-Beijing streptomycin-resistant Mycobacterium tuberculosis isolates from Singapore. Clin Microbiol Infect 16(3):287–289

Sun H, Li S, Xie Z, Yang F, Sun Y, Zhu Y, Zhao X, Jiang S (2012) A novel multidrug resistance plasmid isolated from an Escherichia coli strain resistant to aminoglycosides. J Antimicrob Chemother 67(7):1635–1638

Sun H, Zhang C, Xiang L, Pi R, Guo Z, Zheng C, Li S, Zhao Y, Tang K, Luo M, Rastogi N, Li Y, Sun Q (2016) Characterization of mutations in streptomycin-resistant Mycobacterium tuberculosis isolates in Sichuan, China and the association between Beijing-lineage and dual-mutation in gidB. Tuberculosis (Edinb) 96:102–106

Sun Q, Xiao TY, Liu HC, Zhao XQ, Liu ZG, Li YN, Zeng H, Zhao LL, Wan KL (2018) Mutations within embCAB are associated with variable level of ethambutol resistance in Mycobacterium tuberculosis isolates from China. Antimicrob Agents Chemother 62(1):10–128

Supo-Escalante RR, Médico A, Gushiken E, Olivos-Ramírez GE, Quispe Y, Torres F, Zamudio M, Antiparra R, Amzel LM, Gilman RH, Sheen P, Zimic M (2020) Prediction of Mycobacterium tuberculosis pyrazinamidase function based on structural stability, physicochemical and geometrical descriptors. PLoS One 15(7):e0235643

Suzuki Y, Katsukawa C, Tamaru A, Abe C, Makino M, Mizuguchi Y, Taniguchi H (1998) Detection of kanamycin-resistant Mycobacterium tuberculosis by identifying mutations in the 16S rRNA gene. J Clin Microbiol 36(5):1220–1225

Svetlov MS, Syroegin EA, Aleksandrova EV, Atkinson GC, Gregory ST, Mankin AS et al (2021) Structure of Erm-modified 70S ribosome reveals the mechanism of macrolide resistance. Nat Chem Biol 17(4):412–420. https://doi.org/10.1038/s41589-020-00715-0. PMID: 33462493; PMCID: PMC7990689

Swaney SM, Aoki H, Ganoza MC, Shinabarger DL (1998) The oxazolidinone linezolid inhibits initiation of protein synthesis in bacteria. Antimicrob Agents Chemother 42(12):3251–3255

Szumowski JD, Adams KN, Edelstein PH, Ramakrishnan L (2013) Antimicrobial efflux pumps and mycobacterium tuberculosis drug tolerance: evolutionary considerations. Curr Top Microbiol Immunol 374:81–108. https://doi.org/10.1007/82_2012_300. PMID: 23242857; PMCID: PMC3859842

Tabib-Salazar A, Liu B, Doughty P, Lewis RA, Ghosh S, Parsy ML, Simpson PJ, O'Dwyer K, Matthews SJ, Paget MS (2013) The actinobacterial transcription factor RbpA binds to the principal sigma subunit of RNA polymerase. Nucleic Acids Res 41(11):5679–5691

Takiff HE, Salazar L, Guerrero C, Philipp W, Huang WM, Kreiswirth B, Cole ST, Jacobs WR Jr, Telenti A (1994) Cloning and nucleotide sequence of Mycobacterium tuberculosis gyrA and gyrB genes and detection of quinolone resistance mutations. Antimicrob Agents Chemother 38(4):773–780

Tan Y, Su B, Zheng H, Song Y, Wang Y, Pang Y (2017) Molecular characterization of prothionamide-resistant Mycobacterium tuberculosis isolates in southern China. Front Microbiol 8:2358

Tan YZ, Rodrigues J, Keener JE, Zheng RB, Brunton R, Kloss B, Giacometti SI, Rosário AL, Zhang L, Niederweis M, Clarke OB, Lowary TL, Marty MT, Archer M, Potter CS, Carragher B, Mancia F (2020) Cryo-EM structure of arabinosyltransferase EmbB from Mycobacterium smegmatis. Nat Commun 11(1):3396

Telenti A, Imboden P, Marchesi F, Lowrie D, Cole S, Colston MJ, Matter L, Schopfer K, Bodmer T (1993) Detection of rifampicin-resistance mutations in Mycobacterium tuberculosis. Lancet 341(8846):647–650

Timm J, Bateson A, Solanki P, Paleckyte A, Witney AA, Rofael SAD, Fabiane S, Olugbosi M, McHugh TD, Sun E (2023) Baseline and acquired resistance to bedaquiline, linezolid and pretomanid, and impact on treatment outcomes in four tuberculosis clinical trials containing pretomanid. PLOS Glob Public Health 3(10):e0002283

Todd AE, Orengo CA, Thornton JM (2001) Evolution of function in protein superfamilies, from a structural perspective. J Mol Biol 307(4):1113–1143

Toivonen JM, Boocock MR, Jacobs HT (1999) Modelling in Escherichia coli of mutations in mitoribosomal protein S12: novel mutant phenotypes of rpsL. Mol Microbiol 31(6):1735–1746

Tomas N, Leonelli D, Campoy M, Marthey S, Le NH, Rengel D, Martin V, Pál A, Kordulákóvá J, Eynard N, Guillet V, Mourey L, Daffé M, Lemassu A, André G, Marrakchi H (2022) Bioinformatic Mining and Structure-Activity Profiling of Baeyer-Villiger Monooxygenases from Mycobacterium tuberculosis. mSphere 7(2):e0048221

Tooke CL, Hinchliffe P, Bragginton EC, Colenso CK, Hirvonen VHA, Takebayashi Y, Spencer J (2019) β-Lactamases and β-lactamase inhibitors in the 21st century. J Mol Biol 431(18):3472–3500

Tretter EM, Schoeffler AJ, Weisfield SR, Berger JM (2010) Crystal structure of the DNA gyrase GyrA N-terminal domain from Mycobacterium tuberculosis. Proteins 78(2):492–495

Tsagkalia A, Leontiadou F, Xaplanteri MA, Papadopoulos G, Kalpaxis DL, Choli-Papadopoulou T (2005) Ribosomes containing mutants of L4 ribosomal protein from Thermus thermophilus display multiple defects in ribosomal functions and sensitivity against erythromycin. RNA 11(11):1633–1639

Tsodikov OV, Green KD, Garneau-Tsodikova S (2014) A random sequential mechanism of aminoglycoside acetylation by Mycobacterium tuberculosis Eis protein. PLoS One 9(4):e92370

Tunstall T, Phelan J, Eccleston C, Clark TG, Furnham N (2021) Structural and genomic insights into pyrazinamide resistance in Mycobacterium tuberculosis underlie differences between ancient and modern lineages. Front Mol Biosci 8:619403

Unissa AN, Selvakumar N, Narayanan S, Suganthi C, Hanna LE (2015) Investigation of Ser315 substitutions within katG gene in isoniazid-resistant clinical isolates of Mycobacterium tuberculosis from South India. Biomed Res Int 2015:257983

Ushtanit A, Mikhailova Y, Lyubimova A, Makarova M, Safonova S, Filippov A, Borisov S, Zimenkov D (2021) Genetic profile of linezolid-resistant M. Tuberculosis clinical strains from Moscow. Antibiotics (Basel). 10(10):1243

Valafar SJ (2021) Systematic review of mutations associated with isoniazid resistance points to continuing evolution and subsequent evasion of molecular detection, and potential for emergence of multidrug resistance in clinical strains of Mycobacterium tuberculosis. Antimicrob Agents Chemother 65(3):10–128

Van Bambeke F, Balzi E, Tulkens PM (2000) Antibiotic efflux pumps. Biochem Pharmacol 60(4):457–470

Vannelli TA, Dykman A, Ortiz de Montellano PR (2002) The antituberculosis drug ethionamide is activated by a flavoprotein monooxygenase. J Biol Chem 277(15):12824–12829

Varela MF, Stephen J, Bharti D, Lekshmi M, Kumar S (2023) Inhibition of multidrug efflux pumps belonging to the major facilitator superfamily in bacterial pathogens. Biomedicine 11(5):1448

Velayati AA, Farnia P (2016) Atlas of Mycobacterium tuberculosis, 1st edn. Academic Press, London. Hardback ISBN: 9780128038086. eBook ISBN: 9780128038802

Velayati AA, Farnia P, Masjedi MR (2013a) Totally drug-resistant tuberculosis (TDR-TB): a debate on global health communities. Int J Mycobacteriol. 2(2):71–72

Velayati AA, Farnia P, Masjedi MR (2013b) The totally drug resistant tuberculosis (TDR-TB). Int J Clin Exp Med. 6(4):307–309. PMID: 23641309; PMCID: PMC3631557

Velayati AA, Mitaria S, Farnia P, Farnia P, Ghanavi J, Maleknia M et al (2024) Exploring the genetic landscape of Mycobacterium tuberculosis: unlocking the differences between latent and active tuberculosis. Int J Mycobacteriol. 13(4):410–419

Vemula SL, Gupta MT (2022) Isoniazid Nano-drug delivery systems targeting macrophages for the treatment of tuberculosis. J Prevent Diagnost Treat Strateg Med 1(2):96–104

Vergnolle O, Xu H, Tufariello JM, Favrot L, Malek AA, Jacobs WR Jr, Blanchard JS (2016) Post-translational acetylation of MbtA modulates mycobacterial Siderophore biosynthesis. J Biol Chem 291(42):22315–22326

Verma A, Kinger AK, Tyagi JS (1994) Functional analysis of transcription of the mycobacterium tuberculosis 16S rDNA-encoding gene. Gene 148(1):113-118. https://doi.org/10.1016/0378-1119(94)90243-7. PMID: 7926824

Verma JS, Gupta Y, Nair D, Manzoor N, Rautela RS, Rai A et al (2014) Evaluation of gidB alterations responsible for streptomycin resistance in Mycobacterium tuberculosis. J Antimicrob Chemother 69(11):2935–2941

Verma AK, Kim RQ, Lamprecht DA, Aguilar-Pérez C, Wong S, Veziris N, Aubry A, Bartolomé-Nebreda JM, Carbajo RJ, Wetzel J, Lamers MH (2025) Structural and mechanistic study of a novel inhibitor analogue of Mycobacterium tuberculosis cytochrome bc1:aa3. NPJ Drug Discov 2:–6

Vetting MW, Hegde SS, Fajardo JE, Fiser A, Roderick SL, Takiff HE, Blanchard JS (2006) Pentapeptide repeat proteins. Biochemistry 45(1):1–10

Vilchèze C, Jacobs WR (2014) Resistance to isoniazid and Ethionamide in Mycobacterium tuberculosis: genes, mutations, and causalities. Microbiol Spectr. 2(4):MGM2-0014-2013

Vilchèze C, Jacobs WR (2019) The isoniazid paradigm of killing, resistance, and persistence in Mycobacterium tuberculosis. J Mol Biol 431(18):3450–3461

Vilchèze C, Weinrick B, Leung LW, Jacobs WR Jr (2018) Plasticity of Mycobacterium tuberculosis NADH dehydrogenases and their role in virulence. Proc Natl Acad Sci USA 115(7):1599–1604

Viljoen A, Dubois V, Girard-Misguich F, Blaise M, Herrmann JL, Kremer L (2017) The diverse family of MmpL transporters in mycobacteria: from regulation to antimicrobial developments. Mol Microbiol 104(6):889–904

Villellas C, Aristimuño L, Vitoria MA, Prat C, Blanco S, García de Viedma D, Domínguez J, Samper S, Aínsa JA (2013) Analysis of mutations in streptomycin-resistant strains reveals a simple and reliable genetic marker for identification of the Mycobacterium tuberculosis Beijing genotype. J Clin Microbiol 51(7):2124–2130

Vincent AT, Nyongesa S, Morneau I et al (2018) The mycobacterial cell envelope: a relict from the past or the result of recent evolution? Front Microbiol 9:2341. https://doi.org/10.3389/fmicb.2018.02341. PMID: 30369944; PMCID: PMC6199408

Vögeli B, Rosenthal RG, Stoffel GMM, Wagner T, Kiefer P, Cortina NS, Shima S, Erb TJ (2018) InhA, the enoyl-thioester reductase from mycobacterium tuberculosis, forms a covalent adduct during catalysis. J Biol Chem 293(44):17200–17207. https://doi.org/10.1074/jbc.RA118.005405. PMID: 30217823; PMCID: PMC6222099

Vummidi V (2023) Elucidating the molecular mechanisms underlying mutations in Mycobacterium tuberculosis RNA polymerase that confer resistance to rifampicin and its structural analogues. Biomedicine 43(4):1108–1117

Wade MM, Zhang Y (2004) Mechanisms of drug resistance in mycobacterium tuberculosis. Front Biosci 9:975–994

Wan L, Hu P, Zhang L, Wang ZX, Fleming J, Ni B, Luo J, Guan CX, Bai L, Tan Y, Liu H, Li N, Xiao T, Bai H, Zhang YA, Zhang XE, Wan K, Bi L, Ouyang S, Zhang H (2023) Omics analysis of Mycobacterium tuberculosis isolates uncovers Rv3094c, an ethionamide metabolism-associated gene. Commun Biol 6(1):156

Wang F, Cassidy C, Sacchettini JC (2006) Crystal structure and activity studies of the Mycobacterium tuberculosis beta-lactamase reveal its critical role in resistance to beta-lactam antibiotics. Antimicrob Agents Chemother 50(8):2762–2771

Wang S, Wang K, Song K, Li P, Li D, Sun Y, Mei Y, Xu C, Liao M (2024) Structures of the essential efflux pump EfpA from Mycobacterium tuberculosis reveal the mechanisms of substrate transport and small-molecule inhibition. Res Sq [Preprint]:rs.3.rs-3740027

Warrier T, Kapilashrami K, Argyrou A, Ioerger TR, Little D, Murphy KC, Nandakumar M, Park S, Gold B, Mi J, Zhang T, Meiler E, Rees M, Somersan-Karakaya S, Porras-De Francisco E, Martinez-Hoyos M, Burns-Huang K, Roberts J, Ling Y et al (2016) N-methylation of a bactericidal compound as a resistance mechanism in Mycobacterium tuberculosis. Proc Natl Acad Sci USA 113(31):E4523–E4530

Wasserman S, Louw G, Ramangoaela L, Barber G, Hayes C, Omar SV, Maartens G, Barry C, Song T, Meintjes G (2019) Linezolid resistance in patients with drug-resistant TB and treatment failure in South Africa. J Antimicrob Chemother 74(8):2377–2384

Watanabe S, Matsumura K, Iwai H, Funatogawa K, Haishima Y, Fukui C, Okumura K, Kato-Miyazawa M, Hashimoto M, Teramoto K, Kirikae F, Miyoshi-Akiyama T, Kirikae T (2016) A mutation in the 16S rRNA decoding region attenuates the virulence of Mycobacterium tuberculosis. Infect Immun 84(8):2264–2273

Wei CJ, Lei B, Musser JM, Tu SC (2003) Isoniazid activation defects in recombinant Mycobacterium tuberculosis catalase-peroxidase (KatG) mutants evident in InhA inhibitor production. Antimicrob Agents Chemother 47(2):670–675

Willby MJ, Green KD, Gajadeera CS, Hou C, Tsodikov OV, Posey JE, Garneau-Tsodikova S (2016) Potent inhibitors of acetyltransferase Eis overcome kanamycin resistance in Mycobacterium tuberculosis. ACS Chem Biol 11(6):1639–1646

Willmott CJ, Maxwell A (1993) A single point mutation in the DNA gyrase A protein greatly reduces binding of fluoroquinolones to the gyrase-DNA complex. Antimicrob Agents Chemother 37(1):126–127

Wilson DN (2014) Ribosome-targeting antibiotics and mechanisms of bacterial resistance. Nat Rev Microbiol 12(1):35–48

Winther KS, Brodersen DE, Brown AK, Gerdes K (2013) VapC20 of mycobacterium tuberculosis cleaves the sarcin-ricin loop of 23S rRNA. Nat Commun 4:2796. https://doi.org/10.1038/ncomms3796. PMID: 24225902; PMCID: PMC3847341

Witek MA, Kuiper EG, Minten E, Crispell EK, Conn GL (2017) A novel motif for S-adenosyl-L-methionine binding by the ribosomal RNA methyltransferase TlyA from mycobacterium tuberculosis. J Biol Chem 292(5):1977–1987. https://doi.org/10.1074/jbc.M116.752659. PMID: 28031456; PMCID: PMC5290967

Wivagg CN, Bhattacharyya RP, Hung DT (2014) Mechanisms of β-lactam killing and resistance in the context of Mycobacterium tuberculosis. J Antibiot 67(9):645–654

Wong SY, Lee JS, Kwak HK, Via LE, Boshoff HI, Barry CE 3rd. (2011) Mutations in gidB confer low-level streptomycin resistance in Mycobacterium tuberculosis. Antimicrob Agents Chemother 55(6):2515–2522

Wong SY, Javid B, Addepalli B, Piszczek G, Strader MB, Limbach PA, Barry CE 3rd. (2013) Functional role of methylation of G518 of the 16S rRNA 530 loop by GidB in Mycobacterium tuberculosis. Antimicrob Agents Chemother 57(12):6311–6318

Xu Y, Jia H, Huang H, Sun Z, Zhang Z (2015) Mutations found in embCAB, embR, and ubiA genes of ethambutol-sensitive and -resistant Mycobacterium tuberculosis clinical isolates from China. Biomed Res Int 2015:951706

Xu A, Wen ZX, Wang Y, Wang WB (2022) Prediction of different interventions on the burden of drug-resistant tuberculosis in China: a dynamic modelling study. J Glob Antimicrob Resist 1(29):323–330

Xu H, Kenneson JR, Minton LE, Goodwin DC (2024) A switch and a failsafe: KatG's mechanism for preservation of catalase activity using a conformationally dynamic Arg and an active-site Trp. Front Chem Biol 3:1431412

Yadav S (2024) BPaLM regimen against drug-resistant tuberculosis in India: a need of the hour. Cureus 16(9):e68421. https://doi.org/10.7759/cureus.68421. PMID: 39360094; PMCID: PMC11445631

Yan W, Zheng Y, Dou C, Zhang G, Arnaout T, Cheng W (2022) The pathogenic mechanism of Mycobacterium tuberculosis: implication for new drug development. Mol Biomed 3(1):48

Yang K, Chang JY, Cui Z, Li X, Meng R, Duan L, Thongchol J, Jakana J, Huwe CM, Sacchettini JC, Zhang J (2017) Structural insights into species-specific features of the ribosome from the human pathogen Mycobacterium tuberculosis. Nucleic Acids Res 45(18):10884–10894

Yang S, Xu Y, Wang Y, Ren F, Li S, Ding W, Ma Y, Zhang W (2018) The biological properties and potential interacting proteins of D-alanyl-D-alanine ligase A from Mycobacterium tuberculosis. Molecules 23(2):324

Yaseen I, Kaur P, Nandicoori VK, Khosla S (2015) Mycobacteria modulate host epigenetic machinery by Rv1988 methylation of a non-tail arginine of histone H3. Nature Commun 6(1):8922

Yoon HJ, Kim KH, Yang JK, Suh SW, Kim H, Jang S (2013) A docking study of enhanced intracellular survival protein from Mycobacterium tuberculosis with human DUSP16/MKP-7. J Synchrotron Radiat 20(Pt 6):929–932

Zapata-Pérez R, García-Saura AG, Jebbar M, Golyshin PN, Sánchez-Ferrer Á (2016) Combined whole-cell high-throughput functional screening for identification of new nicotinamidases/pyrazinamidases in metagenomic/polygenomic libraries. Front Microbiol 7:1915

Zengel JM, Jerauld A, Walker A, Wahl MC, Lindahl L (2003) The extended loops of ribosomal proteins L4 and L22 are not required for ribosome assembly or L4-mediated autogenous control. RNA 9(10):1188–1197

Zenkin N, Yuzenkova Y (2015) New insights into the functions of transcription factors that bind the RNA polymerase secondary channel. Biomolecules 5(3):1195–1209. https://doi.org/10.3390/biom5031195. PMID: 26113847; PMCID: PMC4598797

Zhang Y, Yew WW (2015) Mechanisms of drug resistance in Mycobacterium tuberculosis: update 2015. Int J Tuberc Lung Dis 19(11):1276–1289

Zhang Y, Heym B, Allen B, Young D, Cole S (1992) The catalase-peroxidase gene and isoniazid resistance of Mycobacterium tuberculosis. Nature 358(6387):591–593

Zhang Y, Wade MM, Scorpio A, Zhang H, Sun Z (2003a) Mode of action of pyrazinamide: disruption of Mycobacterium tuberculosis membrane transport and energetics by pyrazinoic acid. J Antimicrob Chemother 52(5):790–795

Zhang L, Sato NS, Watanabe K, Suzuki T (2003b) Functional genetic selection of the decoding center in E. coli 16S rRNA. Nucleic Acids Res Suppl:(3):319–320

Zhang H, Deng JY, Bi LJ, Zhou YF, Zhang ZP, Zhang CG, Zhang Y, Zhang XE (2008) Characterization of Mycobacterium tuberculosis nicotinamidase/pyrazinamidase. FEBS J 275(4):753–762

Zhang Y, Feng Y, Chatterjee S, Tuske S, Ho MX, Arnold E, Ebright RH (2012) Structural basis of transcription initiation. Science 338(6110):1076–1080

Zhang Y, Shi W, Zhang W, Mitchison D (2014) Mechanisms of pyrazinamide action and resistance. Microbiology spectrum 2(4):10–128

Zhang HN, Xu ZW, Jiang HW, Wu FL, He X, Liu Y et al (2017) Cyclic di-GMP regulates Mycobacterium tuberculosis resistance to ethionamide. Sci Rep 7:5860

Zhang L, Zhao Y, Gao Y, Wu L, Gao R, Zhang Q, Wang Y, Wu C, Wu F, Gurcha SS, Veerapen N, Batt SM, Zhao W, Qin L, Yang X, Wang M, Zhu Y, Zhang B, Bi L, Zhang X, Yang H, Guddat LW, Xu W, Wang Q, Li J, Besra GS, Rao Z (2020) Structures of cell wall arabinosyltransferases with the anti-tuberculosis drug ethambutol. Science 368(6496):1211–1219

Zhang S, Wang J, Ahn J (2023) Advances in the discovery of efflux pump inhibitors as novel Potentiators to control antimicrobial-resistant pathogens. Antibiotics (Basel). 12(9):1417

Zhang Y, Zhao R, Zhang Z, Liu Q, Zhang A, Ren Q, Li S, Long X, Xu H (2021) Analysis of factors influencing multidrug-resistant tuberculosis and validation of whole-genome sequencing in children with drug-resistant tuberculosis. Infect Drug Resist 14:4375-4393. https://doi.org/10.2147/IDR.S331890. PMID: 34729015; PMCID: PMC8554314

Zhang X, Zhao R, Qi Y, Yan X, Qi G, Peng Q (2024) The progress of Mycobacterium tuberculosis drug targets. Front Med (Lausanne). 11:1455715

Zhang W, Jiang Z, Mi K (2025) The antibiotic-resistant protein MfpA modulates host cell apoptosis and promotes mycobacterial survival by targeting mitochondria and regulating the NF-κB signaling pathway. Cells 14(12):867

Zhao M, Wang L, Zhang H, Dong Y, Gong Y, Zhang L, Wang J (2014) Purification, crystallization and preliminary crystallographic analysis of the 16S rRNA methyltransferase RsmI from Escherichia coli. Acta Crystallogr F Struct Biol Commun 70(Pt 9):1256–1259

Zhao B, Zheng H, Timm J, Song Z, Pei S, Xing R, Guo Y, Ma L, Li F, Li Q, Li Y (2024) Prevalence and genetic basis of Mycobacterium tuberculosis resistance to pretomanid in China. Ann Clin Microbiol Antimicrob 23(1):40

Zhu JH, Wang BW, Pan M, Zeng YN, Rego H, Javid B (2018) Rifampicin can induce antibiotic tolerance in mycobacteria via paradoxical changes in rpoB transcription. Nat Commun 9:4218

Zimenkov D, Ushtanit A, Gordeeva E, Guselnikova E, Schwartz Y, Stavitskaya N (2025) High prevalence of atpE mutations in Bedaquiline-resistant Mycobacterium tuberculosis isolates. Russia Emerg Infect Dis 31(3):525–536

Zwama M, Nishino K (2021) Ever-adapting RND efflux pumps in gram-negative multidrug-resistant pathogens: A race against time. Antibiotics (Basel) 10(7):774

Recent Characterization of Key Proteins to Combat Tuberculosis

10

Abstract

Recent large-scale proteogenomic analyses of *Mycobacterium tuberculosis* (Mtb) have significantly deepened our understanding of its proteome by delineating two major classes of proteins: canonical proteins and microproteins. Canonical proteins, encoded by well-annotated open reading frames (ORFs), perform essential functions in fundamental cellular processes, including DNA replication, metabolism, virulence regulation, and adaptation to stress. Concurrently, growing attention has been directed toward microproteins, small polypeptides translated from previously unannotated small ORFs, which have increasingly been recognized as key modulators of bacterial growth, persistence, and antibiotic response. Microproteins commonly exhibit structural characteristics such as intrinsically disordered regions, signal peptides, and transmembrane domains, enabling their participation in diverse cellular pathways, including membrane dynamics, signal transduction, and transcriptional regulation. These insights highlight the remarkable regulatory complexity and proteomic versatility that underpin Mtb's ability to adapt to host-imposed stress conditions. A significant advancement is the emerging understanding of conserved hypothetical proteins (CHPs), a long-overlooked group previously considered functionally ambiguous. Recent structural and functional characterizations have uncovered critical roles for many CHPs in vital biological processes, such as cell wall biosynthesis, DNA maintenance, metabolic control, and environmental stress responses. Notably, several newly characterized members of both microproteins and CHPs have surfaced as promising vaccine candidates owing to their strong immunogenicity and as attractive drug targets due to their essentiality and amenability to structural studies. Additionally, some exhibit potential as diagnostic biomarkers based on their infection-specific expression and recognition by the host immune system. Collectively, this expanded repertoire, especially the newly characterized proteins, significantly enhances our understanding of Mtb pathogenesis, adaptability, and persistence.

Keywords

Mycobacterium tuberculosis · Proteogenomics · Microproteins · Type II toxin-antitoxin (TA) proteins · PE/PPE family · Stress response microproteins · Subunit vaccines

P. Farnia et al., *Proteins in Mycobacterium Tuberculosis*, Advances in Experimental Medicine and Biology 1484, https://doi.org/10.1007/978-3-031-96883-9_10

10.1 Introduction

Proteogenomics is an innovative approach that integrates protein information, primarily obtained through mass spectrometry, with genomic and transcriptomic data. By combining these diverse sources, researchers can enhance genome annotation accuracy, identify novel proteins, and more effectively validate gene prediction models (Nesvizhskii 2014; Fancello and Burger 2022; Song et al. 2023). This discipline differs from conventional proteomics, which centers on protein detection and quantification, by applying protein-level evidence to refine gene predictions, correct start and stop codons, and discover previously unannotated open reading frames (ORFs), including small ORFs (smORFs) that encode microproteins. An open reading frame (ORF) refers to a continuous sequence of codons in the genome that has the potential to be translated into a polypeptide until a stop codon is encountered, thereby maintaining an uninterrupted translation process (Jungblut et al. 2001; Mumtaz and Couso 2015; Orr et al. 2020; de Souza and Bizarro 2024). ORFs are typically characterized by their length in amino acids; longer ORFs have historically been considered more likely to encode functional proteins, while shorter ORFs were largely disregarded in earlier genome annotations. Small open reading frames (smORFs) are defined as encoding short polypeptides, generally comprising fewer than 100 to 150 amino acids (Lin et al. 2011; Leong et al. 2022). Owing to their small size, smORFs were frequently overlooked by prior annotation pipelines, which presumed such regions were unlikely to produce functional proteins. However, recent proteogenomic investigations have demonstrated that many smORFs do encode biologically significant microproteins, which participate in key cellular processes such as regulation, signaling, and adaptation (Pai et al. 2024).

In Mtb, the identification of smORFs has contributed to an expanded understanding of the proteome, uncovering a previously hidden repertoire of molecular entities essential for bacilli survival, virulence, and drug resistance (Kelkar et al. 2011). Proteogenomic research in Mtb has profoundly advanced our comprehension of its molecular biology by offering direct protein-level validation for a substantial proportion of the predicted proteome, as well as revealing novel proteins, including microproteins. Cole et al. (1998) initially reported 3924 ORFs in the Mtb $H_{37}Rv$ genome, a figure later revised to 3995 in their subsequent annotation (Camus et al. 2002). Comparative studies of gene annotation sets from different sources (e.g., Sanger and TIGR) have revealed that approximately 50% of genes in Mtb $H_{37}Rv$ possess divergent translation start sites (de Souza et al. 2011). The utilization of proteomic data from 449 culture filtrate proteins in the same study enabled the correction of 24 gene annotations. Moreover, it has been proposed that numerous coding sequences (CDSs) in the $H_{37}Rv$ genome remain unannotated (Lew et al. 2013). Through an in-depth proteomic analysis utilizing high-resolution Fourier transform mass spectrometry, Kelkar et al. (2011) identified 3176 Mtb proteins, representing approximately 80% of the total Mtb proteome, and discovered about 250 peptides unmatched by existing annotations. These high-confidence peptide identifications facilitated the delineation of 41 novel genes and the correction of 79 gene models in the $H_{37}Rv$ genome. Subsequent studies have consistently employed proteogenomics to refine genome annotation, delineate gene boundaries, and discover new ORFs previously missed by traditional annotation strategies (Kelkar et al. 2011). Several of these newly identified elements are crucial for bacterial growth, virulence, and antibiotic response, often being co-expressed with functionally important genes within operons. This has unveiled a covert dimension of the proteome, revealing essential molecular players that underlie Mtb's adaptability and persistence (de Souza and Bizarro 2024).

Proteogenomics has also been instrumental in elucidating strain-specific proteomic differences among various Mtb lineages. Comparative proteomic analyses have detected differential protein abundances between lineage 7 (mainly prevalent in Ethiopia and the Horn of Africa) and lineage 4 (the globally prominent Euro-American lineage), findings that align with observed phenotypic disparities in growth, virulence, and drug resistance (Yimer et al. 2020). In this regard, a total of 2867 protein groups were identified, covering 71% of the predicted Mtb proteome, with the abundance of 125 proteins significantly differing between the two lineages. Particularly, proteins implicated in growth and virulence were found to be less abundant in lineage 7 compared to lineage 4. This includes five ABC transporter proteins, three phosphate-binding proteins critical for inorganic phosphate uptake, and six components of the ESX-3 type VII secretion system associated with iron acquisition, all of which were relatively scarce in lineage 7. Such proteogenomic insights provide crucial information regarding the lineage-specific protein profiles, offering clues to the differential properties, such as slower growth, survival fitness, and pathogenicity, of lineage 7 strains (Yimer et al. 2020). Technological advances in proteomics, particularly the advent of shotgun proteomics, have proven pivotal for the high-throughput profiling of complex proteomes exemplified by Mtb. In shotgun proteomics, cellular proteins are enzymatically digested into peptides, which are then separated using liquid chromatography before being analyzed by tandem mass spectrometry. The resulting spectral data are computationally matched against protein databases, enabling the identification and quantification of thousands of proteins in a single experiment, independent of a priori knowledge about the sample contents. This method affords comprehensive coverage of the Mtb proteome, enabling the detection of low-abundance proteins, posttranslational modifications, and novel peptides originating from previously unannotated genes or microproteins. Shotgun proteomics has been particularly instrumental in characterizing the Mtb proteome under diverse physiological conditions that mirror the host environment, such as hypoxia, nutrient deprivation, oxidative stress, and antibiotic exposure (Bespyatykh et al. 2020; Hoffmann et al. 2018; Kirykowicz and Woodward 2020; Nikitushkin et al. 2022). Despite its broad coverage, this approach is inherently semi-quantitative, with run-to-run variability potentially limiting reproducibility and sensitivity, especially for proteins of low abundance or clinical relevance. To reduce such shortcomings, targeted mass spectrometry techniques, including Selected Reaction Monitoring (SRM) and Sequential Window Acquisition of All Theoretical Mass Spectra (SWATH-MS), have emerged as robust alternatives (Sidoli et al. 2015; Kitata et al. 2023). SRM employs a triple quadrupole mass spectrometer to selectively monitor predefined peptide ions and their fragments, facilitating highly sensitive and reproducible quantification of specific proteins across numerous samples. In Mtb research, SRM assays have enabled precise quantification of proteins central to drug resistance, virulence, and metabolism, thereby supporting nuanced investigations into bacterial physiology and responses to treatment (Picotti et al. 2010; Aebersold and Blattmann 2019). SWATH-MS, a data-independent acquisition platform, synthesizes the comprehensive coverage of shotgun proteomics with the quantification accuracy and reproducibility of targeted approaches. This technique sequentially fragments all detectable peptides within selected mass-to-charge windows, creating a comprehensive digital archive of peptide fragments. Subsequent data mining against spectral libraries permits the quantification of thousands of proteins in a single analysis with exceptional precision. Applications of SWATH-MS to Mtb have enabled detailed profiling of proteome dynamics under infection-relevant stresses, supporting the identification of differentially expressed proteins and posttranslational modifications vital for bacterial adaptation (Rosenberger et al. 2017; Banaei-Esfahani et al. 2017; Ludwig et al. 2018; Babele et al. 2023). The comprehensive review by Banaei-Esfahani et al. (2017) highlights the transformative impact of systems proteomics in bacterial pathogen research, particularly with respect to Mtb. The integration of high-resolution mass spectrometry with systems biology approaches facilitates comprehensive, reproducible, and high-throughput analysis of the Mtb proteome. By combining proteomic data with genomic and transcriptomic information, novel proteins can be uncovered, posttranslational modifications characterized, and complex protein interaction networks elucidated, all of which are pivotal to Mtb's survival, virulence, and drug resistance mechanisms. Therefore, advancements in proteomics, from discovery-oriented shotgun methods to highly quantitative techniques such as SRM and SWATH-MS, have fundamentally transformed Mtb research (Sidoli et al. 2015; Shi et al. 2023; Vitorino 2024; Devlin et al. 2025). Each approach offers distinct advantages: shotgun proteomics excels at comprehensive and unbiased protein discovery, whereas SRM and SWATH-MS provide robust, reproducible, and precise quantification, crucial for the analysis of clinically relevant or low-abundance proteins. The ongoing integration of systems proteomics with multi-omics data continues to accelerate our understanding of Mtb pathogenesis and adaptation, facilitating the identification of novel biomarkers, vaccine candidates, and therapeutic targets. These scientific advances are of paramount importance for improving tuberculosis diagnostics and therapeutics, thereby contributing to more effective global TB control strategies.

10.2 Proteogenomic Discovery of Novel Microproteins in *Mycobacterium tuberculosis*

Microproteins represent a distinct class of small, functional proteins typically composed of approximately 50–150 amino acids. These proteins are encoded by small open reading frames (smORFs), which are short DNA sequences historically dismissed as nonfunctional or regarded as genomic noise owing to their limited length and the difficulty in distinguishing them from random short sequences. As a result, microproteins have mainly been overlooked in conventional genome annotation efforts. However, recent advancements in high-throughput transcriptomics and proteogenomics have revealed smORFs to encode a diverse and biologically significant array of microproteins that play essential roles in various cellular processes, including those critical to the physiology and pathogenesis of Mtb (Andrews and Rothnagel 2014; Orr et al. 2020; Leong et al. 2022; Cao et al. 2023). Within the Mtb genome, smORFs contribute to complex regulatory networks and are frequently co-regulated alongside adjacent genes. The transcripts encoding smORFs may be monocistronic, with a single mRNA encoding one microprotein; however, they are more commonly polycistronic, wherein multiple smORFs are transcribed together on a single mRNA molecule. In line with general bacterial transcriptional organization, Mtb predominantly produces polycistronic mRNAs, facilitating the coordinated expression of multiple microproteins from a single transcript. This arrangement enables Mtb to finely regulate the production of functionally related or complementary microproteins, thereby allowing precise modulation of physiological pathways essential for bacterial adaptation, survival, and pathogenesis during infection (Arnvig and Young 2009; Shell et al. 2015; Guerra-Almeida et al. 2021; Moos et al. 2024).

The peptides encoded by smORFs exhibit a wide range of structural and functional properties, underscoring their versatility in cellular roles. A substantial subset of microproteins contains intrinsically disordered regions (IDRs), segments that lack stable three-dimensional structures, which confer structural flexibility. This flexibility enables diverse molecular interactions, allowing microproteins to dynamically participate in activities such as transcriptional regulation, signal transduction, and stress adaptation (van der Lee et al. 2014; Zhang et al. 2013; Uversky 2024). Other microproteins possess signal peptides that direct them to the secretory pathway, targeting them for export or incorporation into the cell envelope. These secreted or membrane-associated microproteins frequently participate in host-pathogen interactions, immune modulation, and remodeling of the bacterial cell surface (Gomez et al. 2000; Stamm et al. 2019; de Souza and Bizarro 2024). Moreover, numerous microproteins contain transmembrane domains, enabling their integration into lipid bilayers where they perform critical functions related to nutrient uptake, ion transport, signal relay, and maintenance of cell envelope integrity, processes especially vital for Mtb's persistence in hostile host environments (Poirier et al. 2014; Gröschel et al. 2016; Abdalla et al. 2020). The emergence of proteogenomic methodologies combining ribosome profiling, RNA sequencing, and mass spectrometry has dramatically advanced the discovery and functional classification of microproteins in Mtb.

Functionally, microproteins are classified into several categories, each fulfilling distinct biological roles. Regulatory microproteins form a specialized group, typically fewer than 100 to 150 amino acids in length, that modulate critical cellular pathways such as gene expression, stress responses, and bacterial dormancy. A well-characterized subset consists of toxins within Type II toxin-antitoxin (TA) systems, which induce reversible growth arrest and promote Mtb persistence under various environmental stressors, such as nutrient limitation, oxidative stress, antibiotic exposure, and immune pressure (Sala et al. 2014; Lee and Lee 2016; Kang et al. 2020). A total of 79 confirmed or putative TA systems have been identified within the Mtb $H_{37}Rv$ strain, the majority of which belong to the Type II class. These systems consist of proteinaceous toxin and antitoxin pairs, where the antitoxin neutralizes the toxin through direct protein-protein interactions (Sala et al. 2014). Among these, 67 Type II toxin-antitoxin (TA) systems belong to six well-characterized families: Virulence-associated protein B and C (VapBC), MazE antitoxin and MazF toxin (MazEF), YefM/YoeB, RelB antitoxin and RelE toxin (RelBE), HigB toxin and HigA antitoxin (HigBA), and ParD antitoxin and ParE toxin (ParDE) (Sala et al. 2014). The VapBC family is the largest, comprising 50 units encoding VapC toxins, which contain PilT N-terminal (PIN) domains with ribonuclease activity, alongside VapB antitoxins that neutralize these toxins, highlighting their significant roles in stress response and persistence (Sala et al. 2014). The MazEF family includes ten units in which MazF toxins cleave mRNA to modulate translation during stress (Tiwari et al. 2015; Hoffer et al. 2017; Thakur et al. 2024). The YefM/YoeB family is represented by a single unit, wherein YoeB toxins inhibit translation by targeting the ribosome (Cherny et al. 2005; McGillick et al. 2021). Both the RelBE and HigBA families comprise two units each; these include ribonuclease toxins implicated in growth regulation and adaptation to stress (Hurley et al. 2011; Sala et al. 2014; Schureck et al. 2014; Sinha et al. 2023). The ParDE family, also consisting of two systems, encodes toxins typically targeting DNA gyrase, thus interfering with DNA replication (Ramage et al. 2009; Sala et al. 2014; Gupta et al. 2016; Ames et al. 2019; Beck et al. 2024).

In addition to these, a tripartite Type II toxin-antitoxin-chaperone (TAC) system, which includes a SecB-like chaperone, provides an extra regulatory layer over the toxin-antitoxin pair (Sala et al. 2014). Furthermore, three putative Type IV toxin-antitoxin (TA) systems have been identified, where antitoxins neutralize toxin effects by stabilizing the toxin's cellular target rather than directly binding the toxin. Additionally, eight uncharacterized TA systems await experimental validation (Yu et al. 2020). The extensive repertoire of TA systems in Mtb underscores their critical role in bacterial physiology, particularly in mediating reversible growth arrest that enables survival under hostile conditions such as nutrient deprivation, oxidative stress, and antibiotic exposure (Ramage et al. 2009; Tiwari et al. 2015). By modulating essential cellular processes, including translation, these systems facilitate persistence and dormancy, promoting the formation of persister cells that complicate tuberculosis treatment (Wang and Wood 2011; Leung and Lévesque 2012; Sala et al. 2014; Korch and Hill 2006). Several TA systems are also implicated in regulating bacterial virulence through adaptation mechanisms within the host environment (Sala et al. 2014; Song et al. 2023). The predominance of VapBC pairs suggests specialized adaptation to the intracellular environment encountered during infection, while novel regulatory elements such as the TAC system highlight the complex, multilayered control of toxin activity (Kang et al. 2020). These toxins are encoded adjacent to their cognate antitoxin genes (vapB) and co-transcribed as bicistronic operons. Mtb contains at least 47 VapBC pairs, reflecting both functional specialization and redundancy (Sala et al. 2014; Ramage et al. 2009).

Noteworthy VapC toxins include VapC30 (Rv0617c), which cleaves specific tRNAs such as tRNA^Ser isoacceptors, resulting in translational inhibition and growth arrest; VapC30 activity is implicated in stress adaptation by modulating protein synthesis under adverse conditions (Ahidjo et al. 2011; Nawrot et al. 2011; Winther et al. 2016). VapC20 (Rv2549c), a PIN domain endoribonuclease, inhibits translation by cleaving the sarcin-ricin loop (SRL) of 23S rRNA, a conserved element crucial for ribosomal A-site function, thereby disrupting translation elongation and inducing dormancy. The SRL is targeted by various ribotoxins, and its cleavage results in ribosome inactivation and cell death (Winther et al. 2013; Deep et al. 2017). VapC21 (Rv2757c) cleaves initiator tRNA^fMet, essential for translation initiation, producing bacteriostatic effects and contributing to persistence during stress (Winther et al. 2016; Sharma et al. 2020). VapC1 (Rv0065) is a PIN-domain ribonuclease that cleaves specific RNA substrates, including tRNAs such as tRNA^Ala and tRNA^Ser, thereby disrupting translation and promoting dormancy or growth arrest under stress (Winther and Gerdes 2011; Winther et al. 2013, 2016). VapC4 (Rv0595c) specifically cleaves tRNA^Cys, causing ribosome stalling at cysteine codons and triggering metabolic reprogramming that enhances survival under oxidative and copper stress encountered within the host. This cleavage activates downstream small open reading frames (sORFs) that regulate gene expression, promoting cysteine biosynthesis and mycothiol production, both critical for maintaining redox homeostasis and virulence during infection (Barth et al. 2021; Winther et al. 2016; Zarin et al. 2024). Chemically, VapC microproteins' PIN domains are approximately 120 to 130 amino acids in length, adopting a compact α/β fold composed of a central five-stranded parallel β-sheet flanked by α-helices (Arcus et al. 2005; Winther et al. 2016). The active site contains four conserved acidic residues (aspartates and glutamates) that coordinate divalent metal ions (Mg^{2+} or Mn^{2+}), essential for catalyzing RNA cleavage (Winther et al. 2016). This enzymatic activity enables VapC toxins to cleave RNA targets site-specifically, resulting in translational arrest and induction of a dormant, nonreplicating state (Ramage et al. 2009; Winther et al. 2016). Toxic effects are tightly regulated by cognate VapB antitoxins, small proteins that bind VapC and occlude its active site to neutralize ribonuclease activity under favorable conditions (Ramage et al. 2009). Under stress, VapB antitoxins are selectively degraded by intracellular proteases such as ClpXP or Lon, releasing active VapC toxins to exert their effects (Ramage et al. 2009; Tiwari et al. 2015). This reversible regulatory mechanism allows Mtb to toggle between active growth and dormancy in response to environmental cues (Sala et al. 2014). Overall, VapC microproteins play a vital role in Mtb's ability to survive hostile host environments by inducing bacteriostasis and dormancy, key features underpinning latent tuberculosis infection. By selectively cleaving transfer RNAs (tRNAs) and ribosomal RNAs (rRNAs), VapC toxins reduce global translation, conserve cellular energy, and reprogram gene expression to help Mtb adapt to oxidative stress, nutrient deprivation, and antibiotic pressure (Winther et al. 2016; Tiwari et al. 2015; Abdalla et al. 2020).

Membrane proteins play a fundamental role in regulating the transport of substances and signals across cellular membranes. While small gases and hydrophobic molecules can diffuse directly through the phospholipid bilayer, larger or charged molecules such as sugars, amino acids, and ions require specialized membrane proteins to facilitate their transport. Beyond transport, membrane proteins mediate physical interactions between cells, controlling processes like cell adhesion, cellular coupling, and membrane fusion. This diverse group includes ion channels, transporters, exchangers, receptors, and enzymes (Makarewich and Olson 2017; Makarewich 2020; Mohsen et al. 2023). Despite extensive study, significant gaps remain in understanding their regulation, assembly, and subcellular localization. The recent discovery of microproteins with potential membrane association offers new perspectives, suggesting that these small proteins may perform critical functions that complement classical membrane proteins. Among membrane-associated microproteins, PE20, a member of the PE/PPE family, is predicted to embed in the mycobacterial envelope, contributing to membrane stability and facilitating immune evasion through antigenic variation. Encoded by gene Rv1806, PE20 is a prototypical PE family microprotein approximately 99 amino acids long (Brennan et al. 2001; Walters et al. 2006; de Souza et al. 2011; Minato et al. 2015). It contains a conserved N-terminal PE domain that mediates protein-protein interactions, particularly with cognate PPE partners such as PPE20, PPE31, and PPE33. PE20 typically forms heterodimeric complexes with these PPE proteins (Chen et al. 2017; Qian et al. 2020; Wang et al. 2020). These complexes likely adopt channel-like structures featuring amphipathic or hydrophobic surfaces, enabling integration into the hydrophobic, lipid-rich outer membrane of Mtb, which acts as a permeability barrier. Biochemical and structural studies have shown that the PE20/PPE20 complex binds calcium ions (Ca^{2+}), with binding sites primarily on the PPE partner, reflecting a functional design tailored for selective ion transport (Boradia et al. 2022). Functionally, PE20 participates in outer membrane-associated transport systems that compensate for Mtb's lack of classical porins. For instance, the PE20/PPE31 complex selectively mediates magnesium ion (Mg^{2+}) uptake, an essential cofactor for numerous enzymatic activities and for stabilizing ribosomes and nucleic acids. Similarly, the PE15/PPE20 complex functions as a specific calcium ion transporter, critical for maintaining intracellular calcium homeostasis, which regulates vital processes such as ATP synthesis, biofilm formation, and intracellular signaling (Boradia et al. 2022). These PE/PPE heterodimers do not act as passive pores; rather, they function as selective, gated ion channels that dynamically respond to environmental stimuli, thereby regulating ion transport in line with the bacterium's metabolic demands. Beyond nutrient uptake, PE family microproteins like PE20 are important for maintaining outer membrane integrity and facilitating immune evasion. By forming dynamic, surface-exposed complexes with variable antigenic profiles, PE/PPE proteins obstruct immune detection and promote antigenic variation, supporting Mtb's persistence within host macrophages (Zhang et al. 2025). Additionally, these complexes remodel the outer membrane architecture, enhancing bacterial survival under hostile conditions such as antibiotic exposure and granuloma-mediated stress. PE20's function depends on specific protein-protein interactions that produce membrane-integrated heterodimers acting as gated or selective channels within the outer membrane. For example, the PE15/PPE20 complex binds Ca^{2+} ions at defined coordination sites, enabling their translocation across the otherwise impermeable lipid-rich outer membrane (Cascioferro et al. 2007; Wang et al. 2020; Burggraaf et al. 2019; Boradia et al. 2022). Likewise, the PE20/PPE31 complex mediates Mg^{2+} import to support enzymatic functions and maintain cell envelope integrity. These transport systems are tightly regulated, likely responding to ionic fluctuations, redox changes, or environmental stresses, thereby ensuring efficient nutrient acquisition and bacterial adaptation in nutrient-limiting or host-imposed stress conditions (Cascioferro et al. 2007; Wang et al. 2020; Burggraaf et al. 2019; Boradia et al. 2022).

Another important category of microproteins includes those that are secreted or surface-exposed, small proteins either released into the extracellular environment or anchored on the cell surface with external accessibility. These proteins play critical roles in cell signaling, host-pathogen interactions, immune modulation, and intercellular communication (Burton et al. 2024; Rodrigues et al. 2024; Stamm et al. 2019; Hassel et al. 2023). Their export occurs via classical signal peptide-dependent pathways or alternative nonclassical mechanisms. Despite their small size, surface-anchored microproteins engage extracellular molecules and function as signaling mediators, toxins, adhesins, or immune modulators, effectively orchestrating intercellular interactions and sensing environmental cues (Burton et al. 2024). In Mtb, these microproteins are essential for manipulating host immune responses, securing critical nutrients, and sustaining long-term survival under hostile conditions. Although generally around 100 amino acids in length, they perform vital roles at the host-pathogen interface by modulating immune signaling, disrupting host membranes, or forming specialized membrane channels (Houben et al. 2014). A well-studied example is EsxW, a 98-amino-acid microprotein encoded by esxW (Rv3620c). EsxW belongs to the WXG100 family of small secreted α-helical proteins characterized by a conserved tryptophan-X-glycine (WXG) motif and helix-turn-helix fold (Damen et al. 2022). EsxW forms a heterodimer with EsxV (Rv3619c), another 98-amino-acid microprotein, and this complex is secreted via the ESX-5 Type VII Secretion System (T7SS), one of five specialized protein export pathways in Mtb (Shah and Briken 2016). Expression and secretion of EsxW are tightly regulated by environmental stress signals and transcription factors such as EspR and the SenX3-RegX3 two-component system (Park et al. 2019). The ESX machinery's core components include EccB, EccC, EccD, and MycP, with EccB and EccD forming transmembrane structures, EccD likely serving as the inner membrane channel, and EccC providing ATPase activity to energize secretion, while MycP processes the substrate (Houben et al. 2014; Roy et al. 2020; Crosskey et al. 2020; Bunduc et al. 2021). The EsxW/EsxV heterodimer is recognized by the ESX-5 translocon and exported across the thick, lipid-rich mycobacterial envelope via ATP hydrolysis (Crosskey et al. 2020). Once secreted or surface-exposed, EsxW modulates the host immune response by suppressing pro-inflammatory cytokine production, disrupting antigen presentation, and altering macrophage activation, thereby facilitating immune evasion and promoting infection persistence (Passos et al. 2024). The ESX systems also secrete two unique mycobacterial protein families, PE and PPE proteins, which are generally secreted as heterodimers composed of one PE and one PPE partner (Cole et al. 1998; Ekiert and Cox 2014). Successful secretion relies on interaction with cognate EspG chaperones that target the complexes to the appropriate ESX machinery (Daleke et al. 2012; Korotkova et al. 2014; Ekiert and Cox 2014). While many PE and PPE proteins remain uncharacterized, several contribute to Mtb virulence and persistence (Sampson 2011). Unlike their larger counterparts, small PE microproteins fit the classical microprotein definition and are surface-exposed. Although structural data are limited, they are predicted to harbor α-helical domains and form heterodimers with PPE partners to aid membrane insertion or association. For example, PE5 (Rv0285) is a 102–103 amino acid protein with a conserved N-terminal PE motif (Brennan and Delogu 2002; Cascioferro et al. 2007; Tiwari et al. 2012; Kumar et al. 2024). PE5 contributes to antigenic variation and immune modulation through multiple mechanisms. Its high sequence variability, especially in C-terminal regions, allows presentation of diverse antigenic profiles, helping Mtb evade adaptive immune responses (Tiwari et al. 2012). Stage-specific and environment-dependent differential expression of PE genes further supports dynamic remodeling of bacterial surface composition, complicating immune targeting. Additionally, PE microproteins may shield immunodominant epitopes or disrupt host receptor signaling to dampen immune responses (Priyanka et al. 2024). Another key microprotein pair in Mtb is EsxA (ESAT-6, Rv3875) and EsxB (CFP-10, Rv3874), both approximately 95 amino acids long. They are secreted as a heterodimer by the ESX-1 Type VII Secretion System, which is essential for Mtb virulence (Cole et al. 1998; Stanley et al. 2003; Champion et al. 2009; Simeone et al. 2009; Gröschel et al. 2016). Recent real-time microscopy studies show that the EsxA/EsxB heterodimer disrupts phagosomal membranes in macrophages via fibril-mediated vesiculation, facilitating bacterial escape into the cytosol. This escape activates host cytosolic surveillance pathways, including inflammasome activation and interleukin-1β (IL-1β) secretion, contributing both to immunopathology and immune evasion (Koiri et al. 2025). The compact structure, surface localization, and sequence plasticity of these secreted microproteins underline their central role in Mtb pathogenesis, making them promising targets for immunotherapy and diagnostics (de Jonge et al. 2007; Passos et al. 2024).

Stress response microproteins play a vital role in enabling Mtb to survive hostile conditions, including oxidative stress, nutrient scarcity, acidic pH, hypoxia, and immune pressures. These small proteins act as key regulators within adaptive and protective networks, mediating environmental sensing and counteracting various stressors (Mohsen et al. 2023; Jain and Vyas 2025). They encompass diverse families such as universal stress proteins (USPs), thioredoxins, and other redox-active enzymes, whose expression is tightly controlled by transcription factors including the DosR dormancy regulon, two-component systems like MprAB, and alternative sigma factors such as σ^E (Hingley-Wilson et al. 2010). The Mtb genome encodes at least ten USPs, many regulated by DosR and highly expressed during hypoxia and nitric oxide exposure, thus implicating them in persistence and intracellular survival within macrophages (Hingley-Wilson et al. 2010; Bespyatykh et al. 2020). Notable USPs include Rv2005c (~146 amino acids), Rv2028c (~279 amino acids), Rv2623 (293–309 amino acids), and Rv1636 (~140 amino acids). Structurally, USPs share a conserved α/β fold that forms a nucleotide-binding cleft for ATP binding, essential for signal transduction and stress sensing (Glass et al. 2017; Mohsen et al. 2023). Functionally, USPs enhance tolerance to oxidative and nitrosative stress, acidic pH, and DNA damage by stabilizing cellular components and activating protective pathways, often showing functional redundancy that ensures survival amid complex stresses (Bespyatykh et al. 2020). Another example is Rv2617c, a stress-responsive microprotein (~120 amino acids) predicted to contain α-helical domains. It is upregulated under oxidative stress and during infection, suggesting a role in redox homeostasis and protection against reactive oxygen species (Moukendza Koundi et al. 2024). Similarly, thioredoxins, including TrxB1 (Rv1471, 123 amino acids), TrxB2 (Rv3913, 335 amino acids), and TrxC (Rv3914, 116 amino acids), are redox-active microproteins essential for oxidative stress resistance (Lee et al. 2013). These proteins share a conserved thioredoxin fold with a catalytic CXXC motif that reduces disulfide bonds in target proteins (Lin et al. 2016). Recent evidence suggests Mtb thioredoxins have evolved specialized functions beyond general redox maintenance, such as repairing oxidatively damaged proteins, supporting DNA synthesis, and modulating redox-sensitive signaling pathways to counteract host-induced oxidative stress (Anjou et al. 2024). Collectively, the thioredoxin system exemplifies the transition from basic bacterial redox maintenance to sophisticated infection survival strategies. Despite their small size, stress response microproteins constitute indispensable components of Mtb's adaptive machinery, utilizing conserved structural motifs to maintain cellular homeostasis and defend against host-imposed stresses (Akif et al. 2008; Attarian et al. 2009; Lin et al. 2016; Moukendza Koundi et al. 2024).

Prophage-associated microproteins in Mtb, typically 50 to 150 amino acids and encoded by small open reading frames (smORFs) within phage-derived genomic regions phiRv1 and phiRv2, also play significant roles in bacterial adaptation (Cole et al. 1998; Bibb and Hatfull 2002). These regions are remnants of temperate bacteriophages integrated into the bacterial genome and remain transcriptionally active despite defective replication potential (Bibb et al. 2005; Fan et al. 2016, 2022). The ~10-kilobase phiRv1 locus spans genes Rv1573 to Rv1584c, encoding proteins involved in phage-related functions such as site-specific recombination and prophage excision regulation. For example, Rv1584c encodes a small, basic, intrinsically disordered protein (78 amino acids) likely involved in recombination directionality control. Other proteins, such as Rv1575 (~150 amino acids), have predicted transmembrane helices and signal peptides, indicating membrane localization and involvement in envelope-associated processes (Bibb et al. 2005). Similarly, the phiRv2 locus (Rv2653c to Rv2662c) encodes a mix of small and larger proteins implicated in stress response and prophage maintenance (Aagaard et al. 2004). Rv2654c produces an 81-amino-acid microprotein with intrinsically disordered regions and a lipoprotein anchor suggestive of membrane association. Rv2656c and Rv2657c (130 and 86 amino acids, respectively) contain transmembrane domains that likely contribute to envelope integrity under oxidative and acidic stress found within macrophage phagosomes (Fan et al. 2016). Larger proteins like Rv2655c (475 amino acids) and Rv2659c (375 amino acids) serve roles in prophage excision and mobility, with Rv2659c functioning as an integrase (Aagaard et al. 2004). Although prophage-encoded proteins do not form a single classical family, they share hallmark features of regulatory or accessory phage proteins, including intrinsic disorder, membrane localization, and structural flexibility, supporting their role as dynamic modulators of bacterial stress responses through interactions with larger protein complexes or membranes, thereby enhancing Mtb survival in the host (Bibb and Hatfull 2002, Bibb et al. 2005). Furthermore, some prophage microproteins aid immune evasion by masking immunogenic surface components or disrupting host signaling pathways, reducing immune recognition and promoting persistent infection (Bibb and Hatfull 2002; Moukendza Koundi et al. 2024). Many contain helix-turn-helix motifs or other regulatory domains, resembling classical phage repressors or activators, modulating transcription of adjacent genes in response to environmental stimuli to coordinate adaptation. Others regulate metabolism or signaling to optimize energy use during stress (Bibb and Hatfull 2002; Mohsen et al. 2023). Expression of prophage microproteins is tightly controlled, with their genes organized into operons co-expressed alongside other prophage elements. Their transcription is induced by environmental stressors such as oxidative and nitrosative stress, hypoxia, nutrient limitation, and acidic pH conditions resembling the granuloma microenvironment in the host (Bespyatykh et al. 2020). Regulation likely involves global stress-response pathways including the DosR/DosS and PhoP/PhoR two-component systems and alternative sigma factors like σ^E and σ^H, which respond to envelope and redox stresses (Manganelli et al. 2001; Park et al. 2003; Rodrigues et al. 2024).

Posttranslational mechanisms further modulate microprotein stability and activity. Their intrinsically disordered regions facilitate rapid, flexible interactions with stress sensors and membrane components, enabling Mtb to mount swift, localized responses essential for survival (Moukendza Koundi et al. 2024).

10.3 Novel *Mycobacterium tuberculosis* Proteins as Promising Targets for Vaccine Design

Vaccine development for Mtb has undergone significant advancement, evolving beyond traditional strategies to include a diverse range of novel antigens and delivery platforms designed to improve protective immunity against both active and latent tuberculosis. The earliest and most widely used vaccine, Bacillus Calmette-Guérin (BCG), is a live attenuated strain of *Mycobacterium bovis* that provides variable efficacy, with notably limited protection against adult pulmonary TB (Li et al. 2022). Subsequent innovations introduced inactivated whole-cell vaccines such as DAR-901 (von Reyn et al. 2017; Munseri et al. 2020), subunit vaccines including H4:IC31, H56:IC31, and ID93-GLA (Srivastava et al. 2023; Borges et al. 2025), and GamTBvac (Vasina et al. 2019; Tkachuk et al. 2020), as well as viral vector-based vaccines like MVA85A (Tameris et al. 2013) and AdHu5Ag85A (Smaill et al. 2013). More recently, advances in nucleic acid vaccine technology, including DNA and self-amplifying RNA vaccines, have enabled precise delivery of antigen-encoding sequences, thereby enhancing targeted immune responses (Li et al. 2019).

Central to recent advances in tuberculosis vaccine development is the rational selection of antigens, driven by computational immunology and proteomic profiling. These approaches enable the identification of proteins expressed across various stages of Mtb infection, including active replication, latency, and reactivation, thereby supporting the design of multistage vaccine strategies. Among the most extensively studied antigens are the early secreted virulence factors ESAT-6 and CFP-10, small proteins approximately 95 and 100 amino acids in length, respectively. These form a stable heterodimer secreted via the ESX-1 Type VII secretion system (An et al. 2025). ESAT-6 features a compact helix-turn-helix structure that disrupts host phagosomal membranes, facilitating bacterial escape into the cytosol (Bates et al. 2024). CFP-10 stabilizes ESAT-6 through hydrophobic interactions, making the complex essential for both secretion and virulence. Both proteins are highly immunogenic, eliciting robust multifunctional CD4+ and CD8+ T cell responses characterized by the secretion of IFN-γ, TNF-α, and IL-2 (Borgström et al. 2020). This potent immunogenicity has motivated their incorporation into chimeric vaccines such as H1 and GamTBvac, which have demonstrated strong immune responses and protective efficacy in preclinical and early-phase clinical trials (Gurmessa et al. 2025). For example, GamTBvac combines multiple Mtb antigens fused into chimeric proteins and incorporates CpG oligodeoxynucleotide adjuvants to enhance antigen presentation and amplify immune activation (Tkachuk et al. 2020). Another key group of vaccine candidates is the antigen 85 (Ag85) complex, comprising Ag85A, Ag85B, and Ag85C (Harris et al. 2014; Metcalfe et al. 2016; 2000). These ~285 amino acid proteins share a conserved α/β hydrolase fold typical of mycolyl transferases, characterized by a catalytic triad composed of serine, glutamate, and histidine residues essential for enzymatic function. The Ag85 complex catalyzes the transfer of mycolic acids to the arabinogalactan layer of the mycobacterial cell wall, a critical process for cell wall integrity and pathogenicity (Johnson 2017; Lesur et al. 2025). Structural analyses reveal a hydrophobic substrate-binding groove and multiple surface-exposed epitopes accessible for immune recognition. Posttranslational modifications, particularly glycosylation, influence immunogenicity by affecting protein folding, stability, and receptor interactions. For instance, glycosylated Ag85B variants exhibit altered binding affinities to host antibodies and antigen-presenting cells, which may impact vaccine efficacy depending on glycan structure and localization (Pillay et al. 2024). Vaccines incorporating Ag85 proteins, such as H4:IC31 and H56:IC31, induce potent Th1-skewed immune responses marked by elevated IFN-γ and TNF-α production, enhancing macrophage activation and bacterial clearance in animal models (Srivastava et al. 2023).

MPT64 (Rv1980c), an immunodominant secreted protein of approximately 228 amino acids, contains multiple epitopes recognized by both CD4+ and CD8+ T cells, particularly within residues 190–198 (Wang et al. 2007). Structurally, MPT64 adopts a predominantly α-helical fold with surface-exposed loops that serve as key sites for immune recognition. Incorporation of MPT64 into recombinant vaccine constructs aims to broaden the spectrum of T cell activation. Empirical studies confirm that MPT64-based vaccines enhance T cell proliferation and promote secretion of critical cytokines such as IFN-γ and TNF-α, essential for effective protection against Mtb infection (Wang et al. 2014; Singh et al. 2022). Peptides derived from MPT64 are processed and presented via MHC class II molecules, stimulating CD4+ T cells, and can also engage CD8+ T cell responses through cross-presentation, thereby augmenting cytotoxic activity.

Latency-associated antigens, including HspX (α-crystallin,~16.3 kDa), Rv2660c, and Rv2031c, are upregulated during bacterial dormancy and represent critical vaccine targets aimed at controlling latent tuberculosis infection. HspX functions as a small heat shock molecular chaperone that stabilizes proteins under stress conditions (Hu et al. 2006; Taylor et al. 2012). Structurally, HspX assembles into oligomeric complexes composed of α-helical domains and β-sandwich motifs, which underpin its chaperone activity (Panda et al. 2017). Vaccination with HspX-containing immunogens elicits both cellular and humoral immune responses, including memory T cell activation essential for long-lasting protection (Kwon et al. 2024). Since the early 2010s, recombinant fusion proteins combining HspX with other antigens such as EsxS have been developed and assessed in preclinical models for their immunogenicity and protective efficacy. These fusion proteins often serve as components of multistage vaccines designed to bolster T cell responses (Khademi et al. 2017; Avarvand et al. 2021). More recently, advanced subunit vaccines like the quadrivalent formulation HSP90-ESAT-6-HspX-RipA (HEHR), evaluated as BCG boosters, have demonstrated enhanced and durable protection against hypervirulent Mtb strains in animal models (Kwon et al. 2024). Clinical trials testing HspX-containing vaccines are currently ongoing, highlighting their promise in TB vaccine development. Proteomic analyses have also identified novel latency-associated proteins such as Rv0232 and Rv1115, both predicted to be membrane-associated with α-helical and β-strand secondary structures. Immunization studies in C57BL/6 mice showed that these proteins triggered robust secretion of key cytokines, including IFN-γ and IL-2, which are central to cellular immunity against Mtb. While Rv0232 stimulated modest CD4+ T cell proliferation, both proteins significantly increased production of inflammatory mediators such as GM-CSF, IL-12p40, IL-17, IP-10, IL-4, IL-6, and TNF-α, supporting their inclusion in multistage vaccine platforms designed to target active and latent TB by enhancing Th1 immune responses (Li et al. 2019).

Emerging research has also focused on hypothetical or previously uncharacterized proteins. One such candidate, NHP1, is a 248-amino acid protein encoded within the RD7 region of Mtb. It is predicted to adopt a flavodoxin-like fold, characterized by a central five-stranded β-sheet flanked by α-helices, a typical structure for proteins involved in electron transfer and ATP-binding functions (Kootery and Sarojini 2022). Structural modeling confirms a stable globular conformation with multiple exposed antigenic epitopes. Immunoinformatics analyses using tools such as the Immune Epitope Database (IEDB) and VaxiJen predict numerous B cell and T cell epitopes, with high antigenicity and immunogenicity scores further indicating NHP1's potential to elicit strong humoral and cellular immune responses. This underscores the growing importance of bioinformatics and structural modeling in identifying and validating novel antigen candidates for tuberculosis vaccine development (Kootery and Sarojini 2022).

Peptide-based vaccines represent a novel strategy against tuberculosis by harnessing immunogenic epitopes derived from Mtb proteins (Gong et al. 2022; Bellini et al. 2023). These epitopes, typically 8 to 15 amino acids long, are selected for their capacity to elicit strong T cell responses crucial for infection control (Bellini et al. 2023). Employing a multistage approach, epitopes are drawn from proteins expressed during various phases of infection, including early secreted proteins (e.g., CFP-10), latency-associated proteins (e.g., HBHA), and cell wall-associated PPE family components (Gong et al. 2022; Bellini et al. 2023). Chemically synthesized epitopes are conjugated via thiol-maleimide chemistry into linear or branched constructs, with N-terminal palmitoylation enhancing antigen uptake, membrane association, and intracellular delivery (Bellini et al. 2023). Structural characterization reveals that palmitoylated conjugates exhibit increased α-helicity and reduced flexibility, improving stability and prolonging antigen presentation. Immunization induces balanced cytokine responses, including IFN-γ, TNF-α, IL-10, and IL-22, through activation of MHC class I and II pathways, supporting potent CD4+ and CD8+ T-cell-mediated immunity essential for effective bacterial clearance and durable protection (Bellini et al. 2023). Peptide vaccines thus offer advantages such as safety, precise design, scalability, and reduced reliance on traditional adjuvants due to intrinsic antigen stabili zation (Gong et al. 2022; Medha et al. 2022; Kumar et al. 2023).

Among modern platforms, the ID91 replicon RNA (repRNA) vaccine encodes a fusion protein comprising four immunodominant Mtb antigens: Rv3619 (EsxV), Rv2389 (RpfD), Rv3478 (PPE60), and Rv1886 (Ag85B). These antigens cover diverse functional classes and infection stages, providing broad immune coverage (Larsen et al. 2022, 2023; Rais et al. 2023). Utilizing an alphavirus-derived backbone, this repRNA features a subgenomic promoter that amplifies intracellular antigen expression within the cytoplasm, thereby avoiding risks of genomic integration while sustaining antigen presentation and stimulating innate immunity (Larsen et al. 2023; Rais et al. 2023). Encapsulated in a nanostructured lipid carrier (NLC), the repRNA is protected from degradation, eliminating the need for viral vectors and circumventing anti-vector immunity (Rais et al. 2023). Preclinical murine studies demonstrated that ID91 repRNA vaccination induces robust CD8+ T cell proliferation and cytokine production; in contrast, protein subunit vaccines formulated with TLR4 agonists preferentially elicit CD4+ Th1 responses. Both platforms significantly reduce pulmonary bacterial loads following challenge. Notably, heterologous prime-boost or simultaneous immunization regimens yield maximal protection and diversified immune responses (Larsen et al. 2023; Rais et al. 2023). Antigens selected for ID91 show no human sequence homology and stimulate immune responses in purified protein derivative (PPD)-positive human peripheral blood mononuclear cells, validating their translational potential (Rais et al. 2023). Western blot analyses confirm in vitro expression of ID91 antigens from the repRNA construct, exemplifying how structure-guided antigen selection combined with cutting-edge RNA delivery enhances vaccine efficacy by eliciting both CD4+ and CD8+ T cell responses critical for TB protection.

In summary, recent advances have significantly broadened the spectrum of vaccine candidates targeting Mtb. These candidates encompass early secreted virulence factors (e.g., ESAT-6, CFP-10, EsxV), enzymes involved in cell wall biosynthesis (Ag85 complex), immunodominant structural proteins (MPT64, PPE18, PPE44, PPE15), latency-associated stress proteins (HspX, Rv2660c, Rv2031c, Rv0232, Rv1115), hypothetical proteins such as NHP1, and synthetic peptide chimeras derived from multiple Mtb antigens. When combined with innovative nucleic acid delivery platforms, including the ID91 replicon RNA vaccine, these candidates form the basis of a comprehensive multistage vaccination strategy aimed at eliciting robust and long-lasting immunity by targeting the intricate biology of Mtb infection and persistence within the host.

10.4 Current Developments in Identifying Therapeutic Protein Targets Against *Mycobacterium tuberculosis*

Over the past year, protein therapeutics have emerged as a prominent pharmaceutical class, effectively addressing previously unmet clinical needs across a broad spectrum of diseases. A primary advantage of protein therapeutics is their exceptional targeting specificity. Glycans (complex carbohydrates attached to proteins through glycosylation) play a crucial role in modulating the biological activity of these therapeutics by influencing their interactions with receptors and molecular targets. Furthermore, glycosylation enhances protein stability by protecting therapeutics from proteolytic degradation, oxidation, precipitation, and denaturation. By shielding hydrophobic regions on the glycoprotein surface, glycosylation prevents aggregation, thereby reducing immunogenicity. Additionally, glycosylation masks immunogenic epitopes, further mitigating adverse immune responses. Despite these benefits, achieving precise and uniform glycosylation patterns remains a significant challenge in therapeutic protein development (Sarkar and Wintrode 2011; O'Flaherty et al. 2018; Garcia-Vilanova et al. 2019; Zhang and Qu 2021; Delobel 2021). Similarly, Mtb utilizes glycosylation as a critical posttranslational modification, attaching sugar moieties to proteins that regulate bacterial physiology and pathogenicity (Birhanu et al. 2019). Posttranslational modifications, particularly glycosylation, contribute significantly to Mtb biology. Unique glycoconjugates in the mycobacterial cell envelope are pivotal in host-pathogen interactions, antigenicity, virulence, and antimicrobial resistance (Jarlier and Nikaido 1994; Herrmann et al. 1996, 2000; Torrelles and Schlesinger 2010; Sonawane et al. 2012; Angala et al. 2014; Calder et al. 2015). Glycosylation also protects sensitive proteolytic cleavage sites, maintains protein attachment to membranes via lipid anchors, and may facilitate protein export

(Herrmann et al. 1996; Kovacs-Simon et al. 2011).

In Mtb, glycosidase enzymes contribute to both glycan degradation and monosaccharide removal, providing substrates for glycosyltransferases involved in glycan biosynthesis. Protein glycosylation often co-occurs with acylation. Notably, O-linked mannosylation, catalyzed by protein O-mannosyltransferases (PMTs), dominates glycosylation, transferring mannose residues to serine/threonine residues, frequently in conjunction with acyl modifications (VanderVen et al. 2005; Tucci et al. 2020). Although the number of identified glycosylated proteins in Mtb is increasing, many aspects remain unclear. Unlike N-glycosylation, which targets asparagine residues within defined motifs, O-glycosylation lacks a well-defined consensus sequence, complicating prediction and study. Comprehensive understanding of O-glycosylation is critical for elucidating its role in Mtb pathogenesis and advancing targeted therapeutics (Birhanu et al. 2019).

Glycoproteomics offers promising avenues for identifying Mtb virulence factors, as glycoproteins on the bacterial envelope mediate crucial interactions that facilitate entry into host macrophages (Graham and Hess 2010). Mannosylated proteins engage host mannose receptors, promoting macrophage invasion (Diaz-Silvestre et al. 2005). Despite their importance, only a limited number of cell wall-associated and secreted glycoproteins have been characterized. Recent glycoproteome mapping revealed numerous membrane-associated glycolipoproteins implicated in pathogenesis, with notable quantitative differences among Mtb lineages that may explain phenotypic diversity. These findings identify attractive targets for drug development and glycoprotein engineering for clinical use (Birhanu et al. 2019).

Modern multi-omics and computational approaches have uncovered novel drug targets in *Mycobacterium tuberculosis* (Mtb). One such target is SufD, an essential component of the sulfur mobilization (SUF) pathway responsible for iron-sulfur (Fe-S) cluster assembly (Gorityala et al. 2024). Fe-S clusters are vital cofactors in numerous cellular processes (Fontecave et al. 2008), and their integrity is crucial for bacterial survival. During infection, Mtb faces multiple hostile conditions within alveolar macrophages and granulomas, including acidic pH, nutrient deprivation, oxidative and nitrosative stress, and often hypoxia (Schnappinger et al. 2003; Via et al. 2008; Parbhoo et al. 2022). Oxidative stress, frequently exacerbated by anti-TB drugs, poses a significant threat to Mtb survival (Bhaskar et al. 2020). To counteract this, Mtb maintains redox homeostasis and relies heavily on Fe-S cluster assembly pathways, which carefully regulate iron utilization and prevent the formation of harmful reactive oxygen species (Elchennawi and Ollagnier de Choudens 2022). Upregulation of Fe-S cluster biogenesis genes enhances resistance to both oxidative and nitrosative stresses (Voskuil et al. 2011), underscoring the critical protective role of these chemically sensitive clusters, which are prime targets for reactive oxygen and nitrogen species. Mtb possesses three evolutionarily conserved Fe-S cluster assembly systems: the iron-sulfur cluster (ISC), sulfur formation (SUF), and nitrogen fixation (NIF) pathways (Ayala-Castro et al. 2008; Garcia et al. 2022). Uniquely, Mtb relies exclusively on the SUF system, encoded by an operon comprising *sufR*, *sufB*, *sufD*, *sufC*, *sufS*, *sufU*, and *sufT* genes (Huet et al. 2005; Pandey et al. 2018). The biogenesis of Fe-S clusters involves their initial assembly on scaffold proteins that receive sulfur, transferred from cysteine via cysteine desulfurase, and iron from as-yet unidentified donors. Subsequently, these clusters are transferred in an ATP-dependent manner to target apoproteins. This tightly regulated and evolutionarily conserved process is essential for maintaining Mtb's metabolic and respiratory functions under stress (Py and Barras 2010; Gorityala et al. 2024). Disruption of Fe-S clusters leads to the release of ferrous ions, which can react with oxygen to produce superoxide and hydrogen peroxide. This process, particularly through the Fenton reaction, generates highly reactive hydroxyl radicals, contributing to bactericidal drug-induced cell death irrespective of the primary drug target (Abe and Miyazawa 2022; Thomas et al. 2009; Pereira et al. 2012).

All *suf* genes in Mtb are essential for bacterial growth, except the negative regulator *sufR* (Sassetti et al. 2003; DeJesus et al. 2017). SufR functions by repressing the expression of the downstream *suf* operon (Willemse et al. 2018). Recent research has illuminated the roles of other key components: the cysteine desulfurase SufS is activated by the cofactor SufU, enhancing sulfur mobilization for Fe-S cluster assembly (Elchennawi et al. 2023), while SufT plays a critical role in cluster maturation and overall cellular homeostasis (Tripathi et al. 2022). In *Escherichia coli*, SufB, SufC, and SufD form a scaffold complex (SufBC_2D) that serves as the platform for Fe-S cluster biogenesis, with specific protein-protein interfaces facilitating cluster formation and transfer (Hirabayashi et al. 2015; Yuda et al. 2017). Although high-resolution structural data on Mtb Suf proteins remain scarce, UniProt annotations for Rv1461, Rv1462, and Rv1463 indicate analogous essential functions in cluster assembly and ATP-dependent transport. Notably, SufD may specifically act to sequester ferrous ions, thereby limiting Fenton chemistry reactions that generate harmful oxidative damage. This crucial function positions SufD as a compelling drug target; its inhibition could disrupt iron homeostasis, significantly increase Mtb's susceptibility to oxidative stress, and exert bactericidal effects. Furthermore, combining SufD inhibitors with reactive oxygen species (ROS)-inducing anti-TB drugs may yield synergistic therapeutic effects, offering a promising strategy to combat resistant Mtb strains (Costa-Gouveia et al. 2017; Gorityala et al. 2024).

Proteomic studies have identified a subset of Mtb enzymes that utilize the rare cofactor F420, a deazaflavin integral to redox metabolism and defense against oxidative stress (Selengut et al. 2007; Greening et al. 2016). F420 functions as a low-potential two-electron carrier facilitating precise electron transfer during catalysis (Greening et al. 2016). Structural analyses of F420-dependent glucose-6-phosphate dehydrogenase reveal active sites optimized for substrate binding, underscoring F420's critical role in maintaining intracellular redox homeostasis (Bashiri et al. 2008). Additionally, hydroxymycolic acid dehydrogenase (Rv0132c) exploits F420 to neutralize reactive oxygen species produced during host immune responses and antibiotic treatment (Purwantini and Mukhopadhyay 2013; Aderemi and Snee 2025; Gurumurthy et al. 2013). Given their key metabolic and protective roles, F420-dependent enzymes represent promising targets to disrupt Mtb physiology and augment bactericidal effects, potentially overcoming drug resistance (Aderemi and Snee 2025). Comparative proteomic analyses of aminoglycoside-resistant versus susceptible Mtb clinical isolates have revealed differential expression of capsule-associated and membrane proteins that contribute to cell wall integrity and regulate drug permeability. These adaptations directly impact resistance mechanisms and highlight attractive targets to sensitize resistant strains to existing therapies. Well-established drug targets remain central to current antitubercular drug development. Among these are cytochrome P450 enzymes, specifically CYP121A1 and CYP124, and KasA, a β-ketoacyl-acyl carrier protein synthase involved in mycolic acid biosynthesis (Padayachee et al. 2024).

CYP121A1 is a heme-containing monooxygenase with a characteristic P450 fold and cysteine-ligated heme prosthetic group. This enzyme catalyzes the formation of mycocyclosin, a critical di-tyrosine crosslink essential for Mtb viability (Ouellet et al. 2010; Campomizzi et al. 2021). The hydrophobic active site accommodates both substrates and inhibitors; azole antifungals such as econazole and clotrimazole inhibit Mtb growth by targeting CYP121A1 in vitro and in vivo (Kumar et al. 2021; Padayachee et al. 2024; Hossam Abdelmonem et al. 2024). Crystallographic studies have unveiled novel inhibitor binding modalities, including azole coordination via water molecules, thereby informing rational drug design strategies (Jasani and Patel 2023).

CYP124 functions as a methyl-branched lipid monooxygenase that metabolizes the promising antitubercular candidate drug SQ109 by hydroxylating its methyl-branched hydrocarbons (Bukhdruker et al. 2020). High-resolution crystal structures of the CYP124-SQ109 complex reveal drug conformations poised for ω-methyl hydroxylation. The hydroxylated SQ109 subsequently forms hydrogen bonds with MmpL3, the presumed drug target, highlighting an important functional interplay. These findings position Mtb CYP enzymes as functional analogs of human drug-metabolizing enzymes that influence drug pharmacodynamics and efficacy (Bukhdruker et al. 2020).

KasA catalyzes key elongation steps in mycolic acid biosynthesis, a process essential for maintaining the integrity and impermeability of the mycobacterial cell wall (Rudraraju et al. 2022). As part of the Type II fatty acid synthase (FAS-II) system, KasA executes sequential Claisen condensation reactions, extending fatty acyl chains by two carbons each cycle (Schaeffer et al. 2001; Bhatt et al. 2007). KasA is a 53 kDa homodimer with a thiolase fold comprising an α/β-hydrolase core and a catalytic triad formed by Cys171, His311, and His345 (Luckner et al. 2009). X-ray crystallography has revealed distinct core and flexible cap domains, with cap dynamics essential for substrate binding and catalysis (Luckner et al. 2009; Kumar et al. 2018; Schiebel et al. 2013). Inhibitor studies with thiolactomycin (TLM) show preferential binding to the acylated form of KasA, occupying the malonyl substrate binding site and interacting with key residues such as Phe404 to block substrate access and elongation (Schiebel et al. 2013). These structural insights have guided the rational design of potent KasA inhibitors, targeting its dynamic active site to disrupt mycolic acid synthesis and compromise bacterial viability. Notably, compounds like GSK3011724A stabilize the open acyl channel conformation of KasA, further underscoring its therapeutic potential (Abrahams et al. 2016; Cunningham et al. 2020).

Another promising protein drug target is the pyrimidine biosynthesis regulator PyrR (Rv1379). The crystal structure of PyrR, resolved at 1.9 Å, reveals a dimeric phosphoribosyltransferase fold featuring binding sites for pyr mRNA as well as allosteric sites for the metabolites UMP and PRPP. These interactions mediate feedback inhibition of pyr gene expression. Disruption of PyrR function impairs pyrimidine nucleotide synthesis, which is essential for DNA and RNA replication, offering a novel therapeutic avenue for tuberculosis treatment (Kantardjieff et al. 2005).

Rv0676, also known as MmpL5, is a pivotal transmembrane efflux pump that plays a central role in Mtb multidrug resistance, particularly against bedaquiline (BDQ) and clofazimine (CFZ), two key drugs used in treating drug-resistant TB (Farnia et al. 2024). Recent research has shown that mutations in the MmpL5 protein, comprising deletions, insertions, and substitutions spanning critical amino acid regions, alter its substrate specificity and enhance drug efflux activity (Farnia et al. 2024). These mutations reduce intracellular drug concentrations, thereby diminishing susceptibility and contributing to treatment failure (Farnia et al. 2024). High-resolution structural studies, including cryo-electron microscopy and computational modeling, have elucidated that MmpL5, often functioning in concert with its partner protein MmpS5, assembles into a trimeric complex featuring gated transmembrane channels that tightly regulate substrate transport (Maharjan et al. 2024). The dynamic assembly of MmpL5 homotrimers facilitated by MmpS5 is essential for efficient efflux function, offering clear mechanistic insights and identifying vulnerable structural elements that could be exploited therapeutically to inhibit efflux and restore drug efficacy (Maharjan et al. 2024; Xiong et al. 2025). In parallel, studies of the PE_PGRS protein family (e.g., Wag22, PE_PGRS31-33) reveal their involvement in antigenic variation and immune evasion via flexible β-strand-rich domains with proline-glutamate repeats and GC-rich motifs (Farnia et al. 2023). Their expression fluctuates between active growth and latency, correlating with bacterial persistence and suggesting that these proteins, alongside efflux systems, represent promising drug targets for addressing latent tuberculosis and drug resistance (Farnia et al. 2023, 2024).

The HutC family transcriptional regulator Rv0792c represents a promising drug target due to its critical role in Mtb's oxidative stress response and virulence. Structural analyses using small-angle X-ray scattering (SAXS) combined with SELEX-derived aptamer assays identified a druggable pocket located at the protein's dimer interface. The small-molecule inhibitor I-OMe-Tyrphostin binds with nanomolar affinity to this site, effectively blocking DNA binding activity, significantly reducing intracellular bacterial growth, and attenuating infection in vivo. These findings highlight the therapeutic potential of targeting Rv0792c to disrupt bacterial stress adaptation mechanisms and combat tuberculosis (Chauhan et al. 2023; Chinnam et al. 2023).

The MurT/GatD complex catalyzes the amidation of peptidoglycan precursors, a critical modification that reinforces the Mtb cell wall, thereby enhancing its resistance to host immune defenses. Structural and biochemical investigations have identified key interaction sites and catalytic residues within this complex, establishing MurT/GatD as a promising drug target. Inhibiting this complex compromises cell wall integrity and significantly impairs bacterial survival (Maitra et al. 2021). Similarly, the putative DNA-alkyltransferase-like protein MtATL is crucial for DNA repair in Mtb, functioning by recruiting nucleotide excision repair proteins to sites of DNA damage. This repair mechanism enables the bacterium to withstand antibiotic-induced genotoxic stress and contributes substantially to the development of antibiotic resistance. Targeting MtATL could sensitize Mtb to DNA-damaging agents and impede resistance evolution, offering a novel anti-evolution therapeutic strategy to more effectively control drug-resistant and persistent tuberculosis (Miggiano et al. 2020; Naz et al. 2021). Overall, integrating structural, biochemical, and functional insights into essential Mtb proteins involved in cell wall biosynthesis, nucleotide metabolism, drug efflux, stress response, and DNA repair establishes a solid foundation for rational drug development. Targeting these critical, druggable proteins presents promising strategies to circumvent drug resistance and enhance the effectiveness of tuberculosis therapies.

10.5 Novel Protein Targets for Enhanced Detection of Tuberculosis

Substantial progress has been achieved in the identification of novel protein and peptide biomarkers for the diagnosis of Mtb. These advancements, arising from proteomic, transcriptomic, peptidomic, immunoinformatics, and immunological studies, offer promising opportunities to develop more accurate, rapid, and noninvasive diagnostic tools. A prominent host-derived biomarker in tuberculosis (TB) is IFI44L (Interferon-Induced Protein 44-Like), an interferon-stimulated gene that is significantly overexpressed in patients with active TB. Early research underscored the influence of host cell responses on bacterial pathogenicity, highlighting the importance of host factors in TB development (Jiang et al. 2018). Building on this, subsequent transcriptomic and functional analyses demonstrated that IFI44L enhances the antimicrobial capacity of human macrophages, correlates strongly with treatment efficacy, and shows potential as both a diagnostic and therapy-monitoring biomarker, thus emphasizing the crucial role of host immune modulation in controlling Mtb infection (Jiang et al. 2021). Further insights into host-pathogen interaction come from transcriptomic studies of IFIT2 in macrophages infected with Mtb. Investigation revealed that IFIT2 regulates gene expression governing macrophage responses to infection, highlighting the intricate interplay between innate immunity and mycobacterial virulence (Ewura-Esi Manful et al. 2025). Collectively, these findings illustrate the potential utility of host-derived biomarkers such as IFI44L and IFIT2, along with proteomic signatures, in improving TB diagnosis, treatment monitoring, and therapeutic guidance. Noninvasive salivary protein biomarkers have also gained attention as effective tools for TB diagnosis and treatment monitoring. Early work demonstrated the diagnostic value of salivary host biomarkers and validated saliva as a medium for immunological assessment. Later, quantitative mass spectrometry-based proteomics has identified a five-protein biosignature in saliva that effectively distinguishes active tuberculosis (TB) patients from healthy individuals and those with other respiratory conditions. Among these proteins, Alpha-1-antitrypsin (AAT; UniProt: P01011) stands out due to its demonstrated association with TB-related lung inflammation and tissue damage, supporting its potential as a diagnostic biomarker. While the other proteins in the signature, Lipocalin-1 (Q8NCW5), Coactosin-like protein (P28072), Mucin-7 (A0A2Q2TTZ9), and Zymogen granule protein 16 (Q99574), have not been individually validated as TB-specific markers, their combined presence contributes to the overall diagnostic accuracy of the panel. This multi-protein saliva biosignature represents a promising, rapid, noninvasive, and accessible point-of-care diagnostic approach, particularly suitable for resource-limited settings (Almeida et al. 2009; Khambati et al. 2021; Guo et al. 2022).

Serum protein analyses have similarly yielded effective diagnostic markers. A panel of four Mtb antigens, Rv1860 (MPT64), Rv3881c, Rv2031c (HspX), and Rv3803c, showed high sensitivity and specificity in differentiating active TB from latent infection (Alonzi et al. 2023; Neudecker et al. 2024). Using the Slow Off-rate Modified Aptamer (SOMAscan) assay, a separate study identified a serum biomarker signature composed of six proteins: tryptophanyl-tRNA synthetase 1 (SYWC/WARS1), kallistatin (SERPINA4), complement component 9 (C9), gelsolin (GSN), testican-2 (SPOCK2), and aldolase C (ALDOC). This protein panel represents diverse host immune mechanisms, including interferon-mediated signaling, inflammatory responses, activation of the complement system, regulation of the cytoskeleton, interactions with the extracellular matrix, and metabolic remodeling. Together, these processes form a comprehensive and reliable biomarker signature with high accuracy for detecting active tuberculosis (De Groote et al. 2017; Guo et al. 2022; Neudecker et al. 2024).

Peptidomic analyses have identified specific Mtb peptides, namely MPT64-2, culture filtrate protein 10–2 (CFP10-2), and phosphate-binding lipoprotein (PstS1-2), that show significantly different abundance levels in tuberculosis (TB) patients compared to controls. These peptides exhibit high sensitivity and specificity, making them valuable diagnostic biomarkers for accurately detecting active TB infection (Chen et al. 2022). Host blood transcriptomic profiling has identified a four-gene signature, Ubiquitin-Conjugating Enzyme E2L6 (UBE2L6), Basic Leucine Zipper ATF-Like Transcription Factor 2 (BATF2), Serpin Family G Member 1 (SERPING1), and Vesicle-Associated Membrane Protein 5 (VAMP5) that reliably differentiates active tuberculosis (TB) from latent infection and healthy individuals. These genes play critical roles in key immune processes, including protein degradation, regulation of inflammatory signaling, complement system activation, and vesicle trafficking involved in immune cell communication. Notably, this gene panel demonstrates superior sensitivity and specificity compared to traditional diagnostic methods, offering improved accuracy for TB detection (Gong et al. 2021). Also, to enhance diagnostic accuracy, multi-epitope recombinant antigens such as HP16118P have been developed. This antigen combines multiple T cell and B cell epitopes derived from several Mtb proteins, eliciting strong immune responses and effectively differentiating latent from active TB (Jiang et al. 2025; Wang et al. 2024). From a broad panel of recombinant Mtb proteins, promising candidates such as E-M, E-C-M, and E-7.7-C are being developed to improve tuberculin skin tests (TST). In these names, "E" refers to early secretory antigenic target 6 kDa

(ESAT-6), a highly specific Mtb antigen absent in Bacillus Calmette-Guérin (BCG) vaccine strains and most nontuberculous mycobacteria (NTM). The E-M antigen, a fusion of ESAT-6 and MPT64, effectively discriminates TB infection with minimal cross-reactivity to BCG or NTM, thereby enhancing the specificity and safety of the test compared to traditional purified protein derivative-based assays (Ou et al. 2025; WHO 2022).

Diagnostic potential has also been identified in newly characterized proteins such as Rv2878c (Mpt53/DsbE) and Rv0792c. Rv2878c is a thioredoxin-like secreted protein involved in oxidative stress protection and elicits detectable antibody responses in TB patients (Goldstone et al. 2016). Rv0792c, a HutC family transcriptional regulator, plays a key role in oxidative stress response and virulence; targeting its druggable pocket inhibits DNA binding and bacterial growth (Chauhan et al. 2023). The universal stress protein Rv2623 regulates Mtb's growth and latency through ATP binding and interaction with transporters, modulating cell wall components. Its deletion leads to hypervirulence and failure to establish latency, making it a valuable biomarker for latent TB (Florczyk et al. 2001; Drumm et al. 2009; Glass et al. 2017).

Lipoarabinomannan (LAM), a key glycolipid component of the Mtb cell wall, plays a significant role in modulating host immune responses and is actively shed during infection. Its presence can be detected noninvasively in urine, serving as the basis for World Health Organization (WHO)-endorsed diagnostic assays such as Alere Determine TB LAM and FujiLAM. These assays demonstrate particularly high effectiveness in HIV-positive patients, where conventional diagnostics often underperform (Mishra et al. 2011; WHO 2019; Stead et al. 2025). Recently, the synthetic multi-epitope antigen CP19128P has been developed, incorporating multiple T cell and B cell epitopes that elicit robust Th1 and Th2 immune responses. This wide-ranging immunogenicity improves the accuracy of immunodiagnostic assays, allowing for more precise differentiation between latent and active tuberculosis infections (Jiang et al. 2023; Wang et al. 2024). Together, these biomarkers and recombinant proteins represent cutting-edge developments in TB diagnostics, integrating biochemical stability, immunological specificity, and noninvasive detection to enhance the precision and accessibility of TB diagnosis and monitoring.

10.6 Significance of Conserved Hypothetical Proteins (CHPs) in *Mycobacterium tuberculosis*

Conserved hypothetical proteins (CHPs) constitute a substantial fraction of the Mtb proteome. These proteins are conserved across mycobacterial species but remain largely uncharacterized regarding their biochemical functions. The landmark Mtb genome sequence published by Cole et al. (1998) identified approximately 4000 genes, of which 606 were classified as proteins of unknown function and 910 as conserved hypothetical proteins. Later re-annotations refined these numbers, decreasing unknown function proteins to 272 while increasing CHPs to 1051 (Camus et al. 2002). Genomic techniques such as transposon site hybridization have been pivotal in determining gene essentiality; Sassetti and Rubin (2003) identified 194 genes essential for Mtb survival in murine models, over half encoding proteins with previously unknown functions. Comparative genomics analyses across diverse mycobacterial species have played a pivotal role in identifying essential gene products. Although many conserved hypothetical proteins (CHPs) remain functionally uncharacterized, their strong evolutionary conservation underscores their likely critical involvement in fundamental processes such as bacterial physiology, virulence, and adaptation to complex host environments. Representing approximately one-third of the *Mycobacterium tuberculosis* genome, CHPs have garnered increasing scientific attention aimed at elucidating their biological functions and exploring their potential as targets for novel therapeutic interventions (Camus et al. 2002; Garg et al. 2015; Yang et al. 2019). Among CHPs, Rv2626c is notably one of the most highly expressed under hypoxic or nitric oxide-induced stress conditions. It is detectable in Mtb culture supernatants, bacterial lysates, and activated macrophages within infected lungs (Jiang et al. 2018; Bashiri et al. 2008; Sun et al. 2017). Rv2626c elicits strong serum antibody responses in patients with active TB, underscoring its diagnostic utility. Functionally, it stimulates pro-inflammatory cytokines such as interleukin-12 (IL-12) and tumor necrosis factor-alpha (TNF-α), enhances costimulatory molecule expression via NF-κB activation, and induces Th1 cytokine production—including interferon-gamma (IFN-γ)—from immune cells of TB patients, suggesting its promise as both a vaccine candidate and diagnostic marker (Bashiri et al. 2008). Structural investigations have significantly advanced our understanding of CHPs. Notably, the crystal structure of Rv0216, determined by X-ray crystallography, revealed a distinctive double hotdog fold, a structural motif commonly associated with fatty acid-metabolizing enzymes (Castell et al. 2005). Although Rv0216 itself lacks catalytic activity against standard substrates, the presence of this characteristic fold has been instrumental in inferring the functions of related essential proteins. The findings suggest that Rv0216 and homologous proteins participate in lipid metabolic processes that are crucial for preserving the integrity and functionality of the Mtb cell envelope (Sassetti and Rubin 2003; Castell et al. 2005). Similarly, Rv2302 has been characterized by nuclear magnetic resonance (NMR) spectroscopy as a compact 80-residue monomeric protein, composed of a five-stranded antiparallel β-sheet core with a C-terminal α-helix. This protein is upregulated during starvation and heat shock, stressors linked to dormancy, and features highly conserved surface elements, including a negatively charged electrostatic potential and a positively charged arginine-rich pocket (R18, R70, R74) (Buchko et al. 2006). These features suggest potential involvement in protein-protein or nucleic acid interactions. Circular dichroism (CD) spectroscopy and heteronuclear Nuclear Overhauser Effect (NOE) experiments reveal that Rv2302 undergoes reversible partial unfolding when exposed to elevated temperatures and possesses a flexible loop region, respectively. Furthermore, structural homology analyses suggest that Rv2302 adopts a novel protein fold, uniquely evolved to facilitate Mtb survival under starvation and heat shock stress conditions (Buchko et al. 2006). The conserved hypothetical protein Rv0574c is crucial for sustaining Mtb cell wall integrity, stress tolerance, and virulence (Garg et al. 2015). It supports bacterial persistence in granulomas, environments rich in carbon dioxide and host-derived stresses, as demonstrated by growth defects and increased sensitivity to host defenses in Rv0574c mutants. Additional CHPs, including Rv2037c and Rv1860, are strongly upregulated in response to environmental stress and elicit robust host immune responses characterized by lymphocyte proliferation and IFN-γ production. Notably, Rv1860 is a glycosylated secreted protein that modulates dendritic cell function and adaptive immunity, potentially influencing vaccine efficacy and protective immunity against TB (Satchidanandam et al. 2014). Another hypothetical protein, CuvA (Rv1422) from H_{37}Rv, has been shown to play a critical role in nutrient utilization, cell wall biosynthesis, and bacterial virulence. It is phosphorylated by the serine/threonine kinases PknA and PknB, which are key regulators of cell growth and peptidoglycan synthesis. Recent progress in comparative functional genomics has improved the characterization of CHPs through the combined use of sequence homology, domain architecture prediction, gene context analysis, and protein-protein interaction mapping (Méheust et al. 2022). These studies highlight the involvement of CHPs in several essential biological pathways, including cell wall biosynthesis, DNA repair, metabolism, and stress response (Galperin and Koonin 2004; Garg et al. 2015). Some CHPs participate in synthesizing the complex lipid layers of the mycobacterial cell envelope, which serve as a protective barrier against host immune defenses and antibiotic penetration (Garg et al. 2015). Others contribute to maintaining genomic integrity by engaging in DNA repair mechanisms that counteract damage caused by reactive oxygen and nitrogen species produced within macrophages (Sassetti and Rubin 2003). Additionally, CHPs play pivotal roles in metabolic adaptability, enabling the bacterium to survive nutrient deprivation and fluctuating oxygen levels through involvement in central carbon metabolism, lipid metabolism, and cofactor biosynthesis. Many CHPs also support adaptation to various environmental stresses, including oxidative, nitrosative, acidic, and drug-induced challenges (Garg et al. 2015). Expression profiling has further revealed that CHPs are differentially regulated under infection-relevant conditions such as hypoxia and exposure to frontline antituberculosis drugs, linking them to critical processes like dormancy, persistence, and drug resistance (Sharma and Bisht 2017; Devlin et al. 2025). Therefore, conserved hypothetical proteins in Mtb constitute a crucial yet largely underexplored category of biomolecules. Advances in immunology, structural biology, and functional genomics are progressively uncovering their fundamental roles in bacterial physiology, modulation of host immune responses, and the mechanisms underlying pathogenesis.

Conflict of Interest The authors declare that they have no conflict of interest related to the content of this chapter.

Ethical Consideration This chapter is a review based entirely on data and information obtained from previously published and publicly available sources, all of which have been appropriately cited. The authors conducted no new experimental work involving human participants or animals. Therefore, no additional ethical approval was required. The authors have ensured proper attribution and compliance with ethical standards in the use of all referenced materials.

References

Aagaard C, Brock I, Olsen A, Ottenhoff THM, Weldingh K, Andersen P (2004) Mapping immune reactivity toward Rv2653 and Rv2654: two novel low-molecular-mass antigens found specifically in the Mycobacterium tuberculosis complex. J Infect Dis 189(5):812–819

Abdalla AE, Yan S, Zeng J, Deng W, Xie L, Xie J (2020) Mycobacterium tuberculosis Rv0341 promotes mycobacterium survival in in vitro hostile environments and within macrophages and induces cytokines expression. Pathogens 9(6):454

Abe C, Miyazawa T (2022) Current use of Fenton reaction in drugs and food. Molecules 27(17):5451

Abrahams KA, Chung CW, Ghidelli-Disse S, Rullas J, Rebollo-López MJ, Gurcha SS, Cox JA, Mendoza A, Jiménez-Navarro E, Martínez-Martínez MS, Neu M (2016) Identification of KasA as the cellular target of an anti-tubercular scaffold. Nat Commun 7(1):12581

Aderemi VA, Snee M (2025) Targeting mycobacterium tuberculosis F420-dependent dehydrogenases for new antitubercular drug discovery. Indian J Tuberc. https://doi.org/10.1016/j.ijtb.2025.03.006

Aebersold R, Blattmann P (2019) Mass spectrometric exploration of the biochemical basis of living systems. Chimia 73(7–8):540

Ahidjo BA, Kuhnert D, McKenzie JL, Machowski EE, Gordhan BG, Arcus V, Abrahams GL, Mizrahi V (2011) VapC toxins from Mycobacterium tuberculosis are ribonucleases that differentially inhibit growth and are neutralized by cognate VapB antitoxins. PLoS One 6(6):e21738

Akif M, Khare G, Tyagi AK, Mande SC, Sardesai AA (2008 Nov) Functional studies of multiple thioredoxins from mycobacterium tuberculosis. J Bacteriol 190(21):7087–7095

Almeida ML, Barbieri MA, Gurgel RQ, Abdurrahman ST, Baba UA, Hart CA et al (2009 Jun) Alpha1-acid glycoprotein and alpha1-antitrypsin as early markers of treatment response in patients receiving the intensive phase of tuberculosis therapy. Trans R Soc Trop Med Hyg 103(6):575–580

Alonzi T, Repele F, Goletti D (2023) Research tests for the diagnosis of tuberculosis infection. Expert Rev Anti-Infect Ther 21(9):783–795

Ames JR, Muthuramalingam M, Murphy T, Najar FZ, Bourne CR (2019 Oct) Expression of different ParE toxins results in conserved phenotypes with distinguishable classes of toxicity. Microbiology 8(10):e902

An Y, Ni R, Zhuang L, Yang L, Ye Z, Li L, Parkkila S, Aspatwar A, Gong W (2025) Tuberculosis vaccines and therapeutic drug: challenges and future directions. Mol Biomed 6:4

Andrews SJ, Rothnagel JA (2014 Mar) Emerging evidence for functional peptides encoded by short open reading frames. Nat Rev Genet 15(3):193–204

Angala SK, Belardinelli JM, Huc-Claustre E, Wheat WH, Jackson M (2014) The cell envelope glycoconjugates of mycobacterium tuberculosis. Crit Rev Biochem Mol Biol 49(5):361–399

Anjou C, Lotoux A, Zhukova A, Royer M, Caulat LC, Capuzzo E, Morvan C, Martin-Verstraete I (2024) The multiplicity of thioredoxin systems meets the specific lifestyles of Clostridia. PLoS Pathog 20(2):e1012001. https://doi.org/10.1371/journal.ppat.1012001. PMID: 38330058; PMCID: PMC10880999

Arcus VL, Rainey PB, Turner SJ (2005) The PIN-domain toxin-antitoxin array in mycobacteria. Trends Microbiol 13(8):360–365. https://doi.org/10.1016/j.tim.2005.06.008

Arnvig KB, Young DB (2009 Aug) Identification of small RNAs in mycobacterium tuberculosis. Mol Microbiol 73(3):397–408

Attarian R, Bennie C, Bach H, Av-Gay Y (2009) Glutathione disulfide and S-nitrosoglutathione detoxification by Mycobacterium tuberculosis thioredoxin system. FEBS Lett 583(19):3215–3220

Avarvand AY, Meshkat Z, Khademi F, Tafaghodi M (2021) Immunogenicity of HspX/EsxS fusion protein of Mycobacterium tuberculosis along with ISCOMATRIX and PLUSCOM nano-adjuvants after subcutaneous administration in animal model. Microb Pathog 154:104842

Ayala-Castro C, Saini A, Outten FW (2008) Fe-S cluster assembly pathways in bacteria. Microbiol Mol Biol Rev 72(1):110–125

Babele P, Midha MK, Rao KVS, Kumar A (2023) Temporal profiling of host proteome against different M. tuberculosis strains reveals delayed epigenetic orchestration. Microorganisms 11(12):2998. . PMID: 38138142; PMCID: PMC10745383 https://doi.org/10.3390/microorganisms11122998

Banaei-Esfahani A, Nicod C, Aebersold R, Collins BC (2017 Oct) Systems proteomics approaches to study bacterial pathogens: application to Mycobacterium tuberculosis. Curr Opin Microbiol 39:64–72

Barth VC, Chauhan U, Zeng J, Su X, Zheng H, Husson RN, Woychik NA (2021) Mycobacterium tuberculosis VapC4 toxin engages small ORFs to initiate an integrated oxidative and copper stress response. Proc Natl Acad Sci USA 118(32):e2022136118

Bashiri G, Squire CJ, Moreland NJ, Baker EN (2008) Crystal structures of F420-dependent glucose-6-phosphate dehydrogenase FGD1 involved in the activation of the anti-tuberculosis drug candidate PA-824 reveal the basis of coenzyme and substrate binding. J Biol Chem 283(25):17531–17541. https://doi.org/10.1074/jbc.M801854200. Epub 2008 Apr 22

Bates TA, Trank-Greene M, Wrynla XH, Anastas A, Gurmessa SK, Merutka IR, Dixon SD, Shumate A, Groncki AR, Parson MAH, Ingram JR, Barklis E, Burke JE, Shinde U, Ploegh HL, Tafesse FG (2024) ESAT-6 undergoes self-association at phagosomal pH and an ESAT-6-specific nanobody restricts M. tuberculosis growth in macrophages. elife 12. RP91930

Beck IN, Arrowsmith TJ, Grobbelaar MJ, Bromley EHC, Marles-Wright J, Blower TR (2024) Toxin release by conditional remodelling of ParDE1 from Mycobacterium tuberculosis leads to gyrase inhibition. Nucleic Acids Res 52(4):1909–1929

Bellini C, Vergara E, Bencs F, Fodor K, Bősze S, Krivić D, Bacsa B, Surguta SE, Tóvári J, Reljic R, Horváti K (2023) Design and characterization of a multistage peptide-based vaccine platform to target Mycobacterium tuberculosis infection. Bioconjug Chem 34(10):1738–1753

Bespyatykh JA, Shitikov EA, Ilina EN (2017) Proteomics for the investigation of mycobacteria. Acta Nat 9(1):15–25. PMID: 28461970; PMCID: PMC5406656

Bespyatykh J, Shitikov E, Bespiatykh D, Guliaev A, Klimina K, Veselovsky V et al (2020) Metabolic changes of mycobacterium tuberculosis during the anti-tuberculosis therapy. Pathogens 9(2):131. https://doi.org/10.3390/pathogens9020131

Bhaskar A, Dwivedi VP, Nandicoori VK (2020) Eliminating mycobacterial persistence: Novel targets for anti-TB therapy. InPathogenicity and drug resistance of human pathogens: mechanisms and novel approaches. Springer Singapore, Singapore, pp 57–79

Bhatt A, Molle V, Besra GS, Jacobs WR Jr, Kremer L (2007 Jun) The Mycobacterium tuberculosis FAS-II condensing enzymes: their role in mycolic acid biosynthesis, acid-fastness, pathogenesis and in future drug development. Mol Microbiol 64(6):1442–1454

Bibb LA, Hatfull GF (2002 Sep) Integration and excision of the Mycobacterium tuberculosis prophage-like element, phiRv1. Mol Microbiol 45(6):1515–1526

Bibb LA, Hancox MI, Hatfull GF (2005 Mar) Integration and excision by the large serine recombinase phiRv1 integrase. Mol Microbiol 55(6):1896–1910

Birhanu AG, Yimer SA, Kalayou S, Riaz T, Zegeye ED, Holm-Hansen C, Norheim G, Aseffa A, Abebe M, Tønjum T (2019) Ample glycosylation in membrane and cell envelope proteins may explain the phenotypic diversity and virulence in the Mycobacterium tuberculosis complex. Sci Rep 9:2927

Boradia V, Frando A, Grundner C (2022) The Mycobacterium tuberculosis PE15/PPE20 complex transports calcium across the outer membrane. PLoS Biol 20(11):e3001906

Borges ÁH, Russell M, Tait D, Scriba TJ, Nemes E, Skallerup P, van Brakel E et al (2025 Jul) Immunogenicity, safety, and efficacy of the vaccine H56:IC31 in reducing the rate of tuberculosis disease recurrence in HIV-negative adults successfully treated for drug-susceptible pulmonary tuberculosis: a double-blind, randomised, placebo-controlled, phase 2b trial. Lancet Infect Dis 25(7):751–763

Borgström EW, Fröberg G, Correia-Neves M, Atterfelt FB, Bellbrant J, Szulkin R, Chryssanthou E, Ängeby K, Tecleab T, Ruhwald M, Andersen P (2020) CD4+ T cell proliferative responses to PPD and CFP-10 associate with recent M. tuberculosis infection. Tuberculosis 123:101959

Brennan MJ, Delogu G (2002) The PE multigene family: a "molecular mantra" for mycobacteria. Trends Microbiol 10(5):246–249. https://doi.org/10.1016/s0966-842x(02)02335-1

Brennan MJ, Delogu G, Chen Y, Bardarov S, Kriakov J, Alavi M, Jacobs WR Jr (2001 Dec) Evidence that mycobacterial PE_PGRS proteins are cell surface constituents that influence interactions with other cells. Infect Immun 69(12):7326–7333

Buchko GW, Kim CY, Terwilliger TC, Kennedy MA (2006 Aug) Solution structure of the conserved hypothetical protein Rv2302 from Mycobacterium tuberculosis. J Bacteriol 188(16):5993–6001

Bukhdruker S, Varaksa T, Grabovec I, Marin E, Shabunya P, Kadukova M, Grudinin S, Kavaleuski A, Gusach A, Gilep A, Borshchevskiy V (2020) Hydroxylation of antitubercular drug candidate, SQ109, by mycobacterial cytochrome P450. Int J Mol Sci 21(20):7683

Bunduc CM, Fahrenkamp D, Wald J, Ummels R, Bitter W, Houben ENG, Marlovits TC (2021 May) Structure and dynamics of a mycobacterial type VII secretion system. Nature 593(7859):445–448

Burggraaf MJ, Speer A, Meijers AS, Ummels R, Van Der Sar AM, Korotkov KV, Bitter W, Kuijl C (2019) Type VII secretion substrates of pathogenic mycobacteria are processed by a surface protease. MBio 10(5):10–128

Burton AT, Zeinert R, Storz G (2024 Nov) Large roles of small proteins. Ann Rev Microbiol 78(1):1–22

Calder B, Soares NC, de Kock E, Blackburn JM (2015) Mycobacterial proteomics: analysis of expressed proteomes and post-translational modifications to identify candidate virulence factors. Expert Rev Proteomics 12(1):21–35

Campomizzi CS, Ghanatios GE, Estrada DF (2021) 19F-NMR reveals substrate specificity of CYP121A1 in Mycobacterium tuberculosis. J Biol Chem 297(5):101287

Camus JC, Pryor MJ, Médigue C, Cole ST (2002) Re-annotation of the genome sequence of mycobacterium tuberculosis H37Rv. Microbiology (Reading) 148(Pt 10):2967–2973. PMID: 12368430 https://doi.org/10.1099/00221287-148-10-2967

Cao K, Heydary YH, Tong G, Martinez TF (2023) Integrated workflow for discovery of microprotein-coding small open reading frames. STAR protocols 4(4):102649

Cascioferro A, Delogu G, Colone M, Sali M, Stringaro A, Arancia G, Fadda G, Palù G, Manganelli R (2007 Dec) PE is a functional domain responsible for protein translocation and localization on mycobacterial cell wall. Mol Microbiol 66(6):1536–1547

Castell A, Johansson P, Unge T, Jones TA, Bäckbro K (2005 Jul) Rv0216, a conserved hypothetical protein from Mycobacterium tuberculosis that is essential for bacterial survival during infection, has a double hotdog fold. Protein Sci 14(7):1850–1862

Champion PA, Champion MM, Manzanillo P, Cox JS (2009 Sep) ESX-1 secreted virulence factors are recognized by multiple cytosolic AAA ATPases in pathogenic mycobacteria. Mol Microbiol 73(5):950–962

Chauhan NK, Anand A, Sharma A, Dhiman K, Gosain TP, Singh P et al (2023) Structural and functional characterization of Rv0792c from mycobacterium tuberculosis: identifying small molecule inhibitor against HutC protein. Microbiol Spectr 11(1):e01973–e01922

Chen X, Cheng HF, Zhou J, Chan CY, Lau KF, Tsui SK, Au SW (2017) Structural basis of the PE-PPE protein interaction in Mycobacterium tuberculosis. J Biol Chem 292(41):16880–16890

Chen H, Li S, Zhao W, Deng J, Yan Z, Zhang T, Wen SA, Guo H, Li L, Yuan J, Zhang H (2022 Jan) A peptidomic approach to identify novel antigen biomarkers for the diagnosis of tuberculosis. Infect Drug Resist 1:4617–4626

Cherny I, Rockah L, Gazit E (2005) The YoeB toxin is a folded protein that forms a physical complex with the unfolded YefM antitoxin: implications for a structural-based differential stability of toxin-antitoxin systems. J Biol Chem 280(34):30063–30072

Chinnam NB, Syed A, Hura GL, Hammel M, Tainer JA, Tsutakawa SE (2023) Combining small angle X-ray scattering (SAXS) with protein structure predictions to characterize conformations in solution. Methods Enzymol 678:351–376. Epub 2022 Oct 31. PMID: 36641214; PMCID: PMC10132260 https://doi.org/10.1016/bs.mie.2022.09.023

Choi JA, Lim YJ, Cho SN, Lee JH, Jeong JA, Kim EJ, Park JB, Kim SH, Park HS, Kim HJ, Song CH (2013) Mycobacterial HBHA induces endoplasmic reticulum stress-mediated apoptosis through the generation of reactive oxygen species and cytosolic Ca2+ in murine macrophage RAW 264.7 cells. Cell Death Dis 4(12):e957

Cole ST, Brosch R, Parkhill J, Garnier T, Churcher C, Harris D, Gordon SV et al (1998) Nature 393(6685):537–544

Costa-Gouveia J, Ainsa J, Brodin P, Lucia A (2017) How can nanoparticles contribute to antituberculosis therapy? Drug Discov Today 22:600

Crosskey TD, Beckham KSH, Wilmanns M (2020) The ATPases of the mycobacterial type VII secretion system: structural and mechanistic insights into secretion. Prog Biophys Mol Biol 152:25–34. https://doi.org/10.1016/j.pbiomolbio.2019.11.008

Cunningham F, Esquivias J, Fernández-Menéndez R, Pérez A, Guardia A, Escribano J, Rivero C, Vimal M, Cacho M, de Dios-Antón P, Martínez-Martínez MS (2020) Exploring the SAR of the β-Ketoacyl-ACP synthase inhibitor GSK3011724A and optimization around a genotoxic metabolite. ACS Infectious Diseases 6(5):1098–1109

de Jonge MI, Pehau-Arnaudet G, Fretz MM, Romain F, Bottai D, Brodin P et al (2007) ESAT-6 from mycobacterium tuberculosis dissociates from its putative chaperone CFP-10 under acidic conditions and exhibits membrane-lysing activity. J Bacteriol 189(16):6028–6034

de Souza GA, Leversen NA, Målen H, Wiker HG (2011) Bacterial proteins with cleaved or uncleaved signal peptides of the general secretory pathway. J Proteome 75(2):502–510. https://doi.org/10.1016/j.jprot.2011.08.016. Epub 2011 Sep 5

Daleke M, Woude A, Parret A, Ummels R, de Groot AM, Watson D, Piersma S, Jimenez C, Bitter W, Houben E (2012) Specific chaperones for the type VII protein secretion pathway. J Biol Chem 287(42):31939–31947

Damen MPM, Meijers AS, Keizer EM, Piersma SR, Jiménez CR, Kuijl CP, Bitter W, Houben ENG (2022) The ESX-1 substrate PPE68 has a key function in ESX-1-mediated secretion in Mycobacterium marinum. MBio 13(6):e0281922

De Groote MA, Sterling DG, Hraha T, Russell TM, Green LS, Wall K, Kraemer S, Ostroff R, Janjic N, Ochsner UA (2017 Oct) Discovery and validation of a six-marker serum protein signature for the diagnosis of active pulmonary tuberculosis. J Clin Microbiol 55(10):3057–3071

de Souza EV, Bizarro CV (2024) Identification of novel bacterial microproteins encoded by small open Reading frames using a computational Proteogenomics workflow. In: Protein bioinformatics. Springer US, New York, pp 19–34

Deep A, Kaundal S, Agarwal S, Singh R, Thakur KG (2017 Dec) Crystal structure of Mycobacterium tuberculosis VapC20 toxin and its interactions with cognate antitoxin, VapB20, suggest a model for toxin-antitoxin assembly. FEBS J 284(23):4066–4082

DeJesus MA, Gerrick ER, Xu W, Park SW, Long JE, Boutte CC, Rubin EJ, Schnappinger D, Ehrt S, Fortune SM, Sassetti CM, Ioerger TR (2017) Comprehensive essentiality analysis of the Mycobacterium tuberculosis genome via saturating transposon mutagenesis. MBio 8(1):e02133–e02116

Delobel A (2021 Apr) Glycosylation of therapeutic proteins: a critical quality attribute. Mass Spectrometry of Glycoproteins: Methods and Protocols 29:1–21

Devlin KL, Leach DT, Stratton KG, Lamichhane G, Lin VS, Beatty KE (2025) Proteomic characterization of *Mycobacterium tuberculosis* subjected to carbon starvation. mSystems. 10(5):e0153024. Epub 2025 Apr 15. PMID: 40231840; PMCID: PMC12090744 https://doi.org/10.1128/msystems.01530-24

Diaz-Silvestre H, Espinosa-Cueto P, Sanchez-Gonzalez A, Esparza-Ceron MA, Pereira-Suarez AL, Bernal-Fernandez G et al (2005) The 19-kDa antigen of mycobacterium tuberculosis is a major adhesin that binds the mannose receptor of THP-1 monocytic cells and promotes phagocytosis of mycobacteria. Microb Pathog 39(3):97–107. https://doi.org/10.1016/j.micpath.2005.06.002

Drumm JE, Mi K, Bilder P, Sun M, Lim J, Bielefeldt-Ohmann H, Basaraba R, So M, Zhu G, Tufariello JM, Izzo AA (2009) Mycobacterium tuberculosis universal stress protein Rv2623 regulates bacillary growth by ATP-binding: requirement for establishing chronic persistent infection. PLoS Pathog 5(5):e1000460

Ekiert DC, Cox JS (2014) Structure of a PE-PPE-EspG complex from Mycobacterium tuberculosis reveals molecular specificity of ESX protein secretion. Proc Natl Acad Sci USA 111(41):14758–14763

Elchennawi I, Ollagnier de Choudens S (2022) Iron–sulfur clusters toward stresses: implication for understanding and fighting tuberculosis. Inorganics 10(10):174

Elchennawi I, Carpentier P, Caux C, Ponge M, Ollagnier de Choudens S (2023) Structural and biochemical characterization of Mycobacterium tuberculosis Zinc SufU-SufS complex. Biomolecules 13(5):732

Fan X, Abd Alla AA, Xie J (2016) Distribution and function of prophage phiRv1 and phiRv2 among Mycobacterium tuberculosis complex. J Biomol Struct Dyn 34(2):233–238

Fan X, Liu Z, Wan Z, Zou H, Ji M, Sun K, Gao R, Li Z, Li W (2022 Jan) Prophage gene Rv2650c enhances intracellular survival of Mycobacterium smegmatis. Front Microbiol 17(12):819837

Fancello L, Burger T (2022) An analysis of proteogenomics and how and when transcriptome-informed reduction of protein databases can enhance eukaryotic proteomics. Genome Biol 23(1):132

Farnia P, Farnia P, Ghanavi J, Besharati S, Velayati AA (2023) Comparison of Proline-glutamate-proline-glutamate-polymorphic GC-rich sequences family protein Wag22 (Rv1759c), PE_PGRS31 (Rv1768), PE_PGRS32 (Rv1803), and PE_PGRS33 gene (Rv1818c) in exponential state and under *in vitro* model of latency in same clinical isolates of *Mycobacterium tuberculosis*: frameshift mutation in extensively drug-resistant and totally drug-resistant tuberculosis bacilli. Biomed Biotechnol Res J 7(4):621–632

Farnia P, Besharati S, Farina P, Ayoubi S, Marjani M, Ghanavi J, Tabarsi P, Velayati AA (2024) The role of efflux pumps transporter in multi-drug resistant tuberculosis: mycobacterial memberane protein (MmpL5). Intl J Mycobacteriol 13(1):7–14

Florczyk MA, McCue L A, Stack RF, Hauer CR, McDonough KA (2001) Identification and characterization of mycobacterial proteins differentially expressed under standing and shaking culture conditions, including Rv2623 from a novel class of putative ATP-binding proteins. Infect Immun 69(9):5777–5785

Flores J, Cancino JC, Chavez-Galan L (2021) Lipoarabinomannan as a point-of-care assay for diagnosis of tuberculosis: how far are we to use it? Front Microbiol 12:638047

Fontecave M, Py B, Ollagnier de Choudens S, Barras F (2008) From iron and cysteine to iron-sulfur clusters: the biogenesis protein machineries. EcoSal Plus 3(1):10–128

Galperin MY, Koonin EV (2004) 'Conserved hypothetical' proteins: prioritization of targets for experimental study. Nucleic Acids Res 32(18):5452–5463

Garcia PS, D'Angelo F, Ollagnier de Choudens S, Dussouchaud M, Bouveret E, Gribaldo S, Barras F

(2022 Oct) An early origin of iron-sulfur cluster biosynthesis machineries before earth oxygenation. Nat Ecol Evol 6(10):1564–1572

Garcia-Vilanova A, Chan J, Torrelles JB (2019) Underestimated manipulative roles of Mycobacterium tuberculosis cell envelope glycolipids during infection. Front Immunol 10:2909

Garg R, Tripathi D, Kant S, Chandra H, Bhatnagar R, Banerjee N (2015 Jan) The conserved hypothetical protein Rv0574c is required for cell wall integrity, stress tolerance, and virulence of Mycobacterium tuberculosis. Infect Immun 83(1):120–129

Glass LN, Swapna G, Chavadi SS, Tufariello JM, Mi K, Drumm JE et al (2017) Mycobacterium tuberculosis universal stress protein Rv2623 interacts with the putative ATP binding cassette (ABC) transporter Rv1747 to regulate mycobacterial growth. PLoS Pathog 13(7):e1006515

Goldstone DC, Metcalf P, Baker EN (2016) Structure of the ectodomain of the electron transporter Rv2874 from Mycobacterium tuberculosis reveals a thioredoxin-like domain combined with a carbohydrate-binding module. Acta Crystallographica: section D. Struct Biol 72(1):40

Gomez M, Johnson S, Gennaro ML (2000 Apr) Identification of secreted proteins of Mycobacterium tuberculosis by a bioinformatic approach. Infect Immun 68(4):2323–2327

Gong Z, Gu Y, Xiong K, Niu J, Zheng R, Su B et al (2021 Jun) The evaluation and validation of blood-derived novel biomarkers for precise and rapid diagnosis of tuberculosis in areas with high-TB burden. Front Microbiol 14(12):650567

Gong W, Pan C, Cheng P, Wang J, Zhao G, Wu X (2022 Jan) Peptide-based vaccines for tuberculosis. Front Immunol 31(13):830497

Gorityala N, Baidya AS, Sagurthi SR (2024) Genome mining of mycobacterium tuberculosis: targeting SufD as a novel drug candidate through in silico characterization and inhibitor screening. Front Microbiol 15:136964

Graham RL, Hess S (2010) Mass spectrometry in the elucidation of the glycoproteome of bacterial pathogens. Curr Proteom 7(1):57–81

Greening C, Ahmed FH, Mohamed AE, Lee BM, Pandey G, Warden AC, Scott C, Oakeshott JG, Taylor MC, Jackson CJ (2016) Physiology, biochemistry, and applications of F420- and Fo-dependent redox reactions. Microbiol Mol Biol Rev 80(2):451–493

Gröschel MI, Sayes F, Simeone R, Majlessi L, Brosch R (2016 Nov) ESX secretion systems: mycobacterial evolution to counter host immunity. Nat Rev Microbiol 14(11):677–691

Guerra-Almeida D, Tschoeke DA, Nunes-da-Fonseca R (2021) Understanding small ORF diversity through a comprehensive transcription feature classification. DNA Res 28(5):dsab007

Guo J, Zhang X, Chen X, Cai Y (2022 Apr) Proteomics in biomarker discovery for tuberculosis: current status and future perspectives. Front Microbiol 26(13):845229

Gupta M, Nayyar N, Chawla M, Sitaraman R, Bhatnagar R, Banerjee N (2016 Jun) The chromosomal parDE2 toxin-antitoxin system of mycobacterium tuberculosis H37Rv: genetic and functional characterization. Front Microbiol 14(7):886

Gurmessa SK, Choi HG, Back YW, Jiang Z, Pham TA, Choi S, Kim HJ (2025) Novel fusion protein REA induces robust prime protection against tuberculosis in mice. npj Vaccines. 10(1):20

Gurumurthy M, Rao M, Mukherjee T, Rao SP, Boshoff HI, Dick T, Barry CE 3rd, Manjunatha UH (2013) A novel F(420) -dependent anti-oxidant mechanism protects Mycobacterium tuberculosis against oxidative stress and bactericidal agents. Mol Microbiol 87(4):744–755

Harris SA, Meyer J, Satti I, Marsay L, Poulton ID, Tanner R et al (2014) Evaluation of a human BCG challenge model to assess antimycobacterial immunity induced by BCG and a candidate tuberculosis vaccine, MVA85A, alone and in combination. J Infect Dis 209:1259–1268

Hassel KR, Brito-Estrada O, Makarewich CA (2023) Microproteins: Overlooked regulators of physiology and disease. iScience 26(6):106781

Herrmann JL, O'Gaora P, Gallagher A, Thole JE, Young DB (1996) Bacterial glycoproteins: a link between glycosylation and proteolytic cleavage of a 19 kDa antigen from Mycobacterium tuberculosis. EMBO J 15(14):3547–3554

Herrmann JL, Delahay R, Gallagher A, Robertson B, Young D (2000) Analysis of post-translational modification of mycobacterial proteins using a cassette expression system. FEBS Lett 473(3):358–362

Hingley-Wilson SM, Lougheed KE, Ferguson K, Leiva S, Williams HD (2010 Jul) Individual Mycobacterium tuberculosis universal stress protein homologues are dispensable in vitro. Tuberculosis (Edinb) 90(4):236–244

Hirabayashi K, Yuda E, Tanaka N, Katayama S, Iwasaki K, Matsumoto T, Kurisu G, Outten FW, Fukuyama K, Takahashi Y, Wada K (2015) Functional dynamics revealed by the structure of the SufBCD complex, a novel ATP-binding cassette (ABC) protein that serves as a scaffold for iron-sulfur cluster biogenesis. J Biol Chem 290(50):29717–29731

Hoffer ED, Miles SJ, Dunham CM (2017) The structure and function of Mycobacterium tuberculosis MazF-mt6 toxin provide insights into conserved features of MazF endonucleases. J Biol Chem 292(19):7718–7726

Hoffmann E, Machelart A, Song OR, Brodin P (2018) Proteomics of mycobacterium infection: moving towards a better understanding of pathogen-driven immunomodulation. Front Immunol 9:86

Hossam Abdelmonem B, Abdelaal NM, Anwer EK, Rashwan AA, Hussein MA, Ahmed YF, Khashana R, Hanna MM, Abdelnaser A (2024) Decoding the role of CYP450 enzymes in metabolism and disease: a comprehensive review. Biomedicine 12(7):1467

Houben EN, Korotkov KV, Bitter W (2014 Aug) Take five - type VII secretion systems of mycobacteria. Biochim Biophys Acta 1843(8):1707–1716

Hu Y, Movahedzadeh F, Stoker NG, Coates AR (2006 Feb) Deletion of the Mycobacterium tuberculosis alpha-crystallin-like hspX gene causes increased bacterial growth in vivo. Infect Immun 74(2):861–868

Huerga H, Mathabire Rucker SC, Bastard M, Mpunga J, Amoros Quiles I, Kabaghe C, Sannino L, Szumilin E (2021) February. Urine lipoarabinomannan testing for all HIV patients hospitalized in medical wards identifies a large proportion of patients with tuberculosis at risk of death. In: Open forum infectious diseases, vol 8, No. 2. US: Oxford University Press, p ofaa639

Huet G, Daffé M, Saves I (2005 Sep) Identification of the Mycobacterium tuberculosis SUF machinery as the exclusive mycobacterial system of [Fe-S] cluster assembly: evidence for its implication in the pathogen's survival. J Bacteriol 187(17):6137–6146

Hurley JM, Cruz JW, Ouyang M, Woychik NA (2011) Bacterial toxin RelE mediates frequent codon-independent mRNA cleavage from the 5′ end of coding regions in vivo. J Biol Chem 286(17):14770–14778

Jain M, Vyas R (2025) Unveiling the silent defenders: mycobacterial stress sensors at the forefront to combat tuberculosis. Crit Rev Biotechnol 29:1–9

Jarlier V, Nikaido H (1994) Mycobacterial cell wall: structure and role in natural resistance to antibiotics. FEMS Microbiol Lett 123(1–2):11–18

Jasani M, Patel L (2023) Design and synthesis of novel substituted pyrazole as small molecule inhibitor of cytochrome P450 CYP121A1. Results Chem 5:100739

Jiang H, Sun J, Chen Y, Chen Z, Wang L, Gao W, Shi Y, Zhang W, Mei Y, Chokkakula S, Vissa V (2018) Landscape of the genome and host cell response of Mycobacterium shigaense reveals pathogenic features. Emerg Microb Infect 7(1):1–3

Jiang H, Tsang L, Wang H, Liu C (2021) IFI44L as a forward regulator enhancing host antituberculosis responses. J Immunol Res 2021:5599408

Jiang F, Liu Y, Li L, Ni R, An Y, Li Y, Zhang L, Gong W (2025 May) Genome-wide expression in human whole blood for diagnosis of latent tuberculosis infection: a multicohort research. Front Microbiol 9(16):1584360

Johnson G (2017) The α/β hydrolase fold proteins of Mycobacterium tuberculosis, with reference to their contribution to virulence. Curr Protein Pept Sci 18(3):190–210

Jungblut PR, Müller EC, Mattow J, Kaufmann SH (2001) Proteomics reveals open reading frames in Mycobacterium tuberculosis H37Rv not predicted by genomics. Infect Immun 69(9):5905–5907

Kang SM, Moon H, Han SW, Kim DH, Kim BM, Lee BJ (2020) Structure-based de novo design of mycobacterium tuberculosis VapC-activating stapled peptides. ACS Chem Biol 15(10):2493–2498

Kantardjieff KA, Vasquez C, Castro P, Warfel NM, Rho BS, Lekin T, Kim CY, Segelke BW, Terwilliger TC, Rupp B (2005 Apr) Structure of pyrR (Rv1379) from Mycobacterium tuberculosis: a persistence gene and protein drug target. Acta Crystallogr D Biol Crystallogr 61(Pt 4):355–364

Keijzer JD (2012) A proteomic portrait of Mycobacterium tuberculosis. Infect Dis 54(4):579–581

Kelkar DS, Kumar D, Kumar P, Balakrishnan L, Muthusamy B, Yadav AK et al (2011) Proteogenomic analysis of Mycobacterium tuberculosis by high resolution mass spectrometry. Mol Cell Proteomics 10(12):M111.011627

Khademi F, Yousefi-Avarvand A, Derakhshan M, Meshkat Z, Tafaghodi M, Ghazvini K et al (2017) Mycobacterium tuberculosis HspX/EsxS fusion protein: gene cloning, protein expression, and purification in Escherichia coli. Rep Biochem Mol Biol 6(1):15–22

Khambati N, Olbrich L, Ellner J, Salgame P, Song R, Bijker EM (2021) Host-based biomarkers in saliva for the diagnosis of pulmonary tuberculosis in children: a mini-review. Front Pediatr 9:756043

Kirykowicz AM, Woodward JD (2020) Shotgun EM of mycobacterial protein complexes during stationary phase stress. Curr Res Struct Biol 2:204–212. PMID: 34235480; PMCID:PMC8244302 https://doi.org/10.1016/j.crstbi.2020.09.002

Kitata RB, Yang JC, Chen YJ (2023) Advances in data-independent acquisition mass spectrometry towards comprehensive digital proteome landscape. Mass Spectrom Rev 42(6):2324–2348

Koiri D, Nandi M, Hameem PMA, Aher JB, Kumar A, Behura A, Meher G, Choudhary V, Choubey S, Saleem M (2025) Real-time visualization reveals Mycobacterium tuberculosis ESAT-6 disrupts phagosome-like compartment via fibril-mediated vesiculation. Cell Rep 44(3):115328

Kootery KP, Sarojini S (2022) Structural and functional characterization of a hypothetical protein in the RD7 region in clinical isolates of mycobacterium tuberculosis—an in silico approach to candidate vaccines. J Genet Eng Biotechnol 20(1):55. PMID: 35394551; PMCID: PMC8993957 https://doi.org/10.1186/s43141-022-00340-5

Korch SB, Hill TM (2006 Jun) Ectopic overexpression of wild-type and mutant hipA genes in Escherichia coli: effects on macromolecular synthesis and persister formation. J Bacteriol 188(11):3826–3836

Korotkova N, Freire D, Phan TH, Ummels R, Creekmore CC, Evans TJ, Wilmanns M, Bitter W, Parret AH, Houben EN, Korotkov KV (2014 Oct) Structure of the Mycobacterium tuberculosis type VII secretion system chaperone EspG5 in complex with PE25-PPE41 dimer. Mol Microbiol 94(2):367–382

Kovacs-Simon A, Titball RW, Michell SL (2011 Feb) Lipoproteins of bacterial pathogens. Infect Immun 79(2):548–561

Kumar P, Capodagli GC, Awasthi D, Shrestha R, Maharaja K, Sukheja P, Li SG, Inoyama D, Zimmerman M, Ho Liang HP, Sarathy J (2018) Synergistic lethality of a binary inhibitor of Mycobacterium tuberculosis KasA. MBio 9(6):10–128

Kumar A, Campomizzi CS, Jay N, Ferguson S, Scheffler EJ, Lioi J, Tu C, Qu J, Simons C, Estrada DF (2021) Surface hydrophobics mediate functional dimerization of CYP121A1 of Mycobacterium tuberculosis. Sci Rep 11(1):394

Kumar A, Sharma P, Arun A, Meena LS (2023 May) Development of peptide vaccine candidate using highly antigenic PE-PGRS family proteins to stimulate the host immune response against Mycobacterium tuberculosis H37Rv: an immuno-informatics approach. J Biomol Struct Dyn 41(8):3382–3404

Kumar A, Kamuju V, Vivekanandan P (2024) RNA G-quadruplexes inhibit translation of the PE/PPE transcripts in mycobacterium tuberculosis. J Biol Chem 300(1). https://doi.org/10.1016/j.jbc.2023.105754

Kwon KW, Choi HG, Kim KS, Park SA, Kim HJ, Shin SJ (2024) BCG-booster vaccination with HSP90-ESAT-6-HspX-RipA multivalent subunit vaccine confers durable protection against hypervirulent Mtb in mice. NPJ Vaccin 8;9(1):55

Larsen SE, Baldwin SL, Coler RN (2023) Tuberculosis vaccines update: is an RNA-based vaccine feasible for tuberculosis? Int J Infect Dis 130(Suppl 1):S47–S51. https://doi.org/10.1016/j.ijid.2023.03.035. Epub 2023 Mar 23. PMID: 36963657; PMCID: PMC10033141

Lee KY, Lee BJ (2016) Structure, biology, and therapeutic application of toxin-antitoxin Systems in Pathogenic Bacteria. Toxins (Basel) 8(10):305

Lee S, Kim SM, Lee RT (2013) Thioredoxin and thioredoxin target proteins: from molecular mechanisms to functional significance. Antioxid Redox Signal 18(10):1165–1207

Lei Z, Wang J, Liu CH (2025) Developing next-generation tuberculosis vaccines based on pathogen–host interactions: towards a holistic perspective. hLife 3(4):164–171

Leong AZ, Lee PY, Mohtar MA, Syafruddin SE, Pung YF, Low TY (2022) Short open reading frames (sORFs) and microproteins: an update on their identification and validation measures. J Biomed Sci 29(1):19

Lesur E, Zhang Y, Dautin N, Dietrich C, Li de la Sierra-Gallay I, Augusto LA, Rollando P, Lazar N, Urban D, Doisneau G, Constantinesco-Becker F, Van Tilbeurgh H, Guianvarc'h D, Bourdreux Y, Bayan N (2025 Mar) Synthetic mycolates derivatives to decipher protein mycoloylation, a unique post-translational modification in bacteria. J Biol Chem 301(3):108243

Leung V, Lévesque CM (2012 May) A stress-inducible quorum-sensing peptide mediates the formation of persister cells with noninherited multidrug tolerance. J Bacteriol 194(9):2265–2274

Lew JM, Mao C, Shukla M, Warren A, Will R, Kuznetsov D, Xenarios I, Robertson BD, Gordon SV, Schnappinger D, Cole ST, Sobral B (2013 Jan) Database resources for the tuberculosis community. Tuberculosis (Edinb) 93(1):12–17

Li H, Liu L, Zhang WJ, Zhang X, Zheng J, Li L, Zhu X, Yang Q, Zhang M, Liu H, Chen X, Jin Q (2019) Analysis of the antigenic properties of membrane proteins of Mycobacterium tuberculosis. Sci Rep 9(1):3042

Lin MF, Jungreis I, Kellis M (2011) PhyloCSF: a comparative genomics method to distinguish protein coding and non-coding regions. Bioinformatics 27(13):i275–i282

Lin K, O'Brien KM, Trujillo C, Wang R, Wallach JB, Schnappinger D, Ehrt S (2016) Mycobacterium tuberculosis Thioredoxin reductase is essential for thiol redox homeostasis but plays a minor role in antioxidant defense. PLoS Pathog 12(6):e1005675

Luckner SR, Machutta CA, Tonge PJ, Kisker C (2009) Crystal structures of Mycobacterium tuberculosis KasA show mode of action within cell wall biosynthesis and its inhibition by thiolactomycin. Structure 17(7):1004–1013

Ludwig C, Gillet L, Rosenberger G, Amon S, Collins BC, Aebersold R (2018) Data-independent acquisition-based SWATH-MS for quantitative proteomics: a tutorial. Mol Syst Biol 14(8):e8126

Maharjan R, Zhang Z, Klenotic PA, Gregor WD, Tringides ML, Cui M, Purdy GE, Yu EW (2024) Structures of the mycobacterial MmpL4 and MmpL5 transporters provide insights into their role in siderophore export and iron acquisition. PLoS Biol 22(10):e3002874

Maitra A, Nukala S, Dickman R, Martin LT, Munshi T, Gupta A, Shepherd AJ, Arnvig KB, Tabor AB, Keep NH, Bhakta S (2021) Characterization of the MurT/GatD complex in Mycobacterium tuberculosis towards validating a novel anti-tubercular drug target. JAC-Antimicrob Resist 3(1):dlab028

Makarewich CA (2020) The hidden world of membrane microproteins. Exp Cell Res 388(2):111853

Makarewich CA, Olson EN (2017 Sep) Mining for Micropeptides. Trends Cell Biol 27(9):685–696

Manful EE, Pietersen RD, Baker B (2025 May) IFIT2-induced transcriptomic changes in Mycobacterium tuberculosis infected macrophages. Front Cell Infect Microbiol 20(15):1536446

Manganelli R, Voskuil MI, Schoolnik GK, Smith I (2001 Jul) The Mycobacterium tuberculosis ECF sigma factor sigmaE: role in global gene expression and survival in macrophages. Mol Microbiol 41(2):423–437

McGillick J, Ames JR, Murphy T, Bourne CR (2021) A YoeB toxin cleaves both RNA and DNA. Sci Rep 11(1):3592

Medha P, Sharma S, Sharma M (2022) Design of a peptide-based vaccine from late stage specific immunogenic cross-reactive antigens of PE/PPE proteins of Mycobacterium tuberculosis. Eur J Pharm Sci 168:106051

Méheust R, Castelle CJ, Jaffe AL, Banfield JF (2022) Conserved and lineage-specific hypothetical proteins may have played a central role in the rise and diversification of major archaeal groups. BMC Biol 20(1):154

Metcalfe HJ, Steinbach S, Jones GJ, Connelley T, Morrison WT, Vordermeier M et al (2016) Protection associated with a TB vaccine is linked to increased frequency of Ag85A-specific CD4(+) T cells but no increase in avidity for Ag85A. Vaccine 34:4520–4525

Miggiano R, Morrone C, Rossi F, Rizzi M (2020) Targeting genome integrity in Mycobacterium tuberculosis: from nucleotide synthesis to DNA replication and repair. Molecules 25(5):1205

Minato Y, Thiede JM, Kordus SL, McKlveen EJ, Turman BJ, Baughn AD (2015) Mycobacterium tuberculosis folate metabolism and the mechanistic basis for Para-aminosalicylic acid susceptibility and resistance. Antimicrob Agents Chemother 59(9):5097–5106. . Epub 2015 Jun 1. PMID: 26033719; PMCID:PMC4538520 https://doi.org/10.1128/AAC.00647-15

Mishra AK, Driessen NN, Appelmelk BJ, Besra GS (2011) Lipoarabinomannan and related glycoconjugates: structure, biogenesis and role in Mycobacterium tuberculosis physiology and host–pathogen interaction. FEMS Microbiol Rev 35(6):1126–1157

Mohsen JJ, Martel AA, Slavoff SA (2023) Microproteins—discovery, structure, and function. Proteomics 23(23–24):e2100211. Epub 2023 Aug 21. PMID: 37603371; PMCID: PMC10841188 https://doi.org/10.1002/pmic.202100211

Moos PJ, Carey AF, Joseph J, Kialo S, Norrie J, Moyarelce JM, Amof A, Nogua H, Lim AL, Barrows LR (2024) Description of bacterial RNA transcripts detected in Mycobacterium tuberculosis – infected cells from peripheral human granulomas using single cell RNA sequencing. bioRxiv [Preprint]:2024.08.20.608852

Moukendza Koundi L, Ekomi Moure UA, Boni FG, Hamdi I, Fan L, Xie J (2024) Mycobacterium tuberculosis Rv2617c is involved in stress response and phage infection resistance. Heliyon 10(5):e27400

Mumtaz MA, Couso JP (2015) Ribosomal profiling adds new coding sequences to the proteome. Biochem Soc Trans 43(6):1271–1276

Munseri P, Said J, Amour M, Magohe A, Matee M, Rees CA, Mackenzie T, Tvaroha S, Bailey-Kellogg C, Maro I, Wieland-Alter W, Adams LV, Horsburgh CR, Nakamura K, Arbeit RD, Pallangyo K, von Reyn CF (2020) DAR-901 vaccine for the prevention of infection with Mycobacterium tuberculosis among BCG-immunized adolescents in Tanzania: a randomized controlled, double-blind phase 2b trial. Vaccine 38(46):7239–7245

Mutavhatsindi H, Calder B, McAnda S, Malherbe ST, Stanley K, Kidd M, Walzl G, Chegou NN (2021) Identification of novel salivary candidate protein biomarkers for tuberculosis diagnosis: a preliminary biomarker discovery study. Tuberculosis 130:102118

Nawrot B, Sochacka E, Düchler M (2011 Dec) tRNA structural and functional changes induced by oxidative stress. Cell Mol Life Sci 68(24):4023–4032

Naz S, Dabral S, Nagarajan SN, Arora D, Singh LV, Kumar P, Singh Y, Kumar D, Varshney U, Nandicoori VK (2021) Compromised base excision repair pathway in Mycobacterium tuberculosis imparts superior adaptability in the host. PLoS Pathog 17(3):e1009452

Nesvizhskii AI (2014) Proteogenomics: concepts, applications and computational strategies. Nat Methods 11(11):1114–1125. PMID: 25357241; PMCID: PMC4392723 https://doi.org/10.1038/nmeth.3144

Neudecker D, Fritschi N, Sutter T, Lu LL, Lu P, Tebruegge M, Santiago-Garcia B, Ritz N (2024) Evaluation of serological assays for the diagnosis of childhood tuberculosis disease: a study protocol. BMC Infect Dis 24(1):481

Nikitushkin V, Shleeva M, Loginov D, Dyčka F, Sterba J, Kaprelyants A (2022) Shotgun proteomic profiling of dormant, 'non-culturable' Mycobacterium tuberculosis. PLoS One 17(8):e0269847

Orr MW, Mao Y, Storz G, Qian SB (2020) Alternative ORFs and small ORFs: shedding light on the dark proteome. Nucleic acids research. 48(3):1029–1042

Ou X, Liu E, Rashid F, Pei S, Zhang G, Anthony R, Xing R, Huang F, Xia H, Zheng Y, Song Y, Zhou Y, Zhao B, Wang S, Lin M, Zhou L, Zhao Y (2025) Screening and preclinical assessment of novel Mycobacterium tuberculosis recombinant antigens based tuberculin skin testing. Front Immunol 16:1498448

Ouellet H, Johnston JB, de Montellano PR (2010) The Mycobacterium tuberculosis cytochrome P450 system. Arch Biochem Biophys 493(1):82–95

O'Flaherty R, Trbojević-Akmačić I, Greville G, Rudd PM, Lauc G (2018) The sweet spot for biologics: recent advances in characterization of biotherapeutic glycoproteins. Expert Rev Proteomics 15(1):13–29

Padayachee T, Lamb DC, Nelson DR, Syed K (2024) Structure–function analysis of the essential Mycobacterium tuberculosis P450 drug target, CYP121A1. Int J Mol Sci 25(9):4886

Pai VJ, Lau CJ, Garcia-Ruiz A, Donaldson C, Vaughan JM, Miller B, De Souza EV, Pinto AM, Diedrich J, Gavva NR, Yu S, DeBoever C, Horman SR, Saghatelian A (2024) Microprotein-encoding RNA regulation in cells treated with pro-inflammatory and pro-fibrotic stimuli. BMC Genomics 25:1034

Panda A, Chakraborty A, Nandi S, Kaushik A, Biswas A (2017) The C-terminal extension of Mycobacterium tuberculosis Hsp16.3 regulates its oligomerization, subunit exchange dynamics and chaperone function. FEBS J 284(2)

Pandey M, Talwar S, Bose S, Pandey AK (2018) Iron homeostasis in Mycobacterium tuberculosis is essential for persistence. Sci Rep 8(1):17359

Parbhoo T, Mouton J, Sampson S (2022) Phenotypic adaptation of Mycobacterium tuberculosis to host-associated stressors that induce persister formation. Front Cell Infect Microbiol 12:956607

Park HD, Guinn KM, Harrell MI, Liao R, Voskuil MI, Tompa M, Schoolnik GK, Sherman DR (2003 May) Rv3133c/dosR is a transcription factor that mediates the hypoxic response of Mycobacterium tuberculosis. Mol Microbiol 48(3):833–843

Park EJ, Kwon YM, Lee JW, Kang HY, Oh JI (2019) Dual control of RegX3 transcriptional activity by SenX3 and PknB. J Biol Chem 294(28):11023–11034

Passos BBS, Araújo-Pereira M, Vinhaes CL, Amaral EP, Andrade BB (2024) The role of ESAT-6 in tuberculosis immunopathology. Front Immunol 15:1383098

Pereira MC, Oliveira LC, Murad E (2012 Sep) Iron oxide catalysts: Fenton and Fentonlike reactions–a review. Clay Miner 47(3):285–302

Picotti P, Rinner O, Stallmach R, Dautel F, Farrah T, Domon B et al (2010) High-throughput generation of selected reaction-monitoring assays for proteins and proteomes. Nat Methods 7(1):43–46

Pillay K, Chiliza TE, Senzani S, Pillay B, Pillay M (2024) In silico design of Mycobacterium tuberculosis multi-epitope adhesin protein vaccines. Heliyon. 10(18):e37536

Poirier V, Bach H, Av-Gay Y (2014) Mycobacterium tuberculosis promotes anti-apoptotic activity of the macrophage by PtpA protein-dependent dephosphorylation of host GSK3. J Biol Chem 289(42):29377–29385. https://doi.org/10.1074/jbc.M114.582502

Prados-Rosales R, Carreño L, Cheng T, Blanc C, Weinrick B, Malek A, Lowary TL, Baena A, Joe M, Bai Y, Kalscheuer R (2017) Enhanced control of Mycobacterium tuberculosis extrapulmonary dissemination in mice by an arabinomannan-protein conjugate vaccine. PLoS Pathog 13(3):e1006250

Priyanka SS, Varma-Basil M, Sharma M (2024 Oct) C-terminal region of Rv1039c (PPE15) protein of Mycobacterium tuberculosis targets host mitochondria to induce macrophage apoptosis. Apoptosis 29(9–10):1757–1779

Purwantini E, Mukhopadhyay B (2013) Rv0132c of Mycobacterium tuberculosis encodes a coenzyme F420-dependent hydroxymycolic acid dehydrogenase. PLoS One 8(12):e81985

Py B, Barras F (2010 Jun) Building Fe-S proteins: bacterial strategies. Nat Rev Microbiol 8(6):436–446

Qian J, Chen R, Wang H, Zhang X (2020) Role of the PE/PPE family in host–pathogen interactions and prospects for anti-tuberculosis vaccine and diagnostic tool design. Front Cell Infect Microbiol 10:594288

Rais M, Abdelaal H, Reese VA, Ferede D, Larsen SE, Pecor T, Erasmus JH, Archer J, Khandhar AP, Cooper SK, Podell BK, Reed SG, Coler RN, Baldwin SL (2023 Jan) Immunogenicity and protection against Mycobacterium avium with a heterologous RNA prime and protein boost vaccine regimen. Tuberculosis (Edinb) 138:102302

Ramage HR, Connolly LE, Cox JS (2009 Dec) Comprehensive functional analysis of Mycobacterium tuberculosis toxin-antitoxin systems: implications for pathogenesis, stress responses, and evolution. PLoS Genet 5(12):e1000767

Rodrigues P, Bangali H, Ali E, Sharma MK, Abdullaev B, Alkhafaji AT et al (2024) Microproteins/micropeptides dysregulation contributes to cancer progression and development: a mechanistic review. Cell Biol Int 48(10):1395–1405. https://doi.org/10.1002/cbin.12219. Epub 2024 Jul 15. PMID: 39010637

Rosenberger G, Liu Y, Röst HL, Ludwig C, Buil A, Bensimon A et al (2017) Inference and quantification of peptidoforms in large sample cohorts by SWATH-MS. Nat Biotechnol 35(8):781–788. https://doi.org/10.1038/nbt.3908. Epub 2017 Jun 12. PMID: 28604659; PMCID: PMC5593115

Roy S, Ghatak D, Das P, BoseDasgupta S (2020) ESX secretion system: the gatekeepers of mycobacterial survivability and pathogenesis. Eur J Microbiol Immunol (Bp) 10(4):202–209

Rudraraju RS, Daher SS, Gallardo-Macias R, Wang X, Neiditch MB, Freundlich JS (2022) Mycobacterium tuberculosis KasA as a drug target: structure-based inhibitor design. Front Cell Infect Microbiol 12:1008213

Sala A, Bordes P, Genevaux P (2014) Multiple toxin-antitoxin systems in Mycobacterium tuberculosis. Toxins (Basel). 6(3):1002–1020

Sali M, Di Sante G, Cascioferro A, Zumbo A, Nicolò C, Donà V, Rocca S, Procoli A, Morandi M, Ria F, Palù G, Fadda G, Manganelli R, Delogu G (2010 Dec) Surface expression of MPT64 as a fusion with the PE domain of PE_PGRS33 enhances Mycobacterium bovis BCG protective activity against Mycobacterium tuberculosis in mice. Infect Immun 78(12):5202–5213

Sampson SL (2011) Mycobacterial PE/PPE proteins at the host-pathogen interface. Clin Dev Immunol 2011:497203

Sarkar A, Wintrode PL (2011 Apr) Effects of glycosylation on the stability and flexibility of a metastable protein: the human serpin α(1)-antitrypsin. Int J Mass Spectrom 302(1–3):69–75

Sassetti CM, Rubin EJ (2003) Genetic requirements for mycobacterial survival during infection. Proc Natl Acad Sci USA 100(22):12989–12994

Sassetti CM, Boyd DH, Rubin EJ (2003 Apr) Genes required for mycobacterial growth defined by high density mutagenesis. Mol Microbiol 48(1):77–84

Satchidanandam V, Kumar N, Jumani RS, Challu V, Elangovan S, Khan NA (2014) The glycosylated Rv1860 protein of Mycobacterium tuberculosis inhibits dendritic cell mediated TH1 and TH17 polarization of T cells and abrogates protective immunity conferred by BCG. PLoS Pathog 10(6):e1004176

Schaeffer ML, Agnihotri G, Volker C, Kallender H, Brennan PJ, Lonsdale JT (2001) Purification and biochemical characterization of the Mycobacterium tuberculosis β-Ketoacyl-acyl carrier protein synthases KasA and KasB. J Biol Chem 276(50):47029–47037

Schiebel J, Kapilashrami K, Fekete A, Bommineni GR, Schaefer CM, Mueller MJ, Tonge PJ, Kisker C (2013) Structural basis for the recognition of mycolic acid precursors by KasA, a condensing enzyme and drug target from Mycobacterium tuberculosis. J Biol Chem 288(47):34190–34204

Schnappinger D, Ehrt S, Voskuil MI, Liu Y, Mangan JA, Monahan IM, Dolganov G, Efron B, Butcher PD, Nathan C, Schoolnik GK (2003) Transcriptional adaptation of Mycobacterium tuberculosis within macrophages: insights into the phagosomal environment. J Exp Med 198(5):693–704

Schureck MA, Maehigashi T, Miles SJ, Marquez J, Cho SE, Erdman R, Dunham CM (2014) Structure of the Proteus vulgaris HigB-(HigA)2-HigB toxin-antitoxin complex. J Biol Chem 289(2):1060–1070

Selengut JD, Haft DH, Davidsen T, Ganapathy A, Gwinn-Giglio M, Nelson WC et al (2007) TIGRFAMs and genome properties: tools for the assignment of molecular function and biological process in prokaryotic genomes. Nucleic Acids Res 35(Database issue):D260–D264. https://doi.org/10.1093/nar/gkl1043. Epub 2006 Dec 6. PMID: 17151080; PMCID: PMC1781115

Shah S, Briken V (2016) Modular organization of the ESX-5 secretion system in Mycobacterium tuberculosis. Front Cell Infect Microbiol 6:49

Sharma D, Bisht DM (2017) tuberculosis hypothetical proteins and proteins of unknown function: hope for exploring novel resistance mechanisms as well as future target of drug resistance. Front Microbiol 8:465

Sharma A, Chattopadhyay G, Chopra P, Bhasin M, Thakur C, Agarwal S, Ahmed S, Chandra N, Varadarajan R, Singh R (2020 Sep) VapC21 toxin contributes to drug-tolerance and interacts with non-cognate VapB32 antitoxin in Mycobacterium tuberculosis. Front Microbiol 11(11):2037

Shell SS, Wang J, Lapierre P, Mir M, Chase MR, Pyle MM, Gawande R, Ahmad R, Sarracino DA, Ioerger TR, Fortune SM, Derbyshire KM, Wade JT, Gray TA (2015) Leaderless transcripts and small proteins are common features of the mycobacterial translational landscape. PLoS Genet 11(11):e1005641

Shi T, Shou F, He Y, Zhou K, Gao W, Nie X, Han M, Liao C, Li T (2023) Whole genome sequencing of drug resistance mycobacterium tuberculosis from extra-pulmonary sites. Life Sci Alliance 6(11):e202302076. https://doi.org/10.26508/lsa.202302076. PMID: 37591723; PMCID: PMC10435967

Sidoli S, Lin S, Xiong L, Bhanu NV, Karch KR, Johansen E, Hunter C, Mollah S, Garcia BA (2015) SWATH analysis for characterization and quantification of histone post-translational modifications. Mol Cell Proteomics 14:2420–2428

Simeone R, Bottai D, Brosch R (2009 Feb) ESX/type VII secretion systems and their role in host-pathogen interaction. Curr Opin Microbiol 12(1):4–10

Singh S, Maurya SK, Aqdas M, Bashir H, Arora A, Bhalla V, Agrewala JN (2022) Mycobacterium tuberculosis exploits MPT64 to generate myeloid-derived suppressor cells to evade the immune system. Cell Mol Life Sci 79(11):567

Sinha SK, Rs N, Devarakonda Y, Rathi A, Reddy Regatti P, Batra S, Syal K (2023) Tale of twin Bifunctional second messenger (p)ppGpp Synthetases and their function in mycobacteria. ACS Omega 8(36):32258–32270

Smaill F, Jeyanathan M, Smieja M, Medina MF, Thanthrige-Don N, Zganiacz A, Yin C, Heriazon A, Damjanovic D, Puri L, Hamid J, Xie F, Foley R, Bramson J, Gauldie J, Xing Z (2013) A human type 5 adenovirus-based tuberculosis vaccine induces robust T cell responses in humans despite preexisting anti-adenovirus immunity. Sci Transl Med 5(205):205ra134

Sonawane A, Mohanty S, Jagannathan L, Bekolay A, Banerjee S (2012) Role of glycans and glycoproteins in disease development by Mycobacterium tuberculosis. Crit Rev Microbiol 38(3):250–266

Song Y-C, Das D, Zhang Y, Chen M-X, Fernie AR, Zhu F-Y, Han J (2023) Proteogenomics-based functional genome research: approaches, applications, and perspectives in plants. Trends Biotechnol 41(12):1532–1548

Srivastava S, Dey S, Mukhopadhyay S (2023) Vaccines against tuberculosis: where are we now? Vaccines (Basel). 11(5):1013

Stamm CE, Pasko BL, Chaisavaneeyakorn S, Franco LH, Nair VR, Weigele BA, Alto NM, Shiloh MU (2019) Screening Mycobacterium tuberculosis secreted proteins identifies Mpt64 as a eukaryotic membrane-binding bacterial effector. mSphere 4(3):e00354–e00319

Stanley SA, Raghavan S, Hwang WW, Cox JS (2003) Acute infection and macrophage subversion by Mycobacterium tuberculosis require a specialized secretion system. Proc Natl Acad Sci USA 100(22):13001–13006

Stead D, Wasserman S, Steenkamp E, Parrish A, Barr D, Meintjes G (2025) Comparative performance of urine lipoarabinomannan and urine Xpert MTB/RIF ultra for diagnosing tuberculosis in adult inpatients with human immunodeficiency virus in East London, South Africa. Clin Infect Dis:ciaf080

Sun C, Yang G, Yuan J, Peng X, Zhang C, Zhai X et al (2017 Jul) *Mycobacterium tuberculosis* hypoxic response protein 1 (Hrp1) augments the pro-inflammatory response and enhances the survival of *Mycobacterium smegmatis* in murine macrophages. J Med Microbiol 66(7):1033–1044

Tameris MD, Hatherill M, Landry BS, Scriba TJ, Snowden MA, Lockhart S, Shea JE, McClain JB, Hussey GD, Hanekom WA, Mahomed H, McShane H (2013) MVA85A 020 trial study team. Safety and efficacy of MVA85A, a new tuberculosis vaccine, in infants previously vaccinated with BCG: a randomised, placebo-controlled phase 2b trial. Lancet 381(9871):1021–1028

Taylor JL, Wieczorek A, Keyser AR, Grover A, Flinkstrom R, Karls RK, Bielefeldt-Ohmann H, Dobos KM, Izzo AA (2012 Nov) HspX-mediated protection against tuberculosis depends on its chaperoning of a mycobacterial molecule. Immunol Cell Biol 90(9):945–954

Thakur Z, Chaudhary R, Mehta PK (2024 Nov) MazEF toxin-antitoxin systems: their role in Mycobacterium tuberculosis stress response and drug resistance. Expert Rev Anti-Infect Ther 22(11):921–924

Thomas C, Mackey MM, Diaz AA, Cox DP (2009) Hydroxyl radical is produced via the Fenton reaction in submitochondrial particles under oxidative stress: implications for diseases associated with iron accumulation. Redox Rep 14(3):102–108

Tiwari BM, Kannan N, Vemu L, Raghunand TR (2012) The Mycobacterium tuberculosis PE proteins Rv0285 and Rv1386 modulate innate immunity and mediate bacillary survival in macrophages. PLoS One 7(12):e51686

Tiwari P, Arora G, Singh M, Kidwai S, Narayan OP, Singh R (2015 Jan) MazF ribonucleases promote Mycobacterium tuberculosis drug tolerance and virulence in Guinea pigs. Nat Commun 22(6):6059

Tkachuk AP, Bykonia EN, Popova LI, Kleymenov DA, Semashko MA, Chulanov VP, Fitilev SB, Maksimov SL, Smolyarchuk EA, Manuylov VA, Vasina DV, Gushchin VA, Gintsburg AL (2020) Safety and immunogenicity of the GamTBvac, the recombinant subunit tuberculosis vaccine candidate: a phase II, multi-center, double-blind, randomized, placebo-controlled study. Vaccines (Basel). 8(4):652

Torrelles JB, Schlesinger LS (2010) Diversity in mycobacterium tuberculosis mannosylated cell wall determinants impacts adaptation to the host. Tuberculosis 90(2):84–93

Tripathi A, Anand K, Das M, O'Niel RA, Thakur C, Rajmani RS, Chandra N, Laxman S, Singh A (2022) Mycobacterium tuberculosis requires SufT for Fe-S cluster maturation, metabolism, and survival in vivo. PLoS Pathog 18(4):e1010475

Tucci P, Portela M, Chetto CR, Gonzalez-Sapienza G, Marín M (2020) Integrative proteomic and glycoproteomic profiling of Mycobacterium tuberculosis culture filtrate. PLoS One 15(3):e0221837

Uversky VN (2024) Functional unfoldomics: roles of intrinsic disorder in protein (multi)functionality. Adv Protein Chem Struct Biol 138:179–210

VanderVen BC, Harder JD, Crick DC, Belisle JT (2005) Export-mediated assembly of mycobacterial glycoproteins parallels eukaryotic pathways. Science 309(5736):941–943

Vasina DV, Kleymenov DA, Manuylov VA, Mazunina EP, Koptev EY, Tukhovskaya EA, Murashev AN, Gintsburg AL, Gushchin VA, Tkachuk AP (2019) First-in-human trials of GamTBvac, a recombinant subunit tuberculosis vaccine candidate: safety and immunogenicity assessment. Vaccines (Basel) 7(4):166

Via LE, Lin PL, Ray SM, Carrillo J, Allen SS, Eum SY, Taylor K, Klein E, Manjunatha U, Gonzales J, Lee EG (2008 Jun) Tuberculous granulomas are hypoxic in Guinea pigs, rabbits, and nonhuman primates. Infect Immun 76(6):2333–2340

Vitorino R (2024) Transforming clinical research: the power of high-throughput omics integration. Proteomes 12(3):25

Voskuil MI, Bartek IL, Visconti K, Schoolnik GK (2011 May) The response of Mycobacterium tuberculosis to reactive oxygen and nitrogen species. Front Microbiol 13(2):105

van der Lee R, Buljan M, Lang B, Weatheritt RJ, Daughdrill GW, Dunker AK et al (2014) Classification of intrinsically disordered regions and proteins. Chem Rev 114(13):6589–6631. https://doi.org/10.1021/cr400525m

von Reyn CF, Lahey T, Arbeit RD, Landry B, Kailani L, Adams LV, Haynes BC, Mackenzie T, Wieland-Alter W, Connor RI, Tvaroha S, Hokey DA, Ginsberg AM, Waddell R (2017) Safety and immunogenicity of an inactivated whole cell tuberculosis vaccine booster in adults primed with BCG: A randomized, controlled trial of DAR-901. PLoS One 12(5):e0175215. https://doi.org/10.1371/journal.pone.0175215. PMID: 28498853; PMCID: PMC5429024

Walters SB, Dubnau E, Kolesnikova I, Laval F, Daffe M, Smith I (2006 Apr) The Mycobacterium tuberculosis PhoPR two-component system regulates genes essential for virulence and complex lipid biosynthesis. Mol Microbiol 60(2):312–330

Wang X, Wood TK (2011) Toxin-antitoxin systems influence biofilm and persister cell formation and the general stress response. Appl Environ Microbiol 77(16):5577–5583

Wang Z, Potter BM, Gray AM, Sacksteder KA, Geisbrecht BV, Laity JH (2007) The solution structure of antigen MPT64 from Mycobacterium tuberculosis defines a new family of beta-grasp proteins. J Mol Biol 366(2):375–381

Wang Q, Liu S, Tang Y, Liu Q, Yao Y (2014) MPT64 protein from Mycobacterium tuberculosis inhibits apoptosis of macrophages through NF-κB–miRNA21–Bcl-2 pathway. PLoS One 9(7):e100949

Wang Q, Boshoff HI, Harrison JR, Ray PC, Green SR, Wyatt PG, Barry CE III (2020) PE/PPE proteins mediate nutrient transport across the outer membrane of Mycobacterium tuberculosis. Science 367(6482):1147–1151

Wang L, He Y, Wang P, Lou H, Liu H, Sha W (2024) Single-cell transcriptome sequencing reveals altered peripheral blood immune cells in patients with severe tuberculosis. Eur J Med Res 29(1):434

Willemse D, Weber B, Masino L, Warren RM, Adinolfi S, Pastore A, Williams MJ (2018) Rv1460, a SufR homologue, is a repressor of the suf operon in Mycobacterium tuberculosis. PLoS One 13(7):e0200145

Winther KS, Gerdes K (2011) Enteric virulence associated protein VapC inhibits translation by cleavage of initiator tRNA. Proc Natl Acad Sci USA 108(18):7403–7407. https://doi.org/10.1073/pnas.1019587108. Epub 2011 Apr 18.PMID: 21502523; PMCID: PMC3088637.

Winther KS, Brodersen DE, Brown AK, Gerdes K (2013) VapC20 of mycobacterium tuberculosis cleaves the sarcinricin loop of 23S rRNA. Nat Commun 4:2796. https://doi.org/10.1038/ncomms3796

Winther K, Tree JJ, Tollervey D, Gerdes K (2016) VapCs of mycobacterium tuberculosis cleave RNAs essential for translation. Nucleic Acids Res 44(20):9860–9871. https://doi.org/10.1093/nar/gkw781. Epub 2016 Sep 5. PMID:27599842; PMCID: PMC5175351

World Health Organization (2019) Lateral flow urine lipoarabinomannan assay (LF-LAM) for the diagnosis of active tuberculosis in people living with HIV: 2019 update. WHO, Geneva. Available from: https://www.who.int/publications-detail-redirect/9789241550604

World Health Organization (WHO) (2022) WHO announces updates on new TB antigen-based skin tests for the diagnosis of TB infection. WHO Newsroom. Available from: https://www.who.int/news/item/04--04-2022-who-announces-updates-on-new-tb-antigen--based-skin-tests-for-the-diagnosis-of-tb-infection

Xiong Z, Yang X, Wang S, Smart CJ, Sisson HM, Lin Z et al (2025) Structure and assembly of the MmpL5/

MmpS5 efflux transporter from *Mycobacterium tuberculosis*. Nat Commun 29;16(1):4976

Yang Z, Zeng X, Tsui SK (2019) Investigating function roles of hypothetical proteins encoded by the Mycobacterium tuberculosis H37Rv genome. BMC Genomics 20(1):394

Yimer SA, Kalayou S, Homberset H, Birhanu AG, Riaz T, Zegeye ED et al (2020) Lineage-specific proteomic signatures in the mycobacterium tuberculosis complex reveal differential abundance of proteins involved in virulence, DNA repair, CRISPR-Cas, bioenergetics and lipid metabolism. Front Microbiol 11:550760. https://doi.org/10.3389/fmicb.2020.550760. PMID: 33072011; PMCID: PMC7536270

Yu X, Gao X, Zhu K, Yin H, Mao X, Wojdyla JA, Qin B, Huang H, Wang M, Sun YC, Cui S (2020) Characterization of a toxin-antitoxin system in Mycobacterium tuberculosis suggests neutralization by phosphorylation as the antitoxicity mechanism. Commun Biol 3(1):216

Yuda E, Tanaka N, Fujishiro T, Yokoyama N, Hirabayashi K, Fukuyama K, Wada K, Takahashi Y (2017) Mapping the key residues of SufB and SufD essential for biosynthesis of iron-sulfur clusters. Sci Rep 7:9387

Zarin S, Shariq M, Rastogi N, Ahuja Y, Manjunath P, Alam A, Hasnain SE, Ehtesham NZ (2024) Rv2231c, a unique histidinol phosphate aminotransferase from Mycobacterium tuberculosis, supports virulence by inhibiting host-directed defense. Cell Mol Life Sci 81(1):203

Zhang XL, Qu H (2021) The role of glycosylation in infectious diseases. The role of glycosylation in health and disease. 9:219–237

Zhang T, Faraggi E, Li Z, Zhou Y (2013) Intrinsically semi-disordered state and its role in induced folding and protein aggregation. Cell Biochem Biophys 67(3):1193–1205. https://doi.org/10.1007/s12013-013-9638-0. PMID: 23723000; PMCID: PMC3838602

Zhang Z, Dong L, Li X, Deng T, Wang Q (2025) The PE/PPE family proteins of Mycobacterium tuberculosis: evolution, function, and prospects for tuberculosis control. Front Immunol 16:1606311

Future Research Directions on *Mycobacterium tuberculosis* Proteins

11

Abstract

Future research on *Mycobacterium tuberculosis* (Mtb) proteins is essential for advancing tuberculosis (TB) diagnosis, treatment, and control amid rising drug resistance and global health challenges. Recent studies focus on elucidating protein structures and interactions critical to bacterial survival, virulence, and drug resistance, enabling the identification of novel therapeutic targets. Cutting-edge technologies such as single-cell proteomics, cryo-electron microscopy, and spatial proteomics provide unprecedented resolution of protein localization, dynamics, and host-pathogen interactions. Posttranslational modifications and dynamic proteomic profiling reveal bacterial adaptive mechanisms, while integrative multi-omics combined with artificial intelligence (AI) and machine learning accelerate functional annotation and predictive modeling. Clustered Regularly Interspaced Short Palindromic Repeats (CRISPR)-based functional genomics complements these approaches by enabling precise gene regulation studies. Advances in AI-powered diagnostics and genomic surveillance are transforming TB detection and drug resistance profiling, improving patient outcomes, especially in resource-limited settings. Overall, integrating experimental and computational innovations promises to deepen understanding of Mtb biology, accelerate drug discovery, enhance diagnostics, and ultimately contribute to global efforts in combating tuberculosis.

Keywords

Mycobacterium tuberculosis · Single-cell proteomics · Spatial proteomics · Posttranslational modification (PTM) · Clustered Regularly Interspaced Short Palindromic Repeats (CRISPR) · Artificial intelligence (AI)

11.1 Introduction

Future research on *Mycobacterium tuberculosis* (Mtb) proteins is critically important, given the pathogen's complex biology, its exceptional capacity for dormancy, and the increasing challenge posed by antibiotic resistance (Kelly et al. 2020; Kelly and McLoughlin 2020). The ongoing global challenge of tuberculosis (TB) control is significantly impeded by an incomplete understanding of the molecular mechanisms that govern Mtb's survival, virulence, and drug resistance pathways (Slavov 2023). Proteins drive nearly all cellular processes in Mtb, from metabolism and replication to immune evasion and antimicrobial resistance, highlighting the urgent need for a comprehensive understanding of the Mtb proteome. Such knowledge is essential for identify-

P. Farnia et al., *Proteins in Mycobacterium Tuberculosis*, Advances in Experimental Medicine and Biology 1484, https://doi.org/10.1007/978-3-031-96883-9_11

ing novel therapeutic targets, developing more sensitive diagnostic tools, and uncovering the pathogen's adaptive strategies within the host environment (Kelly and McLoughlin 2020). The field is currently advancing rapidly, propelled by synergistic innovations in proteomics, structural biology, and computational analysis. These approaches enable unprecedented resolution in characterizing protein functions, molecular interactions, subcellular localization, and dynamic behaviors within the bacterial cell (Slavov 2023; Dubochet et al. 1988). A particularly promising area is single-cell proteomics, which surmounts limitations of bulk proteomic analyses by profiling protein expression at the level of individual bacterial cells. This method reveals heterogeneity within Mtb populations that is critical for understanding phenotypic diversity, especially under stress or antibiotic exposure (Slavov 2023). For example, the Single-Cell Proteomics by Mass Spectrometry (SCoPE-MS) technique employs tandem mass tags to multiplex protein detection in single cells, while the Nanodroplet Processing in One pot for Trace Samples (nanoPOTS) platform reduces sample loss and enables deep proteomic profiling from minimal biological material. These technologies shed light on the molecular basis of persistence phenotypes and treatment failures by dissecting responses at the subpopulation level (Kelly and McLoughlin 2020; Slavov 2023). In structural biology, cryo-electron microscopy (cryo-EM) has revolutionized the determination of protein structures at near-atomic resolution, a breakthrough honored by the 2017 Nobel Prize in Chemistry (Dubochet et al. 1988). Cryo-EM facilitates detailed visualization of key Mtb macromolecular assemblies, such as the RNA polymerase (RNAP) complex, providing essential structural blueprints for rational drug design (Bendre et al. 2021). More recently, cryo-electron tomography (cryo-ET) has risen in prominence, enabling in situ examination of native cellular architectures that preserve physiological context and reveal the spatial organization of protein complexes within intact Mtb cells (Dubochet et al. 1988). Spatial proteomics is another cutting-edge frontier focused on precisely mapping the intracellular localization of Mtb proteins within bacterial and host cells, thereby inferring their functions (Roux et al. 2012; Branon et al. 2018). Proximity-labeling techniques such as biotin identification (BioID), Turbo biotin identification (TurboID), and engineered ascorbate peroxidase 2 (APEX2) enable mapping of protein interaction networks in living cells, with TurboID particularly suited for rapid labeling to capture transient interactions (Branon et al. 2018; Lam et al. 2015). When combined with correlative light and electron microscopy (CLEM), these methods offer high-resolution contextualization of bacterial proteins during infection and reveal their interplay with host cellular components (Lam et al. 2015). Posttranslational modifications (PTMs), including phosphorylation, acetylation, and pupylation, play critical roles in regulating Mtb protein function and reflect adaptive responses to environmental cues (Pearce et al. 2008). Since the early 2000s, phosphoproteomics has mapped phosphorylation patterns that regulate essential signaling cascades (Olsen et al. 2006). In the mid-2010s, redox proteomics emerged to identify oxidative modifications of cysteine residues, integral to Mtb's defense against oxidative stress (Leichert et al. 2008). Pupylation, a prokaryotic ubiquitin-like modification first described by Pearce et al. (2008), has expanded understanding of bacterial protein turnover, quality control, and proteostasis. Temporal proteomic techniques such as pulsed Stable Isotope Labeling by Amino acids in Cell culture (pulsed SILAC) provide dynamic insights into protein synthesis and degradation rates, key parameters for unraveling Mtb's stress adaptations (Schwanhäusser et al. 2011). Complementary tools like auxin-inducible degron systems enable conditional, temporal control of protein abundance, facilitating precise functional analyses (Nishimura et al. 2009). The adoption of integrative multi-omics approaches, combining proteomic, genomic, transcriptomic, and metabolomic data, has become vital for dissecting complex regulatory networks. Utilizing network visualization platforms such as Cytoscape (Shannon et al. 2003) alongside artificial intelligence (AI)-driven analytic frameworks like DeepOmics (Zhou et al. 2024) empowers researchers to decode Mtb pathogenesis and predict resistance evolution, with promising applications in precision medicine. Proteogenomics, which links proteomic data with genomic annotations, has proven powerful in identifying previously unannotated coding regions, small open reading frames, and alternative isoforms. This approach also enables correction of genomic misannotations and the discovery of novel proteins that may serve as drug targets or vaccine antigens (Nesvizhskii 2014). The advent of Clustered Regularly Interspaced Short Palindromic Repeats (CRISPR)-based functional genomics has transformed *Mycobacterium tuberculosis* genetics. CRISPR interference (CRISPRi), specifically adapted for bacterial systems, enables precise gene silencing to assess gene essentiality and related phenotypic effects. When combined with proteomic profiling, CRISPRi enhances understanding of gene-protein interactions and regulatory networks. Furthermore, advanced base editing technologies (Komor et al. 2016) allow targeted nucleotide modifications, thereby facilitating in-depth studies of antibiotic resistance mechanisms. Artificial intelligence and machine learning (ML) have dramatically improved functional annotation of Mtb proteins, especially those lacking experimental characterization. State-of-the-art models such as AlphaFold (DeepMind, 2020) and ESMFold (Meta AI, 2022) generate highly accurate protein structure predictions from sequence data, while models like ProtT5 and ESM-2 infer protein functions using large-scale protein databases (Jumper et al. 2021; Callaway 2022; Lin et al. 2023). These AI-driven tools accelerate hypothesis generation and prioritization for experimental validation. Finally, host-pathogen interaction proteomics, employing dual proteomics and proximity labeling, illuminates molecular crosstalk during infection, identifying bacterial effectors and host factors that influence susceptibility and immune response (Roux et al. 2012; Branon et al. 2018). These insights are foundational for the development of host-directed therapies designed to enhance immune control of TB. Thereby, the future of Mtb protein research lies in integrating advanced experimental methodologies with computational innovations. This multidimensional approach promises to deepen understanding of Mtb biology and pathogenesis, accelerate the discovery of novel therapeutics and diagnostics, and ultimately contribute to reducing the global burden of tuberculosis (Bendre et al. 2021).

11.2 Application of Single-Cell Methodologies to Study *Mycobacterium tuberculosis* Heterogeneity

Single-cell methodologies have emerged as invaluable tools in microbiological research, overcoming limitations of conventional population-level analyses by revealing cellular heterogeneity obscured in bulk measurements. Unlike traditional approaches that average signals from millions of bacterial cells, single-cell studies provide detailed insights into genetic, biochemical, and phenotypic diversity within microbial populations (Dahlberg et al. 1998; Benoit 2002; Brehm-Stecher and Johnson 2004; Potter 2018; Larsson et al. 2021; Homberger et al. 2025; Lyu et al. 2025). This is particularly relevant for complex pathogens such as Mtb, where subpopulation dynamics critically influence pathogenesis, antibiotic tolerance, and disease progression. A diverse array of cutting-edge tools facilitates single-cell interrogation. Fluorescence-based techniques, including green fluorescent protein (GFP) tagging and fluorescence in situ hybridization (FISH), enable precise visualization of gene expression and protein localization within individual bacterial cells (Webb et al. 1995; Kain 1999; Yang et al. 2016; MacGilvary and Tan 2018; Li et al. 2017; Barbosa et al. 2023; Kairišs et al. 2024). Complementing these, atomic force microscopy (AFM) provides nanoscale imaging of cellular mechanics (elasticity, adhesion) and enzymatic activity, exposing functional heterogeneity inaccessible by bulk biochemical assays (Müller and Dufrêne 2011; Dufrêne 2014; Zhavnerko et al. 2017; Matias et al. 2025). Spectroscopic methods such as Raman micro-spectroscopy and fluorescence microspectroscopy offer detailed metabolic and elemental mapping on a single-cell basis, elucidating metabolic states and heterogeneity within populations (Buijtels et al. 2008; Stöckel et al. 2015; Verma et al. 2019; Lee et al. 2021). Microfluidic platforms now enable unparalleled isolation, manipulation, and longitudinal observation of individual bacterial cells, allowing dynamic monitoring of cellular responses to environmental stress and antimicrobials with exceptional temporal precision (Baron et al. 2020; Molloy et al. 2021; Mistretta et al. 2022, 2024). Significantly, recent advances in single-cell proteomics have transformed our ability to quantify protein expression and posttranslational modifications in individual bacteria. Two key techniques exemplify this progress: Single-Cell Proteomics by Mass Spectrometry (SCoPE-MS) and Nanodroplet Processing in One Pot for Trace Samples (nanoPOTS). SCoPE-MS uses tandem mass tags to multiplex peptides from multiple single cells, pooling them for simultaneous mass spectrometric analysis. This multiplexing substantially increases throughput and sensitivity, allowing quantification of thousands of proteins per bacterium and revealing proteomic heterogeneity important for understanding variable drug susceptibility (Budnik et al. 2018). Concurrently, nanoPOTS minimizes sample loss by performing all processing within nanoliter volumes on microfabricated chips, maximizing protein recovery and digestion efficiency from trace clinical or environmental samples such as sputum and tissue biopsies (Kelly et al. 2018). Together, these innovations enable direct measurement of protein-level variations underlying bacterial physiology and adaptive behavior, complementing single-cell genomic and transcriptomic data. Applying single-cell technologies to Mtb is particularly transformative, as this pathogen exhibits notable phenotypic diversity even among genetically identical bacilli due to stochastic gene expression, environmental modulation, and cell cycle stages (Pisu et al. 2021; Chung et al. 2024). Conventional population averages mask such variability, which has profound implications for drug tolerance, persistence, and immune evasion. Critical to these analyses is the isolation of single Mtb cells from complex biological samples. Fluorescence-activated cell sorting (FACS) enables high-throughput selection based on fluorescent labels, enriching subpopulations by size, viability, or marker expression (Ambriz-Aviña et al. 2014; Rothchild et al. 2020; Barr et al. 2021). Microfluidic sorting offers gentle, precise manipulation in defined fluidic environments. Other complementary isolation methods include laser capture microdissection, manual micromanipulation, and dielectrophoretic sorting, especially useful for rare or specific Mtb cell populations (Moore et al. 2023; Cañadas-Ortega et al. 2023). On the genomic front, whole-genome amplification techniques now permit sequencing of individual bacterial genomes, uncovering mutations and structural variants relevant to pathogenesis and drug resistance (McNerney et al. 2018; Gessese et al. 2025). Advanced single-cell transcriptomic workflows, such as single-cell RNA sequencing (scRNA-seq) and combinatorial barcoding methods like SPLiT-seq, provide comprehensive gene expression profiles for individual bacilli, illuminating transcriptional heterogeneity and subpopulation dynamics within infected tissues (Pisu et al. 2021; Pan et al. 2023; Ma et al. 2025). Despite challenges due to limited bacterial protein content, advances in single-cell proteomics, including sensitive mass spectrometry and antibody-based assays, have enabled detailed proteome landscapes. Platforms like nanoPOTS have enhanced proteomic coverage and facilitated the identification of rare bacterial phenotypes such as persister cells, crucial to understanding therapeutic failure (Zhu et al. 2024). Furthermore, single-cell metabolomics reveal intracellular metabolite distributions, shedding light on metabolic adaptations and states influencing survival under antimicrobial or immune pressures (Mashabela et al. 2019; Zhang et al. 2025).

Integrated single-cell multi-omics approaches empower researchers to track Mtb molecular states within the host, revealing bacterial subpopulations driving persistence, antibiotic tolerance, and intracellular survival strategies. Large-scale proteomic profiling has expanded biomarker discovery efforts. For instance, Schiff et al. (2024) used an advanced non-depletion tandem mass spectrometry (MS/MS) approach on plasma from tuberculosis (TB) patients to identify a six-protein signature, fetuin-B (FETUB), Fc gamma receptor IIIb (FCGR3B), leucine-rich alpha-2-glycoprotein 1 (LRG1), selectin-L (SELL), cluster of differentiation 14 (CD14), and adenosine deaminase 2 (ADA2), with high diagnostic accuracy. Similarly, integrated proteomic and network biology analyses defined a three-protein urinary signature capable of differentiating active TB from other pulmonary diseases (Isa et al. 2018). Beyond molecular profiling, integrating single-cell proteomics with spatially resolved techniques, such as proximity labeling tools, Turbo Biotin Identification (TurboID), and Engineered Ascorbate Peroxidase 2 (APEX2), has enabled mapping of protein interactions and subcellular localizations within infected host cells, enhancing understanding of host-pathogen interplay (Branon et al. 2018; Lam et al. 2015; Jaisinghani et al. 2024). Coupled with single-cell transcriptomics and high-resolution imaging, these approaches have revolutionized insight into Mtb biology within the host microenvironment. Notably, Pisu et al. (2021) simultaneously profiled Mtb bacilli and host macrophages in infected lungs via single-cell RNA sequencing combined with imaging, uncovering bacterial heterogeneity alongside diverse immune cell phenotypes. Their study revealed distinct macrophage subsets interacting differentially with Mtb and demonstrated how bacterial effectors, including the ESX-1 secretion system, create an anti-inflammatory niche promoting bacterial survival. Single-cell imaging paired with mathematical modeling (Chung et al. 2024) showed that Mtb growth follows asymmetric and linear dynamics, driving phenotypic diversity with direct implications for antibiotic susceptibility. Other single-cell studies dissected macrophage heterogeneity during infection, exposing differences in bacterial control capacity among alveolar and interstitial macrophages (Pisu et al. 2021). Additionally, single-cell profiling of Mtb-specific CD4+ T cells revealed subsets with distinct cytokine profiles, informing vaccine design efforts aimed at modulating protective immunity (Sakai et al. 2025). Studies further identified that only a minority of resting macrophages produce tumor necrosis factor (TNF) protein during infection, underscoring innate immune response heterogeneity and potential Mtb evasion tactics (Yuk et al. 2024). Functional perturbations in natural killer (NK) cells during active TB were also characterized at the single-cell level, highlighting their critical role in immune surveillance and suggesting NK cell phenotypes as promising biomarkers for disease status and treatment efficacy (Garand et al. 2018; Lyu et al. 2024). Collectively, these advances, from foundational single-cell isolation and multi-omic profiling to integrative spatial proteomics and immune phenotyping, are transforming tuberculosis research. They facilitate precise biomarker discovery, deepen molecular understanding of host-pathogen interactions, and lay the groundwork for novel diagnostics and targeted therapies. Moving forward, rigorous clinical validation and translation of these findings into accessible point-of-care applications remain essential to mitigate the global tuberculosis burden. Ultimately, single-cell investigations reaffirm that Mtb's persistence and pathogenicity depend on complex interactions among bacterial subpopulations and the heterogeneous host immune landscape (Mittal et al. 2023; Chung et al. 2024; Feng et al. 2025).

11.3 Advances and Prospects in Structural Biology for *Mycobacterium tuberculosis* Research

Recent years have seen transformative progress in structural biology, largely driven by advances in cryo-electron microscopy (cryo-EM). Cryo-EM is a sophisticated imaging technique that uses transmission electron microscopy to visualize biological specimens rapidly frozen at cryogenic temperatures. This vitrification process preserves biomolecules in their native conformations by immobilizing them in glass-like ice, protecting them from radiation damage while maintaining physiological structures. The technique involves capturing thousands of two-dimensional (2D) projection images of macromolecules from multiple orientations. Advanced computational algorithms then align and integrate these images to reconstruct detailed three-dimensional (3D) molecular architectures with near-atomic resolution (Yin 2018). The mathematical principle underpinning this 3D reconstruction is the central section theorem, formulated by De Rosier and Klug in 1968. This theorem states that the Fourier transform of a projection corresponds to a slice through the origin of the Fourier transform of the full 3D object, oriented perpendicular to the projection direction (Rossmann and Tao 1999). Utilizing this principle, cryo-EM translates large datasets of 2D projections into precise 3D models of complex biomolecules, effectively overcoming challenges associated with studying large, flexible, or non-crystallizable assemblies (Nogales and Scheres 2015). The significance of cryo-EM was recognized by the 2017 Nobel Prize in Chemistry, awarded to Jacques Dubochet, Joachim Frank, and Richard Henderson for their pioneering work enabling near-atomic resolution imaging of biomolecules without the need for crystallization. This breakthrough revolutionized structural biology by allowing visualization of biomolecules in states that closely emulate their natural environments, shifting away from traditional X-ray crystallography approaches. Cryo-EM has since delivered landmark structures of complex biomolecules, including ribosomes, RNA polymerases, and membrane protein complexes, providing profound insights into enzymatic mechanisms, protein interactions, and conformational dynamics essential for cellular function (Murata and Wolf 2018; Quentin and Raunser 2018; Benjin and Ling 2020; Kühlbrandt 2022; de la Cruz and Eng 2023). In mycobacterial research, particularly involving Mtb and *Mycobacterium smegmatis*, cryo-EM has dramatically advanced knowledge of integral membrane proteins and complexes. Numerous high-resolution structures now available in public repositories such as the Protein Data Bank have deepened our understanding of bacterial persistence and antibiotic action (Bendre et al. 2021). For instance, the Mtb 70S ribosome structure bound to the antibiotic capreomycin revealed key molecular interactions that inhibit protein synthesis, providing a foundation for understanding mechanisms of antibiotic resistance (Yang et al. 2022). Similarly, multiple functional states of the *Mycobacterium smegmatis* 70S ribosome have been characterized, elucidating mechanisms of translation regulation and bacterial dormancy, critical in pathogenesis and persistence (Mishra et al. 2018). Further structural studies resolved ribosome complexes with regulatory proteins such as MpY, highlighting potential targets to disrupt bacterial survival under stress. The structural biology of bacterial RNA polymerase (RNAP), central to transcription, has also greatly benefited from cryo-EM. High-resolution structures of Mtb RNAP in open-promoter and DNA-unwinding states, bound to transcription factors RbpA and CarD, have illuminated transcription initiation and revealed novel antibiotic target sites (Boyaci et al. 2020). Structures of RNAP in complex with the antibiotic fidaxomicin have provided critical insights for drug development (Boyaci et al. 2020; Lin et al. 2018). Key metabolic enzymes have been examined by cryo-EM as well. The respiratory supercomplex from *Mycobacterium smegmatis* has been structurally characterized, creating a framework for designing inhibitors targeting bacterial respiration (Wiseman et al. 2018). ATP synthase, essential for energy production, has been imaged in multiple conformations, with and without the antituberculosis drug bedaquiline, clarifying the drug's mechanism of action and informing therapeutic optimization (Guo et al. 2021). Cell wall biosynthesis enzymes, vital drug targets in mycobacteria, have also been structurally defined. For example, arabinosyltransferase complexes EmbA-EmbB-AcpM2 bound to ethambutol or arabinose substrates have revealed insights into substrate recognition and drug resistance mechanisms (Zhang et al. 2020). Further studies on arabinofuranosyltransferase EmbB2-AcpM2 in resting and active states detailed essential conformational changes needed for enzymatic function (Zhang et al. 2020), while investigations into arabinofuranosyltransferase AftD enhanced understanding of cell wall assembly (Alderwick et al. 2018). High-resolution cryo-EM structures of secretion systems integral to mycobacterial virulence have elucidated functional mechanisms. For example, structures of the ESX-3 secretion system and its EccB3 dimer revealed critical details of secretion pathways and pathogenesis, identifying promising drug targets (Famelis et al. 2019). Additionally, proteasome complexes involved in protein quality control, including bacterial activator Bpa and PafE-capped proteasomes, have been structurally characterized to understand survival strategies under stress conditions (Bolten et al. 2016; von Rosen et al. 2023). Virulence factors such as EspB, which forms oligomeric ring structures, have been examined to clarify how mycobacteria interact with host defenses (Korotkova et al. 2015; Solomonson et al. 2015). Investigations into fatty acid synthase complexes, key for lipid metabolism and cell wall biosynthesis, provide templates vital for rational drug design (Elad et al. 2018; Ciccarelli et al. 2013). Further structural studies have focused on bacterial stress response and DNA repair mechanisms. Detailed cryo-EM characterization of the ClpB chaperone and AdnAB helicase-nuclease complexes offers insights into protein homeostasis and genome maintenance (Yu et al. 2018; Alam et al. 2021). Transport proteins like the ABC transporter Rv1819c have been structurally resolved, shedding light on substrate translocation and drug resistance mechanisms (Rempel et al. 2020). A particularly noteworthy study by Munir et al. (2021) resolved the structure of KatG, a catalase-peroxidase from Mtb responsible for activating the frontline anti-TB drug isoniazid. Comparison of wild-type and clinically resistant KatG variants revealed how mutations such as S315T introduce steric hindrance in substrate access channels, reducing drug activation and driving resistance. These structural results were corroborated by biochemical assays showing decreased enzymatic activity caused by the mutations. Bendre et al. (2021) comprehensively reviewed the impact of cryo-EM on understanding the conformational diversity of mycobacterial membrane proteins, including respiratory complexes, secretion systems, and cell wall enzymes, underscoring how high-resolution structures deepen insight into bacterial physiology, virulence, and resistance, and guide rational therapeutic design. Thereby, cryo-electron microscopy has ushered in a new era of molecular exploration for Mtb research. By delivering near-atomic resolution structures of large, flexible protein assemblies once inaccessible by other methods, cryo-EM fundamentally enhances our understanding of Mtb biology and pathogenesis. These structural insights drive the targeted development of improved drugs and vaccines, accelerating efforts to control and ultimately eliminate tuberculosis. As cryo-EM technologies continue to evolve, they will remain indispensable tools for unraveling mycobacterial molecular machinery, fostering innovation in antituberculosis interventions, and contributing significantly to global health.

11.4 Spatial Proteomics and Interaction Mapping in the Future Research of *Mycobacterium tuberculosis*

Spatial proteomics, an advanced subfield of proteomics, focuses on precisely mapping protein localization and distribution within cells and tissues while preserving their native spatial context (Horvath and Coscia 2025). Unlike conventional proteomics, which measures protein abundance in homogenized bulk samples, spatial proteomics maintains the intricate intracellular and tissue architecture. This preservation enables a deeper understanding of how subcellular protein positioning influences cellular processes, signaling pathways, and molecular interactions, an aspect especially critical for bacterial pathogens like Mtb, where spatial organization regulates metabolic activity, virulence, and host-pathogen interactions (Mohammadnabi et al. 2024; Magoulopoulou et al. 2022; Nargan et al. 2024). The growing importance of spatial proteomics is highlighted by its applications in elucidating disease mechanisms, mapping molecular pathways, and identifying biomarkers. Technological advances have propelled global efforts to build high-resolution spatial atlases of organs and tissues, explore complex microenvironments, reveal cellular crosstalk, and discover diagnostic and therapeutic targets (Gengenbacher et al. 2014; Flores-Villalva et al. 2015; Huang et al. 2020; Wu et al. 2024; Qiu et al. 2024). Despite ongoing challenges in sample preparation, data acquisition, and bioinformatic analysis, spatial proteomics has established itself as a foundational tool in biomedical and infectious disease research. Two complementary techniques have become central to spatial proteomics and protein interaction mapping in bacterial studies: Turbo Biotin Identification (TurboID) proximity labeling and correlative light and electron microscopy (CLEM) (Solomon 2023; Fay et al. 2025; Jiang and Dickson 2025). TurboID is an engineered biotin ligase that rapidly biotinylates proteins in close proximity, capturing transient and weak interactions critical for bacterial survival and virulence (Branon et al. 2018; Cho et al. 2020). CLEM combines the molecular specificity of fluorescence microscopy with the ultrastructural resolution of electron microscopy, enabling precise protein localization within bacterial compartments and at pathogen-host interfaces (Russell et al. 2017; Finin et al. 2023). Together, these techniques provide a powerful toolkit for in situ protein network mapping and elucidate spatial relationships essential for understanding bacterial physiology and pathogenesis.

In tuberculosis research, proteomics has become indispensable for biomarker discovery aimed at improving early diagnosis, monitoring infection progression, and evaluating treatment. Protein biomarkers are especially valuable as they actively participate in biological processes and can be detected in accessible fluids such as blood, urine, and sputum (Guo et al. 2022; Leo et al. 2024). Early in vivo proteomic studies characterized the Mtb proteome during infection, revealing host-induced changes in protein expression critical for pathogenesis and identifying antigenic targets with diagnostic and vaccine potential (Kruh et al. 2010). Building on this foundation, Deng et al. (2014) developed a high-throughput proteome microarray that greatly expedited antigen discovery. Label-free quantitative proteomics further unveiled proteins enabling Mtb's adaptation to host-imposed stresses (Borah et al. 2021). Comparative proteomics under host granuloma-like conditions, particularly contrasting aerobic and hypoxic growth, has identified hypoxia-responsive proteins falling into two categories: "over-expressed" proteins with increased levels under low oxygen, and "newly expressed" proteins induced exclusively during hypoxia. These proteins mediate virulence, cell wall remodeling, and metabolic adaptation, supporting persistence in hostile granulomatous niches (Devasundaram et al. 2016).

Hoffmann et al. (2018) utilized high-resolution mass spectrometry to simultaneously profile both host and *Mycobacterium tuberculosis* proteomes during infection. Their analysis revealed dynamic modulation of bacterial secretion systems and immunomodulatory proteins, including members of the Esx and proline-proline-glutamic acid (PPE) families, highlighting Mtb's ability to adapt its phenotype to evade immune detection within diverse host cell types such as macrophages and dendritic cells. Comprehensive profiling under diverse stress conditions revealed the induction of stress-responsive proteins and pathways critical for Mtb survival and persistence, uncovering promising new therapeutic targets (Birhanu et al. 2022). Additionally, Choudhary et al. (2022) utilized liquid chromatography-tandem mass spectrometry (LC-MS/MS) to catalog thousands of Mtb proteins across physiologically relevant stressors, defining a core proteome essential for bacterial viability under varying environmental challenges.

Proximity labeling of proteins within the cytosol and cell wall of live bacteria was achieved using the engineered peroxidase APEX2, effectively overcoming the permeability challenges posed by the complex mycobacterial envelope. This approach was further extended to infected lung tissue, revealing novel macrophage polarization states during Mtb infection and delineating the spatial and functional heterogeneity of host immune responses in situ. Recent "surface shaving" proteomics methods have defined the mycobacterial surfaceome, identifying key surface-exposed proteins involved in host-pathogen interactions with potential as vaccine and drug targets. Furthermore, the functions of Type VII ESX secretion systems, particularly ESX-1 and ESX-3, were clarified, demonstrating that substrates such as EsxB (CFP-10) and EsxG transiently localize in the periplasmic space, thereby elucidating how virulence factors navigate the complex multilayered mycobacterial envelope to reach extracellular or host-directed destinations (Jaisinghani et al. 2024; Seto et al. 2022; Lepe et al. 2025). Therefore, spatial proteomics is rapidly emerging as a transformative discipline in Mtb research. By offering unprecedented resolution of protein localization, interaction mapping, and dynamic profiling within bacterial cells and infected tissues, it significantly enhances our molecular understanding of tuberculosis pathogenesis (Wu et al. 2024; Horvath and Coscia 2025). Looking forward, the integration of spatial proteomics with advanced multi-omic platforms and high-resolution imaging is poised to accelerate the discovery of novel drug targets, vaccine candidates, and diagnostic biomarkers. These evolving technologies hold the potential to enable real-time visualization of protein interactions during infection, facilitate precision medicine tailored to individual patients, and reveal previously inaccessible aspects of tuberculosis biology and treatment.

11.5 Post-Translational Modifications (PTMs) in the Future Research of *Mycobacterium tuberculosis*

Posttranslational modification (PTM) of proteins by ubiquitin-like modifiers has recently gained considerable attention due to its essential role in regulating a wide array of biological processes. While ubiquitin conjugation was initially identified as a mechanism targeting proteins for proteasomal degradation, it is now recognized as a versatile regulator of numerous cellular functions and physiological activities (Hershko and Ciechanover 1986, 1998; Crosas et al. 2006; Grabbe and Dikic 2009). PTMs profoundly influence protein function, stability, localization, and interaction networks, enabling bacteria to rapidly adapt to fluctuating environmental conditions and stressors. Common bacterial PTMs include phosphorylation, acetylation, methylation, glycosylation, lipidation, and proteolytic cleavage. Each modification modulates enzymatic activity, signaling pathways, or protein turnover, thereby supporting bacterial physiology, virulence, and survival. Among bacterial PTMs, pupylation and redox modifications are unique prokaryotic systems with distinct biological roles.

Mtb encodes a small ubiquitin-like protein called Pup, encoded by the gene Rv2111c, representing the first known bacterial ubiquitin-like system (Pearce et al. 2008; Burns et al. 2009). Pupylation genes are largely restricted to Actinobacteria and often co-localize with proteasome components (Iyer et al. 2008). Pup exists in two forms based on its C-terminal residue: Pup^Gln and Pup^Glu. Mycobacteria synthesize Pup^Gln, which is deamidated to Pup^Glu by the deamidase Dop before conjugation to substrate lysines (Striebel et al. 2009; Barandun et al. 2012). The proteasome accessory factor A (PafA) attaches Pup^Glu to target proteins via an isopeptide bond, marking them for proteasomal degradation (Burns et al. 2009; Pearce et al. 2008; Guth et al. 2011). This process is reversible as Dop also acts as a depupylase, hydrolyzing the Pup-substrate bond to recycle Pup (Burns et al. 2010; Imkamp et al. 2010). Although depupylation is not essential for proteasomal degradation in vitro (Striebel et al. 2010), it facilitates Pup recycling and may enhance degradation efficiency of specific substrates in vivo (Burns et al. 2010; Cerda-Maira et al. 2010). Key Pup substrates characterized in Mtb include malonyl CoA-acyl carrier protein transacylase (FabD), important for cell wall lipid biosynthesis (Kremer et al. 2002); myo-inositol-1-phosphate synthase (MIPS), catalyzing myo-inositol biosynthesis (Stieglitz et al. 2005); and superoxide dismutase (SOD), vital for detoxifying reactive oxygen species (ROS) (Kusunose et al. 1976). Proteomic analyses have identified hundreds of pupylation targets spanning diverse functions such as metabolism, respiration, lipid metabolism, virulence, and detoxification (Festa et al. 2010; Poulsen et al. 2010). Pupylation is a dynamic modification varying with growth conditions, signifying an adaptable regulatory system. Significantly, several pupylated proteins linked to pathogenesis—including Isocitrate Lyase (Icl), Inositol-1-Phosphate Synthase (Ino1), Response Regulator MtrA, and Phosphate Response Regulator PhoP—demonstrate regulation beyond degradation, underscoring pupylation's complex role (Festa et al. 2010). Reconstitution of the Mtb pupylation system in *Escherichia coli*, which lacks Pup and related enzymes, showed that Pup and PafA alone are sufficient for substrate pupylation. This indicates substrate recognition is inherent to the target protein, independent of species-specific cofactors (Cerda-Maira et al. 2011). Functionally, the Pup-proteasome system (PPS) is essential for regulating nitrate metabolism and overall protein quality control in Mtb. PPS mutants show impaired nitrate utilization, accumulate misfolded proteins, and suffer compromised enzymatic activities, especially under stress requiring stringent proteostasis (Becker et al. 2019; von Rosen et al. 2021). Recently, a pseudopantothenate kinase named CoaX was discovered as a novel regulator of depupylation. CoaX modulates the stability of PanB, a key enzyme in pantothenate (vitamin B5) biosynthesis, by enhancing its depupylation; this activity is feedback-inhibited by pantothenate, suggesting metabolic control over protein turnover (Kahne et al. 2024). Findings highlight the therapeutic potential of targeting pupylation components, which are absent in humans, as a strategy to disrupt Mtb survival. Future research should focus on proteome-wide substrate mapping, mechanistic insights into regulation, and combinatorial therapies pairing pupylation inhibitors with existing antimicrobials to combat drug resistance.

Redox proteomics, an important branch of posttranslational modification (PTM) research, focuses on characterizing oxidative and nitrosative modifications of cysteine residues caused by reactive oxygen species (ROS) and reactive nitrogen species (RNS) (Petushkova and Zamyatnin 2020). Endogenous ROS and RNS include molecules such as superoxide anion ($O_2^{-}\cdot$), hydrogen peroxide (H_2O_2), hydroxyl radicals ($\cdot OH$), nitric oxide (NO), and peroxynitrite ($ONOO^-$). While these species play essential roles in physiological signaling, their dysregulation can lead to cellular damage (Hemnani and Parihar 1998; Cumming et al. 2014). Cells maintain tight control over ROS and RNS levels through sophisticated antioxidant systems to prevent molecular injury; however, oxidative stress can still arise during inflammatory responses, mitochondrial respiration, and nicotinamide adenine dinucleotide phosphate (NADPH) oxidase activity (Cumming et al. 2014). Excess ROS/RNS preferentially target cysteine thiol groups because of their unique chemical reactivity and key catalytic roles in proteins (Ulrich and Jakob 2019; Khan et al. 2021). Under physiological pH, most cysteine thiols are protonated with relatively high pKa values (above 8.0), rendering them less reactive. However, their reactivity is enhanced when local protein environments stabilize the thiolate anion form, thus increasing susceptibility to oxidation (Giles et al. 2001; Poole 2017). Reversible cysteine modifications, including sulfenic acid formation, S-nitrosylation, S-glutathionylation, and disulfide bond formation, act as molecular switches that regulate protein activity and facilitate redox-based signaling pathways (Paulsen and Carroll 2013). Irreversible oxidation to sulfonic acid, occurring under excessive oxidative stress, often leads to loss of protein function and degradation. Standard redox proteomics workflows begin by blocking free thiols to prevent artificial oxidation, followed by selective reduction of oxidized cysteines and labeling with mass tags or fluorescent probes. These samples are then analyzed using liquid chromatography coupled with tandem mass spectrometry (LC-MS/MS) to identify and quantify modifications (Lu et al. 2023). Advanced isotope labeling approaches, such as Selective Proteomic Experimental Analysis of Redox (SPEAR) and Oxidation-dependent Isotope-coded Affinity Tagging (OxiCAT), allow simultaneous and precise quantification of both protein abundance and cysteine oxidation levels (Zhou and Yin 2019). These cysteine modifications influence protein charge, structure, and interactions, thereby regulating redox-sensitive pathways, including responses to hypoxia mediated by hypoxia-inducible factors (HIFs). Despite technical complexities, redox proteomics provides powerful insights into how proteins dynamically respond to oxidative stimuli (Khan et al. 2021). In Mtb, redox proteomics has shed light on mechanisms that maintain intracellular redox homeostasis essential for survival amid oxidative and nitrosative stresses encountered within host macrophages. The Mtb genome encodes over 200 proteins containing conserved redox-active motifs such as CXXC and TXXC, which participate in thiol-disulfide exchange reactions critical for metabolism, DNA repair, cell wall integrity, and overall redox balance (Kumar et al. 2011). Central to this system are thioredoxins (TrxA, TrxB, TrxC) and thioredoxin reductase (TrxR), which regenerate reduced thiols using electrons from NADPH-dependent reactions (Kumar et al. 2011; Sugandhi et al. 2023). Members of the peroxiredoxin family, including alkylhydroperoxide reductase (AhpC) and thiol peroxidase (Tpx), detoxify peroxides to protect against oxidative damage (Jaeger 2007). Proteins such as Rv0991c (Ruc), a redox-activated chaperone, interact with the heat shock protein Hsp70 system to preserve protein folding and function under oxidative stress conditions (Becker et al. 2020). Moreover, modulation of mycothione reductase (Mtr), a unique thiol-disulfide oxidoreductase in Mtb, affects antibiotic susceptibility, acid resistance, and the bacterium's ability to withstand oxidative stress inside macrophages (Lin et al. 2016). Genetic disruption of key redox regulators like melH markedly diminishes Mtb's ability to neutralize ROS, resulting in increased oxidative damage, shifts in intracellular redox buffers, such as mycothiol and ergothioneine, plus altered energy metabolism. These changes cumulatively reduce bacterial survival and virulence (Pacl et al. 2018). Hence, posttranslational modifications provide critical regulatory layers beyond gene expression, significantly influencing metabolic adaptation, modulation of virulence factors, and drug resistance in Mtb. Comprehensive mapping of PTMs, including phosphorylation, acetylation, ubiquitination, and pupylation, through advanced proteomic technologies uncovers dynamic molecular changes throughout infection. These insights are vital for deciphering bacterial mechanisms of adaptation, persistence, and survival within the hostile host environment, paving the way for innovative therapeutic strategies to combat tuberculosis.

11.6 CRISPR and Its Emerging Impact in Tuberculosis Genetics and Diagnostics

Clustered Regularly Interspaced Short Palindromic Repeats (CRISPR) form a sophisticated, naturally occurring adaptive immune system in bacteria and archaea, providing defense against invading bacteriophages and foreign genetic elements (Haurwitz et al. 2010; Westra et al. 2012; Heussler et al. 2015; Ai et al. 2019; Zhang et al. 2023). Mechanistically, CRISPR loci are transcribed into long precursor RNAs that are processed into mature CRISPR RNAs (crRNAs), each containing a unique spacer sequence flanked by partial repeat sequences. These crRNAs associate with CRISPR-associated (Cas) proteins to assemble ribonucleoprotein complexes crucial for sequence-specific recognition and cleavage of foreign nucleic acids (Reeks et al. 2013; Charpentier and Marraffini 2014; van der Oost et al. 2014). CRISPR-Cas systems are widespread, found in over 84% of sequenced archaeal genomes and nearly 47% of bacterial genomes (Grissa et al. 2007; McDonald et al. 2019). These systems consist of arrays of CRISPR repeats and spacers transcribed into crRNA and trans-activating crRNA (tracrRNA), along with diverse cas genes encoding nucleases and other proteins necessary for function. Based on cas gene composition and effector complexes, CRISPR-Cas systems are classified into two main classes: Class 1 systems, characterized by multi-protein effector complexes (Types I, III, and IV), and Class 2 systems, employing a single effector protein (Types II, V, and VI) (Makarova et al. 2018). Despite this diversity, all CRISPR-Cas systems operate on a conserved principle: mature crRNAs contain invader-derived spacers that guide Cas proteins to complementary nucleic acid targets for sequence-specific cleavage, providing effective immunity (van der Oost et al. 2014). The advent of CRISPR-Cas technology has revolutionized genetic engineering, particularly in microbiology, by enabling precise, programmable genome editing. The CRISPR-Cas9 platform is the most commonly used, featuring a Cas9 endonuclease guided by a synthetic single-guide RNA (sgRNA) that induces targeted double-strand breaks in DNA. This facilitates gene knockout or precise sequence editing via endogenous DNA repair pathways, making CRISPR-Cas9 an indispensable tool in bacterial genetics. Beyond gene editing, CRISPR technologies have been adapted for transcriptional control without DNA cleavage through CRISPR interference (CRISPRi). This method employs a catalytically inactive Cas9 (dCas9) that sterically blocks transcription machinery at target DNA sites, allowing reversible, specific gene repression (Cheng et al. 2021). CRISPRi is particularly valuable for dissecting essential gene functions and advancing microbial functional genomics and proteomics.

Within the *Mycobacterium tuberculosis* complex (MTBC), the CRISPR locus, defined by conserved direct repeat regions interspersed with unique spacer sequences, is highly preserved across strains. This locus forms the basis for spoligotyping, a genotyping technique that differentiates strains by assessing the presence or absence of 43 canonical spacers, aiding in tracking transmission dynamics. Genomic analyses show that the ancestral MTBC genome contained approximately 68 spacers, with CRISPR evolution primarily shaped by deletions, IS6110 transposon insertions, and single-nucleotide polymorphisms rather than the acquisition of new spacers (Zhang et al. 2010; Sola 2015; Refrégier et al. 2020; Zein-Eddine et al. 2023). The CRISPR-Cas region remains highly conserved among MTBC lineages, with deletions serving as markers for particular strain families, such as the Beijing lineage, ensuring spoligotyping's continued utility in global TB surveillance (Kato-Maeda et al. 2011). Although the adaptive immunity role of CRISPR-Cas in Mtb appears partially retained, key components like Cas1 and Cas2, involved in new spacer acquisition, are inactive, limiting adaptation to novel phages or plasmids (Jiao et al. 2021). Nonetheless, other Cas proteins remain functional and contribute to defense via interference against invading DNA. MTBC strains can degrade plasmids homologous to their CRISPR spacers, primarily through ancillary RNase activities of crRNA-protein complexes rather than direct nuclease cleavage. Certain Cas proteins also exert immunomodulatory effects and influence antibiotic susceptibility, revealing multifaceted roles for CRISPR-Cas beyond classical immunity in Mtb (Zein-Eddine et al. 2023). CRISPR-Cas systems have further transformed molecular diagnostics, especially for infectious diseases like tuberculosis. Unlike traditional nucleic acid amplification tests (NAATs), which suffer from reduced sensitivity in challenging clinical scenarios, such as pediatric, HIV-positive, or extrapulmonary TB cases, CRISPR-based assays offer ultra-sensitive and highly specific detection of Mtb DNA. These diagnostics utilize programmable Cas nucleases like Cas12 and Cas13, which, upon target recognition, activate collateral cleavage of reporter molecules, amplifying detection signals with high speed and accuracy (Zein-Eddine et al. 2023; Zu et al. 2025). When combined with isothermal amplification methods such as recombinase polymerase amplification (RPA) or loop-mediated isothermal amplification (LAMP), CRISPR diagnostics enable point-of-care testing without complex equipment, a critical advantage in resource-limited settings (Mao et al. 2023). Additionally, CRISPR-based platforms have successfully detected Mtb DNA in paucibacillary and extrapulmonary infections where conventional assays often fail. They facilitate early diagnosis and treatment monitoring through noninvasive samples like blood (Huang et al. 2023; Zein-Eddine et al. 2023; Ahmed et al. 2024). These technologies also hold promise for identifying drug-resistant Mtb strains by targeting resistance-conferring mutations, though challenges in multiplexing and sgRNA design remain (Qi et al. 2013). Overall, CRISPR-Cas-based diagnostics offer superior sensitivity, specificity, rapid turnaround, and greater accessibility, representing a significant leap forward in TB diagnosis and patient management worldwide.

11.7 The Role of AI and Machine Learning in Advancing Tuberculosis Control and Treatment

Artificial intelligence (AI) and machine learning (ML) technologies are rapidly transforming medical diagnostics and disease prediction, offering unprecedented capabilities to understand and anticipate drug resistance, disease progression, and treatment outcomes. By enabling precise, data-driven insights, these tools facilitate earlier clinical interventions and the development of personalized therapeutic strategies, significantly improving healthcare delivery (Zhai et al. 2021; Sadr et al. 2025). Traditional ML algorithms, such as Naive Bayes (NB), k-nearest neighbor (kNN), support vector machines (SVMs), and artificial neural networks (ANN), have been widely used to predict mutations linked to drug resistance. These methods analyze complex genomic and clinical datasets to identify specific genetic variations that confer resistance to antimicrobial and anticancer agents (Obaido et al. 2024). For example, SVMs excel at handling high-dimensional genomic data by finding optimal hyperplanes that separate resistant from susceptible mutations. The kNN algorithm classifies based on proximity to known cases, while Naive Bayes, valued for computational efficiency and interpretability, offers rapid screening across large mutation datasets. Deep learning, a powerful subset of ML, has emerged especially for interpreting complex unstructured data, including medical images and genomic sequences (LeCun et al. 2015; Esteva et al. 2019). Convolutional neural networks (CNNs) have become standard for identifying subtle patterns in radiological images, enabling early disease detection, tumor characterization, and monitoring metastatic progression (Litjens et al. 2017). Complementarily, recurrent neural networks (RNNs) and transformer-based models analyze sequential and longitudinal data from health records and genomic sequences, supporting dynamic modeling of disease trajectories and personalized treatment responses. Ensemble learning techniques, combining multiple models like decision trees, random forests, gradient boosting, and neural networks, further enhance predictive accuracy by leveraging complementary strengths, an advantage crucial for clinical decision-making where precision impacts patient outcomes (Zhai et al. 2021). Radiomics, an emerging AI-driven field, involves extracting and quantitatively analyzing high-dimensional imaging features, often imperceptible to human observers, to reveal tumor phenotypes, heterogeneity, and biological behavior (Koçak et al. 2019). Predictive analytics integrates diverse data sources—including structured electronic health records (EHRs), laboratory results, imaging, and continuous wearable device streams—to model disease progression and forecast complications. Through advanced time-series analyses, survival models, and AI-enabled pattern recognition, clinicians gain the ability to anticipate deterioration, optimize treatments, and allocate healthcare resources more effectively.

Tuberculosis (TB) research exemplifies AI/ML's transformative impact. The explosion of biological data from high-throughput technologies demands sophisticated computational approaches to glean meaningful insights from complex datasets. AI-driven methods are indispensable for predicting drug targets, improving diagnostics, elucidating microbial interactions, and understanding antimicrobial resistance mechanisms. Drug-resistant TB, including multidrug-resistant (MDR-TB) and extensively drug-resistant TB (XDR-TB), remains a critical global health challenge, per the World Health Organization (WHO) reports. AI and ML models such as SVM, Random Forests, Deep Neural Networks (DNN), and Graph Neural Networks (GNN) have been trained on large whole-genome sequencing (WGS) datasets to accurately predict mutations conferring drug resistance (Jamal et al. 2020; Liang et al. 2022).

Jamal et al. (2020) applied such algorithms to classify single-nucleotide variations in genes like *rpoB*, *inhA*, *katG*, *pncA*, *gyrA*, and *gyrB*, responsible for resistance to key TB drugs, including rifampicin, isoniazid, pyrazinamide, and fluoroquinolones. Integrating sequence and structural features, their framework enables rapid, accurate resistance detection, complementing traditional diagnostics and facilitating timely, tailored TB treatment (Phelan et al. 2019). These computational tools significantly expedite resistance profiling in a cost-effective manner, accelerating the initiation of appropriate therapies. AI-powered diagnostics using deep learning, particularly CNNs, are transforming TB detection from chest X-rays (CXRs) and computed tomography (CT) images. CNNs can identify subtle radiographic abnormalities indicative of TB, differentiate it from other pulmonary diseases such as pneumonia and lung cancer, and support early diagnosis (Lakhani and Sundaram 2017; Hansun et al. 2023). Radiomics enhances assessment of lesion heterogeneity, severity, and treatment response beyond human perception. Combining radiomics with deep learning shows promise in early detection of drug-resistant TB and monitoring therapeutic efficacy, especially in resource-limited settings lacking expert radiologists (Song et al. 2025). For instance, Jaeger et al. (2014) analyzed over 5000 CXR images with CNNs, achieving an Area under the curve (AUC) of 0.85 for MDR-TB detection, while Gao and Qian (2018) reported over 90% accuracy in MDR-TB prediction using hybrid CNN-SVM models on CT data. ML models are also employed to analyze TB transmission dynamics and identify high-risk populations or geographic hotspots through integrated datasets from EHRs, contact tracing, and social determinants of health (Zhao et al. 2024). AI-driven predictive analytics further combine clinical, demographic, and genomic data to forecast individual patient outcomes, pinpointing those at high risk of treatment failure or relapse (Dixon et al. 2024). This enables personalized management with optimized drug dosages, effective drug combinations, and adherence support, ultimately reducing the emergence of resistance. Recent clinical studies demonstrate that AI-based decision support can predict treatment duration and effectiveness, enhancing TB management efficiency (Zhang et al. 2024). Collectively, AI and ML are foundational to advancing the WHO End TB Strategy (WHO 2015). Integrated into TB control programs, they promise to revolutionize diagnosis, surveillance, therapy optimization, and prevention, particularly in resource-constrained, high-burden settings. One vital application is the development of rapid, point-of-care diagnostics. AI-driven mobile imaging platforms combined with molecular detection methods enable swift, accurate TB diagnosis, even in remote locations with limited healthcare infrastructure. Deep learning algorithms analyzing chest radiographs autonomously detect TB-related abnormalities with high sensitivity and specificity, empowering frontline healthcare workers to initiate timely treatment and reduce diagnostic delays (Lakhani and Sundaram 2017; Palanivel et al. 2024). This democratization of diagnostic capabilities stands to substantially improve global patient outcomes. Beyond diagnostics, AI-powered genomic surveillance platforms facilitate near-real-time detection of TB outbreaks and emerging drug resistance by leveraging extensive whole-genome sequencing data. These platforms enable precise tracking of transmission and identification of novel resistance mutations, supporting targeted containment measures essential to controlling MDR-TB spread (Yang et al. 2019; Du et al. 2024). AI and ML also accelerate vaccine development by guiding antigen discovery and deciphering complex host-pathogen interactions. Machine learning algorithms analyze comprehensive immunological and genomic data to prioritize vaccine candidates and predict immunogenicity, potentially shortening development timelines for next-generation vaccines critical to TB elimination (Li et al. 2023). Finally, AI-integrated digital health solutions, including wearable sensors and mobile apps, enable continuous, real-time monitoring of patient symptoms and treatment adherence. These technologies can provide personalized reminders, detect early signs of treatment failure or relapse, and facilitate remote clinician oversight, ultimately improving compliance and reducing resistance emergence (Rajkomar et al. 2018). The convergence of AI with patient monitoring technologies promises to enhance healthcare quality and sustain treatment success. Thereby, artificial intelligence and machine learning are poised to become indispensable components of global tuberculosis control. By enabling rapid diagnostics, real-time epidemiological surveillance, accelerated vaccine research, and personalized patient management, these technologies will be critical to achieving the WHO End TB Strategy targets and reducing the global burden of tuberculosis.

Conflict of Interest The authors declare that they have no conflict of interest related to the content of this chapter.

Ethical Consideration This chapter is a review based entirely on data and information obtained from previously published and publicly available sources, all of which have been appropriately cited. The authors conducted no new experimental work involving human participants or animals. Therefore, no additional ethical approval was required. The authors have ensured proper attribution and compliance with ethical standards in the use of all referenced materials.

References

Ahmed MM, Kayode HH, Okesanya OJ, Ukoaka BM, Eshun G, Mourid MR, Adigun OA, Ogaya JB, Mohamed ZO, Lucero-Prisno DE III (2024 Dec) CRISPR-Cas systems in the fight against antimicrobial resistance: current status, potentials, and future directions. Infect Drug Resist 31:5229–5245

Ai JW, Zhou X, Xu T, Yang M, Chen Y, He GQ, Pan N, Cai Y, Li Y, Wang X, Su H (2019) CRISPR-based rapid and ultra-sensitive diagnostic test for Mycobacterium tuberculosis. Emerging Microb Infect 8(1):1361–1369

Alam A, Bröms JE, Kumar R, Sjöstedt A (2021) The role of ClpB in bacterial stress responses and virulence. Front Mol Biosci 8:668910. https://doi.org/10.3389/fmolb.2021.668910. PMID: 33968993; PMCID: PMC8100447

Alderwick LJ, Birch HL, Krumbach K, Bott M, Eggeling L, Besra GS (2018) AftD functions as an α1 → 5 arabinofuranosyltransferase involved in the biosynthesis of the mycobacterial cell wall core. Cell Surf 1:2–14. https://doi.org/10.1016/j.tcsw.2017.10.001. PMID: 29998212; PMCID: PMC6034362

Ambriz-Aviña V, Contreras-Garduño JA, Pedraza-Reyes M (2014) Applications of flow cytometry to characterize bacterial physiological responses. Biomed Res Int 2014(1):461941

Barandun J, Delley CL, Weber-Ban E (2012) The pupylation pathway and its role in mycobacteria. BMC Biol 10:95. https://doi.org/10.1186/1741-7007-10-95. PMID: 23198822; PMCID: PMC3511204

Barbosa A, Miranda S, Azevedo NF, Cerqueira L, Azevedo AS (2023) Imaging biofilms using fluorescence *in situ* hybridization: seeing is believing. Front Cell Infect Microbiol 13:1195803. https://doi.org/10.3389/fcimb.2023.1195803. PMID: 37284501; PMCID: PMC10239779

Baron VO, Chen M, Hammarstrom B, Hammond RJ, Glynne-Jones P, Gillespie SH, Dholakia K (2020) Real-time monitoring of live mycobacteria with a microfluidic acoustic-Raman platform. Commun Biol 3(1):236

Barr DA, Omollo C, Mason M, Koch A, Wilkinson RJ, Lalloo DG, Meintjes G, Mizrahi V, Warner DF, Davies G (2021) Flow cytometry method for absolute counting and single-cell phenotyping of mycobacteria. Sci Rep 11(1):18661

Becker SH, Jastrab JB, Dhabaria A, Chaton CT, Rush JS, Korotkov KV, Ueberheide B, Darwin KH (2019) The *Mycobacterium tuberculosis* Pup-proteasome system regulates nitrate metabolism through an essential protein quality control pathway. Proc Natl Acad Sci U S A 116(8):3202–3210. https://doi.org/10.1073/pnas.1819468116. Epub 2019 Feb 5. PMID: 30723150; PMCID: PMC6386731

Becker SH, Ulrich K, Dhabaria A, Ueberheide B, Beavers W, Skaar EP, Iyer LM, Aravind L, Jakob U, Darwin KH (2020) Mycobacterium tuberculosis Rv0991c is a redox-regulated molecular chaperone. MBio 11(4):e01545–e01520. https://doi.org/10.1128/mBio.01545-20. PMID: 32843553; PMCID: PMC7448276

Bendre AD, Peters PJ, Kumar J (2021 Jun) Recent insights into the structure and function of mycobacterial membrane proteins facilitated by Cryo-EM. J Membr Biol 254(3):321–341

Benjin X, Ling L (2020 Apr) Developments, applications, and prospects of cryo-electron microscopy. Protein Sci 29(4):872–882

Benoit M (2002) Cell adhesion measured by force spectroscopy on living cells. In: Methods in cell biology, vol 68. Academic, pp 91–114

Birhanu AG, Gómez-Muñoz M, Kalayou S, Riaz T, Lutter T, Yimer SA, Abebe M, Tønjum T (2022) Proteome profiling of *Mycobacterium tuberculosis* cells exposed to Nitrosative stress. ACS Omega 7(4):3470–3482. https://doi.org/10.1021/acsomega.1c05923. PMID: 35128256; PMCID: PMC8811941

Bolten M, Delley CL, Leibundgut M, Boehringer D, Ban N, Weber-Ban E (2016) Structural analysis of the bacterial proteasome activator Bpa in complex with the 20S proteasome. Structure 24(12):2138–2151

Borah K, Xu Y, McFadden J (2021) Dissecting host-pathogen interactions in TB using systems-based Omic approaches. Front Immunol 12:762315. https://doi.org/10.3389/fimmu.2021.762315. PMID: 34795672; PMCID: PMC8593131

Boyaci H, Saecker RM, Campbell EA (2020) Transcription initiation in mycobacteria: a biophysical perspective. Transcription 11(2):53–65

Branon TC, Bosch JA, Sanchez AD, Udeshi ND, Svinkina T, Carr SA, Feldman JL, Perrimon N, Ting AY (2018) Efficient proximity labeling in living cells and organisms with TurboID. Nat Biotechnol 36(9):880–887. https://doi.org/10.1038/nbt.4201. Epub 2018 Aug 20. Erratum in: Nat Biotechnol. 2020 Jan;38(1):108. doi: 10.1038/s41587-019-0355-0. PMID: 30125270; PMCID: PMC6126969

Brehm-Stecher BF, Johnson EA (2004) Single-cell microbiology: tools, technologies, and applications. Microbiol Mol Biol Rev 68(3):538–559

Budnik B, Levy E, Harmange G, Slavov N (2018) SCoPE-MS: mass spectrometry of single mammalian cells quantifies proteome heterogeneity during cell differentiation. Genome Biol 19:161

Buijtels PC, Willemse-Erix HF, Petit PL, Endtz HP, Puppels GJ, Verbrugh HA, van Belkum A, van Soolingen D, Maquelin K (2008 Mar) Rapid identification of mycobacteria by Raman spectroscopy. J Clin Microbiol 46(3):961–965

Burns KE, Liu WT, Boshoff HI, Dorrestein PC, Barry CE III (2009) Proteasomal protein degradation in *mycobacteria* is dependent upon a prokaryotic ubiquitin-like protein. J Biol Chem 284:3069–3075

Burns KE, Cerda-Maira FA, Wang T, Li H, Bishai WR, Darwin KH (2010) 'Depupylation' of prokaryotic ubiquitin-like protein from mycobacterial proteasome substrates. Mol Cell 39:821–827

Callaway E (2022) Alphafold's new rival? Meta AI predicts shape of 600 million proteins. Nature 611:211–212

Cañadas-Ortega M, Gómez-Cruz C, Vaquero JJ, Muñoz-Barrutia A (2023) The role of diagnostic microdevices in the fight against tuberculosis. In: Tuberculosis: integrated studies for a complex disease. Springer, Cham, pp 117–137

Cerda-Maira FA, Pearce MJ, Fuortes M, Bishai WR, Hubbard SR, Darwin KH (2010) Molecular analysis of the prokaryotic ubiquitin-like protein (pup) conjugation pathway in *Mycobacterium tuberculosis*. Mol Microbiol 77:1123–1135

Cerda-Maira FA, McAllister F, Bode NJ, Burns KE, Gygi SP, Darwin KH (2011) Reconstitution of the mycobacterium tuberculosis pupylation pathway in Escherichia coli. EMBO Rep 12(8):863–870. https://doi.org/10.1038/embor.2011.109. PMID: 21738222; PMCID: PMC3147258

Charpentier E, Marraffini LA (2014) Harnessing CRISPR-Cas9 immunity for genetic engineering. Curr Opin Microbiol 19:114–119

Cheng H, Zhang F, Ding Y (2021) CRISPR/Cas9 delivery system engineering for genome editing in therapeutic applications. Pharmaceutics 13(10):1649. https://doi.org/10.3390/pharmaceutics13101649. PMID: 34683943; PMCID: PMC8538656

Cho KF, Branon TC, Udeshi ND, Myers SA, Carr SA, Ting AY (2020 Dec) Proximity labeling in mammalian cells with TurboID and split-TurboID. Nat Protoc 15(12):3971–3999

Choudhary E, Sharma R, Pal P, Agarwal N (2022) Deciphering the proteomic landscape of *Mycobacterium tuberculosis* in response to acid and oxidative stresses. ACS Omega 7(30):26749–26766. https://doi.org/10.1021/acsomega.2c03092. PMID: 35936415; PMCID: PMC9352160

Chung ES, Kar P, Kamkaew M, Amir A, Aldridge BB (2024) Single-cell imaging of the Mycobacterium tuberculosis cell cycle reveals linear and heterogenous growth. Nat Microbiol. 9(12):3332–3344. https://doi.org/10.1038/s41564-024-01846-z. Epub 2024 Nov 15. PMID: 39548343; PMCID: PMC11602732

Ciccarelli L, Connell SR, Enderle M, Mills DJ, Vonck J, Grininger M (2013) Structure and conformational variability of the Mycobacterium tuberculosis fatty acid synthase multienzyme complex. Structure 21(7):1251–1257

Crosas B, Hanna J, Kirkpatrick DS, Zhang DP, Tone Y, Hathaway NA, Buecker C, Leggett DS, Schmidt M, King RW, Gygi SP, Finley D (2006) Ubiquitin chains are remodeled at the proteasome by opposing ubiquitin ligase and deubiquitinating activities. Cell 127:1401–1413

Cumming BM, Lamprecht DA, Wells RM, Saini V, Mazorodze JH, Steyn AJ (2014 Sep) The physiology and genetics of oxidative stress in mycobacteria. Mol Genet Mycobact 10:297–322

Dahlberg C, Bergström M, Hermansson M (1998) In situ detection of high levels of horizontal plasmid transfer in marine bacterial communities. Appl Environ Microbiol 64(7):2670–2675

de la Cruz MJ, Eng ET (2023) Scaling up cryo-EM for biology and chemistry: the journey from niche technology to mainstream method. Structure 31(12):1487–1498

Deng J, Bi L, Zhou L, Guo S, Fleming J, Jiang H, Zhou Y, Gu J, Zhong Q, Wang Z, Liu Z, Deng R, Gao J, Chen T, Li W, Wang J, Wang X, Li H, Ge F, Zhu G, Zhang H, Gu J, Wu F, Zhang Z, Wang D, Hang H, Li Y, Cheng L, He X, Tao S, Zhang X (2014) *Mycobacterium tuberculosis* proteome microarray for global studies of protein function and immunogenicity. Cell Rep 9(6):2317–2329. https://doi.org/10.1016/j.celrep.2014.11.023

Devasundaram S, Gopalan A, Das SD, Raja A (2016) Proteomics analysis of three different strains of Mycobacterium tuberculosis under in vitro hypoxia and evaluation of hypoxia associated antigen's specific memory T cells in healthy household contacts. Front Microbiol 7:1275. https://doi.org/10.3389/fmicb.2016.01275. PMID: 27667981; PMCID: PMC5017210

Dixon D, Sattar H, Moros N, Kesireddy SR, Ahsan H, Lakkimsetti M, Fatima M, Doshi D, Sadhu K, Junaid HM (2024) Unveiling the influence of AI predictive analytics on patient outcomes: a comprehensive narrative review. Cureus 16(5):e59954. https://doi.org/10.7759/cureus.59954. PMID: 38854327; PMCID: PMC11161909

Du J, Su Y, Qiao J, Gao S, Dong E, Wang R, Nie Y, Ji J, Wang Z, Liang J, Gong W (2024) Application of artificial intelligence in diagnosis of pulmonary tuberculosis. Chin Med J 137(05):559–561

Dubochet J, Adrian M, Chang JJ, Homo JC, Lepault J, McDowall AW, Schultz P (1988 May) Cryo-electron microscopy of vitrified specimens. Q Rev Biophys 21(2):129–228. https://doi.org/10.1017/s0033583500004297

Dufrêne YF (2014) Atomic force microscopy in microbiology: new structural and functional insights into the microbial cell surface. MBio 5(4):10–128

Elad N, Baron S, Peleg Y, Albeck S, Grunwald J, Raviv G, Shakked Z, Zimhony O, Diskin R (2018) Structure of type-I Mycobacterium tuberculosis fatty acid synthase at 3.3 Å resolution. Nat Commun 9(1):3886

Esteva A, Robicquet A, Ramsundar B, Kuleshov V, DePristo M, Chou K, Cui C, Corrado G, Thrun S, Dean J (2019) A guide to deep learning in healthcare. Nat Med 25(1):24–29. https://doi.org/10.1038/s41591--018-0316-z. Epub 2019 Jan 7. PMID: 30617335

Famelis N, Rivera-Calzada A, Degliesposti G, Wingender M, Mietrach N, Skehel JM, Fernandez-Leiro R, Böttcher B, Schlosser A, Llorca O, Geibel S (2019) Architecture of the mycobacterial type VII secretion system. Nature 576(7786):321–325

Fay A, Kurland AP, Li Z, Monetti M, Johnson JR, Glickman MS (2025) A split ALFA tag-nanobody system for protein localization and proximity proteomics in mycobacteria. MBio:e00971-25

Feng Y, Chen Y, Zhang W, Shen X, Yan J, Yao L, Zhang L, Niu Y, Zhang J, Tang P, Ling C (2025) Single-cell analysis of peripheral blood and pleural effusion reveals functional diversity of $\gamma\delta$ T cells in tuberculosis infection. Front Immunol 16:1605827. https://doi.org/10.3389/fimmu.2025.1605827. PMID: 40607390; PMCID: PMC12213264

Festa RA, McAllister F, Pearce MJ, Mintseris J, Burns KE, Gygi SP, Darwin KH (2010) Prokayrotic ubiquitin-like protein (pup) proteome of Mycobacterium tuberculosis. PLoS One 5(1):e8589

Finin P, Khan RMN, Oh S, Boshoff HIM, Barry CE (2023) Chemical approaches to unraveling the biology of mycobacteria. Cell Chem Biol 30(5):420–435. https://doi.org/10.1016/j.chembiol.2023.04.014

Flores-Villalva S, Rogríguez-Hernández E, Rubio-Venegas Y, Cantó-Alarcón JG, Milián-Suazo F (2015) What can proteomics tell us about tuberculosis? J Microbiol Biotechnol 25(8):1181–1194

Gao XW, Qian Y (2018) Prediction of multidrug-resistant TB from CT pulmonary images based on deep learning techniques. Mol Pharm 15(10):4326–4335. https://doi.org/10.1021/acs.molpharmaceut.7b00875. Epub 2018 Jan 29 PMID: 29257894

Garand M, Goodier M, Owolabi O, Donkor S, Kampmann B, Sutherland JS (2018) Functional and phenotypic changes of natural killer cells in whole blood during Mycobacterium tuberculosis infection and disease. Front Immunol 9:257

Gengenbacher M, Mouritsen J, Schubert OT, Aebersold R, Kaufmann SH (2014 Sep) Mycobacterium tuberculosis in the proteomics era. Mol Genet Mycobacter 10:239–260

Gessese AT, Kinde MZ, Eshetu T, Kerisew B (2025) Whole-genome sequencing analysis to identify antimicrobial resistance regions and virulence factors in Mycobacterium tuberculosis isolates from the Amhara region, Ethiopia. Sci Rep 15(1):16076

Giles GI, Tasker KM, Jacob C (2001) Hypothesis: the role of reactive sulfur species in oxidative stress. Free Radic Biol Med 31(10):1279–1283

Grabbe C, Dikic I (2009) Functional roles of ubiquitin-like domain (ULD) and ubiquitin-binding domain (UBD) containing proteins. Chem Rev 109:1481–1494

Grissa I, Vergnaud G, Pourcel C (2007) The CRISPRdb database and tools to display CRISPRs and to generate dictionaries of spacers and repeats. BMC Bioinform 8(1):172

Guo H, Courbon GM, Bueler SA, Mai J, Liu J, Rubinstein JL (2021) Structure of mycobacterial ATP synthase bound to the tuberculosis drug bedaquiline. Nature 589:143–147

Guo J, Zhang X, Chen X, Cai Y (2022) Proteomics in biomarker discovery for tuberculosis: current status and future perspectives. Front Microbiol 13:845229. https://doi.org/10.3389/fmicb.2022.845229. PMCID: PMC9087271 PMID: 35558124

Guth E, Thommen M, Weber-Ban E (2011) Mycobacterial ubiquitin-like protein ligase PafA follows a two-step reaction pathway with a phosphorylated pup intermediate. J Biol Chem 286:4412–4419

Hansun S, Argha A, Liaw ST, Celler BG, Marks GB (2023) Machine and deep learning for tuberculosis detection on chest X-rays: systematic literature review. J Med Internet Res 3;25:e43154. https://doi.org/10.2196/43154. PMID: 37399055; PMCID: PMC10365622

Haurwitz RE, Jinek M, Wiedenheft B, Zhou K, Doudna JA (2010) Sequence-and structure-specific RNA processing by a CRISPR endonuclease. Science 329(5997):1355–1358

Hemnani T, Parihar MS (1998) Reactive oxygen species and oxidative DNA damage. Indian J Physiol Pharmacol 42(4):440–452. PMID: 10874342

Hershko A, Ciechanover A (1986) The ubiquitin pathway for the degradation of intracellular proteins. Prog Nucleic Acid Res Mol Biol 33(19–56):301. https://doi.org/10.1016/s0079-6603(08)60019-7. PMID: 3025922

Hershko A, Ciechanover A (1998) The ubiquitin system. Annu Rev Biochem 67:425–479

Heussler GE, Cady KC, Koeppen K, Bhuju S, Stanton BA, O'Toole GA (2015) Clustered regularly interspaced short palindromic repeat-dependent, biofilm-specific death of Pseudomonas aeruginosa mediated by increased expression of phage-related genes. MBio 6(3):10–128

Hoffmann E, Machelart A, Song OR, Brodin P (2018) Proteomics of *Mycobacterium* infection: moving towards a better understanding of pathogen-driven immunomodulation. Front Immunol 9:86. https://doi.org/10.3389/fimmu.2018.00086. PMID: 29441067; PMCID: PMC5797607

Homberger C, Imdahl F, Hayward RJ, Barquist L, Saliba AE, Vogel J (2025 Apr) Transcriptomic profiling of individual bacteria by MATQ-seq. Nat Protoc 9:1–30

Horvath P, Coscia F (2025) Spatial proteomics in translational and clinical research. Mol Syst Biol 21(6):526–530. https://doi.org/10.1038/s44320--025-00101-9. Epub 2025 Apr 14. PMID: 40229558; PMCID: PMC12130312

Huang H, Wang C, Rubelt F, Scriba TJ, Davis MM (2020 Oct) Analyzing the Mycobacterium tuberculosis immune response by T-cell receptor clustering with GLIPH2 and genome-wide antigen screening. Nat Biotechnol 38(10):1194–1202

Huang Z, Zhang G, Lyon CJ, Hu TY, Lu S (2023) Outlook for CRISPR-based tuberculosis assays now in their infancy. Front Immunol 14:1172035. https://doi.org/10.3389/fimmu.2023.1172035. PMID: 37600797; PMCID: PMC10436990

Imkamp F, Striebel F, Sutter M, Ozcelik D, Zimmermann N, Sander P, Weber-Ban E (2010) Dop functions as a depupylase in the prokaryotic ubiquitin-like modification pathway. EMBO Rep 11:791–797

Isa F, Collins S, Lee MH, Decome D, Dorvil N, Joseph P, Smith L, Salerno S, Wells MT, Fischer S, Bean JM, Pape JW, Johnson WD, Fitzgerald DW, Rhee KY (2018 May) Mass spectrometric identification of urinary biomarkers of pulmonary tuberculosis. EBioMedicine 31:157–165

Iyer LM, Burroughs AM, Aravind L (2008) Unraveling the biochemistry and provenance of pupylation: a prokaryotic analog of ubiquitination. Biol Direct 3:45. https://doi.org/10.1186/1745-6150-3-45. PMID: 18980670; PMCID: PMC2588565

Jaeger T (2007) Peroxiredoxin systems in mycobacteria. Subcell Biochem 44:207–217. https://doi.org/10.1007/978-1-4020-6051-9_9. PMID: 18084895

Jaeger S, Candemir S, Antani S, Wáng YX, Lu PX, Thoma G (2014) Two public chest X-ray datasets for computer-aided screening of pulmonary diseases. Quant Imaging Med Surg 4(6):475–477. https://doi.org/10.3978/j.issn.2223-4292.2014.11.20. PMID: 25525580; PMCID: PMC4256233

Jaisinghani N, Previti ML, Andrade J, Askenazi M, Ueberheide B, Seeliger JC (2024) Proteomics from compartment-specific APEX2 labeling in Mycobacterium tuberculosis reveals type VII secretion substrates in the cell wall. Cell Chem Biol 31(3):523–533

Jamal S, Khubaib M, Gangwar R, Grover S, Grover A, Hasnain SE (2020) Artificial intelligence and machine learning based prediction of resistant and susceptible mutations in Mycobacterium tuberculosis. Sci Rep 10(1):5487

Jiang P, Dickson DW (2025) Correlative light and electron microscopy imaging of proteinaceous deposits in cell cultures and brain tissues. Acta Neuropathol Commun 13(1):53. https://doi.org/10.1186/s40478-025-01969-2. PMID: 40057802; PMCID: PMC11889819

Jiao J, Zheng N, Wei W, Fleming J, Wang X, Li Z, Zhang L, Liu Y, Zhang Z, Shen A, Chuanyou L. M. Tuberculosis CRISPR/Cas proteins are secreted virulence factors that trigger cellular immune responses. Virulence 2021 (1):3032–3044, 12

Jumper J, Evans R, Pritzel A, Green T, Figurnov M, Ronneberger O, Tunyasuvunakool K et al (2021) Highly accurate protein structure prediction with AlphaFold. Nature 596(7873):583–589. https://doi.org/10.1038/s41586-021-03819-2. Epub 2021 Jul 15. PMID: 34265844; PMCID: PMC8371605

Kahne SC, Yoo JH, Chen J, Nakedi K, Iyer LM, Putzel G, Samhadaneh NM, Pironti A, Aravind L, Ekiert DC, Bhabha G (2024) Identification of a depupylation regulator for an essential enzyme in Mycobacterium tuberculosis. Proc Natl Acad Sci 121(49):e2407239121

Kain SR (1999) Green fluorescent protein (GFP): applications in cell-based assays for drug discovery. Drug Discov Today 4(7):304–312

Kairišs K, Sokolova N, Zilova L, Schlagheck C, Reinhardt R, Baumbach T, Faragó T, van de Kamp T, Wittbrodt J, Weinhardt V (2024) Visualisation of gene expression within the context of tissues using an X-ray computed tomography-based multimodal approach. Sci Rep 14:8543

Kato-Maeda M, Metcalfe JZ, Flores L (2011) Genotyping of Mycobacterium tuberculosis: application in epidemiologic studies. Future Microbiol 6(2):203–216

Kelly RT (2020) Single-cell proteomics: progress and prospects. Mol Cell Proteomics 19(11):1739–1748. https://doi.org/10.1074/mcp.R120.002234

Kelly AM, McLoughlin RM (2020) Target the host, kill the bug; targeting host respiratory immunosuppressive responses as a novel strategy to improve bacterial clearance during lung infection. Front Immunol 11:767

Khan MZ, Singha B, Ali MF, Taunk K, Rapole S, Gourinath S, Nandicoori VK (2021) Redox homeostasis in Mycobacterium tuberculosis is modulated by a novel actinomycete-specific transcription factor. EMBO J 40(14):e106111. https://doi.org/10.15252/embj.2020106111. Epub 2021 May 21. PMID: 34018220; PMCID: PMC8280819

Koçak B, Durmaz EŞ, Ateş E, Kılıçkesmez Ö (2019) Radiomics with artificial intelligence: a practical guide for beginners. Diagn Interv Radiol 25(6):485–495. https://doi.org/10.5152/dir.2019.19321. PMID: 31650960; PMCID: PMC6837295

Komor AC, Kim YB, Packer MS, Zuris JA, Liu DR (2016) Programmable editing of a target base in genomic

DNA without double-stranded DNA cleavage. Nature 533:420–424. https://doi.org/10.1038/nature17946

Korotkova N, Piton J, Wagner JM, Boy-Röttger S, Japaridze A, Evans TJ, Cole ST, Pojer F, Korotkov KV (2015) Structure of EspB, a secreted substrate of the ESX-1 secretion system of Mycobacterium tuberculosis. J Struct Biol 191(2):236–244

Kremer L, Dover LG, Carrere S, Nampoothiri KM, Lesjean S, Brown AK, Brennan PJ, Minnikin DE, Locht C, Besra GS (2002) Mycolic acid biosynthesis and enzymic characterization of the beta-ketoacyl-ACP synthase A-condensing enzyme from *Mycobacterium tuberculosis*. Biochem J 364:423–430

Kruh NA, Troudt J, Izzo A, Prenni J, Dobos KM (2010) Portrait of a pathogen: the Mycobacterium tuberculosis proteome in vivo. PLoS One 5(11):e13938. https://doi.org/10.1371/journal.pone.0013938. PMID: 21085642; PMCID: PMC2978697

Kühlbrandt W (2022) Concluding remarks: challenges and future developments in biological electron cryo-microscopy. Faraday Discuss 240:323–335

Kumar A, Farhana A, Guidry L, Saini V, Hondalus M, Steyn AJ (2011) Redox homeostasis in mycobacteria: the key to tuberculosis control? Expert Rev Mol Med 13:e39. https://doi.org/10.1017/S1462399411002079. PMID: 22172201; PMCID: PMC3241215

Kusunose E, Ichihara K, Noda Y, Kusunose M (1976) Superoxide dismutase from *Mycobacterium tuberculosis*. J Biochem 80:1343–1352

Lakhani P, Sundaram B (2017 Aug) Deep learning at chest radiography: automated classification of pulmonary tuberculosis by using convolutional neural networks. Radiology 284(2):574–582

Lam SS, Martell JD, Kamer KJ, Deerinck TJ, Ellisman MH, Mootha VK, Ting AY (2015) Directed evolution of APEX2 for electron microscopy and proximity labeling. Nat Methods 12(1):51–54. https://doi.org/10.1038/nmeth.3179. Epub 2014 Nov 24. PMID: 25419960; PMCID: PMC4296904

Larsson I, Dalmo E, Elgendy R, Niklasson M, Doroszko M, Segerman A, Jörnsten R, Westermark B, Nelander S (2021 Sep) Modeling glioblastoma heterogeneity as a dynamic network of cell states. Mol Syst Biol 17(9):e10105

LeCun Y, Bengio Y, Hinton G (2015) Deep learning. Nature 521(7553):436–444. https://doi.org/10.1038/nature14539. PMID: 26017442

Lee KS, Landry Z, Pereira FC, Wagner M, Berry D, Huang WE, Taylor GT, Kneipp J, Popp J, Zhang M, Cheng JX (2021) Raman microspectroscopy for microbiology. Nature Rev Methods Primers 1(1):80

Leichert LI, Gehrke F, Gudiseva HV, Blackwell T, Ilbert M, Walker AK, Strahler JR, Andrews PC, Jakob U (2008) Quantifying changes in the thiol redox proteome upon oxidative stress in vivo. Proc Natl Acad Sci U S A 105(24):8197–8202. https://doi.org/10.1073/pnas.0707723105. Epub 2008 Feb 14. PMID: 18287020; PMCID: PMC2448814

Leo S, Narasimhan M, Rathinam S, Banerjee A (2024) Biomarkers in diagnosing and therapeutic monitoring of tuberculosis: a review. Ann Med 56(1):2386030. https://doi.org/10.1080/07853890.2024.2386030. Epub 2024 Aug 3. PMID: 39097795; PMCID: PMC11299445

Lepe BA, Zheng CR, Leddy OK, Allsup BL, Solomon SL, Bryson BD (2025) Protease shaving of *Mycobacterium tuberculosis* facilitates vaccine antigen discovery and delivery of novel cargoes to the Mtb surface. Microbiol Spectr 13(2):e0227724. https://doi.org/10.1128/spectrum.02277-24. Epub 2024 Dec 17. PMID: 39688428; PMCID: PMC11792546

Li J, Zhang H, Menguy N, Benzerara K, Wang F, Lin X, Chen Z, Pan Y (2017) Single-cell resolution of uncultured magnetotactic bacteria via fluorescence-coupled electron microscopy. Appl Environ Microbiol 83(12):e00409–e00417

Li LS, Yang L, Zhuang L, Ye ZY, Zhao WG, Gong WP (2023) From immunology to artificial intelligence: revolutionizing latent tuberculosis infection diagnosis with machine learning. Mil Med Res 10(1):58

Liang S, Ma J, Wang G, Shao J, Li J, Deng H, Wang C, Li W (2022) The application of artificial intelligence in the diagnosis and drug resistance prediction of pulmonary tuberculosis. Front Med 9:935080

Lin K, O'Brien KM, Trujillo C, Wang R, Wallach JB, Schnappinger D, Ehrt S (2016) Mycobacterium tuberculosis Thioredoxin reductase is essential for thiol redox homeostasis but plays a minor role in antioxidant defense. PLoS Pathog 12(6):e1005675. https://doi.org/10.1371/journal.ppat.1005675. PMID: 27249779; PMCID: PMC4889078

Lin W, Das K, Degen D, Mazumder A, Duchi D, Wang D, Ebright YW, Ebright RY, Sineva E, Gigliotti M, Srivastava A (2018) Structural basis of transcription inhibition by fidaxomicin (lipiarmycin A3). Mol Cell 70(1):60–71

Lin Z, Akin H, Rao R, Hie B, Zhu Z, Lu W, Smetanin N, Verkuil R, Kabeli O, Shmueli Y, dos Santos CA (2023) Evolutionary-scale prediction of atomic-level protein structure with a language model. Science 379(6637):1123–1130

Litjens G, Kooi T, Bejnordi BE, Setio AAA, Ciompi F, Ghafoorian M, van der Laak JAWM, van Ginneken B, Sánchez CI (2017) A survey on deep learning in medical image analysis. Med Image Anal 42:60–88. https://doi.org/10.1016/j.media.2017.07.005. Epub 2017 Jul 26. PMID: 28778026

Lu Y, Chen H, Wang P, Pang J, Lu X, Li G, Hu X, Wang X, Yang X, Li C, Lu Y, You X (2023) Identification and quantification of S-Sulfenylation proteome of Mycobacterium tuberculosis under oxidative stress. Microbiol Spectr 11(2):e0338622. https://doi.org/10.1128/spectrum.03386-22. Epub ahead of print. PMID: 36943050; PMCID: PMC10101118

Lyu J, Narum DE, Baldwin SL, Larsen SE, Bai X, Griffith DE, Dartois V, Naidoo T, Steyn AJ, Coler RN, Chan ED (2024) Understanding the development of tuberculous granulomas: insights into host protection and pathogenesis, a review in humans and animals. Front Immunol 15:1427559

Lyu M, Liu Y, Zhou J, Lai H, Ruan H, Wu D, Zhu S, Zhou X, Ma W, Huang Y, Lei S (2025) Advances in single-cell sequencing for infectious diseases: progress and perspectives. Adv Science:e15678

Ma R, Yang W, Guo W, Zhang H, Wang Z, Ge Z (2025) Single-cell transcriptome analysis reveals the dysregulated monocyte state associated with tuberculosis progression. BMC Infect Dis 25(1):210

MacGilvary NJ, Tan S (2018) Fluorescent Mycobacterium tuberculosis reporters: illuminating host–pathogen interactions. Pathogens Dis 76(3):fty017

Magoulopoulou A, Qian X, Pediatama Setiabudiawan T, Marco Salas S, Yokota C, Rottenberg ME, Nilsson M, Carow B (2022) Spatial resolution of *Mycobacterium tuberculosis* bacteria and their surrounding immune environments based on selected key transcripts in mouse lungs. Front Immunol 13:876321. https://doi.org/10.3389/fimmu.2022.876321. PMID: 35663950; PMCID: PMC9157500

Makarova KS, Wolf YI, Koonin EV (2018) Classification and nomenclature of CRISPR-Cas systems: where from here? CRISPR J 1(5):325–336. https://doi.org/10.1089/crispr.2018.0033. PMID: 31021272; PMCID: PMC6636873

Mao X, Xu M, Luo S, Yang Y, Zhong J, Zhou J, Fan H, Li X, Chen Z (2023) Advancements in the synergy of isothermal amplification and CRISPR-cas technologies for pathogen detection. Front Bioeng Biotechnol 11:1273988. https://doi.org/10.3389/fbioe.2023.1273988. PMID: 37885449; PMCID: PMC10598474

Mashabela GT, de Wet TJ, Warner DF (2019) *Mycobacterium tuberculosis* Metabolism. Microbiol Spectr 7:4. https://doi.org/10.1128/microbiolspec.GPP3-0067-2019. PMID: 31350832; PMCID: PMC10957194

Matias Z, Lopes CS, Santos NC, Carvalho FA (2025 Apr) Nanotechnology meets medicine: applications of atomic force microscopy in disease. Biophys Rev 3:1–26

McDonald ND, Regmi A, Morreale DP, Borowski JD, Boyd EF (2019) CRISPR-Cas systems are present predominantly on mobile genetic elements in *vibrio* species. BMC Genomics 20:105. https://doi.org/10.1186/s12864-019-5439-1

McNerney R, Zignol M, Clark TG (2018) Use of whole genome sequencing in surveillance of drug resistant tuberculosis. Expert Rev Anti-Infect Ther 16(5):433–442

Mishra S, Ahmed T, Tyagi A, Shi J, Bhushan S (2018) Structures of Mycobacterium smegmatis 70S ribosomes in complex with HPF, tmRNA, and P-tRNA. Sci Rep 8(1):13587

Mistretta M, Gangneux N, Manina G (2022) Microfluidic dose–response platform to track the dynamics of drug response in single mycobacterial cells. Sci Rep 12(1):19578

Mistretta M, Cimino M, Campagne P, Volant S, Kornobis E, Hebert O, Rochais C, Dallemagne P, Lecoutey C, Tisnerat C, Lepailleur A (2024) Dynamic microfluidic single-cell screening identifies pheno-tuning compounds to potentiate tuberculosis therapy. Nat Commun 15(1):4175

Mittal E, Roth AT, Seth A, Singamaneni S, Beatty W, Philips JA (2023) Single cell preparations of *Mycobacterium tuberculosis* damage the mycobacterial envelope and disrupt macrophage interactions. elife 12:e85416

Mohammadnabi N, Shamseddin J, Emadi M, Bodaghi AB, Varseh M, Shariati A, Rezaei M, Dastranj M, Farahani A (2024) Mycobacterium tuberculosis: the mechanism of pathogenicity, immune responses, and diagnostic challenges. J Clin Lab Anal 38(23):e25122. https://doi.org/10.1002/jcla.25122. Epub 2024 Nov 26. PMID: 39593272; PMCID: PMC11632860

Molloy A, Harrison J, McGrath JS, Owen Z, Smith C, Liu X, Li X, Cox JA (2021) Microfluidics as a novel technique for tuberculosis: from diagnostics to drug discovery. Microorganisms 9(11):2330

Moore CM, Dhillon J, Flynn R, Gizynski K, Adams C, Morgan G, McGurk D, Boada E, Shabestary S, Peat J, O'Halloran J, Stoker NG, Butcher PD, Murton H (2023) A novel microfluidic Dielectrophoresis technology to enable rapid diagnosis of Mycobacteria tuberculosis in clinical samples. J Mol Diagn 25(7):513–523. https://doi.org/10.1016/j.jmoldx.2023.04.005. Erratum in: J Mol Diagn 2023 Oct;25(10):782. doi: 10.1016/j.jmoldx.2023.07.002. PMID: 37355278; PMCID: PMC12179504

Müller DJ, Dufrêne YF (2011) Atomic force microscopy: a nanoscopic window on the cell surface. Trends Cell Biol 21(8):461–469

Munir A, Wilson MT, Hardwick SW, Chirgadze DY, Worrall JA, Blundell TL, Chaplin AK (2021) Using cryo-EM to understand antimycobacterial resistance in the catalase-peroxidase (KatG) from Mycobacterium tuberculosis. Structure 29(8):899–912

Murata K, Wolf M (2018) Cryo-electron microscopy for structural analysis of dynamic biological macromolecules. Biochimica et Biophysica Acta (BBA) Gen Sub 1862(2):324–334

Nargan K, Glasgow JN, Nadeem S, Naidoo T, Wells G, Hunter RL, Hutton A, Lumamba K, Msimang M, Benson PV, Steyn AJC (2024) Spatial distribution of Mycobacterium tuberculosis mRNA and secreted antigens in acid-fast negative human antemortem and resected tissue. EBioMedicine 105:105196. https://doi.org/10.1016/j.ebiom.2024.105196. Epub 2024 Jun 15. PMID: 38880068; PMCID: PMC11233921

Nesvizhskii AI (2014) Proteogenomics: concepts, applications and computational strategies. Nat Methods 11:1114–1125. https://doi.org/10.1038/nmeth.3144. PMID: 25357241; PMCID: PMC4392723

Nishimura K, Fukagawa T, Takisawa H, Kakimoto T, Kanemaki M (2009) An auxin-based degron system for the rapid depletion of proteins in nonplant cells. Nat Methods 6(12):917–922. https://doi.org/10.1038/nmeth.1401. Epub 2009 Nov 15. PMID: 19915560

Nogales E, Scheres SH (2015) Cryo-EM: a unique tool for the visualization of macromolecular complexity. Mol Cell 58(4):677–689

Obaido G, Mienye ID, Egbelowo OF, Emmanuel ID, Ogunleye A, Ogbuokiri B, Mienye P, Aruleba K (2024) Supervised machine learning in drug discovery and development: algorithms, applications, challenges, and prospects. Machine Learn Appl 17:100576. https://doi.org/10.1016/j.mlwa.2024.100576

Olsen JV, Blagoev B, Gnad F, Macek B, Kumar C, Mortensen P, Mann M (2006) Global, in vivo, and site-specific phosphorylation dynamics in signaling networks. Cell 127(3):635–648. https://doi.org/10.1016/j.cell.2006.09.026

Pacl HT, Reddy VP, Saini V, Chinta KC, Steyn AJC (2018 Jul) Host-pathogen redox dynamics modulate Mycobacterium tuberculosis pathogenesis. Pathog Dis 76(5):fty036. https://doi.org/10.1093/femspd/fty036. PMID: 29873719; PMCID: PMC5989597

Palanivel J, Manikkam R, Sounderrajan V, Jayaraj S, Rao SS, Thangam T, Parthasarathy K (2024) An emerging artificial intelligence tool for the advancement of modern health care in Tuberculosis. In: Translational research in biomedical sciences: recent progress and future prospects. Springer Nature Singapore, Singapore, pp 111–120

Pan J, Chang Z, Zhang X, Dong Q, Zhao H, Shi J, Wang G (2023) Research progress of single-cell sequencing in tuberculosis. Front Immunol 14:1276194

Paulsen CE, Carroll KS (2013) Cysteine-mediated redox signaling: chemistry, biology, and tools for discovery. Chem Rev 113(7):4633–4679. https://doi.org/10.1021/cr300163e. Epub 2013 Mar 20. PMID: 23514336; PMCID: PMC4303468

Pearce MJ, Mintseris J, Ferreyra J, Gygi SP, Darwin KH (2008) Ubiquitin-like protein involved in the proteasome pathway of Mycobacterium tuberculosis. Science 322(5904):1104–1107

Petushkova AI, Zamyatnin AA (2020) Redox-mediated post-translational modifications of proteolytic enzymes and their role in protease functioning. Biomolecules 10(4):650. https://doi.org/10.3390/biom10040650. PMID: 32340246; PMCID: PMC7226053

Phelan JE, O'Sullivan DM, Machado D, Ramos J, Oppong YEA, Campino S, O'Grady J, McNerney R, Hibberd ML, Viveiros M, Huggett JF, Clark TG (2019) Integrating informatics tools and portable sequencing technology for rapid detection of resistance to anti-tuberculous drugs. Genome Med 11:41. https://doi.org/10.1186/s13073-019-0650-x

Pisu D, Huang L, Narang V, Theriault M, Lê-Bury G, Lee B, Lakudzala AE, Mzinza DT, Mhango DV, Mitini-Nkhoma SC, Jambo KC, Singhal A, Mwandumba HC, Russell DG (2021) Single cell analysis of M. tuberculosis phenotype and macrophage lineages in the infected lung. J Exp Med 218(9):e20210615

Poole T (2017) Understanding sulfur based redox biology through advancements in chemical biology. Wake Forest University

Potter SS (2018) Single-cell RNA sequencing for the study of development, physiology and disease. Nat Rev Nephrol 14(8):479–492. https://doi.org/10.1038/s41581-018-0021-7. PMID: 29789704; PMCID: PMC6070143

Poulsen C, Akhter Y, Jeon AH, Schmitt-Ulms G, Meyer HE, Stefanski A, Stühler K, Wilmanns M, Song YH (2010) Proteome-wide identification of mycobacterial pupylation targets. Mol Syst Biol 6:386. https://doi.org/10.1038/msb.2010.39. PMID: 20631680; PMCID: PMC2925521

Qi LS, Larson MH, Gilbert LA, Doudna JA, Weissman JS, Arkin AP, Lim WA (2013) Repurposing CRISPR as an RNA-guided platform for sequence-specific control of gene expression. Cell 152(5):1173–1183. https://doi.org/10.1016/j.cell.2013.02.022. Erratum in: Cell 2021 Feb 4;184(3):844. doi: 10.1016/j.cell.2021.01.019. PMID: 23452860; PMCID: PMC3664290

Qiu X, Zhong P, Yue L, Li C, Yun Z, Si G, Li M, Chen Z, Tan Y, Bao P (2024) Spatial transcriptomic sequencing reveals immune microenvironment features of Mycobacterium tuberculosis granulomas in lung and omentum. Theranostics 14(16):6185

Quentin D, Raunser S (2018 Jun) Electron cryomicroscopy as a powerful tool in biomedical research. J Mol Med 96(6):483–493

Rajkomar A, Oren E, Chen K, Dai AM, Hajaj N, Hardt M, Liu PJ et al (2018) Scalable and accurate deep learning with electronic health records. NPJ Digit Med 1:18. https://doi.org/10.1038/s41746-018-0029-1. PMID: 31304302; PMCID: PMC6550175

Reeks J, Naismith JH, White MF (2013) CRISPR interference: a structural perspective. Biochem J 453(2):155–166

Refrégier G, Sola C, Guyeux C (2020) Unexpected diversity of CRISPR unveils some evolutionary patterns of repeated sequences in Mycobacterium tuberculosis. BMC Genomics 21(1):841

Rempel S, Gati C, Nijland M, Thangaratnarajah C, Karyolaimos A, de Gier JW, Guskov A, Slotboom DJ (2020) A mycobacterial ABC transporter mediates the uptake of hydrophilic compounds. Nature 580(7803):409–412

Rossmann MG, Tao Y (1999) Cryo-electron-microscopy reconstruction of partially symmetric objects. J Struct Biol 125(2–3):196–208

Rothchild AC, Mai D, Aderem A, Diercks AH (2020) Flow cytometry analysis and fluorescence-activated cell sorting of myeloid cells from lung and bronchoalveolar lavage samples from Mycobacterium tuberculosis-infected mice. Bio-protocol 10(10):e3630

Roux KJ, Kim DI, Raida M, Burke B (2012) A promiscuous biotin ligase fusion protein identifies proximal and interacting proteins in mammalian cells. J Cell Biol 196(6):801–810. https://doi.org/10.1083/jcb.201112098. Epub 2012 Mar 12. PMID: 22412018; PMCID: PMC3308701

Russell MR, Lerner TR, Burden JJ, Nkwe DO, Pelchen-Matthews A, Domart MC, Durgan J, Weston A, Jones

ML, Peddie CJ, Carzaniga R, Florey O, Marsh M, Gutierrez MG, Collinson LM (2017) 3D correlative light and electron microscopy of cultured cells using serial blockface scanning electron microscopy. J Cell Sci 130(1):278–291. https://doi.org/10.1242/jcs.188433. Epub 2016 Jul 21. PMID: 27445312; PMCID: PMC5394779

Sadr H, Nazari M, Khodaverdian Z, Farzan R, Yousefzadeh-Chabok S, Ashoobi MT, Hemmati H, Hendi A, Ashraf A, Pedram MM, Hasannejad-Bibalan M, Yamaghani MR (2025) Unveiling the potential of artificial intelligence in revolutionizing disease diagnosis and prediction: a comprehensive review of machine learning and deep learning approaches. Eur J Med Res 30:418. https://doi.org/10.1186/s40001-025-02680-7

Sakai Y, Asa M, Hirose M, Kusuhara W, Fujiwara N, Tamashima H, Ikazaki T, Oka S, Kuraba K, Tanaka K, Yoshiyama T (2025) A conserved human CD4+ T cell subset recognizing the mycobacterial adjuvant trehalose monomycolate. J Clin Invest 135(6)

Schiff HF, Walker NF, Ugarte-Gil C, Tebruegge M, Manousopoulou A, Garbis SD, Mansour S, Wong PHM, Rockett G, Piazza P, Niranjan M, Vallejo AF, Woelk CH, Wilkinson RJ, Tezera LB, Garay-Baquero D, Elkington P (2024) Integrated plasma proteomics identifies tuberculosis-specific diagnostic biomarkers. JCI Insight 9(8):e173273. https://doi.org/10.1172/jci.insight.173273. PMID: 38512356; PMCID: PMC11141874

Schwanhäusser B, Busse D, Li N, Dittmar G, Schuchhardt J, Wolf J, Chen W, Selbach M (2011) Global quantification of mammalian gene expression control. Nature 473(7347):337–342. https://doi.org/10.1038/nature10098. Erratum in: Nature 2013 Mar 07;495(7439):126–7. doi: 10.1038/nature11848. PMID: 21593866

Seto S, Nakamura H, Guo TC, Hikichi H, Wakabayashi K, Miyabayashi A, Nagata T, Hijikata M, Keicho N (2022) Spatial multiomic profiling reveals the novel polarization of foamy macrophages within necrotic granulomatous lesions developed in lungs of C3HeB/FeJ mice infected with *Mycobacterium tuberculosis*. Front Cell Infect Microbiol 12:968543. https://doi.org/10.3389/fcimb.2022.968543. PMID: 36237431; PMCID: PMC9551193

Shannon P, Markiel A, Ozier O, Baliga NS, Wang JT, Ramage D, Amin N, Schwikowski B, Ideker T (2003) Cytoscape: a software environment for integrated models of biomolecular interaction networks. Genome Res 13(11):2498–2504. https://doi.org/10.1101/gr.1239303. PMID: 14597658; PMCID: PMC403769

Slavov N (2023 Jul) Single-cell proteomics: quantifying post-transcriptional regulation during development with mass-spectrometry. Development 150(13):dev201492

Sola C (2015) Clustured regularly interspersed short palindromic repeats (CRISPR) genetic diversity studies as a mean to reconstruct the evolution of the Mycobacterium tuberculosis complex. Tuberculosis 95:S159–S166

Solomon SL (2023) Development of tools to quantitatively probe Mycobacterium tuberculosis-induced host Phagosomal and signaling responses. Massachusetts Institute of Technology

Solomonson M, Setiaputra D, Makepeace KA, Lameignere E, Petrotchenko EV, Conrady DG, Bergeron JR, Vuckovic M, DiMaio F, Borchers CH, Yip CK (2015) Structure of EspB from the ESX-1 type VII secretion system and insights into its export mechanism. Structure 23(3):571–583

Song C, Huang A, Zhao C, Wen L, Song S, Lin Y, Xu C, Qiang H, Zhu Q (2025) Multimodal radiomics integrating deep learning and clinical features for diagnosing multidrug-resistant tuberculosis in HIV/AIDS patients. J Glob Antimicrob Resist 43:134–142. https://doi.org/10.1016/j.jgar.2025.04.013. Epub 2025 May 3. PMID: 40324593

Stieglitz KA, Yang H, Roberts MF, Stec B (2005) Reaching for mechanistic consensus across life kingdoms: structure and insights into catalysis of the myo-inositol-1-phosphate synthase (mIPS) from *Archaeoglobus fulgidus*. Biochemistry 44:213–224

Stöckel S, Stanca AS, Helbig J, Rösch P, Popp J (2015 Nov) Raman spectroscopic monitoring of the growth of pigmented and non-pigmented mycobacteria. Anal Bioanal Chem 407(29):8919–8923

Striebel F, Imkamp F, Sutter M, Steiner M, Mamedov A, Weber-Ban E (2009) Bacterial ubiquitin-like modifier pup is deamidated and conjugated to substrates by distinct but homologous enzymes. Nat Struct Mol Biol 16:647–651

Striebel F, Hunkeler M, Summer H, Weber-Ban E (2010) The mycobacterial Mpa-proteasome unfolds and degrades pupylated substrates by engaging pup's N-terminus. EMBO J 29:1262–1271

Sugandhi S, Rajmane V, Taunk K, Jadhav S, Nema V, Rapole S, Mande SC (2023) The role of thioredoxin proteins in *Mycobacterium tuberculosis* probed by proteome-wide target profiling. Biochem Biophys Rep 35:101512. https://doi.org/10.1016/j.bbrep.2023.101512. PMID: 37521372; PMCID: PMC10371808

Ulrich K, Jakob U (2019) The role of thiols in antioxidant systems. Free Radic Biol Med 140:14–27. https://doi.org/10.1016/j.freeradbiomed.2019.05.035. Epub 2019 Jun 13. PMID: 31201851; PMCID: PMC7041647

van der Oost J, Westra ER, Jackson RN, Wiedenheft B (2014) Unravelling the structural and mechanistic basis of CRISPR-Cas systems. Nat Rev Microbiol 12(7):479–492. https://doi.org/10.1038/nrmicro3279. Epub 2014 Jun 9. PMID: 24909109; PMCID: PMC4225775

Verma T, Podder S, Mehta M, Singh S, Singh A, Umapathy S, Nandi D (2019 Dec) Raman spectroscopy reveals distinct differences between two closely related bacterial strains, Mycobacterium indicus pranii and Mycobacterium intracellulare. Anal Bioanal Chem 411(30):7997–8009

von Rosen T, Keller LML, Weber-Ban E (2021 Jun) Survival in hostile conditions: pupylation and the proteasome in actinobacterial stress response pathways. Front Mol Biosci 7(8):685757. https://doi.org/10.3389/fmolb.2021.685757

von Rosen T, Pepelnjak M, Quast JP, Picotti P, Weber-Ban E (2023) ATP-independent substrate recruitment to proteasomal degradation in mycobacteria. Life Sci Alliance 6(10):e202301923. https://doi.org/10.26508/lsa.202301923. PMID: 37562848; PMCID: PMC10415612

Webb CD, Decatur A, Teleman A, Losick R (1995 Oct) Use of green fluorescent protein for visualization of cell-specific gene expression and subcellular protein localization during sporulation in Bacillus subtilis. J Bacteriol 177(20):5906–5911

Westra ER, van Erp PB, Künne T, Wong SP, Staals RH, Seegers CL, Bollen S, Jore MM, Semenova E, Severinov K, de Vos WM (2012) CRISPR immunity relies on the consecutive binding and degradation of negatively supercoiled invader DNA by Cascade and Cas3. Mol Cell 46(5):595–605

Wiseman B, Nitharwal RG, Fedotovskaya O, Schäfer J, Guo H, Kuang Q, Benlekbir S, Sjöstrand D, Ädelroth P, Rubinstein JL, Brzezinski P (2018 Dec) Structure of a functional obligate complex III2IV2 respiratory supercomplex from Mycobacterium smegmatis. Nat Struct Mol Biol 25(12):1128–1136

World Health Organization (2015) Digital health for the end TB strategy: an agenda for action [internet]. WHO, Geneva. [cited 2025 Jul 19]. Technical document. Available from: [URL]

Wu M, Tao H, Xu T, Zheng X, Wen C, Wang G, Peng Y, Dai Y (2024) Spatial proteomics: unveiling the multidimensional landscape of protein localization in human diseases. Proteome Sci 22(1):7. https://doi.org/10.1186/s12953-024-00231-2. PMID: 39304896; PMCID: PMC11416001

Yang D, Ding F, Mitachi K, Kurosu M, Lee RE, Kong Y (2016) A fluorescent probe for detecting Mycobacterium tuberculosis and identifying genes critical for cell entry. Front Microbiol 7:2021

Yang Y, Walker TM, Walker AS, Wilson DJ, Peto TE, Crook DW, Shamout F, Zhu T, Clifton DA (2019) DeepAMR for predicting co-occurrent resistance of Mycobacterium tuberculosis. Bioinformatics 35(18):3240–3249

Yang Y, Arseni D, Zhang W, Huang M, Lövestam S, Schweighauser M, Kotecha A, Murzin AG, Peak-Chew SY, Macdonald J, Lavenir I (2022) Cryo-EM structures of amyloid-β 42 filaments from human brains. Science 375(6577):167–172

Yin CC (2018) Structural biology revolution led by technical breakthroughs in cryo-electron microscopy. Chinese Physics B 27(5):058703

Yu H, Lupoli TJ, Kovach A, Meng X, Zhao G, Nathan CF, Li H (2018) ATP hydrolysis-coupled peptide translocation mechanism of Mycobacterium tuberculosis ClpB. Proc Natl Acad Sci 115(41):E9560-9

Yuk JM, Kim JK, Kim IS, Jo EK (2024) TNF in human tuberculosis: a double-edged sword. Immune Netw 24(1):e4. https://doi.org/10.4110/in.2024.24.e4. PMID: 38455468; PMCID: PMC10917576

Zein-Eddine R, Refrégier G, Cervantes J, Yokobori NK (2023) The future of CRISPR in Mycobacterium tuberculosis infection. J Biomed Sci 30(1):34

Zhai X, Chu X, Chai CS, Jong MS, Istenic A, Spector M, Liu JB, Yuan J, Li Y (2021) A review of artificial intelligence (AI) in education from 2010 to 2020. Complexity 2021(1):8812542

Zhang J, Abadia E, Refregier G, Tafaj S, Boschiroli ML, Guillard B, Andremont A, Ruimy R, Sola C (2010 Mar) Mycobacterium tuberculosis complex CRISPR genotyping: improving efficiency, throughput and discriminative power of 'spoligotyping' with new spacers and a microbead-based hybridization assay. J Med Microbiol 59(3):285–294

Zhang L, Zhao Y, Gao Y, Wu L, Gao R, Zhang Q, Wang Y, Wu C, Wu F, Gurcha SS, Veerapen N, Batt SM, Zhao W, Qin L, Yang X, Wang M, Zhu Y, Zhang B, Bi L, Zhang X, Yang H, Guddat LW, Xu W, Wang Q, Li J, Besra GS, Rao Z (2020) Structures of cell wall arabinosyltransferases with the anti-tuberculosis drug ethambutol. Science 368(6496):1211–1219. https://doi.org/10.1126/science.aba9102

Zhang X, He X, Zhang Y, Chen L, Pan Z, Huang Y, Li H (2023) A new method for the detection of Mycobacterium tuberculosis based on the CRISPR/Cas system. BMC Infect Dis 23(1):680

Zhang F, Zhang F, Li L, Pang Y (2024) Clinical utilization of artificial intelligence in predicting therapeutic efficacy in pulmonary tuberculosis. J Infect Public Health 17(4):632–641

Zhang Y, Shi M, Li M, Qin S, Miao D, Bai Y (2025) Dynamic single-cell metabolomics reveals cell-cell interaction between tumor cells and macrophages. Nat Commun 16(1):4582. https://doi.org/10.1038/s41467-025-59878-w. PMID: 40379657; PMCID: PMC12084531

Zhao AP, Li S, Cao Z, Hu PJ, Wang J, Xiang Y, Xie D, Lu X (2024) AI for science: predicting infectious diseases. J Saf Sci Resil 5(2):130–146. https://doi.org/10.1016/j.jnlssr.2024.02.002

Zhavnerko G, Farnia P, Poleschuk N (2017) Brief reports on the use of atomic force microscopy in visualization of Mycobacterium tuberculosis. Biomed Biotechnol Res J (BBRJ) 1(1):9–13

Zhou J, Yin Y (2019) Use of liquid chromatography-mass spectrometry-based metabolomics to identify biomarkers of tuberculosis. Methods Mol Biol 1859:241–251. https://doi.org/10.1007/978-1-4939--8757-3_13. PMID: 30421233

Zhou L, Wang N, Zhu Z, Gao H, Lu N, Su H, Wang X (2024) Multi-omics fusion based on attention mechanism for survival and drug response prediction in digestive system tumors. Neurocomputing 572:127168. https://doi.org/10.1016/j.neucom.2023.127168

Zhu J, Wang H, Chen L (2024) Recent advances in nanomaterials for the detection of mycobacterium tuberculosis. Int J Mol Med 55(3):36

Zu Y, Chang H, Cui Z (2025) Molecular point-of-care testing technologies: current status and challenges. Nexus 2(2):100059. https://doi.org/10.1016/j.ynexs.2025.100059

MIX
Papier aus verantwortungsvollen Quellen
Paper from responsible sources
FSC® C105338

If you have any concerns about our products,
you can contact us on
ProductSafety@springernature.com

In case Publisher is established outside the EU,
the EU authorized representative is:
Springer Nature Customer Service Center GmbH
Europaplatz 3, 69115 Heidelberg, Germany

Printed by Libri Plureos GmbH
in Hamburg, Germany